The Building Environment: Active and Passive Control Systems

THIRD EDITION

Vaughn Bradshaw, P.E.

WILEY

John Wiley & Sons, Inc.

Published by John Wiley & Sons, Inc., Hoboken, New Jersey
Published simultaneously in Canada

For general information on our other products and services or for technical support, please contact our Customer Care Department within the United States at (800) 762-2974, outside the United States at (317) 572-3993 or fax (317) 572-4002.

Wiley also publishes its books in a variety of electronic formats. Some content that appears in print may not be available in electronic books. For more information about Wiley products, visit our web site at www.wiley.com.

Library of Congress Cataloging-in-Publication Data:

Bradshaw, Vaughn.
    The building environment : active and passive control systems /
    Vaughn Bradshaw.– 3rd ed.
      p. cm.
    Rev. ed. of: Building control systems. 1993.
    Includes bibliographical references and index.
    ISBN-13 978-0-471-68965-2
    ISBN-10 0-471-68965-3
        1. Intelligent buildings. 2. Buildings–Electric equipment.
    3. Buildings–Mechanical equipment. I. Bradshaw, Vaughn.
    Building control systems. II. Title.

    TH6012.B73 2006
    696–dc22

                                                2005051670

Printed in the United States of America

10 9 8 7 6 5 4 3 2

Contents

Preface   ix

Introduction   xiii

PART 1   THERMAL CONTROL CONCEPTS   1

Chapter 1   Human Comfort and Health Requirements   3

Human Physiology   5
Heat Balance   7
Metabolism   13
Clothing   14
Environmental Factors   16
Thermal Comfort Standards   21
Temperature and Humidity Extremes   28
Air Quality and Quantity   30

Chapter 2   Thermodynamic Principles   39

Heat Transfer   40
Psychrometrics   47

Chapter 3   Thermal Dynamics of Buildings   57

Heat Gain and Loss   58
Thermal Mass Dynamics   69
Moisture Problems   72

Chapter 4   Load Calculations   79

Heating and Cooling Loads   80
Selecting Design Temperature and Humidity Conditions   85
Solar Gain Through Fenestration   87

Heat Transmission Through the Building Envelope	95
Internal Loads	100
Outside Air	104
Annual Energy Use Calculations	106

PART 2   THERMAL CONTROL SYSTEMS   **117**

Chapter 5   A Building's Impact on the Environment   **119**

Ozone Depletion	119
Global Warming	120
Energy Conservation	122
Green Design/Sustainable Design	122
Case Study	134

Chapter 6   Active HVAC Systems   **139**

Central Heat Sources	145
Central Cooling Sources	161
Distribution Media	171
Delivery Devices	184
Localized Exhaust Systems	197
Heat Recovery	199
Thermal Storage	200
Automatic Controls	203
Noise and Vibration	205
District Heating and Cooling	206

Chapter 7   Passive Methods   **213**

Load Reductions	214
Passive Solar Heating	235
Passive Cooling	241
Case Study	246

PART 3   ELECTRICAL SYSTEMS   **251**

Chapter 8   Lighting   **253**

Principles of Illumination	257
Electrical Lighting	261
Daylighting	276
Emergency Lighting	285
Outdoor Lighting	285

Chapter 9   Normal Electrical Service   289

Loads   292
Service Entrance   296
Interior Distribution   302
Battery Power Supply   308
Utility Demand Management   310

Chapter 10   On-Site Power Generation   315

Thermal-Source Generation   316
Wind Turbines   323
Solar Photovoltaic Cells   326
Hydropower   328

Chapter 11   Special Systems   335

Signal Systems   335
Conveying Systems   340

PART 4   OTHER PIPED SYSTEMS   353

Chapter 12   Plumbing Systems   355

Water Supply   356
Plumbing Fixtures   360
Sanitary Piping   370
Stormwater Drainage   375

Chapter 13   Fire Protection   383

Detection and Alarm Systems   384
Exits: Means of Egress   389
Compartmentation   391
Smoke Control   393
Fire Suppression   399

PART 5   RELATED SUPPORT SERVICES   409

Chapter 14   Architectural Acoustics   411

Sound Theory   412
Room Acoustics   415
Sound Isolation   421

Chapter 15 **Building Commissioning** **433**

The Commissioning Plan 434
Applicable Systems and Equipment 435

Chapter 16 **Design Economics** **437**

Payback Period 438
Life-Cycle Cost 440
Rate of Return 446
Comparative Value Analysis 447
Other Considerations 448

Appendix **453**
Index **565**

Preface

This book is an introductory survey of the broad field of mechanical/electrical/plumbing (MEP) systems in buildings. It is intended to provide students of architecture with the information necessary to make proper provisions and allowances for these systems in their building designs.

Architects used to be responsible for designing entire buildings, including the heating and ventilation (usually steam radiators and operable windows), lighting, and power systems. These systems were much simpler than their sophisticated modern counterparts. With the advent of more and more complex systems, specialist consultants (structural, mechanical, electrical, civil engineering, fire protection, acoustical, lighting, elevator, etc.) are now usually required to design and handle the details of these technologies. Architects today find themselves more in the position of managing a team of engineers and other consultants than of designing entire buildings themselves.

Textbooks on building technologies, which have traditionally taught architects to be proficient in all the disciplines, have become unwieldy and cumbersome in their attempts to remain comprehensive. They contain much more detail and depth than is necessary for today's architects, and as a result, many students lose the essential overview.

This book, a departure from that tradition, provides a nontechnical treatment of the physical principles and equipment related to building control systems. Only the pertinent aspects of HVAC, electrical, lighting, plumbing, and acoustical systems are covered. The higher mathematics used by specialist consultants in each field are, for the most part, avoided, and the text concentrates on concepts in order to give students an appreciation of the larger picture without getting bogged down in details.

The general purpose of this book is to provide only the information that an architect, as generalist and "orchestrator," needs to communicate and coordinate with the consultants in all of the many disciplines. Detailed information and procedures for sizing and specifying mechanical and electrical systems are not included. Rather, the book emphasizes the conceptual understanding of the functions of the various systems and how they interact with the other building components. Items of equipment are regarded simply as "black boxes" having a given function.

The reader should come away with (1) an understanding of the environmental requirements of buildings, (2) a general sense of the systems and equipment typically used to satisfy those requirements, (3) an awareness of passive alternatives, (4) an awareness of the need to allow space and access for the necessary systems and equipment during the conceptual stages of design development, and (5) enough of a working knowledge of the terminology and requirements (spatial and structural) to ask the right questions of the consultants and, on occasion, to direct a mechanical or electrical contractor on projects too small to justify hiring an engineer.

Those interested in designing building control systems are referred to the bibliography at the end of each chapter. The references listed provide more detailed literature on specific topics. In addition, the end of each chapter contains a chapter summary, a list of key words, and study questions intended to stimulate thinking about, and retention of, the information presented.

ORGANIZATION

The book is organized into five parts. Part One presents the theoretical bases for thermal control, including the concepts of comfort, heat transfer, psychrometrics, building heat gains and losses, thermal mass, and condensation problems. It concludes with a chapter on load and energy-use calculations. The procedures presented in that chapter, based on the theory presented in the preceding chapters, are intended to demonstrate the magnitude of the energy consequences of design decisions and material selections. All the necessary tables and charts from ASHRAE and other sources are included in the appendix at the back of the book for convenient reference.

Part Two describes the systems, both active and passive, used to control the thermal environment within buildings. It also addresses the environmental impact from active systems and how passive systems can be used to mitigate some of the environmental problems. Passive environmental control is dealt with only in a cursory way, not because it is unimportant (quite the contrary), but because this subject can take up many volumes in itself. Good reference works on the subject, listed at the end of Chapter 7, are readily available. The objective here is to provide a survey of the passive alternatives to active building control systems.

Part Three concentrates on electrical systems, including power and lighting, on-site generating systems, signal and communication systems, and conveying systems. Natural daylighting is included in the chapter on lighting.

Part Four covers the other piped systems in buildings (outside of HVAC): plumbing services and fire protection.

Part Five presents support services that can apply to all of the systems described in Parts Two, Three, and Four. These are noise control, building commissioning, and methods for making design decisions based on economics. With today's high cost of maintenance and energy, the owning and operating costs of a building often overshadow the initial costs. The procedures introduced in Chapter 16 show how to make simple comparisons of life-cycle costs.

RECOMMENDATIONS FOR CLASSROOM USE

This book is intended to be a source of basic background information, which should be supplemented with classroom discussions and examples appropriate to the instructor's emphasis. I have deliberately avoided filling up the book with specific examples of design solutions because it is impossible to include the number of examples necessary to cover all aspects of each subject. Instead, I have kept information general and inclusive. Class time can be used to answer questions, add specific information from the instructor's own experience, go over examples of actual installations to clarify the concepts, and provide students with an opportunity to practice what they have learned.

The scope and depth of this book are appropriate for a well-rounded survey curriculum of one or two semesters (one to three quarters). Supplementary textbooks are recommended for more in-depth coverage of passive systems, lighting, acoustics, and energy where these subjects are especially emphasized or taught as separate courses. Appropriate specialized books may be selected from the bibliographies at the ends of the chapters.

OTHER USES

Although this book is directed to students of architecture and construction technology, it may also prove useful for building owners, designers, financiers, and others interested in how building control systems work. It is written primarily for instructional purposes, but the information is organized so that it can also serve the reader as a practical reference. Those who might benefit from it include architects, builders, developers, contractors, beginning HVAC designers and engineering students, building managers, and homeowners. While the book does not contain enough technical detail to be used as an engineering design manual on all subjects, it is useful as an overview for engineers and designers just entering the field.

THIRD EDITION

The cost of fossil fuels is of increasing concern to building users. It is also becoming increasingly evident that human activities are adversely affecting the global environment in significant ways. A new chapter accordingly has been added on the effect that buildings have on the environment. It includes discussions of ozone depletion, global warming, energy conservation, and green design or sustainable design.

Another new chapter has been added that addresses the subject of building commissioning, which helps to ensure the effectiveness of energy-conserving features included in a building's design.

In addition, most of the photographs of fixtures and equipment have been updated, and the bibliographies at the end of each chapter also have been updated. Finally, improvements have been incorporated in response to a number of very useful comments from users of the book.

A NOTE ABOUT SI UNITS

The metric system originated in France and was adopted there during the French Revolution. It gradually gained acceptance throughout Europe and in the French and Spanish colonies, and in 1875 the United States joined 16 other countries in signing the Treaty of the Meter. A revised metric system was introduced in 1960, named Système Internationale d'Unités (SI). Almost every country in the world has now committed to this universal system of units, which reduces the risk of mathematical errors.

In the United States, the Metric Conversion Act of 1975 prescribed a gradual voluntary conversion to SI units. U.S. industry, while reluctant to change, has begun shifting to SI units as the emphasis on foreign trade has increased. Several states already record surveys and plat plans in metric. The major organizations for building professionals, including ASHRAE and AMCA, are in the process of gradually shifting to metric units in their standards and publications.

The momentum of conversion to SI units in the United States is growing, but the English or inch-pound system of units is still in common usage. For that reason, all units in this book are still presented in English units with their equivalence in SI units following in parentheses.

The American Institute of Architects has produced the *AIA Metric Building and Construction Guide* to assist those making the transition. Appendix A of *Architectural Graphic Standards* (10th ed., John Wiley & Sons, Inc., 2000) also includes a useful discussion of SI units.

The following table lists the numerical value of the standard metric prefixes. Appendix Table A1 lists equivalents of common English and SI units.

SI PREFIXES

Multiplying Factor	Prefix	Symbol
1 000 000 000 000 = 10^{12}	tera	T
1 000 000 000 = 10^{9}	giga	G
1 000 000 = 10^{6}	mega	M
1 000 = 10^{3}	kilo	k
100 = 10^{2}	hecto[a]	h
10 = 10^{1}	deka[a]	da
0.1 = 10^{-1}	deci[a]	d
0.01 = 10^{-2}	centi	c
0.001 = 10^{-3}	milli	m
0.000 001 = 10^{-6}	micro	μ
0.000 000 001 = 10^{-9}	nano	n
0.000 000 000 001 = 10^{-12}	pico	p

[a] These prefixes are not normally used.

Introduction

A building's form is an expression of the creativity and inspiration of the architect in response to the client's objectives, under the limitations of the constraints imposed. The initial objectives typically include a projected building size and the intended functions of the facility. Limitations have traditionally been imposed by the local site characteristics of topography, orientation, and climate; the budget; owner preferences; and the legal conditions of building codes and zoning restrictions in the area. Today energy efficiency must be added as another strong influence on the design. During the design process, careful consideration must be given to the selection of building and system components with regard to energy efficiency.

Another related issue that must now be addressed is the selection of the energy source(s) for a building. In general, the energy resources we currently have available are solar, geothermal, fossil fuels, tidal, and nuclear. Solar energy accounts for 99.9776 percent of the total available. Wind, wood, alcohol (from plant biomass), and hydropower are all created by solar energy, and are considered renewable because they are continually being replenished.

Renewable fuel resources are available indefinitely but at a finite rate of replenishment. While the influx of solar energy varies from day to day at any given location on the planet, it supplies the earth as a whole at a relatively steady rate, and will continue to do so for the conceivable future. For example, a wood lot produces a limited amount of wood fuel each year, but it will repeat its production year after year if properly managed. Renewable fuels are analogous to a fixed but steady monetary income, such as a salary with no raises. By contrast, nonrenewable fuels are like savings accounts that draw no interest; once spent, they are gone.

The sun is also the ultimate source of fossil fuels, but the difference is that coal, oil, and natural gas are fossilized solar energy, collected and stored over hundreds of millions of years. Once they are used up, it will take a similar length of time for them to be replenished. For this reason, fossil fuels are called *nonrenewable resources.*

Technically, nuclear fission fuels and geothermal reserves are also exhaustible, but at the rate of current use, the supply is for all intents and purposes as limitless as solar energy itself.

Electricity can be generated from virtually any of these forms of energy. In the United States today, the vast majority is produced from coal, oil, and natural gas. Hydropower and nuclear power—although predominant in some regions—each carry only a minor share of the load nationwide. There are also a few solar thermal, solar photovoltaic, wind power, and geothermal power plants in operation.

Historically, the dominant fuel resource has shifted. Initially, human beings probably relied exclusively on renewable fuel resources: the sun directly for heat and light, supplemented by burning wood at night and on cold days. More sophisticated technology brought the use of the wind for transportation, processing (grinding grains), and other industrial purposes. The power in free-flowing streams and rivers was also tapped for stationary industries.

Before the nineteenth century, wood was used for most fuel needs. With mineral discoveries and mining, the vast reserves of fossil fuels (coal, petroleum, natural gas) became available, offering the advantages of portability, convenience, and reliability.

The 1800s saw the infancy of the coal industry. By the 1900s, coal had taken over as the predominant fuel. At the same time, hydropower and natural gas first began to be used on a significant scale. By 1950, petroleum and natural gas (cleaner-burning fuels) had taken the lead. Natural gas and petroleum presently have about even shares of the market, while coal use in buildings has been declining. Most of the coal currently consumed is for electricity generation and heavy industrial processes, where the problems of fuel storage and air pollution can be centrally treated. Notwithstanding modern techniques for scrubbing and filtering out sulfur ash from coal combustion emissions, many large cities currently place limits on coal use in individual buildings.

Nuclear power has entered the fuel mix since the 1950s. A number of factors, however, render the future of nuclear power questionable. These include (1) the economics relating to long construction periods and the need to contain extremely high temperature, pressure, and radioactivity levels during operation, as well as (2) some serious public safety issues related to the normal release of low-level radiation over long periods of time and the risks of accidental high-level releases. The cost and safety considerations inherent in nuclear power limit its civilian use to research and electricity generation by utility companies.

As recently as 1950, the United States was completely energy self-sufficient; our power needs were easily met with relatively cheap and abundant domestic fuels—first with coal, and then increasingly with oil and natural gas—along with hydroelectric power. Since then, as demand has grown, our reliance on imports of crude oil and petroleum products has steadily risen.

Since 1973, the price of oil has fluctuated widely in response to geopolitical forces and increasing demand around the world. When the price is high, energy conservation and alternative energy resources are on people's minds, and when the price drops, those interests are displaced by other matters.

But the instability of petroleum prices makes fiscal planning for building operations difficult, and when energy costs rise sharply, it has a significant impact on cash flow. Most building owners and developers have responded by placing a priority on energy efficiency to minimize the effect of energy cost increases. In addition, energy conservation standards are now included in virtually all building codes in the United States.

When fossil fuels began to be used the reserves seemed limitless, so no attempts were made to manage them for the most intelligent long-term benefit. Instead, their use was simply dictated by immediate economic factors such as the labor cost of mining, distribution costs, and the supply and demand forces. In fact, it was only in the latter half of the twentieth century that scientists tried to draw attention to the limitations in the fuels we rely upon so heavily.

The fact is that the worldwide supply of nonrenewable resources is dwindling, and energy costs will assuredly tend to rise over the long term. The exact amount of reserves of nonrenewable fuel resources is not known for sure. Two things, however, are certain: (1) there is a fixed limit to the nonrenewable fuel reserves, and (2) we are using them up at

an ever-increasing rate. And regardless of how much is still in the ground, it will become increasingly expensive, and in some cases environmentally objectionable, to extract the remaining amounts.

The obvious ramification to this is that buildings built today with a 50-year functional life and 100-year structural life may very well outlast the supplies of fossil fuels. Designers are thus faced with the following challenges:

1. Use the minimum energy that can be justified by current economic conditions.
2. Design buildings so that they can eventually be weaned away from dependence upon nonrenewable fuels.
3. Use only a "fair share" of renewable fuels, recognizing that such resources are limited even if they continue to be available.

Building designs emphasizing energy conservation and the substitution of renewable for nonrenewable energy sources can have a significant impact. About 35 percent of all energy used in the United States is consumed directly in buildings; another 6 percent is consumed off-site in support facilities such as those for sewage treatment, water supply, and solid waste management; and approximately 7 percent more is used to process, produce, and transport materials used for building construction. Altogether, about 48 percent of all energy used is in and for buildings.

There are many ways of improving energy efficiency in buildings. It is the responsibility of a building's occupants to operate it in an energy-conserving way, but the occupants can only operate within the capabilities provided by the building's designers. It is ultimately up to the designers to provide the building's owners and occupants with the advantage of the most energy-efficient building feasible. Not only is this a service from an economic standpoint, it will prevent the building from becoming obsolete in the future due to high energy costs.

The purpose of constructing buildings is to provide an artificial environment that is aesthetically appropriate and more conducive to the intended process or human occupancy than the natural environment. This implies, among other things, the maintenance of thermal and atmospheric conditions in enclosed spaces that differ from those existing concurrently outdoors. These conditions are controlled by the characteristics of the enclosure's envelope (as conceived by the architect) and the equipment serving it (as devised by the engineers).

Traditionally, the building envelope has been regarded as a "barrier" separating the interior from the outdoor environment. The architect provided an isolated environment, and the engineers equipped it with energy-using devices to control the conditions. Now the need to conserve energy has brought a reevaluation of the envelope's role, and it is conceived increasingly as a dynamic boundary, interacting with both external natural energy forces and the internal building environment. The concept of the envelope is evolving into that of an *energy mediator*—sensitively attuned to the indigenous natural resources of sun, wind, and water. These are viewed as resources to be manipulated in order to balance the energy flows between inside and outside, rather than as environmental intruders. The architect has the first crack at providing the proper thermal and lighting conditions by using the structure itself. Any remaining needs must be satisfied by the mechanical/electrical systems.

An important aspect of mechanical and electrical systems is that they use energy; the greater the dependency upon them, the more energy is consumed by the building. By displacing them with passive control mechanisms in the envelope itself, energy consumption can be minimized.

Architects are being called upon to provide new solutions to these new design challenges. Energy-conscious design requires that the total energy use of the building be thought out from its inception. The superior design is one that meets aesthetic, cost, and performance requirements in an elegant, energy-efficient manner.

The sophistication involved in energy-conscious design requires a team effort. The best solution to the control problem is arrived at through the close coordination of the entire design team from the earliest stage. The design process must be a cooperative effort between architects and engineers. Engineers can provide efficient equipment and systems, but in a sense, the control systems can only respond to the architectural decisions. While the mechanical system designer determines the efficiency of the equipment that consumes the energy, the building designer influences the extent to which that equipment must be operated. Thus, a finely tuned,

energy-efficient structure requires close cooperation among all the disciplines.

But energy efficiency is not the only reason that engineers should be represented in the preliminary stage of design. Many building mistakes can be avoided by early communication. Just as a building must work architecturally, it must work mechanically, and the two are inextricably linked. Even though mechanical equipment is generally concealed and unnoticed until it *doesn't* work, attention must be paid to the requirements that are crucial for proper operation. A beautiful building with a reputation for poor comfort conditions is less marketable.

In other words, rather than being considered as additional equipment to make an architectural design work, the mechanical and electrical control systems of a building should be considered integrally with the initial planning. They can have a strong bearing on the optimum form, location on a site, and orientation of a building. Any design can be made to work in terms of the thermal environment, lighting, plumbing, and acoustics—given enough money, and sometimes with a compromise of the original architecture. But by integrating their needs from the beginning, the most economical and elegant solution can be obtained.

Altogether, the design team should expect to spend increasing amounts of time considering strategies to reduce energy consumption in their buildings. The more scarce nonrenewable fuels get, the more expensive they will get. It is to everyone's advantage to create buildings that use fuels as efficiently as possible. The most efficient and self-sufficient buildings will only increase in value in the future.

Many clients today come to an architect with energy efficiency as a criterion. Others may not have considered it or don't believe they can afford it. Considering the importance of energy in every building's future, it is imperative that creative design make each building as energy efficient as possible within the program and budget constraints.

PART ONE

THERMAL CONTROL CONCEPTS

Energy conservation should not be regarded as cutting back on comfort conditioning to the limit of what the occupants can bear. The technology is available for maintaining perfectly acceptable comfort conditions at energy consumption levels much lower than those currently experienced. Sometimes, by taking advantage of innovative technology and new design concepts, and by applying a measure of creativity and imagination, this can be accomplished at little or no additional cost, and in some cases, it may actually be less expensive. But first, some basic information is needed to creatively apply energy-efficient innovations.

Chapter 1 covers occupant comfort, including thermal comfort, air quality, and air quantity. The chapter is intended to instill a sensitivity to comfort issues. Under most circum-

stances, standard design practices are sufficient, but it is important to be aware of what conditions need further attention.

Since heat loss plays a crucial role in maintaining the body's thermal equilibrium, Chapter 2 elaborates on the physical principles of this interactive process in order to provide a basis for designing an environment that satisfies the objectives of comfort and health. A procedure is introduced for calculating the rate of heat flow. This knowledge of the fundamentals of heat transfer theory will be applied to buildings in Chapter 3, and will be used as the basis for learning how to perform building heat loss, heat gain, and energy use calculations in Chapter 4. Radiation heat transfer, in particular, will be important in the subject of solar energy utilization in Chapters 6 and 7.

1

Chapter 2 also includes an elaboration of psychrometrics and an introduction to the psychrometric chart. This information will be useful in understanding the sources of moisture problems discussed in Chapter 3. It will also be important background for the load calculations in Chapter 4 and the mechanical systems for heating, ventilation, and air conditioning in Chapter 6.

Chapter 4, the only one that is extensively numerical, is intended to give the reader a quantitative sensitivity to the various factors involved. It is more detailed than necessary for estimating building loads in standard practice. Readers are encouraged to develop their own rules of thumb and simplified procedures based on their own experience. The simplifications previously used are, for the most part, no longer accurate, since changing insulation standards, increasing attention to sealing air leaks, higher internal gains, use of passive solar heating, and increasing energy costs have invalidated old assumptions.

Architects typically do not need to perform these kinds of calculations except for optimizing insulation levels and the number of panes in windows, for occasional sizing of residential heating and cooling systems, or for special energy-integrated designs. It is important to go through the exercise, however, in order to develop a general understanding of how design decisions influence energy consumption.

Chapter 1

Human Comfort and Health Requirements

COMFORT CONDITIONS

HUMAN PHYSIOLOGY

HEAT VS. TEMPERATURE

BODY TEMPERATURE CONTROL

HEAT BALANCE

EVAPORATION

RADIATION

CONVECTION-CONDUCTION

COMBINED EFFECTS

METABOLISM

CLOTHING

ENVIRONMENTAL FACTORS

DRY-BULB TEMPERATURE

HUMIDITY

MEAN RADIANT TEMPERATURE

AIR MOVEMENT

THERMAL COMFORT STANDARDS

THERMAL INDICES

THE COMFORT CHART

ASHRAE'S THERMAL COMFORT STANDARD

DESIGN CONSIDERATIONS

INDIVIDUAL VARIABILITY

TEMPERATURE AND HUMIDITY EXTREMES

HEAT STRESS

RESPONSE TO EXTREME COLD

AIR QUALITY AND QUANTITY

AIR CONTAMINANTS

ODORS

VENTILATION

Thermal and atmospheric conditions in an enclosed space are usually controlled in order to ensure (1) the health and comfort of the occupants or (2) the proper functioning of sensitive electronic equipment, such as computers, or certain manufacturing processes that have a limited range of temperature and humidity tolerance. The former is referred to as *comfort conditioning,* and the latter is called *process air conditioning.* The conditions required for optimum operation of machinery may or may not coincide with those conducive to human comfort.

Process air conditioning requirements are highly specific to the equipment or operation involved. Specifications are generally available from the producer or manufacturer, and the ASHRAE[a] *Handbook of Applications* provides a description of acceptable conditions for a number of generic industrial processes.

Once the necessary conditions for process or machinery operation are established, attention must be paid to providing acceptable comfort, or at least relief from discomfort or physiological stress, for any people also occupying the space.

Although human beings can be considered very versatile "machines" having the capacity to adapt to wide variations in their working environment while continuing to function, their productivity does vary according to the conditions in their immediate environment. Benefits associated with improvements in thermal environment and lighting quality include:

- Increased attentiveness and fewer errors
- Increased productivity and improved quality of products and services
- Lower rates of absenteeism and employee turnover
- Fewer accidents
- Reduced health hazards such as respiratory illnesses

Indeed, in many cases, air conditioning costs can be justified on the basis of increased profits. The widespread availability of air conditioning has also enabled many U.S. companies to expand into the Sun Belt, which was previously impractical.

Air conditioning and electric lights have eliminated the

need for large windows, which provided light and ventilation in older commercial and institutional buildings. Although windows are still important for aesthetics, daylighting, and natural ventilation, windowless interior spaces may now be used to a much greater extent. Air conditioning allows for more compact designs with lower ceilings, fewer windows, less exterior wall areas, and less land space for a given enclosed area. Conditioned air, which is cleaner and humidity-controlled, contributes to reduced maintenance of the space. As a testament to the importance placed on air conditioning, over one-third of the entire U.S. population presently spends a substantial amount of time in air-conditioned environments. And all of this represents growth since the commercialization of refrigeration cooling in the early 1950s.

On the other hand, this improvement in comfort has come about at the expense of greater equipment installation, maintenance, and energy costs. A substantial portion of the energy consumed in buildings is related to the maintenance of comfortable environmental conditions. In fact, approximately 20 percent of the *total* U.S. energy consumption is directed toward this task.

But this doesn't have to continue to be the case. With an understanding of the factors that determine comfort in relation to climate conditions, designers may select design strategies that provide human comfort more economically. Thus, prior to investigating the energy-consuming mechanical systems in buildings, we will begin by discussing the concepts of human comfort.

Comfort Conditions

Besides being aesthetically pleasing, the human environment must provide light, air, and thermal comfort. In addition, proper acoustics and hygiene are important. Air requirements and thermal comfort are covered in this chapter, while illumination and acoustical considerations will be presented in later chapters.

Comfort is best defined as the absence of discomfort. People feel uncomfortable when they are too hot or too cold, or when the air is odorous and stale. Positive comfort conditions are those that do not distract by causing unpleasant sensations of temperature, drafts, humidity, or other aspects of the environment. Ideally, in a properly conditioned space, people should not be aware of equipment noise, heat, or air motion.

The feeling of comfort—or, more accurately, discomfort—is based on a network of sense organs: the eyes, ears, nose, tactile sensors, heat sensors, and brain. *Thermal comfort* is that state of mind that is satisfied with the thermal environment; it is thus the condition of minimal stimulation

[a] The American Society of Heating, Refrigerating, and Air-Conditioning Engineers, Inc. (ASHRAE), a professional organization of engineers, conducts basic research of importance to the progress of the industry. Its handbooks are recognized references and data from them are widely used in this text. A major function of ASHRAE is to promulgate national voluntary consensus standards relating to the work of its members. Based on authoritative research and assembled with great care, ASHRAE standards are accepted throughout the industry and usually serve as the basis for state and local building codes.

of the skin's heat sensors and of the heat-sensing portion of the brain.

The environmental conditions conducive to thermal comfort are not absolute, but rather vary with the individual's metabolism, the nature of the activity engaged in, and the body's ability to adjust to a wider or narrower range of ambients.

For comfort and efficiency, the human body requires a fairly narrow range of environmental conditions compared with the full scope of those found in nature. The factors that affect humans pleasantly or adversely include:

1. Temperature of the surrounding air
2. Radiant temperatures of the surrounding surfaces
3. Humidity of the air
4. Air motion
5. Odors
6. Dust
7. Aesthetics
8. Acoustics
9. Lighting

Of these, the first four relate to thermal interactions between people and their immediate environment. In order to illustrate how thermal interactions affect human comfort, the explanation below describes the body temperature control mechanisms and how environmental conditions affect them.

HUMAN PHYSIOLOGY

Heat vs. Temperature

The sense of touch tells whether objects are hot or cold, but it can be misleading in telling just how hot or cold they are. The sense of touch is influenced more by the rapidity with which objects conduct heat to or from the body than by the actual temperature of the objects. Thus, steel feels colder than wood at the same temperature because heat is conducted away from the fingers more quickly by steel than by wood.

As another example, consider the act of removing a pan of biscuits from an oven. Our early childhood training would tell us to avoid touching the hot pan, but at the same time, we would have no trouble picking up the biscuits themselves. The pan and biscuits are at the same temperature, but the metal is a better conductor of heat and may burn us. As this example illustrates, the sensors on our skin are poor gauges of temperature, but rather are designed to sense the degree of heat flow.

Heat

By definition, *heat* is a form of energy that flows from a point at one temperature to another point at a lower temperature. There are two forms of heat of concern in planning for comfort: (1) sensible heat and (2) latent heat. The first is the one we usually have in mind when we speak of heat.

Sensible Heat. *Sensible heat* is an expression of the degree of molecular excitation of a given mass. Such excitation can be caused by a variety of sources, such as exposure to radiation, friction between two objects, chemical reaction, or contact with a hotter object.

When the temperature of a substance changes, it is the heat content of the object that is changing. Every material has a property called its *specific heat,* which identifies how much its temperature changes due to a given input of sensible heat.

The three means of transferring sensible heat are radiation, convection, and conduction. All bodies emit *thermal radiation.* The net exchange of radiant heat between two bodies is a function of the difference in temperature between the two bodies. When radiation encounters a mass, one of three things happens: (1) the radiation continues its journey unaffected (in which case it is said to be transmitted), (2) it is deflected from its course (in which case it is said to be reflected), or (3) its journey comes to an end (and it is said to be absorbed). Usually, the response of radiation to a material is some combination of transmission, reflection, and absorption. The radiation characteristics of a material are determined by its temperature, emissivity (emitting characteristics), absorptivity, reflectivity, and transmissivity.

Conduction is the process whereby molecular excitation spreads through a substance or from one substance to another by direct contact. *Convection* occurs in fluids and is the process of carrying heat stored in a particle of the fluid to another location where the heat can conduct away. The heat transfer mechanisms of radiation, conduction, and convection are elaborated on in Chapter 2.

Latent Heat. Heat that changes the state of matter from solid to liquid or liquid to gas is called *latent heat. The latent heat of fusion* is that which is needed to melt a solid object into a liquid. A property of the material, it is expressed per unit mass (per pound or per kilogram). The *latent heat of vaporization* is the heat required to change a liquid to a gas. When a gas liquefies (condenses) or when a liquid solidifies, it releases its latent heat.

Enthalpy. *Enthalpy* is the sum of the sensible and latent heat of a substance. For example, the air in our ambient environment is actually a mixture of air and water vapor. If the total heat content or enthalpy of air is known, and the enthalpy of the desired comfort condition is also known, the difference between them is the enthalpy or heat that must be added (by heating and humidification) or removed (by cooling and dehumidification).

Units.   The common measure of quantity of heat energy in the English system of units is the *British thermal unit* (*Btu*). It is that heat energy required to raise 1 pound of water 1 degree Fahrenheit. The rate of flow of heat in these units is expressed in Btu per hour (Btuh).

In the International system of units (SI units), the corresponding measure is the *joule.* The rate of heat flow in SI units is joules per second, or *watts* (*W*).

Temperature

Temperature is a measure of the degree of heat intensity. The temperature of a body is an expression of its molecular excitation. The temperature difference between two points indicates a potential for heat to move from the warmer to the colder point.

The English system of units uses the Fahrenheit degree scale, while in SI units the Celsius degree scale is used. Note that temperature is a measure of heat *intensity,* whereas a Btu or joule is a measure of the *amount* of heat energy.

The *dry-bulb temperature* of a gas or mixture of gases is the temperature taken with an unwetted bulb that is shielded from radiant exchange. The familiar wall thermometer registers the dry-bulb temperature of the air.

If a thermometer bulb is moistened, any evaporation of water extracts sensible heat from the air surrounding the bulb. The sensible heat vaporizes the water and becomes latent heat. The exchange of sensible for latent heat in the air does not change the total heat content, but the air temperature is lowered. Thus, a thermometer with a wetted bulb (such as that shown in Figure 1.9) indicates a lower temperature than a dry-bulb thermometer. The drier the air, the greater the exchange of latent for sensible heat, making the wet-bulb temperature correspondingly lower.

The *wet-bulb temperature* is therefore a means of expressing the humidity of air. Dry- and wet-bulb temperatures that are the same indicate that the air has already absorbed all the water vapor it can hold, no evaporation can take place, and the percentage humidity is 100 percent. By comparing dry- and wet-bulb temperature readings, one can determine the level of humidity in the air. The larger the temperature difference, the lower the humidity.

The capability of air to hold moisture depends on the dry-bulb temperature. The higher the temperature, the

more moisture the air can hold. Thus, as a mixture of air and water vapor is cooled, it becomes relatively more humid. At some temperature, the air becomes saturated with the given water content. In other words, the quantity of water vapor present is all that the air can hold at that temperature. Any further lowering of temperature will cause condensation of some of the water vapor. The temperature at which condensation begins is known as the *dew-point temperature.*

The temperatures of the surfaces surrounding an enclosed space in relation to the temperature of a body within the space determine the rate and direction of radiant heat flow between the body and the surrounding surfaces. The comfort of a person in a space is affected by this radiant exchange of heat. Therefore, it is useful to know the radiant surface temperatures or the average (mean) value of them. The equivalent uniform temperature of an enclosure causing the same radiant exchange as the given real conditions is known as the *mean radiant temperature* (*MRT*).

These concepts will be reviewed in Chapter 2 and are merely introduced here in order to provide an understanding of the following discussion of human physiology.

Body Temperature Control

Human beings are essentially constant-temperature animals with a normal internal body temperature of about 98.6°F (37.0°C). Heat is produced in the body as a result of metabolic activity, so its production can be controlled, to some extent, by controlling metabolism. Given a set metabolic rate, however, the body must reject heat at the proper rate in order to maintain thermal equilibrium.

If the internal temperature rises or falls beyond its normal range, mental and physical operation is curtailed, and if the temperature deviation is extreme, serious physiological disorders or even death can result. Sometimes the body's own immunological system initiates a body temperature rise in order to kill infections or viruses.

The importance of maintaining a fairly precise internal temperature is illustrated in Figure 1.1, which shows the consequences of deep-body temperature deviations. When body temperature falls, respiratory activity—particularly in muscle tissue—automatically increases and generates more heat. Shivering is the extreme manifestation of this form of body temperature control.

An extremely sensitive portion of the brain called the *hypothalmus* constantly registers the temperature of the blood and seems to be stimulated by minute changes in blood temperature originating anywhere in the body (this could result from drinking a hot beverage or a change in skin

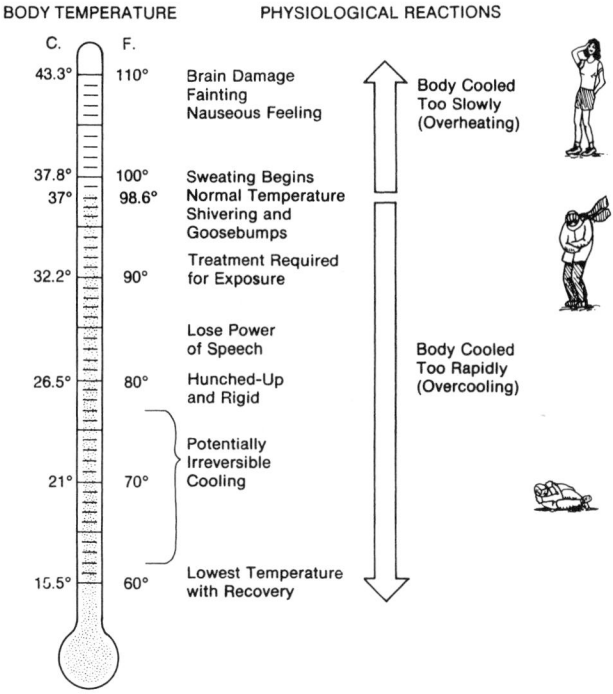

FIGURE 1.1. Physiological reactions to body temperature.

surface temperature). The skin also has sensors that signal to the brain the level of heat gain or loss at the skin.

It is the hypothalmus that appears to trigger the heat control mechanisms to either increase or decrease heat loss. This is accomplished by controlling the flow of blood to the skin by constricting or dilating the blood vessels within it. Since blood has high thermal conductivity, this is a very effective means of rapid thermal control of the body. By controlling peripheral blood flow, the body is able to (1) increase skin temperature to speed up elimination of body heat, (2) support sweating, or (3) reduce heat loss in the cold.

When body temperature rises above normal, the blood vessels in the skin dilate, bringing more heat-carrying blood to the surface. This results in a higher skin temperature and, consequently, increased heat loss. At the same time, sweat glands are stimulated, opening the pores of the skin to the passage of body fluids which evaporate on the surface of the skin and thereby cool the body. This evaporating perspiration is responsible for a great deal of heat loss. A minor amount of heat is also lost continuously by evaporation of water from the lungs and respiratory tracts.

When body heat loss is high, people experience a feeling of lassitude and mental dullness brought about by the fact that an increased amount of the blood pumped by the heart goes directly from the heart to the skin and back to the heart, bypassing the brain and other organs. A hot environment also increases strain on the heart, since it has to beat more rapidly to pump more blood to the periphery of the body.

When the body loses more heat to a cold environment than it produces, it decreases heat loss by constricting the outer blood vessels, thereby reducing blood flow to the outer surface of the skin. This converts the skin surface to a layer of insulation between the interior of the body and the environment. It has about the same effect as putting on a light sweater. If the body is still losing too much heat, the control device increases heat production by calling for involuntary muscular activity or shivering. When heat loss is too great, the body tends to hunch up and undergo muscular tension, resulting in a strained posture and physical exhaustion if the condition persists for any length of time.

Within limits, the body can acclimate itself to thermal environmental change. Such limits are not large, especially when the change is abrupt, such as when passing from indoors to outdoors. The slower seasonal changes are accommodated more easily. Changes in clothing assist the acclimatization. Whenever the body cannot adjust itself to the thermal environment, heat stroke or freezing to death is inevitable.

The physiological interpretation of comfort is the achievement of thermal equilibrium at our normal body temperature with the minimum amount of bodily regulation. We feel uncomfortable when our body has to work too hard to maintain thermal equilibrium. Under conditions of comfort, heat production equals heat loss without any action necessary by the heat control mechanisms. When the comfort condition exists, the mind is alert and the body operates at maximum efficiency.

It has been found that maximum productivity occurs under this condition and that industrial accidents increase at higher and lower temperatures. Postural awkwardness due to a cold feeling results in just as many accidents as does mental dullness caused by a too warm environment.

HEAT BALANCE

Like all mammals, humans "burn" food for energy and must discard the excess heat. This is accomplished by evaporation coupled with the three modes of sensible heat transfer: conduction, convection, and radiation. For a person to remain healthy, the heat must not be lost too fast or too slowly, and a very narrow range of body temperature must be maintained.

The body is in a state of thermal equilibrium with its environment when it loses heat at exactly the same rate as it gains heat. Mathematically, the relationship between the body's heat production and all its other heat gains and losses is:

$$\text{Heat production} = \text{heat loss} \qquad (1.1)$$

or

$$M = E \pm R \pm C \pm S \qquad (1.2)$$

where:

M = metabolic rate
E = rate of heat loss by evaporation, respiration, and elimination
R = radiation rate
C = conduction and convection rate
S = body heat storage rate

Equation 1.2 is illustrated in Figure 1.2.

The body always produces heat, so the metabolic rate (M) is always positive, varying with the degree of exertion. If environmental conditions are such that the combined heat loss from radiation, conduction, convection, and evaporation is less than the body's rate of heat production, the excess heat must be stored in body tissue. But body heat storage (S) is always small because the body has a limited thermal storage capacity. Therefore, as its interior becomes

warmer, the body reacts to correct the situation by increasing blood flow to the skin surface and increasing perspiration. As a result, body heat loss is increased, thereby maintaining the desired body temperature and the balance expressed by Equation 1.2.

The converse condition—where heat loss is greater than body heat production—causes a reversal of the above process and, if necessary, shivering. This increased activity raises the metabolic rate.

Table 1.1 indicates the environmental and human factors that influence each of the major terms in Equation 1.2. Metabolism is discussed at greater length later in this chapter, while the other major factors—evaporation, radiation, conduction, and convection—are discussed below.

Evaporation

The body can either gain or lose heat by radiation (R) and conductive-convective heat transfer (C), depending on the temperature of the surrounding objects and ambient air. By contrast, evaporation (E) is exclusively a cooling mechanism.

Evaporative losses usually play an insignificant role in the body's heat balance at cool temperatures. They become the predominant factor, however, when ambient temperatures are so high that radiant or convective heat losses cannot occur.

At comfortable temperatures, there is a steady flow of sensible heat from the skin to the surrounding air. The amount of this sensible heat depends upon the temperature difference between the skin and air. Although the deep body temperature remains relatively constant, the skin temperature may vary from 40° to 105°F (4° to 41°C) according to the surrounding temperature, humidity, and air velocity. During the heating season, the average surface temperature of an adult indoors wearing comfortable clothing is approximately 80°F (27°C). At lower surrounding temperatures, the skin temperature is correspondingly lower.

When the surrounding environment is about 70°F (21°C), most people lose sensible heat at a rate that makes them feel comfortable. If the ambient temperature rises to the skin temperature, the sensible heat loss drops to zero. If the ambient temperature continues to rise, the body gains heat from the environment, and the only way it can lose heat is by increasing evaporation.

Evaporative heat losses also increase at high activity levels, when the metabolic heat production rises. A person engaged in strenuous physical work may sweat as much as a quart of fluid in an hour.

The rate of evaporation and evaporative heat loss is deter-

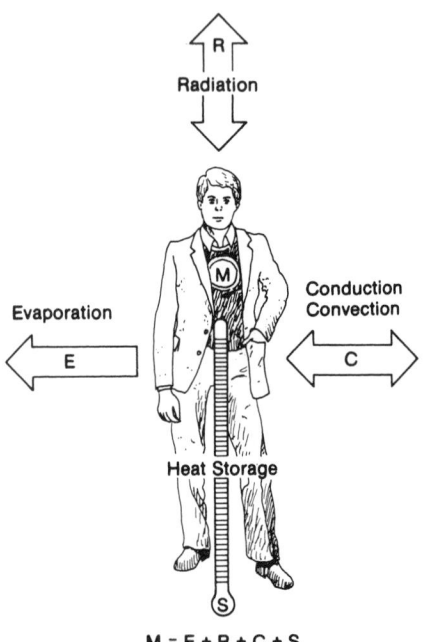

$$M = E \pm R \pm C \pm S$$

FIGURE 1.2. Heat balance of the human body interacting with its environment.

TABLE 1.1 FACTORS INFLUENCING THE HEAT BALANCE EQUATION

Factor	Environment	Human
Metabolism (M)	Little effect	Activity Weight Surface area Age Sex
Evaporation (E)	Wet-bulb temperature Dry-bulb temperature Velocity	Ability to produce sweat Surface area Clothing
Radiation (R)	Temperature difference between bodies Emissivity of surfaces	Surface area Clothing
Convection (C)	Dry-bulb temperature Velocity	Clothing Mean body surface temperature Surface area

Source: John Blankenbaker, "Ventilating Systems for Hot Industries," *Heating/Piping/Air Conditioning,* Vol. 54, No. 2, February 1982, p. 61. (Reproduced from the original: *Industrial Ventilation.* American Conference of Governmental Industrial Hygienists, Committee on Industrial Ventilation, Lansing, Michigan, 1976, p. 3-2.) Used by permission of Reinhold Publishing, a Division of Penton/IPC.

mined by the evaporation potential of the air. It is dependent to a minor degree on the relative humidity of the surrounding air and, to a much greater extent, on the velocity of air motion. Moisture, which is evaporated from the skin surface, is carried away by the passing air stream.

Sufficient heat must be added to the perspiration to vaporize it, and this heat is drawn from the body. This heat loss equals the latent heat of vaporization of all the moisture evaporated. It is thus commonly known as the latent heat component of the total heat rejected by the body.

While the skin sweats only at moderate to high temperatures, evaporative losses of water from the respiratory passages and lungs occur continuously. The breath "seen" when exhaled in frosty weather is evidence that the air leaving the lungs has a high moisture content. We generally exhale air that is saturated (100 percent RH), and even at rest, the body requires about 100 Btuh (30 W) of heat to evaporate this moisture from the lungs into the inhaled air. Since it takes a considerable amount of heat to convert water into vapor, the evaporative heat loss from our lungs and skin plays an important role in disposing of body heat.

Radiation

Radiation is the net exchange of radiant energy between two bodies across an open space. The human body gains or loses radiant heat, for example, when exposed to an open fire, the sun, or a window on a cold winter day.

Each body—the earth, the sun, a human body, a wall, a window, or a piece of furniture—interacts with every other body in a direct line of sight with it. Radiation affects two bodies only when they are in sight of each other. This means that the energy cannot go around corners or be affected by air motion. For example, when we are uncomfortably hot in the direct light of the sun, we can cut off the radiant energy coming directly from the sun by stepping into the shade of a tree. Since air is a poor absorber of radiant heat, nearly all radiant exchanges are with solid surfaces to which we are exposed.

Radiant heat may travel toward or away from a human body, depending on whether the radiating temperatures of surrounding surfaces are higher or lower than the body's temperature. In a cold room, the warmer body or its clothing transmits radiant heat to all cooler surfaces such as walls, glass, and other construction within view. If there is a cold window in sight, it will typically have the largest impact in terms of draining heat away and making the body feel colder. By closing the drapes, a person can block the radiant transfer in the same way that a person can cut off the radiant energy from the sun by stepping into the shade of a tree.

The rate of radiant transfer depends on the temperature differential, the thermal absorptivity of the surfaces, and the distance between the surfaces. The body gains or loses heat by radiation according to the difference between the body surface (bare skin and clothing) temperature, and the MRT of the surrounding surfaces.

The MRT is a weighted average of the temperatures of all the surfaces in direct line of sight of the body (see Figure

1.3). Although the MRT tends to stabilize near the room air temperature, it is also affected by large glass areas, degree of insulation, hot lights, and so on. The inside surface temperature of an insulated wall will be much closer to the room air temperature than will that of an uninsulated wall.

If the MRT is below the body temperature, the radiant heat term, R, in Equation 1.2 is a positive number, and the body is losing radiant heat. If the MRT is above the body temperature, R is negative, and the body is gaining radiant heat. This could be a benefit if the room air temperature is cool, causing excess body heat loss, while it would be detrimental if the ambient conditions are hot and humid, and the body is already having trouble rejecting heat.

It should be kept in mind that the body loses radiant heat according to its surface temperature. For a comfortable, normally dressed adult, the weighted average temperature of the bare skin and clothed surfaces is about 80°F (27°C). In still air at a temperature near skin temperature, radiant exchange is the principal form of heat exchange between the body and its environment.

To illustrate the body's radiant interaction with surrounding surfaces, consider a person during the heating season

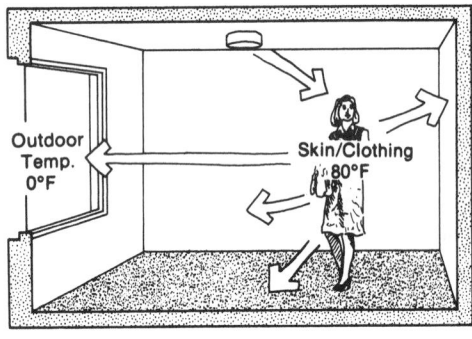

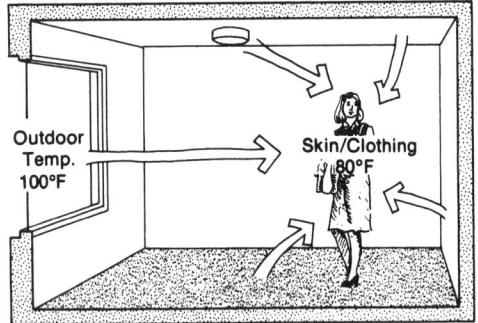

FIGURE 1.3. Radiant heat transfer with surrounding surfaces.

working at a desk facing the center of an office with his or her back 5 feet (1.5 m) from an outside wall (see Figure 1.4a). The wall surface temperature is 59°F (15.0°C). If the room air temperature is 74°F (23.3°C) at the ceiling, 72°F (22.2°C) at the floor, and a uniform 73°F (22.8°C) in between, including the space between the person's back and the wall, will he or she be comfortable? Probably not, because the radiant heat loss to the cold wall is so high that the office worker will feel chilly. (As a rule of thumb, if the MRT is 10°F (5°C) hotter or colder than comfortable room air conditions, an occupant will feel uncomfortable.) What can be done to correct this situation?

1. The wall surface temperature could be changed by adding insulation to the wall construction, or by hanging an insulative tapestry or wall hanging over the wall, as was done in medieval castles (Figure 1.4b).
2. The position of the desk might be changed, moving the person closer to an inside wall (Figure 1.4c). The radiant exchange would then be predominantly influenced by the surface temperature of the inside wall, which would be near the air temperature of 73°F (22.8°C). Thus, the radiant heat loss would be one-third of what it was: 80° − 73° = 7°F instead of 80° − 59° = 21°F (27° − 23° = 4°C instead of 27° − 15° = 14°C).
3. If the desk cannot be moved, the temperature of the air might be increased by turning up the thermostat. Increasing the air temperature would decrease the convective heat loss from the body. Suppose that setting the air temperature at 77°F (25°C) would decrease the convective heat loss by the same amount that the radiant heat loss would be decreased by moving the desk away from the outside wall. This would balance the heat loss from the body. The trouble is that everyone else in the room not sitting near an outside wall would be too warm.

Exactly the same thing might be true during the cooling season. A person might be too warm because of the radiant heat the body gains from a warm outside wall or window. In this case, the sensible heat loss from the body could be *increased* by *decreasing* the air temperature. This puts one person's body heat loss in balance, but everyone else in the room would be too cool.

Thus, not only is good, properly operated heating and cooling equipment important for maintaining comfort, but the building construction itself can also have a strong influence. Poorly insulated walls and windows should be flagged as comfort problems. Furthermore, the type of occupancy must be borne in mind when analyzing the intended comfort conditions.

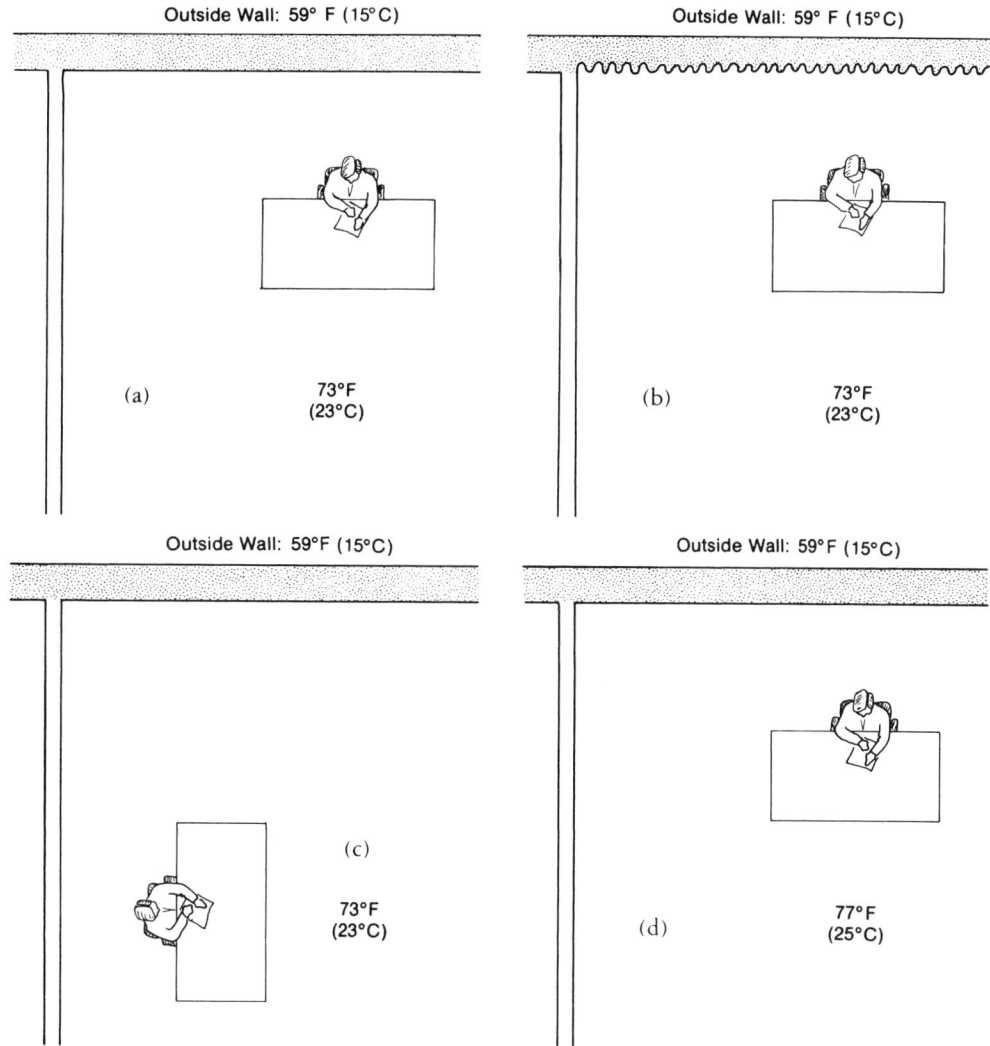

FIGURE 1.4. Alternative methods of achieving radiant comfort.

Convection-Conduction

Air passing over the skin surface is instrumental not only to the evaporation of moisture, but also to the transference of sensible heat to or from the body. The faster the rate of air movement, the larger the temperature difference between the body and surrounding air, and the larger the body surface area, the greater the rate of heat transfer.

When the air temperature is lower than the skin (and clothing) temperature, the convective heat term in equation 1.2 is "plus," and the body loses heat to the air. If the air is warmer than the skin temperature, the convective heat term is "minus," and the body gains heat from the air. Convection becomes increasingly effective at dissipating heat as air temperature decreases and air movement increases.

The conduction heat loss or gain occurs through contact of the body with physical objects such as the floor and chairs. If two chairs—one with a metal seat and the other with a fabric seat—have been in a 70°F (21°C) room for a period of time, they will both have a temperature of 70°F (21°C), but the metal one will feel colder than the one with the woven seat.

There are two reasons for this. First, metal is a good conductor, and it is the rate at which heat is conducted away—not the temperature—that we feel. Also, the metal chair has a smoother surface, which makes a good contact and thus facilitates better conduction. Clothing also plays an important role in conductive heat transfer, insulating us from the warm or cold surface, just as a pot holder protects us from a hot pot.

Combined Effects

The physiological basis of comfort was previously stated as the achievement of thermal equilibrium with a minimal amount of body regulation of M, E, R, and C. Figure 1.5 shows the relation between all these factors for lightly clothed and unclothed subjects at rest. Note that convective and radiant heat loss is greater for the lightly clothed subjects. Also, the heat loss by convection and radiation decreases with increasing air temperature, while evaporative heat loss increases with increasing air temperature.

Heat loss by evaporation is relatively constant below certain air temperatures—approximately 75°F (24°C) for the heavily clothed subject and 85°F (29°C) for the lightly clothed subject. The metabolic rate at a given activity level is stable when the temperature ranges from about 70° to 90°F (21° to 32°C).

To illustrate the various modes of heat loss operating in conjunction, consider a person outdoors in 100°F (38°C) air temperature. Referring to Equation 1.2, the convective heat loss is "minus" because the body is gaining heat from the air.

The MRT is much higher than the body surface temperature—the sidewalk, street, building walls, sunny sky, and everything else in the range of view of the body is warmer than the body surface temperature. Thus, the radiant heat term is also "minus" because the body is gaining radiant heat.

But as the person walks down the sidewalk, the metabolism produces about 700 Btuh (200 W), and all that heat must be lost in addition to that gained by convection and radiation in order to maintain the heat balance. The total the body must lose may be over 1,000 Btuh (300 W), all by evaporation. The sweat glands automatically open, and the resultant moisture emitted onto the body surface then evaporates. The heat drawn from the body to evaporate the moisture keeps the skin cool as long as the surrounding air will carry away the water vapor so that more can be evaporated. This in turn keeps the deep-body temperature close to 98.6°F (37.0°C).

As the dry-bulb temperature of the surrounding air rises from the comfortable 70s to the 80s and 90s, less sensible heat (convective and radiative) is lost by the body, while the latent heat (evaporation) loss increases. Thus, if a body at rest produces 400 Btuh (117 W), it may lose 290 Btuh (85 W) of sensible heat and 110 Btuh (32 W) of latent heat at 70°F (21°C). At higher temperatures, the sensible component drops to nearly zero, and the latent heat must increase to almost the full 400 Btuh (117 W) in order to lose the same amount of heat.

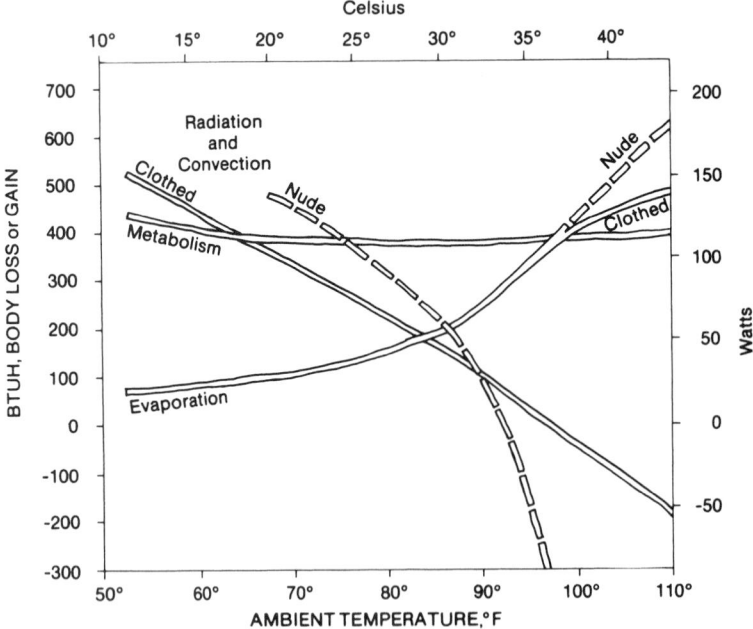

FIGURE 1.5. Relationship between metabolism, evaporation, radiation, convection, and temperature.

When people work under conditions of high temperature and extremely high humidity, both the sensible heat loss and the evaporation of moisture from their skins are reduced. Under these conditions, the rate of evaporation must be increased by blowing air rapidly over the body.

METABOLISM

As part of the process of being alive, people metabolize (oxidize) the food taken into the body, converting it into electrochemical energy. This energy is used for growth, regeneration, and operation of the body's organs, such as muscle contraction, blood circulation, and breathing. It enables us to carry out our normal bodily functions and to perform work upon objects around us.

As with all conversions from one form of energy to another, there is a certain conversion efficiency. Only about 20 percent of all the potential energy stored in the food is available for useful work. The other 80 percent takes the form of heat as a by-product of the conversion. This results in the continuous generation of heat within the body, which must be rejected by means of sensible heat flow (radiation, convection, or conduction) to the surrounding environment or by evaporating body fluids. If more food energy is ingested than is needed, it is stored as fat tissue for later use.

There is a continuous draw of energy for the operation of life-sustaining organs such as the heart. This is the idle level of bodily activity corresponding to the state of rest. It requires minimum energy conversion, and thus a minimum amount of heat is released as a by-product. When the body is engaged in additional mental or physical activity, metabolism increases to provide the necessary energy. At the same time, by-product heat generation also increases. The fuel for this is drawn from food currently being digested or, if necessary, from the fat stores.

When the body heat loss increases and the internal temperature begins to drop, metabolism increases in an effort to stabilize the temperature even though there is no additional mental or physical activity. In this case, all of the additional energy metabolized is converted into heat.

In general, the metabolic rate is proportional to body weight, and is also dependent upon the individual's activity level, body surface area, health, sex, age, amount of clothing, and surrounding thermal and atmospheric conditions. Metabolism rises to peak production at around 10 years of age and drops off to minimum values at old age. It increases due to a fever, continuous activity, or cold environmental conditions if the body is not thermally protected.

To determine the optimum environmental conditions for comfort and health, one must ascertain the metabolic level during the course of routine physical activities, since body heat production increases in proportion to the level of exercise. When the activity level shifts from sleeping to heavy work, the metabolism varies accordingly, as shown in Figure 1.6.

Table 1.2 presents average metabolic rates for a variety of steady activities in *met* units. One met is defined in terms of body surface area as

$$18.4 \text{ Btuh/ft}^2 = 58.2 \text{ W/m}^2 = 50 \text{ kcal/m}^2 \cdot \text{hr}$$

For an average-size man, the met unit corresponds to approximately

$$360 \text{ Btuh} = 100 \text{ W} = 90 \text{ kcal/hr}$$

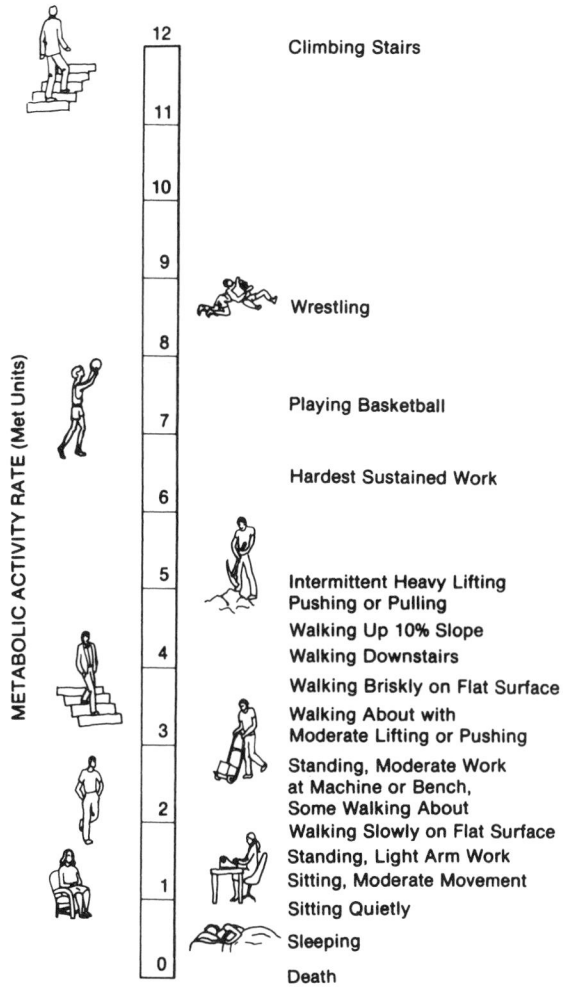

FIGURE 1.6. Scale of activity level variations (in met units).

TABLE 1.2 METABOLIC RATE AT DIFFERENT TYPICAL ACTIVITIES

Activity	Metabolic Rate in Met Units[a]	Activity	Metabolic Rate in Met Units[a]
Resting		**Miscellaneous Work**	
Sleeping	0.7	Watch-repairing, seated	1.1
Reclining	0.8	Lifting/packing	1.2 to 2.4
Seated, reading	0.9	Garage work (e.g., replacing tires, raising cars by jack)	2.2 to 3.0
Office Work		**Vehicle Driving**	
Seated, writing	1.0	Car	1.5
Seated, typing or talking	1.2 to 1.4	Motorcycle	2.0
Seated, filing	1.2	Heavy vehicle	3.2
Standing, talking	1.2	Aircraft flying, routine	1.4
Drafting	1.1 to 1.3	Instrument landing	1.8
Miscellaneous office work	1.1 to 1.3	Combat flying	2.4
Standing, filing	1.4	**Leisure Activities**	
Walking (on Level Ground)		Stream fishing	1.2 to 2.0
2 mph (0.89 m/s)	2.0	Golf, swinging and walking	1.4 to 2.6
3 mph (1.34 m/s)	2.6	Golf, swinging and with golf cart	1.4 to 1.8
4 mph (1.79 m/s)	3.8	Dancing	2.4 to 4.4
Domestic Work		Calisthenics exercise	3.0 to 4.0
Shopping	1.4 to 1.8	Tennis, singles	3.6 to 4.6
Cooking	1.6 to 2.0	Squash, singles	5.0 to 7.2
House cleaning	2.0 to 3.4	Basketball, half court	5.0 to 7.6
Washing by hand and ironing	2.0 to 3.6	Wrestling, competitive or intensive	7.0 to 8.7
Carpentry			
Machine sawing, table	1.8 to 2.2		
Sawing by hand	4.0 to 4.8		
Planing by hand	5.6 to 6.4		

[a] Ranges are for activities that may vary considerably from place to place and from person to person. 1 met = 18.4 Btuh/ft^2 = 58.2 W/m^2 = 50 kcal/hr·m^2. Some activities are difficult to evaluate because of differences in exercise intensity and body position among people.

Source: Reprinted from ASHRAE *Handbook of Fundamentals 1989* by permission of the American Society of Heating, Refrigerating, and Air-Conditioning Engineers, Inc.

A met is the average amount of heat produced by a sedentary man, and any metabolic rate can be expressed in multiples of this standard unit.

For the more intense activities listed in Table 1.2, actual metabolic rates depend on the relation between the intensity of the given activity and the individual's peak capacity. Another factor affecting metabolic rate is the heavy, protective clothing worn in cold weather, which may add 10 to 15 percent to the rate. Pregnancy and lactation may increase values by 10 percent.

CLOTHING

Another determinant of thermal comfort is clothing. In the majority of cases, building occupants are sedentary or slightly active and wear typical indoor clothing. Clothing, through its insulation properties, is an important modifier of body heat loss and comfort.

The insulation properties of clothing are a result of the small air pockets separated from each other to prevent air from migrating through the material. Newspaper, for example, can serve as good insulation if several sheets are separated so that there are layers of air between the layers of paper; this can be used as a crude, but effective, emergency blanket to cover the body. Similarly, the fine, soft down of ducks is a poor conductor and traps air in small, confined spaces. In general, all clothing makes use of this principle of trapped air within the layers of cloth fabric.

Clothing insulation can be described in terms of its *clo* value. The clo value is a numerical representation of a clothing ensemble's thermal resistance. 1 clo = 0.88 ft^2·hr·°F/Btu

(0.155 m²·°C/W). A heavy two-piece business suit and accessories have an insulation value of about 1 clo, while a pair of shorts is about 0.05 clo. Clo values for common articles of clothing are listed in Table 1.3. The total insulation value of a clothing ensemble can be estimated as the sum of the individual garment clo values.

The relationship between clothing insulation and room temperature necessary for a neutral thermal sensation is presented in Figure 1.7 for sedentary occupants, and specified air speed and humidity. Comfortable clothing levels are expressed as a function of *operative temperature*, which is based on both air and mean radiant temperatures. At air speeds of 8 fpm (0.4 m/s) or less and MRT less than 120°F (50°C), the operative temperature is approximately the average of the air and mean radiant temperatures and is equal to the adjusted dry-bulb temperature.

There is no combination of conditions that would satisfy all people all of the time. The *optimum operative temperature,* represented by the middle line in Figure 1.7, is the temperature that satisfies the greatest number of people with a given amount of clothing and specified activity level. The upper and lower *thermal acceptability limits* demarcate a room environment that at least 80 percent of the occupants would find thermally acceptable.

From the 1920s to the early 1970s, energy was abundant and inexpensive. During this period, the preferred amount of clothing worn by building occupants decreased, and correspondingly the preferred temperatures increased from about 68°F (20°C) for winter to the year-round range of 72° to 78°F (22° to 25.5°C). Present conditions, however, make it desirable to minimize energy consumption for providing thermal comfort.

TABLE 1.3   CLO VALUES FOR INDIVIDUAL ITEMS OF CLOTHING

Men		Women	
Clothing	clo	Clothing	clo
Underwear		Underwear	
Sleeveless	0.06	Girdle	0.04
T-shirt	0.09	Bra and panties	0.05
Briefs	0.05	Half slip	0.13
Long underwear, upper	0.10	Full slip	0.19
Long underwear, lower	0.10	Long underwear, upper	0.10
		Long underwear, lower	0.10
Shirt		Blouse	
Light, short sleeve	0.14	Light, long sleeve	0.20
long sleeve	0.22	Heavy, long sleeve	0.29
Heavy, short sleeve	0.25	Dress, light	0.22
long sleeve	0.29	Dress, heavy	0.70
(Plus 5% for tie or turtleneck)			
Vest, light	0.15	Skirt, light	0.10
Vest, heavy	0.29	Skirt, heavy	0.22
Trousers, light	0.26	Slacks, light	0.10
Trousers, heavy	0.32	Slacks, heavy	0.44
		Sweater	
Sweater, light	0.20	Light, sleeveless	0.17
Sweater, heavy	0.37	Heavy, long sleeve	0.37
Jacket, light	0.22	Jacket, light	0.17
Jacket, heavy	0.49	Jacket, heavy	0.37
Socks		Stockings	
Ankle length, thin	0.03	Any length	0.01
thick	0.04	Panty hose	0.01
Knee high	0.10		
Shoes		Shoes	
Sandals	0.02	Sandals	0.02
Oxfords	0.04	Pumps	0.04
Boots	0.08	Boots	0.08
Hat and overcoat	2.00	Hat and overcoat	2.00

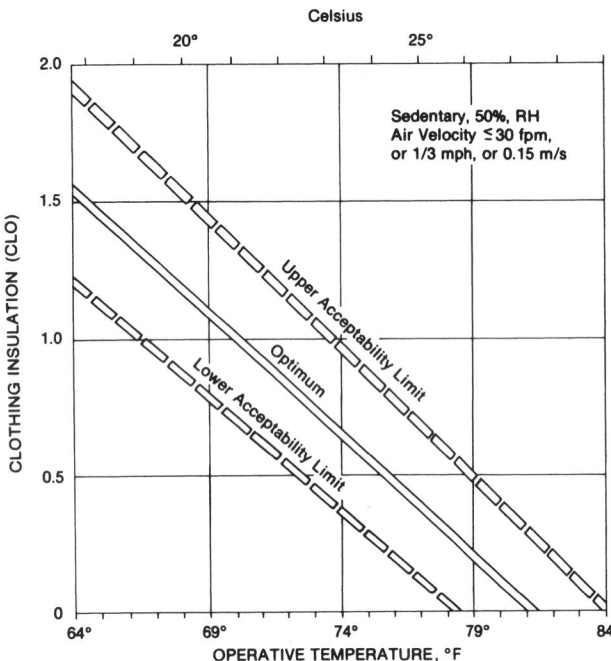

FIGURE 1.7. Clothing level (in clo units) necessary for comfort at different operative temperatures. (Reprinted from *Standard 55* by permission of the American Society of Heating, Refrigerating and Air-Conditioning Engineers, Inc.)

Conditions that are thermally acceptable to at least 80 percent of normally clothed occupants are presented in Figure 1.7. By adjusting clothing as desired, the remaining occupants can satisfy their own comfort requirements. Energy savings can be achieved if the insulation value of clothing worn by people indoors is appropriate to the season and outside weather conditions.

During the summer months, suitable clothing in commercial establishments consists of lightweight dresses, lightweight slacks, short-sleeved shirts or blouses, stockings, shoes, underwear, accessories, and sometimes a thin jacket. These ensembles have insulation values ranging from 0.35 to 0.6 clo.

The winter heating season brings a change to thicker, heavier clothing. A typical winter ensemble—including heavy slacks or skirt, long-sleeved shirt or blouse, warm sweater or jacket, and appropriately warm accessories—would have an insulation value ranging from 0.8 to 1.2 clo. During more temperate seasons, the clothing would likely consist of medium-weight slacks or skirt, long-sleeved shirt or blouse, and so on, having a combined insulation value of 0.6 to 0.8 clo. Figure 1.8 illustrates various clo values.

These seasonal clothing variations of building occupants allow indoor temperature ranges to be higher in the summer than in the winter and yet remain comfortable. In the wintertime, additional clothing lowers the ambient temperature necessary for comfort and for thermal neutrality. Adding 1 clo of insulation permits a reduction in air temperature of approximately 13°F (7.2°C) without changing the thermal sensation. At lower temperatures, however, comfort requires a fairly uniform level of clothing insulation over the entire body. For sedentary occupancy of more than an hour, the operative temperature should not be less than 65°F (18°C).

The insulation of a given clothing ensemble can be estimated by adding up the clo values of the individual items worn, as listed in Table 1.3, and multiplying the sum by 0.82. A rough approximation of the clo value may also be estimated by multiplying each pound of clothing by 0.15 clo (or each kilogram by 0.35 clo).

ENVIRONMENTAL FACTORS

Satisfaction with the thermal environment is a complex, subjective response to many interacting variables. Our perception of comfort is influenced by these variables, which include the characteristics of the physical environment, amount of clothing, activity level, and the demographic character of the subject (age, sex, health, etc.). Researchers have identified the seven major determinants of thermal comfort response:

1. Air (dry-bulb) temperature
2. Humidity
3. Mean radiant temperature
4. Air movement
5. Clothing
6. Activity level
7. Rate of change of any of the above

As any one of these variables changes, the others need to be adjusted to maintain the thermal equilibrium between heat gain and heat loss in order for a person to continue to feel comfortable. The important environmental parameters are temperature, humidity, radiation, and air movement, while the important personal parameters are clothing and activity. The personal parameters have already been covered, so the following discussion concentrates on the environmental parameters.

Dry-Bulb Temperature

Dry-bulb temperature affects the rate of convective and evaporative body heat loss. It is perhaps the most important determinant of comfort, since a narrow range of comfortable

FIGURE 1.8. Illustration of a range of clo values.

temperatures can be established almost independently of the other variables. There is actually a fairly wide range of temperatures that can provide comfort when combined with the proper combination of relative humidity, MRT, and air flow. But as any one of these conditions varies, the dry-bulb temperature must be adjusted in order to maintain comfort conditions.

Temperature drifts or ramps are gradual temperature changes over time. *Drifts* refer to passive temperature changes, while *ramps* are actively controlled temperature changes. People may feel comfortable with temperatures that rise or fall like a ramp over the course of time, even though they would be uncomfortable if some of the temperatures were held constant. Ideal comfort standards call for a change of no more than 1°F/hr (0.6°C/hr) during occupancy, provided that the

temperature excursion doesn't extend far beyond the specified comfort conditions and for very long.

Air temperature in an enclosed space generally increases from floor to ceiling. If this variation is sufficiently large, discomfort could result from the temperature being overly warm at the head and/or overly cold at the feet, even though the body as a whole is thermally neutral. Therefore, to prevent local discomfort, the vertical air temperature difference within the occupied zone should not exceed 5°F (3°C). The *occupied zone* within a space is the region normally occupied by people. It is generally considered to be the first 6 feet (1.8 m) above the floor and 2 feet (0.6 m) or more away from walls or fixed air conditioning equipment. The floor temperature should be between 65° and 84°F (18° and 29°C) to minimize discomfort for people wearing appropriate indoor footwear.

Humidity

Humidity is the amount of water vapor in a given space. The density of water vapor per unit volume of air is called *absolute humidity*. It is expressed in units of pounds (of water) per cubic foot (of dry air). The *humidity ratio* or *specific humidity* is the weight of water vapor per unit weight of dry air; it is given in either grains per pound or pound per pound (kg/kg).

The amount of moisture that air can hold is a function of the temperature. The warmer the air, the more moisture it can hold. The amount of water present in the air relative to the maximum amount it can hold at a given temperature without causing condensation (water present ÷ maximum water holding capability) is known as the *degree of saturation*. This ratio multiplied by 100 is the *percentage humidity*. This percentage is a measure of the dryness of air. Low percentages indicate relative dryness, and high percentages indicate high moisture.

Percentage humidity is often mistakenly called relative humidity. *Relative humidity* (*RH*) is the ratio of the actual vapor pressure of the air-vapor mixture to the pressure of saturated water vapor at the same dry-bulb temperature times 100. Percentage and relative humidity are numerically close to each other but are not identical. The concept of vapor pressure will be discussed later; the intent here is simply to point out the distinction between percentage humidity and RH.

Although human tolerance to humidity variations is much greater than tolerance to temperature variations, humidity control is also important. High humidity can cause condensation problems on cold surfaces and retards human heat loss by evaporative cooling (sweating and respiration). Air already laden with moisture cannot absorb much more from the skin.

The drier and warmer the air, the greater the rate of heat flow from the skin by evaporation of perspiration into the air. However, low humidity tends to dry throat and nasal passages. Low humidity can cause annoying static electric sparks, which can be hazardous in the presence of explosive gases. Carpeting is commercially available with a conductive material such as copper or stainless steel woven into the pile and backing to reduce voltage buildup and help alleviate static electricity problems.

The thermal effect of humidity on the comfort of sedentary persons is small, that is, comfort is maintained over a wide range of humidity conditions. In winter, the body feels no discomfort over a range of RH from 50 percent down to 20 percent. In summer, the tolerance range extends even higher, up to 60 percent RH when the temperature is 75°F; above that, the skin feels sweaty.

Nevertheless, some types of industrial applications, such as textile manufacturing, optical lens grinding, and food storage, maintain an RH above 60 percent because of equipment, manufacturing processes, or product storage requirements. At the other extreme, certain pharmaceutical products, plywood cold pressing, and some other processes require an RH below 20 percent. Hospitals also must carefully control humidity since the level of bacteria propagation is lowest between 50 and 55 percent RH.

Humidity can be expressed as dew point, RH, wet-bulb temperature, or vapor pressure. None of these, however, by itself defines the amount of moisture present without knowledge of one of the others or the coincident dry-bulb temperature. In general, any of these five parameters can be found by means of tables or a psychrometric chart (Figure 2.11) if any two of them are already known.

A common and simple instrument for determining humidity is the sling psychrometer shown in Figure 1.9. It consists of two mercury-filled glass thermometers mounted side by side on a frame fitted with a handle by which the device can be whirled through the air. One of the thermometers has a cloth sock that is wetted. As moisture from the wet sock evaporates into the moving air, the wet-bulb temperature drops.

The drier the air surrounding the sling psychrometer, the more moisture that can evaporate from the sock. This evaporation lowers the wet-bulb temperature accordingly. The greater the difference between the wet-bulb and dry-bulb temperatures (called the *wet-bulb depression*), the lower the RH. A table is normally provided with the device for correlating dry- and wet-bulb temperatures with RH.

Mean Radiant Temperature

As an illustration of the importance of radiant temperature, experiments have been conducted in rooms in which the surface temperatures were controlled. Subjects were surprised to find out that they were warm at air temperatures of 50°F (10°C) when the room surfaces were sufficiently heated and that they were cool at air temperatures of 120°F (49°C) when room surfaces were cooled.

The MRT affects the rate of radiant heat loss from the body. Since the surrounding surface temperatures may vary widely, the MRT is a weighted average of all radiating surface temperatures within line of sight. Two-dimensionally, it can be calculated as follows:

$$MRT = \frac{\Sigma T\theta}{360} = \frac{T_1\theta_1 + T_2\theta_2 + \cdots + T_n\theta_n}{360} \qquad (1.3)$$

where

T = surface temperature

θ = surface exposure angle (relative to occupant) in degrees

Example

The office in Figure 1.10 has insulated walls and glass walls. Regarding the human occupants as cylinders, and ignoring the radiant contributions from the floor and ceiling:

1. What is the MRT for occupant A on a cold winter day when it is 0°F (−18°C) outside? The inside surface temperature of the exterior wall is 67°F (19°C), and the glass temperature is 48°F (9°C). Interior partitions are at 72°F (22°C).

2. What is the MRT for occupant B under the same conditions?

3. What is the MRT for occupant A on a warm summer day when it is 95°F (35°C) outside? The inside surface temperature of the exterior wall is 80°F (27°C), and the shaded glass temperature is 85°F (29°C). Interior partitions are at 77°F (25°C).

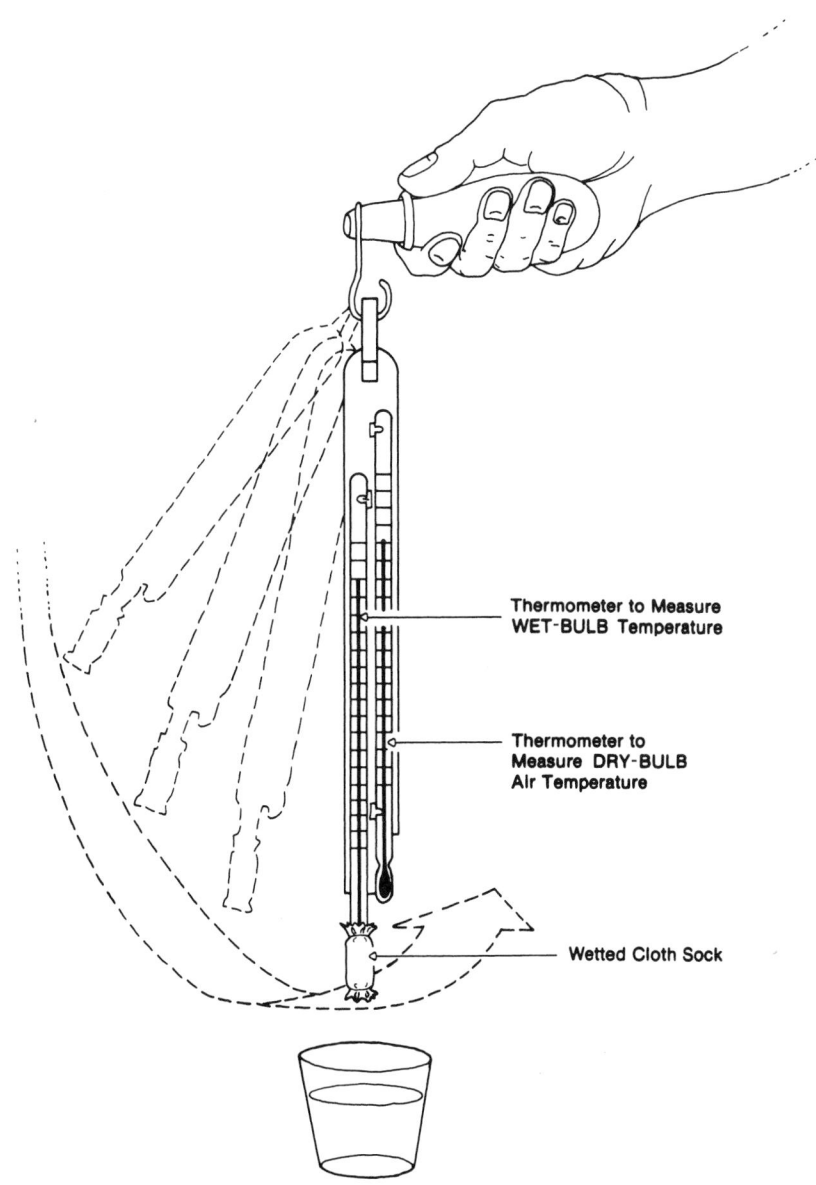

Thermometer to Measure
WET-BULB Temperature

Thermometer to
Measure DRY-BULB
Air Temperature

Wetted Cloth Sock

FIGURE 1.9. Sling psychrometer.

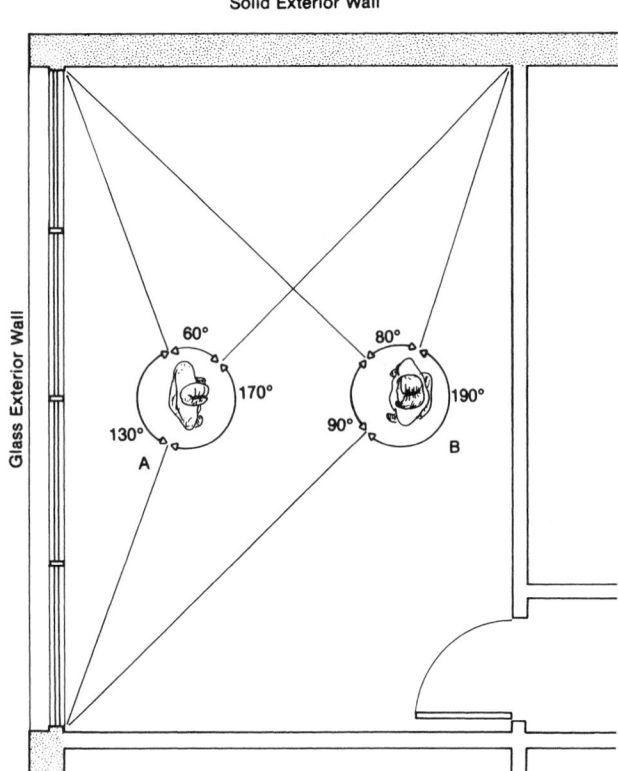

Solid Exterior Wall

Glass Exterior Wall

60°
170°
130°
A

80°
190°
90°
B

FIGURE 1.10. Example of MRT calculation.

Answers (English Units)

1. $\text{MRT} = \dfrac{\Sigma T\theta}{360}$

$= \dfrac{(48 \times 130) + (67 \times 60) + (72 \times 170)}{360}$

$= 62°\text{F}$

2. $\text{MRT} = \dfrac{\Sigma T\theta}{360}$

$= \dfrac{(48 \times 90) + (67 \times 80) + (72 \times 190)}{360}$

$= 65°\text{F}$

3. $\text{MRT} = \dfrac{\Sigma T\theta}{360}$

$= \dfrac{(85 \times 130) + (80 \times 60) + (77 \times 170)}{360}$

$= 80°\text{F}$

Answers (SI Units)

1. $\text{MRT} = \dfrac{\Sigma T\theta}{360}$

$= \dfrac{(9 \times 130) + (19 \times 60) + (22 \times 170)}{360}$

$= 17°\text{C}$

2. $\text{MRT} = \dfrac{\Sigma T\theta}{360}$

$= \dfrac{(9 \times 90) + (19 \times 80) + (22 \times 190)}{360}$

$= 18°\text{C}$

3. $\text{MRT} = \dfrac{\Sigma T\theta}{360}$

$= \dfrac{(29 \times 130) + (27 \times 60) + (25 \times 170)}{360}$

$= 27°\text{C}$

The MRT for office workers should be in the range of 65° to 80°F (18° to 27°C), depending on the clothing worn and the activity. In winter, levels of wall, roof, and floor insulation together with window treatments such as double glazing, blinds, and drapes in accordance with good design practice and current energy codes should generally result in indoor surface temperatures that are no more than 5°F (2.8°C) below the indoor air temperature.

Air Movement

Air motion significantly affects body heat transfer by convection and evaporation. Air movement results from free (natural) and forced convection as well as from the occupants' bodily movements. The faster the motion, the greater the rate of heat flow by both convection and evaporation.

When ambient temperatures are within acceptable limits, there is no minimum air movement that must be provided for thermal comfort. The natural convection of air over the surface of the body allows for the continuous dissipation of body heat. When ambient temperatures rise, however, natural air flow velocity is no longer sufficient and must be artificially increased, such as by the use of fans. Typical human responses to air motion are shown in Table 1.4.

In general, insufficient air motion promotes stuffiness and air stratification. Stratification causes air temperatures to vary from floor to ceiling. When air motion is too rapid, unpleasant drafts are felt by the room occupants. The exact limits to acceptable air motion in the occupied zone are a function of the overall room conditions of temperature, humidity, and MRT, along with the temperature and humidity conditions of the moving air stream.

A noticeable air movement across the body when there is perspiration on the skin may be regarded as a pleasant cooling breeze. When the surrounding surface and room air temperatures are cool, however, it will probably be considered a chilly draft. The neck, upper back, and ankles are most sensitive to drafts, particularly when the entering cool air is 3°F (1.5°C) or more below normal room temperature.

Every 15 fpm increase in air movement above a velocity of 30 fpm is sensed by the body as a 1° temperature drop. Air systems are usually designed for a maximum of 50 fpm in the occupied zone, but that is typically exceeded at the outlet of air registers.

Cool air can impinge on an occupant in two general cases: Air that is warm when introduced into the room may cool off before reaching the occupant, or the air is intended to cool the occupant under overly warm ambient conditions. In either case, when the temperature of the air impinging on an occupant is below the ambient temperature, the individual becomes more sensitive to air motion and may complain of drafts. Therefore, careful attention must be given to air distribution as well as velocity.

The tendency of warm air to rise can greatly affect occupant comfort due to convective air motion, and thus influ-ences the correct placement of the heat source in a room. The consideration of proper air distribution discussed in Chapter 6 affects the placement of outside air openings for natural ventilation, of radiation devices, and of air registers. Air outlet design is determined by the air distribution pattern it is intended to create.

Besides the removal of heat and humidity, another function of air motion in alleviating stuffiness is the dispersion of body odors and air contaminants. The subject of air quality is addressed separately later in this chapter.

THERMAL COMFORT STANDARDS

Thermal Indices

Thermal sensations can be described as feelings of being hot, warm, neutral, cool, cold, and a range of classifications in between. There have been numerous attempts to find a single index—integrating some or all of the environmental factors—that could be used to determine thermal comfort conditions (for a given metabolic rate and amount of clothing). The following are the most common of these indices still in use.

Dry- and Wet-Bulb Temperatures

The simplest practical index of cold and warmth is the reading obtained with an ordinary dry-bulb thermometer. This long-established gauge is fairly effective in judging comfort for average humidity (40 to 60 percent RH), especially in cold conditions.

In the heat, when humidity greatly affects the efficiency of body temperature regulation by sweating, the significance

TABLE 1.4 SUBJECTIVE RESPONSE TO AIR MOTION

Air Velocity		Occupant Reaction
fpm	m/s	
0 to 10	0 to 0.05	Complaints about stagnation
10 to 50	0.05 to 0.25	Generally favorable (air outlet devices normally designed for 50 fpm in the occupied zone)
50 to 100	0.25 to 0.51	Awareness of air motion, but may be comfortable, depending on moving air temperature and room conditions
100 to 200	0.51 to 1.02	Constant awareness of air motion, but can be acceptable (e.g., in some factories) if air supply is intermittent and if moving air temperature and room conditions are acceptable
200 (about 2 mph) and above	1.02 and above	Complaints about blowing of papers and hair, and other annoyances

of the dry-bulb temperature is limited. The wet-bulb temperature represents an improvement over the simple dry-bulb temperature by taking humidity into account.

Globe Thermometer Temperature

The *globe thermometer temperature* combines the effects of radiation and air movement. It uses a 6-inch (150-mm)-diameter black globe. The equilibrium temperature of the globe is a single temperature index describing the combined physical effect of dry-bulb temperature, air movement, and net radiant heat received from the surrounding surfaces.

The globe temperature is an approximate measure of operative temperature. It is usually used as a simple device for determining MRT.

Operative Temperature

Operative temperature is the uniform temperature of an imaginary enclosure in which the occupant would exchange the same heat by radiation and convection as in the actual environment. An alternative definition of operative temperature is an average of MRT and dry-bulb temperatures weighted by the respective radiation and convection heat transfer coefficients.

Humid operative temperature is the uniform temperature of an imaginary environment at 100 percent RH with which the occupant would exchange the same heat by radiation, convection, conductance through clothing, and evaporation as in the actual environment.

New Effective Temperature

Effective temperature is not an actual temperature in the sense that it can be measured by a thermometer. It is an experimentally determined index of the various combinations of dry-bulb temperature, humidity, radiant conditions, and air movement that induce the same thermal sensation. Those combinations that induce the same feeling of warmth or cold are called thermo-equivalent conditions.

The *new effective temperature* (*ET**) of a given space is defined as the dry-bulb temperature of a thermo-equivalent environment at 50 percent RH and a specific uniform radiation condition. The thermo-equivalent heat exchange is based on clothing at 0.6 clo, still air (40 fpm = 0.2 m/s or less), 1-hour exposure time, and a sedentary activity level (approximately 1 met). Thus, any space has an ET* of 70°F (21°C) when it induces a sensation of warmth like that experienced in still air at 70°F (21°C), 50 percent RH, and the proper radiant conditions.

ET* is, in general, a reliable indicator of discomfort or dissatisfaction with the thermal environment. If ET* could be envisioned as a thermometer scale, it would appear as in Figure 1.11.

The Comfort Chart

The comfort chart, shown in Figure 1.12, correlates the perception of comfort with the various environmental factors known to influence it. The dry-bulb temperature is indicated along the bottom. The right side of the chart contains a dew-point scale, and the left side a wet-bulb temperature scale indicating guide marks for imaginary lines sloping diagonally down from left to right. The lines curving upward from left to right represent RHs.

ET* lines are also drawn. These are the sloping dashed lines that cross the RH lines and are labeled in increments of 5°F. At any point along any one of these lines, an individual will experience the same thermal sensation and will have the same amount of skin wetness due to regulatory sweating. Clo levels at which 94 percent of occupants will find acceptable comfort are also indicated.

Notice that the comfort chart in Figure 1.12 is derived from what is called the psychrometric chart. A description of the psychrometric chart and its importance is addressed in Chapter 2.

Two *comfort envelopes* or zones are defined by the shaded regions on the comfort chart—one for winter and one for summer. The thermal conditions within these envelopes are estimated to be acceptable to 80 percent of the occupants when wearing the clothing ensemble indicated. To satisfy 90 percent of the people, the limits of the acceptable comfort zone are sharply reduced to one-third of the above ranges. The zones overlap in the 73° to 75°F (23° to 24°C) range. In this region, people in summer dress tend to be slightly cool, while those in winter clothing feel a slightly warm sensation.

Figure 1.12 generally applies when altitudes range from sea level to 7,000 feet (2,134 m), MRT is nearly equal to dry-bulb air temperature, and air velocity is less than 40 fpm (0.2 m/s). Under these conditions, thermal comfort can be defined in terms of two variables: dry-bulb air temperature and humidity.

Mean radiant temperature is actually just as important as air temperature in affecting comfort. When air movement in an indoor environment is low, the operative temperature is approximately the average of air temperature and MRT. When the MRT in the occupied zone significantly differs from the air temperature, the operative temperature should be substituted for the dry-bulb temperature scale along the bottom of Figure 1.12.

The comfort chart is primarily useful for occupants with

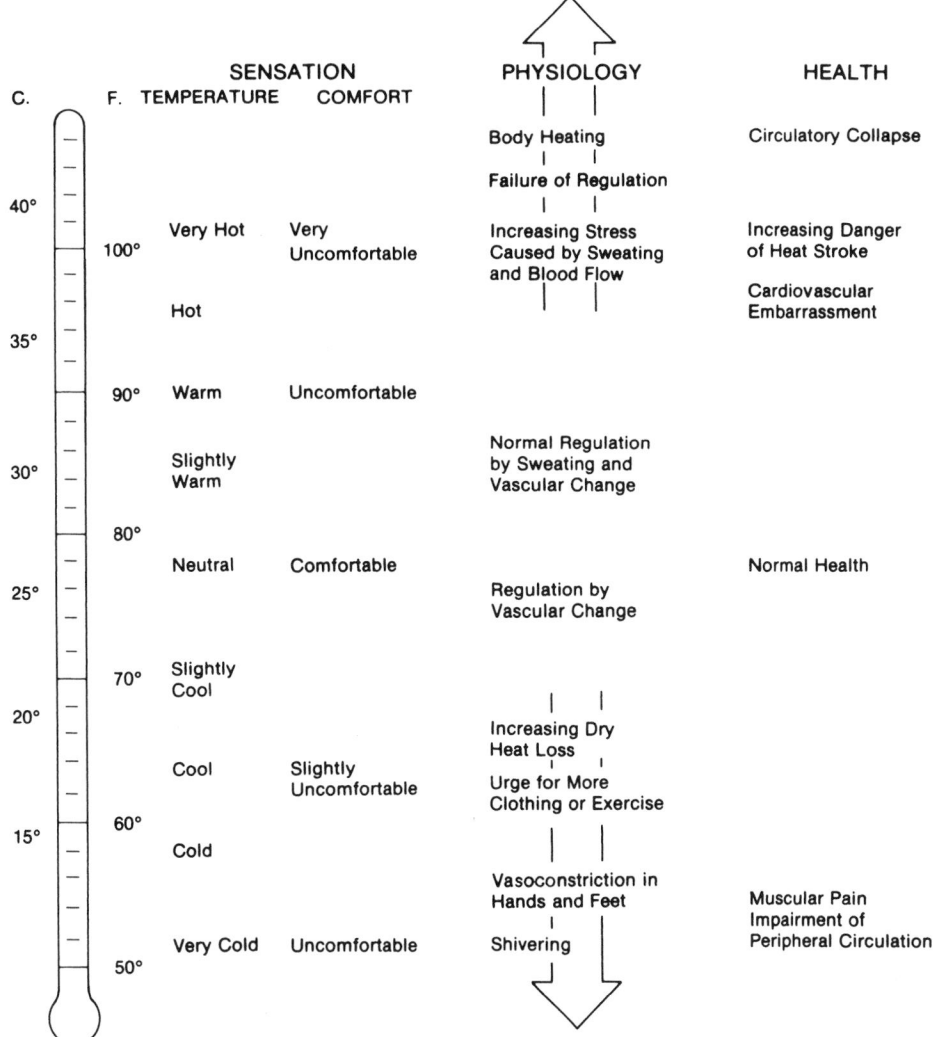

FIGURE 1.11. The ET* scale correlated to physiological reactions, comfort, and health.

a minimum of 1 hour of exposure, 0.6 clo (standard shirt-sleeve indoor office clothing), and a 1-met (seated or sedentary) activity level. It is secondarily useful at higher temperatures to identify when there is a risk of sedentary heat stress. Although the ET* scale is based on 1-hour exposure, data show no significant changes in response with longer exposures unless the limits of heat stress—ET* greater than 90°F (32°C)—are approached. As the ET* lines show, humidity between about 20 and 55 percent RH has only a small effect on thermal comfort. Its effect on discomfort increases with both temperature and the degree of regulatory sweating. Evaporative heat loss near the comfort range is only about 25 percent of the total heat loss. As the temperature increases, this percentage grows until the ambient temperature equals skin (and clothes) temperature, at which point evaporation accounts for 100 percent of the heat loss.

The upper and lower humidity limits on the comfort envelope of Figure 1.12 are based on considerations of respiratory health, mold growth, and other moisture-related phenomena in addition to comfort. Humidification in winter must be limited at times to prevent condensation on cold building surfaces such as windows.

The environmental parameters of temperature, radiation,

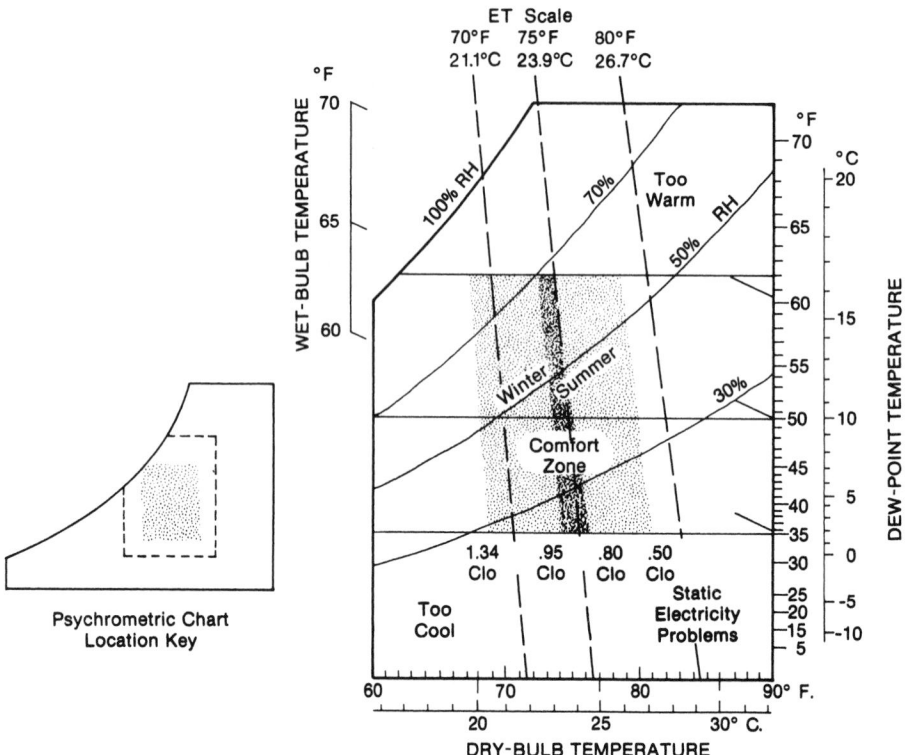

FIGURE 1.12. The comfort chart.

humidity, and air movement necessary for thermal comfort depend upon the occupant's clothing and activity level. The comfort chart was developed from ASHRAE research, which has usually been limited to lightly clothed occupants (0.5 to 0.6 clo) engaged in sedentary activities. The reasoning behind this approach is that 90 percent of people's indoor work and leisure time is spent at or near the sedentary activity level. In line with this rationale, the comfort envelope defined in Figure 1.12 strictly applies only to sedentary and slightly active, normally clothed persons at low air velocities, when the MRT is equal to air temperature. For other conditions, the comfort zone must be adjusted accordingly.

For example, comfort can be maintained at temperatures as low as 68°F (20°C) for an individual wearing a clothing ensemble measuring less than 1.34 clo if he or she gets up and moves around for at least 10 minutes out of every half hour. On the other end of the scale, comfort conditions may be extended upward to 82°F (28°C) with a fan-induced air velocity of 160 fpm (0.8 m/s).

Within the comfort envelope of Figure 1.12, there is no minimum air movement necessary for thermal comfort. However, the maximum allowable air motion is lower in the winter than in the summer. In the wintertime, the average air

movement within the occupied zone should not exceed 30 fpm (0.15 m/s). If the temperature is less than the optimum or neutral sensation temperature, slight increases in air velocity or irregularity of air movement can cause uncomfortable localized drafts. While possibly of little consequence in an active factory, this can create significant problems in professional buildings, religious buildings, and other places where people are seated and wearing light indoor clothing.

In the summer, the average air movement in the occupied zone can go as high as 50 fpm (0.25 m/s) under standard temperature and humidity conditions. Above 79°F (26°C), comfort can be maintained by increasing the average air motion 30 fpm for each °F (0.275 m/s for each °C) of increased temperature up to a maximum of 160 fpm (0.8 m/s). At that point, loose paper, hair, and other light objects start blowing around.

As average steady-state activity increases above the sedentary or slightly active (1.2-met) levels, sweating increases. To maintain comfort, clothing must be adjusted as indicated in Table 1.5, the air motion must be increased, or the operative temperature must be decreased.

The development of reliable thermal comfort indices has

TABLE 1.5 CLOTHING ADJUSTMENTS FOR DIFFERENT ACTIVITY LEVELS

Activity Level (met)	Adjusted Clothing Level[a] (clo)
1	0.6
2	0.4
3	0.3

[a] These clothing adjustments will permit the comfort chart (Figure 1.12) to remain valid for the given activity level.

been very important for the effective control of the thermal environment using no more energy and equipment than is necessary. The five variables affecting comfort for a given activity or room function are dry-bulb temperature, humidity, MRT, air movement, and clothing. Sometimes when one of these conditions is out of the comfort range, adjusting one or more of the other conditions will restore comfort with the addition of little or no additional energy.

The most commonly recommended design conditions for comfort are:

ET* = 75°F (24°C)
Dry-bulb air temperature = MRT
Relative humidity = 40% (20 to 60% range)
Air velocity less than 40 fpm (0.2 m/s)

ASHRAE's Thermal Comfort Standard

ASHRAE's Standard 55, *Thermal Environmental Conditions for Human Occupancy,* describes the combinations of indoor space conditions and personal factors necessary to provide comfort. It addresses the interactions between temperature, thermal radiation, humidity, air speed, personal activity level, and clothing.

The standard recommends conditions that have been found experimentally to be acceptable to at least 80 percent of the occupants within a space. The operative temperature range for building occupants in typical winter clothing (0.8 to 1.2 clo) is specified as 68° to 74°F (20° to 23.5°C). The preferred temperature range for occupants dressed in summer clothes (0.35 to 0.6 clo) is 73° to 79°F (22.5° to 26°C).

These values are based on 60 percent RH, an activity level of approximately 1.2 met, and an air speed low enough to avoid drafts. The standard includes a chart that relates the allowable air speed to room air temperature and the turbulence of the air. For each 0.1 clo of increased clothing insulation, the acceptable temperature range is lowered by 1°F (0.6°C). However, as the temperature decreases, comfort depends more and more on maintaining a uniform distribution of clothing insulation over the entire body, especially the hands and feet. For sedentary occupancy of more than an hour, the operative temperature should not drop below 65°F (18°C).

Design Considerations

The comfort chart is useful for determining design conditions to be met by a building envelope and its *heating, ventilation, and air conditioning* (HVAC) equipment. But considerable judgment must still be exercised if this chart is to be employed as a guide. For example, noticeably uneven radiation from hot and cold surfaces, temperature stratification in the air, a wide disparity between air temperature and MRT, a chilly draft, contact with a warm or cool floor, and other factors can cause local discomfort and reduce the thermal acceptability level of the space. Although a person may feel thermally neutral in general—preferring neither a warmer nor a cooler environment—thermal discomfort may exist if one part of the body is warm and another is cold.

This relates to both the building envelope design and the system for HVAC. A higher air temperature may be necessary to compensate for extensive cold glass areas. Or if large radiant heating panels are contemplated, a lower air temperature might be allowable, since radiant gain to the body will permit greater convective and evaporative losses. Direct sunlight from large windows or skylights *requires* lower room air temperature in order to compensate for the high radiant gain. In order to avoid having to lower the air temperature in the summer, the windows or skylights could be shaded. When radiant surfaces are too cool for comfort, the air temperature must be increased from 0.3° to 1° for every 1° reduction in MRT, depending on room conditions.

Since the comfort chart is the result of observations of healthy, clothed, sedentary subjects, spaces to be conditioned for very active, ill, or nude persons may require considerably different conditions for comfort than those indicated. Table 1.6 suggests some guidelines for winter heating and summer cooling design temperatures in various types of spaces. These should serve as points of departure according to the following discussion. The condition, clothing, and activity level of the occupants, as well as the humidity, MRT, and air motion conditions in the space, must also be taken into consideration.

People engaged in physical work need a lower effective temperature for comfort than do sedentary ones. The greater the activity and the more clothing worn, the lower the effective temperature must be for comfort. Although the latent heat liberated by people engaged in any physical activity

TABLE 1.6 GUIDELINE ROOM AIR TEMPERATURES

Type of Space	°F		°C	
	Summer	Winter	Summer	Winter
Residences, apartments, hotel and motel guest rooms, convalescent homes, offices, conference rooms, classrooms, courtrooms, and hospital patient rooms	74–78	68–72	23–26	20–22
Theaters, auditoriums, churches, chapels, synagogues, assembly halls, lobbies, and lounges	76–80	70–72	24–27	21–22
Restaurants, cafeterias, and bars	72–78	68–70	22–26	20–21
School dining and lunch rooms	75–78	65–70	24–26	18–21
Ballrooms and dance halls	70–72	65–70	21–22	18–21
Retail shops and supermarkets	74–80	65–68	23–27	18–20
Medical operating rooms[a]	68–76	68–76	20–24	20–24
Medical delivery rooms[a]	70–76	70–76	21–24	21–24
Medical recovery rooms and nursery units	75	75	24	24
Medical intensive care rooms[a]	72–78	72–78	22–26	22–26
Special medical care nursery units[a]	75–80	75–80	24–27	24–27
Kitchens and laundries	76–80	65–68	24–27	18–20
Toilet rooms, service rooms, and corridors	80	68	27	20
Bathrooms and shower areas	75–80	70–75	24–27	21–24
Steam baths	110	110	43	43
Warm air baths	120	120	49	49
Gymnasiums and exercise rooms	68–72	55–65	20–22	13–18
Swimming pools	75 or above	75	24	24
Locker rooms	75–80	65–68	24–27	18–20
Children's play rooms	75–78	60–65	24–26	16–18
Factories and industrial shops	80–85	65–68	27–29	18–20
Machinery spaces, foundries, boiler shops, and garages	—	50–60	—	10–16
Industrial paint shops	—	75–80	—	24–27

[a] Variable temperature range required with individual room control.

rises sharply, the liberated sensible heat changes very little. For example, an average man seated at rest in a room at 80°F DB gives off 180 Btuh of sensible heat and 150 Btuh of latent heat, or a total of 330 Btuh (27°C, 53 W, 44 W, and 97 W, respectively). Now, if he engages in light bench work, he will liberate 220 Btuh of sensible heat and 530 Btuh of latent heat (64 W and 155 W). This represents a sensible heat increase of about 20 percent but a latent increase of about 250 percent.

When air temperature is low, convective heat loss increases with air motion associated with increased activity, thereby decreasing the heat load on the body evaporative system and resulting in a wider range of activity before discomfort is felt. The maximum range of activity in which people feel comfortable is therefore achieved by minimizing dry-bulb temperature and RH while compensating with an MRT sufficient to maintain comfort.

Short-term acclimation plays an important part in deter-

mining the best conditions for comfort. In summer, the body's heat-controlling mechanisms become adjusted to the higher outdoor temperature, so a period of time is required for their readjustment to the lower temperature and humidity of a conditioned indoor environment. During acclimatization to the cooler interior, the body experiences a greatly lessened rate of perspiration and the blood vessels recede from the skin surface. A conditioned space maintained to be comfortable for longer periods of occupancy may be too cool for a person who has just left the oppressive heat of the outdoors. If one passes quickly to the hot, humid outside conditions again, a greater sensation of discomfort results than was originally experienced outside because perspiration cannot increase immediately and blood vessels do not move quickly to the skin surface to promptly balance the body heat production and loss.

In spaces of short-time occupancy, such as stores or public lobbies, it is best to maintain a relatively warm, dry climate in which the perspiration rate will change only a small amount. A rough guideline is to reduce the inside-to-outside temperature difference at least 5°F (2.8°C) for occupancy times of less than 1 hour. In this way, the body's thermal control system can easily revert to meet the outside condition by increasing moisture loss and heat loss from the blood vessels already at the surface.

Where the periods of occupancy are long, such as in offices, the body becomes accustomed to the conditioned environment, and the comfort chart may be used to establish appropriate conditions. In many commercial facilities, the temperatures and humidities to be maintained must be a compromise between those necessary to ensure the comfort of employees and those necessary to avoid too great a contrast between indoor and outdoor temperatures for the customers.

Even though restaurants can be classified as short-occupancy environments, the digestion of food increases the metabolic rate, requiring cooler temperatures and/or humidities. In some air-conditioned factories, half-hour lunch periods, lunch rooms, and cafeteria service discourage employees from leaving the air-conditioned building, thus avoiding the need for physiological readjustment.

The economical selection and operation of HVAC equipment require that indoor design conditions involve the smallest change possible from the outside environment while maintaining comfort. In winter, the problem is usually one of raising air temperature and humidity. In summer, the reverse problem exists: Cooling and dehumidification must be accomplished, and the smallest change that can be accommodated in terms of comfort should be sought. Therefore, in winter, conditions near the lower left-hand corner of the comfort zone should be selected, while in summer those near the upper right-hand corner should be chosen.

Other distinctions between winter and summer that must be taken into consideration in selecting design conditions are:

- The MRT of perimeter and top-floor spaces are lower in winter than in summer due to the influence of outdoor air conditions on the building shell.
- Conventions of clothing insulation are different in winter and summer.
- Expectations (psychological and physiological acclimatization) of indoor thermal conditions vary according to season.

Individual Variability

Thermal comfort standards (ASHRAE or any other) usually give a particular value or range of conditions for general application and make no provision for individual differences. However, there are no conditions that will provide comfort for all people. Under the same conditions of temperature and humidity, a healthy young man may be slightly warm, while an elderly woman may be cool. A pedestrian stepping into an air-conditioned store experiences a welcome sense of relief from the blazing summer heat, while the active clerk who has been in the store for several hours may be a bit too warm for comfort. Dancers out on the dance floor may feel quite warm, while their friends seated at a nearby table are comfortable or even slightly cool.

Men generally feel warmer than women on initial exposure to a given temperature but later feel cooler, approaching women's thermal sensations after 1 to 2 hours in the environment. Elderly subjects exposed to conditions of the comfort envelope seem to have responses nearly identical to those of college-age subjects.

Comfort conditions seem independent of the time of day or night. Workers prefer the same thermal environment during a night shift as during the day. Individuals are normally consistent in their thermal preference from day to day, but preferences differ considerably between individuals.

Different countries may have different comfort standards as a result of particular climate extremes and of the relative economics of providing and running heating and cooling systems. And different clothing customs can also be a dominant factor. Even within a country, different conventions of dress for men and women or style preferences among individuals can lead to greatly different comfort requirements.

While different countries have different standards, and geographical location may account for a spread of a few degrees in the most desirable ET*, variations in sensation among individuals within a particular environment tend to be greater than variations due to a difference in geographical location. Apart from general categorical differences, individuals vary greatly in their physiological reaction to their thermal environment. Under the same thermal conditions, some individuals may feel too hot, while others wearing identical clothing feel too cold.

The psychology of *expectancy*—the level of comfort occupants are accustomed to—plays an important part in attitudes towards ambient conditions. People are also sensitive to different aspects of their environment. Moreover, people who believe they are uncomfortable are just as uncomfortable as they would be were they physiologically uncomfortable.

There is no one set of conditions that will satisfy all occupants. Each person has a distinct perception of too hot, too cold, and comfortable. The objective in designing a common thermal environment is to satisfy a majority of occupants and to minimize the number of people who will inevitably be dissatisfied.

TEMPERATURE AND HUMIDITY EXTREMES

As conditions become warmer or colder than the comfort zone, thermal sensations gradually increase and are accompanied by increasing discomfort and strains on the cardiovascular system, respiratory system, and other internal systems involved in bodily thermal regulation. When the thermal exposure is sufficiently intense, pain occurs in conjunction with failure of the body's thermal regulation capability, which can eventually lead to death.

Figure 1.11 relates the various human responses of temperature sensation, comfort, health, and physiology for sedentary subjects over a wide range of thermal conditions represented by the ET* scale. The physiological consequences signify that regulation of the thermal environment is more important than just for providing comfort; there are serious health purposes, too. At the high end of the scale, there is heat stress to contend with and, in extremely cold conditions, a variety of respiratory ailments, incapacity, and heart failure.

Heat Stress

Heat stress occurs when one gains heat faster than one can lose it. When this condition persists without relief, there is the danger that workers, such as those in hot industries, can experience heat prostration.

A person's tolerance to high temperature may be limited if he or she cannot (1) sense temperature, (2) lose heat by regulatory sweating, and (3) move heat by blood flow from the body core to the skin surface where cooling can occur.

Pain receptors in the skin are normally triggered by a skin surface temperature of 115°F (46°C). Although direct contact with a metal surface at this temperature is painful, much higher dry air temperatures can be tolerated since the layer of air at the skin surface provides some thermal insulation. Tolerance times for lightly clothed men at rest in environments with dew points lower than 85°F (30°C) are listed in Table 1.7.

Many individuals find exposure to dry air at 180°F (85°C) for brief periods in a Finnish sauna bath stimulating. Cooling by evaporation of sweat makes short exposures to such extremely hot environments tolerable. However, temperatures of 122°F (50°C) may well be intolerable when the dew-point temperature is greater than 77°F (25°C).

TABLE 1.7   HIGH AMBIENT TEMPERATURE TOLERANCE TIMES

Air DB Temperature	Tolerance Time
180°F (82°C)	almost 50 minutes
200°F (93°C)	33 minutes
220°F (104°C)	26 minutes
239°F (115°C)	24 minutes

Source: Data from ASHRAE *Handbook of Fundamentals 1989.*

The tolerance limit actually represents the body heat-storage limit. The voluntary tolerance limit for an average-sized man is about 2.5°F (1.4°C). Individuals who remain in the heat and continue to work beyond this point increase their risk of heat exhaustion. Collapse can occur at a 5°F (2.8°C) rise in internal body temperature.

The tolerance limit to extreme heat is also affected by the cardiovascular system. In a normal, healthy subject, heart rate and cardiac output increase in response to a hot environment in an attempt to maintain blood pressure and supply blood to the brain. At a heart rate as high as 180 beats per minute, there may not be enough time between contractions to fill the chambers of the heart as completely as required. As the heart rate increases further, cardiac output may drop lower, providing an insufficient amount of blood to the skin for heat loss and, perhaps more importantly, an inadequate blood supply to the brain as less blood is pumped. At some point, the individual will faint or black out from heat exhaustion.

An accelerated heart rate may also result from inadequate blood return to the heart caused by the accumulation of blood in the skin and lower extremities. In this case, cardiac output is limited because not enough blood returns to fill the heart between beats. This frequently occurs when an overheated individual, having worked hard in the heat, suddenly stops working. The muscles suddenly are no longer massaging the blood past the valves in the veins back toward the heart. Dehydration from sweating compounds the problem by reducing the fluid volume in the vascular system.

If the body core temperature goes too high—above 106°F (41°C)—proteins in the delicate nerve tissues in the hypothalmus of the brain, which helps regulate body temperature, may be damaged. Inappropriate vascular constriction, cessation of sweating, increased heat production by shivering, or some combination of these may result. Such *heat stroke* damage is frequently irreversible and carries a high risk of fatality.

A final problem is hyperventilation, or overbreathing, which occurs predominantly in hot-humid conditions. This exhaling of more carbon dioxide from the blood than is desirable can lead to tingling sensations and skin numbness, possibly resulting in vascular constriction in the brain with occasional loss of consciousness.

The Heat Stress Index (HSI) was developed to provide an indication of the severity of the ambient environment on workers. It consists of a ratio of the evaporative heat loss required to maintain thermal equilibrium (E_{req}) divided by the maximum possible evaporation rate in the given environment (E_{max}). When HSI is greater than 100, body heating occurs; when it is less than 0 (negative), body cooling occurs.

The upper limit reported for E_{max} is

$$6 \text{ mets} = 110 \text{ Btuh/ft}^2 = 350 \text{ W/m}^2.$$

For an average-sized man, this corresponds to

$$2,388 \text{ Btuh} = 700 \text{ W} = 602 \text{ kcal/hr},$$

or approximately 17 g/min of sweating.

The physiological and health implications associated with HSI values are described in Table 1.8. HSI serves as a reliable reference for planning worker environments in hot industries. Graphic correlations between dry-bulb tempera-ture, humidity, MRT, air movement, and HSI are available for this purpose.

Response to Extreme Cold

The effect of exposure to extreme cold depends on the main-tenance of the thermal balance. People exposed to cold can lose heat faster than they produce it for only a limited time. Such a "heat debt" results in a drop in body temperature, which is sensed as "acutely uncomfortable" when it reaches 4.7°F (2.6°C). An adequate level of clothing insulation for any given cold environment reduces body heat losses enough to maintain the thermal balance. The extent to which cloth-ing adjustment is practical is dictated by clothing customs and the limits of restricted mobility for the intended activity.

When vascular constriction is unable to prevent body heat loss, a second, automatic, more efficient defense against cold is shivering. It may be triggered by low deep-body tem-perature, low skin temperature, rapid change of skin temper-ature, or some combination of all three. Shivering is usually preceded by an imperceptible increase in muscle tension and by noticeable "goose bumps" produced by muscle contrac-tion in the skin. It begins slowly in small muscle groups, and

TABLE 1.8   PHYSIOLOGICAL AND HEALTH IMPLICATIONS OF 8-HR EXPOSURES TO VARIOUS HEAT STRESSES

Heat Stress Index	Effect on Male Subjects
−20	
−10	Mild cold strain.
0	No thermal strain.
+10	Mild to moderate heat strain. Where a job involves higher intellectual functions, dexterity,
20	or alertness, subtle to substantial decrements in performance may be expected. Little
30	decrement in performing heavy physical work.
40	Severe heat strain involving a threat to health unless subject is physically fit. Break-in
50	period required for men not previously acclimatized. Some decrement in performance
60	of physical work is to be expected. Medical selection of personnel is desirable because these conditions are unsuitable for people with cardiovascular or respiratory impairment, or with chronic dermatitis. These working conditions are also unsuitable for activities requiring sustained mental efforts.
70	Very severe heat strain. Only a small percentage of the population may be expected to
80	qualify for this work. Personnel should be selected: (a) by medical examination, and
90	(b) by trial on the job (after acclimatization). Special measures are needed to ensure adequate water and salt intake. Amelioration of working conditions by any feasible means is highly desirable, and may be expected to decrease the health hazard while increasing job efficiency. Slight "indispositions," which in most jobs would be insufficient to affect performance, may render workers unfit for this exposure.
100	The maximum strain tolerated daily by fit, acclimatized young men.

Source: John Blankenbaker, "Ventilating Systems for Hot Industries," *Heating/Piping/Air Conditioning,* Vol. 54, No. 2, February 1982, p. 58. (Reproduced from the original: H.S. Belding and T.F. Hatch, "Index for Evaluating Heat Stress in Terms of Resulting Physiological Strains," *Heating/Piping/Air Condi-tioning,* August 1955, pp. 129–136.) Used by permission of Reinhold Publishing, a Division of Penton/IPC.

may initially increase total heat production 50 to 100 percent from resting levels. As body cooling increases, additional body segments are involved. Ultimately, violent whole-body shivering may result in a maximum heat production of about six times resting levels, rendering the individual totally ineffective.

There are two means at our disposal for adapting to cold: accustomization and acclimatization. Accustomizing is learning how to survive in cold environments. Acclimatizing is a physiological process in response to long exposure to cold. The physiological changes involve (1) hormonal changes that cause the metabolism of free fatty acids released from fat tissue, (2) maintaining circulatory heat flow to the skin, resulting in a greater sensation of comfort, and (3) improved regulation of heat to the extremities, reducing the risk of cold injury. These physiological changes are generally small and are induced only by repeated uncomfortable exposures to cold. Nonphysiological factors, such as the selection of adequate protective clothing, are generally more useful than dependence on these physiological changes.

Proper protection against low temperatures may be attained either by maintaining high metabolic heat production through activity, or by reducing heat loss with clothing or some other means of controlling the body's microclimate. Spot radiant heat, an "envelope" of hot air for work at a fixed location, or heated clothing are all practical possibilities. The extremities, such as fingers and toes, pose more of a problem than the torso, because, as thin cylinders, they lose heat much more rapidly and, especially in the case of fingers, are difficult to insulate without hampering mobility. Vascular constriction reduces circulatory heat input to extremities by over 90 percent.

As far as insulating material for protective clothing is concerned, radiation-reflective materials can be quite effective, especially if they seal the body from cold air currents at the same time. Otherwise, insulation is primarily a function of clothing thickness. The greater the fiber thickness, the greater the thickness of the insulating trapped air.

AIR QUALITY AND QUANTITY

Besides the thermal conditions of an environment, comfort and health also depend on the composition of the air itself. For example, people feel uncomfortable when the air is odorous or stale.

The quality of air in a space can even seriously affect its ability to support life. Under heavy occupancy of a space, the concentration of carbon dioxide can rise to deleterious levels. In addition, excessive accumulations of some air contaminants become hazardous to both plants and animals.

Air Contaminants

Air normally contains both oxygen and small amounts of carbon dioxide (0.03 percent), along with varying amounts of particulate materials referred to as *permanent atmospheric impurities*. These materials arise from such natural processes as wind erosion, evaporation of sea spray, and volcanic eruption. The concentrations of these materials in the air vary considerably but are usually below the concentrations caused by human activity.

Air composition can change drastically. In sewers, sewage treatment plants, tunnels, and mines, the oxygen content of air may become so low that it cannot support human life. Concentrations of people in confined spaces, such as theaters, require the removal of carbon dioxide given off by respiration and replacement with oxygen.

At atmospheric pressure, oxygen concentrations of less than 12 percent or carbon dioxide concentrations greater than 5 percent are dangerous even for short periods. Smaller deviations from normal concentrations can be hazardous under prolonged exposures.

Artificial contaminants are numerous, originating from a variety of human activities. Contaminants in the indoor environment, of which tobacco is a prime example, are of particular concern to building designers.

Air contaminants can be particulate or gaseous, organic or inorganic, visible or invisible, toxic or harmless. Loose classifications are (1) dust, fumes, and smoke, which are chiefly *solid* particulates (although smoke often contains liquid particles); (2) mist and fog, which are *liquid* particulates; and (3) vapors and gases, which are *nonparticulates*.

Dust consists of solid particles projected into the air by natural forces such as wind, volcanic eruption, or earthquakes, or by human activities. *Fumes* are solid airborne particles usually 100 times smaller than dust particles, commonly formed by condensation of vapors of normally solid materials. Fumes that are permitted to age tend to agglomerate into larger clusters. *Smoke* is made up of solid or liquid particles about the same size as fumes, produced by the incomplete combustion of organic substances such as tobacco, wood, coal, and oil. This class also encompasses *airborne living organisms* which range in size from submicroscopic viruses to the larger pollen grains. Included are bacteria and fungus spores, but not the smallest insects.

Mist is defined as very small airborne droplets of a liquid that are formed by atomizing, spraying, mixing, violent chemical reactions, evaporation, or escape of a dissolved gas when pressure is released. Sneezing expels or atomizes very small droplets containing microorganisms that become air contaminants. *Fog* is very fine airborne droplets usually formed by condensation of vapor. Fogs are composed of droplets that are smaller than those in mists, but the distinction is insignificant, and both terms are commonly used to indicate the same condition.

Tobacco Smoke

Tobacco smoke is the most common indoor pollutant. A growing percentage of the public is finding it extremely objectionable. Smokers and nonsmokers are segregated in public restaurants, on airplanes, and on many other modes of public transportation. But isolation of the smoker does not solve the problem as long as the nonsmoker breathes the same air circulated by a common air handling system. It only results in better dilution.

Tobacco smoke is an extremely complex mixture of combustion products that consist of particulate matter (visible smoke) and gaseous contaminants. The gas constituents of tobacco smoke include nitrogen dioxide, formaldehyde, hydrogen sulfide, hydrogen cyanide, ammonia, and nicotine. All of these gases are irritants, carcinogens, or toxic substances in sufficient concentration. These gases combine to form the "odor" portion of tobacco smoke and can be extremely noxious at high concentrations. The particulate component of tobacco smoke contains tens of trillions of fine particles of tar and nicotine per cigarette.

Formaldehyde

Formaldehyde, a colorless, strong-smelling gas, is used in the manufacture of synthetic resins and dyes, and as a preservative and disinfectant. Carpeting and panelboard in newly constructed or renovated buildings may give off small quantities of formaldehyde gas for many years.

At full strength, formaldehyde gas is lethal. In buildings, it can reach concentrations that may cause irritation, discomfort, and with long exposure, more severe effects.

Aeroallergens

Hay fever, asthma, eczema, and contact dermatitis are allergic disorders. An allergic person has an altered capacity to react to substances such as foods, dust, pollens, bacteria, fungi, medicines, and other chemicals, and exposure to them may result in adverse symptoms. But environmental conditions such as dust, irritating gases, or changes of temperature and humidity can precipitate asthmatic attacks in allergic individuals, even without exposure to specific allergens.

Fortunately, pollen grains discharged by weeds, grasses, and trees, which are responsible for hay fever, are even larger than ordinary dust particles and can be readily filtered out of the air. Most grains are hygroscopic, varying in size and weight with the humidity.

Airborne Microorganisms

Bacteria and other airborne microorganisms can cause infections and diseases in humans. Yeasts and molds in the air can contaminate many food products and can cause expensive damage in the food industry. Microorganisms often become airborne by attaching themselves to dust particles which are then suspended in the air by nearby activity.

The most successful methods of controlling airborne microorganisms are *dust control, air sterilization,* and carefully designed *ventilation.* In critical areas such as operating rooms, special down-draft ventilation is employed. Air sterilization by in-duct radiation with ultraviolet light is sometimes used in premature nurseries or laboratories but is difficult to maintain. In general, wherever the risk of airborne infection is high, such as medical-treatment and research areas, air movement must be designed so as to avoid moving potentially infected air into uninfected areas. This is accomplished by using outdoor air, avoiding air movement from one room to another, and special cleaning of recirculated air.

Radioactive Air Contaminants

Radioactive contaminants may be particulate or gaseous and are physically similar to any other air contaminants. Many radioactive materials would be chemically toxic if present in high concentrations, but it is their radioactivity that generally necessitates limiting their concentration in the air.

A distinction should be made between radioactive materials and the radiation emitted from them. Radioactive particulates and gases can be removed from the air by filters and other devices, whereas the gamma radiation from such material is able to penetrate solid walls. The hazardous effects of most radioactive air contaminants occur when they are taken into and retained inside the body.

Special problems make radioactive materials distinctive among contaminants. For example, radon and other radioactive gases may decay, producing radioactive particulates; in some materials the opposite occurs. The contaminants may generate enough heat to damage filtration equipment, or they may spontaneously ignite. The concentrations at which most radioactive materials are hazardous are far lower than those of other materials, so special electronic instruments must be used to detect them.

For sensitive industrial plants, such as those in the photographic industry, contaminants must be prevented from entering the plant. If radioactive materials are handled inside a plant, the contaminant should be removed from the air as close to the source as possible before the air is released outdoors. Where X-ray and radiation therapy equipment is used, the room itself and the air ducts to and from the room must be lined with lead to contain the radiation.

Odors

Odors are important in the enjoyment of food, in the attraction of one person to another, and in a person's evaluation of whether the surroundings are clean and well maintained. Malodors may signal poor maintenance, if not actually unsafe conditions, such as spoiled food or the presence of natural gas or other toxic substances.

Any given odor is not always desirable. What is pleasant in one context may be objectionable in another; what appeals to one person may nauseate another. Although one odor may seem inherently more pleasant than another, all odors become unpleasant at high levels of perceived intensity. Odor control must therefore be directed at the general attenuation of odor levels.

Our olfactory sensitivity often makes it possible to detect and thus eliminate potentially harmful substances at concentrations below dangerous levels. While foul-smelling air is not necessarily unhealthy, the sheer unpleasantness of an odor can initiate symptoms such as nausea, headache, and loss of appetite, even if the air is not toxic. In such cases, the stronger the odor, the more intense the symptoms. Even a mild but recognizable odor may arouse uneasiness among a room's occupants.

Discomfort in occupied areas may result from the intake of outdoor air containing automobile exhaust, furnace effluents, industrial effluents, or smog. Industrial spaces may have odors from chemical products, such as printing ink, dyes, and synthetics, as well as from manufactured products. Offices, assembly rooms, and other enclosed, densely occupied spaces can contain objectionable body odors and tobacco smoke. Smoking produces a large variety of odorous compounds and irritants, and it also impairs visibility. Odors can result from wetted air-conditioning coils or certain metals and coatings used on the coils. Odors may be caused by linoleum, paint, upholstery, rugs, drapes, or other room furnishings. Food, cooking, and decomposition of animal and vegetable matter are also frequent sources.

Spaces frequently retain occupancy odors from people long after occupancy has ceased, since odors accumulate on the furnishings during occupancy and are later released to the space. Cotton, wool, rayon, and fir wood have each been found to have odor-absorbing capabilities, and each gives off odors at a varying rate, depending on temperature and RH.

Odors emanating from these sources can be decreased by lowering temperature and RH. Where the odor sources are the materials themselves, as in the case of linoleum, paint, rubber, and upholstery, a reduction in RH decreases the rate of odor release.

Odor *perception* is also affected by temperature and humidity, but in the opposite direction. The perception of smoke, cooking, body odors, and many other vapors decreases as humidity *increases*. This effect is more pronounced for some odorants than for others. An increase in temperature slightly reduces some odors, such as cigarette smoke.

Adaptation to odors takes place rapidly during the initial stages of exposure. While the perceptible odor level of cigarette smoke decreases with exposure time, irritation to the eyes and nose generally increases. The irritation is greatest at low RH. In order to keep odor perception and irritation at a minimum, the optimum RH for a conditioned space is 45 to 60 percent.

When the concentration of an odorous vapor is so low as to be imperceptible, the air is said to be *odor-free*. Therefore, to abate an odor condition, it is necessary to (1) remove the offending gases or vapors or adequately reduce their concentration below perceptible limits, or (2) interfere with the perception of the odor.

Odor may be removed from the air by physical or chemical means. The available methods include ventilation with clean outdoor air; air washing or scrubbing; adsorption by activated charcoal or other materials; chemical reaction; odor modification; oxidation; and combustion. Equipment is available to accomplish any of these methods.

Ventilation is the exhausting of space air containing objectionable gaseous odors, irritants, particulates that obscure vision, and toxic matter and replacing this air with clean outdoor air. Except in mild climates, introducing outdoor air adds substantially to heating and cooling needs, so it is desirable to minimize such ventilation. Moreover, increasing air pollution has reduced outdoor air quality below the minimum acceptable limits in some areas, and thus the outdoor air cannot be used directly for ventilation.

Adsorption is the physical attaching of a gas or vapor onto an activated solid substrate. While *ab*sorption can be visualized as the action of a sponge, *ad*sorption is more like the action of a magnet upon iron filings.

Theoretically, odors can be eliminated by oxidation. However, while oxidizing gases such as ozone and chlorine can neutralize odors in water, the concentrations of these gases required for air deodorization would be so high that they would be toxic to human occupants. The primary effect of ozone generators as used for deodorizing is to reduce the sensitivity of the occupants' sense of smell rather than reduce the actual odor concentration.

An alternative to removal or destruction of objectionable odorants is to introduce other chemicals that will (1) modify the perceived odor quality to make it more acceptable or (2) reduce the perceived malodor intensity to an acceptable level. The first alternative is known as *odor masking;* the second is called *counteraction.* A mixture of odorants—malodorant and "counteracting agent"—will generally smell less intense than the sum of the intensities of the unmixed components. Given the correct proportions, the mixture may, in fact, smell less intense than the malodorant alone. This is the objective of counteraction.

General guidelines on the use of counteractants are:

1. They should not be used as a substitute for ventilation.
2. They usually work best against weak malodors and should be used only after ventilation or some other procedure has reduced the malodor to a low level.
3. They are usually quite odorous themselves, and are formulated to mask as well as to counteract.
4. Since quality masking is one of their functions, the perceived quality of the counteractant should be chosen with care.
5. The counteractant should not be permitted to mask or otherwise interfere with warning odors, such as the odor of leaking natural gas.

Ventilation

The concentration of indoor air contaminants and odors can be maintained below levels known to impair health or cause discomfort, by the controlled introduction of fresh air to exchange with room air. This is known as ventilation.

Humans require fresh air for an adequate supply of oxygen, which is necessary for metabolism of food to sustain life. Carbon and hydrogen in foods are oxidized to carbon dioxide and water, which are eliminated by the body as waste products. The rate at which oxygen is consumed and carbon dioxide generated depends on physical activity, and the ratio of carbohydrates, fats, and protein eaten.

It was once thought that the carbon dioxide content of the air from respiration was responsible for the condition of stale air experienced in places of concentrated occupancy. Actually, the sense of staleness is primarily a result of the buildup of heat, moisture, and unpleasant odors given off by the body.

While high carbon dioxide levels are responsible for headaches and loss of judgment, acute discomfort from odors, and health problems from other sources of air conta-

mination usually occurs long before the carbon dioxide concentration rises that high. The generally accepted safe limit is a 0.5 percent concentration for healthy, sedentary occupants eating a normal diet. This corresponds to 2.25 CFM (cubic feet per minute) of outdoor air per person where the outdoor air contains a normal proportion of carbon dioxide.

To allow for individual variations in health, eating habits, and activity level, and the presence of other air contaminants, with a margin of safety, ASHRAE Standard 62.1, *Ventilation for Acceptable Indoor Air Quality,* specifies a minimum of 15 CFM (8 L/s) of outdoor air per person.

Every building, without exception, contains one or more of the following contaminants: asbestos, benzene, carbon monoxide, chlordane, formaldehyde, lead, mercury, nitrogen dioxide, ozone, particulates, radon, and sulfur dioxide. For example, carpeting, draperies, upholstery fabrics, some insulation products, shelving, particle board, and laminated woods are manufactured in part from formaldehyde-based materials. Formaldehyde then outgases from these building materials and other products. Other objectionable and even dangerous contaminants are tobacco smoke and ammonia fumes emitted from blueprint machines. In addition, human occupants naturally generate carbon dioxide, water vapor, particulates, biological aerosols containing infectious and allergenic organisms, and other contaminants.

Outdoor air ventilation requirements for various indoor spaces are summarized in ASHRAE Standard 62.1, and reprinted in Table B6.3 in Appendix B. In most cases, the predominant contamination is presumed to be in proportion to the number of persons in the space, and for these applications, ventilation rates are presented in CFM (L/s) per person. Where ventilating rates are listed as CFM/ft^2 ($L/s·m^2$), contamination is primarily due to other factors. The table also lists suggested occupant densities for use in determining ventilation rates when the actual number of occupants is not known.

These ventilation rates are believed to provide a generally acceptable level of carbon dioxide, particulates, odors, and other contaminants common to those spaces when the outdoor air is of an acceptable quality. Where human carcinogens or other harmful contaminants are suspected of being present in the occupied space, other relevant standards such as those of the Occupational Safety and Health Administration (OSHA) or the Environmental Protection Agency (EPA) must supersede the rates presented in Table B6.3.

The ventilation rate actually needed to provide satisfactory air quality depends on ventilation effectiveness. Ventilation effectiveness depends on the design, performance, and location of the supply outlet and return inlet. When some of the supply air flows directly to the return inlet without passing through the occupied zone of a room, the effectiveness of the ventilation is reduced. When outdoor air passes through the system without ever being used to dilute contaminants in the occupied zone, the energy used to heat, cool, and circulate the air is also wasted.

When spaces are unoccupied, ventilation is generally not required unless it is necessary to prevent an accumulation within the space of contaminants which would be hazardous to returning occupants or injurious to the contents or structure.

In some cases, outdoor air is so polluted as to be unacceptable as ventilation air. If the concentration of contaminants in the outdoor air exceeds acceptable levels, the air must be cleaned or treated before being introduced into the space. Elaborate ventilating systems with recirculation and treatment may considerably reduce pollution below prevailing ambient levels. Except for certain critical areas like hospital operating rooms, cleaning and recirculation of air within a building is permitted as long as the concentration of all contaminants of concern is maintained within specified acceptable levels.

Standard 62.1 includes an alternative performance method of providing acceptable indoor air quality. It specifies maximum allowable contaminant concentrations which can be tested for and corrected by an air-cleaning system. Air-cleaning systems can reduce both particulate and gaseous contaminants. This approach may be taken if, for example, the introduction of outdoor air must be curtailed to conserve energy or because of unacceptable outside air quality.

Normally, pollution levels inside buildings are slightly lower than those concurrently found outdoors. Peak concentrations indoors are lower and occur somewhat later than those outdoors. However, indoor production of carbon dioxide, formaldehyde, radon, tobacco smoke, and other contaminants can make indoor levels higher than ambient outdoor levels.

In recent years, illnesses due to high concentrations of indoor contaminants have received increased attention and emphasis. Some of these pollutants, such as asbestos dust, radon gas, benzines, chlorinated hydrocarbons, vapors of formaldehyde and mercury, paper copying and ink fumes, outgasing from plastics, and fire-retardant chemicals, are generated by the building, its contents, and its site. Some are produced by unvented indoor combustion or sewer gases. Some are released into the indoor environment by applications of adhesives, paint, cleaning compounds, and maintenance activities. Still others are the products of tobacco smoking, substances given off by human bodies, and fumes arising from food preparation. These contaminants are commonly found in most buildings, but the problem arose because of inadequate ventilation.

Prior to 1973, inexpensive energy for heating and cooling enabled designers to introduce high quantities of outdoor air into buildings to replace or dilute dangerous or simply odorous gases. Since then, building designers and code officials have responded to higher fuel costs by tightening up structures, that is, increasing insulation, reducing air leakage through the building envelope, and decreasing mechanical ventilation rates.

Air quality was initially ignored, and the objective was simply to minimize air exchange between indoors and outdoors. This led to the point where the concentration of contaminants rose in some cases to levels harmful to public health.

This became known as the *indoor air quality (IAQ) problem* or *sick building syndrome*. In response, research studies and experiments established levels of outdoor air ventilation needed to achieve acceptable indoor air quality.

Designers are now trying to achieve a balance between energy conservation and acceptable IAQ. Some means of reducing energy consumption while still providing adequate ventilation are:

- Heat recovery between exhaust and make-up outside air
- Tracking occupancy (providing only the ventilation necessary for the current number of people in the building)
- Opening outside air dampers 1 hour after occupancy (where permitted) to take advantage of the dilution capacity of large volumes of room air, and the natural dissipation of contaminants during long vacant periods
- Segregating smokers (on a separate, higher-ventilated HVAC system), preferably located in perimeter areas

When ventilation air is brought in from outdoors, it may be by either mechanical (active) or natural (passive) means. In spaces with low-density occupancy and exterior walls, sufficient outside air may be introduced by leakage through doors and windows. Interior zones and heavily populated areas, however, require the introduction of ventilated air by mechanical equipment. Also, if the outside air needs to be conditioned, it should be passed through the conditioning equipment first and then delivered to the space.

Whether ventilation air is brought in from outside or is predominantly recirculated, it must still be introduced at a rate sufficient to remove objectionable odors and contaminants from the space. With proper air distribution, the motion of the ventilation air blends with the room air, creating a unified thermal condition. And as the air gently passes by the occupants, it carries away heat, humidity, and odors given off by the body. This should result in a feeling of freshness.

SUMMARY

The human body is essentially a constant-temperature device. Heat is continuously produced by bodily processes and dissipated in an automatically regulated manner to maintain the body temperature at its correct level despite variations in ambient conditions. In terms of physiology, the experience of comfort is the achievement of thermal equilibrium with the minimum amount of body regulation.

The human body normally rejects heat to the environment using evaporative cooling (sweating) and the heat transfer mechanisms of radiation, convection, and conduction (see Chapter 2 for further discussion of these heat transfer mechanisms).

The relative roles of these heat transfer mechanisms are determined by the individual's metabolism, clothing, and activity level, as well as by the surrounding environmental conditions of radiation, humidity, air temperature, and air motion. The acceptable value of each of these features is not fixed, but can vary in conjunction with one or more of the others. It is possible for the body to vary its own balance of losses, for example, through increased sweating; or the insulating value of the clothing worn can be varied to a limited degree to compensate for conditions beyond the body's ability to make its own adequate adjustment.

The comfort of a given individual is affected by many variables. Health, age, activity, clothing, gender, food, and acclimatization are all determining factors of the comfort conditions for any particular person. Since these factors will not be identical for all people, room conditions are provided under which a majority of the expected occupants will feel comfortable.

In addition to its thermal climate, the air quality of each indoor environment affects the sense of comfort. Air may contain a variety of possible contaminants that may or may not be harmful to human occupants. Along with possible toxicity, contaminants can impart odors to the space, and the toxicity and odor intensity are often related to the concentration of impurities. To reduce health hazards and eliminate objectionable odors, concentrations of impurities are controlled either by dilution from outside air ventilation or by treatment of the air in an air conditioning system, or by both. Proper fresh air distribution throughout a space is important for mixing the air in order to achieve acceptable overall quality, and

for keeping the air steadily moving around the occupants to carry away heat, moisture, and odors generated by them.

KEY TERMS

Comfort conditioning
Process air conditioning
Comfort
Body heat balance
Evaporative cooling
Thermal comfort
Heat
Sensible heat
Specific heat
Thermal radiation
Conduction
Convection
Latent heat
Latent heat of fusion
Latent heat of vaporization
Radiant heat
Enthalpy
British thermal unit (Btu)
Btuh (Btu/hr)
Joule
Watt (W)
Dry-bulb (DB) temperature
Wet-bulb (WB) temperature

Dew-point temperature
Mean radiant temperature (MRT)
Radiation
Met
Clo
Operative temperature
Optimum operative temperature
Thermal acceptability limit
Drift
Ramp
Occupied zone
Absolute humidity
Humidity ratio (specific humidity)
Degree of saturation
Percentage humidity
Relative humidity (RH)
Wet-bulb depression
Globe thermometer temperature

Operative temperature
New effective temperature (ET*)
Comfort envelope
Heating, ventilation, and air conditioning (HVAC)
Expectancy
Heat stroke
Permanent atmospheric impurities
Dust
Fumes
Smoke
Airborn living organisms
Mist
Fog
Dust control
Air sterilization
Ventilation
Odor masking
Counteraction
Indoor air quality (IAQ) problem (sick building syndrome)
ASHRAE

STUDY QUESTIONS

1. Explain the difference between heat and temperature. What are the conventional (U.S.) and SI units for each?

2. What are the numerical temperature and humidity limits to the ASHRAE comfort zone for sedentary, lightly clothed occupants? What are the air motion and MRT presumptions for these limits?

3. What is the MRT in the horizontal plane for an occupant in a space with all glass on one side at 45°F (7°C), and symmetrical interior partitions on the other side at 70°F (21°C), disregarding the effect of lighting and other surfaces in the space?

4. At what ET* is there a risk of sedentary heat stress?

5. What are the three principal functions of ventilation relating to carbon dioxide concentration; odorous and noxious contaminants; and body heat, moisture, and odors?

6. Name five common interior finish products that outgas formaldehyde.

7. What effect does smoking have on ventilation requirements for comfort and health?

8. List two reasons for restricting the introduction of outdoor air for ventilation.

BIBLIOGRAPHY

American Society of Heating, Refrigerating, and Air-Conditioning Engineers, Inc. *ASHRAE Handbook of Fundamentals.*

American Society of Heating, Refrigerating, and Air-Conditioning Engineers, Inc. *Standard 55—Thermal Environmental Conditions for Human Occupancy,* (*ANSI Approved*).

American Society of Heating, Refrigerating, and Air-Conditioning Engineers, Inc. *Standard 62.1—Ventilation for Acceptable Indoor Air Quality,* (*ANSI Approved*).

American Society of Heating, Refrigerating, and Air-Conditioning Engineers, Inc. *Standard 62.2—Ventilation and Acceptable Indoor Air Quality in Low-Rise Residential Buildings* (*ANSI Approved*).

Bardana, Emil J., Jr., and Anthony Montanaro (editors). *Indoor Air Pollution and Health.* Marcel Dekker, Inc., 1997.

Bas, Ed. *Indoor Air Quality,* 2nd ed. Marcel Dekker, Inc., 2003.

Bearg, David W. *Indoor Air Quality and HVAC Systems.* CRC Press, 1993.

Brooks, Bradford O., and William F. Davis. *Understanding Indoor Air Quality.* CRC Press, 1991.

Fourt, Lyman Edwin. *Clothing Comfort and Function.* Marcel Dekker, Inc., 1970.

Gammage, Richard B., and Barry A. Berven (editors). *Indoor Air and Human Health,* 2nd ed. CRC Press, 1996.

Godish, Thad. *Indoor Air Pollution Control.* CRC Press, 1989.

Givoni, Baruch. *Passive Low Energy Cooling of Buildings.* John Wiley & Sons, Inc., 1994.

Godish, Thad. *Sick Buildings: Definition, Diagnosis and Mitigation.* CRC Press, 1994.

Givoni, Baruch. *Climate Considerations in Building and Urban Design.* Wiley, 1998.

Godish, Thad. *Indoor Environmental Quality.* CRC Press, 2000.

Hansen, Shirley J., and H. E. Burroughs. *Managing Indoor Air Quality,* 3rd ed. 2004.

Hyde, Richard. *Climate Responsive Design: A Study of Buildings in Moderate and Hot Humid Climates.* Brunner-Routledge, 2000.

Kay, Jack G., George E. Keller, and Jay F. Miller. *Indoor Air Pollution: Radon, Bioaerosols, and VOCs.* CRC Press, 1991.

Koch-Neilsen, Holger. *Stay Cool: A Design Guide for the Built Environment in Hot Climates.* Earthscan Publications, 2002.

Kroger, Detlev G. *Air-Cooled Heat Exchangers and Cooling Towers: Thermal-Flow Performance Evaluation and Design, Vol. 1.* Penwell Corporation, 2004.

Kroger, Detlev. *Air-Cooled Heat Exchangers and Cooling Towers: Thermal-Flower Performance Evaluation and Design, Vol. 2.* Pennwell Books, 2004.

Lechner, Norbert. *Heating, Cooling, Lighting: Design Methods for Architects,* 2nd ed. John Wiley & Sons, Inc., 2001.

Leslie, G. B., and F. W. Lunau (editors). *Indoor Air Pollution: Problems and Priorities.* Cambridge University Press, 1994.

Lstiburek, Joseph. *Builder's Guide to Mixed Climates: Details for Design and Construction.* Taunton Press, 2000.

Lstiburek, Joseph. *Builder's Guide to Cold Climates: Details for Design and Construction.* Taunton Press, 2000.

Mahnke, Frank H., and Rudolf H. Mahnke. *Color and Light in Man-made Environments.* John Wiley & Sons, Inc., 1993.

Moore, Suzi, Nora Burba Trulsson, Suzi Moore McGregor (Photographer), and Terrence Moore. *Living*

Homes: Sustainable Architecture and Design. Chronicle Books, 2001.

Odom, J. David (editor), and George Dubose. *Commissioning Buildings in Hot Humid Climates: Design and Construction Guidelines.* 2000.

O'Reilly, James T., Philip Hagan, and Ronald Gots. *Keeping Buildings Healthy.* John Wiley & Sons, Inc., 1998.

Reynolds, John S. *Courtyards: Aesthetic, Social, and Thermal Delight.* John Wiley & Sons, Inc., 2001.

Roaf, Sue, David Crichton, and Fergus Nicol. *Adapting Buildings and Cities for Climate Change: A 21st Century Survival Guide.* Architectural Press, 2004.

Salmon, Cleveland. *Architectural Design for Tropical Regions.* John Wiley & Sons, Inc., 1999.

Samet, Jonathan M., and John D. Spengler. *Indoor Air Pollution: A Health Perspective* (*The Johns Hopkins Series in Environmental Toxicology*). Johns Hopkins University Press, 1991.

Sanders, Mark S., and Ernest J. McCormick. *Human Factors in Engineering and Design,* 7th ed. McGraw-Hill Book Company, 1993.

Shurcliff, William A. *Air-to-Air Heat Exchangers for Houses: How to Bring Fresh Air into Your Home and Expel Polluted Air, While Recovering Valuable Heat.* Brick House Publishing Company, 1982.

Shurcliff, William A. *Superinsulated Houses and Air-to-Air Heat Exchangers.* Brick House Publishing Company, 1988.

Wadden, Richard A., and Peter A. Scheff. *Indoor Air Pollution: Characterization, Prediction, and Control.* John Wiley & Sons, Inc., 1982.

Woodson, Wesley E. *Human Factors Design Handbook: Information and Guidelines for the Design of Systems, Facilities, Equipment, and Products for Human Use.* McGraw-Hill Book Company, 1981.

Woodson, Wesley E., and Peggy Tillman. *Human Factors Design Handbook,* 2nd ed. McGraw-Hill Book Company, 1992.

Chapter 2

Thermodynamic Principles

HEAT TRANSFER

GENERAL THEORY

STEADY-STATE HEAT TRANSFER

PSYCHROMETRICS

DEFINITION OF TERMS

THE PSYCHROMETRIC CHART

AIR-CONDITIONING PROCESSES

EVAPORATIVE COOLING

VAPOR PRESSURE

Thermodynamics is a science dealing with the relationship between heat and other forms of energy. It includes the subjects of heat transfer and psychrometrics, which are the focuses of this chapter.

The discipline of architecture typically relies on graphic and visual models to analyze architectural space. A knowledge of the physical principles that explain the thermal environment is essential, for by understanding these basic principles, the thermal space can be created as an active part of the designer's imagination.

Weather has a tremendous impact on both occupant comfort and energy use within buildings. A building is essentially a shelter from the elements. The building envelope acts as a mediator between the external and internal conditions. The mechanism by which the building interior thermally interacts with its surroundings, and by which mechanical heating and cooling equipment delivers thermal comfort, is heat transfer.

Using heat transfer theory as a tool, the building envelope can be analyzed for its heat flow characteristics, including its ability to control heat gain and loss by its construction, orientation, and use of particular building materials. By comparing the desired indoor temperature with the average climate conditions, one can estimate the envelope's ability to control thermal transfer and regulate interior conditions for thermal comfort.

There was a time when heat transmission was of almost no concern in the selection of envelope construction because heating and cooling energy was cheap and represented only a minor portion of a building's operating expenses. As long as air leaks didn't create drafts and compromise occupant comfort, no one cared. Special effects were created with all-glass buildings, and first costs were minimized by omitting roof and wall insulation; it was simply up to the HVAC designer to make the building comfortable by the use of as much mechanical equipment as was necessary.

As fuel costs have increased, however, energy has become one of the largest expenses in a building's operating budget. Owners and operators are beginning to appreciate the need to spend more on construction to save that much in energy bills in a few years' time.

The significance of the designer's choices is dramatically illustrated by the fact that every square foot of the envelope has a permanent energy liability attached to it for the life of the building. The cost of that energy liability is likely to continue growing until the occupants can no longer afford to heat and cool the structure. Improvements can be retrofitted at some later time, but at considerable expense. The magnitude of the energy liability may determine how long a building will remain in service before becoming obsolete.

On the other hand, an integrated awareness of heat flow characteristics allows the designer to conceptualize an energy-efficient building from the start. When building heat gain and heat loss are minimized, less reliance need be placed on mechanical systems which are costly to install, can break down, and use expensive energy.

Ever since energy costs began surging, cutting energy costs has overshadowed comfort in some designers' minds. But, in fact, the better isolated a building interior is from severe outside conditions, the better control there is over occupant comfort. Nevertheless, some energy conservation strategies are indeed at the expense of comfort.

So, where is the balance between initial costs, energy costs, and comfort? How much insulation is enough? An understanding of heat transfer theory, together with the mathematical tools presented in Chapter 4 and the economic considerations described in Chapter 16, will provide the reader with the ability to answer these questions.

HEAT TRANSFER

General Theory

The flow of heat is defined as energy transfer between two regions due to a difference in temperature. Heat transfer is actually the transfer of energy on a microscopic scale. This is envisioned as the excitation of molecules, in which the level of excitation corresponds to the amount of energy contained within a group of molecules (see Figure 2.1). Heat transfer is the transfer of motion from a more excited group of molecules to a less excited group, as shown in Figure 2.2. A cold region is simply one containing less thermal energy, and a

warm region is one containing more thermal energy. Heat transfer always occurs from a region of higher temperature to a region of lower temperature. As long as there is a temperature difference, there will be a tendency for heat to flow from the region with the higher temperature to that with the lower temperature, thereby decreasing the temperature of the former and increasing the temperature of the latter.

When heat transfer no longer occurs between two regions that are not thermally isolated from one another, they have the same temperature and are said to be in *thermal equilibrium*. Whenever the temperatures are not the same, energy transfer as heat will tend to occur until thermal equilibrium is achieved. The rate of heat transfer will be a function of (1) the magnitude of the temperature difference, (2) the area perpendicular to the heat flow path, and (3) the resistance to heat transfer of whatever medium separates the two areas (see Figure 2.3). Consider the indoor environment as an example of one region and the outdoor ambient as another.

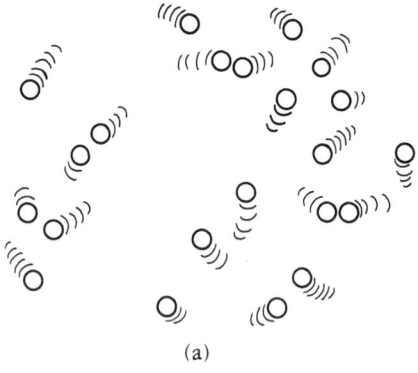

(a)

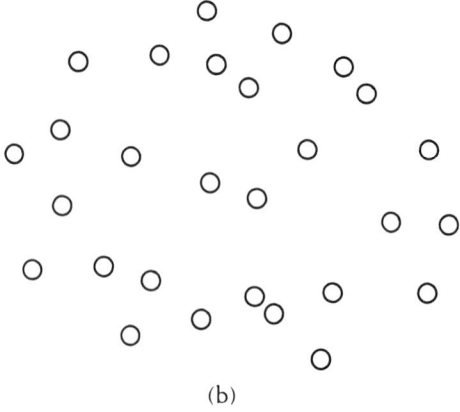

(b)

FIGURE 2.1. Thermal energy (heat) on a molecular level. (a) Warm molecules; (b) molecules at absolute zero.

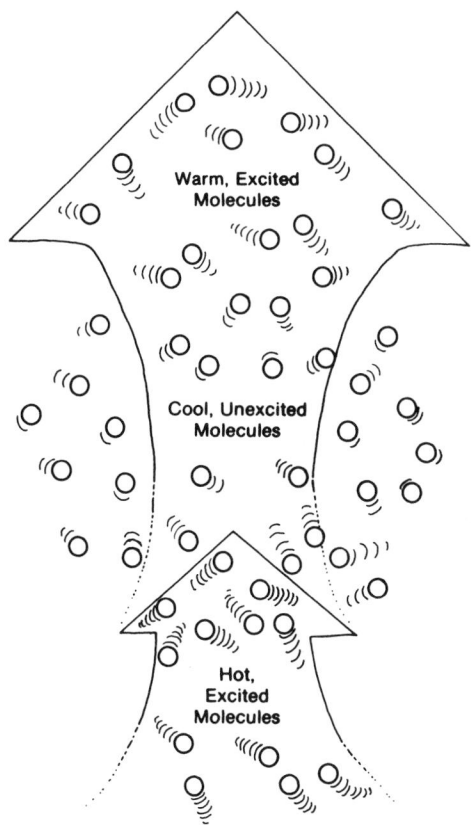

FIGURE 2.2. Heat transfer on a molecular level. When hot, excited molecules are blended with cool, unexcited molecules; the result is a level of molecular excitation (temperature) between the two extremes.

In thermodynamic terms, cold is not the opposite of heat, but simply a quantity that is always relative to something warmer. The colder something is, the lower its temperature and the less thermal energy contained within it. At the condition of absolute zero, there is no thermal energy or, in other words, no molecular motion. At all temperatures above

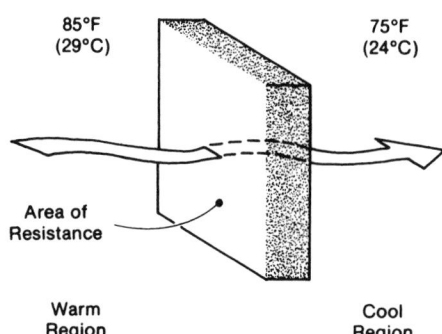

FIGURE 2.3. Heat transfer path through a medium.

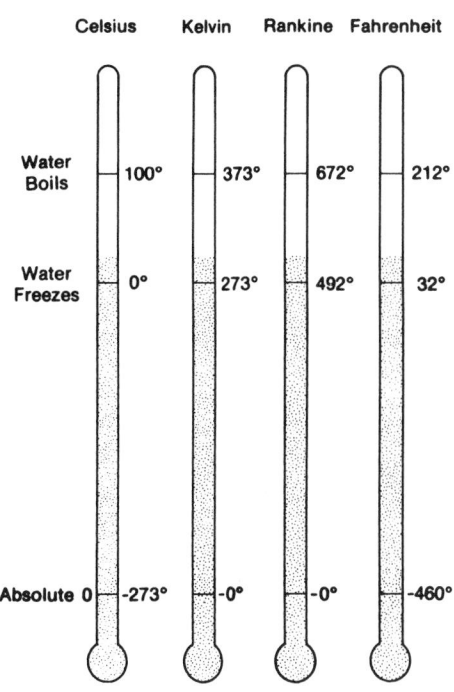

FIGURE 2.4. Relationship between the four temperature scales.

absolute zero, there is some degree of thermal energy which is in proportion to temperature.

There are two *absolute temperature* scales in which 0° represents absolute zero—*Rankine* and *Kelvin*. Rankine corresponds to the *Fahrenheit* scale (0° Rankine = −460°F), and Kelvin to the *Celsius* scale (0° Kelvin = −273°C). The relationship between all four temperature scales is illustrated in Figure 2.4.

Steady-State Heat Transfer

Heat or thermal energy transfer from one region to another occurs by three modes, as illustrated in Figure 2.5. They are:

1. Conduction, which is heat transfer from one particle to another in direct contact with it
2. Convection, which is heat transfer by fluid motion
3. Radiation, which is heat transfer by electromagnetic waves

Conduction

Conduction is the heat transfer through solid materials. It is caused by molecules vibrating at a faster rate (higher temperature) bumping into and thus transferring energy to molecules vibrating at a slower rate (lower temperature). This transfer occurs when there is a temperature difference across

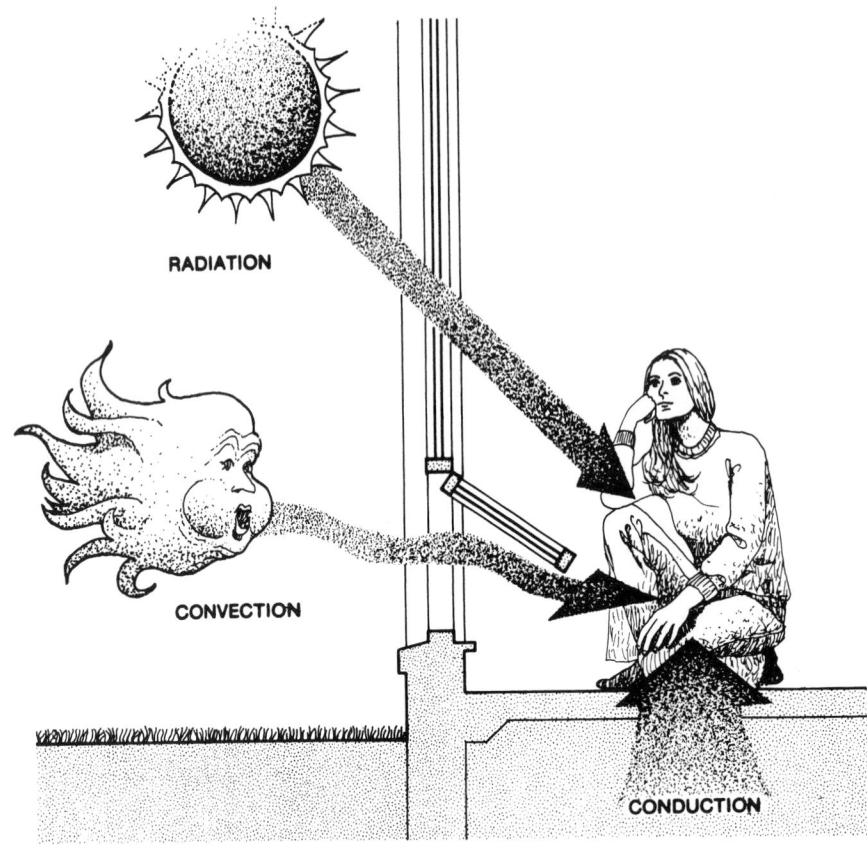

FIGURE 2.5. The three modes of heat transfer.

the material. The rate of transfer is directly proportional to the temperature difference, $T_1 - T_2$ (°F or °C), the surface area, A (ft^2 or m^2), and the material *conductance*, C (Btu/hr·ft^2·°F, or W/m^2·°C). Thus conductive heat transfer is expressed as:

$$Q = C \times A \times (T_1 - T_2) \qquad (2.1)$$

A familiar example of conduction is a pan on a stove. Heat is applied at the bottom, but you will get burned if you touch the metal on the top or sides because the heat has conducted throughout the metal. In fact, the heat conducted through the pan is what boils water and cooks the food inside.

It is common knowledge that the ability of various materials to conduct heat differs considerably. The best conductors of heat are metals. The poorer conductors (wood, plastics, gases, and ceramics) are called *insulators*. The ability of a substance to transmit heat by conduction is a physical property of the particular material. It is called *thermal conductivity*. This is generally shortened to *conductivity*, represented by the symbol "k."

The relationship between conductance and conductivity is as follows:

$$C = \frac{k}{x} \quad \frac{Btu}{hr \cdot ft^2 \cdot °F} \left(\frac{W}{m^2 \cdot °C} \right) \qquad (2.2)$$

Notice that for a material with a given conductivity, k, the conductance, C, decreases as the thickness, x, increases.

The reciprocal of conductivity (1/k) is *resistivity* (r), and the reciprocal of conductance (1/C) is *resistance* (R). Resistance and resistivity measure the insulating quality of a material in the same way that conductance and conductivity measure the ability of a material to conduct heat. The relationships between all four quantities are as follows:

$$C = \frac{1}{R} = \frac{1}{rx} = \frac{k}{x} \qquad (2.3)$$

$$R = \frac{1}{C} = rx = \frac{x}{k} \qquad (2.4)$$

The resistance, R, increases in direct proportion to an increase in thickness.

Notice that C and R are functions of thickness, and that r and k must have a thickness specified for them to be applied. Homogeneous materials such as insulating board or concrete have uniform properties and the same conductivity throughout. These substances are typically identified by their conductivity or resistivity property. The total conductance or resistance is then determined on the basis of thickness. Many materials used in building construction, however, come in a standard thickness (⅜-in. plywood sheathing, ½-in. gypsum board, etc.) or are composite materials in which the heat flow through each inch of thickness is not necessarily identical with that through the preceding inch (glass blocks, hollow clay tile, concrete blocks, etc.). The heat transfer characteristic of these products is typically given in terms of overall conductance or resistance.

Conductivity is established by empirical tests and is a basic rating for a material. Appendix B4.2 provides a reference of insulating and conducting characteristics of common building materials. Where conductance, C, is referred to instead of conductivity, k, the material is either nonhomogeneous or is a standard thickness other than 1 inch.

Notice the distinction between conductance and conductivities in the following two examples:

Sand and gravel or stone aggregate concrete poured in place may have a *conductivity* of 12 Btuh/ft^2· °F/in. (173 W/m^2·°C/cm) thickness. Two inches would have a total conductance of 6 Btuh/ft^2·°F (34 W/m^2·°C).

Gypsum wall board (½ inch thick) has a *conductance* of 2.22 Btuh/ft^2·°F (12.6 W/m^2·°C).

Thermal conductance can be thought of as the rate of heat flow through a unit area of a body per unit of temperature difference. This is important for a conceptual understanding of the heat flow process. But in practice the resistance, or *R-value,* of a material is most commonly used for computational purposes. The R-value can be thought of as the number of hours (seconds) required for 1 Btu (joule) to penetrate 1 square foot (square meter) of a material for each degree of temperature difference between the two sides. Thus, the higher the R-value, the slower (more hours required for) the heat transfer through a material.

When heat flows through a series of several materials, such as in a typical wall section (Figure 2.6), the resistances are additive, so the total thermal resistance, R, is

$$R = R_1 + R_2 + R_3 + \cdots \qquad (2.5)$$

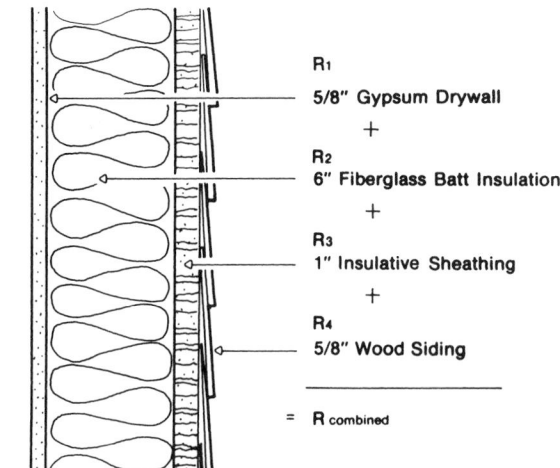

FIGURE 2.6. Section through a typical cavity wall.

In Figure 2.6, the successive layers of drywall, fiberglass batt insulation, insulative sheathing, and wood siding yield the following combined R-value:

| | | Resistance | |
| | | ft^2·°F | m^2·°C |
Construction Element		Btuh	Watt
R_1	drywall	.45	0.079
R_2	6-inch fiberglass batt	19.00	3.346
R_3	1-inch insulative sheathing	1.32	0.232
R_4	wood siding	0.81	0.143
		21.58	3.800

Notice that the dominant portion of the total resistance is the fiberglass batt.

Convection

Convection is thermal energy carried by the flow of a liquid or gas. It is the transport of heat by fluid motion and is normally associated with the transfer of heat between a surface and a fluid, such as air or water, moving across it. An example of convection is room air giving up heat to a window on a cold day. The glass conducts the heat outside. If it is a double-pane window (see Figure 2.7), the heat will be convected across the air space to the outer pane and then conducted through to the outside.

In detail, convection is a combination of conduction and fluid flow. Heat is transferred by conduction between the region very close to a solid boundary and the solid itself. The fluid flow is responsible for the rapid transport of heat to or

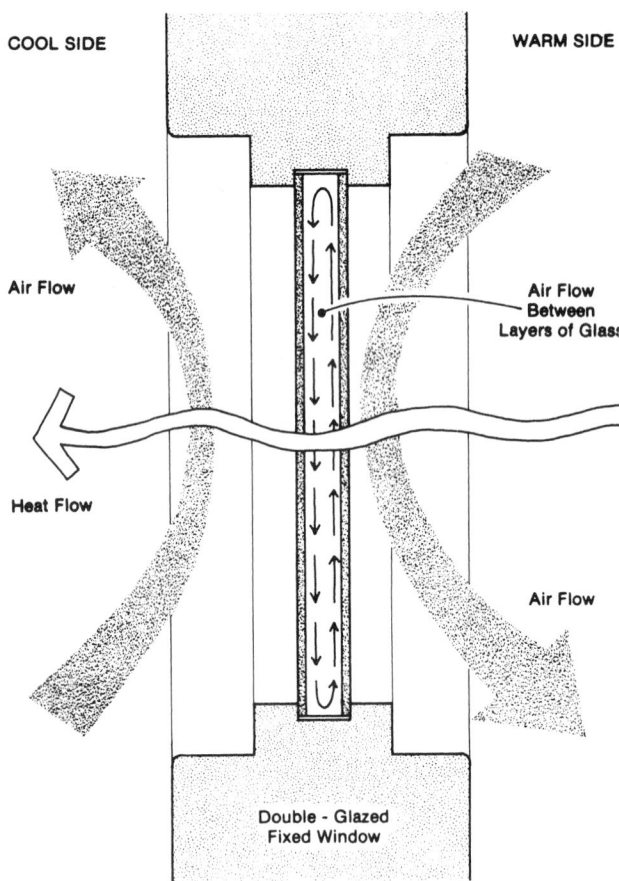

COOL SIDE   WARM SIDE

Air Flow

Air Flow
Between
Layers of Glass

Heat Flow

Air Flow

Double - Glazed
Fixed Window

FIGURE 2.7. Heat transfer through a double-pane window. Heat convects from the warm airflow to the first layer of glass, conducts through the glass, convects from the first layer of glass to the second layer, conducts through the glass, and then convects into the air stream on the cool side.

sity. As a result, the warmer region is more buoyant and will be displaced upward by the colder region, which is being pulled downward by gravity. This phenomenon is usually not as powerful as any forced air flow present and is therefore overridden by it.

The transfer of heat from air to a surface or from a surface to air is called *surface conductance.* It is also referred to as *film conductance* or *film coefficient.* As implied by the term "conductance," this coefficient is used in the same way material conductance is in Equation 2.1. Using "f" as the surface conductance, it becomes:

$$Q = f \times A \times (T_1 - T_2) \tag{2.6}$$

The rate of convection is a function of (1) the surface roughness and orientation, (2) the direction of flow, (3) the type of fluid, and (4) whether the process is free or forced. It is often temperature-dependent and must be obtained experimentally. The inverse of surface conductance is a resistance to heat flow. The resistances of materials constituting a given heat flow path are additive.

A special case of convection from a surface to air is that of heat transfer across an air space where the heat travels from a surface to air and then across to another surface. The heat transfer across such an air space is dependent upon (1) the distance between the surfaces, (2) the areas of the two surfaces, and (3) the smoothness of the surfaces. The smaller the cross-sectional area of the air space, the more constricted the natural convection currents and the lower the heat transfer.

If a material contains many small air spaces, thermal conduction will be broken up by the poorly conducting air, and convection will be inhibited by the smallness of the air pockets. Consequently, overall heat transfer through the material will be low. This is the principle behind most insulations.

Other special cases of convection are infiltration and ventilation. *Infiltration* is the exchange between conditioned room air and outdoor air through cracks and leaks in the building skin. It is called *ventilation* when the exchange is intentional, for example, when caused by window openings or a fan. The magnitude of this exchange is measured in cubic feet per minute (CFM) or liters per second (L/s). The resultant heat transfer into or out of the building is calculated by multiplying the volume of air exchanged (CFM or L/s) by its storage capacity (Btu/ft³·°F or joules/m³·°C) and by the temperature difference (°F or °C) between the incoming and outgoing air:

$$Q = CFM \times 1.08^a \times (T_2 - T_1) \tag{2.7}$$

$$Q = L/s \times 1.2^a \times (T_2 - T_1) \tag{2.8}$$

---

[a] This term (called the *O.A. factor*) is explained in Chapter 4.

from the boundary region. Conduction also occurs between molecules throughout the fluid, but the faster and usually dominant mechanism is the physical motion of the fluid carrying thermal energy. As with conduction, the rate of convective heat transfer is directly proportional to the area of the solid surface, A, and the temperature difference between the surface and the fluid, $T_1 - T_2$.

Convection is broadly divided into two categories: natural and forced convection. *Forced convection* occurs when the fluid motion is caused by a fan or pump or the wind. *Natural (or free) convection* occurs when the fluid motion is primarily caused by the temperature difference between the solid surface and the fluid in contact with it. In a warmer region of the fluid, the molecules are vibrating more excitedly than in a colder region. The warmer molecules will thus take up more volume per unit weight, causing a difference in den-

Radiation

For conduction and convection, the transfer of heat takes place through matter, either solid or fluid. In the case of radiant heat transfer, the internal energy or molecular vibrations set up electromagnetic waves that emanate from the warm object and carry the energy to all bodies within a direct line of site. Then the electromagnetic waves excite the molecules of the receiving bodies and thereby increase their internal energy. Familiar examples of radiant heat transfer are the sun's rays and the hot coals of a fire.

By making use of electromagnetic waves, radiation does not need any medium for transmission and can occur through a vacuum. A very important implication of this is that radiant energy from the sun is able to travel across the expansive vacuum of space to provide the source of nearly all the energy we use here on earth.

When radiant energy comes in contact with an object or medium, it can be (1) reflected from the surface, (2) absorbed by the material, or (3) transmitted through. When no radiant energy is transmitted, the material is said to be *opaque*. When radiant energy passes through the material, it is said to be *transparent*. One way we experience radiation is when sunshine travels through space, through our atmosphere, and through a car window on a cold day. Until then, it has passed through transparent materials, but it now strikes the contents of the car and warms them up even though the air all around the car is cold.

But not all of the sunlight passes through the glass. The ultraviolet rays, for example, are reflected. Generally, materials selectively transmit electromagnetic energy of some wavelengths while reflecting or absorbing others. Materials that are shiny, silvered, or mirrored (that is, materials that reflect visible light) will also reflect radiant heat. Glass and other inorganic crystals are transparent to visible light and to radiant energy from the hot sun, but are opaque to wavelengths of thermal radiation released by objects at normal terrestrial temperatures. Most materials that are opaque to the visible wavelengths will not stop X-rays from passing through. For reference, the electromagnetic spectrum is presented in Figure 2.8.

All radiation travels in a straight line at the speed of light (186,000 miles/s), but at different wavelengths. The amount of energy transmitted by radiation is inversely proportional to its wavelength (that is, the shorter the wavelength, the higher the energy content). The wavelength of radiation is inversely proportional to the temperature of the source.

Thus, both wavelength and the rate of radiant heat transfer depend on temperature. Whereas conduction and convection rates are directly proportional to the temperature difference, radiant heat transfer increases at an exponential

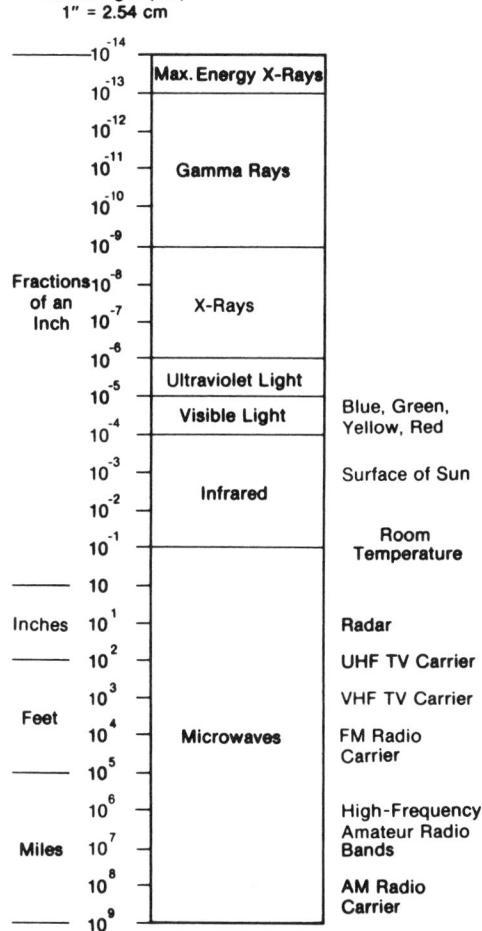

FIGURE 2.8. The electromagnetic spectrum.

rate with increasing temperature. Radiant heat transfer is proportional to $(T_1^4 - T_2^4)$, where the temperatures are based on the absolute scale, that is, Rankine or Kelvin.

All the objects and surfaces in a room radiate thermal energy according to the fourth power of their absolute temperature. The only net transfer of heat, however, occurs from warm bodies to cooler ones. If two objects are at the same temperature, they will each radiate to the other, but no net exchange will take place. Like conduction and convection, radiant heat transfer is influenced by the exposed surface areas (ft^2 or m^2) of the warmer and cooler objects.

Sunshine is a special case of thermal radiation. The sun's temperature is so high (10,000° R) that the entire exchange is virtually one way.[b] Solar radiation arrives at the outer edge

[b] The earth does not continually warm up, however, because it radiates heat to the much colder outer space. A balance between incoming and outgoing heat is thus maintained.

of the atmosphere as a nearly constant energy flow (429 Btuh/ft^2). As it travels to the earth's surface, it is considerably weakened from reflection and absorption by the atmosphere. The amount of decrease in intensity is determined by such factors as angle of incidence, time of day, time of year, and atmospheric conditions. The radiation that does make it through the atmosphere is then further attenuated by the transmissivity of window glass and is absorbed according to the absorptivity of the receiving object.

Multiple Modes

Heat loss or heat gain through the building envelope involves all three basic modes of heat transfer: conduction, convection, and radiation. For convenience, the composite effect is assigned a single heat transfer coefficient that is determined experimentally for common types of building construction under specific conditions. Some representative samples of these empirical values are listed in Appendix B4. The composite coefficient, known as the *U-value* of an assembly of materials, is used to correlate the overall rate of heat transfer with surface area and temperature difference in the following formula, which takes the same simple form as Equation 2.1:

$$Q = U \times A \times (T_1 - T_2) \tag{2.9}$$

The reciprocal of the U-value is the R-value, which represents the degree of resistance to heat flow, or insulating ability, offered by an element of the building envelope. The lower the U-value, the higher the R-value or insulating value.

As heat flows through the envelope, each element that it passes through in turn plays a part in retarding the heat transfer:

1. The convective heat transfer between the outside air and the exterior surface varies with wind velocity and with the physical character of the surface. The resistance to heat flow is determined by the value of the surface conductance.
2. Each layer of material contributes some resistance to heat flow, depending on the material properties and thickness. Heavy, compact materials usually have less resistance than light ones.
3. Each air space adds a measurable amount of resistance. The value of the resistance varies with the dimensions of the space and the character of the surfaces facing the space.
4. The inside surface also has a film conductance that represents a resistance to heat flow.

Each of the above elements has a numerical R-value associated with it. When added together as in Equation 2.5, they collectively represent the total thermal resistance of the given envelope section. Thus, the overall coefficient of heat transmission or thermal transmittance for a building envelope section is the reciprocal of R, where R is the summation of the resistances of the air films, materials, and air spaces that make up the assembly:

$$U = \frac{1}{R}$$

$$= \frac{1}{(R_1 + R_2 + R_3 + \cdots)} \frac{Btu}{hr \cdot ft^2 \cdot {}^\circ F} \left(\frac{W}{m^2 \cdot {}^\circ C} \right) \tag{2.10}$$

The U-value is usually less than 1. A table of R-values with their reciprocals is presented in Appendix B4.5 for convenience in evaluating U-values.

The surface conductance of building materials is a function of both the convective and radiative properties and is thus dependent on the color, smoothness, temperature, area, and position of the surface, as well as on the temperature and velocity of the surrounding air. For the inside surface (still air) of a vertical wall, the surface conductance is usually given as 1.47 Btuh/ft^2·°F (8.35 W/m^2·°C). The standard value for a wind velocity of 15 mph (6.71 m/s) is 6.00 Btuh/ft^2·°F (34.07 W/m^2·°C); a surface conductance of 4.00 Btuh/ft^2·°F (22.7 W/m^2·°C) is used for a wind velocity of 7½ mph (3.35 m/s).

The U-value of an envelope section can be reduced (improved) by introducing additional air spaces or insulation. Unlike most insulation materials, an air space has an R-value that is not related primarily to thickness but is determined by many factors, including the position of the space, its dimensions and shape, the texture and color of the surfaces facing the space, the mean temperature of the space, and the direction of heat flow.

Heat passes into the space by conduction from one face to air, then by convection through the space, and finally by conduction to the opposite face. Heat also crosses the space by radiation from the warm face to the colder face. Radiant transmission across the space is greatly reduced by the use of shiny, reflective material such as aluminum foil. Vacuum bottles are mirrored to inhibit radiation across the vacuum that is there to prevent convection. In buildings, a vacuum is usually impractical, but reflective foils are effective in reducing the radiant component of heat transfer across air spaces. The foil is needed only on one of the sides because using it on the opposite side adds very little improvement. When heat flow is upward, the convective transfer is much greater than the radiant transfer, and the relative importance of reflective foil is low. The reverse is true for heat flow downward. Reflective foil is extremely useful for reducing transmission of summer solar heat across the air spaces between roof and ceiling and for reduction of winter heat losses in floors over unheated areas.

Appendix Table B4.1 provides values for air space conductance under various conditions of thickness, surface reflectivity, and direction of heat flow (up, down, horizontal). Examples in Figures 2.9 and 2.10 demonstrate how to calculate overall R-values and U-values using the information in Appendix B4.

PSYCHROMETRICS

The properties of air that affect comfort were introduced in Chapter 1. They are reviewed here because this section will explain in greater depth how they are related, and why they are important in the systems and equipment that condition the air to provide comfortable and healthy indoor environments.

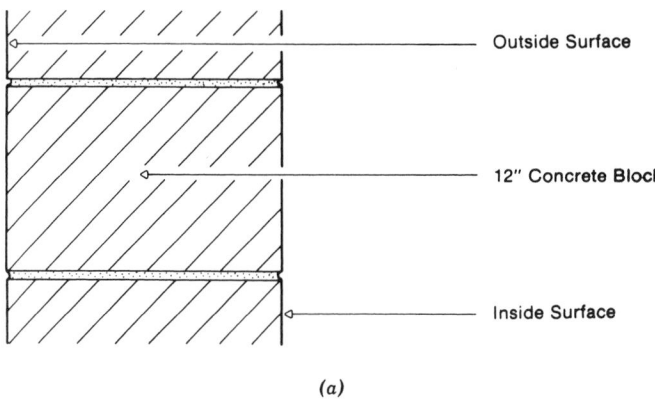

(a)

Construction (a)	R	U 1/R
1. Outside surface (15 mph wind)	0.17	
2. 12-inch concrete block	2.27	
3. Inside surface (still air)	0.68	
	3.12	0.32

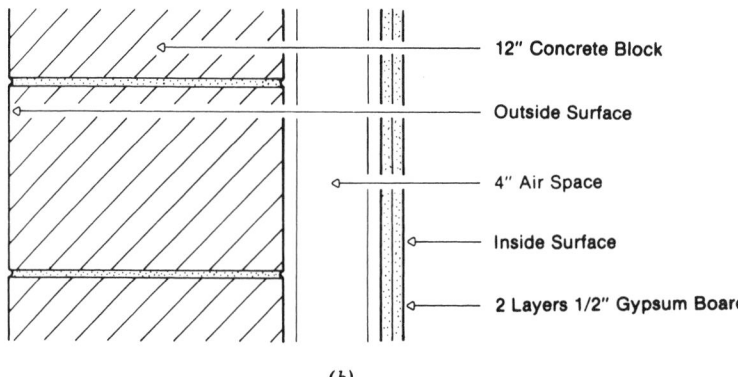

(b)

Construction (b)	R	U 1/R
1. Outside surface (15 mph wind)	0.17	
2. 12-inch concrete block	2.27	
3. 4-inch air space (winter conditions, both surfaces nonreflective)	1.01	
4. Two layers ½-inch gypsum board (2 × 0.45)	0.90	
5. Inside surface (still air)	0.68	
	5.03	0.20

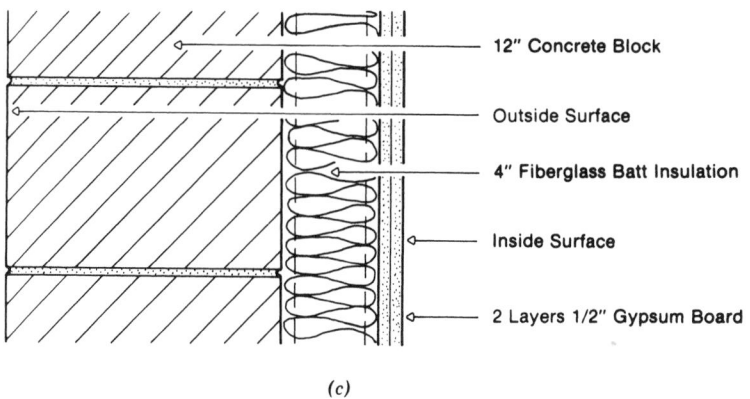

(c)

Construction (c)	R	U 1/R
1. Outside surface (15 mph wind)	0.17	
2. 12-inch concrete block	2.27	
3. 4-inch fiberglass insulation	11.00	
4. Two layers ½-inch gypsum board (2 × 0.45)	0.90	
5. Inside surface (still air)	0.68	
	15.02	0.07

FIGURE 2.9. Sample calculations of U-value for wall sections.

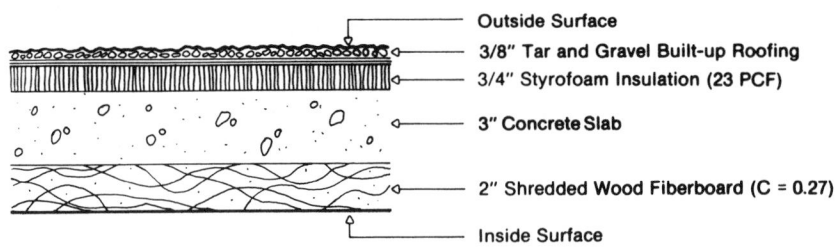

Construction (a)	R	U 1/R
1. Outside surface (15 mph wind)	0.17	
2. ⅜-inch tar and gravel built-up roofing	0.33	
3. ¾-inch styrofoam insulation	3.38	
4. 3-inch concrete slab (3 × 0.08)	0.24	
5. 2-inch shredded-wood fiberboard	3.70	
6. Inside surface (still air)	0.61	
	8.43	0.12

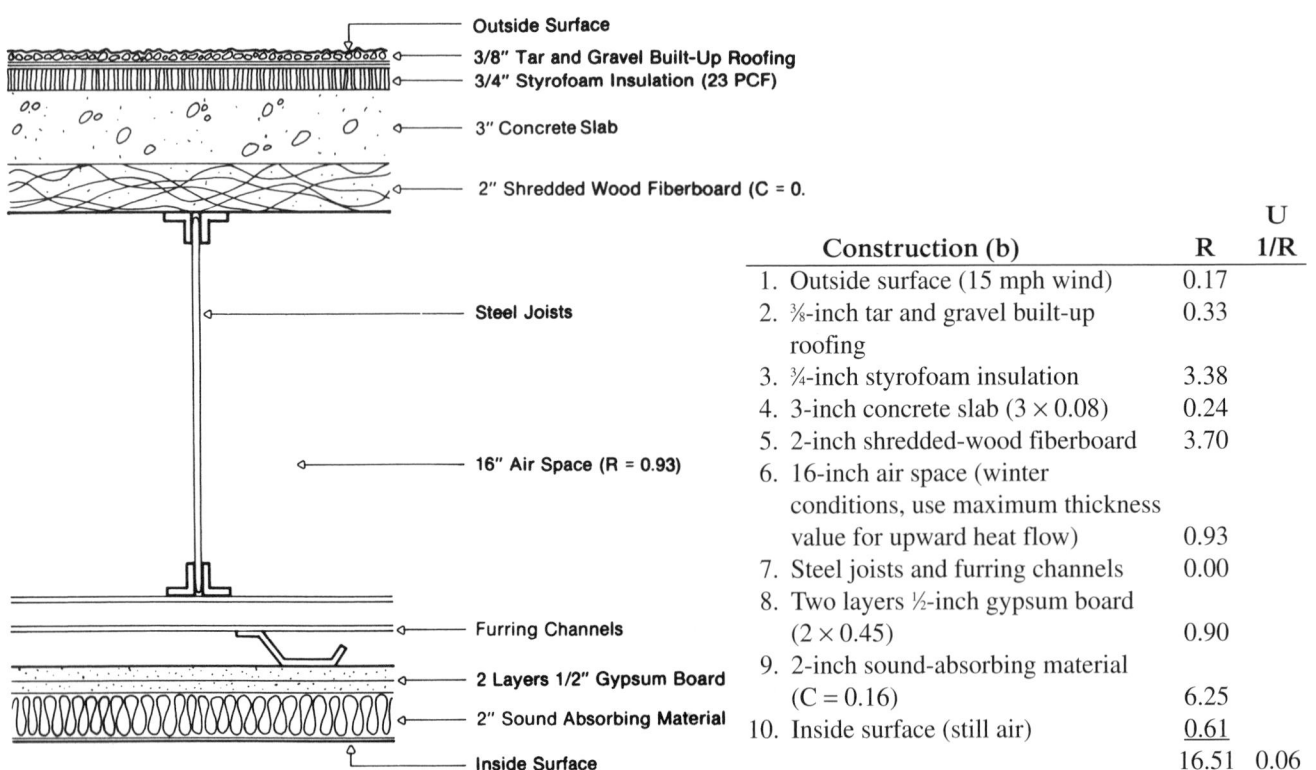

Construction (b)	R	U 1/R
1. Outside surface (15 mph wind)	0.17	
2. ⅜-inch tar and gravel built-up roofing	0.33	
3. ¾-inch styrofoam insulation	3.38	
4. 3-inch concrete slab (3 × 0.08)	0.24	
5. 2-inch shredded-wood fiberboard	3.70	
6. 16-inch air space (winter conditions, use maximum thickness value for upward heat flow)	0.93	
7. Steel joists and furring channels	0.00	
8. Two layers ½-inch gypsum board (2 × 0.45)	0.90	
9. 2-inch sound-absorbing material (C = 0.16)	6.25	
10. Inside surface (still air)	0.61	
	16.51	0.06

FIGURE 2.10. Sample calculations of U-value for roof sections.

Definition of Terms

Dry-bulb Temperature (DB).   The temperature of air as registered by an ordinary thermometer.

Wet-bulb Temperature (WB).   The temperature registered by a thermometer whose bulb is covered by a wetted wick and exposed to a current of rapidly moving air.

Dew-Point Temperature.   The temperature at which condensation of moisture begins when the air is cooled.

Relative Humidity (RH).   Ratio of the actual water vapor pressure of the air to the saturated water vapor pressure of the air at the same temperature.

Specific Humidity or Moisture Content.   The weight of water vapor in grains or pounds of moisture per pound of dry air.

Dry Air.   Air that contains no water vapor.

Saturated Air.   Air that contains all the moisture it can hold. If saturated air is cooled or if any more moisture is added to it, moisture will begin condensing out of it (100 percent RH). When air is saturated with moisture, dry-bulb, wet-bulb, and dew-point temperatures are all equal.

Sensible Heat.   Heat associated with the change in dry-bulb temperature of air.

Latent Heat.   Heat associated with the change of state of the moisture in air. When moisture is condensed from air, the air is reducing its latent heat (dehumidification). When moisture is evaporated into the air (from a pool of water, human perspiration, or the release of steam), the air is increasing its latent heat (humidification).

Enthalpy.   A thermal property indicating the quantity of heat in the air expressed in British thermal units per pound of dry air (joules per kilogram of dry air). Enthalpy represents the sum of the latent heat and sensible heat of air.

The Psychrometric Chart

Psychrometrics is a branch of physics concerned with the study of atmospheric conditions and, in particular, the relationship between air and moisture when mixed together.

Air is a mixture of water vapor, nitrogen, oxygen, carbon dioxide, and traces of other gases. Although its water vapor content is often less than 1 percent of the total, it is a major factor in determining the condition of the air mixture, due to the great energy content of water when in vapor form. The latent heat in water vapor (the energy in the form of heat required to change water from liquid to vapor) is the largest of any common liquid. As a result, the small amount of water vapor in the air mixture often contains the major part of the total heat energy of the mixture.

A *psychrometric chart* is a graphic representation of the thermodynamics and thermal properties of moist air. The dry-bulb, wet-bulb, and dew-point temperatures and the RH are related so that if any two properties are known, all other properties may then be determined from the psychrometric chart (Figure 2.11). Values determined from the chart can be converted to SI units using Table 2.1.

On the chart, the vertical lines having numbers along the bottom are dry-bulb temperatures and are those air temperatures measured with an ordinary thermometer. All points falling on a given vertical line will be at the same dry-bulb temperature. The horizontal lines are water-content lines, and the numbers on the right side designate the water vapor content of air in terms of grains or pounds of moisture per pound of dry air. All points on a given horizontal line contain the same amount of water vapor. (There are 7,000 grains in 1 pound of water, and 1 pound of air at normal temperatures occupies about 13.5 cubic feet. The water vapor content may be expressed in grains instead of pounds in order to avoid using very small fractions.) Any point on the chart thus indicates the dry-bulb temperature and water vapor content for a given pound of air.

The curved lines that radiate from the lower left of the chart to the upper right are the RH lines. The horizontal line at the bottom of the chart is the 0 percent RH line and coincides with the 0 grains of water content line. The uppermost curved line is the 100 percent RH, or saturation, line. The water content line terminating at any point on this saturation line designates the maximum grains of water vapor that saturated air can hold at that intersecting dry-bulb temperature. This intersecting point is known as the *dew point.*

As an example, on the 100 percent RH line, locate the terminus of the 60°F dry-bulb line; note that at this condition a pound of air contains 77 grains of water vapor. Follow the 60°F dry-bulb line down to the 50 percent RH line. The horizontal water vapor content line shows that air at this latter point contains only half as much, or approximately 38 grains of water vapor. The air now is only 50 percent full and has an RH of 50 percent. RH therefore approximately states, as a fraction in terms of percent, the amount of water vapor in air compared to the amount it could actually hold at that temperature.

The temperature of air greatly affects its capacity to hold water vapor. In the previous example, a pound of air at 60°F dry-bulb and 100 percent RH contained 77 grains of water vapor. If this air is warmed up 20°F to a temperature of 80°F (locate 60°F and 100 percent RH, and trace the line horizontally to the right to 80°F), the RH decreases to 50 percent. Air at 80°F could hold 156 grains of water vapor when saturated (follow the 80°F line up to the saturation line), but it actually contains only 77 grains, and thus is only half full. If

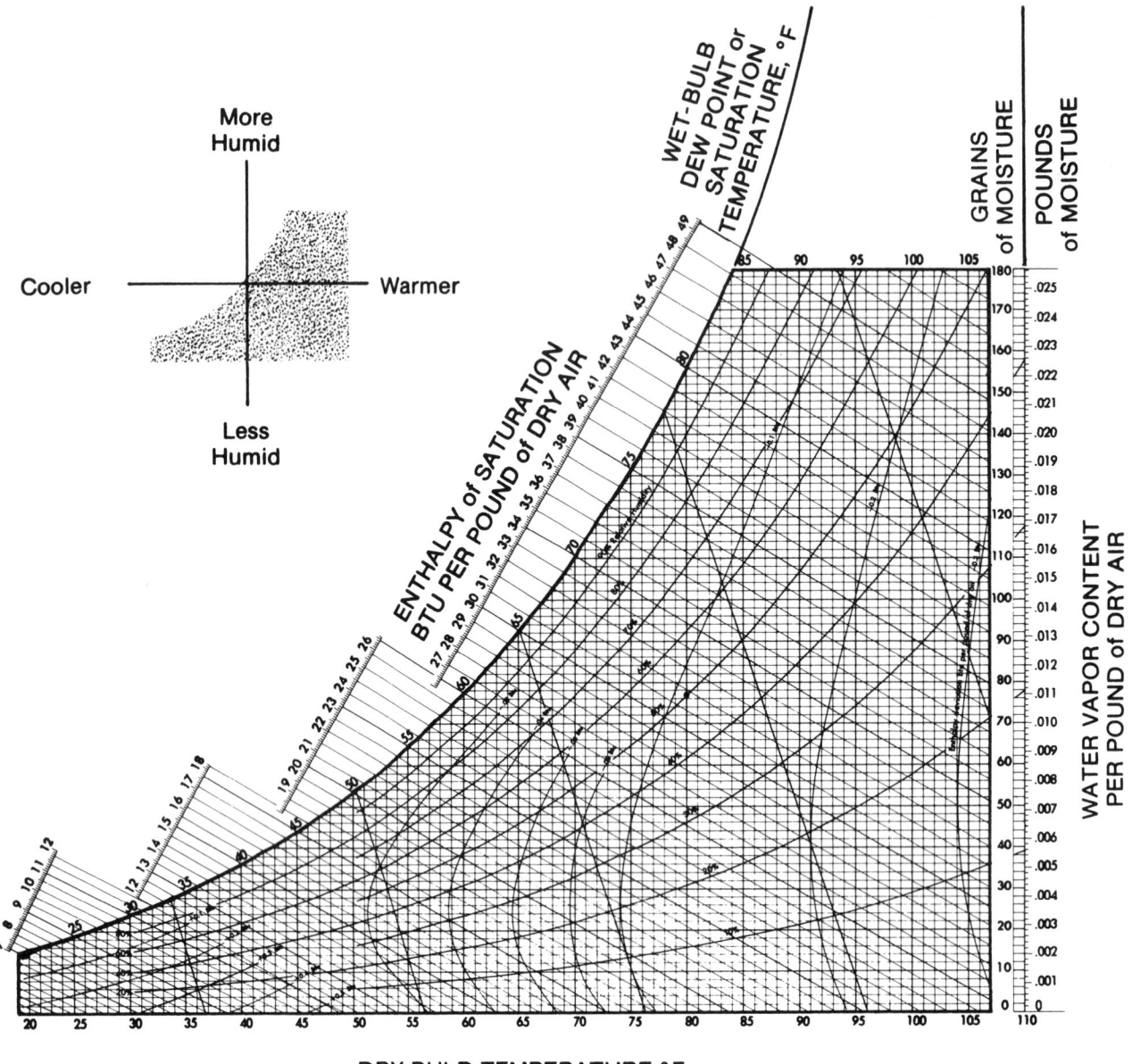

FIGURE 2.11. Psychrometric chart.

this air is warmed up another 20°F to 100°, its RH drops to about 25 percent. In this way, when room air is warmed up, its RH sometimes drops below specified limits. Humidification would then need to accompany the heating process in order to maintain comfort, health, or process requirements.

When the reverse occurs and air becomes colder, its capacity to hold water vapor is reduced. If conditions cause air temperature to drop so that the RH reaches 100 percent, any further cooling will cause condensation. Referring back

to the earlier example, the pound of air at 100 percent RH and 60°F contained 77 grains of water vapor. If it is cooled down to 40°F (follow the curved 100 percent RH line down to the 40°F dry-bulb line), it can contain only 38 grains. The other 39 grains of water vapor it contained must condense out as liquid drops of water.

Figure 2.12 shows a typical air-conditioning process traced on a psychrometric chart. Outdoor air at condition (B) is mixed with return air from the room at condition (A) and

TABLE 2.1   UNIT CONVERSION FOR THE PSYCHROMETRIC CHART

To Convert from	to	Multiply by
Btu per pound	Joule per kilogram	2,326
Pounds per pound	Kilograms per kilogram	1
Degree Fahrenheit	Degree Celsius	$°C = (°F - 32)/1.8$

enters the apparatus at mixed condition (C). Air flows through the conditioning apparatus and is cooled and dehumidified from condition (C) to condition (D) and is then supplied to the space at condition (D). As the air picks up heat and moisture from the room, it moves along the line from condition (D) back to condition (A), and the cycle then is repeated.

The psychrometric chart also indicates the energy content of the air/water vapor mixture. The energy or heat content of dry air is completely determined by, and is proportional to, its dry-bulb temperature. It is called *sensible heat* because it can be felt.

The energy content of the water vapor is due to its latent heat of vaporization and is proportional to the amount of water vapor in the moist air mixture. *Latent heat* is the energy, in the form of heat, that is required to convert water from liquid to vapor without any change in temperature. It requires approximately 1,060 Btu (1.12 million joules) of heat to convert 1 pound (0.45 kg) of water from a liquid to a vapor.

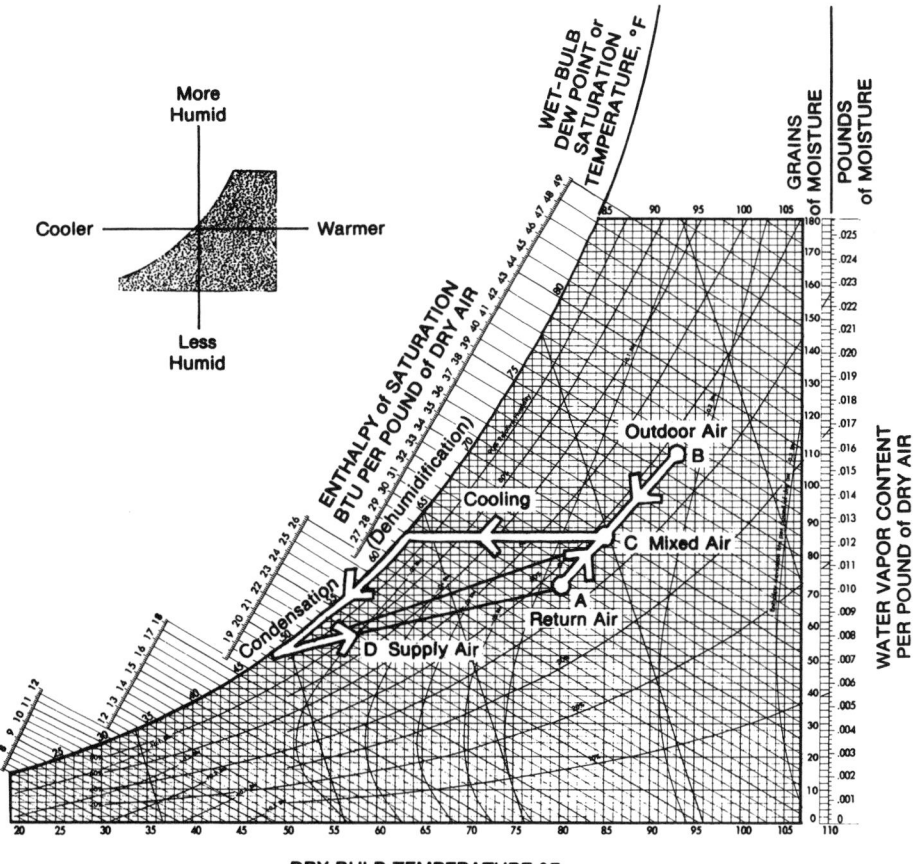

FIGURE 2.12. A typical air-conditioning process on the psychrometric chart.

The sum of these two energies is called *total heat,* or *enthalpy.* Typically, psychrometric charts are not extended down to 0°F (−18°C) dry-bulb temperature. However, at that point, the dry air is considered to contain no heat or moisture; therefore, the enthalpy at any higher dry-bulb temperature represents the total heat content of the water/air mixture required to raise it from 0°F to the particular temperature and moisture level.

Enthalpy is proportional to the wet-bulb temperature of air. The wet-bulb temperatures are designated by the diagonal lines that slope from the upper left to the lower right of the psychrometric chart. At 100 percent RH, the dry-bulb and wet-bulb temperatures of air coincide, but at any other condition less than saturation, the wet-bulb temperature is always less than the dry-bulb temperature.

For air with a constant moisture content, a falling dry-bulb temperature causes a drop in total heat content. This is a sensible heat loss from the air because water vapor is being neither added nor subtracted to cause a latent heat change. A drop in wet-bulb temperature accompanies the heat loss, but the wet-bulb temperature does not drop as fast as the dry-bulb temperature. The air is getting wetter, and indicatively, the RH is rising.

When the temperature and humidity conditions of a quantity of air are known, it is possible to find out the amount of energy needed to be added to or extracted from the air to obtain some other desired condition. First, locate the two points corresponding to the initial and desired conditions (for example, points [C] and [A], respectively, in Figure 2.12). Then follow a line from each point diagonally upward to the left. Where each line intersects the enthalpy scale is the total heat content of each condition, respectively. The difference between them represents the net heat energy to be added or extracted.

Notice, however, that in Figure 2.12, in order to get from condition (C) to condition (A), the air must first be brought to condition (D). So, while the *net* energy difference is just enthalpy (C) (33.9 Btu/lb) minus enthalpy (A) (30.2 Btu/lb), or 3.7 Btu/lb, in actuality, the air must be cooled all the way to condition (D) (20.8 Btu/lb). Thus, the actual amount of heat that must be removed is 33.9 − 20.8 = 13.1 Btu/lb. Additional heat will be added to bring the air to condition (A) by mixing it with the room air before it reaches the occupied zone.

When two streams of air are mixed together, the combined properties will be in between the individual properties of the two streams before mixing. This is illustrated in Figure 2.12 by the air mixture of point (C) in between its constituent air conditions: points (A) and (B). This sort of mixing occurs in the ductwork when outside ventilation air is combined with return air from the room, as in Figure 2.12, or when conditioned air is blended with room air before reaching the occupied zone of a room.

The relationships among thermodynamic properties vary with pressure, which is affected by geographic elevation. The psychrometric chart shown in Figure 2.11 is for standard atmospheric pressure. However, little error is introduced for elevations up to 2,000 feet above sea level.

The psychrometric relationships discussed here are the basis for the heating and cooling load calculations described in Chapter 4. They will also be important in understanding the phenomenon of condensation in preparation for the discussion of moisture problems within buildings in Chapter 3.

Air-Conditioning Processes

There are basically four processes of thermal air conditioning: heating, cooling, humidification, and dehumidification. Humidification, when employed, normally accompanies heating, and dehumidification is associated with cooling.

Referring to the psychrometric chart, the heating process—or the adding of sensible heat—is represented by moving horizontally to the right. Cooling is the opposite of heating. Moving horizontally to the left on the psychrometric chart represents the removal of sensible heat.

Latent heat is associated with the vertical direction on the psychrometric chart. Humidification (the addition of moisture to the air) increases the latent heat and is represented by moving upward. Dehumidification decreases the latent heat and is shown as moving downward on the psychrometric chart.

A normal air conditioner cools by removing sensible heat from the air. From our knowledge of heat transfer, we know that a transfer of sensible heat requires a temperature difference, that is, a warmer region and a cooler one. In this case, a surface, such as a coil or cooling panel, is placed in contact with the air and is kept at a temperature below that of the air by a continuous extraction of heat.

Now, if the temperature of this surface is below the dew-point temperature of the air, condensation will begin forming on it. The air will be losing moisture and thereby decreasing in latent heat. Typically, air conditioners both cool and dehumidify in this way. The area and temperature of the surface and, in the case of coils in a duct, the air velocity are carefully selected to achieve the intended balance between cooling and dehumidification (sensible and latent heat removal, respectively).

One implication of this is that whenever air is mechanically cooled, it almost always involves some condensation that will require some means of draining it away. *Cooling equipment generally must have some provision for draining condensate.* Heating, on the other hand, can result in a reduction of RH below acceptable levels, necessitating some humidification of the air as a separate step. *Humidification equipment must be provided with a continuous supply of water.*

One departure from these general rules is evaporative cooling. Evaporative cooling systems do not need a condensate drain, but instead, like humidifiers, need a continuous supply of water.

Evaporative Cooling

Perfectly dry (0 percent RH) air at 98°F (37°C) dry-bulb temperature has a wet-bulb temperature of 56°F (13°C) and contains only sensible heat, having no latent heat due to the complete absence of moisture. However, if moisture is evaporated into this dry air until it becomes 100 percent humid, its dry-bulb temperature will drop to 56°F (13°C) (locate 98°F/37°C dry-bulb at 0 percent RH and follow the diagonal 56°F/37°C wet-bulb line upward to the 100 percent RH line). This decrease in dry-bulb temperature is caused by using up some of the sensible heat of the air to evaporate the water into the air. This loss in sensible heat energy (reduction of the dry-bulb temperature to 56°F/37°C is exactly balanced by an increase in the latent heat energy, and consequently the total energy content of the air mixture remains unchanged. This process follows along any of the wet-bulb lines on the psychrometric chart and is the basis for *evaporative cooling.* The lowest temperature to which air can be cooled by the evaporation of water is the wet-bulb temperature.

An example of wet-bulb temperature is the temperature a person feels upon stepping out of a swimming pool into a blowing wind. As long as the skin is wet while in a strong breeze, its temperature will drop to the prevailing wet-bulb temperature.

An evaporative cooling system precools outside air before passing it through a space. It does this by first passing the air through a wet pad, from which water is evaporated, increasing the air's water content (latent heat), and reducing its dry-bulb temperature (sensible heat) almost to the wet-bulb temperature.

Thus, the wet-bulb temperature limits the extent to which air can be cooled by evaporative cooling. In actual equipment, the wet-bulb temperature cannot quite be reached, but the air can typically be cooled down to within about 3°F (1.7°C) of the wet-bulb temperature.

Since the wet-bulb temperature, not the RH, determines the temperature to which air can be cooled by evaporation of water, it is of importance in determining the comfort potential of evaporative cooling in a given locale. Wet-bulb temperatures, compiled for many years from meteorological data, are available from government sources for most localities. Peak values can also be found in the ASHRAE *Handbook of Fundamentals.* For convenience, maps of peak dry- and wet-bulb temperatures throughout the United States are presented in Figure 2.13. Notice that the maximum wet-bulb temperature for 95 percent of the four hottest months, June through September, ranges from 70° to 80°F (21° to 27°C). At high elevations, the wet-bulb temperatures are even lower. As previously explained, air can be cooled down in commercial evaporative cooling equipment to within about 3°F (1.7°C) of these wet-bulb temperatures.

As can be seen from these maps, evaporative cooling will work in most parts of the country as long as high humidities and rapid outside air flow rates can be tolerated. While the dry-bulb temperature fluctuates greatly over a 24-hour day (typically 20°F/11°C), the wet-bulb temperature changes only about a third as much. Thus, during the hottest time of the day, when both the dry-bulb and wet-bulb temperatures are at their peak, the difference between them is greatest. Consequently, the greatest potential for evaporative cooling occurs during the part of the day when it is needed most.

Vapor Pressure

Water vapor is a gas that occupies the same space as the other gases that collectively constitute air. The atmosphere is an air/vapor mixture, so when air is moved, the water vapor is carried along with it. The overall conditions of the air are just as dependent on the moisture content as on the temperature of the dry air itself.

But in some ways, water vapor acts independently of the air. For any given temperature and degree of saturation, water vapor in the air exerts its own vapor pressure. It flows or migrates from areas of higher vapor pressure towards areas of lower vapor pressure (in air or in other materials) regardless of the direction of any slight, steady air motion. Moisture, driven by vapor pressure, can even travel through porous materials through which air cannot go.

RH is approximately the ratio, expressed in percent, of the actual vapor pressure to the vapor pressure at saturation for the same temperature. This means that at a given temperature, the higher the RH, the higher the vapor pressure. There is always a tendency to equalize the vapor pressures between any two regions where moisture transfer can occur. So in general, moisture will migrate from high moisture-content areas to low moisture-content areas.

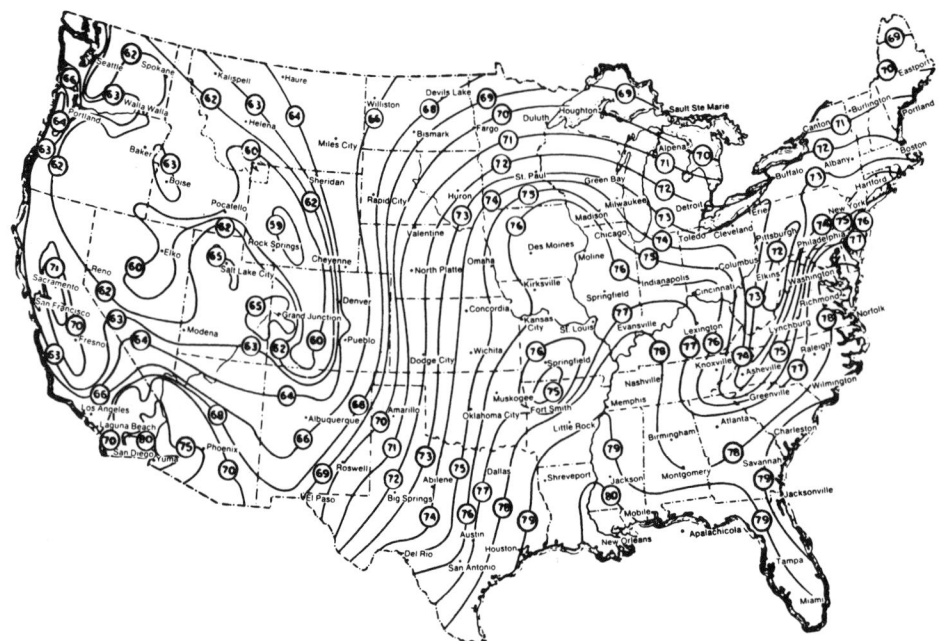

Summer wet-bulb temperature data. The wet-bulb temperatures shown will be exceeded not more than 5 percent of the total hours, during June to September inclusive, of a normal summer.

(a)

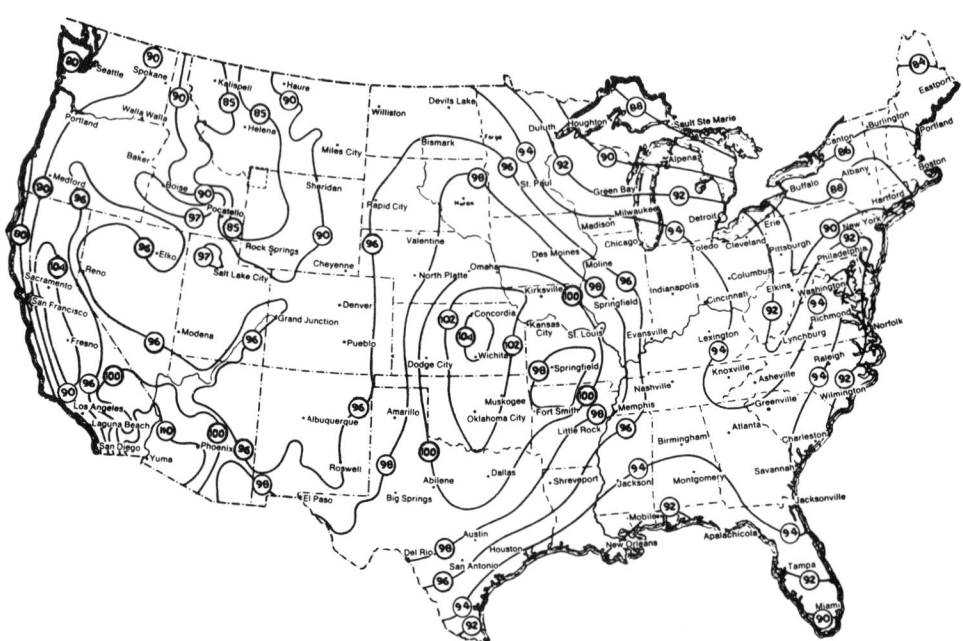

Summer dry-bulb temperature data. The dry-bulb temperatures shown will be exceeded not more than 5 percent of the 12 hours during the middle of the day in June to September inclusive, of a normal summer.

(b)

FIGURE 2.13. (a) Peak wet-bulb temperatures in the United States. (b) Peak dry-bulb temperatures in the United States. (Copyright 1983, Acme Engineering and Manufacturing Corporation.)

Beyond this generality, the subject of moisture transfer is an extremely complicated one because it is interdependent with heat flow, and all other factors can be overridden by a sufficiently high air flow carrying the moisture along with it. One subtlety is that liquid and vapor can flow in opposite directions through permeable materials. In particular, liquid moisture may flow downward by gravity, while vapor flows upward under the influence of vapor pressure differences, or by buoyancy, or both.

In porous materials partially saturated with water, moisture is known to migrate to the cold side by a process of evaporation, vapor flow, and condensation within the material. The transfer of latent heat along with the vapor can be substantial. This phenomenon can be explained as follows:

When moist air comes in contact with a surface below its dew-point temperature, condensation will occur. Once the moisture is released from the air, the vapor pressure at the condensing surface is reduced, thereby creating a vapor-pressure "vacuum" which draws nearby vapor. In this way, a pattern of continuous migration of moisture to the (cold) condensing surface is established, and will continue as long as there is a source of vapor to replenish that which is condensed, and the surface temperature remains below the dew point of the vapor. This is the cause of the concealed condensation problems that will be discussed in Chapter 3.

SUMMARY

Heat transfer is the flow of energy from a region of higher temperature to a region of lower temperature as a result of the temperature difference. A temperature difference between two points indicates that heat will move from the warmer to the colder point. Any two points will always tend to achieve thermal equilibrium (equalization) by heat transfer.

The relationship between the temperatures surrounding an enclosed space and the interior temperature determines the rate and direction of heat flow between the space and its environment. Heat absorbed by or transmitted through a building envelope is a combination of heat transfer by radiation, convection, and conduction. All three modes of heat transfer are a direct function of a temperature difference.

Thermal conduction is the mechanism whereby energy is transferred through solid materials. The transfer of heat occurs when heated molecules bump into adjacent molecules and pass their thermal energy to them. Convection heat transfer takes advantage of fluid motion (air, water, etc.) to transport heat more rapidly. Radiation is the transmission of heat energy by electromagnetic waves through either a trans-

parent medium or a vacuum. When the energy encounters an opaque object, it is absorbed, causing the object's temperature to increase.

In actuality, most heat transfer within buildings occurs as a composite effect of two or three modes. For this reason, heat transfer through buildings is for the most part viewed overall, rather than treated separately as conduction, convection, or radiation. Overall thermal resistances (R-values) are assigned to standard materials on the basis of empirical tests. The total R-value of a given construction element, composed of a series of layers of materials and air spaces, is the sum of the component R-values. The reciprocal of the total R-value is the overall coefficient of heat transmission (U-value). The U-value is used to calculate the rate of heat flow through the element by Equation 2.9.

Air spaces, insulating materials, and reflective linings in air spaces can be used to reduce building heat loss or gain. The resistance, or R-value, of air spaces is treated the same as the resistance of any material in series with it along the heat flow path.

Thermal air conditioning refers to the processes of heating, cooling, humidification, and dehumidification. Dehumidification can be seen on the psychrometric chart (Figure 2.11) as simply cooling air to below its dew point. During the process of humidification or dehumidification, the total enthalpy—sensible plus latent heat—remains the same. Evaporative cooling takes advantage of this phenomenon by evaporating moisture into a moving stream of air without adding more heat. The latent heat of vaporization necessary for the evaporation is drawn from the sensible heat, and as a result, the air (dry-bulb) temperature is decreased.

Condensation occurs when a saturated vapor contacts a surface at a lower temperature than the saturation temperature (dew point) of the vapor. The vapor loses its latent heat of vaporization and thus condenses on the surface.

Vapor pressure is an independent condition of water vapor contained within air. At a given temperature, the higher the RH, the higher the vapor pressure. Moisture tends to migrate from a high vapor pressure to a low vapor pressure, which usually means from high to low moisture-content areas.

Architects and designers conventionally use graphic and visual models to analyze buildings. Although a departure from this practice, a conceptual understanding of these thermodynamic principles is very important in evaluating the energy efficiency of a building shell (that is, how fast heat will enter or leave the building). Designs today are expected to conform to energy codes and standards, or to special energy criteria established by some clients, and the designer must be in a position to evaluate the performance of the envelope. Furthermore, new concepts are being proposed, and new products intended to save energy are being introduced in the market. The successful designer needs to analyze these claims in order to separate the innovations from the unsound gimmicks.

KEY TERMS

Thermal equilibrium	Surface conductance	Dew-point temperature
Absolute temperature	Film conductance	Relative humidity (RH)
Rankine	Film coefficient	Specific humidity (moisture content)
Kelvin	Infiltration	
Fahrenheit	Ventilation	
Celsius	CFM	
Conduction	Radiation (radiant heat transfer)	Dry air
Conductance		Saturated air
Insulation	Emissivity	Saturation line
Thermal conductivity (k)	Absorptivity	Sensible heat
Resistivity (r)	Opaque material	Latent heat
Resistance (R-value)	Transparent material	Enthalpy (total heat)
Convection	U-value	Grains
Forced convection	Dry-bulb temperature	Dew point
Natural (free) convection	Wet-bulb temperature	Evaporative cooling
		Vapor pressure

STUDY QUESTIONS

1. When heat flows through a series of materials, which of the following is directly additive?
   Conductance
   Conductivity
   Resistance
   Resistivity
2. What is the distinction between the four terms above?
3. If 1,000 CFM (1,700 m³/hr) of 0°F (−18°C) outside air is introduced into a heated room at 70°F (21°C) either by infiltration or intentional ventilation, how much energy will be needed to heat up the air to room temperature?
4. If outside air is at 95°F dry-bulb and 78°F wet-bulb temperature (35°C and 26°C, respectively),

a. What is the RH?
b. What is the moisture content?
c. What is the dew point?
d. How cool would an object have to be (what temperature) in order for condensation to begin forming on its surface?
e. How much heat must be removed per pound (or kilogram) in order to introduce it at 55°F (13°C) into a room?
f. What is the moisture content of the 55°F (13°C) air as it enters the room?
g. If the outdoor temperature increased to 105°F (41°C) dry-bulb but the wet-bulb temperature remained the same, what would the RH be?

BIBLIOGRAPHY

American Society of Heating, Refrigerating, and Air-Conditioning Engineers, Inc. *ASHRAE Handbook of Fundamentals.*

Bobenhausen, William. *Simplified Design of HVAC Systems.* Wiley-Interscience, 1994.

Burghardt, M.D. *Engineering Thermodynamics with Applications,* 3rd ed. HarperCollins College Division, 1986.

Carrier Air Conditioning Co. *Handbook of Air-Conditioning System Design.* McGraw-Hill Book Company, 1966.

Holman, J.P. *Heat Transfer,* 9th ed. McGraw-Hill Book Company, 2001.

Incropera, Frank P., and David P. DeWitt. *Introduction to Heat Transfer,* 4th ed. John Wiley & Sons, Inc., 2001.

Kreith, Frank, and Mark S. Bohn. *Principles of Heat Transfer,* 6th ed. Thomson-Engineering, 2000.

Lechner, Norbert, *Heating, Cooling, Lighting: Design Methods for Architects,* 2nd ed. John Wiley & Sons, Inc., 2001.

Rohsenow, W.W. *Handbook of Heat Transfer Applications,* 2nd ed. Bergano Book Company, 1986.

Rohsenow, W.W., and J.P. Hartnett. *Handbook of Heat Transfer Fundamentals,* 2nd ed. McGraw-Hill Book Company, 1985.

Rohsenow, Warren M., James P. Hartnett, and Young I. Cho. *Handbook of Heat Transfer,* 3rd ed. McGraw-Hill Professional, 1998.

Shurcliff, William A. *Superinsulated Houses and Air-to-Air Heat Exchangers.* Brick House Publishing Company, 1988.

Chapter 3

Thermal Dynamics of Buildings

HEAT GAIN AND LOSS

SITE RESOURCES AND MICROCLIMATES

THERMAL LOADS

HEAT TRANSFER THROUGH THE BUILDING ENVELOPE

HEAT GENERATION WITHIN BUILDINGS

OUTSIDE AIR

THERMAL MASS DYNAMICS

MOISTURE PROBLEMS

VISIBLE CONDENSATION

CONCEALED CONDENSATION

EXTREMELY LOW HUMIDITY

In this chapter, we will begin to apply the knowledge of thermodynamic principles to an understanding of the thermal behavior of buildings. The principles of thermodynamics are abstracted in the two *Laws of Thermodynamics*.

The *First Law of Thermodynamics* states that energy can be neither created nor destroyed. The total amount of energy in an environment remains constant and can be accounted for as it transfers from one place to another and as it transforms from one form to another. There is not a problem of energy shortage, but rather of energy not being well managed. It is possible to reuse the same energy over and over again for different purposes in buildings, thus realizing great energy efficiency. How buildings respond to energy is the subject of this chapter, while Chapter 7 suggests some techniques for designing energy-efficient responses into buildings.

A quality of energy can be defined whereby electrical energy has a higher quality than thermal energy, and the higher the temperature of thermal energy, the higher the quality. Notice that energy *quantity* pertains to how many Btu or joules of energy there are, while energy *quality* refers to the form of energy and its temperature. The *Second Law of Thermodynamics* states that as time goes by, energy will tend toward a lower and lower quality. The quality of energy never increases unless a penalty (conversion efficiency) is exacted. This means that in order to increase the quality of energy, you must add more quantity, so that as you increase the quality of some energy you decrease the quality of the rest, with the net effect being an overall decrease in quality. For example, thermal energy (heat) can be converted to electricity by a turbine generator. There is a practical limit of roughly one-third of the initial thermal energy that can be converted to electricity, while the remaining two-thirds must be exhausted from the process at a lower quality (temperature) for it to work. When the electricity is then used to illuminate a light bulb or energize a motor, all of the energy eventually ends up as heat when the electromagnetic radiation from the light is absorbed by some opaque material or when the motion created by the motor ends up as friction.

Ideally, a higher-quality energy should be used only for high-quality requirements (for example, electricity used for lighting and motors, and not for electric heating), and once it has degraded to lower-quality energy, it should be used for space heating or water heating. In actual practice, occasions arise when it is appropriate to use electricity for space heating, for instance, because it does not need a source of fuel within the building or a flue to exhaust combustion gases. But proper energy management would avoid resorting to electric heating due to the inherent discordance in quality between electricity as a resource and heating as a requirement. Furthermore, it takes more energy to produce electric heating than to produce direct combustion heating. Remember that the production of each unit of electrical energy takes 3 units of heat energy. By comparison, combustion efficiencies are typically 60 to 80 percent, with some newer heating equipment reaching above 90 percent efficiency. This means that every unit of usable heat requires $\frac{1}{90}$ to $\frac{1}{60}$ percent, or 1.1 to 1.7 units of combustion fuel (oil, coal, gas, wood, etc.). Of course, the best method is to heat with energy that has already been used once for some other purpose, such as lighting, or with electromagnetic radiation coming from the sun.

One of a designer's first concerns is to recognize the resources and the relevant environmental factors that exist on and around a site and to decide how best to integrate them into the final design while successfully integrating the design into the surrounding environment. Methods of integrating external environmental factors into building envelope designs for passive control of the interior are discussed in Chapter 7. For now, we will simply introduce some ways in which site conditions influence the thermal gains and losses of a building. Then we will look at how the envelope dynamically responds to these external conditions and the effect of internal sources of heat. And finally, using the psychrometric principles discussed in the previous chapter, we will address the subject of moisture migration and the potential problems resulting from it.

HEAT GAIN AND LOSS

Site Resources and Microclimates

A *microclimate* is the set of weather conditions of the environment immediately surrounding a building or group of buildings, differing from the general regional climate. Microclimates are created by such modifying elements as the presence or absence of plants that provide summer shading, windbreaks, and transpiration cooling; thermal storage from massive geological formations; and nearby buildings or other structures.

A building is subject to its microclimate, but it may also influence it. For example, redirecting rainwater away from its natural path into areas where it nourishes plant life can change vegetation patterns. The type of vegetation is different on the shaded side of a building than on the sunny side. Buildings and other structures alter and redirect wind patterns by blocking or channeling the wind between structures, creating areas of diminished and accelerated wind speed. The thermal storage effect of massive building materials, which will be explained later in this chapter, can have a moderating effect on temperatures in the immediate vicinity. It can also maintain uncomfortably hot temperatures in an area long after the air would otherwise have cooled down somewhat.

Cities

When many buildings are clustered, the number of local climate effects thereby created is multiplied, and many of the effects are magnified. In large cities, the following climatic changes are generally experienced:

- Higher temperatures
- Lower wind speeds in most areas, with higher wind speeds where channeled by tall buildings
- More cloudiness
- More rain instead of snow
- More air contaminants
- Less solar radiation

One reason for higher year-round temperatures in a city is its concentration of heat sources: heating and air-conditioning equipment, people, automobile engines, lights, and other machinery. Rainfall can be a very effective cooling mechanism, especially as water evaporates from wet surfaces. But streets and buildings are usually constructed of nonabsorbent materials and designed to drain water away quickly and thoroughly, so there is a minimum of evaporative cooling.

By blocking or channeling wind down narrow streets, cities also diminish the overall cooling effect from the wind. The geometry of high vertical walls and narrow streets adds to the summer heat by reflecting the sun's rays downward to be trapped and absorbed by the streets and building surfaces. In winter, this same geometry causes neighboring structures to shade all but the uppermost portions of south-facing walls from the sun's rays.

Another influence on sunshine is air pollution. Small particles in most urban air block some sunlight from ever reaching the city, while preventing the city's internally generated heat from radiating outward to dissipate. These particles also form the nucleus of fog droplets, which obscure even more sunlight. Up to twice as much fog occurs in the city in winter, when sunlight is most needed, as in the surrounding countryside.

Besides shading, another nuisance for city dwellers is the solar reflections from the highly reflective (or "mirror") glass that is used to reduce heat gain in office buildings, and from solar collectors on walls and sloped roofs. Although the reflected heat could be welcome in winter, the glare is intense. In summer, the reflected sunshine adds to the cooling load of the neighboring buildings. Since the most intense reflections occur when the sun's rays are nearly parallel to the wall, such reflections can be easily eliminated by external projections around windows or by strategically placed foliage.

Topography, Wind, and Sun

The inclination and orientation of a sloped site affect both the amount of sunshine received and the resulting air temperature on the site. Because of the position of the sun, southern slopes are generally warmer in the winter in the Northern Hemisphere. In the Southern Hemisphere, the solar configuration is reversed, so that the northern slopes are warmer in the winter.

Topography affects wind patterns by constricting the wind (Figure 3.1a), increasing its velocity in some areas, while sheltering other areas. The velocity increases because the same amount of air is being squeezed through a smaller area, so the leading air must hurry through to allow the air behind it to pass. Imagine a group of 10 people walking abreast toward a turnstile. If the entire group is to continue at the same overall speed, they must hurry through the turnstile in single file at 10 times the group rate. For the same reason, wind speeds increase at corners of large buildings (Figure 3.1b and 3.1c).

Slopes to the leeward side of winter winds offer a more protected building site (Figure 3.2). Hill crests, where wind velocities are increased, should be avoided in all but hot, humid climates. The bottoms of valleys and ravines, along with other topographic depressions, may channel and trap cold air masses during the night and in winter. Figure 3.3 shows how the distance that is effectively sheltered behind a windbreak is affected by the shape of a building's roof.

Surface Water

Because of their ability to absorb heat during warm periods and release it during cold periods, large water bodies such as

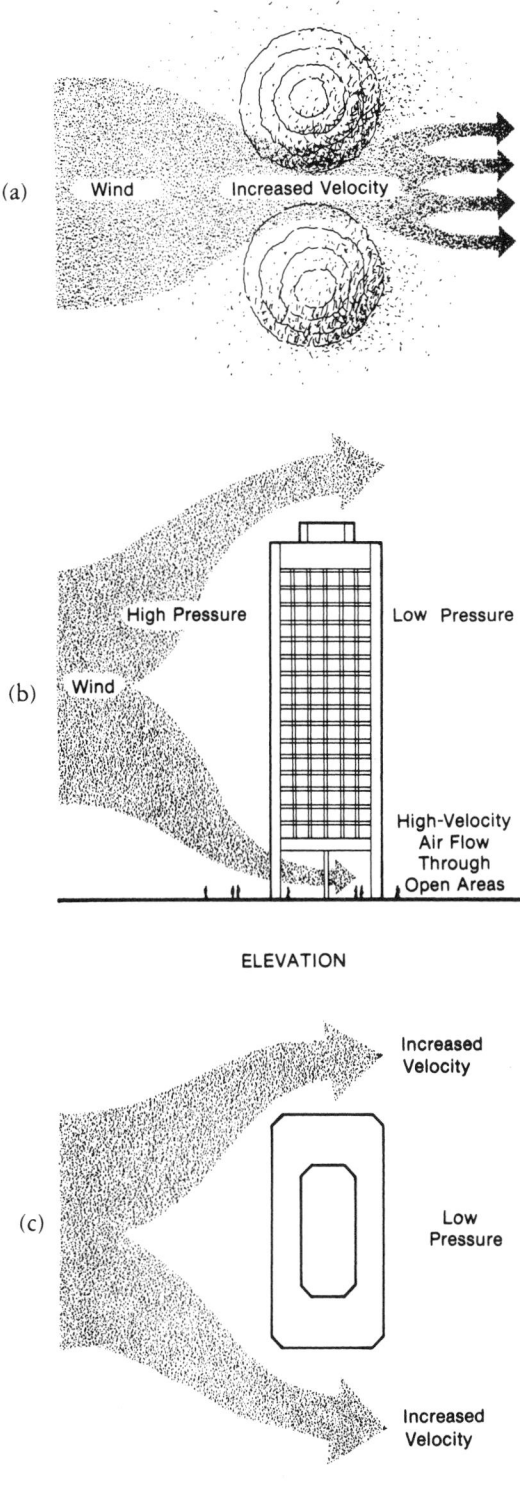

FIGURE 3.1. Wind patterns. (a) Constricted by topography. (b) Above and below tall buildings. (c) Around large buildings.

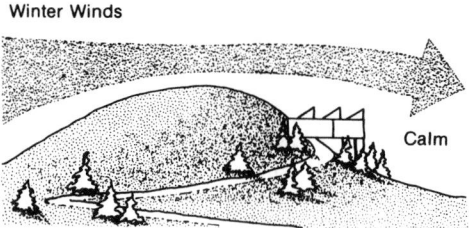

FIGURE 3.2. Sheltered building site.

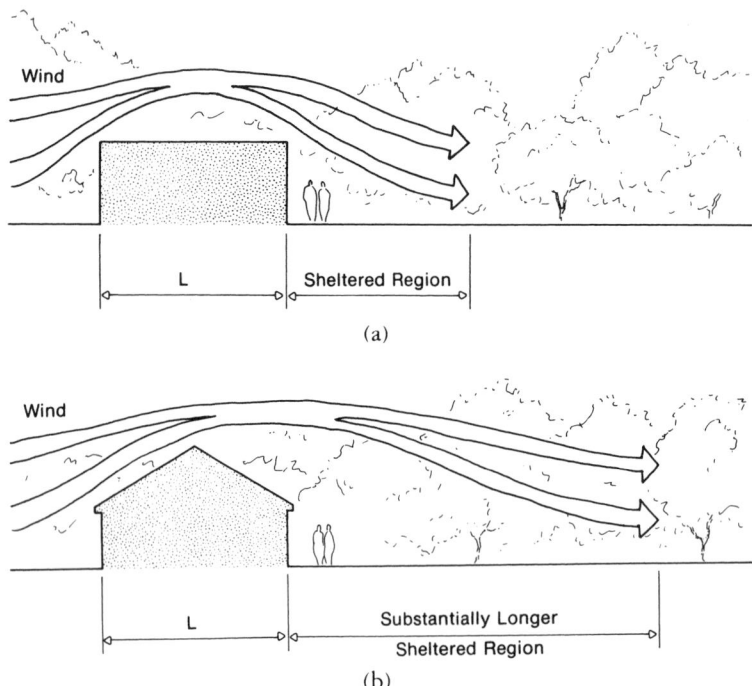

FIGURE 3.3. Shaping of the wind by roof configuration. (a) Flat roof. (b) Sloped roof.

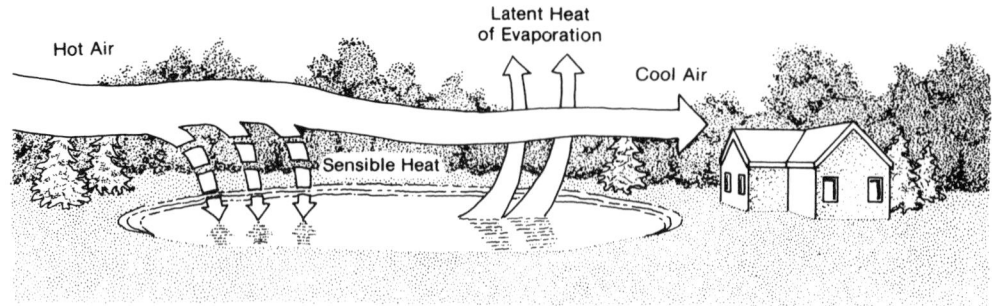

FIGURE 3.4. Evaporative cooling by water bodies.

lakes or the ocean exert a moderating influence on air temperatures, from day to night and throughout the year. Even small streams and ponds, in the process of evaporation, cool air temperatures in the summer (Figure 3.4). In addition, the large, flat surface of a lake or a river valley makes an excellent low-resistance path for winds to sweep across.

Vegetation

Vegetation has numerous influences on the microclimate around a building. It provides:

- Absorption of moisture during the day, and release of moisture at night
- Conversion of carbon dioxide into oxygen by day
- A source of food
- Privacy
- Seasonal variation
- Barrier to cold winter winds
- Shading from the summer sun
- Reduction of glare
- A psychological sense of life and growth

Healthy trees and shrubs provide shade, reduce glare, and, in dense configurations, are effective windbreaks. They also have a cooling effect in the summer, absorbing heat in the process of evapotranspiration in their foliage. They reduce the albedo (reflectivity) of the land surface, thereby reducing heat gain from outdoor reflection. Dense growths effectively absorb sound, enhance privacy, and are remarkably efficient air filters for dust and other particles.

Natural Convection Air Patterns

When air is heated it expands. It thus becomes less dense and more *buoyant* than the air around it that is not heated. The result is a tendency for warm air to rise, displacing colder air downward. Figure 3.5 illustrates the phenomenon known as the *stack effect,* which occurs because of warm air buoyancy and differential air pressures.

In high-rise buildings, the surrounding air at the upper stories is less dense than that surrounding the lower stories. Naturally, the less dense air will rise higher than the more dense air. The high-density air will tend to exert pressure against the building at its base, causing infiltration. When this infiltrating air is colder than room temperature, it will be warmed by room surfaces, internal sources of heat, or solar radiation entering through fenestration. As the air is warmed, it will tend to rise up through the building and exit near the top level under the pressure of the warm air behind it, overcoming the lower pressure exerted against the building by the low-density surrounding air.

In contemporary high-rise buildings, fire protection codes

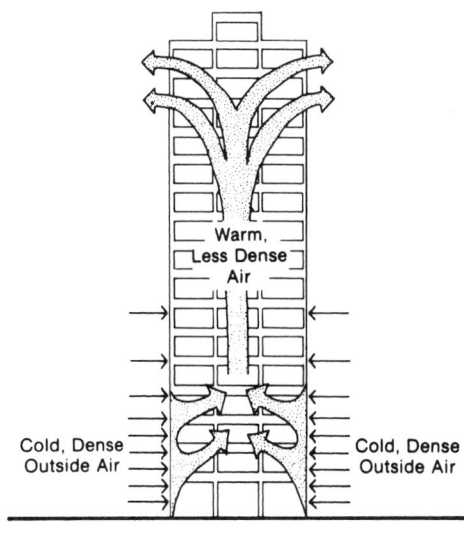

FIGURE 3.5. Stack effect in tall buildings.

are now restricting air interaction between floors. The purpose is to contain smoke, which also tends to rise, and prevent it from spreading throughout the building. In these buildings, the stack effect is very effectively eliminated.

Thermal Loads

An analogy can be made between the heat gain of a building and the leaking of a ship. If the leaks are big and numerous, water leaks rapidly into the boat just as heat leaks rapidly into a building. In the case of the building, mechanical air-conditioning equipment may be used to draw the heat back out of the building, and in the boat, a bilge pump is used to remove the water. The capacity of the bilge pump to remove water is dependent on its flow rate and how long it runs, and this must be well matched to the rate at which water is leaking in. If the pump is sized too small, it will not be able to keep up with the water leaking in, causing the boat to sink. If it is sized too large, the pump will turn on and off with excessive frequency and will wear out sooner. The same is true with air-conditioning equipment. If it is too small, adequate comfort (or process conditions) will not be maintained, and if it is too large, it will cycle on and off too frequently and wear out faster. In addition, with some equipment, excessive cycling will decrease its efficiency and cause more energy to be used. The rate at which heat needs to be removed is referred to as the *cooling load,* and the ability of the equipment to remove heat is called its *capacity.*

In the case of an occupied building, outside air is normally introduced intentionally for ventilation. Air-conditioning systems must therefore be sized to eliminate both the heat and humidity that unintentionally leak into the building (or are generated within it), as well as the heat and humidity introduced deliberately for ventilation.

Another name for the total load on the cooling system is *heat gain*. Actually, these terms (*cooling load* and *heat gain*) are not exactly synonymous. The distinction will be explained in Chapter 4, but for the time being, they will be used interchangeably.

In the winter, heat flows in the opposite direction, from the warm indoors to the colder outdoors. This is referred to as *heat loss,* or *heating load.* In this case, the analogy is to a water tank with a leak and an incoming water supply line refilling it. The larger the leak, the more water that must be added to overcome the loss and keep the amount of water at its proper level. In the same way, a heating system must replenish the heat in a space as it leaks out through the cracks and poorly insulated areas in order to maintain comfort conditions.

If the temperature inside a building is lower than that outdoors, there will be a continuous flow of heat inward (heat gain). If the inside temperature is higher than the outdoor temperature, there will be a continuous flow of heat outward (heat loss).

Whenever one object loses heat, another object or the surrounding environment must receive that heat by getting warmer, increasing its humidity, or increasing in some other form of energy. In the winter, a building loses heat either to the cold surrounding air as it passes by or to the exposed sky as radiant heat. The rise in temperature of the outside environment is imperceptible because the heat is spread out over a large area.

When air is cooled by means of chilled water circulating in a pipe, the heat lost by the air is transferred to the cooling water. Correspondingly, there is a rise in the temperature of the water. This warmer water is then returned to a piece of equipment called a *chiller,* which cools the water back down by removing some heat and rejecting it outside. The cooled water is then resupplied to the place where it can pick up some more heat from the inside room air, and the cycle repeats.

Building Envelope

When there is a climate difference between interior and exterior air, the building envelope acts to block the transfer of air and heat. While building skins can approach a perfect air seal, there is no such thing as a perfect thermal insulator. As long as a material has a finite thickness, it will allow some

heat transfer through it. By using enough thickness of a sufficiently insulative material, any degree of insulation is achievable.

Heat transfer through the building envelope encompasses all three modes: convection, conduction, and radiation. Air currents within a room bring air molecules into contact with the various surfaces. Surface conductance occurs, transferring heat between the air and the room surfaces. If heat is transferred to the air, the film of air next to the wall will rise in temperature. The air film will then rise because of its greater buoyancy. A convective air current is set up in this way, continuously bringing cooler air into contact with the warmer wall. If the heat transfer occurs from the air to the surface, the air circulation pattern will be just the reverse due to the falling of the heavier-density cooler air as it gives up its heat.

Once inside the solid material of the wall, the heat continues through by means of ordinary conduction. If there are air spaces within the wall, such as in a hollow stud wall, heat must flow either by a free convection current or by radiation.

When the heat finally passes through to the outside surface, the same convection process occurs between the surface and the air. The inside air is normally still except for convective currents and air motion caused by people and moving objects, while the outside air varies from still to whatever the wind speed is.

This process of heat transfer through the building envelope results in heat losses and sensible heat gains within the building. When the outside ambient conditions are severe enough, heating or sensible cooling loads are imposed on the mechanical climate-control systems. When all windows, doors, and other openings are closed tightly, the building skin allows only a minimal amount of air to enter, and thus the latent cooling load associated with the moisture in infiltrating air is also minimal. However, there are two other general sources of latent load: people and mechanically controlled ventilation to support those people. People continually add moisture to the air in varying amounts, depending on their activity level and the room temperature and humidity. Ventilation air may or may not contribute moisture to the interior, depending on the outside humidity. Under low outdoor moisture conditions, ventilation air will actually decrease the latent cooling load, and it is sometimes used for that purpose. However, when the outside moisture level is high, ventilation will increase the latent cooling load.

Loads

The various heating and cooling loads impinging upon an interior climate are as follows:

	Sensible	Latent
Heating	heat loss through envelope	
	outside air (infiltration/ ventilation)	
Cooling	heat gain through envelope	
	outside air (infiltration/ ventilation)	outside air
	people	people
	cooking	cooking
	lights	showers
	appliances and equipment	special processes

Some of these loads are distributed throughout a building, while others occur only at the envelope boundary. The last four cooling loads (both sensible and latent) are internally generated and are therefore referred to as *internal gains* or *internal loads*. Along with any load due to controlled ventilation, these four loads are distributed throughout a building wherever the particular source is located. In contrast, the heat gains and losses through the envelope occur only at the perimeter and in rooms exposed to the roof or ground floor. Similarly, infiltration loads occur only at the envelope. The heat gain and loss through the envelope include both transmission of heat through solid surfaces and heat gain from solar radiation through fenestration.

In multistory buildings, interior spaces on intermediate floors only have ventilation air as a source of heating load. These interior zones in office buildings and other buildings with high internal gains are served only with cooling, since the level of electrical and biological activity and other heat-generating sources within the space usually outweigh the cooling effect of minimum required outdoor air even in winter. Even in summer, the bulk of interior loads is internally generated heat gains. The perimeter areas, on the other hand, are much more weather-sensitive. This discrepancy calls for a different treatment, and as a result, interior and perimeter zones are often served by different systems, or at least different branches on the same system.

A building has the same type of heat balance equation as was defined for people in Chapter 1. It can be represented as follows:

$$\pm M = I + S \pm T \pm O \qquad (3.1)$$

where

$\quad$ M = mechanical heating (−) or cooling (+)
$\quad$ I = internal gains

$\quad$ S = solar heat gains
$\quad$ T = transmission of heat through envelope; loss (−) or gain (+)
$\quad$ O = outside air load; heat loss (−) or gain (+)

This relationship is illustrated in Figure 3.6.

The internal gains come from people, lighting, and any other energy-using appliances or equipment. Each person gives off about the same amount of heat as a single incandescent light bulb—between 100 and 350 W, the exact amount depending on sex, age, and activity level. The total amount of internal heat gain from occupants is thus determined by their number, the output per person, and the length of occupation. Each watt of electricity used for lighting contributes 1 W (3.413 Btuh) of heat to the building interior. Similarly, each watt used for any equipment or process inside a space contributes 1 W of heat to the space.

Solar gains through windows and skylights vary from zero (at night) to 335 Btuh/ft^2 (1,058 W/m^2). The exact amount of heat gain depends upon the time of day, time of year, latitude, cloudiness, orientation of the glass, the tilt angle, and the type and number of layers of glazing. It is also affected by any shading devices, inside or outside.

The transmission of heat through the envelope (glass, walls, roof, and floor) is a function of conductive characteristics, surface area, and the difference in dry-bulb temperature between the inside and outside. The outside air load is determined by the flow rate of outside air into the building (either by infiltration or ventilation) and by the difference in temperature and humidity between the inside and outside air. Air flow is expressed in cubic feet per minute (CFM) or liters per second (L/s).

It is worth mentioning that in the winter in the Northern Hemisphere, the maximum solar gain occurs through south-facing windows. Thus, while windows have an extremely high rate of transmission heat loss, south-facing windows may actually contribute a net heat gain over the winter because of the solar gain. In fact, many buildings with large glass areas are in the cooling mode year round, even at 0°F outside temperature. This is an example of overheating with passive solar methods.

Referring to Equation 3.1, when I + S does not equal ± T ± O, additional heat gain or removal must be provided by M in order to maintain a specified temperature and humidity level. Normally, there is some internal gain in almost every building. The building design can have considerable influence over the solar gain. Transmission is set by envelope material selections, area, and the indoor temperature specified; outside air can be minimized by paying attention to

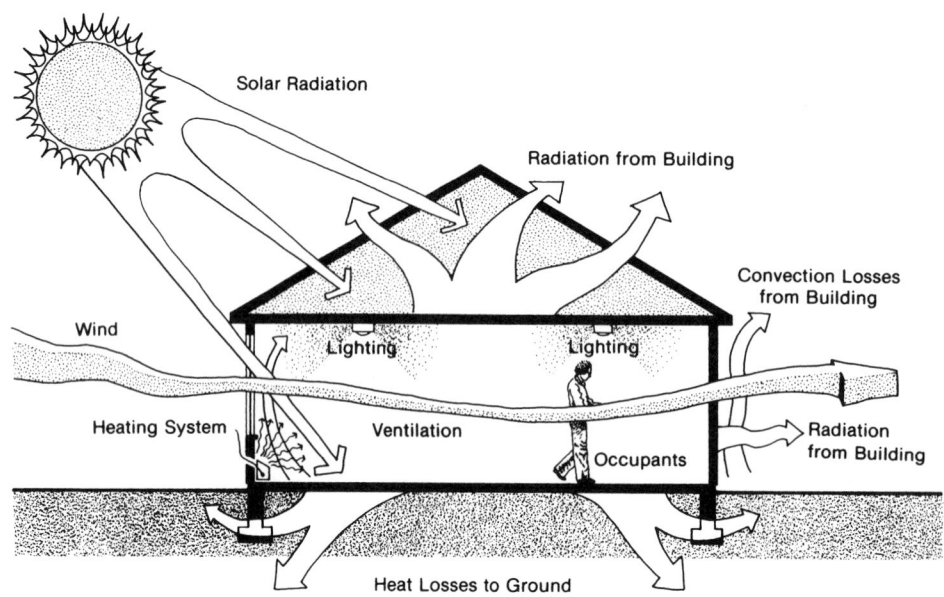

FIGURE 3.6. Heat gains and losses in buildings.

construction details and keeping ventilation rates as low as possible. Given the above parameters, every building has an outside balance-point temperature above which $I + S = \pm T \pm O$. Below this point, $-M$ is required to maintain specified conditions. There is also a cooling balance-point temperature below which $I + S = \pm T \pm O$ and above which M is required to maintain specified conditions. In between the heating and cooling balance points, the amount of outside air can usually be modulated to maintain a balance between $I + S$ and $\pm T \pm O$. Chapter 4 will show how to quantify the various loads, allowing the designer to select appropriate balance points.

When mechanical heating or cooling is required, there are a wide variety of systems that can be used to fulfill those functions. The various systems, each of which has an appropriate application, will be described in detail in Chapter 6.

For heating purposes, either heat is introduced directly into the room or warm air is supplied at a temperature higher than is comfortable so that it will offset the losses to the outside. In either case, the heat is applied at the envelope where the losses occur. There is also some heating or tempering of cold outside air, even when it is applied in the interior core of a building.

When cooling is required, air is supplied to the room at a temperature and humidity lower than desired so that it will "soak up" the excess heat and moisture. There are alternative methods of cooling that do not rely on air circulation, but cooling systems usually transport coolness in the form of cool air to the occupied space, where it will mix with the room air.

Outside air under cold winter conditions sometimes requires humidification as well as heating. Unless the outside air already contains enough moisture, or there is a source of adequate moisture as part of some process within the space, the heating of the air will reduce the RH to annoying or even dangerous levels. Low humidity is a problem in that it causes wood to shrink, resulting in cracks and the loosening of furniture joints. Also, the drying out of glue causes delamination. Low humidity, when extreme, can also cause discomfort and even health problems—from rough dry skin to respiratory illness. It is important, however, not to overhumidify, which can result in other problems during cold weather, as explained later in this chapter.

A dynamic analysis of the thermal loads on a building control system takes account of the following: the external factors of climate and topography; the internal factors of people, lights, and equipment; the envelope materials and construction techniques; any seasonal adjustments to the envelope (such as movable awnings); and the organization of inside spaces relative to orientation, function, and comfort requirements. There are an almost infinite number of energy exchanges taking place throughout a building at any given time. The thermal performance and energy efficiency of an occupied space are determined just as much by the design and construction of the space as by its heating, cooling, and ventilating systems and their controls.

Heat Transfer Through the Building Envelope

The building envelope is a mediator between the external and internal environments. Its components are fenestration, doors, walls, roof, and floor. It is made up of all the surfaces that serve as a border between the inside and outside. An envelope is erected primarily to enclose and shelter space. Providing a barrier to rain is practically a universal feature of all buildings. Another function of the envelope is to provide varying degrees of protection from the sun, wind, and harsh temperatures.

While buildings are generally regarded as fixed structures, the envelope may be dynamic and sensitive to changing needs and conditions, sometimes letting in heat, light, air, and sound, and sometimes closing these resources out. Each element of the envelope can serve as a static barrier (such as a conventional wall), an on–off switch (such as a door or window), or a modulating regulator (such as venetian blinds).

If a climate is consistently harsh, an appropriate envelope would be a static barrier. Where there is an intermittent resource, such as the sun, that is desirable at times, a dynamic and controllable element would be called for. When external conditions are close to the desired interior qualities, the envelope can be simply an open structural frame. In most inhabited areas, both harsh and pleasant conditions prevail at various times, so the envelope should be somewhat dynamic, allowing for a modulation of the degree of separation between the inside and outside. At certain times the envelope would resemble a protective shell, while at other times openings in the shell would allow in some or all of the outdoor resources.

Building envelopes are becoming increasingly dynamic, and Chapter 7 discusses some methods that are now used to accomplish proper control. The exact architectural solution in any given case will depend on the dynamic range required, local material availability, and the influence of local style preferences.

With dynamic envelopes, it is vital that occupants (or at least a designated operator) understand how, why, and when to make adjustments, and it is the responsibility of the designer to convey this information to appropriate personnel or automated controls. An additional benefit of this sort of flexibility is that the occupants may feel more in control of their environment.

Fenestration

Fenestration is the term used for any light-transmitting opening in a building wall or roof. Examples are windows, skylights, and clerestories. It includes the glazing material (glass or plastic); framing, mullions, muntins, and dividers; shading devices, internal, external, or integral (in between layers of glazing); and curtains, drapes, or other forms of adjustable barriers. Control elements such as screens, shutters, drapes, blinds, diffusing glass, and reflecting glass can make fenestration even more dynamic than a mere operable window. They make it possible to change the transmissability of heat, light, air, and sound.

The purposes of fenestration are (1) visual communication with the outside world (awareness of weather and changing events) and relief from claustrophobia and monotony; (2) admittance of solar radiation to provide light and heat; (3) admittance of outside air when desired; (4) when near ground level, a means of entrance or egress (exit) in case of fire or other emergencies; and (5) enhancement of the exterior and interior appearance of the envelope itself.

From the point of view of energy and comfort, the most important considerations in the use of fenestration are adequate control of sunshine and heat transmission. By its very nature, fenestration is intended to permit the entrance of light and, along with it, the sun's thermal energy. To prevent excessive glare and radiant heat buildup through fenestration that is exposed to direct rays or reflections from neighboring surfaces, some form of solar control is required. On the other hand, sometimes it is desirable to collect solar heat gain within a space.

Sun control devices that admit heat gain but inhibit glare are typically located on the interior side of the glazing so that sunshine can be converted to heat within the space. Wherever the sun's heat is not desired, barriers or light-filtering devices should be located outside the glazing so that the heat from solar radiation can be intercepted before entering the envelope and be carried away by the outside air.

Exterior shading rejects about 80 percent of the incident solar radiation, while internal shading accepts all that is transmitted through the glazing, reradiating much of it into the space. Outside shading devices are continuously cooled by ambient breezes carrying away the absorbed heat, but inside shades intercept the radiation only after it is trapped inside the envelope. The exterior location of shading mechanisms also provides additional design options for the facade. However, exterior mechanisms are subject to problems of dirt accumulation, wind damage, and weathering.

Old-fashioned cast-iron radiators give off about 240 Btuh/ft^2 (757 W/m^2). By comparison, the heat gain through a square foot of unshaded glass facing south at a latitude of 40° north at noon on a sunny day in January is about 250 Btuh (790 W/m^2). In August, it is about half that amount.

Imagine every square foot of window facing east, south, or west that is unshaded in the middle of the summer as thermally equal to half of that area in cast-iron radiators at full output. Unshaded windows can require tremendous unnecessary expenditures of energy and money for mechanical cooling, and in many cases, comfort is virtually impossible to achieve through mechanical systems, no matter how large they are.

There are situations in which that amount of "free" heat contribution to a space in the winter could be very helpful. In that event, some sort of dynamic control is necessary to prevent overheating during the summer. Chapter 7 will discuss some methods for dynamically controlling solar gain for its benefit in winter while blocking it out when not desired.

The other important thermal aspect of fenestration is transmission of heat between the inside and outside. Fenestration is typically less thermally resistant than insulated opaque walls and roofs, and when not supplemented by operable insulation, it is *much* less thermally resistant.

This relatively high thermal conductivity of fenestration causes two types of comfort problems. First of all, perimeter glazed areas cool the adjacent interior air in the winter. The vertical layer of cooled, denser air thus created along the glass then drops to the floor, blanketing it like a chilly carpet. If kept in motion by a pattern of air currents, it can even result in uncomfortable drafts. The second problem is the cold radiant temperature of glazings. The high conductivity keeps the inside and outside surfaces at close to the same temperature, about halfway between the indoor and outdoor temperatures. Thus, the radiant temperature of the internal surface is very much influenced by the external temperature: Large fenestration areas will contribute to an MRT approaching the exterior temperature, having an adverse effect on comfort when the exterior temperature is much higher or lower than desired. These comfort problems can be offset by proper treatment with mechanical systems, but at the expense of higher energy and equipment costs.

Aside from the use of operable thermal barriers, the thermal resistance value of fenestration can also be increased by adding a second or third layer of glazing to the opening. Multiple layers of glass with air spaces between them are sometimes referred to as *insulating glass*. Figure 3.7 illustrates how double and triple glazing is normally affected. A separate sash unit (known as a *storm window*) added to a single-glazed window cuts thermal conduction and infiltration nearly in half (b). A single sash containing insulating glass (c) results in a comparable reduction in transmission of heat but does not affect the infiltration rate. A sash containing insulating glass combined with a storm window (d) provides a reduction in transmission to one-third and infiltration to

one-half. Economic limitations often do not permit both multiple glazing and operable thermal barriers. In that case, a choice must be made on the basis of economics, aesthetics, and relative solar benefits. (A single layer of glazing allows more solar radiation to enter, but in that case, some form of nighttime insulation should be used.)

Fenestration can be highly beneficial to a space but can also contribute to problems of excessive heat gain, heat loss, and infiltration. The proportion of glazing to opaque insulated area should be studied for its impact on energy economy and human comfort. Fenestration can contribute to energy conservation by providing solar thermal energy, natural ventilation cooling, and illumination as a substitute for electric lighting. An evaluation of whether to have fenestration and, if so, what kind, how much area, and so on, should be based on (1) architectural considerations, (2) thermal control capability, (3) economics (first costs versus energy costs and overall life-cycle costs), and (4) human needs, such as the psychological desire or physical need for windows.

Opaque Barriers

The opaque portions of the envelope typically act as a fixed barrier to heat, light, air, and sound. The greatest variability is in their ability to block heat transmission.

The relative ability of an envelope component to limit heat gain and loss is a function of its construction, orientation, and selection of materials. The orientation has a bearing on absorption of solar radiation and convective heat transfer from strong winds. The construction methods and selection of materials will determine the R-value and U-value as defined in Chapter 2. An envelope's ability to resist thermal transfer and regulate interior conditions for thermal comfort can be estimated with a knowledge of the R-value or U-value, the desired indoor temperature, and the outdoor climate conditions. The procedure for this estimation is the subject of Chapter 4.

Heat is transferred through an envelope in the following manner: The gentle motion of nominally still air within a room, caused by activity and the buoyancy effects of warmer and cooler air, brings a flow of room air in contact with the inside surface of the envelope, causing convection transfer. Heat conducts through solid layers and is transmitted by radiation and convection through air pockets within the envelope. The motion of the wind against the outside surfaces convects heat to or from the envelope. The more energetic the wind, the greater the convective heat transfer. There are also radiant interactions both between the envelope's inside surfaces and the contents of a room, and between the outside surfaces and their surroundings.

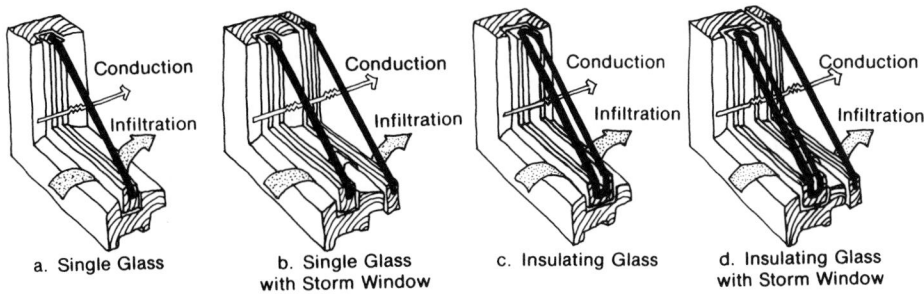

FIGURE 3.7. Effect of multiple glazings on conduction and infiltration through windows.

The link in the chain of heat transfers most readily controllable is that within the envelope itself. The thermal resistance can be greatly increased by adding more material (insulation or reflective sheets) or more air spaces. The use of insulating materials having slow conduction rates and thus high thermal resistance can dramatically reduce the rate of heat transmission. The important characteristic of reflective linings is a very low emissivity: 0.05 compared to an emissivity of 0.82 for common building materials. Air spaces are not as effective in reducing heat transfer as insulating materials, but they do add some benefit. Generally, the thickness of the air space is not critical.

High levels of insulation are desirable because they (1) maintain a comfortable interior MRT, (2) control condensation and moisture problems, and (3) reduce heat transmission through the envelope, which determines the energy requirements for both heating and cooling.

A poorly insulated wall, roof, or floor will allow the exterior temperature to have a greater influence on the interior MRT. Increased thermal resistance of the envelope will result in interior surface temperatures (and therefore MRT) that are closer to the controlled indoor temperature and thus supply greater comfort.

Heat Generation Within Buildings

Heat transfer in a building can be thought of as similar to that of a human body. In this analogy, the process of metabolism corresponds to activities of the occupants and the operation of equipment, lights, and heating systems. These internal activities produce heat.

Heat is generated by the occupants in proportion to their quantity and collective activity level. While engaged in sedentary activity, the human body loses heat at about the same rate as a 100-W light bulb. It drops to about half this level during sleep and increases to about 10 times the sleeping rate for the hardest sustained work. Heat is also introduced by lights and all equipment that uses electricity, such as appliances, heavy machinery, and computers. In addition, high levels of lighting create a high MRT environment, much as stage lights do.

Regardless of usage (lighting or machinery), all electrical energy entering a space is eventually converted into heat energy and is considered internal gain. The energy conversion rate is simply:

$$1 \text{ W (electricity)} = 3.4 \text{ Btuh (heat)} \qquad (3.2)$$

Other internal gains come from such sources as cooking and laundry equipment and water evaporation during bathing.

The thermal performance of a building is a function of the balance between the heat generated internally and the heat exchange with the external environment. Just as the human body must maintain a constant internal temperature, a building must maintain a constant specified temperature. That means that whatever heat is produced inside must be transmitted outside—no more and no less.

When it is colder outside than inside, the internal heat gain can be dissipated to the outside. If the temperature is cold enough outside, and the conduction rate of the envelope combined with the air exchange rate is high enough, all the heat produced is transmitted out. If the combined envelope conduction and air change rates are too high for a given climate condition, too much heat is transmitted out, and either the heat loss rate needs to be reduced somehow or the internal heat generation must be increased. Usually, the latter is done by activating a space-heating system. On the other hand, if the outside ambient temperature is higher than the inside, or if there is too much gain from solar radiation, the envelope cannot lose heat and may gain additional heat. When this happens, a special mechanism must be employed to remove the heat actively, in the same way that the bilge pump is used to remove the water from a leaky boat. In the case of the human body, that mechanism is sweating; for a building, it is a mechanical cooling system.

One of the differences between residential and nonresidential buildings is that at the same outside temperature a residence may require additional heat because it has little internal gain, while an office or factory may not require heating because of its concentration of internal heat sources (people, lights, and equipment).

The need for heating in a nonresidential space is dependent on the level of insulation in the envelope. Even in residences, the new "superinsulated" homes eliminate the need for a central heating system; the little heat necessary to offset the low heat loss is supplied by cooking, appliances, lighting, and body heat. But many commercial buildings have such a high internal heat generation—especially modern offices with computing equipment and a high density of people and bright, even illumination—that they need no regular heating source even with only a moderate amount of envelope insulation. The reason many conventional commercial buildings have traditionally needed heat at all is that they have had virtually no insulation.

If there is sufficient insulation and thermal storage mass within a building, it can hold the prescribed temperature overnight and needs no heating at all. In most cases, however, some heating is required to warm up the space in the morning after being unoccupied over a cold winter night.

Due to the high internal gain in nonresidential buildings, the space inside the larger, partitioned ones is usually divided into at least two zones for thermal control:

- Interior zones: cooling only (plus the tempering of ventilation air)
- Perimeter zones: cooling and heating at the envelope as required

The perimeter zone is usually the first 15 to 20 feet inside the building's envelope, so some small commercial buildings may have no interior zone.

In interior zones, there is little or no direct heat transfer with the outside environment, so all the internal heat produced must be mechanically removed by a cooling system. In perimeter zones, the internal heat gain can be dissipated directly outdoors if the outside temperature is low enough and if the envelope is able to transmit it. Often, the heat loss at the envelope is greater than the adjacent internal heat gain. At those times, a perimeter heating system is required.

In interior zones, and at the perimeter enclosed by a well-insulated envelope, the amount of heat removed in cold weather can be modulated by increasing or decreasing the amount of ventilation air. Using cool outside air for neutralizing internal heat gain instead of mechanical cooling is known as the *economizer cycle.* Buildings with high internal heat gains normally have some sort of mechanical cooling

system, and they can be fitted with economizer cycle controls that will automatically open dampers to draw a large amount of outside air into the ventilation system, which is normally set for a minimum of outside air.

Outside Air

Outside air must be brought in to continuously replace indoor air that has become contaminated with odors, carbon dioxide, and the other harmful substances discussed in Chapter 1. It is necessary to remove these contaminants in order to sustain human occupancy, and while it is allowable to filter and recirculate room air to remove contaminants under some circumstances, a minimum amount of air exchange with the outdoors is always required to dilute the carbon dioxide buildup.

There are two ways to introduce outside air into a building: infiltration and ventilation. The latter may be either mechanically controlled or natural ventilation (discussed in Chapter 7). Ventilation air is controlled and desired. Infiltration is uncontrolled and generally undesired. Prior to the advent of mechanical ventilation, the common architectural style of high ceilings provided for a larger volume of air available to the occupants for dilution of odors and carbon dioxide. Operable windows and infiltration were relied upon for the exchange of air with the outdoors. The high ceilings in older assembly spaces such as auditoriums allow a reserve of fresh air to build up in between periods of dense occupancy.

Infiltration is air leakage into buildings through cracks around windows and doors and through walls; it is caused by outdoor wind pressure. Winds on the windward side of a building blow outdoor air into indoor spaces, displacing conditioned air, which leaves through similar cracks on the opposite side of the building where the lower outside air pressure creates a suction. The outside air that has entered the space must then be conditioned (heated, cooled, humidified, dehumidified, or cleaned).

Infiltration is reduced by using weather stripping or gasketing around doors and windows and caulking for fixed cracks or penetrations in walls. The life of these materials is usually 10 years or less, and they need to be replaced within that time span to maintain control over infiltration. A tightly sealed space will still have one-half to one air change per hour from infiltration. A very leaky space can have two to three air changes or more per hour.

To alleviate the discomfort of drafts and uneven temperatures caused by infiltration, it is frequently the practice to introduce more air into a space than is being mechanically removed. A slightly higher than atmospheric pressure is established, and the infiltration is overcome by *exfiltration,* which is the leakage of air out of the building. A deliberate positive pressure in a building is known as *pressurization.*

In order to introduce more air into a space than that being removed, the central air-handling system must bring in more outside air than it exhausts. The same amount of outside air ends up in the space with either infiltration or pressurization, and thus the same amount of heating or cooling is required ultimately to maintain the prescribed conditions, but more uniform control and better comfort are achieved if the outside air is conditioned first before entering a room.

In addition to exfiltration, room air is directly removed to the outside through exhaust hoods in kitchens and over certain industrial processes that give off air contaminants. In some cases, outside air can be introduced at or near the exhaust hood with little or no conditioning required, which reduces energy requirements for heating and cooling.

In those spaces where overall room exhaust fans are used, such as bathrooms, storage rooms, janitors closets, and dark rooms, the outdoor air quantity supplied is slightly less than the exhausted air, creating a negative pressure condition that draws air in from surrounding areas and prevents odors and contaminants from migrating to those surrounding spaces.

THERMAL MASS DYNAMICS

The sources of heat gains to a space are the sun, lights, people, transmission through the envelope (walls, roof, glass), outside air, and electrical equipment (machinery, appliances, office equipment). A large portion of it is radiant heat that does not immediately heat up the space, since it must first strike a solid surface and be absorbed before it can be reradiated back into the space or be convected into the room air.

As radiant heat from the above-mentioned sources strikes a solid surface such as a wall, floor, or ceiling, some of it is absorbed, raising the temperature at the surface above that inside the material and above that of the air adjacent to the surface. This temperature difference causes heat to flow into the material by conduction and into the air by convection until the temperatures equalize, as illustrated in Figure 3.8.

Heat conducted into the material is stored, and heat convected into the air becomes heat gain to the space. The proportion of the radiant heat being stored depends on (1) the ratio of the thermal resistance into the material to the thermal resistance into the air film and (2) the temperature difference between the surface and the rest of the solid material compared to the temperature difference between the surface and the surrounding air. The resistance to heat flow into most construction materials is much lower than the air film resistance, so when the temperatures of the air and the solid material start out equal, the majority of the absorbed heat is stored in the material. However, as the process of storing radiant heat continues, the inside of the material becomes warmer, which slows down the heat transfer and the storing of heat. On the other hand, if the space temperature rises, less heat is convected from the surface and more heat is stored.

Besides being warmed by radiant heat, a surface can also be warmed by convection from adjacent air at a higher temperature, and this heat may then conduct into the material and be stored in it. However, since the thermal conductance through the air film is relatively low, the effect of any incident radiation tends to dominate. The air temperature must be considerably higher than the surface temperature to enable appreciable storage of heat from the air.

As the intensity of solar radiation fluctuates in its diurnal (day-night) cycle, the incoming radiant heat increases much faster than it can be dissipated into the room air. The intensity then drops just as rapidly after peaking at a fairly high level. As a result, much of the solar heat is first stored as it enters a room and then is gradually released into the space. This delays and diminishes the peak solar heat gain, when thermal storage in solid objects is available.

In contrast, artificial lights contribute relatively constant heat, but a large portion of it is stored just after the lights are turned on, with a decreasing amount being stored the longer the lights are on. As the storing material becomes saturated with heat, less and less of the absorbed radiant heat conducts into the material and increasingly more convects into the air. The heat released to the air will tend to increase the air temperature. If the air temperature is kept constant by removing the heat with an air-conditioning system, the heat convected into the air will become part of the load on the air-conditioning equipment.

Heat is stored in building materials and furnishings much the same way that a sponge soaks up water. The capacity of a material to store heat depends upon the quantity (*mass*) present and its characteristic *specific heat* (Btu/lb or joule/kg). As a mass gets warmer, it has a decreasing capacity to store more, just as a sponge becomes saturated with water.

In an air-conditioned space, the accumulation of heat during the day is stored in the solids present. When the cooling system is shut off overnight, transmission of heat through the envelope continues either into or out of the building, depending on the outside temperature. If the nighttime ambient temperature is cooler than the indoor temperature, heat may be lost to the surrounding environment, so the heat stored in the mass will gradually be released. If the outside temperature

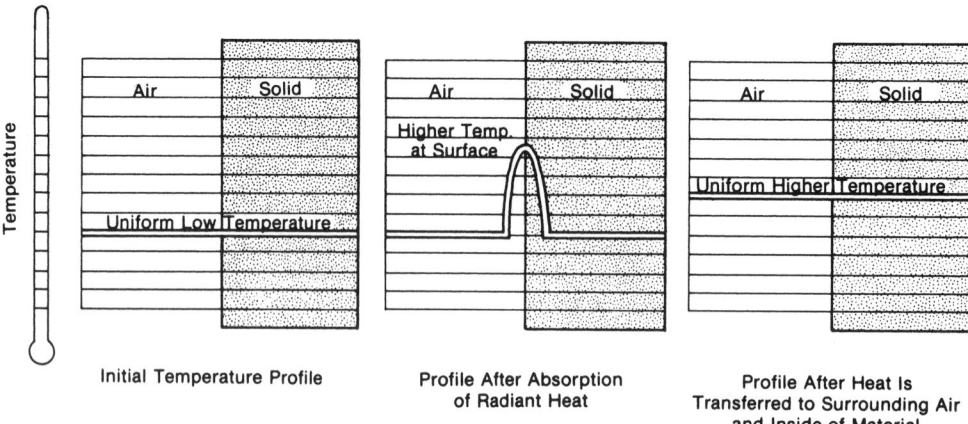

FIGURE 3.8. Radiant heat penetration into a solid object.

remains warmer than the indoor temperature, heat will continue to build up in the space, some of it flowing into the solid materials, where it is stored. This represents an initial cooling load for the air-conditioning equipment when it is turned on again.

Similarly, during the heating season, heat discharges from the collective mass overnight when the thermostat is set back. The mass must then absorb heat during the warmup period to raise its temperature. The bulk of the heating load during the warmup period is attributable to warming the collective mass.

The magnitude of the storage potential depends on the *thermal capacity*—or heat-holding capacity—of all the materials within and surrounding a space. The thermal capacity of a material is its weight times its specific heat, and thus is directly proportional to its weight. Since the specific heat of most common construction materials is approximately 0.2 Btu/lb·°F (830 joules/kg·°C), the thermal capacity can be roughly determined by the weight of the materials.

The most massive materials are those with the highest density (lb/ft^3 or kg/m^3): brick, stone, concrete, masonry blocks, baked clay, adobe, rammed earth, and so on. The table of thermal properties in Appendix Table B4.2 compares the densities of common construction materials. The specific heat of each material is also given.

A large amount of mass within the envelope enables a building to store excess thermal energy from the sun or internal gain for later use. By delaying the heat gain to the room air, *thermal storage mass* helps to stabilize interior temperatures in spite of widely varying exterior conditions and internal heat generation. Figure 3.9 illustrates this moderating effect in a building subjected to large fluctuations in outdoor temperature.

The ability of materials to store heat and release it at a later time is frequently compared to the ability of a flywheel to store and release energy as momentum. There are various means of increasing thermal storage mass. For example, high-mass materials may be integral parts of the envelope itself, or they may be incorporated into the furnishings or decor of the interior space. The important point is that the mass must be within the insulated boundary of the envelope for the maximum benefit.

If the envelope contains a large amount of mass, it will store the heat as it passes through, causing a delay in transmission.[a] This *thermal lag* can be a matter of several hours or several days. The more mass there is, the longer the delay.

The magnitude and configuration of the mass can be arranged in order to store heat from the sun or from occupants so that it can be used to provide warmth at night. In commercial buildings, the thermal lag can be designed to last overnight until morning so that the building doesn't have to be warmed back up after a nighttime thermostat setback. The heat storage and subsequent rate of release into or out of the building depend on the thermal storage capacity and the severity of the inside-outside temperature difference. Misjudgments in the use of thermal storage mass can result in excessively high temperatures or cooling loads on sunny days or insufficient storage overnight.

The desirability of high or low thermal storage mass depends on the climate, site, interior design condition, and operating patterns. For example, a high thermal storage mass

---

[a] This should not be confused with the low transmission rate resulting from thermal resistance. Thermal storage mass and thermal resistance are distinct properties. It is possible to have a building with a high thermal resistance and a low thermal storage mass, or vice versa.

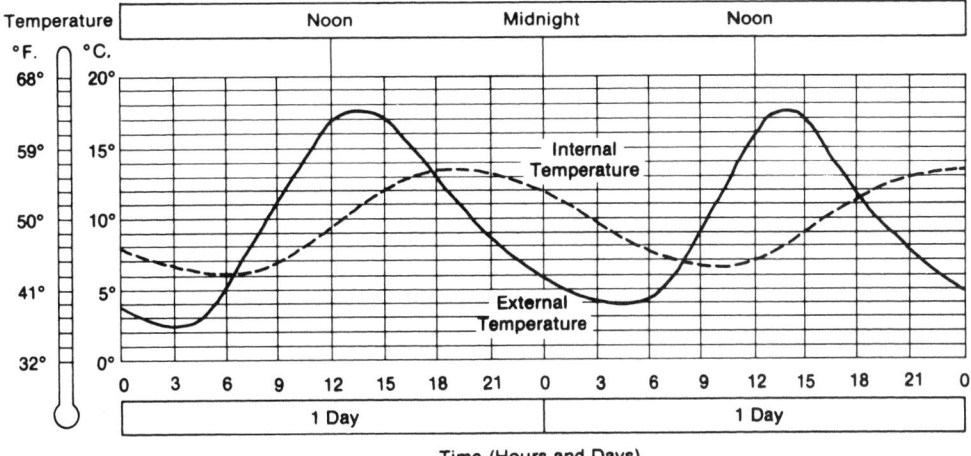

FIGURE 3.9. Moderating effect of thermal mass. Note the time lag and decreased amplitude.

is desirable when the outside temperature swings widely above and below the desired interior temperature. Low thermal storage mass is better when the outside temperature remains consistently above or below the acceptable indoor temperature. Or, if solar heat gain is to be used, the storage potential allows the heat to be stored for nighttime and cloudy-day use.

The experiential wisdom of indigenous cultures is demonstrated in the application of the thermal storage mass principle in the thick adobe houses of the southwestern United States and in the heavy masonry buildings of the desert areas in the Middle East. The dry air in those locations permits large diurnal swings of outside air temperature. When the morning sun begins to penetrate the wall exterior, the heat is absorbed and is not released at the inside surface until early evening. The gradual release of heat from the walls then helps keep the occupants warm all night. As a result, there is no need for mechanical cooling during the day or for mechanical heating at night.

Consider a climate where the days are warmer than comfortable and the nights are colder than desired. If the walls are not massive, an air conditioner would be needed to provide comfort during the day, and a heating system would have to run at night. With proper placement and quantity of mass, heat entering the building during the day would be absorbed and stored so that there would be little or no rise in the space temperature. At night, as the outside temperature drops and heat is lost through the envelope to the outside, the heat in the mass is released to keep the inside temperature comfortable. Both heating and cooling requirements are reduced or eliminated.

Now consider a different climate where both nights and days are uncomfortably cold. During the day, massive walls exposed to the sun store a portion of the solar heat, while lighter-weight walls transmit more of it into the space. The greater solar heat gain benefits the lighter building by decreasing the load on the heating system. At night, however, the sun's heat stops, and the heat stored in the massive walls of a building is released so that less heat is required from the heating system. The net effect of massive or low-mass walls may be roughly the same, resulting in an equivalent average load on the heating systems. The heat savings in the massive-walled building at night are canceled by the added heating necessary during the day.

The occupancy program and the control procedure can also have a major impact on the optimum amount of mass. How much temperature drift is allowed and night setback are two examples of such variables. A lightweight wall will result in direct energy savings during the night setback period. With a massive wall, nighttime savings may be even higher, since the heat released to the space will offset some of the losses. But the heat released may keep the space warmer than necessary, causing a larger heat loss. And in the morning, the heating system on its warmup cycle will need to reheat not only the air, but also the mass.

Another important factor is the placement of the mass. The mass can be in the form of (1) masonry walls on the outside with insulation on the inside, (2) massive components on the inside with insulation on the exterior, or (3) heavy objects within the interior of the building. Each placement of the mass has a different influence on the overall thermal behavior of the building.

As mentioned before, a storage mass exchanges (that is, stores or releases) heat by convection with the inside or outside air and by radiation. The outside of a thermal storage wall exchanges radiation with the surrounding environment, the sun, and the open sky. On the inside, the wall interacts with other walls, partitions, furnishings, floors, ceilings, and any occupants.

From the occupants' point of view, the thermal storage mass constitutes a portion of the MRT of a room. When a room air temperature normally at 70°F (21°C) is allowed to drift down to 50°F (10°C) during the unoccupied period, the MRT gradually approaches the lower value. When the air temperature is rapidly brought back up to 70°F (21°C), the MRT gradually approaches the higher value. The response time, or lag, in heating or cooling depends on the amount of mass and its heat transfer properties, and may be a few hours or a week or more. In order to provide adequate comfort, the radiant chill during the warmup period must be compensated for by warmer-than-normal air temperatures or some other means. The same applies in a cooling situation: A slightly cooler than normal air temperature or some other compensating condition must be provided while the storage mass temperature and MRT drop back down to their desired point.

One strategy for reducing the required size of the cooling system for a building with relatively short, but intense, peak load periods is called *precooling*. Precooling a space, its contents, and the surrounding mass to the desired temperature prior to occupancy can increase the heat storage capacity available later when the heat gain is at its peak. The result is less net heat released to the room air during that peak period.

When the space is precooled to a temperature lower than desired and the thermostat is reset upward to the desired condition when the occupants arrive, no additional storage occurs. Instead, the cooling equipment shuts off, and there is no cooling during the time the mass warms back up. When the cooling system begins to supply cooling again, the load is about what it would have been without any precooling. An exception to this is when the peak load occurs right away or when there is sufficient mass to delay its warming back up until the peak load period occurs.

Precooling is especially helpful when there is a high thermal storage capacity, or when the cooling system operates much more efficiently during cooler nighttime periods, or in buildings such as churches and theaters where peak loads from concentrated occupancy occur for a relatively short duration.

Heat storage effects are generally neglected in sizing heating plants, since peak heat loss conditions can persist for a period of days after the thermal storage has been depleted. On the other hand, heat storage effects in cooling operation should not be overlooked, since cooling loads generally vary with solar time and, if properly evaluated, may result in substantial savings both in initial investment and in operating cost.

The interdependency of thermal mass dynamics, building occupancy, and mechanical system controls is an excellent example of why it is important to include participation from the full team of architects and engineering consultants at the earliest stages of a project when material selection and construction-type decisions are being made.

MOISTURE PROBLEMS

Problems involving moisture may arise from changes in moisture content, the presence of too little or too much moisture, or the effects associated with its *changes of state*. Of particular interest is the change from the vapor to either the liquid or solid state, known as *condensation*. Specifically, there are three types of moisture problems in building construction:

Visible condensation
Concealed condensation
Extremely low humidity

Each of these will be addressed in turn in this section.

Moisture problems in residences are becoming more prevalent as smaller, tighter homes are being built. Water vapor originates from such necessary activities as cooking, laundering, bathing, and the breathing and perspiring of people. In a typical family of four, the average rate of production of water vapor from these sources is about 25 pounds per day, and may be much greater if humidifiers and automatic dryers are used. But moisture problems are by no means limited to residences. Water vapor is released by many commercial and industrial processes as well as by people. Another large source of water vapor may be the bare earth in a crawl space or basement. Also, in new construction, moisture is added by the drying of poured concrete slabs, masonry construction, and new plaster. All of this water vapor must somehow escape from the building.

Referring to the psychrometric chart (Figure 2.11) for a brief review, the saturation line (the curve indicating 100 percent RH) represents the maximum concentration of water vapor that can exist as vapor at any given temperature. As air is cooled—represented by moving horizontally to the left on the psychrometric chart—the RH increases. When the RH reaches 100 percent, the air is said to be saturated. The temperature at which this particular air/vapor mixture, upon cooling, becomes saturated is its dew-point temperature. Upon further cooling, the air can no longer hold all the moisture, and some will condense out.

(The dew observed on the ground or on objects in the morning following a clear night is the result of the surfaces being exposed to the open sky. With no cloud cover to reflect radiant heat back to earth at night, the surfaces radiate their heat out to the cold emptiness of outer space. They consequently become colder than the surrounding air. As the surfaces cool the air adjacent to them, the air temperature approaches the dew point. The dew finally forms on the surfaces when they cool the air to the dew point. The same phenomenon of condensation is observed on the outer surface of a glass of an iced beverage on a warm, humid day.)

The process of cooling beyond the dew point occurs when warm indoor air comes into contact with a cool window surface. The result of the air giving up some of its moisture is visible condensation on the glass surface. On windows with aluminum or steel frames that don't have a thermal break, condensation occurs on the metal frames as well. This condition is sometimes referred to as *sweating*. If the surface temperature is below 32°F (0°C), the condensation takes the form of frost.

Some of the deleterious effects of water on building materials are:

1. Significant changes take place in the dimensions of many building materials when there is a change in moisture content. The most familiar case is wood, but almost all plant and animal fibers show appreciable moisture changes with changing RH, and undergo changes of dimension of the same order as those occurring to wood. It is less well known that dimensional changes can also occur in masonry materials as a result of changes in moisture content.

2. Water is either an essential or contributory factor in chemical changes (such as the rusting of steel), physical changes (such as the spalling of masonry by frost action), or biological processes (such as the rotting of wood). Surface condensation commonly damages decorative finishes and window sashes (both wood and metal) as well as structural members.

3. Water is a good conductor of heat. Moisture in a material increases the thermal conductivity by providing a "short-circuit" path for heat flow.

Visible Condensation

Visible condensation occurs when any interior surface is colder than the dew point of the nearby air. Once the air temperature drops below its dew point, moisture is released onto the surface. The loss of moisture from the air causes a reduction in vapor pressure which then draws moisture in from the surrounding air. This moisture also condenses, causing a continuous migration of moisture toward the cold surface, as referred to in Chapter 2. This pattern will continue as long as the surface temperature remains below the dew point associated with the remaining concentration of water vapor in the air. Allowing condensation to occur sometimes results in extremely low humidity conditions.

In winter, visible condensation may collect on cold closet walls, attic roofs, and windows, especially single-pane windows. The temperature of any envelope surface—fenestration, wall, or roof—depends on the inside and outside air temperatures and on the heat transfer coefficient (U) of the construction. Note that U-values are commonly used as an average for large areas, within which there may be portions—such as at studs in an insulated frame wall—where the heat transfer is greater. The inside surface temperature of a wall will generally be lower at the bottom due to such things as inside air stratification, air leakage, and convection in walls with air spaces. The lines in Figure 3.10 represent the RH at which visible condensation will appear on an inside envelope surface. At a given indoor design RH and outdoor temperature, the U-value must be greater than that indicated by Figure 3.10. Alternatively, at a given U-value and outdoor design temperature, the RH must be kept below that in Figure 3.10.

Summer condensation occurs on concrete basement walls and floors cooled by the earth and on concrete floor slabs on grade (if not part of the heating system). Being massive, these structures tend to maintain a constant temperature from day to day while the dew point rises. When no additional water vapor is released into the space, the dew point tends to equal that of the outdoors (although when ventilation is low, there will probably be a lag). Rugs on such floors or interior insulation on basement walls contribute to the problem by inhibiting the rise in concrete slab temperature. In addition, condensation or very high RH may damage rugs and insulation. Insulation on the exterior or below the slab along with well-drained gravel helps alleviate the problem.

Another summer problem is condensation on cold water pipes and cool ducts from warm, moist room air. Water then drips down onto any objects below, causing considerable damage. The solution is adequate insulation levels and vapor barriers.

Wherever exposed surfaces are at a lower temperature than the air, condensation must be considered a likelihood. Even when a surface isn't quite cold enough to produce vis-

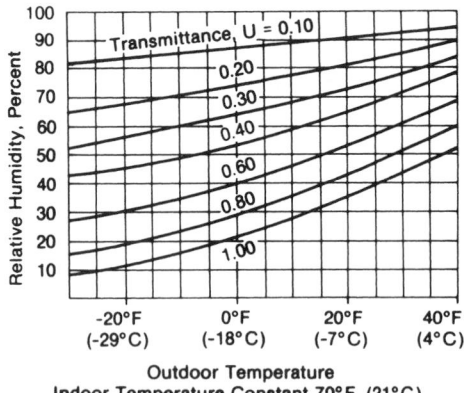

FIGURE 3.10. Maximum RH to avoid visible condensation.

ible condensation, moisture damage such as swelling, mold, and discoloration of the surface may still occur.

Possible solutions to visible condensation problems are:

1. Insulate interior surfaces from the cold outdoors (winter); also, insulate cold water pipes and ductwork (summer).
2. Provide enough air motion to keep condensation from settling on cold surfaces (winter).
3. Reduce the amount of moisture introduced into the air (summer and winter).
4. Ventilate moist air out of the space to keep vapor concentrations low (summer or winter).
5. Reduce ventilation rates at times when the dew point is higher outside than inside (summer).
6. Artificially warm the cold surfaces; for example, blow warm air across perimeter windows (winter).
7. Avoid designing room arrangements with cold out-of-the-way surfaces that are shielded from radiant room heat and air movement (winter).
8. Mechanically dehumidify the air to keep vapor concentrations low.

It is desirable to use methods that do not rely on energy consumption (such as fans, heaters, and dehumidifiers) and do not increase the heat loss in winter or heat gain in summer. In other words, before resorting to additional heating or dehumidifying, use insulation and select proper room arrangements that avoid pockets of still air and surfaces shielded from the radiant heat in the rest of the room.

As a note of caution, operable insulation over windows can actually contribute to condensation problems. The interior surface of the window is thus shielded from the heating source in the room and becomes a cold surface. If warm room air can pass around or through the insulating barrier, its moisture will condense on the window. The problem is alle-

viated by properly gasketing or sealing the edges (sides, top, and bottom) of the insulation to prevent moist room air from entering the space between the insulation and the glass, and by making sure that the insulation construction is impermeable to moisture.

Concealed Condensation

When there is a higher concentration of water vapor inside than outside, the *vapor pressure* is also higher on the inside. This represents a driving force causing a diffusion of moisture through the envelope. The buildup of vapor pressure within a building depends on the amount of vapor produced, its inability to escape, and the air temperature.

An uninsulated envelope will be at a fairly uniform temperature all the way through. Moisture passes through any porous materials even if air cannot go through, and it may travel all the way to the outside before encountering a surface as cold as its dew point.

When this same envelope is insulated, its inside surface temperature rises while its exterior surface temperature drops, causing a wider variation in temperature through the envelope. A temperature profile through an insulated wall is shown in Figure 3.11. Along with a lower exterior surface temperature comes an increased likelihood that the moisture traveling through the structure from the warm, moist side to the cold side will at some point reach a surface cold enough to cause condensation. Condensation, which normally requires a solid surface, generally does not occur in fibrous insulation itself, except when frost forms on sheathing and gradually builds back into the fibers. Nonetheless, insulation may become wet as a result of the above condition or from liquid condensation seeping down from a surface above.

Condensation within the envelope construction leads to deterioration of materials, paint peeling, and insulation saturation. Saturation with moisture destroys the value of the insulation. Besides leading to increased energy use because of the higher heat loss, this may mean that the heating system is no longer capable of keeping the building contents warm. Not only is heat conducted through the envelope at higher rates, but porous materials—particularly fibrous insulation—partially saturated with water allow a migration of moisture to the cold side. This occurs through evaporation, vapor flow, and condensation within the material. In the process, a substantial amount of heat is transferred as latent heat of the vapor. Other deleterious effects are material damage due to freeze-thaw cycles and spalling of masonry by frost action.

While ventilating or mechanically dehumidifying the interior space will relieve some of the vapor pressure and

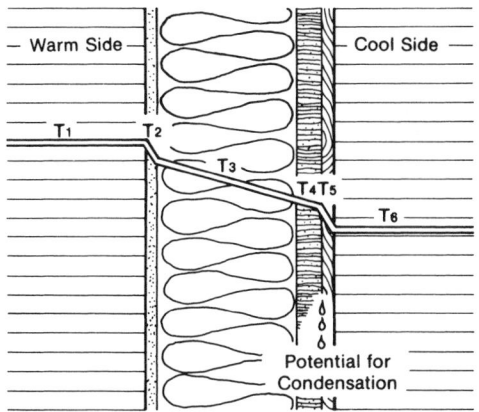

FIGURE 3.11. Temperature profile through an insulated wall.

reduce the amount of moisture passing through the envelope, the primary method of concealed condensation control is to provide tightly sealed *vapor barriers* that prevent moisture from entering the building skin. Since fibrous or porous insulations offer little resistance to vapor migration, some impermeable material is required to block it. Moisture migration will normally occur from the inside to the outside because the inside vapor pressure is greater. The vapor barrier should therefore be installed on the interior side. In general, the vapor barrier should be applied to the heated side.

(In the summer, the vapor pressure may be higher outside than inside, and in some circumstances, the vapor barrier may be most useful if placed on the exterior side of the insulation. However, the problem is usually more severe in the winter season, and in the summer the cooler interior temperature may never be as low as the dew point. In no case should a vapor barrier be located on both the interior and exterior sides of insulation. Any vapor leaking into the insulation would be trapped and the liquid condensation would be unable to escape.)

Another means of controlling concealed condensation is by ventilating the wall cavity to remove any vapor before it can condense on the cold surface. Chapter 7 of Ramsey and Sleeper's *Architectural Graphic Standards* (10th ed., 2000) provides some specific details on thermal and moisture protection.

Extremely Low Humidity

The minimum RH that should normally be maintained is 20 percent. An RH below this level can result in human discomfort, rough, dry skin, and respiratory problems. Beyond these effects, extremely low RH causes shrinking of porous materials and delamination of furniture, paneling, and other wood products. The shrinkage of wood lateral to the direction of the grain can cause unsightly cracks and the loosening of furniture joints.

But even at an RH of 20 percent, condensation can be a problem. Single-glazed windows will condense moisture at 15 percent RH when the temperature is 0°F (−18°C). That is why, in cold climates, double- or triple-glazed windows should be used. At 0°F (−18°C), a double-glazed window will not condense moisture out of the air until the RH gets as high as 40 percent.

SUMMARY

Chapter 3 continues the discussion of thermodynamics by explaining how heat energy is transferred and transformed in and around buildings. It begins with a statement of the First and Second Laws of Thermodynamics. It then goes on to describe the site as a collection of resources (sun, wind, water, earth, vegetation, and other structures). These should be utilized as much as possible to passively maintain an acceptable interior climate with minimum additional energy use. Outdoor spaces can complement the indoor ones. The integration of the building within the site and the utilization of site resources should be considered at the initial planning stage.

Heating and cooling loads for a building consist of:

- Solar gain
- Transmission through the envelope
- Internal gains
- Outside air

At any given time, if the sum of the gains equals the sum of the losses, then no mechanical heating or cooling is required. Otherwise, systems must be employed to actively add heat to or remove it from the space. Portions of the envelope may be designed, in some cases, with the ability to vary their transmission of heat, light, and air. Providing these necessities naturally, when available and desired, minimizes the use of mechanical systems.

Time itself can have a great influence on thermal loads, since massive materials that are part of the envelope or contained within a space can absorb excess heat which will be released later. Temperature extremes are thus moderated, and by taking advantage of the cyclical nature of some loads, heat gains can be delayed by using thermal storage mass to offset heat losses at a later time. The net requirements for mechanical heating and cooling can sometimes be reduced or eliminated altogether in this way.

Chapter 4 will show how the various thermal loads can be quantified, and Chapter 6 will describe the mechanical systems and equipment available for internal climate control.

Since air contains moisture, it is imperative that cold spots below the dew-point temperature be avoided within the space or within the envelope construction; they can result in visible or concealed condensation. In general, the presence of water can cause swelling in all plant and animal fibers and in masonry materials, rotting of wood, rusting or corrosion of metals, damage to decorative finishes, and a weakening of products and materials that are not designed for contact with water. Solutions involve removing some of the humidity from the air, maintaining all visible surfaces above the dew point, and locating a vapor barrier on the warm, humid side of all insulations to block moisture from passing through to a cold surface and condensing where it is concealed.

Another moisture problem is too little humidity. Health and the maintenance of finishes and furnishings dictate at least a 20 percent RH. In order to maintain that level, the envelope must have a certain minimum thermal resistance—and, in particular, fenestration must be double-glazed—in cold climates in order to avoid condensation.

KEY TERMS

Laws of Thermo-dynamics	Storm window	Thermal capacity
Microclimate	Economizer cycle	Thermal storage mass
Buoyancy	Infiltration	Thermal lag
Stack effect	Exfiltration	Precooling
Cooling load	Positive pressure	Condensation
Capacity	Negative pressure	Sweating
Heat gain	Pressurization	Visible condensation
Heat loss	Diurnal (day-night) cycle	Concealed condensation
Heating load		Dew point
Chiller	Mass	Vapor pressure
Internal gain	Specific heat	Vapor barrier
Internal load		
Fenestration		
Insulating glass		

STUDY QUESTIONS

1. State in your own words the First and Second Laws of Thermodynamics.
2. What is the difference between energy quality and quantity?
3. Explain why warm air tends to rise and stratify above the layers of cooler air.
4. Which of the following loads are perimeter loads? Which are internal loads?
   Transmission through the envelope
   People
   Lights
   Solar gain
   Ventilation
   Office electrical equipment
   Infiltration
5. What determines the thermal capacity of a given thermal storage mass? What determines how fast it will release the heat?
6. At 50 percent RH and 0°F outside air temperature, what is the required U-value to prevent visible condensation on the interior surface?

BIBLIOGRAPHY

American Society of Heating, Refrigerating, and Air-Conditioning Engineers, Inc. *ASHRAE Handbook of Fundamentals.*

Burghardt, M.D. *Engineering Thermodynamics with Applications,* 3rd ed. HarperCollins College Division, 1986.

Climatic Atlas of the United States CD-ROM version 2.0. NOAA National Climatic Data Center, Asheville, NC. US Government On-line store: http://nndc .noaa.gov/?http://ols.nndc.noaa.gov/plolstore/plsql/ olstore.prodspecific?prodnum=C00519-CDR-A0001. (This version, which includes Alaska and Hawaii, replaces the paper atlas last published in 1968 and supersedes version 1.0 of the *Climate Atlas of the Contiguous United States* published on CD in 2000. The new CD atlas contains 2023 color maps of climatic elements such as temperature, precipitation, snow, wind, pressure, etc. The period of record of the data for most of the maps is 1961–1990. There is an online version of the *Climate Atlas* at http://www.ncdc.noaa .gov/olca/ having restricted GIS capabilities, but viewable at no charge.)

Lechner, Norbert. *Heating, Cooling, Lighting: Design Methods for Architects,* 2nd ed. John Wiley & Sons, Inc., 2001.

Ramsey, C.G., H.R. Sleeper, and J. R. Hoke, Jr. *Architectural Graphic Standards,* 9th ed. John Wiley & Sons, Inc., 2000.

Simiu, Emil, and Robert H. Scanlan. *Wind Effects on Structures: An Introduction to Wind Engineering,* 2nd ed. John Wiley & Sons, Inc., 1986. (Includes the dynamic response of tall buildings, the performance of glass curtain walls in high-rise structures, the effect of wind on pedestrians near buildings, and the estimation of the probability that extreme winds will occur at a given site.)

Simiu, Emil, and Robert H. Scanlan. *Wind Effects on Structures: Fundamentals and Applications to Design,* 3rd ed. Wiley-Interscience, 1996.

Stein, Benjamin, and John S. Reynolds. *Mechanical and Electrical Equipment for Buildings,* 10th ed. John Wiley & Sons, Inc., 2006.

United States Environmental Data Service (corporate author) and John L. Baldwin. *Weather Atlas of the United States: Originally Titled Climatic Atlas of the United States.* Gale Group, 1975.

Load Calculations

HEATING AND COOLING LOADS

HEATING LOAD CALCULATIONS

COOLING LOAD CALCULATIONS

SELECTING DESIGN TEMPERATURE AND HUMIDITY CONDITIONS

INDOORS

OUTDOORS

SOLAR GAIN THROUGH FENESTRATION

SOLAR RADIATION

BUILDING SOLAR HEAT LOAD

SHGF

SHADING COEFFICIENTS

GLASS BLOCKS

SHADING

THERMAL LAG FACTOR

HEAT TRANSMISSION THROUGH THE BUILDING ENVELOPE

ENERGY CODE COMPLIANCE

SOL-AIR TEMPERATURE

TEMPERATURE IN ADJACENT UNCONDITIONED SPACES

HEAT LOSS THROUGH GROUND

THERMAL LAG FACTOR

INTERNAL LOADS

PEOPLE

LIGHTING

APPLIANCES, EQUIPMENT, AND MISCELLANEOUS SOURCES

DIVERSITY

OUTSIDE AIR

INFILTRATION

VENTILATION

CALCULATION OF HEATING OR COOLING LOAD DUE TO OUTSIDE AIR

ANNUAL ENERGY USE CALCULATIONS

DEGREE DAY METHOD

COOLING ENERGY USE COMPUTATIONS

BIN METHOD

HOURLY COMPUTER SIMULATIONS

A numerical analysis of the heating and cooling loads of a building takes into account a number of variables: the external factors of climate and topography; the internal factors of people, lights, and equipment; the materials and construction techniques used for the envelope itself; the opportunities to adjust the envelope seasonally to regulate comfort; the variability possible in the envelope, depending on the orientation of each elevation; and the organization of inside spaces relative to orientation, function, and comfort requirements.

Load calculations are performed for a number of different reasons. Mechanical engineers and designers use them to determine the appropriate size of heating and cooling equipment, air flow rates, and duct and pipe sizes. Architects calculate loads to make sure that envelope insulation meets the pertinent criteria or applicable energy conservation standards; to compare the relative energy efficiency of alternative envelope treatments; to estimate preliminary mechanical system costs; or to evaluate solar energy benefits. These same calculations serve as the basis for estimates of a building's annual energy use.

These calculations can become very complicated when detailed cost comparisons of alternative systems are called for. The degree of sophistication sometimes requires that heating and cooling energy use be calculated in hourly increments for a year's time. The hour-by-hour calculations can account precisely for the heat storage effects of the building structure. Because of the number of calculations involved, computer processing is necessary. This also provides the capability of calculating hourly solar angles and solar intensities and of analyzing the various shading effects. The computer can accumulate all the load components each hour over the entire year and determine when the peak load will occur.

But even with this powerful tool and its high degree of precision, the results are no better than a load *estimate* based on the average weather conditions, uncertain construction quality, and other approximations of reality. In the end, the process is still as much an art as it is a science, depending on the judgment of the person selecting the input data.

With that in mind, this chapter presents a simplified calculation approach to familiarize the reader with the concepts and to allow him or her to become accustomed to performing the calculations without getting bogged down in details. For readers who wish to refine the procedure, more detailed calculations can be found in some of the references cited in the bibliography at the end of this chapter, most notably the *ASHRAE Handbook of Fundamentals.* This source also contains the algorithms (equations) necessary to computerize the process.

There are many computer programs that will perform the detailed hour-by-hour calculations as well as the more simplified versions such as that presented here. The user only has to follow the directions to produce a quick calculation. However, without an understanding of the process, a designer cannot fully and creatively use the tool and is likely to fall prey to the long-standing computer adage "garbage in, garbage out."

While a simplification, the process described in this chapter is based on currently accepted technical information and can prove quite useful. It can be employed to estimate annual fuel costs for life-cycle cost/benefit analyses and for evaluation of energy conservation opportunities, as explained in Chapter 16.

HEATING AND COOLING LOADS

Chapter 3 described conceptually the various building heat gains and losses. This chapter is concerned with quantifying those thermal interactions.

Before getting into the numbers, there is another important concept to understand. There are actually three distinct heat flow rates related to buildings:

1. Heat gain or loss
2. Cooling load or heating load
3. Heat extraction or heat addition rate

A *heat gain* is the rate at which heat enters or is generated within a space at a given instant. Heat gain is classified by the manner in which it enters the space and by whether it is sensible or latent heat. The manner in which heat enters a space is typically indicated as follows:

- Solar radiation through fenestration
- Heat conduction through the envelope
- Heat generated within the space by people, lights, electrical equipment or appliances, or any other electrical, mechanical, or thermal processes within the space
- The exchange of cool indoor air for warmer outside air by infiltration and/or ventilation

Sensible heat gain is the direct addition of heat to an enclosure—apart from any change in the moisture content—by any or all of the mechanisms of conduction, convection, and radiation. When moisture is added to the space—for example, by vapor emitted by the occupants—there is an energy quantity associated with it that is referred to as *latent heat.* The sensible and latent heat gains are separated here because different principles and equations are used to calculate each one. They must also be identified separately in order to design the cooling and dehumidifying system, but we will not go into that here. For our purposes, we will simply calculate them separately and then add them together for a total value.

A *heat loss* is the rate at which heat flows out of a space to the surrounding environment. Heat losses are classified as:

- Heat conduction through the envelope
- The exchange of warm indoor air for cold outside air by infiltration or ventilation

Heat gains and losses can be considered the source of *cooling* and *heating loads,* respectively. A *load* is the rate at which heat must be added or removed to offset the heat losses or gains in order to maintain the interior air temperature and humidity at the desired levels.

In the case of heating, the heat loss and heating load are for all practical purposes the same. The distinction is more pronounced in the case of cooling, since the heat gain by radiation is partially absorbed by the surfaces and contents of the space and does not affect the room air until some time later. The sensible cooling load at a given time is therefore generally considerably below the instantaneous heat gain at that time. The latent heat gain, however, is essentially an instantaneous cooling load. As a result, the total cooling load is lower than the total heat gain at any instant, although they will approach the same value for a constant heat source that has had time to heat up the entire mass within the space. When heat gains fluctuate—such as due to the sun's intensity, lights being turned on and off, and intermittent occupancy—this thermal inertia or storage effect can be quite significant and is accounted for in the calculation of cooling loads.

The third category of building heat flow is referred to as the *heat extraction* or *heat addition rate.* It is also known as *equipment load* because it is the amount of heating, cooling, humidifying, or dehumidifying that needs to be provided by the fuel-consuming equipment (boilers, furnaces, chillers, air conditioners, etc.). It includes heat gains and losses of the distribution equipment. For mechanical design purposes, the outside air load due to ventilation is excluded from the space load and is accounted for in this third category of equipment loads since it actually is a load imposed on the equipment without ever impinging on the space. This allows the main fuel-consuming equipment and the distribution network to be sized properly.

For the purpose of simplification, the outside air loads (whether infiltration or ventilation) will here be considered as integral parts of the heating or cooling load, and the distribution losses will be accounted for along with energy conversion efficiency as a single system efficiency factor.

Heating and cooling loads are calculated in much the same manner, but calculating the cooling load is more complicated and includes many additional considerations. Heating and cooling loads will thus be presented separately as follows:

Heating Load Calculations

The components of a building heating load are:

Transmission Heat Loss Through

- Fenestration
- Opaque walls
- The roof
- The floor, below-grade walls, slab edge, and so on.

Outside Air Heat Losses

- Due to infiltration and/or ventilation

Figure 4.1 shows a sample form for tabulating the calculations and recording the results.

This chapter refers extensively to the tables in Appendix B. The reader may wish to pause here to become familiar with them. These reference data are provided for instructional purposes. For actual design applications, it is recommended that the latest edition of the ASHRAE handbooks, manufacturers' product specifications, and other project specific data be obtained.

The general procedure for calculating a heating load is as follows:

	Appendix Reference

Step 1

Take off net areas of fenestration, opaque walls, roof, and floor (area or perimeter) from building plans or from the actual building (inside dimensions).

Step 2

Determine design criteria.
A. Select the design outdoor temperature, wind speed, and wind direction.   B2
B. Select the design indoor temperature.

Step 3

Determine the *coefficient of transmission* (U-factor) for all elements of the building envelope—fenestration, walls, roof, floors, and so on.   B4

Step 4

Calculate transmission heat loss through each exterior surface: Load = net area × U-factor × ΔT.

HEATING LOAD CALCULATIONS

PROJECT _____   PREPARED BY _____

LOCATION _____   DATE _____

	INSIDE DESIGN TEMP _____
	OUTSIDE DESIGN TEMP _____
	DIFFERENCE _____

ITEM	QUANTITY	X	U-VALUE	X	ΔT	=	BTUH (W)
GLASS	FT² (M²)				°F (°C)		
NET WALL (ABOVE GRADE)	FT² (M²)				°F (°C)		
NET WALL (BELOW GRADE)	FT² (M²)				°F (°C)		
ROOF	FT² (M²)				°F (°C)		
OUTSIDE AIR	CFM (L/S)		1.08 (1.20)		°F (°C)		
TOTAL							

FIGURE 4.1.

Step 5

Determine outside air load due to ventilation, infiltration, or special exhaust: Load = CFM (liters/sec) × O.A. factor[a] × ΔT. **B6**

Step 6

Add up all heat losses.

Cooling Load Calculations

The components of a building cooling load are:

Solar Heat Gain Through Fenestration

Transmission Heat Gain Through

- Fenestration
- Opaque walls
- The roof

Internal Heat Gain From

- People
- Lights
- Electrical appliances and equipment

Outside Air Heat Gain

- Due to the exchange of cool indoor air for hot outside air by infiltration and/or ventilation

Figure 4.2 shows a sample form for tabulating the calculations and recording the results.

The first two components listed above, solar heat gain and transmission heat gain, are strictly sensible heat gains, that is, they raise the temperature of the space but do not affect the amount of moisture present. The second two components, internal heat gain and outside air heat gain, include both sensible and latent (moisture) heat gains.

The calculation of a building's cooling load requires detailed building design information and weather data at selected design conditions. The general procedure is as follows:

[a] O.A. factor = density of air × specific heat of air.

$$1.08 \frac{Btuh}{CFM \cdot °F} = 0.075 \text{ #/ft}^3 × 0.240 \text{ Btu/#·°F} × 60 \frac{min}{hr}$$

$$0.018 \frac{Btuh}{CFH \cdot °F} = 0.075 \text{ #/ft}^3 × 0.240 \text{ Btu/#·°F}$$

$$1.20 \frac{W}{(L/s)·°C} = 1.2 \text{ kg/m}^3 × 1,000 \text{ joules/kg·°C} × 0.001 \text{ m}^3/L$$

Appendix Reference

Step 1

Take off net areas of fenestration, opaque walls, and roof from building plans or from the actual building (inside dimensions). Areas should be tabulated separately for each orientation (north, south, east, west, etc.).

Step 2

Determine design criteria:
A. Select the design *outdoor* temperature and humidity conditions. **B2**
B. Select the design *indoor* temperature and humidity conditions (Chapter 1).
C. Select the hour of peak load by finding the dominant load (south glass, west glass, roof, internal gains, etc.) and then determining by visual inspection of the appendix tables or by knowledge of the building operation schedule **B3** when the dominant load will peak. **B4**

Step 3

Determine the solar heat gain factors and shading coefficients of the fenestration on each exposure (north, south, east, west, and horizontal) and calculate the solar load. **B3**

Step 4

Determine the U-factor of fenestration, opaque walls, and roof. Calculate transmission heat gains through each exterior surface: Load = net area × U-factor × ΔT. **B4**
A. Fenestration ΔT = outside DB temperature minus inside DB temperature.
B. Opaque walls and roof ΔT = respective sol-air temperature minus inside DB temperature.

Step 5

Determine the sensible internal heat gain due to people, lights, equipment, etc. **B5**

Step 6

Determine the sensible heat gain from outside air: Sensible load = CFM (L/s) − O.A. factor − ΔT **B6**

Step 7

Add all loads from Steps 3 through 6 to obtain the total sensible load.

Step 8

Determine the latent internal loads due to people and other sources of moisture. **B5**

Step 9

Determine the latent load due to outside air. **B6**

Step 10

Add up all loads from Steps 8 and 9 to obtain the total latent load.

PROJECT _____   PREPARED BY _____
LOCATION _____   DATE _____

COOLING LOAD CALCULATIONS

SPACE USE _____
FLOOR AREA _____   VOLUME _____
PEAK LOAD DATE _____   TIME _____   HRS/DAY OF OPERATION _____
GLAZING _____   SHADING _____
WALL COLOR: LT ☐ DK ☐   ROOF COLOR: LT ☐ DK ☐   LATITUDE _____

CONDITIONS	DB	WB	% RH	DP	HUM RATIO GR/LB (KG/KG)
OUTDOOR	_____	_____	_____	_____	_____
ROOM	_____	_____	_____	_____	_____
DIFFERENCE	_____	_____	_____	_____	_____

SENSIBLE LOADS

SOLAR

EXPOSURE	AREA	X	SHGF	X	SC	X	TLF	=	BTUH (W)	SUBTOTAL	TOTALS
_____	_____		_____		_____		_____		_____	_____	
_____	_____		_____		_____		_____		_____	_____	
_____	_____		_____		_____		_____		_____	_____	
_____	_____		_____		_____		_____		_____	_____	

TRANSMISSION

ITEM	EXP	AREA	X	ΔT	X	U-VALUE	X	TLF	=	BTUH (W)	SUBTOTAL
GLASS											
NET WALL		_____		_____		_____		_____		_____	_____
		_____		_____		_____		_____		_____	_____
		_____		_____		_____		_____		_____	_____
		_____		_____		_____		_____		_____	_____
ROOF		_____		_____		_____		_____		_____	_____
		_____		_____		_____		_____		_____	_____

O.A.

INFILTRATION
_____ CFM (L/S) X **1.08 (1.20)** X _____ °F (°C) = _____

VENTILATION
_____ CFM (L/S) X **1.08 (1.20)** X _____ °F (°C) = _____

INTERNAL HEAT

								BTUH (W)	SUBTOTAL
PEOPLE:	_____ PEOPLE	X	_____ BTUH(W) EA	X	_____ BSF	X	_____ DIVERSITY	=	_____
LIGHTS:	_____ WATTS	X **3.4*** X	_____ BALLAST	X	_____ BSF	X	_____ DIVERS	=	_____
EQUIPT:	_____ WATTS	X **3.4*** X	_____	X	_____ BSF	X	_____ DIVERSITY	=	_____
APPLIANCES:	_____ WATTS	X **3.4*** X	_____	X	_____ BSF	X	_____ DIVERSITY	=	_____
OTHER:								=	_____

* Omit for SI Units

TOTAL SENSIBLE LOAD _____

LATENT LOADS

						BTUH (W)
PEOPLE:	_____ PEOPLE	X	_____ BTUH(W)/PERSON	X	_____ DIVERSITY	= _____

APPLIANCES: _____
OUTSIDE AIR: _____ CFM (L/S) X _____ GR/LB (KG/KG) X **0.68 (2.808)** = _____

TOTAL LATENT LOAD _____

TOTAL LOAD (SENSIBLE & LATENT) _____

FIGURE 4.2.

Step 11

Determine the total load by adding the results of Steps 7 and 10.

The details of each step will be explained in the next sections: Selecting Design Temperature and Humidity Conditions, Solar Gain Through Fenestration, Heat Transmission Through the Building Envelope, Internal Loads, and Outside Air.

Prior to calculating a heating or cooling load, it is useful to survey the pertinent conditions of the building and assemble all the necessary data. The summary at the end of this chapter provides a checklist for this purpose. The more exact the information that can be obtained about the building characteristics, heat sources, weather data, and so on, the more accurate the load estimate. Additional information can then be obtained from the tables in Appendix B. Once the necessary input data are in place, the loads can be calculated and totaled conveniently on forms such as those in Figures 4.1 and 4.2.

It is a good idea to check the results against averages relating square feet of floor area with various loads. A reference table of such check figures is included in Appendix B1. They should be used with discretion and only as a rough approximation to make sure that an erroneous assumption or an arithmetical error doesn't lead to a gross mistake.

SELECTING DESIGN TEMPERATURE AND HUMIDITY CONDITIONS

Indoors

The indoor design temperature for comfort should be based on the discussions in Chapter 1 and, in particular, Table 1.6. One must be alert, however, to unusual circumstances that may require special conditions. For example if occupants are to be engaged in strenuous activity or must wear heavy protective clothing, a lower indoor design temperature would be required.

Generally, an RH of 25 to 70 percent will provide comfort in conjunction with appropriate dry-bulb temperatures. For cooling-load estimating purposes, 50 percent is commonly used. If humidification is to be provided in the winter, it should be designed for a maximum of 30 percent RH.

Some industrial processes and sensitive computer equipment require a fairly narrow range of temperature and humidity conditions. These should be obtained from the client, equipment manufacturer, or engineering consultant.

Storage areas also often have specific requirements for temperature and humidity. Perishable food storage conditions range over a wide spectrum, from below freezing to 55°F (13°C). The best temperature and humidity range for books generally falls within the human comfort range—70° to 80°F (21° to 27°C) and 40 to 55 percent RH—and rapid changes in temperature and humidity are more harmful than the maintenance of marginal conditions at all times. The precise temperatures and humidities that should be maintained for various types of paper, magnetic tape, disks, and film can usually be obtained from manufacturers.

Outdoors

The table in Appendix B2 contains outdoor design conditions for both summer cooling and winter heating in various locales in the United States, Canada, and other countries. Data for an area not represented in the table can be interpolated from listed nearby locations with a similar elevation and topography.

When design requirements must be more precisely established, it may be advisable to consult with a meteorologist. If this is not feasible, the following guidelines can be used to adjust the design data for nearby locations supplied in Appendix Table B2:

1. *Elevation.* Increase the design values for lower elevations and decrease them for higher elevations as follows:
   Dry-bulb temperature   1°F per 200 feet[a]
   Wet-bulb temperature   1°F per 500 feet[a]
   These adjustments do not apply to narrow valleys in the winter or to locations where considerable radiation cooling occurs at night.
2. *Proximity of large bodies of water.* Daily temperature ranges are smaller near large bodies of water. In summer, the dry-bulb temperature increases with distance from the oceans and large lakes. Wet-bulb temperatures generally tend to be higher near large water bodies.
3. *Urban areas.* Urban locations tend to be slightly warmer than the surrounding countryside.

Heating

Heating load calculations are normally based on near-peak winter conditions. The maximum heat loss usually occurs at night, when there is no benefit of sunlight and when the building may not be occupied. No credit is taken for heat given off by internal sources such as people, lights, and appliances. Because the most adverse conditions may occur during sustained periods of very cold weather, the heat loss must be considered relatively constant, so the load is not reduced by time lag and thermal inertia effects.

[a] Degree Fahrenheit can be converted to degree Celsius by multiplying by ⁵⁄₉. Feet can be converted to meters by multiplying by 0.3048.

The effect of wind on heating requirements should be considered for two reasons:

1. Wind movement increases heat transmission through walls, glass, and roof (affecting poorly insulated walls to a much greater extent than well-insulated walls).
2. Wind increases the infiltration of cold air through cracks around doors and windows, and even through building materials themselves.

The *design conditions* for heating thus are the coldest temperature and the highest wind speed. Minimum temperatures usually occur in the early morning hours on clear nights with relatively wide-ranging 24-hour temperature swings. But the *most* severe weather conditions do not repeat every year. The most extreme condition is so infrequent that load estimates are based instead on design temperatures that are only rarely surpassed.

Two levels of design conditions are provided in Appendix Table B2: normal and maximum. The normal level represents the lowest temperature 97.5 percent of the time during December, January, and February (a total of 2,160 hours) in the Northern Hemisphere and during June, July, and August in the Southern Hemisphere (a total of 2,208 hours). The maximum level represents the lowest temperature 99 percent of the time during those months. In a normal winter, there would be approximately 22 hours when temperatures are at or below the 99 percent value and 54 hours when temperatures are at or below the 97.5 percent value. For the Canadian stations, the 99 and 97.5 percent values are based only on the month of January.

The maximum value should be used only if the structure is constructed of lightweight material (low heat capacity), is poorly insulated, and has large glazed areas, or if space temperature control is critical. Otherwise, use the normal value.

Cooling

Cooling load calculations are based on peak summer conditions: the hottest temperature, the highest likely coincident solar heat gain, maximum occupancy, and the highest simultaneous equipment on line.

The design day is one in which:

- Dry- and wet-bulb temperatures are at their coincident peak.
- There is a minimum of haze in the air to reduce the solar heat.
- The building is occupied and in full operation.

The month in which the peak load occurs will depend upon the changes in *solar load* each month and the changes in weather conditions. Any seasonal variation of internal loads, such as people, is also important.

The outside dry-bulb temperature is lower in winter than in the summer months. But if the combination of solar load on the southern elevations and a seasonal increase in internal loads is greater than the effects of lower outdoor conditions, then the peak load may occur in September, October, or even December.

In Birmingham, Alabama, for example, the solar load on south-facing glass is three times greater in October than in July. The solar loads on east- and west-facing glass remains almost constant. The solar load on the north-facing glass is cut in half but is relatively small to begin with. The outside peak dry-bulb temperature drops from 94° to 84°F (34° to 29°C). Consequently, a building in Birmingham with its largest glass area facing south may well have its maximum cooling load during October.

The time of day when the peak load occurs depends upon the balance between the following loads:

1. Solar heat gain on each exposure
2. Transmission gains
3. Internal gains
4. Ventilation load

Maximum outside temperatures usually occur between 2:00 p.m. and 4:00 p.m. sun time, but time lag can cause the peak transmission gain to occur later than that.

Most residential and other low internal-load buildings are more sensitive to the envelope loads. They will generally peak when the solar gain through the fenestration and the load through the roof are at their highest, usually in the late afternoon.

The time of peak load can usually be established by inspection of Tables B3 and B4 in the appendix, together with a knowledge of the orientation of the building and the internal gains. In some cases, estimates must be made for several different times of the day to determine the maximum.

When the dominant glass area faces west and there is little exposed roof, the load will peak in the late afternoon or early evening. East-facing glazing will contribute to a morning peak. When the roof load dominates, the total peak will be when the roof load peaks. In some cases, the cooling load will be dominated by high internal loads, resulting in an almost uniform load throughout the day.

Most conditioned areas with more than one glass exposure and normal daytime occupancies will peak between 1:00 p.m. and 6:00 p.m. solar time. Exceptions are generally caused by glass exposures with northeast to southeast orientations, or by occupancy schedules such as those for restaurants or theaters.

Table B2 in the appendix provides three levels of design conditions for summer cooling: maximum, normal, and minimum. The table lists peak dry-bulb temperatures for each level. For the United States and Canada, the corresponding coincident wet-bulb temperatures are also given. The coincident wet-bulb temperature listed with each design dry-bulb temperature is the mean of all wet-bulb temperatures occurring at that specific dry-bulb temperature.

The table also presents peak wet-bulb temperatures that were selected independently; these are not necessarily coincident with the design dry-bulb temperatures at each level. The peak dry bulb and peak wet-bulb temperatures tend to be coincident in maritime areas but are usually not coincident throughout inland regions. The erroneous assumption of simultaneous peaking of dry-bulb and wet-bulb temperatures can lead to an overestimation of weather-dependent loads by up to one-third. When a building load is dominated by outside ventilation, it is more sensitive to the wet-bulb temperature, so the independent peak wet-bulb temperature should be used.

The maximum, normal, and minimum levels correspond respectively to conditions that are equaled or exceeded 1, 2.5, and 5 percent of the time during June through September (a total of 2,928 hours) in the Northern Hemisphere and during December through March in the Southern Hemisphere (a total of 2,904 hours). In a normal summer, approximately 30 hours would be at or above the 1 percent level and 150 hours at or above the 5 percent level. For Canadian stations, the 1, 2.5, and 5 percent levels are based only on the month of July.

The maximum level for summer design conditions is recommended for laboratories, industrial applications, or hospitals, where exceeding the room design conditions for even brief periods of time can be detrimental to a product or process or could jeopardize patient recovery. Hospitals and some industrial processes also require large quantities of outside air. Since the impact of outside conditions has a direct effect on the outside air load without any time lag, these applications are especially sensitive to outside air conditions.

Another consideration in critical control cases is that when the design temperature difference is relatively small—for instance, 2° to 4°—an outside temperature exceeding the design value by a few degrees can have a dramatic impact on the load. In locations where the design temperature difference is 15° to 20°, on the other hand, an outside temperature over the design value by a few degrees would be much less significant. Furthermore, when the dry-bulb temperature

does exceed the design level, it is usually accompanied by wet-bulb temperatures lower than the design level.

The mean daily range shown in Table B2 is the difference between the average daily maximum and average daily minimum temperatures in the warmest month. For countries other than the United States and Canada, the daily range is the long-term average.

SOLAR GAIN THROUGH FENESTRATION

There are two sources of heat flow through fenestration: (1) the conduction heat gain (transmission) due to a difference between outdoor and indoor air temperatures and (2) the solar heat gain due to radiation from the sun that is transmitted through the opening and absorbed within the space.

Fenestration load = solar + transmission

The first component, solar heat gain, is present only during the day when the fenestration is exposed to the solar radiation and is therefore related to the intensity of that radiation. The second component, the transmission gain, occurs whenever the temperature difference exists, whether or not the sun is shining.

These two components will be considered separately since they require different methods for computing them. Solar heat gain is the subject of this section, while the transmission through fenestration will be addressed in the next section, along with transmission through the opaque portions of the envelope.

Solar Radiation

Radiant heat from the sun arrives at the earth directly from the sun. Since radiation tends to spread out as it travels from its source, its intensity drops off with distance. The earth moves in a slightly elliptical orbit around the sun, causing a variation in the earth-sun distance. As a result, the radiation intensity ranges from a maximum of 444 Btuh/ft^2 (1,400 W/m^2) on January 3, when the earth is closest to the sun, to a minimum of 416 Btuh/ft^2 (1,310 W/m^2) on July 6, when the earth is farthest away.

The earth's orbital velocity also varies throughout its annual cycle, causing *solar time* as measured by a sun dial to vary slightly from the *mean time* kept by a clock running at a uniform rate synchronized to a 24-hour day.

As it passes through the atmosphere, a portion of the sun's radiation is reflected, scattered, and absorbed by dust, smoke, gas molecules, ozone, carbon dioxide, and water vapor. The familiar blue color of the sky is a result of the scattering of some of the shorter wavelengths from the visible portion of the spectrum. The familiar red at sunset results from the scattering of longer wavelengths by dust or cloud particles near the earth. Some radiation, particularly ultraviolet radiation, may be absorbed by ozone in the upper atmosphere, and other radiation is absorbed by water vapor near the earth's surface. That part of the radiation that is not scattered or absorbed and does reach the earth's surface is called *direct radiation.* It is accompanied by *diffuse radiation,* which is radiation that has been scattered or re-emitted.

The extent of depletion of the sun's rays by the earth's atmosphere is determined by the composition of the atmosphere (cloudiness, dust and pollutants, atmospheric pressure, and humidity), and by the length of the path of the rays through the atmosphere. In the morning or evening, for example, the sun's rays must travel along a much longer path through the atmosphere than they do at noontime. Similarly, the rays that hit the polar regions at midday have passed through a longer atmospheric path than those that hit the tropical regions at midday.

As the distance traveled or the amount of haze increases, the diffuse radiation component increases, while the direct component decreases. The net effect is a reduction in the total quantity of heat reaching the earth's surface. The total quantity of solar heat falling on any surface consists of:

1. Unshaded direct radiation
2. Unobstructed diffuse radiation from the sky
3. Reflected solar radiation from adjacent surfaces, particularly the ground to the south of the surface

Reflected radiation is classified as either *specular* or *diffuse.* Diffuse reflection from surrounding terrain is known as albedo.

Building Solar Heat Load

In order to determine the solar component of heat gain through fenestration, it is necessary to compile certain data from which the load can be estimated. This information includes:

- Compass point orientation of the fenestration
- Area of and number of layers of glazing on each exposure
- Possible shading effects of nearby permanent structures
- Intended shading devices integral with the building
- Reflective surfaces, such as water, sand, parking lots, and so on to the south, east, or west

Once these data are accumulated, the tables in Appendix B3 can be consulted for the peak solar load per square foot (square meter) of fenestration area. Notice that the solar radiation varies considerably throughout the day, and that the peak total cooling load hour must be selected in order to use these tables.

The formula for the *solar heat gain* (SHG) through fenestration is as follows:

$$SHG = A \times SHGF \times SC \qquad (4.1)$$

where

A = area of the fenestration in ft^2 (m^2)
$SHGF$ = solar heat gain factor in $Btuh/ft^2$ (W/m^2)
SC = shading coefficient

The solar cooling load, q, which lags behind the solar heat gain, is obtained by the use of a correction factor called the *thermal lag factor* (TLF) in the following formula:

$$q = A \times SHGF \times SC \times TLF \qquad (4.2)$$

The above procedure is based on a method developed by ASHRAE. The method establishes a *solar heat gain factor* (SHGF) in $Btuh/ft^2$ (W/m^2) through fenestration in various orientations for the daylight hours of the 21st day of each month, based on a reference glazing material of double-strength (DSA) sheet glass 0.125 inches (3.175 mm) thick. To account for different types of fenestration and shading devices that might be used, the *shading coefficient* (SC) is used to correct the SHGF. The shading coefficient is defined as:

$$SC = \frac{\text{solar heat gain of actual fenestration}}{\text{solar heat gain of unshaded double-strength glass}} \qquad (4.3)$$

The time lag difference for converting heat gain to a cooling load is represented by a multiplier called the thermal lag factor, which is a modification of ASHRAE's cooling load factor (CLF).

The tables in Appendix B3 that pertain to the calculation of solar loads are listed below:

Table	Title
B3.1	Solar Intensity and Heat Gain Factors
B3.2	Solar Reflectances of Various Foreground Surfaces
B3.3	Shading Coefficients (SC) for Glass

B3.4	Shading Coefficients (SC) for Plastic
B3.5	Shading Coefficients (SC) for Glass with Exterior Shading
B3.6	Shading Coefficients (SC) for Single and Insulating Glass with Draperies
B3.7	Solar Heat Gain Factors for Glass Block with and without Shading Devices
B3.8	Maximum Solar Heat Gain Factor for Externally Shaded Glass
B3.9	Shadow Lengths and Shadow Widths for Building Exterior Projections
B3.10	Thermal Lag Factors for Glass Solar Load without Interior Shading
B3.11	Thermal Lag Factors for Glass Solar Load with Interior Shading

SHGF

The SHG through fenestration depends on the window's location on the earth's surface (latitude), time of day, time of year, and the direction it faces. The direct radiation component results in a heat gain to the conditioned space only when the window is in the direct rays of the sun, whereas the diffuse radiation component results in a heat gain even when the window is not facing the sun.

Table B3.1 lists SHGF values for north latitudes at 0°, 8°, 16°, 24°, 32°, 40°, 48°, 56°, and 64°; for each month of the year; and for each hour of daylight. For dates, times, and latitudes other than those provided in Table B3.1, interpolation may be used. Note well that the headings for glass orientation and time of day for the morning hours are read across the top and down the left column. *For afternoon hours, they are read across the bottom and up the right column.*

This table includes the direct and diffuse radiation, as well as that portion of the heat that is absorbed within the glass and flows into the space. Notice that even those surfaces receiving no direct sun at all may receive some diffuse sky radiation and reflected radiation. The large increase in values that occurs when direct radiation is added indicates the significant contribution this radiation makes and underscores the value of proper shading devices in reducing solar heat gains.

The underlined values in Table B3.1 indicate the maximum SHG for the month for each exposure. The values that are boxed indicate the yearly maximums for each exposure.

Clearness Correction

The SHGF values in Table B3.1 represent the conditions on average cloudless days. For high elevations and very clear atmospheres, the maximum SHGF can be as much as 15 percent above the tabulated values. For very dusty industrial atmospheres and exceptionally humid locations, they may be from 20 to 30 percent lower than the tabulated values. The clearness factors in Figure B3.1 correct these values for regional variations in the normal atmospheric conditions.

Foreground Reflectance Correction

The SHGF values in Table B3.1 were computed using a ground reflectance of 0.2. Using Table B3.2, the SHGF may be adjusted for other known foreground reflectances. In order to correct for a different foreground reflectance, the following procedure may be used:

1. Determine the angle between incident solar rays and the vertical from the data in Table B3.9.
2. Using Table B3.2, find the solar reflectance of the specific foreground surface at that incident angle.
3. Subtract the standard 0.2 from the actual reflectance value (step 2).
4. Multiply that sum by the SHGF for the horizontal at the proper latitude, month, and hour from Table B3.1.
5. Multiply the above by 0.575.
6. Add the resulting positive or negative number to the SHGF for the vertical glazing concerned.

Example

Given: Steel casement windows on the west exposure in Albuquerque, New Mexico.

Find: The peak solar heat gain through the window per square foot.

Solution:
From Table B2,
  Elevation = 5,310 ft
  Latitude = 35°
  Design dry-bulb and mean coincident wet-bulb = 94/61.

From Table B3.1, the boxed value indicating the peak solar heat gain at latitude 32°[a] on the west exposure is 227 Btuh/ft^2 (717 W/m^2) and occurs at 4:00 p.m. in April.

From Figure B3.1, the clearness number for Albuquerque in the winter is about 1.02.

---

[a] Note that interpolating to the actual 35° latitude would provide even greater precision if desired.

From the psychrometric chart, Figure 2.11, the design dew point corresponding to 94° dry-bulb/61° wet-bulb is 34°F (1°C).

From the footnotes to Table B3.1, the correction factors are:
Steel sash: 1.17
Haze: (−)15% due to the high air pollution in the city.
Altitude: (+)0.7%/1,000 ft × 5,310 ft = (+)3.7%
Dew point: (67° − 34°) × (+)7%/10° = (+)23%

Therefore,

Peak solar heat gain =
$(227)(1.02)(1.17)(1 − .15)(1 + .037)(1 + .23) =$
    294 Btuh/ft^2
$(717)(1.02)(1.17)(1 − .15)(1 + .037)(1 + .23) =$
    928 W/m^2

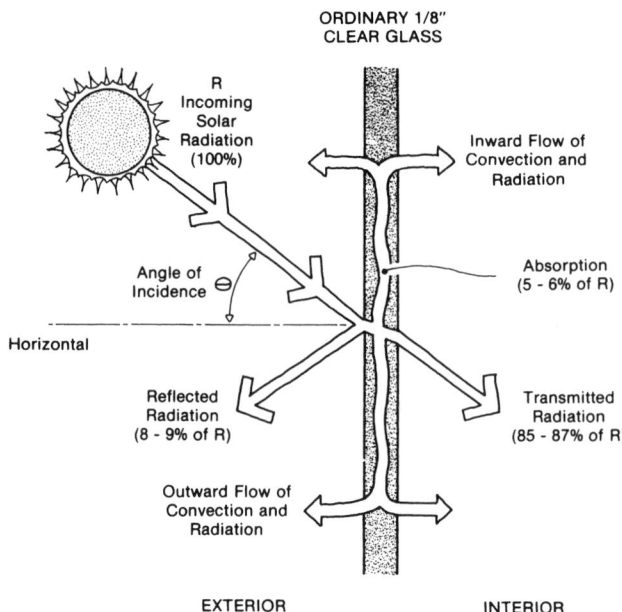

FIGURE 4.3. Solar reflection, transmission, and absorption at fenestration.

Shading Coefficients

Most fenestration has some type of interior shading to provide privacy and aesthetic effects, as well as to give varying degrees of sun control. There are a variety of types of blinds and shades that may be used to reduce solar gain. However, the type of glazing itself, due to its transmission, reflection, and absorption characteristics, can also have a great effect on solar heat gain.

When solar radiation strikes an unshaded window of ordinary glass (see Figure 4.3), a small portion of the solar heat (5 to 6 percent) is absorbed, and the rest is either reflected or transmitted directly indoors to become part of the cooling load. The amount reflected or transmitted depends on the *angle of incidence,* which is the angle between the sun's rays and an imaginary line perpendicular to the window surface. At low angles of incidence, about 86 or 87 percent is transmitted and 8 to 9 percent is reflected back outdoors. As the angle of incidence increases, more solar heat is reflected and less is transmitted.

Special heat-absorbing glazing materials increase the portion of solar heat that is absorbed to up to 50 percent, depending on their composition and thickness. Reflective coatings or films on glazings will cause most of the radiation to be reflected no matter what the angle of incidence. The total solar heat gain to the space is the directly transmitted heat plus the portion of the absorbed radiation that flows inward (about 40 percent).

The values for SHGF in Table B3.1 are based on the attenuation of solar radiation by a single pane of ⅛-inch (3-mm) glass with no shading. Under standard peak summer conditions, about 87 percent of the solar energy incident on the outside of the glass gets through.

The SCs in Tables B3.3 through B3.6 can be used as simple multipliers to correct the SHGF for some common types of glazing alternatives and shading devices. The first two columns in Table B3.3 refer to the exterior and interior panes of double-glazed (or insulating glass) units. Single-glazed units are noted in the exterior column only. The tabulated values for venetian blinds apply specifically to horizontal louvered blinds. For vertical louvered blinds with opaque white or beige louvers that are tightly closed, the SC is approximately the same as for opaque white roller shades. However, unless it is known for certain that the blinds will be tightly closed during the time of peak solar gain, it is better to disregard them and calculate the SHG as if they weren't there.

The effectiveness of an internal shading device depends on its ability to reflect incoming solar radiation back through the fenestration before it is absorbed and converted into heat within the building. Thus, light-colored, reflective shades are more effective than dark ones at keeping out unwanted solar heat. Notice how this is illustrated by the SC values for light versus dark interior shades.

In general, exterior shading devices are more effective than interior ones. Table B3.5 in the appendix lists SC values for several types of commercially available exterior sun screens and for outside awnings. The sun screens are divided into four groups, for each of which the SC is given as a function of the solar altitude angle (see Figure 4.4). Note that the SC decreases as the altitude angle increases. Commercially available sun screens will completely exclude direct solar radiation, but not reflected radiation when the solar altitude angle exceeds 20° for groups 1 and 2 or when it exceeds 40° for groups 3 and 4. The manufacturer of a given glazing or shading device can usually provide data on its SC or solar transmittance.

For fenestration shaded indoors by a fabric drape, the SC can be found in Table B3.6. Drapery fabrics may be classified in terms of their solar-optical properties of fabric transmittance and reflectance. Fabric reflectance is the major factor in determining the ability of a fabric to reduce solar heat gain. The properties of solar transmittance and reflectance can be described in terms of yarn color and openness of weave. The apparent color of a fabric is determined by the reflectance of the yarn itself.

The reflectance value and *openness factor* (ratio of the open area between the fibers to the total area of the fabric) can be accurately measured by laboratory tests; manufacturers can usually supply these. However, for the purpose of approximate load estimations, a visual inspection of the fabric properties will be sufficient. Figure B3.2 presents nine classifications in terms of openness factor and yarn reflectance. The openness factor is designated by Roman numerals as follows:

 I Open-weave (loosely woven) materials that allow light to pass freely
 II Semi-open
III Closed-weave (tightly woven) fabrics that permit little or no light to pass between the fibers

Yarn reflectance is designated by the capital letters: D for dark, M for medium, and L for light.

The nomagraph leading to Table B3.6 in the appendix provides coordinates in terms of yarn reflectance and openness factor, as well as the more precise fabric transmittance and fabric reflectance values which can be obtained from the manufacturer. The SC can be determined by locating the coordinates on the graph associated with a given fabric and then following the heavy line up to Table B3.6.

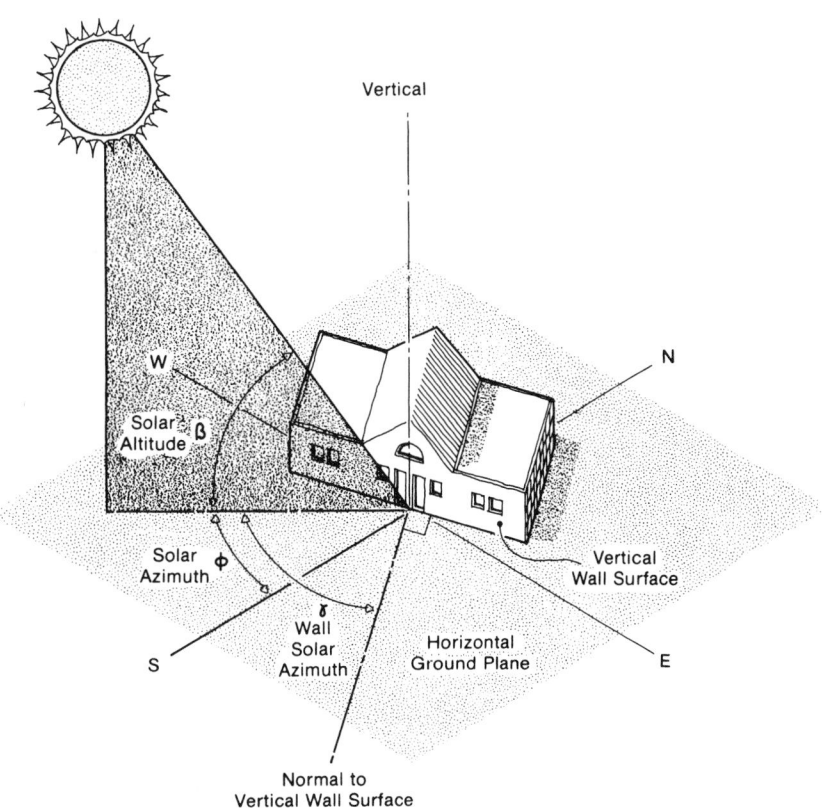

FIGURE 4.4. Solar altitude and azimuth angles.

Note that Table B3.6 applies to ⅛-inch clear glass and to single-drape arrangements where the drapery is hung with 100 percent fullness (drapery width twice the width of the draped area). If the drapery is hung flat like a window shade, a different SC will result. As an extreme example, a flat opaque drape having a reflective (aluminized or similar) lining with a reflectance of 0.80, in combination with ¼-inch (6-mm) clear glass, has an SC of 0.18, as compared with 0.32 extrapolated from Table B3.6 for that material in draped form.

Example

Given: A 10-ft² (0.93-m²) double-glazed (insulating glass) window with each pane ⅛ inch thick on a south elevation at 10:00 a.m., on July 21, at 40°N latitude.

Find: The SHG
   a. For an all-grass foreground
   b. If light-colored venetian blinds are used
   c. If a louvered sun screen, dark colored and 17 louvers per inch, is used
   d. If the outer pane of glass is reflective-coated (0.30 SC), and draperies with a semi-open weave and medium color are used

Solutions:
   a. The SHGF for the horizontal surface, from Table B3.1, is 231 Btuh/ft² (729 W/m²). The ground-reflected contribution to the SHGF is determined as follows:

   From Table B3.9, the solar altitude at 10:00 a.m., on July 21, at 40°N latitude is 57°. The geometry of the solar altitude angle indicates that the angle between the incident solar rays and the vertical = (90° − the solar altitude). Thus the angle desired = 90° − 57° = 33°.

   From Table B3.2, the solar reflectance of grass at 33° angle of incidence = 0.22.

   The formula for the correction factor is: horizontal SHGF × (actual reflectance − 0.2) × 0.575:

   $231 \times (0.22 - 0.2) \times 0.575 = 2.66$ Btuh/ft²
   7.29 W/m² $\times (0.22 - 0.2) \times 0.575 = 8.4$ W/m².

   The standard SHGF, found in Table B3.1, is 81 Btuh/ft² (255 W/m²), and the SC, found in Table B3.3, is 0.88. So the actual SHG =

   $(81 + 2.66) \times 0.88 \times 10$ ft² = 736 Btuh
   $(255 + 8.4) \times 0.88 \times 0.93$ m² = 216 W.

   b. From Table B3.3, the SC for light-colored venetian blinds with double glazing is 0.51. The SHG is then

81 Btuh/ft² $\times 0.51 \times 10$ ft² = 413 Btuh
255 W/m² $\times 0.51 \times 0.93$ m² = 121 W.

   c. According to the footnotes of Table B3.5, a dark-colored sun screen with 17 louvers per inch would fall into Group 3. The solar altitude angle was found above to be 57°. Since this is above 40°, the SC at 40° applies. From Table B3.5, the SC for single-pane clear glass is 0.18. The SC for unshaded double-pane glass was found above to be 0.88, so the SHG =

81 Btuh/ft² $\times 0.18 \times 0.88 \times 10$ ft² = 128 Btuh
255 W/m² $\times 0.18 \times 0.88 \times 0.93$ m² = 38 W.

   d. From Table B3.6, the SC for insulating glass with the outer pane reflective-coated combined with semi-open weave medium-color draperies is 0.26. The SHG =

81 Btuh/ft² $\times 0.26 \times 10$ ft² = 211 Btuh
255 W/m² $\times 0.26 \times 0.93$ m² = 62 W.

Glass Blocks

Glass blocks differ from sheet glass in that there is an appreciable absorption of solar heat as it passes through them. The large thermal storage capacity of the blocks results in a fairly long time lag (about 3 hours) before the heat reaches the building interior.

For this reason, SHG through glass block is estimated differently than that through sheet glass. Instead of a single SC, glass blocks have two factors as presented in Table B3.7. The first factor, SC_i, is multiplied by the SHGF, found in Table B3.1, for the proper wall orientation at the selected peak time. This is the rate of heat gain due to the instantaneous transmission of solar radiation through the glass. The second factor, SC_a, is multiplied by the SHGF for 3 hours earlier. This represents the rate of heat gain due to solar radiation that has been absorbed by the glass and delayed before passing through to the inside. All of it doesn't come through 3 hours later because the heat, upon being absorbed in the glass, is conducted in both directions so that a portion of it is transmitted back outside.

Shading devices on the exterior of glass blocks are nearly as effective as those used with any other type of glass since they block the solar radiation from encountering the fenestration. Shading devices on the inside, however, are not as effective in reducing heat gain through glass block because most of the radiant heat reflected by the shading device is absorbed within the glass block and partially reintroduced to the interior. Also, much of the heat is convected from the warm block directly into the air, and thus would not be affected by an indoor shade unless it is tightly sealed around the window casing. The footnotes to Table B3.7 give appropriate factors for heat gain with outdoor and indoor shading devices.

Example

Given: A 10-ft^2 (0.93-m^2) glass block opening on a north wall at 32°S latitude in January.

Find: The solar load through it at 4:00 p.m.

Solution:
From Table B3.1, the SHGF is as follows:
    4:00 p.m.   28 Btuh/ft^2 (89 W/m^2)
    1:00 p.m.   66 Btuh/ft^2 (210 W/m^2)

According to the footnote at the bottom of Table B3.1, the correction for south latitudes is to add 7 percent. Thus the SHGF becomes:
    4:00 p.m.   30 Btuh/ft^2 (95 W/m^2) = SHGF$_i$
    1:00 p.m.   71 Btuh/ft^2 (225 W/m^2) = SHGF$_a$

The glass block factors from Table B3.7 are:
    Instantaneous transmission (B$_i$) = 0.27
    Absorption transmission (B$_a$) = 0.24

According to the footnotes to Table B3.7, the solar load = (B$_i$ × SHGF$_i$) + (B$_a$ × SHGF$_a$):

$\{(0.27 \times 30 \text{ Btuh/ft}^2) + (0.24 \times 71 \text{ Btuh/ft}^2)\} \times 10 \text{ ft}^2 = 251$ Btuh

$\{(0.27 \times 95 \text{ W/m}^2) + (0.24 \times 225 \text{ W/m}^2)\} \times 0.93 \text{ m}^2 = 74$ W

Shading

When exterior shading is uniform over the entire fenestration, as is that provided by sun screens, the SC may be obtained directly from Table B3.5 and used to adjust the unshaded SHGF from Table B3.1. However, nonuniform shading from the surrounding landscape, tall vegetation, another structure, or some feature of the building itself is not associated with a particular SC. Instead, the solar loads are calculated separately for the shaded and for the unshaded areas.

External shading of an area of glazing cuts off the direct rays of the sun from that surface, allowing only the diffuse component through. This can reduce the SHG to a space up to 80 percent. Since the north[a] orientation at a latitude greater than 24° is exposed only to predominantly diffuse solar radiation, the SHGF (Table B3.1) for northern expo-

[a] Note: In the Southern Hemisphere, north and south orientations must be reversed for the purpose of this discussion.

sures is to be used for shaded fenestration at those latitudes. At latitudes of 24° or less, the northern surfaces receive direct solar radiation during part of the year, so they cannot be substituted for shaded surfaces in those locations. For shaded fenestration at latitudes of 24° or less, the peak SHGF is obtained from Table B3.8 in the appendix. For shaded skylights at all latitudes, use the values for horizontal glass in Table B3.8.

Before the correct values for solar gain can be applied, the extent of shading must be determined. If there is any doubt about the exterior shading, or if the fenestration is shaded for only short durations during the day, the SHGF for sunlit glazing should be used to be on the safe side. The areas shaded and unshaded can be determined graphically using silhouette charts (cylindrical sun charts) as presented in Edward Mazria's *Passive Solar Energy Book*. Otherwise, the shadow line must be located from the dimensional information on the design drawings and the angles defined in Figure 4.4.

The two angles β and ϕ define the location of the sun. The *solar altitude,* β, is the angle in a vertical plane between a ray from the sun and the horizontal ground plane. The *solar azimuth,* ϕ, is the angle in the horizontal plane between the north-south line and the vertical plane of β. The location of the sun with respect to a particular wall orientation is defined by β and the *wall solar azimuth.* The wall solar azimuth, γ, is the angle in the horizontal plane between the perpendicular to the wall in question and the vertical plane of β.

The amount of shading produced by any given configuration of overhangs, vertical projections, recessed windows, or adjacent buildings at any time of the day and year can be determined from the geometry when the above angles are known. Table B3.9 in the appendix lists solar altitude and azimuth angles for each month, each hour of the day, and at north latitudes of 0°, 8°, 16°, 24°, 32°, 40°, 48°, 56°, and 64°. Table B3.9 also lists shadow heights and widths per foot of projection for the common case of vertical walls. Blank spaces in the table denote complete shading; values of zero denote a full sunlit surface.

Example 1 Shading by Adjacent Building

Given: Buildings located as shown in Figure 4.5 at 40°N latitude.

Find: Shadow lines at 3:00 p.m. in July on the building to be air-conditioned.

Solution:
From Table B3.9, shadow height = 1.1 ft/ft
                shadow width = 0.2 ft/ft

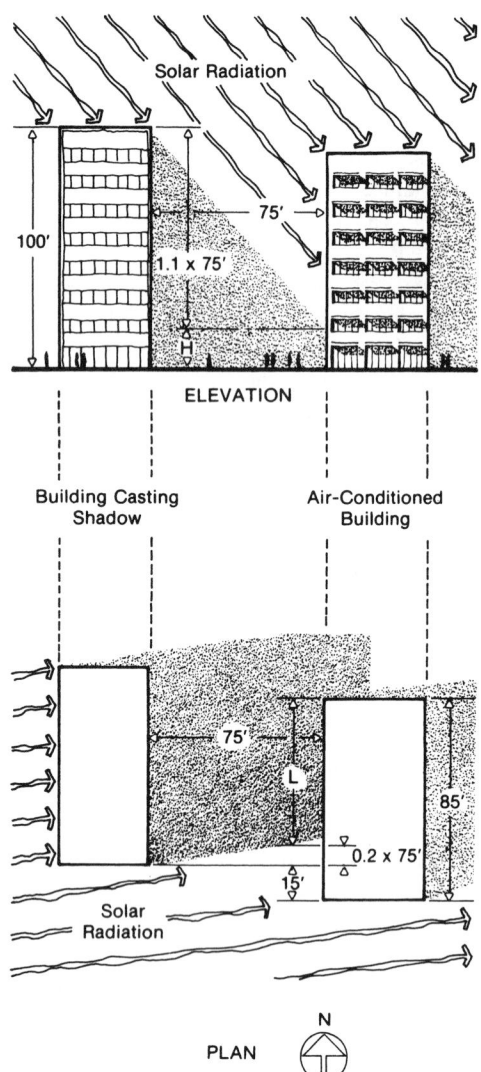

FIGURE 4.5. Example of shading by adjacent building.

Shadow height, H, on the shaded building = 100 − (75 × 1.1) = 17.5 ft

Shadow length, L, on the shaded building = 85 − 15 − (0.2 × 75) = 55 ft

The air-conditioned building is shaded to a height of 17.5 feet and a width of 55 feet along the face at 3:00 p.m. in July.

Example 2 Shading by Overhang and Reveal

Given: A recessed window on a west elevation with an 8-inch reveal and a 2-foot overhang 6 inches above the window at 40°N latitude.

Find: Shadow lines at 2:00 p.m. in July.

Solution:
Referring to Figure 4.6, the total overhang projection is 2 feet 8 inches, or 32 inches.

From Table B3.9, shadow height = 1.8 in./in.
           shadow width = 0.6 in./in.

Shadow height = 1.8 × 32 = 57.6 in.

Since the overhang is 6 inches above the window, the portion of the window shaded is the top 57.6 − 6 = 51.6 inches.

Shading from the side reveal = 0.6 × 8 = 4.8 in.

It is usually helpful to sketch to scale the plan and elevation views of a shading configuration with the approximate location of the sun in order to visualize the shadow lines. Table B3.9 can also be used to design shading schemes that will allow solar heat to enter in winter and be excluded in summer.

Thermal Lag Factor

Chapter 3 described how radiant heat gain does not immediately become a load on the cooling system. Instead, it is first absorbed and partially stored when it strikes a solid surface. Only when the solid surfaces become appreciably warmer than the room air does this energy become part of the cooling load.

The lag between the time when the heat enters the space and when it is received by the room air is determined by the heat storage characteristics of the structure and its contents, the presence of shading, and whether the shading is interior or exterior. This delaying of the heat gain can account for a substantial reduction in the peak cooling load.

The magnitude of the storage effect is largely a function of the thermal capacity or heat-holding capacity of the materials surrounding and within the space. The *thermal capacity* of a material is defined as the weight times the specific heat of the material. Since the specific heat of most construction materials is approximately 0.20 Btu/lb·°F (837 joules/kg·°C), their thermal capacity is directly proportional to their weight.

Tables B3.10 and B3.11 in the appendix provide thermal lag factors that account for the thermal storage effect on the solar cooling load. The thermal lag factors were developed as a function of the maximum SHGF at any given latitude, month, and exposure, and take into account the variation in solar heat intensity with time, the massiveness of the structure, and the presence of internal shading.

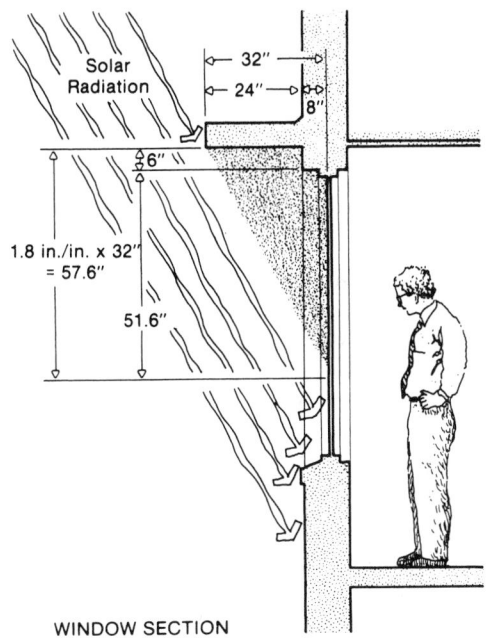

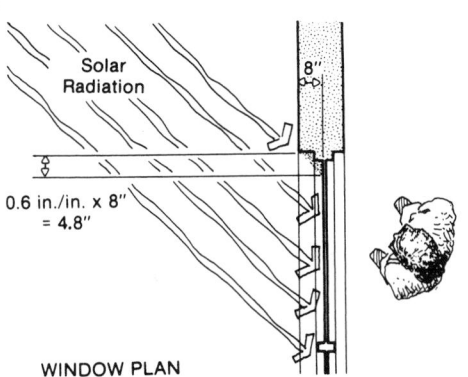

FIGURE 4.6. Example of shading by overhang and reveal.

The thermal lag factors in Tables B3.10 and B3.11 are intended to be multiplied by (1) the maximum SHGF underlined in Table B3.1 or in Table B3.8 and (2) the appropriate SC from Table B3.3, 3.4, 3.5, or 3.6, all in the appendix. Table B3.10 lists thermal lag factors when there is no interior shading. Table B3.11 lists thermal lag factors when there is some form of interior shading. For externally shaded fenestration that is not covered in Table B3.8, use the thermal lag factor for the north orientation in the Northern Hemisphere and the south orientation in the Southern Hemisphere, regardless of the actual orientation.

Example

Given: The window in Figure 4.6.

Find: The solar load at 2:00 p.m. in July if the window is double-glazed with ⅛-inch clear glass and the room is of medium-weight construction. The window is 4 feet wide and 6 feet high.

Solution:
From Table B3.1, the *maximum* SHGF in July for the unshaded area is 216 Btuh/ft^2 (681 W/m^2). The *maximum* SHGF in July for the shaded area is 38 Btuh/ft^2 (120 W/m^2).

From Table B3.3, the SC is 0.88.

From Table B3.10, the thermal lag factor is 0.29 for the unshaded area and 0.75 for the shaded area.

From the dimensions in Figure 4.6 and the solution to the previous example, the shaded area is 51.6 in. × 4.8 in. − 248 in.^{2} = 1.7 ft^2 (0.16 m^2), and the unshaded area is (72 − 51.6) in. × (48 − 4.8) in. = 881 in.^{2} = 6 ft^2 (0.57 m^2).

Finally, the solar load =
$(216 \times 0.88 \times 0.29 \times 6)$
$+ (38 \times 0.88 \times 0.75 \times 1.7)$
$= 373$ Btuh

$(681 \times 0.88 \times 0.29 \times 0.57)$
$+ (120 \times 0.88 \times 0.75 \times 0.16)$
$= 112$ W.

HEAT TRANSMISSION THROUGH THE BUILDING ENVELOPE

The factors that affect heat transfer through a given element of the building envelope are its surface area, construction materials, thickness, position with respect to the sun, exterior color, shading by adjacent structures, as well as the temperature of surrounding spaces or environment and temperature of the interior space. In the case of windows, the sash material (wood, metal, thermal break), the number of panes of glazing, and the glazing material all have an effect.

Heat is conducted through the envelope whenever there is a temperature difference between the inside and the outside. The heat thus transferred through walls, roof, floor, doors, and fenestration is all sensible heat and is referred to as *transmission.* The formula for calculating transmission, q, is Equation 2.9, which is

$$q = UA (T_1 - T_2) \qquad (4.4)$$

where

U = overall heat transfer coefficient
A = area
$(T_1 - T_2)$ = temperature difference between inside and outside

The area is determined from building plans or direct measurement, and the temperature difference is determined by the procedure for selecting design temperature presented earlier in this chapter. The overall heat transfer coefficient, or *U-value,* is determined as outlined in Chapter 2. In summary, the steps are:

1. Determine the resistance of each component of a given structure and also of the inside and outdoor air surface films.
2. Add these resistances together:

$$R = r_1 + r_2 + r_3 + \ldots r_n$$

3. Take the reciprocal, $U = 1/R$.

The following tables will be useful in determining the individual resistances, r:

B4.1 R-Values of Air
B4.2 R-Values of Typical Building Materials
B4.3A Overall Coefficients of Heat Transmission (*U*-Factor) of Windows and Skylights
B4.3B Adjustment Factors for Various Window and Sliding Patio Door Types
B4.4 Coefficients of Transmission (*U*) for Slab Doors
B4.5 Table of Reciprocals
B4.6 Conversion Table for Wall Coefficient *U* for Various Wind Velocities

Using these tables, the overall U-values are calculated as demonstrated in Figures 2.9 and 2.10.

Notice in Table B4.1 that surface air films are more insulative under lower wind velocities than under higher ones. The wind speed at the interior wall surface is normally taken as zero (still air). Unless specific conditions are known, winter calculations (heat loss) are usually based on 15 mph (6.7 m/s). The wind velocity for summer conditions is usually taken to be half the winter design value, but not less than about 7½ mph (3.4 m/s). Actually, in well-insulated construction, one needn't analyze air film insulation values very carefully, since they represent only a very small portion of the overall U-value.

Another point worth noting is that even very small areas of highly conductive materials can serve as a *thermal bridge* and "short-circuit" the insulative effect of an envelope. For example, steel beams that penetrate the envelope to the outside can seriously undermine the conservation efforts of thick wall insulation. This concept will be discussed more fully later in conjunction with floor heat loss.

Energy Code Compliance

ASHRAE has developed over the years a set of building energy conservation standards which have been almost universally adopted in whole or in part by building codes and local building departments throughout the United States. The latest versions for new and existing buildings are listed in the Bibliography at the end of this chapter.

The provisions of the current ASHRAE energy conservation standards are too lengthy to include here, but the reader is cautioned to refer to the applicable local energy code, which establishes the minimum thermal requirements for exterior envelopes, before developing outside wall and roof sections. These requirements will affect the U-values used in an actual load calculation.

For flexibility, the ASHRAE energy conservation standards allow the U-value of any one assembly—such as a roof, wall, or floor—to be increased above the prescribed maximum, as long as the U-values for other components are decreased so that the overall heat gain and/or loss for the entire building envelope does not exceed the total that would result from conformance to the prescribed U-values.

Sol-Air Temperature

The heat that flows into a space through the above-grade portions of the envelope in the summer is a result of the sun rays striking the walls and roof and of the high outdoor air temperature. In order to simplify calculations, the solar and convective sources of heat at the outside surface are combined into one imaginary term called *sol-air temperature.* The sol-air temperature is defined as the equivalent outside temperature that would, in the absence of solar radiation, cause the same amount of heat gain as actually exists due to both the real outside air temperature and the solar radiation. In other words, the sol-air temperature is a fictitious higher temperature that takes into account the effect of the solar heat absorbed by the opaque wall and roof surfaces.

In the calculation of heat gain through walls and roof, the appropriate sol-air temperature is simply substituted for the summer design temperature found in Table B2 in the appendix. Table B4.7 lists sol-air temperatures over 24 hours under the conditions of July 21 and 40°N latitude. Values are given for light-colored surfaces and dark-colored surfaces.

Sol-air temperatures can be estimated for other dates and latitudes by using the data in Table B3.1. The portion of the sol-air temperature attributable to the solar heat is found by subtracting the air temperature given in the second column of Table B4.7 from the sol-air temperature. By multiplying that portion by the ratio of solar heat gain factors from Table B3.1 (actual SHGF:SHGF on July 21, 40°N latitude), the actual portion attributable to solar heat can be approximated. The actual temperature range of the specific locality should also be used. The procedure for adjusting the sol-air temperature can perhaps best be illustrated in the following example:

Example

Calculate: the sol-air temperature on a flat roof at 2:00 p.m. on August 21 in Tokyo, Japan.

Solution:
From Table B2, the daily range in Tokyo is 14°F (8°C); the latitude is 36°N; the 1 percent level of the summer design dry-bulb temperature is 91°F (33°C).

From Table B3.1, the SHGF at 2:00 p.m. on the horizontal is

August 21, 36°N: 222 Btuh/ft^2 (698 W/m^2)^a
July 21, 40°N: 231 Btuh/ft^2 (729 W/m^2)

From Table B4.7, the following information is gathered for 2:00 p.m. (hour 14):

	°F	°C
Light-colored roof		
Sol-air temperature	126	52.2
Air temperature	94	34.4
Dark-colored roof		
Sol-air temperature	166	74.4
Air temperature	94	34.4

The component of the actual sol-air temperature due to solar heat therefore is, for a light-colored roof:

$$\frac{222}{231} \times (126 - 94) = 31°F$$

$$\frac{698}{729} \times (52.2 - 34.4) = 17°C$$

^{a} By interpolation between the 32°N and 40°N tables.

For a dark-colored roof:

$$\frac{222}{231} \times (166 - 94) = 69°F$$

$$\frac{698}{729} \times (74.4 - 34.4) = 38°C$$

According to Table B4.8, the percentage of the daily range at 2:00 p.m. is 3 percent. The component of the actual sol-air temperature due to the outside air temperature is thus:

Design dry-bulb – (daily range × percentage)

$$= 91°F - (14°F \times 3\%) = 90.6°F$$
$$= 33°C - (8°C \times 3\%) = 32.8°C$$

Putting the two components together, the total actual sol-air temperature is, for a light-colored roof:

$$31°F + 90.6°F = 121.6°F$$
$$17°C + 32.8°C = 49.8°C$$

For a dark-colored roof:

$$69°F + 90.6°F = 159.6°F$$
$$38°C + 32.8°C = 70.8°C$$

Depending on whether the roof is light- or dark-colored and whether the calculations are in conventional or SI units, the indoor room temperature would be subtracted from one of the above sol-air temperatures, and the difference would be used as the ΔT in the U × A × ΔT calculation of transmission heat gain into a space through the roof. A comparison between the results for light- and dark-colored roofs may influence the designer's choice of roof color and/or materials. The list of materials and their respective solar absorptances in Table 4.1 indicates the difference it can make.

Temperature in Adjacent Unconditioned Spaces

The sol-air temperature provides a means for evaluating an effective outdoor temperature when the envelope exterior surface is exposed to solar radiation. Another special case is when the exterior surface of the envelope of the space under consideration is bordered by a space that is not heated or cooled. Such is the case, for example, when a building tenant is adjoining a vacant space. In the wintertime, the vacant space, if unheated, can approach the outside temperature; and in the summer, it can become very hot under some circumstances. Some estimate must be made of the adjoining space temperature in order to calculate heat loss and gain of the conditioned space.

TABLE 4.1 SOLAR ABSORPTANCE OF VARIOUS SURFACES

Material	Total Incident Heat Absorbed (%)
Tar and gravel, asphalt, etc.	93
Slate, dark soil, etc.	85
Grass, dry	70
Copper	
Tarnished	64
New	25
Paint	
Light gray	75
Red	74
Aluminum	54
Light green	50
Light cream	35
White	25
Whitewash, fresh snow cover, etc.	20
Aluminum foil	5

Normally, this temperature lies in the range between the indoor and outdoor temperatures. If the envelope of the adjacent space has a larger surface area exposed to the outside than to the conditioned space, the temperature in the space will be more influenced by the outside than by the conditioned space. Likewise, the surface of that envelope with the greatest U-value will have the greatest influence on the space temperature. For a rough approximation of the temperature of the adjacent space, the effect of area and U-value can be proportioned between the outside and conditioned temperatures.

If, however, air is allowed to pass between the adjacent space and either the outside or the conditioned space, the situation becomes more complicated. In general, unconditioned spaces with large openings or glass areas exposed to the outdoors can be assumed to be at the same temperature as the outdoors. (However, in an airtight unconditioned space with large glass areas exposed to the sun, solar heat buildup can raise temperatures well above the outside air.)

Another example of unconditioned adjacent spaces is crawl spaces. These need to be vented in order to prevent moisture condensation, but venting brings the crawl space temperature close to that of the outdoors. Crawl spaces can be insulated either at the ceiling (which is the floor of the space above) or on the outside or inside of the perimeter wall. If the floor above is insulated, the crawl space vents should be kept open because the crawl space temperature may drop below the dew point. If the perimeter wall is insulated, the vents should be kept closed during the heating season and open during the remainder of the year.

Heat Loss Through Ground

The problem in calculating the heat transfer through below-grade walls and floors and through slab floors is that the heat capacity of the soil causes it to maintain a relatively stable temperature while the temperature of the outside air fluctuates from hour to hour and day to day. Furthermore, the ground temperature varies with the depth below the surface. It is the ground temperature that will directly influence heat loss through below-grade surfaces. Summer heat gain through below-grade portions of the envelope is normally negligible compared to other loads.

While ground temperature does fluctuate gradually over time and is related to the heat capacity and conductivity of the earth, the relationship is an extremely complicated one. These properties of the earth also vary with local conditions and are usually unknown. Based on laboratory tests, a commonly used average value for winter heat loss is 2 Btuh/ft^2 through an uninsulated concrete basement floor. Test results indicate a higher heat loss through below-grade walls since they're closer to the surface; the heat loss normally used is twice that of the floor, or 4 Btuh/ft^2. For calculating heat loss through insulated below-grade walls and floors, or for estimating the economically optimum depth below grade to extend insulation, Table B4.9 may be used. The heat loss factor obtained from Table B4.9 must be multiplied by the wall or floor area and by the temperature difference between indoor and outdoor air. (For a heated slab, the average temperature of the pipe, duct, or electric element is used for the indoor temperature.)

It turns out that for concrete slab floors at grade level, the heat loss is proportional to the length of exposed edge rather than to the total area. If the slab edge is not insulated, the heat loss is about 0.81 Btuh per linear foot of exposed edge per degree Fahrenheit difference between the indoor temperature and the average outdoor air temperature. This heat loss may, however, be greatly reduced by insulating the slab edge.

A concrete slab floor may be unheated, or it may be heated by heated pipes, ducts, or electric elements embedded within it. A heated floor constituting a large radiant slab is sometimes used as a complete or partial space heating system. In fact, in severe climates (above 4,000 degree days), slab floors should not be used unless they are heated. Perimeter insulation of a slab-on-grade floor is important for comfort and energy conservation in either case, but for a heated slab floor it is crucial to prevent excessive heat loss.

Since heat loss from a concrete slab floor occurs mostly through the perimeter rather than down through the floor and into the ground, no floor loss other than perimeter loss needs to be considered if it is calculated accurately. Heat loss through an insulated perimeter can be estimated by the following equation:

$$q = F \times P \qquad (4.5)$$

where

q = perimeter heat loss, Btuh (W)
F = heat loss coefficient, Btuh/°F per foot
    (W/°C per meter) of perimeter (Table B4.10)
P = length of perimeter or exposed edge of floor, ft (m)

Since water rapidly absorbs heat and carries it away, contact with groundwater drastically increases heat loss and should therefore be avoided. Accordingly, concrete floor slabs are usually placed on gravel at least 4 inches thick both to insulate the floor from the earth and to inhibit the rise of groundwater by capillary action. A waterproof membrane should be installed over the gravel, and the slab itself should be laid several inches above grade level and be provided with effective subsoil drainage to avoid rain or melting snow from soaking the slab.

In addition to the above precautions, edge insulation is necessary. Insulation should be of a water-resistant material with a minimum of R-2.5. The insulation should be placed vertically next to the exposed slab edge, extending downward to some distance below grade, or vertically to cover the exposed slab edge and the above-grade portion of the foundation wall, and then extended horizontally in an L shape either outward below grade or inward under the floor.

The thermal effect is approximately the same for both configurations if the same thickness and total length of insulation are used. The extra advantage of the L-shaped arrangement is that it also prevents freezing adjacent to the foundation, and eliminates the potential for frost heave damage. The important point is to avoid breaks or joints when the insulation is installed; otherwise, local thermal bridges, which reduce the overall efficiency of the insulation, can be formed.

One final note on below-grade insulation: The thermal performance degrades somewhat over time, largely due to moisture absorbed by the insulation. Thus, the aged and reduced R-value of any below-grade insulation should be used in calculations. The following field test results[a] indicate the percent of R-value retained by various commercially

[a] Reported by Dow Chemical Co. in *Specifying Engineer,* February 1983.

available, rigid, foam plastic insulations after an 18-month earth exposure:

Material	Retained R-Value (%)
Styrofoam	98
Molded-bead polystyrene	60–75
Aluminum-foil-faced polyisocyanurate	64
Polyurethane (faced and unfaced)	85

Example

Given: A below-grade structure $30 \times 28 \times 10$ ft ($9.1 \times 8.5 \times 3.0$ m) with insulation R-8 (R-1.47) in the walls and roof. The roof is 1 foot (0.3 m) below the ground surface.

Find: Total heat loss from the structure when it is 70°F (21°C) inside and 20°F (−7°C) outside.

Solution:
Wall (Using Table B4.9)
(Conventional Units)
Perimeter = $2(30 + 28) = 116$ ft

Area for each 1-ft band around the structure =
    116 ft × 1 ft = 116 ft²

2nd ft below grade:
    0.079 Btuh/ft²·°F × 116 ft² = 9.16 Btuh/°F

3rd ft below grade:
    0.068 Btuh/ft²·°F × 116 ft² = 7.89 Btuh/°F

4th ft below grade:
    0.060 Btuh/ft²·°F × 116 ft² = 6.96 Btuh/°F

5th ft below grade:
    0.053 Btuh/ft²·°F × 116 ft² = 6.15 Btuh/°F

6th ft below grade:
    0.048 Btuh/ft²·°F × 116 ft² = 5.57 Btuh/°F

7th ft below grade:
    0.044 Btuh/ft²·°F × 116 ft² = 5.10 Btuh/°F

8th ft below grade:
    0.044 Btuh/ft²·°F × 116 ft² = 5.10 Btuh/°F

9th ft below grade:
    0.044 Btuh/ft²·°F × 116 ft² = 5.10 Btuh/°F

10th ft below grade:
    0.044 Btuh/ft²·°F × 116 ft² = 5.10 Btuh/°F

11th ft below grade:
    0.044 Btuh/ft²·°F × 116 ft² = <u>5.10 Btuh/°F</u>
Total wall heat loss   61.23 Btuh/°F

(SI units)

Perimeter $= 2(9.1 + 8.5) = 35.2$

Area for each 0.3-m band around the structure $=$
　$35.2 \text{ m} \times 0.3 \text{ m} = 10.56 \text{ m}^2$

0.6 m below grade:
　$0.45 \text{ W/m}^2 \cdot °\text{C} \times 10.56 \text{ m}^2 = 4.75 \text{ W/}°\text{C}$

0.9 m below grade:
　$0.38 \text{ W/m}^2 \cdot °\text{C} \times 10.56 \text{ m}^2 = 4.01 \text{ W/}°\text{C}$

1.2 m below grade:
　$0.34 \text{ W/m}^2 \cdot °\text{C} \times 10.56 \text{ m}^2 = 3.59 \text{ W/}°\text{C}$

1.5 m below grade:
　$0.30 \text{ W/m}^2 \cdot °\text{C} \times 10.56 \text{ m}^2 = 3.17 \text{ W/}°\text{C}$

1.8 m below grade:
　$0.27 \text{ W/m}^2 \cdot °\text{C} \times 10.56 \text{ m}^2 = 2.85 \text{ W/}°\text{C}$

2.1 m below grade:
　$0.25 \text{ W/m}^2 \cdot °\text{C} \times 10.56 \text{ m}^2 = 2.64 \text{ W/}°\text{C}$

2.4 m below grade:
　$0.25 \text{ W/m}^2 \cdot °\text{C} \times 10.56 \text{ m}^2 = 2.64 \text{ W/}°\text{C}$

2.7 m below grade:
　$0.25 \text{ W/m}^2 \cdot °\text{C} \times 10.56 \text{ m}^2 = 2.64 \text{ W/}°\text{C}$

3.0 m below grade:
　$0.25 \text{ W/m}^2 \cdot °\text{C} \times 10.56 \text{ m}^2 = 2.64 \text{ W/}°\text{C}$

3.3 m below grade:
　$0.25 \text{ W/m}^2 \cdot °\text{C} \times 10.56 \text{ m}^2 = \underline{2.64 \text{ W/}°\text{C}}$
　Total wall heat loss　　　31.57 W/°C

Floor (Using Table B4.10a)

Average heat loss per square foot (m^2):
0.023 Btuh/ft$^2 \cdot °$F
(0.131 W/m$^2 \cdot °$C) (greater than 7 feet below grade = 7 feet below grade)

Floor area: 28×30 (8.5×9.1) = 840 ft^2 (77.4 m^2)

Total floor heat loss
　$= 0.023 \times 840 = 19.32 \text{ Btuh/}°\text{F}$
　$= (0.131 \times 77.4) = (10.14 \text{ W/}°\text{C})$

Roof (Using Table B4.9)

Average heat loss per square feet (m^2): Interpolate between 0.68 and 2.27 feet (0.20 and 0.69 m) for a 1.0-foot (0.30-m) path length through soil. A horizontal roof surface can be approximated by a vertical basement wall surface, since the only difference will be in the inside air film coefficient

(R-0.61 for the roof, R-0.68 for the wall), which contributes a very small portion to the R-8 roof.

0.68 ft (0.20 m)	0.093 Btuh/ft$^2 \cdot °$F (0.53 W/m$^2 \cdot °$C)
1.0 ft (0.30 m)	0.090 Btuh/ft$^2 \cdot °$F (0.51 W/m$^2 \cdot °$C)
2.27 ft (0.69 m)	0.079 Btuh/ft$^2 \cdot °$F (0.45 W/m$^2 \cdot °$C)

Roof area　840 ft^2 (77.4 m^2)

Total roof heat loss $= 0.090 \times 840 = 75.60$ Btuh/°F
　　　　　　　　　$= (0.51 \times 77.4) = (39.47 \text{ W/}°\text{C})$

	Total	
	Btuh/°F	W/°C
Heat loss per degree	61.23	(31.57)
Fahrenheit (°C) =	19.32	(10.14)
	75.60	(39.47)
	<u>156.15</u>	<u>(81.18)</u>

Temperature difference $= 70 - 20 = 50°$F
　　　　　　　　　$= (21 - (-7)) = (28°\text{C})$

Total heat loss $= 156.15 \times 50 = 7,808$ Btuh
　　　　　　$= (81.18 \times 28) = (2,273 \text{ W})$

Thermal Lag Factor

The thermal lag factor for heat transmission through the opaque envelope is similar to the factor for SHG through fenestration, except that it represents a delay from the time heat enters the exterior surface until it exits the envelope and enters the space as cooling load.

Table B4.11 lists thermal lag factors for opaque walls and roofs at various latitudes, compass orientations, and times of the year. These factors are to be multiplied by the area, U-value, and sol-air temperature difference to calculate cooling loads through walls and roofs. The thermal lag effect through sheet glazings is negligible, so it is not taken into account for transmission through fenestration. In the case of glass blocks, the thermal lag effect is accounted for in the SHG calculation.

INTERNAL LOADS

Internal gains are components of the cooling load that arise from sources internal to the conditioned space. They include both the sensible and latent heat released by the following components:

- People
- Lighting

- Appliances and equipment (e.g., electric, gas, or steam appliances for cooking, drying, or humidifying; electric calculators, typewriters, duplicators, data processing equipment, and motors; laboratory equipment)
- Miscellaneous sources (e.g., escaping steam, air circulation fans, piping, ductwork, and other sources not included in the above categories)

Normally, all electricity that enters a space for lighting, equipment, electronics, and so on, ends up as internal heat gain in the space.

This section contains the data required to determine the instantaneous heat gain from these sources. It also outlines procedures for estimating reduced internal cooling loads, when heat gains are subject to diversity, as well as radiant heat absorption and storage in the building structure (as described in Chapter 3).

Internal gain is highly dependent on the use of a space, that is, whether it is used as an office, hospital, department store, specialty shop, theater, assembly hall, machine shop, factory, assembly plant, and so on. The type of use determines occupancy density, lighting levels required, and the number and type of appliances and heat-producing equipment.

Internal gain is normally considered only a part of the cooling load; no credit is taken for internal heat gains to offset any heat loss when calculating the peak heating load. Although the internal heat gains will compensate for some winter heat loss and decrease the heating requirements at times, the peak heating load is calculated for the night, weekend, or other unoccupied periods. Exceptions are some industrial plants where large internal gains are always present during occupancy. In this case, the continuous heat sources may offset a portion of the heating load. In fact, there are situations where so much heat is available that outdoor winter air must be used to cool the space. However, sufficient heating equipment must still be provided to prevent freezing pipes during periods of unoccupancy. The heating load would be based on maintaining a temperature of at least 40°F (4.4°C) in the building.

The basic equation for any form of internal load is:

$$\text{Cooling load} = (\text{heat gain}) \times (\text{building storage factor}) \times (\text{diversity factor}) \qquad (4.6)$$

People

The human body generates heat within itself through the process of metabolism, and releases it by radiation, convection, and evaporation from the skin or clothing, and by convection and evaporation in the respiratory tract. While the portion of the heat gain due to evaporation of moisture is latent, the rest is sensible. The proportion of heat released in the form of sensible versus latent heat depends on the surrounding temperature. The total amount of heat gain to the space from people depends on their number, duration of occupancy, and activity level. Table B5.1 in the appendix lists the rate of heat gain from people at various levels of activity. The heat gain listed for restaurant applications takes into account the heat from the food served as well as the raised metabolic level for digestion. The data in Table B5.1 are for continuous occupancy of at least an hour. Where occupancies are brief (under 15 min), the excess heat and moisture brought in by people may increase the heat gain by as much as 10 percent above that listed in Table B5.1.

The occupant density must be estimated from knowledge of the specific circumstances and the space use. Private offices usually have one individual per office. Conference and seminar rooms may have an occupancy load of 10 ft² per person. Auditoriums, religious buildings, theaters, opera houses, concert halls, enclosed stadiums, community centers and sports arenas typically have a capacity for about 5 ft² per person in the seating areas, which are generally occupied for 1 to 4 hours. Department stores contain 25 to 100 ft² per person. Often, the building program will specify a desired occupant density or the total number of people to be accommodated within a given space. Life safety regulations may limit the number of occupants of a space. In residences, occupancy can be sporadic and variable. Living rooms and recreation rooms can be used for entertaining large groups. If there is no indication of the need for special entertainment provisions, and the exact number of occupants in the residence is unknown, a rule of thumb is to figure twice as many occupants as bedrooms.

The heat gain from people has two components: sensible and latent. Moisture from people goes directly into the air, and this latent heat becomes part of the space cooling load with no delay. However, part of the sensible component is first absorbed and stored by the surrounding structure and furnishings and then convected into the air at some later time. The *building storage factor* is used to account for this delay effect. The building storage factor depends on the total hours the occupants are in the space and varies according to the time elapsed since they first entered. Values for a continuously cooled space (during the 24-hour period) are listed in Table B5.2 in the appendix.

If the cooling system is shut down overnight, a building storage factor of 1.0 should be used, since any energy stored in the building at closing time may still be there the next morning, so there is no more storage capacity available to reduce the load. When the cooling equipment is operated for several hours before or after occupancy, the building storage factors are about the same as for continuous equipment operation.

In densely occupied areas such as theaters and auditoriums, a building storage factor of 1.0 should be used. Since very little body surface area is exposed to radiate to the walls and furnishings of such structures, the bulk of the body heat given off must be in the form of latent heat.

Lighting

Electric lights, whether incandescent, fluorescent, or high intensity discharge (H.I.D.), convert electrical power into light and heat. Even the light is eventually reduced to its equivalent heat energy in the space, so, in effect, all the electrical power entering a lighting fixture ends up as heat in the space.

Some of the heat is convected from the fixture to the surrounding air and immediately becomes part of the cooling load. The rest is radiated to surrounding surfaces, except for a small portion that is conducted to adjacent materials. These amounts are then convected to the air to become part of the cooling load at a later time. All of this heat is sensible heat.

It is extremely important to accurately estimate the load imposed by lighting, since it is often the largest component of the total cooling load. The instantaneous rate of heat gain from electric lighting, q, is:

$$q = \text{total light wattage} \times \text{ballast factor} \qquad (4.7)$$

The *total light wattage* is obtained from the ratings of all fixtures or lamps installed in the space. For incandescent lights, the rated wattage is the actual power consumption, and the ballast factor is 1.0. For all other types of lighting, the power consumption is greater than their rated wattage due to the electricity required for the lamp ballast; hence the term ballast factor. *Ballast factors* for various sizes of fluorescent fixtures are presented in Table B5.3 in the appendix. For other H.I.D. fixtures, such as mercury vapor, low-pressure sodium, high-pressure sodium, and metal halide, ballast factors vary from 1.04 to 1.37. Manufacturers' specifications should be consulted whenever possible.

For conventional units, multiply the value of q found in Equation 4.7 by 3.4 Btuh/W. For SI units, leave q in terms of watts as calculated in Equation 4.7.

When actual lighting layouts are not available and the total light wattage of a space is not known, it must be estimated from the approximate wattage per square foot. Some average rules of thumb are presented in Table B5.4 in the appendix.

As mentioned above, there is a time lag between the heat gain from lights and the resulting space cooling load; the cooling load may then be lower than the instantaneous heat gain. Several studies have indicated the effects on the cooling load of the type of light fixture, the manner in which it is installed, the type of air supply and return, the furnishings in the space, and the mass of the structure.

A recessed light fixture will tend to transfer heat to the surrounding structure, whereas a hanging fixture will transfer more heat directly to the air. Some light fixtures are designed so that air returns through them and absorbs heat that would otherwise go into the space.

The above parameters have been formulated into a set of building storage factors for lighting, which are presented in Table B5.5 in the appendix. The "a" and "b" classifications necessary to obtain the building storage factors are defined in the notes at the bottom of the table. If the lights are left on for 24 hours a day, an equilibrium condition is approached in which the cooling load equals the power input to the lights, so that the building storage factor is 1.0. If the cooling system operates only during the occupied hours (when the lights are on), the building storage factor is also 1.0. Where one area of lighting is on a different schedule from another, treat each separately.

Appliances, Equipment, and Miscellaneous Sources

There is an almost infinite variety of equipment and appliances that contribute a heat gain to the conditioned space. The key is to be thorough in identifying all the heat-producing sources in a given space. Table B5.6 in the appendix provides recommended heat gain values for some typical sources.

Restaurants, hospitals, laboratories, and some specialty shops (e.g., beauty shops) have electrical, gas, or steam appliances that release heat into the space. For most appliances, the total rated energy input is not the same as the input required on an hourly average to maintain the desired appliance surface temperatures; in actuality, the appliance cycles on and off over time. That is why the recommended heat gain values differ from the appliance input ratings listed in Table B5.6.

Kitchen appliances sometimes have a hood over them to exhaust air, which reduces the heat gain appreciably. Remember, however, that the exhausted air must be replaced by outdoor air, and that cooling that air must be accounted for in the overall load.

When the actual equipment in a space is unidentified, the approximate heat dissipation rates per unit area in Table 4.2 may be used.

TABLE 4.2 APPROXIMATE HEAT DISSIPATION RATES PER UNIT AREA

Equipment	Btuh/ft^2	W/m^2
Mainframe computers		
Digital	55–130	180–415
Analog	40–115	120–355
Laboratory equipment	15–70	45–220
Manufacturing equipment		
General assembly, and stamping	20	65
Planting, foaming, and curing	150	475
Office equipment		
General offices	3–4	9.5–12.6
Purchasing and accounting departments	6–7	19–22
With computer display units	15	47

Another interior source of heat may be steam or hot water pipes running through the air-conditioned space or hot water tanks within a space. In many industrial processes, tanks are open to the air, allowing water to evaporate into the space and adding to the latent load. Another source of heat and moisture gain within a space is escaping steam from industrial equipment such as cleaning and pressing machines. These sources are difficult to estimate, so an engineer should be consulted.

The effect of these appliances and equipment on the sensible portion of the cooling load is delayed in the same manner as the other load components already discussed. Table B5.7 in the appendix provides the building storage factors for appliances and equipment. The building storage factor should be applied only to the sensible portion of the heat gain, and the sensible and latent cooling load components should then be figured separately.

The total residential appliance loads can be approximated by the major appliances in the kitchen. A value of 1,200 Btuh (350 W) of sensible heat gain released by kitchen appliances is sufficient under most circumstances. Although this is not as much as the heat given off by even one top burner or element of a domestic range, such factors as the intermittent use of appliances, building heat storage effects, and kitchen ventilating fans make this a reasonable value to use.

Diversity

The third component of Equation 4.6 is the *diversity factor*, sometimes referred to as a *usage factor*. Diversity is taken into account when all of the internal loads do not occur at the same time. It is usually unlikely that all of the people are present in a building and that all of the lights are operating at the time of the peak cooling load. Some people may be away

on other business, and the lighting arrangement should allow some lights to be off when not required.

All business machines in a given space may not be used simultaneously. On the other hand, electronic computer equipment is typically in use or at least in a standby mode nearly all the time. In a kitchen, a toaster or waffle iron may not be used in the evening, and a fry kettle may not be used during the morning. Electric motors in industrial buildings may operate at a continuous overload, at less than the rated capacity, or intermittently. Electric motors are a major portion of the cooling load in industrial buildings, so their operating time and usage characteristics should be evaluated carefully and thoroughly.

The size of the diversity factor depends on the location, type, and size of the building, and is ultimately dependent on the judgment and experience of the person performing the load estimate. For example, the diversity factor for a single small office with one or two people is 1.0 (no reduction). In contrast, a whole floor of an office building with 50 to 100 occupants may have 5 to 10 percent of its people absent at the time of the peak load; and a 20- or 40-story building may have 10 to 20 percent absent during the peak. An area predominantly of sales offices may have many people absent in the normal course of business, or it may be full of customers in addition to the regular employees.

In apartments and hotels, very few people are normally present at the time that the solar and transmission loads are peaking, and the lights are typically turned on only after sundown. In these cases, the diversity factor can be much greater than that for office buildings.

The diversity factors listed in Table B5.8 in the appendix are intended only to be a guide. The diversity factor for any particular application must be based on a judgment of the effects of all of the many variables involved.

OUTSIDE AIR

Outside air (O.A.) in the form of infiltration and ventilation imposes a heating or cooling load on the conditioned space and on the mechanical systems that control the temperature and humidity conditions. Ventilation is supplied to control air purity and odor levels, while infiltration arises from controlled or uncontrolled leakage around doors and windows or through walls.

Infiltration

The rate of infiltration is one of the most difficult quantities to estimate accurately. The difficulty lies in the wide variation in type, quality of construction, shape, and location of a building; the type of heating system; and the design variation in window and door construction.

Outdoor air infiltrates into the indoor space through cracks and crevices around doors and windows, through open doors and windows, through penetrations of the envelope, through joints between walls and floor, and through the building skin, itself. The causes of infiltration are the stack effect and the force of the wind. The amount depends on the area of opening and the pressure difference across the opening. Two methods are used to calculate the quantity of infiltration: the *crack length method* and the *air change method.*

Crack Length Method

The crack length method is the more precise means of evaluating infiltration, but it requires specific information about the dimensions and construction details of windows, doors, and other openings. When this information is not available, the air change method may be used.

The crack length method is a calculation of the air flow produced by a pressure difference acting on each leakage path. The air leakage is expressed as air flow rate per unit length of crack. These unit values depend on the kind or width of crack and the pressure difference (due to wind speed or stack effect). It is common to use a wind speed of 15 mph (6.7 m/s) for normal design conditions unless local experience suggests another, more appropriate speed. Even when the prevailing winds are not expected to reach that speed, the air velocity increases as it flows around or over a building.

An alternative variation of the crack length method expresses the air flow rate per unit area of window or door. For fixed, sealed glass, the infiltration rate is normally taken as about 0.25 CFM/ft^2 of glass surface; for movable sash windows it is about 0.5 CFM/ft^2. The crack method is more accurate than the area method, but the area method is useful when it is difficult to evaluate the exact crack dimensions.

Whenever air enters a building, the same amount must also leave at the same rate; otherwise, an ever-increasing pressure would build up. Unless a mechanical exhaust system is operating, air entering through cracks on the windward side is assumed to flow through openings in interior partitions to cracks within other rooms on the leeward side of the building, where it is expelled. Thus, only half of the total crack length is used to estimate the leakage rate. The same idea applies to conditions of stack effect. Air that enters on one floor may exit on another floor; half of the crack length is for incoming air and half for outgoing air. Where negative pressure is created by a power exhaust, however, air could infiltrate through 100 percent of the crack area.

Table B6.1 in the appendix provides air leakage rates through windows and doors of various construction tightnesses and at various wind pressures (speeds). Air flow rates are given in CFM per linear foot, CFM per square foot, liters per second per linear meter, and liters per second per square meter.

How a window is closed or fits when closed, and how well it is crafted in general, are factors that have considerable influence on air leakage. Storm windows added to prime windows reduce infiltration and provide an air space to reduce heat transmission and help prevent frosting. Adding a storm window with the same leakage characteristic as a well-fitted prime window reduces prime-window air leakage about 35 percent. Applying tight storm window units to loosely fitted windows reduces air leakage approximately 50 percent, which is roughly equivalent to the effect of weatherstripping.

Door infiltration depends on the type of door, room, and building. For residences and small buildings where doors are used infrequently, infiltration can be estimated on the basis of air leakage through the cracks between the door and frame. Door fit varies greatly and is affected by warping. For a well-fitted door, leakage approximates that of a poorly fitted double-hung window; for a poorly fitted door, the figure may be doubled. If the door is weatherstripped, the value may be halved.

A frequently opened single door, such as that in a small retail store, may be assigned an air leakage value of three or more times that of a well-fitted door in order to allow for losses due to opening and closing. Figures B6.1, B6.2, and B6.3 in the appendix provide air leakage rates through heavily trafficked swinging doors, door cracks of unused swinging doors, and revolving doors, respectively. Leakage through door cracks is significant only in periods of very low traffic. The pressure differential needed in Figure B6.2 can be determined from Figure B6.1. Vestibule-type doors are two doors in series that form an air lock between them. These doors often appear as two *pairs* of doors in series, which should be considered as two vestibule-type doors.

Fireplaces, chimneys, and flues account for additional infiltration. When the fireplace is not in use, large quantities of air will rise through the chimney, drawing outdoor air through cracks in the envelope. If the chimney damper is closed, the problem is reduced, but fireplace dampers usually don't fit tightly and still allow some air flow even when closed. The recommended allowance is at least 50 CFM (24 L/s) infiltration for a fireplace not in use.

When a fireplace is in use, air loss through the chimney increases, causing even more infiltration throughout the structure. The contribution of heat from the fireplace can be disregarded because it is limited to the immediate vicinity of the fireplace and usually does not compensate for the added heating load from the induced infiltration. Many modern fireplaces contain an integral combustion air intake that is intended to eliminate the need for the fireplace to draw room air up the chimney. That, combined with a glass door on the fireplace, significantly reduces the infiltration caused by fireplaces. It is still advisable to figure an added 50 CFM (24 L/s) infiltration due to a fireplace.

Direct-fired warm-air furnaces are sometimes installed within a conditioned space. If combustion air is not brought in from outdoors through a closed duct, conditioned air from the space will be drawn in and exhausted through the flue. In this case, 10 ft^3 (0.28 m^3) of air should be allowed for per 1,000 Btu (1.06×10^6 joules) input of fuel into the furnace. For example, if the furnace is rated at 10 MBH (2,940 W) input, 10 ft$^3 \times$ 10 (1,000 Btu)/hr, or 1.67 CFM (0.78 L/s) of infiltration should be included.

Stack Effect

Buildings are divided into two categories for calculation of infiltration: (1) those that are low-rise, with fewer than five stories (less than about 100 feet tall), and (2) those with more than five stories (more than about 100 feet tall). Unless the inside-outside temperature difference is extreme, the *stack effect* is generally negligible in low-rise buildings, and the only infiltration that needs to be considered is that due to the effect of wind blowing through cracks in the envelope. In high-rise buildings, the stack effect may be dominant, causing air to leak through walls and around fixed window panels as well as regular window and door leakage.[a] Figures

B6.4, B6.5, and B6.6 may be used to determine air infiltration through curtain walls.

The increased air infiltration through windows caused by the stack effect is evaluated by converting the stack effect force to an equivalent wind velocity, and then looking up the flow rates in Table B6.1 with the corrected wind velocity. In high-rise buildings, the equivalent wind velocity may be calculated from the following formulas based on a neutral pressure point at the mid-height of the building and a typical winter inside-outside temperature difference of 70°F (21°C).[b]

$$V_e = \sqrt{V^2 - 1.75a} \text{ (for upper section of}$$
$$\text{tall buildings—winter)} \quad (4.8a)$$
$$V_e = \sqrt{V^2 + 2.75b} \text{ (for lower section of}$$
$$\text{tall buildings—winter)} \quad (4.8b)$$

where

V_e = equivalent wind velocity (mph)
V = wind velocity normally calculated for location (mph)
a = distance window is above mid-height (feet)
b = distance window is below mid-height (feet)

Note: The total crackage is considered when calculating infiltration from stack effect.

Air Change Method

When construction quality and crack length cannot be accurately evaluated or when wind conditions are not known, the air change method may be used as an alternative. Table B6.2 in the appendix lists the number of air changes per hour in rooms with varying exposures. These are rates of air leakage. The table lists rates for a medium construction quality at different design outdoor temperatures and wind speeds. The notes at the bottom of the table give adjustment factors for different construction qualities.

The total infiltration rate is taken to be one-half of the values in Table B6.2 so as to account for the fact that half of the cracks allow for infiltration and the other half for exfiltration. The procedure for using Table B6.2 is to multiply the volume of each room by the appropriate air change rate. The sum of the leakage rates of all the rooms in a building is then divided by 2 to get the estimated infiltration rate.

Some judgment must be exercised in using the air change method. A large room with few small windows will have a lower air exchange rate than a small room with many large windows. The infiltration for buildings with large, open industrial-type doors can be as much as five air changes per

---

[a] In areas of extremely high winds, infiltration should be considered through walls and fixed fenestration.

[b] Stack effect is not normally significant during the summer because the temperature and density difference is relatively slight.

hour. The tightness of construction and any unusual wind conditions will need to be judged from experience. The air change method should be considered only a gross approximation.

Ventilation

The reasons for ventilation and the distinction between ventilation and infiltration were discussed in Chapters 1 and 3. The rate of ventilation air introduced into a building is determined by fresh air requirements and by building pressurization to control infiltration. Typically, when mechanical ventilation is present, enough pressure is built up in the enclosed spaces to offset the pressure that would otherwise cause infiltration. Thus, infiltration can be disregarded in ventilated buildings, since the only outside air load is due to the controlled ventilation.

Recommended ventilation levels in CFM per person or CFM per square foot are listed for various applications in Table B6.3 in the appendix. In cases where the specific occupancy is not known, the number of people per square foot (square meter) from Table B6.3 can be used with the floor area under consideration to determine the total CFM (L/s). Where local codes call for greater levels of ventilation, those levels should of course be used.

Mechanical ventilation may be provided in residential air-conditioning systems, although it is not generally required. Under normal circumstances, infiltration will be sufficient to furnish enough outside air exchange. Situations in which the introduction of outside air through a central air-conditioning system might occur in a residence are where large-scale entertaining is anticipated, in very tightly built or underground homes, or where requested by the owner. In larger homes, ventilation is often provided. Residential ventilation can be figured at about one air change per hour.

Calculation of Heating or Cooling Load Due to Outside Air

Once the amount of outside air due to infiltration and/or ventilation is established, the resulting heating or cooling load may be calculated. The heat gain or loss due to outside air has a sensible and a latent component. The sensible load is the energy quantity associated with raising or lowering the temperature of outdoor air to indoor room conditions. The latent load is the energy quantity associated with the addition or removal of moisture in the outdoor air to bring it to the proper RH condition.

The *sensible load,* either heating or cooling, is:

$$q = 1.08 \times CFM \times \Delta T(°F) \qquad (4.9a)$$
$$= 1.20 \times L/s \times \Delta T(°C) \qquad (4.9b)$$

where

q	= sensible load, Btuh or W
CFM and L/s	= total outside air
ΔT	= temperature difference between indoor and outdoor air

The *latent load* is normally a component of the cooling load, but it is considered in the heating load only when winter humidification is used. The formula is similar to that for sensible load:

$$q = 0.68 \times CFM \times \Delta W \text{ (Gr/lb)} \qquad (4.10a)$$
$$= 4,840 \times CFM \times \Delta W \text{ (lb/lb)} \qquad (4.10b)$$
$$= 3,012 \times L/s \times \Delta W \text{ (kg/kg)} \qquad (4.10c)$$

where

q	= latent load, Btuh or W
CFM and L/s	= total outside air
ΔW	= difference in moisture content between indoor and outdoor air

The procedure for establishing the total quantity of outside air is as follows: First, determine the infiltration rate due to wind pressure and stack effect, if applicable. Add the CFM (L/s) capacity of all exhaust fans in the conditioned space. Then determine the ventilation rate required for people.

The required ventilation rate will usually be greater than the infiltration rate in nonresidential buildings. All of the infiltration would then be offset by air being introduced through the mechanical system for ventilation. The positive pressure resulting from more air being supplied to the space than exhausted from it can prevent infiltration. In any case, the larger outside air rate (infiltration or ventilation) should be used.

ANNUAL ENERGY USE CALCULATIONS

It is sometimes necessary to estimate the energy requirements and fuel consumption of HVAC systems for long periods of operation. These quantities can be much more difficult to calculate than peak heat loss and gain, since they involve the combined influence of many factors, such as weather variables (sunshine, wind, temperature, humidity), the building operating patterns, and thermal characteristics that may vary with time. It is difficult to foresee accurately how these factors will vary and the way in which they will interact. The calculations required to take all such variations into account become very complex. For these and other reasons, records of past operating experience, when available, provide the most reliable basis for accurately predicting future energy requirements.

When past energy records are not available for an existing or proposed building, some form of calculation is necessary. This section discusses the available methods for calculating projected energy use.

The space load calculations presented earlier in this chapter determine the amount of energy that must be added to or removed from a space to maintain the desired interior conditions. Energy consumption over time can be thought of simply as the summation of instantaneous loads over the period considered. But since the result depends entirely on weather, which is unpredictable, an energy consumption projection can only be an estimate of likely or average conditions, and will be entirely accurate only by coincidence.

There are two ways to approach the necessary computations: single-step and multiple-step.

The *single-step method* requires only one basic calculation, which relies on averages, approximations, and simplifications. The result is a monthly or seasonal energy requirement. The justification for this method is that over a long period of time, short-term deviations and fluctuations tend to cancel out each other. Examples of single-step calculations are heating degree days, equivalent full-load hours of cooling, and cooling degree days.

Multiple-step methods involve many repetitive calculations of instantaneous loads at periodic intervals (hourly, daily, etc.) at specific conditions of weather, HVAC system operation, structural response to thermal exposure (sun, exterior and interior temperatures and humidities), and internal loads. These methods depend on the availability of weather data for the time increments used. This is usually a determining factor. Due to the time-consuming and repetitive nature of these calculations, computer programs are commonly used. Hourly profiles of weather conditions for an entire year are available for most U.S. weather station sites.

It is worth noting that performing calculations that are more detailed than is appropriate for the least detailed of the input data is useless and sometimes introduces additional error. An analogy can be made with a chain that will break at the weakest link no matter how strong the rest of the links are. In fact, if the other links are made excessively strong, they may become heavier and more cumbersome, thus causing the weak link to break sooner. In the same way, excessive detail disproportionate to the weakest link in a data chain will cause a distortion in the overall calculation and a reduction of accuracy.

Whatever calculating method is employed, always compare results with any actual records available and try to reconcile any major discrepancies. For existing buildings, past fuel bills provide the most accurate and reliable check on calculations of future projections. For new buildings, data from similar structures under similar use patterns in the same locality, or at least the same climate, may be applied with caution to the planned design.

Degree Day Method

The simplest method for computing fuel requirements is the traditional *degree day method.* It presupposes that the energy required to maintain comfort is a function of a single parameter: the outdoor dry-bulb temperature.

The degree day method was devised over 50 years ago as a means of determining how much heating oil had been used by residences and therefore when to deliver more fuel as a function of weather severity. It is based on average solar and internal gains offsetting any heat loss when the outdoor temperature is above a *balance point temperature.* Below that point, the energy requirements for heating are proportional to the difference between the balance point temperature and the outside temperature.

Monthly heating degree days, as a measure of weather severity, are defined as the number of degrees the average monthly temperatures are below the balance point temperature multiplied by the number of days in the month. Annual degree days are an accumulation of the monthly degree days. Table B7.1 in the appendix presents monthly and annual heating degree days based on a 65°F (18.3°C) balance point for weather stations in the United States and Canada.

Studies in the 1930s by the American Gas Association and the National District Heating Association substantiated the validity of the 65°F (18.3°C) base temperature. Since that time, insulation practices have improved from virtually none in 1930 to between R-11 and R-19 for walls and R-30 to R-38 for roofs today. Also, internal gains from appliances and equipment have increased dramatically during the same period. Both of these trends decrease the balance point temperature. Accordingly, calculations using 65°F-based degree days overestimate heating energy consumption by modern buildings. Recent research suggests that the heating balance point has indeed shifted and may now be closer to 55°F (13°C). But until a different base temperature is established or a more valid, equally simple procedure for estimating heating loads is developed, the standard will continue to be the degree day method with a 65°F (18.3°C) base in conjunction with a modification factor, M, as listed in Table B7.2 in the appendix.

The national Climatic Center has published degree days based on a wide range of balance point temperatures. If the actual balance point of a building is known, these data may be used in Equation 4.11 below without the modification factor, M.

Another factor is used with the degree day method to account for inefficiencies in part-load operation of heating equipment over a heating season and for oversize equipment. This factor, P, is presented in Table B7.3 in the appendix.

Energy Consumption

In summary, the procedure for calculating heating energy consumption is as follows. First, determine values for the following variables:

Variable	Source
Peak heat loss in Btuh (H) at a specified ΔT	Earlier in Chapter 4
Degree days (DD)	Table B7.1
Modification factor (M)	Table B7.2
Part-load correction factor (P)	Table B7.3
Rated full-load efficiency of heating unit (n)	Product literature or Table B7.4
Fuel heating value (V)	Table B7.4

Then the energy consumption in Btu is given by:

$$E = \frac{H \times 24 \times DD \times M}{\Delta T} \qquad (4.11)$$

Fuel Consumption

The amount of fuel required for the above energy consumption is:

$$F = \frac{E \times P}{n \times V} \qquad (4.12)$$

Fuel Cost

The cost of the use above is simply:

$$\$ = F \times \text{cost per unit} \qquad (4.13)$$
$$\text{(gallon, therm, kilowatt-hour, etc.)}$$

Limitations

The degree day method is not known for great accuracy. Rather, its advantages are its simplicity and its suitability for quick estimates. But discrepancies between these calculations and actual fuel consumption can occur for a variety of reasons.

The degree day method assumes that fuel consumption is directly proportional to the number of degree days, regardless of whether they occur within a brief period or over a long duration. By not distinguishing a short, intense cold spell from a longer, milder condition, there's no way to account for the different equipment operating efficiencies under those differing circumstances.

Since it uses average temperatures, the procedure also does not account for the temperature swings within each day. This omission can result in underestimation of energy use in mild climates. For example, a building may require heat during parts of a day when the average temperature is above the balance point.

Other reasons for discrepancies are variations in actual occupancy and operating patterns from those anticipated; abnormal losses or gains through the envelope; and inefficient controls or control arrangements.

The efficiency of fuel utilization in an individual case can vary widely from average values. For example, a building with its chimney in the interior may recover as much as 35 percent of the heat normally lost up the chimney by being conducted through the chimney walls directly to the outdoors. On the other hand, when solar and internal gains contribute substantially to the heating of a building, the heating unit sized for the maximum load will in fact frequently operate at part load. Consequently, the heating efficiency may be lower than expected. Typically the efficiency of a boiler or furnace varies as the load varies with the time of year.

A building and its HVAC system can be modeled exactly if all the parameters are known in sufficient detail and a large number of computations are made. Most computerized energy use simulations take into account many of these details, although each one is streamlined with its own set of simplifying assumptions. The beauty of the degree day method is that it allows quick and easy estimates without such detail.

The degree day procedure, Equations 4.11 and 4.12, are applicable to estimates of oil and gas consumption as well as to electrical resistance-type devices such as electric baseboard units, electric furnaces, and electric boilers. The technique may be used for heat pumps by means of the addition of a "seasonal performance factor." The estimation of a seasonal performance factor, however, requires considerable experience and judgment. For heat pumps, the bin method calculation discussed later is recommended.

Cooling Energy Use Computations

Cooling loads, correlated more to such factors as solar gain and internal loads than to temperature, are more complex than heating loads. As a result of these various factors that determine cooling loads, energy usage is more dependent on certain building features than on a single climate parameter.

Despite this difficulty, two single-step procedures are popularly used. These are the *equivalent full-load hours method* and the *cooling degree day method*. It is the author's opinion that the equivalent full-load hours method is generally more reliable, so that is the one presented here. A description of the cooling degree day method may be found in the *ASHRAE Handbook of Fundamentals.*

Equivalent Full-Load (EFL) Hours

This procedure consists of determining the following values and then applying them in Equation 4.14:

Variable	Source
Peak cooling load in tons or thermal kw	Earlier in Chapter 4
kw/ton or kw (electric)/ kw (thermal)	Table B7.5 (or manufacturer's literature)
EFL hours	Table B7.6

Peak cooling load (tons) × kw/ton × EFL hours

$$= \text{seasonal kwh for cooling} \qquad (4.14a)$$

Peak cooling load (thermal kw)

$$\times \frac{\text{kw (electric)}}{\text{kw (thermal)}} \times \text{EFL hours}$$

$$= \text{seasonal kwh for cooling} \qquad (4.14b)$$

Since Table B7.6 in the appendix provides a range of values, and the correct number for any given application could even be above or below those averages, some judgment is required. That judgment should take into account the amount of sunshine, the number of abnormally hot or cool days anticipated, the local humidity conditions, the efficiency of the equipment, and such human factors as patterns of equipment operations, use of an economizer cycle, and preferred indoor temperatures. Local experience of climate and use patterns is extremely important.

The EFL hour values in Table B7.6 in the appendix are estimated ranges based on a survey of electric utility companies and are based on an indoor temperature of 75°F (23.9°C). In general, residential applications will be toward the lower end of the range and light commercial ones toward the higher end. Buildings with heavy loads, such as hospitals, may exceed the values listed.

Bin Method

The *bin method* is used for commercial and industrial buildings with unusual operations, and houses utilizing passive heating/cooling designs with high mass thermal storage. It is a multiple-step method allowing consideration of more details than the single-step methods, but it is still simple enough that the calculations can be made manually.

The idea is that 5° temperature ranges, or bins, are estab-

lished for a given location. The heating or cooling load is then calculated at each temperature bin occurring at that location. Once the load is determined in Btuh (W), it is multiplied by the number of hours during which the associated bin condition occurs. That results in annual Btu (Wh) at each bin. Then all the Btu (Wh) are added together to arrive at the total Btu (Wh) energy use for the year.

Table B7.7 in the appendix lists hourly occurrences at various bins 72°F (22°C) and below for selected sites in the United States. More complete tables of temperature occurrences above and below 72°F (22°C) at U.S. Air Force weather stations are contained in Air Force Manual 88-29 entitled *Engineering Weather Data,* available from the Superintendent of Documents, Government Printing Office, Washington, DC 26402.

While the bin method involves more detail than a single-step process, and is in fact the preferred method for calculations relating to heat pump operations, it does not account for solar heat gain, RH, wind conditions, or the time of day the temperature occurs. For this reason, the bin method is not as accurate as the hour-by-hour methods.

Hourly Computer Simulations

Computerized hourly calculations are necessary for accurately analyzing large commercial or industrial buildings but are not normally warranted for simple residences. Passive solar-heated homes, however, require hourly computer calculations to properly simulate their energy use.

A wide variety of energy programs based on hour-by-hour simulation have been developed by governmental organizations, utilities, manufacturers, and private consultants. Many of these programs are commercially available. In general, they involve the inputting of hourly weather data, building envelope characteristics, system descriptions, and operating patterns. Based on this input, a load calculation is performed for each hour of the year. The loads are totaled to arrive at the annual energy consumption.

The advantage of the computer is that it rapidly performs many calculations that would otherwise be very time-consuming and likely to result in computational errors.

However, although computers compute flawlessly, their results are only as reliable as the input data. Often, if an answer is printed out by a computer, it is automatically considered to be correct. It is important, however, to check a computer's answers against known realities and rough estimates. There is always a possibility that the program itself contains errors, that the input data are either incorrect or a mistake has been made in inputting them, or that the program is inappropriate for the intended application. Before using a computer simulation, the limitations of the program should always be checked. Perhaps the most reliable use for these programs is to analyze quickly the results of an architectural or engineering design change to determine exact changes in energy consumption.

Example

Calculate the heating and cooling loads and annual energy use for heating and cooling of the five-story square office building described below.

Location: St. Louis, Missouri.

Dimensions: 50,000 ft^2 (4,645 m^2) total floor area. 8 ft (2.44-m) floor-to-ceiling height. 11-ft (3.35-m) overall floor-to-floor height. Fenestration areas as a percentage of gross wall areas for each exposure are: north 5 percent, east 30 percent, south 20 percent, and west 5 percent.

Windows: Operable, well-fitted, weatherstripped casement type with a crack length of 1 linear foot per square foot (3.28 linear m per m^2) and glazed with two panes of ¼-inch (6-mm) clear glass. There are no interior window coverings or protrusions on the exterior to provide shading, so the designer has decided to use an exterior window shading device similar to group 1 in Table B3.5. The building is surrounded by grass and other vegetation.

Construction: Light-colored metal curtain walls with 6-inch (15-cm) metal studs, ½-inch (12-mm) drywall on the inside, and 6 inches (15 cm) of fiberglass blanket insulation in the wall cavity; flat built-up roof with a 4-inch (10-cm) heavy-weight concrete slab surfaced with a white material; suspended ceilings with a 9-inch (22-cm) fiberglass blanket laid above the ceiling on the top floor; 2-in. (5-cm) concrete floor slabs; and a full basement 8-ft (2.4-m) tall insulated on the exterior with R-8 (R-1.5) to a depth of 6 ft (1.8 m).

Occupancy and operation: An average of 100 ft^2/person (9.3 m^2/person); 2.5 W of fluorescent lighting per square foot (26.9 W/m^2); 1 W/ft^2 (10.8 W/m^2) for office automation equipment; a diversity of 80 percent for occupancy, lighting, and equipment; and operation of 9 hours per day.

Solution

The solution to this example, illustrated in Figures 4.7 and 4.8, is as follows:

The SHGFs are interpolated between those for latitude 32° and 40° and multiplied by the clearness number from Figure B3.1. There is no correction for foreground reflectance because green grass has a reflectance of close to 0.20, which is the standard on which the SHGF tables are based.

Heat loss through below-grade basement walls is:

0–1 ft: 400 S.F. × .093 × 66 = 2,455
1–2 ft: 400 S.F. × .079 × 66 = 2,086
2–3 ft: 400 S.F. × .068 × 66 = 1,795
3–4 ft: 400 S.F. × .060 × 66 = 1,584
4–5 ft: 400 S.F. × .053 × 66 = 1,399
5–6 ft: 400 S.F. × .048 × 66 = 1,267
6–7 ft: 400 S.F. × .069 × 66 = 1,822
7–8 ft: 400 S.F. × .069 × 66 = 1,822
                                14,230 Btuh

Annual heating energy is:

Peak heat loss   920,410
Degree days   4,900
M factor   0.76

$$E = \frac{920,410 \times 24 \times 4,900 \times 0.76}{66}$$

$$= 1,246 \text{ MMBtu}$$

The example does not indicate which fuel is to be used, so each one is figured below:

Electricity: $\dfrac{1,246,000,000 \times 1.0}{0.95 \times 3,413} = 384 \text{ kWh}$

Natural gas: $\dfrac{1,246,000 \times 1.36}{0.65 \times 1.03} = 2,531,083 \text{ ft}^3$

#2 Oil: $\dfrac{1,246,000 \times 1.36}{0.65 \times 138.69} = 18,797 \text{ gal}$

Annual cooling energy:

Peak cooling load: 94 tons
kW/ton: 1.18 + 0.21 = 1.39   (air-cooled)
        0.94 + 0.17 = 1.11   (water-cooled)
EFL hours: 1,500
        94 × 1.39 × 1,500 = 195,990 kWh
        94 × 1.11 × 1,500 = 156,510 kWh

SUMMARY

To calculate a heating or cooling load, detailed building design information, schedules of operation, and a knowledge of design indoor and outdoor temperature and humidity conditions are required. A checklist of the data needed is as follows:

PROJECT _Example_

PREPARED BY _VB_

LOCATION _St. Louis, Missouri_

DATE _8/29/05_

HEATING LOAD CALCULATIONS

INSIDE DESIGN TEMP ___72___

OUTSIDE DESIGN TEMP ___6___

DIFFERENCE ___66___

ITEM	QUANTITY X	U-VALUE X	ΔT	= BTUH (W)
GLASS	3,300 FT² (M²)	.45	66 °F (°C)	98,010
NET WALL (ABOVE GRADE)	18,700 FT² (M²)	.05	66 °F (°C)	61,710
NET WALL (BELOW GRADE)	3,200 FT² (M²)		 °F (°C)	14,230
ROOF	10,000 FT² (M²)	.03	66 °F (°C)	19,800
OUTSIDE AIR	10,000 CFM (L/S)	1.08 (1.20)	66 °F (°C)	712,800
Basement Floor	10,000 ft²	.021	66	13,860
TOTAL				920,410

FIGURE 4.7. Example of heating load calculations.

PROJECT *Example*
LOCATION *St. Louis, Missouri*

PREPARED BY *VB*
DATE *8/29/05*

COOLING LOAD CALCULATIONS

SPACE USE *Office*
FLOOR AREA *50,000 ft²*   VOLUME *400,000 ft³*
PEAK LOAD DATE *June*   TIME *2:00*   HRS/DAY OF OPERATION *9*
GLAZING *double glass*   SHADING *group I exterior shading*
WALL COLOR: LT [X] DK [ ]   ROOF COLOR: LT [X] DK [ ]   LATITUDE *38°*

CONDITIONS	DB	WB	% RH	DP	HUM RATIO GR/LB (KG/KG)
OUTDOOR	94	75			100
ROOM	74		50		62
DIFFERENCE	20				38

SENSIBLE LOADS

SOLAR

EXPOSURE	AREA X	SHGF X	SC X	TLF =	BTUH (W)	SUBTOTAL	TOTALS
N	275	35	.10	.75	722		
E	1650	34	.10	.31	1739		
S	1100	61	.10	.58	3892		
W	275	140	.10	.29	1116		
						7,469	

TRANSMISSION

ITEM	EXP	AREA X	ΔT X	U-VALUE X	TLF =	BTUH (W)	SUBTOTAL
GLASS		3300	20	.50	1.0	33,000	
NET WALL	N	5225	25	.05	.60	3,919	
	E	3850	25	.05	.37	1,781	
	S	4400	32	.05	.33	2,323	
	W	5225	45	.05	.37	4,350	
ROOF		10,000	51	.03	.37	5,661	
							51,034

O.A.

INFILTRATION
_495_ CFM (L/S) X **1.08 (1.20)** X _____ °F (°C) = _____

VENTILATION
_10,000_ CFM (L/S) X **1.08 (1.20)** X _20_ °F (°C) = _216,000_

INTERNAL HEAT

						BTUH (W)	SUBTOTAL
PEOPLE:	500 PEOPLE	X 230 BTUH(W) EA	X .80 BSF	X .80 DIVERSITY	=	73,600	
LIGHTS:	125,000 WATTS	X **3.4***	X 1.25 BALLAST	X .77 BSF	X .80 DIVERS	= 327,250	
EQUIPT:	50,000 WATTS	X **3.4***	X .82 BSF	X .80 DIVERSITY	=	111,520	
APPLIANCES:	____ WATTS	X **3.4***	X ____ BSF	X ____ DIVERSITY	=		
OTHER:					=		512,370

* Omit for SI Units

TOTAL SENSIBLE LOAD   786,873

LATENT LOADS

				BTUH (W)
PEOPLE:	500 PEOPLE	X 190 BTUH(W)/PERSON	X .80 DIVERSITY	= 76,000
APPLIANCES:				
OUTSIDE AIR:	10,000 CFM (L/S)	X 38 GR/LB (KG/KG)	X **0.68 (2.808)**	= 258,400

TOTAL LATENT LOAD   334,400

TOTAL LOAD (SENSIBLE & LATENT)   (94 Tons)   1,121,273

FIGURE 4.8. Example of cooling load calculations.

Data	Applicable to: (H = heating C = cooling)
Building Design Characteristics	
Project location (city, state)	H, C
Latitude	C
Elevation	C
Use of conditioned space	H, C
Number of stories	H, C
Area take-offs and dimensions	
Conditioned floor area	H, C
Floor-to-ceiling height	H, C
Floor-to-floor height	H, C
Fenestration area (for each compass orientation or horizontal)	H, C
Wall area (for each orientation or below grade)	H, C
Roof area	H, C
Construction materials (U- or R-value)	
Fenestration	H, C
Walls	H, C
Roof	H, C
Floor	H
Darkness of exterior surface	
Walls	C
Roof	C
Window shading or shading from other structures	C
Window and door crack length	H, C
or	
Approximate air changes per hour infiltration	H, C
or	
Ventilation rate (in CFM or L/s)	H, C
Operating Schedules	
Number of occupants and their activity level	C
Schedule of occupancy	C
Lighting wattage	C
Schedule of lighting use	C
Equipment and appliance usage	C
Indoor and Outdoor Conditions	
Indoor dry-bulb temperature and RH desired	H, C
Outdoor temperature and humidity at the selected design conditions	H, C

The building characteristics are normally found from building plans and/or by a site inspection of an existing building. Shading from adjacent structures can sometimes be determined by a site plan or, more reliably, by visiting the site. Surrounding landscaping such as water, sand, or parking lots may cause abnormal ground-reflected solar radiation and should be noted. Any equipment, appliance, or process that may contribute to the internal thermal load should also be noted. Weather data may be obtained from Table B2 in the appendix, local weather stations, or the National Climatic Center (Asheville, NC 28801).

Once the above data have been gathered, the analytical work begins. The heating load can be calculated using the above information and Figure 4.1 as a guide. For cooling load calculations, it is first necessary to select the month and time of day that will result in the highest calculated load. Most often, the day and month are dictated by the peak solar conditions for the building exposure most sensitive to them (that is, with the most fenestration). Sometimes the peak load time and date can be readily determined by simple inspection of Table B3.1, and sometimes several different date/time combinations will need to be tried. In southern zones at latitudes south of 32°N, the peak cooling load in buildings with large fenestration areas usually occurs in January or February rather than in summer. Under that condition, the warmest temperature to be expected in the winter months must be known. This information is available from the *Climatic Atlas of the United States* (see Bibliography).

Once the design month and time of day have been selected, the cooling load can be calculated using Figure 4.2 as a guide. The tables and figures in Appendix B may be used in the analysis in the absence of exact data more specific to a given project.

From the design peak heating and cooling loads, the amount of energy consumed in a building over the year can be calculated using the procedures outlined: heating degree day method (heating), equivalent full-load hours method (cooling), and bin method or hourly computer simulation method (heating or cooling).

In estimating energy use, it should be realized that any estimating method will produce more reliable results over a long period of time than over a short period. For one thing, weather conditions in a single year will never be the same as the average conditions on which the calculations are based. Thus, a projection for a period of years will be more accurate on the average than one for only a single year. In addition, all the methods take advantage of certain average conditions and are designed for use over a full annual heating or cooling season. Estimates for shorter periods, such as a month, may produce inaccurate results, because as the period of the estimate is shortened, there is more chance that some factor only indirectly taken into account in the estimating method will deviate from its long-term average value. This may lead to an error in the predicted energy requirement.

Energy calculations are very often part of an economic analysis intended to establish the cost effectiveness of an energy conservation measure. A thorough energy analysis provides additional information, such as the time of maximum demand for energy, and helps to estimate utility charges more precisely. An economic analysis also involves estimating the capital costs of the equipment to accomplish the energy conservation. Such economic analyses are described in Chapter 16.

KEY TERMS

Heat gain	Shading	Usage factor
Sensible heat gain	coefficient (SC)	Outside air loads
Latent heat gain	Clearness correction	Crack length method
Heat loss	Foreground	Air change method
Cooling load	reflectance correction	Stack effect
Heating load	Angle of incidence	Sensible load
Load	Openness factor	Latent load
Heat extraction rate	Solar altitude angle	Mean coincident wet-bulb temperature
Heat addition rate	Solar azimuth angle	Single-step method
Equipment load	Wall solar azimuth	Multiple-step methods
Design conditions	Transmission	Degree day method
Solar load	U-value	Balance point temperature
Solar time	R-value	Equivalent full-load hours
Mean time	Thermal bridge	
Direct solar radiation	Sol-air temperature	Cooling degree day method
Diffuse solar radiation	Building storage factor	Bin method
Specular reflection	Mean daily temperature range	Hourly computer simulations
Diffuse reflection	Specific heat	
Interpolation	Internal loads	
Solar heat gain (SHG)	Total light wattage	
Time lag	Ballast factor	
Thermal lag factor (TLF)	Diversity factor	
Solar heat gain factor (SHGF)		

1. How would it affect the results of the example at the end of this chapter if the building were in Atlanta, Georgia, instead of St. Louis? If it were in London, England? Recalculate the example to find out.
2. What would be the effect of deleting the window shades in that example?

BIBLIOGRAPHY

American Society of Heating, Refrigerating, and Air-Conditioning Engineers, Inc. *Standard 62.1—Ventilation for Acceptable Indoor Air Quality (ANSI Approved).*

American Society of Heating, Refrigerating, and Air-Conditioning Engineers, Inc. *Standard 62.2—Ventilation and Acceptable Indoor Air Quality in Low-Rise Residential Buildings (ANSI Approved).*

American Society of Heating, Refrigerating, and Air-Conditioning Engineers, Inc. *Standard 90.1—Energy Code for Commercial and High-Rise Residential Buildings.*

American Society of Heating, Refrigerating, and Air-Conditioning Engineers, Inc. *Standard 90.1—User's Manual.*

American Society of Heating, Refrigerating, and Air-Conditioning Engineers, Inc. *Standard 90.1—Energy Standard for Buildings Except Low-Rise Residential Buildings (ANSI Approved; IESNA Cosponsored).*

American Society of Heating, Refrigerating, and Air-Conditioning Engineers, Inc. *Standard 90.2—Energy-Efficient Design of Low-Rise Residential Buildings (ANSI Approved).*

American Society of Heating, Refrigerating, and Air-Conditioning Engineers, Inc. *Standard 90.2—Energy Code for New Low-Rise Residential Buildings Based on ASHRAE 90.2.*

American Society of Heating, Refrigerating, and Air-Conditioning Engineers, Inc. *ASHRAE Standard 90.2—Energy Efficient Design of Low-Rise Residential Buildings.*

American Society of Heating, Refrigerating, and Air-Conditioning Engineers, Inc. *Standard 100—Energy Conservation in Existing Buildings (IESNA Cosponsored, ANSI Approved).*

American Society of Heating, Refrigerating, and Air-

Conditioning Engineers, Inc. *ANSI/ASHRAE/IES Standard 100—Energy Conservation in Existing Buildings.*

American Society of Heating, Refrigerating, and Air-Conditioning Engineers, Inc. *Standard 105—Standard Methods of Measuring and Expressing Building Energy Performance (ANSI Approved).*

American Society of Heating, Refrigerating, and Air-Conditioning Engineers, Inc. *Standard 119—Air Leakage Performance for Detached Single-Family Residential Buildings (ANSI Approved).*

American Society of Heating, Refrigerating, and Air-Conditioning Engineers, Inc. *Standard 136—A Method of Determining Air Change Rates in Detached Dwellings (ANSI Approved).*

American Society of Heating, Refrigerating, and Air-Conditioning Engineers, Inc. *ASHRAE Handbook of Fundamentals.*

Carrier Air Conditioning Company. *Handbook of Air-Conditioning System Design.* McGraw-Hill Book Company, 1966.

Clarke, Joseph. *Energy Simulation in Building Design.* Butterworth-Heinemann, 2001.

Climatic Atlas of the United States CD-ROM version 2.0. NOAA National Climatic Data Center, Asheville, NC. US Government On-line store: http://nndc.noaa.gov/?http://ols.nndc.noaa.gov/plolstore/plsql/olstore.prodspecific?prodnum=C00519-CDR-A0001.

(This version, which includes Alaska and Hawaii, replaces the paper atlas last published in 1968 and supersedes version 1.0 of the *Climate Atlas of the Contiguous United States* published on CD in 2000. The new CD atlas contains 2023 color maps of climatic elements such as temperature, precipitation, snow, wind, pressure, etc. The period of record of the data for most of the maps is 1961–1990. There is an online version of the *Climate Atlas* at http://www.ncdc.noaa.gov/olca/ having restricted GIS capabilities, but viewable at no charge.)

Mazria, Edward. *The Passive Solar Energy Book: A Complete Guide to Passive Solar Home, Greenhouse and Building Design.* Rodale Press, 1979.

Thumann, Albert, P.E., C.E.M. *Optimizing HVAC Systems.* The Fairmont Press, Inc. 1988.

United States Environmental Data Service (corporate author), and John L. Baldwin. *Weather Atlas of the United States: Originally Titled Climatic Atlas of the United States.* Gale Group, 1975.

PART TWO

THERMAL CONTROL SYSTEMS

Buildings affect the environment in many ways—during their construction, throughout their operation, and from their disposition, when they are no longer viable. They impact their immediate surroundings and global resources by how they make use of land, energy, water, and lighting, and the selection of construction materials.

Mechanical systems—when required for controlling the thermal environment—use energy, add heat to the environment, and contribute to air pollution, global warming, and depletion of the ozone layer around the planet. It is important to understand how mechanical systems work as well as what they are and are not capable of doing.

Mechanical systems also have an impact on the economics of building construction and use, and on aesthetics. The exterior and interior of a building must accommodate rooftop equipment, grilles, ductwork, and the interior machinery. It is desirable, for many reasons, to minimize the amount of active control, and whatever equipment is necessary should be highly efficient.

Passive heating, cooling, and ventilating methods may have an even greater impact on aesthetics and often dictate the building's form. An elegant passive solution often creates the most energy-efficient, economical, and pleasing environment.

Chapter 5

A Building's Impact on the Environment

Treat the Earth well.
It was not given to you by your parents.
It was loaned to you by your children
KENYAN PROVERB

OZONE DEPLETION

GLOBAL WARMING

ENERGY CONSERVATION

GREEN DESIGN/SUSTAINABLE DESIGN
THE DEFINITION OF GREEN DESIGN
POLLUTION
MATERIALS
LAND AS A RESOURCE
BUILDING SITING
LIGHT TRESPASS AND LIGHT POLLUTION
HEAT ISLANDS
GREEN ROOFS
VERTICAL LANDSCAPING
WATER RESOURCES
ECOSYSTEMS
CONSTRUCTION
PLANNING FOR BUILDING DISPOSAL/RECYCLING

CASE STUDY

OZONE DEPLETION

The earth is enveloped by a blanket of ozone gas, which screens us from harmful ultraviolet radiation from the sun. A series of measurements taken over the years show this stratospheric ozone layer to be diminishing. The resulting increase in ultraviolet burning increases the risk of skin cancers and eye cataracts, and may impair crop production in many parts of the world. The number of deaths from skin cancer is expected to increase significantly due to ozone depletion.

It has been determined that the ozone layer is being destroyed by chlorine and bromine released into the air (Figure 5.1). The chlorofluorocarbon (CFC) refrigerants that have been used in building and automobile air conditioning, and in refrigeration for transporting and storing foods, are one class of chlorine-containing chemicals. If allowed to escape, the CFC molecules eventually rise to the upper atmosphere, where they release their chlorine atoms that then break down the ozone.

To exacerbate the problem, CFC molecules have long atmospheric lives—up to an estimated 145 years. This means that chlorine atoms from refrigerant released today will be circulating in the upper atmosphere well into, and in some cases beyond, the next century.

Because they contribute to the depletion of the protective ozone layer, CFC production is being phased out. In 1987, the United Nations assembled representatives from the majority of CFC user and producer countries, who agreed to the Montreal Protocol on Substances That Deplete the Ozone Layer.

In the years following, scientific evidence revealed that the rate of ozone depletion was accelerating. In response, the Montreal Protocol was amended in 1990 to call for terminat-

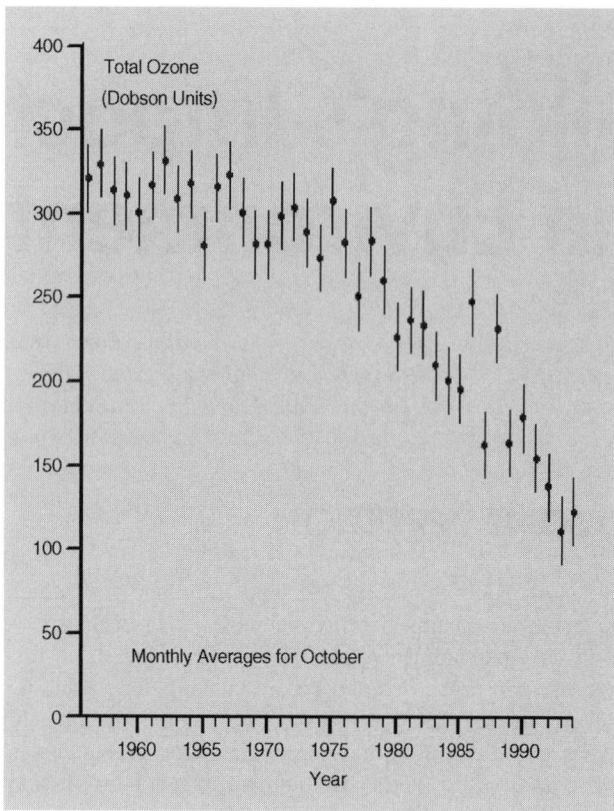

FIGURE 5.1. Depletion of the earth's ozone layer over time. (Graph courtesy of the Centre for Atmospheric Science, University of Cambridge, UK. http://www.atm.ch.cam.ac.uk/tour/tour_images/total_ozone.gif.)

ing all production of CFCs, halons, and other ozone-destructive chemicals by the year 2000 for developed nations and by 2010 for developing nations.

However, these substances are still in use, and the ozone layer is still thinning due to previously released chemicals. Thinning of the ozone layer is observed in the middle and high latitudes of both the Northern and Southern Hemispheres. This means that ozone depletion now exists over populated regions including almost all of North America, Western and Eastern Europe, Australia, New Zealand, and a sizable part of Latin America.

This is a serious problem, but we can hardly do without refrigerants. The refrigeration process is essential for food preservation and many modern industrial processes. All hospitals require air conditioning and refrigeration to reduce the spread of bacteria and infections, to provide a healthful environment for patients, and to store blood, vaccines, medicines, body tissues, and organs. Comfort air conditioning at home, at work, and in transit—whether by car, bus, subway, train, airplane, or ship—has become an integral part of our

standard of living, particularly in America's Sun Belt. Commerce and industry have come to depend on comfort cooling to such an extent that our economy could collapse without it.

To meet these needs, substitute refrigerants have been developed for use in some applications where CFCs had previously been used. Hydrofluorocarbons (HFCs), such as HFC-134a, contain no chlorine, so they have no impact on stratospheric ozone. Hydrochlorofluorocarbons (HCFCs), like HCFC-123 and HCFC-22, do contain chlorine, but they also have an atom of hydrogen that causes them to break down much more quickly in the atmosphere than CFCs. HCFC-123 has an atmospheric life of only 2 years, and the life of HCFC-22 is estimated to range from 15 to 19 years.

These shorter atmospheric lives give the HCFCs lower ozone depletion potentials (ODP) than CFCs. CFC-11 was assigned an ODP of 1. The ODP for HCFC-22 is .05, while for HCFC-123, it is .02. HFC-134a has an ODP of 0.

GLOBAL WARMING

For all of its harmful effects, ozone depletion has been found to help cool the atmosphere, which somewhat counters another environmental problem: global warming caused by the greenhouse effect. Certain gases in our atmosphere allow energy from the sun in the form of ultraviolet radiation to pass through, but not the infrared radiation emitted by sources on earth. This traps the sun's heat within our atmosphere much as glass does in a greenhouse, hence the term *greenhouse effect.*

A number of gases contribute to the greenhouse effect and have done so since the earth's atmosphere was formed. The most common of these is carbon dioxide. Burning hydrocarbons (wood, coal, oil, and natural gas) releases not only energy, but also carbon dioxide.

The greater the quantity of these gases, the more heat is trapped. The problem for us is that the production of some of these gases, principally carbon dioxide, is increasing, resulting in a global warming trend (Figure 5.2). The earth's current average temperature is about 60°F. Increasing this temperature by a few degrees is expected to (1) raise sea levels (as more of the polar ice caps melt), causing coastal flooding, and (2) change climates, thereby affecting agricultural production and the habitability of some areas.

To complicate matters, some refrigerants are also greenhouse gases. CFCs, HCFCs, and HFCs contribute to global warming in two ways. When released, they trap heat in the atmosphere in the same way that carbon dioxide does. That phenomenon, known as the *direct effect,* is measured by the

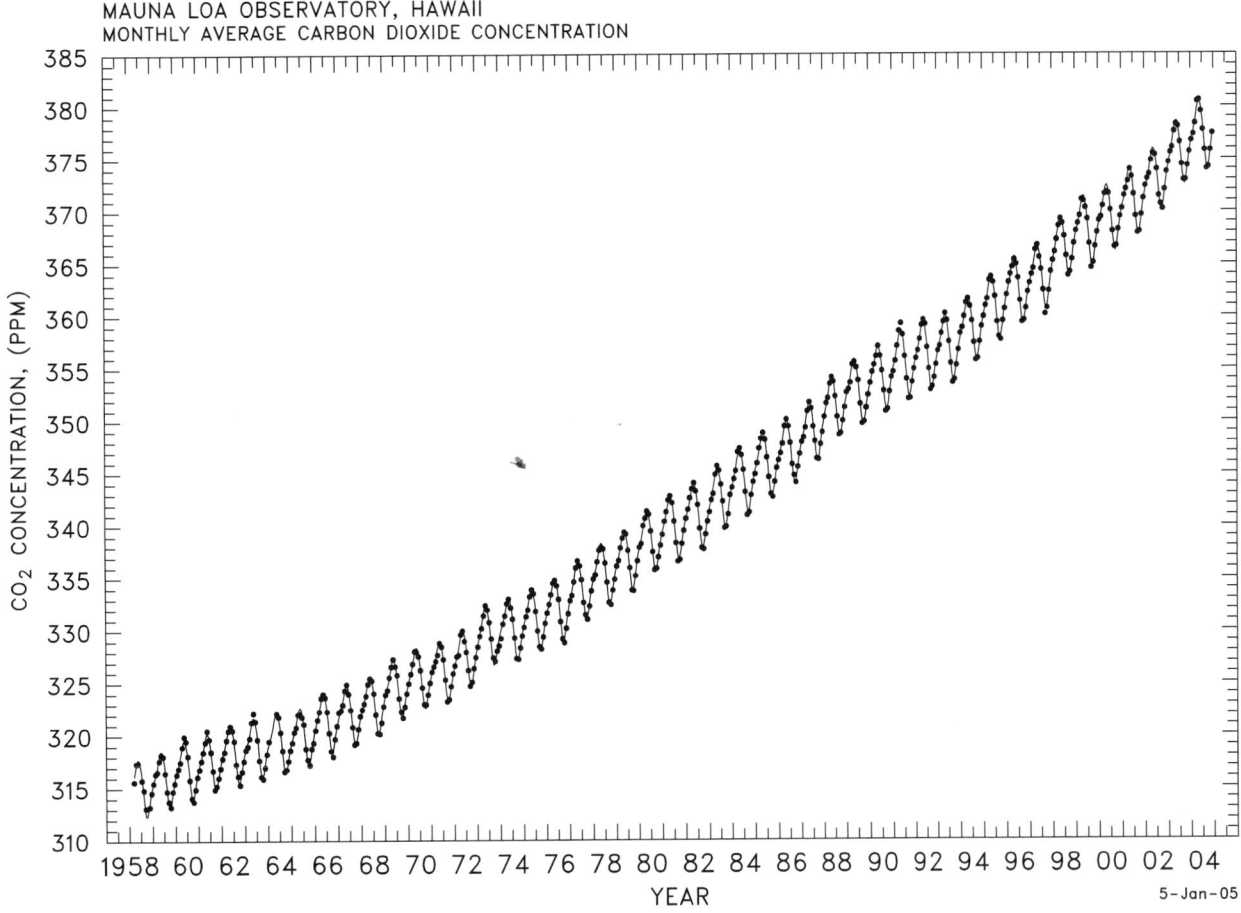

MAUNA LOA OBSERVATORY, HAWAII
MONTHLY AVERAGE CARBON DIOXIDE CONCENTRATION

5-Jan-05

FIGURE 5.2. The trend in global temperature increase is represented by the rising atmospheric concentration of carbon dioxide in the Keeling Curve. (Graph created by Tim Whorf and provided courtesy of CD Keeling, Scripps Institution of Oceanography.)

global warming potential (GWP). The baseline for GWP is carbon dioxide, which has a GWP of 1. CFC-11, for example, had a GWP of 2,000, which means that a molecule of the refrigerant would have 2,000 times the effect of a molecule of carbon dioxide in preventing heat from escaping from the atmosphere. The GWP for other refrigerants ranges from 38 for HCFC-123 to 6,200 for CFC-12.

But the process of refrigeration and cooling can add to the greenhouse effect even if no refrigerant is released into the environment. Much of the electricity used to run these systems is produced by the burning of fossil fuels, which releases large amounts of carbon dioxide into the air. The efficiency of the cooling or refrigeration system therefore has an additional indirect effect on global warming.

Refrigerants cause no direct global warming until they are released into the atmosphere. Manufacturers are building more tightly sealed cooling and refrigeration machines than

they once did, and refrigerant recovery and recycling methods are now commonly employed. Over the lifetime of a piece of equipment, the indirect effect is likely to considerably outweigh the direct effect.

The research and development process to find acceptable alternatives to CFCs in all applications has been difficult, costly, and time-consuming. So far, no perfect refrigerant has been found. Ones that have no ozone depletion potential contribute to direct global warming and, perhaps more importantly, are less efficient refrigerants and therefore contribute more to indirect global warming. With the exception of absorption refrigeration systems, the only known viable alternatives to HCFCs and HFCs are toxic and/or flammable.

The amended Montreal Protocol included a nonbinding resolution recommending, but not mandating, that nations eliminate their production of HCFCs no later than 2040, and by 2020 if possible. The United States took this action a step further on its own. The Clean Air Act of 1990 allows manufacturers to sell new refrigeration equipment that uses HCFCs until 2020. After that, only service of existing systems is permitted. The act freezes production levels of HCFCs beginning in 2015 and bans all production in 2030.

Emissions of carbon dioxide are a result of combustion. Carbon dioxide is eliminated in the photosynthetic process by green plants in the presence of sunlight. Thus, reversing the global warming trend means reducing the use of fossil fuels (coal, oil, and gas) by generally conserving energy and not cutting down forests for firewood.

ENERGY CONSERVATION

Energy conservation in building design is now more important than ever. As Figure 5.3 shows, we are currently at or fast approaching the peak of worldwide oil production. Scientific opinions about when approximately 50 percent of the world's oil supply will have been depleted vary between the year 2005 and 2020. In any case, there is no doubt that over the coming decades it will become harder and harder to extract the diminishing amount of oil left, and at some point, the energy needed to draw oil out of the ground and refine it will become greater than its energy value (the amount of energy into which the oil can be converted).

The fall in production will lead to increasingly scarce supplies of oil. If the demand continues to remain high, it will inevitably begin to exceed the supply, and the competition between the United States and other major world economies will force the price of gasoline and other fossil fuel–derived forms of energy to skyrocket in the coming decades.

At present, there are no readily available technologies to substitute for the world's rapidly depleting oil supplies. Potential alternatives such as hydrogen or fuels derived from coal and tar sands are considered prohibitively expensive. However, once the price of oil rises considerably higher, they will eventually become competitive.

Heating, refrigeration, and air conditioning account for approximately one-third of the global nonrenewable energy consumed annually. The expense of operating buildings heated and cooled with fuel oil and natural gas, or with electricity generated from fossil fuels, will increase. So, too, will the cost of fossil fuel–dependent industrial production, transportation, and building construction. A shift toward more efficient buildings and transportation cannot begin soon enough.

There are also concerns about the contribution of fossil fuel consumption to climate change. Worldwide, over 70 percent of electricity is produced by the combustion of fossil fuels. For every additional kilowatt-hour consumed, more carbon dioxide is discharged up the electrical utility's smokestack to the atmosphere. Designing for energy conservation, passive HVAC strategies, and renewable energy technologies minimizes the need for mechanical heating, ventilation, and cooling, and therefore helps to reduce the contribution to both ozone depletion and global warming.

Reducing electricity and fossil fuel use by designing energy-efficient buildings (1) lowers fuel and electricity costs, (2) reduces peak power demand, (3) reduces the demand for additional energy infrastructure such as power plants and transmission lines, and (4) reduces air pollution, carbon dioxide emissions, and adverse environmental impacts from the production and distribution of fossil fuels.

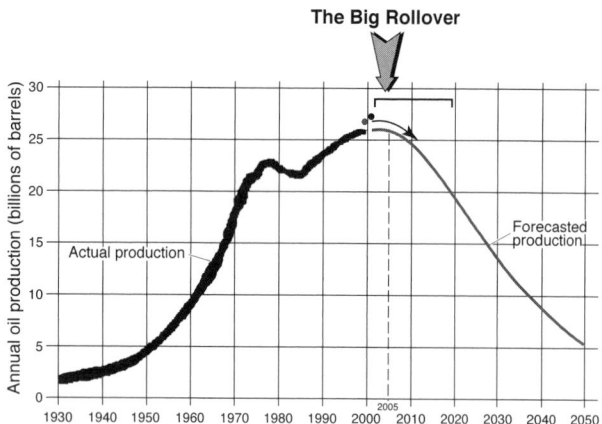

FIGURE 5.3. Worldwide oil production historically and projected into the future. The peak is estimated to occur sometime between 2005 and 2020. After that point, oil production will begin to decline. (Drawing by Bilge Celik.)

GREEN DESIGN/ SUSTAINABLE DESIGN

The concept of environmental sustainability is to leave the earth in as good as—or better—shape for future generations than we found it in ourselves. Human activity is environmentally sustainable only when it can be performed or maintained indefinitely without depleting the natural resources or degrading the natural environment on which it depends. A sustainable system is one that meets the needs of the present while allowing the needs of the future to also be fulfilled.

The terms *green design* and *sustainable design* have been

used interchangeably to represent this concept. *Green design* will be primarily used here because it is the designation adopted by the U.S. Green Building Council (USGBC), the organization that created the preeminent building assessment standard in the United States (and perhaps worldwide) relating to such designs. That standard, called *LEED (Leadership in Environmental and Energy Design)*, is an independent rating system used to quantitatively evaluate a building's resource efficiency and environmental impacts. (There are similar standards such as BREEAM [Building Research Establishment's Environmental Assessment Method] in the United Kingdom, Hong Kong, and Canada, and others in countries throughout the world.)

The term "green" in this context is a reference to the greenery of nature (grass, trees, leaves). It serves as a symbol of the concepts and practices of green design that integrate a building with nature.

The Definition of Green Design

Although green design has been defined in various ways, a good working description is that it is a design incorporating an awareness of—and respect for—nature and the natural order of things. A green design results in a healthy built environment that allows humans to live and prosper in accord with nature. It minimizes any adverse impact on the natural surroundings, materials, resources, and the earth's natural processes in order to avoid inflicting irreversible damage on those processes and ensuring the continuing habitability of the planet.

Ultimately, sustaining life on the planet depends on the net effect of the various activities and processes continually taking place around the globe. In addition to human activities, there are many natural events such as earthquakes, volcanic eruptions, forest fires, floods, and tornados that impact life. With so many variables, there is no known way to determine exactly what contribution toward sustainability any one of them in particular needs to make to assure our ongoing survival. In addition, every building has its own unique set of program, environmental, and budgetary circumstances.

Green design cannot, therefore, be universally prescribed (as life safety codes are), and there is no absolute formula that can be applied to all buildings. Rather, it is a useful paradigm for thinking about a building's rightful relationship with the environment. The idea is to make designers of buildings, building systems, and equipment aware of this philosophy and why it is important, and to induce them to employ practical green building design techniques. The resulting buildings should make a significant contribution toward maintaining the earth in a healthy state indefinitely.

This is not to suggest that green design is purely optional. There are numerous economic and practical advantages to the building owner, to the surrounding local community, and to society at large. Furthermore, some areas and some clients now require a LEED (or similar) certification. And due to increasing petroleum shortages in the near future and the rapidly deteriorating environment, LEED standards or something comparable may soon become a widespread requirement in building codes.

The principles of green design are intended to apply during every phase of a building's life from initial planning to final disposal. They inform decisions about the resources (such as land, energy, water, raw materials, and ecosystems) needed to design, construct, operate, demolish (deconstruct), and dispose of buildings and their components.

Green architecture is concerned not only with the traditional aesthetics of good design (massing, proportion, scale, texture, shadow, and light), but also with the long-term environmental, economic, and human costs of a building. The facility design team should pursue all possible measures to achieve an environmentally friendly, energy-efficient, long-lasting building or development project by effectively managing natural resources. There should be an elegant relationship between use areas, circulation, building form, mechanical and electrical systems, and construction technology.

There are three basic conditions that need to be met for sustainability: (1) Nonrenewable resources should not be consumed when sustainable renewable substitutes are available or can be developed, (2) the rate at which renewable resources are used must not exceed their rate of regeneration, and (3) the rate at which pollution is released must not exceed the capacity of the environment to absorb and transform it.

Green design thus seeks to:

1. *Minimize impacts on the earth from constructing buildings and producing the materials used in construction.* These impacts include land pattern disruption, mining and harvesting operations, embodied energy, and pollution from manufacturing. All possible measures should be taken to relate the form and plan of the design to the site, the region, and the climate. The form of the building should foster a harmonious relationship between its inhabitants and nature. Measures should be taken to minimize negative impacts on site ecosystems, and to heal and augment the natural environment of the site.

   Embodied energy is the quantity of energy required by all the activities associated with a production process. This includes the energy used to acquire natural resources and to manufacture and fabricate products, the energy used in making manufacturing equipment, and the energy used for other supporting functions. (The energy input required to quarry, transport, and manufacture building materials, plus the energy used in the construction process, can amount to a quarter of the energy used by a very energy-efficient building over its lifetime.)

2. *Minimize operating costs and environmental impacts that occur during occupancy.* These impacts include energy use, water use, and resources for operating and maintenance activities throughout the building's life. Energy efficiency is of paramount importance. HVAC, lighting, hot water heating, and process systems should incorporate techniques, technologies, and products that reduce or eliminate energy use. (Chapters 6 and 7 discuss energy-efficient strategies for heating, ventilation, and cooling. Chapter 8 addresses energy-efficient lighting.) Water use should be minimized for sanitation and potable water, food preparation, landscape irrigation, and process requirements. (Chapter 12 describes point-of-use water heaters and presents some techniques for reducing domestic water demand.) Harmful emissions to land, water, and air should be minimized. Buildings should be designed to be easily maintainable.

3. *Reduce the impact of the structure at the end of its life* (whether it is left to decay in place or be dismantled and disposed of). Demolition resources include energy for deconstruction and energy for transportation of materials to the disposal site.

4. *Create a more desirable human experience.* To provide a healthful interior environment, all possible measures should be taken to prevent materials and building systems from emitting toxic substances and gases into the interior atmosphere. Additional measures should be taken to clean and revitalize interior air with filtration and plantings.

Pollution

A green design should minimize the release of substances that have a negative impact on the environment during a building's entire life cycle. This includes emissions into the atmosphere, such as greenhouse gases that contribute to global warming, ozone depletion gases, particulates, chemicals that result in acid rain, and so on. The discharge of harmful liquid effluents and solid waste should also be minimized:

1. During construction
2. Throughout the building's operation
3. During demolition

Materials

Assuming that a particular building is in fact needed for a given function, the use of materials for construction of the building and its infrastructure should be reduced as much as possible, and the materials needed should be utilized as efficiently as possible. Building materials and products that minimize destruction of the global environment should be selected where possible. Toxic waste output during the production of materials and products should be considered. The following sources for materials are listed in the order of priority for reducing their embodied energy and minimizing the environmental impacts:

1. *Reuse existing buildings and structures* wherever possible (provided that their operating energy costs can be reduced to an acceptable level). Demolish and rebuild only when it is not economical or practical to reuse, adapt, or extend an existing structure. By modifying an existing building and reusing as much of its structure and systems as possible, the need to employ new materials is minimized. This reduces their accompanying impacts of resource extraction, transportation, and processing energy and waste. Reuse also reduces the need for transportation energy during demolition and construction. Clearly, trade-offs must be made when considering a building for reuse. For example, a building that, historically, has been inefficient and would need significant changes to its envelope and mechanical/electrical systems might incur significant waste, as well as require enormous quantities of new material in order for the original structure to be retrofitted for its new use. On the other hand, there may be historical value in preserving certain buildings by updating them to meet current codes and energy efficiency standards.

2. *Reduce materials use.* Using the minimal amount of materials required for a building project also lowers the environmental impact of introducing products produced from virgin sources. Waste of materials resulting from handling and conventional construction processes should also be minimized.

3. *Use materials created from renewable resources.* Avoid using materials from nonrenewable sources or materials which cannot be reused or recycled, especially in structures that are expected to have a short life. Wood should be selected based on nondestructive forestry practices.

4. *Reuse building components.* Reusing intact materials and building components salvaged from deconstructed buildings reduces the environmental impacts of building materials, because these items generally require fewer resources for reprocessing.

5. *Use recyclable and recycled content material.*
6. *Use locally produced and low-energy materials.* Using locally produced materials and locally manufactured products minimizes transportation distances and can greatly reduce the overall environmental impacts of materials.

The materials selection process may be summarized as the 3Rs: reduce, reuse, and recycle (the meaning of *recycle* includes use of products and materials with recycled content or from renewable resources).

Land as a Resource

The philosophy of sustainable land use is based upon the principle that land is a precious finite resource and that the development of greenfields should be minimized. *Greenfields* are undeveloped natural properties that have experienced little or no impact from human activities. In order to support biodiversity, encroachment on animal habitats should be avoided. The definition of greenfields can also include agricultural land that has had no activity other than farming. Loss of prime farmland should be especially avoided.

Effective planning can create efficient, environmentally sustainable urban forms and minimize urban sprawl. Proximity to mass transit is a highly desirable green measure. This avoids overdependence on automobiles for transportation, unnecessary fossil fuel consumption, and higher pollution levels. A building designed and constructed to the most exacting green building standards will be seriously compromised if the users must all drive long distances to the facility.

Like other resources, land is recyclable and should be restored to productive use whenever possible. Recycling of disturbed land is an important objective in green design. *Land recycling* refers to reusing land already impacted by human activities instead of developing greenfields. Bringing such land back to productive use facilitates land conservation, and can promote economic and social revitalization in distressed areas. There are at least three identifiable categories of potentially recyclable land: brownfields, grayfields, and blackfields.

The U.S. EPA defines *brownfields* as abandoned, idled, or underused industrial and commercial facilities for which expansion or redevelopment is complicated by real or perceived environmental contamination. Reuse of such land may be complicated by the presence—or potential presence—of a hazardous substance, pollutant, or contaminant. In many U.S. cities, brownfields have become valuable real estate because of their proximity to considerable preexisting

infrastructure and a potential workforce. Placing brownfields back into productive use in a building project has the additional advantage of improving the local environment.

Grayfields, another form of urban property, is defined as a blighted or obsolete building sitting on land that is not necessarily contaminated. Like brownfields, grayfields can be valuable properties because of the scarcity of available urban land; the presence of preexisting infrastructure in urban areas; and tax and other incentives offered by local and state governments. While the developer of such property may receive ready access to infrastructure and a willing workforce, the cities in which these sites are located generally receive greater tax revenues, creating a true win-win scenario.

Another category of blighted land is *blackfields*. These properties are abandoned coal strip mines and subsurface mines. Surface water in these zones can have a very low pH value and can be contaminated with iron, aluminum, manganese, and sulfates.

The very act of constructing a building and providing the infrastructure needed to support operation of the facility (highways, roads, bridges, and sidewalks required to access the facility; connections to power, water, sewer, and communication utilities) invokes tremendous changes to the site. Reusing sites that have already been disturbed by human activities minimizes adverse impacts caused by the building project.

On the other hand, when developing greenfields, the footprint of construction on the site should be minimized to preserve as much of the original site function and its ecological systems as possible. In a green building project, the construction manager plans the construction process to minimize soil compaction and the destruction of plants and animal habitat. Detailed planning to control erosion and sedimentation, and to minimize soil flows during construction, is part of the green building delivery process.

Building Siting

The building owner has usually selected the building site before retaining an architect, but sometimes the members of the project team have input. When possible, it is wise for the architect and all consultants to be involved early in the site selection process.

Careful site selection is essential to minimize the negative environmental impacts that may accompany a project. The implications of nearby pollution sources, ambient air quality, groundwater levels, site drainage, availability of or access to various energy sources (including renewable sources), and other site characteristics can significantly affect the success of a design. Prudent site selection can lower the first cost, operating and maintenance costs, environmental cost, and human cost.

Energy resources that may be available at a given site include:

- Geothermal/groundwater
- Surface water
- Wind
- Solar
- District heating/cooling

Site location, building orientation and geometry, and local climatic characteristics are all factors the design team should address. These considerations can have a significant impact on the quality of the occupants' environment and the efficiency of the building. Buildings that effectively use the attributes of their surroundings generally provide psychological and physiological benefits for the occupants and tend to be more efficient.

Land use and landscape design are closely coupled. For green designs, the role of the landscape architect should be redefined from that of simply providing exterior amenities for the project to serving as an integrator of ecology and nature within the built environment. Because they are typically the most knowledgeable professionals on the project team about the ecology around buildings and the flow of matter and energy across the human-nature interface, landscape architects are usually the best equipped to deal with natural systems. They can also provide expertise to the rest of the project team on the relationship between buildings and natural systems. Landscape design can address the topics of stormwater uptake, wastewater treatment, and integrated food production, and can assist in developing passive heating and cooling strategies.

Close coordination is needed among architects, landscape architects, civil engineers, structural engineers, and construction managers. Careful interdisciplinary consideration of the site's geology, hydrology, topography, and local *microclimates* can produce a building that optimizes the use of the site, that is highly integrated with the local ecosystems, that minimizes impacts during construction and operation, and that employs landscaping as a powerful adjunct to its technical systems.

Climatic factors include ambient temperature and humidity patterns, ambient air quality, potential pollution sources, solar availability and intensity, weather patterns, wind patterns, soil conditions, freshwater availability and quality, and site drainage. The climatic characteristics of a site obviously have an impact on how the building performs, especially its energy performance and the impact on its surroundings. The entire design team should be aware of the key characteristics so that each member can determine the affects on his or her discipline and how they might be utilized or accommodated in some way to further the goal of green design for the building.

Building orientation affects many aspects of green design, ranging from energy performance to visual stimulation for the building occupants. Solar orientation; prevailing breezes; availability of natural light; shading created by natural vegetation, topography, or adjacent structures; and views all impact the choice of how to orient the building on a site. Site orientation can also affect landscaping choices and irrigation water requirements. The benefits, drawbacks, and trade-offs need to be weighed when choosing the orientation.

Buildings that minimize east and west exposures, especially where a lot of glass is used, are generally more energy efficient because of the huge solar heat gains associated with east- and west-facing elevations during cooling months. If natural ventilation is planned to help meet cooling requirements, the building needs to be equipped with operable windows and oriented with the dominant elevations perpendicular to the prevailing breezes so that outside air will be drawn through the building by the windward/leeward effects.

Light Trespass and Light Pollution

Light trespass (unwanted illumination spilling over into off-site areas) poses a number of problems ranging from being a mere nuisance to causing safety problems when it temporarily blinds pedestrians and automobile drivers. It can also negatively impact wildlife—as it can the health of humans—because it can interrupt normal daily light cycles that are needed for the average person's well-being. *Light pollution* is illumination that interferes with the ability of astronomers and the general population to view the night sky constellations.

Minimizing light trespass and light pollution, and ensuring that lighting energy is used efficiently, requires the electrical engineer to carefully design exterior lighting systems to avoid unnecessary illumination of the building's surroundings. The location, mounting height, and aim of exterior luminaires must all be correctly designed to ensure that lighting energy is used efficiently and for its intended purposes. Specifically:

- Parking lot and street lighting must be designed to minimize upward transmission of light.
- Exterior building and sign lighting must be reduced or turned off when not needed.
- Exterior lighting systems must be professionally

designed to provide exactly the level and quality of lighting needed to meet the project's criteria without straying off-site and causing undesirable conditions.

Heat Islands

Temperatures in cities are generally higher than those in surrounding rural areas. Urban environments are typically 2° to 10°F hotter (Figure 5.4). This results in higher cooling requirements for buildings in urban areas than those located in rural settings. These higher cooling loads require greater energy use, which results in more air pollution, greater resource consumption, and higher costs.

The so-called urban *heat island* effect is caused by the removal of vegetation and its replacement with energy-storing materials in the building and on the site. These include asphalt and concrete roads, as well as other buildings and structures. The shading effect of trees and the evapo-transpiration (natural cooling effect) from other vegetation are replaced by human-made structures that absorb, store, and then release solar energy. Additional heat is also generated by electricity consumption for lighting and process use, and for air conditioning, and by combustion of fuel in automobiles. The more heat is added, the higher the cooling loads, which results in even more heat.

Another negative impact of heat islands is that they contribute to global warming by increasing fossil fuel consumption at power plants to operate the additional air conditioning. Heat islands also increase ground-level ozone (smog) pollution by increasing the reaction rate between nitrogen oxides and volatile organic compounds (VOCs). The increased temperatures and ground-level ozone in heat islands adversely affect human health—especially that of children and older people.

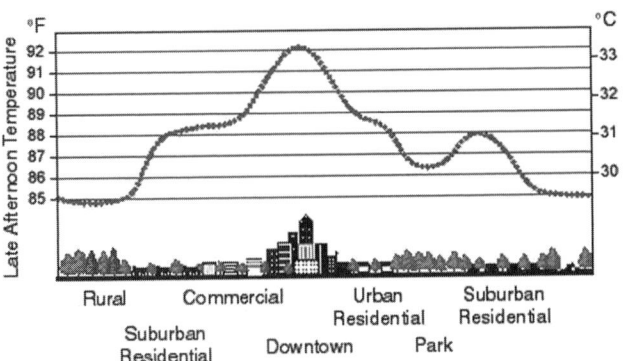

FIGURE 5.4. Urban heat island effect. (Graphic courtesy of USEPA.)

Heat island effects can be mitigated by (1) installing highly reflective and highly emissive roofs that reflect and re-radiate solar energy back into space, (2) planting shade trees near buildings to reduce surface and ambient air temperatures, and (3) using light-colored construction materials where possible to reflect rather than absorb solar radiation.

Green Roofs

A *green*—or living—*roof* (Figure 5.5) is simply a modern application of the ancient sod roof used in Europe. An alternative term for this practice is *eco-roof*. It can be manifested as a grass roof or a rooftop garden. Living roofs have been found to reduce building energy costs by 10 percent and to decrease summer roof temperatures by 70°F. These roofs can reduce stormwater runoff by 90 percent and delay the flow of stormwater for several hours. This can reduce stormwater and sewer system overflow in areas that experience heavy rainfall. Living roofs can also filter pollution and heavy metals from rainwater and help protect the regional water supply. Green roofs can serve as an aesthetic feature, to help blend a building into its environment, and to help support climatic stabilization. Green roofs protect the waterproofing from damage by ultraviolet radiation and eliminate the need for tiles, shingles, or other surfacing.

Green roofs must be built on a sufficiently strong frame with carefully applied waterproofing, because it is very difficult to locate leaks once the growing medium is in place. A living roof requires ongoing tending. Green roofs generally cost twice as much as conventional roofs, but can pay back that investment rapidly with energy savings.

Vertical Landscaping

Skyscrapers can utilize *vertical landscaping* by introducing plants and ecosystem components on the facade at each floor. These create windbreaks, provide shade and cooling by evapotranspiration, absorb carbon dioxide and generate oxygen, and improve the sense of well-being of the occupants by providing greenery to counterbalance the mass of concrete, glass, and steel. A vertical landscape that is well integrated with the building can provide visual relief from otherwise uninteresting, nondescript surfaces.

The use of trellises also allows for vertical growth and landscaping from ground level to the roof. But because wind speeds at roof level can be twice those at ground level, plants at upper levels may need protection, which can be provided by side louvers that allow the landscape to be seen, yet deflect the wind from around the plants.

FIGURE 5.5. Green roof on the Seattle City Hall. (Photo by Vaughn Bradshaw. Special thanks to Jackie Campbell and Andy Worline for access.)

Water Resources

Water is essential at nearly every building site, but over the past several decades there has been a growing realization that it is fast becoming a precious resource. While the total amount of water (in its various forms) on the planet is constant, the amount of fresh water—of a quality suitable for the intended purposes—is not uniformly distributed (e.g., 20 percent of the world's fresh water is in the U.S.' Great Lakes, while elsewhere it is often nonexistent or in very meager supply).

Although direct consumption by people in buildings is not a large fraction of total water use in the United States (buildings account for about 12 percent of freshwater use), water shortages in many areas of the country are interfering with development and construction projects. Therefore we not only need to use water prudently, we also need to avoid

contaminating water supplies. While many of the measures to protect and preserve the world's freshwater supplies are beyond the influence of building designers, there are a number of simple strategies relating to building sites that can be included as part of a green design.

Conventional building plumbing systems use high-quality potable water for all purposes and then discharge the contaminated water down the drain. High-quality potable water is also used to irrigate landscaping. This is inefficient and wasteful. Potable water depends on elaborate systems to extract it from surface water and groundwater sources, and then pump it to facilities for treatment and distribution. This requires large quantities of energy. Similarly, wastewater must be pumped through an extensive system of sanitary sewers and lift stations to central wastewater treatment plants, consuming relatively large amounts of energy in the process. Reducing water consumption, therefore, not only reduces flows through the system, but also lowers overall energy consumption and associated pollution from energy sources.

Increasing water efficiency reduces a building's potable water use, which in turn reduces the demand on the municipal water supply. It also reduces a building's overall waste generation, thus putting fewer burdens on the sewage system. This not only reduces annual water and wastewater utility charges, it also lowers the energy use associated with water. Furthermore, municipal water and wastewater treatment plants save operating and capital costs for new facilities. Potentially, fewer water and wastewater treatment plants may need to be built. Other benefits are reduced discharge to waterways, preservation of water resources for agricultural and recreational uses, preservation of water resources for wildlife, and less strain on aquatic ecosystems.

Prior to discussing a green design strategy for building water use, some terms need to be defined:

Blackwater is water containing human waste from such sources as toilets and urinals. Wastewater from kitchen sinks and dishwashers is sometimes considered blackwater because it may contain oil, grease, and food scraps.

Graywater is water from showers, bathtubs, bathroom sinks, washing machines, and drinking fountains. It may also include condensation pan water from refrigeration equipment and air conditioners, hot tub drainwater, pond and fountain drainwater, and cistern drainwater. Graywater contains a minimum amount of contamination and can be reused for some landscape applications. Although its acceptability is still being debated by public health officials, no case of illness has ever been traced to graywater reuse.

Both graywater and blackwater contain pathogens—humans should avoid contact with both—but blackwater is considered a much higher risk for the transmission of waterborne diseases. Even though they are not blackwater, the following water sources should not be included in graywater that is to be used for irrigation: garden and greenhouse sinks, water softener backflush, floor drains, and swimming pool water. In buildings served exclusively by composting toilets and thus producing no true blackwater, kitchen wastewater can be included in the graywater by eliminating organic matter. A graywater system reduces a building's potable water use, in turn reducing demand on the municipal water supply and lowering costs associated with water use. It also reduces a building's wastewater generation, thus putting less demand on the associated sewage system.

Reclaimed water is water from a wastewater treatment plant that has been treated and can be used for nonpotable uses. In some areas of the United States, reclaimed water may be referred to as *irrigation-quality* (IQ) water, even though applications are not limited to irrigation.

Rainwater harvesting is the collection, storage, and use of rainwater. After purification, rainwater is usually very safe and in some cases can be of higher quality than conventional sources of water. Rainwater harvesting offers several important environmental benefits, including reducing demand for limited water supplies and reducing stormwater runoff and flooding. An example is shown in Figure 5.6.

Xeriscape is a sustainable landscaping strategy that focuses on using drought-tolerant, native, and adapted species that require minimal or no water for their maintenance. The term is derived from the Greek *xeros,* meaning "dry."

A Green Building Water Strategy

The following is a series of logical steps that can be used to develop a water resource strategy for green buildings:

1. Select appropriate water sources for each purpose. Potable water must be used for all applications that involve human consumption or ingestion. Other water sources—including rainwater, graywater, and reclaimed water—can be used for landscape irrigation, fire protection, cooling tower make-up, chilled and heating water, toilet and urinal flushing, industrial process uses, and other applications not requiring valuable potable water. The availability of each type of alternative water source should be analyzed to determine the optimum mix for a particular project based on its specific water requirements.

2. Employ technologies that minimize water consumption.

3. Evaluate the potential for a dual wastewater system that separates lightly contaminated graywater from human waste–contaminated blackwater sources. This dual piping system allows for water recycling within the building.

4. Analyze the potential for innovative wastewater treatment strategies such as the use of constructed wetlands or living machines to process effluent. These approaches use nature in symbiosis with the building.

5. Apply life-cycle costing (LCC) to analyze the cost/benefit ratio of proposed techniques to reduce water flow through the building and its landscape. A simple life-cycle cost analysis that examines nothing more than the cost of potable water will generally indicate a long time frame for payback of the initial investment—perhaps 10 to 20 years. The payback time may be shortened by including reductions in wastewater generation and the costs associated with its treatment; the actual energy cost of moving water and wastewater; emissions associated with energy generation; worker productivity improve-

(a)

FIGURE 5.6. Rainwater harvesting system for the Seattle City Hall. The nonpotable rainwater is used for irrigation and flushing of water closets. During times of the year when there is insufficient rainwater, potable water is used as a backup. (a) Day tank and pumping system. (b) Controls for water treatment as needed. (Photos by Vaughn Bradshaw. Special thanks to Jackie Campbell and Andy Worline for access.)

ments; and general environmental benefits. In addition, it can be reasonably expected that the price of potable water in most regions will increase at a greater rate than the general inflation rate. Including this in the life-cycle cost evaluation should bring the payback into the same range as a good energy conservation measure—7 years or less.

Water Conservation

Water conservation strategies can include a combination of low-flow fixtures (toilets, urinals, faucets, and shower heads), no-flow fixtures (composting toilets, waterless urinals), and water-saving controls (infrared sensors). This can sometimes result in a reduction in capital costs, since some

fixtures—such as waterless urinals and low-flow lavatories—are in some cases less expensive to install.

For landscaping, highly efficient drip irrigation systems use far less water and deliver the water to the plant roots at greater than 90 percent efficiency. Often xeriscape strategies can be employed in the landscape scheme to eliminate the need for an irrigation system.

Rainwater Harvesting

Rainwater has been a crucial source of water for human survival throughout human existence. For buildings, rain was typically collected from the roof and conducted into a storage tank or cistern. The development of centralized potable

(b)

FIGURE 5.6. *(Continued)*

water systems displaced rainwater systems until the emergence of the modern green building movement.

A rainwater harvesting system generally consists of the following key components:

- *Catchment (collection) area.* Most rainwater harvesting systems use the building's roof as the catchment area. The roof surface for rainwater harvesting should not support biological growth (e.g., algae, mold, moss), should be fairly smooth so that pollutants deposited on the roof are readily removed by the roof-wash system, and should have minimal overhanging tree branches above it. Galvanized metal is the most commonly used roofing material for rainwater harvesting, although copper is also suitable. Composite three-tab roofing should not be used if the water is to be used for landscape irrigation.
- *Roof-wash system.* This is a system for keeping dust and pollutants that have settled on the roof out of the cistern. It is a requirement for systems used as a source of potable water, but is also advisable for other systems in order to keep contaminants out of the tank. A roof-wash system is designed to purge the initial water flowing off of a roof during rainfall.
- *Prestorage filtration.* In order to keep leaves, and other debris out of the cistern, a domed stainless steel screen is secured over each inlet to the cistern. In addition, leaf guards can be added over gutters in areas with significant windblown debris or overhanging trees.
- *Rainwater conveyance.* This is the system of gutters, downspouts, and piping used to carry water from the roof to the cistern.
- *Storage cistern.* This is usually the most expensive component of a rainwater harvesting system. It is typically made of galvanized steel, concrete, ferro-cement, fiberglass, polyethylene, or durable wood (e.g., red-

wood or cypress). Costs and expected lifetimes vary considerably among these options. Tanks may be located in a basement, buried outdoors, or located above ground outdoors. Light should be kept out to prevent algae growth. Cistern capacity should be sized to meet the expected demand. For systems designed as the sole water supply, sizing should be based on 30-year precipitation records, and with sufficient storage to meet demand during times of the year having little or no rainfall.

- *Water delivery.* A pump is generally required to deliver water from the cistern to its point of use, because gravity-fed systems require a substantially elevated tank.
- *Water treatment system.* When the water is to be used solely for landscape irrigation, only sediment filtration is typically required. To protect plumbing and irrigation lines (especially with drip irrigation), water should be filtered through sediment cartridges to remove particulates—preferably down to 5 microns. For systems providing water for potable uses, additional treatment is required to purify it sufficiently to ensure a safe water supply. This can be provided by microfiltration, ultraviolet sterilization, reverse osmosis, or ozonation (or a combination of these methods). One option is to provide higher levels of treatment only at selected faucets where potable water is drawn.

If a building design is to include a graywater system or landscape irrigation—and space for storage can be found—rainwater harvesting is a simple addition to those systems. In some cases, rainwater harvesting may be more cost-effective than creating stormwater conveyance systems. Rainwater harvesting is most attractive where a municipal water supply is either nonexistent or unreliable, such as in rural regions and developing countries.

Rainwater is soft and does not cause scale buildup in piping, equipment, and appliances. It can thus extend the life of some systems.

The advantages of a rainwater harvesting system must be weighed against the added first cost associated with the cistern and treatment system. The additional materials and their associated embodied energy costs should be taken into consideration. Space must be made available to accommodate the storage vessel(s). Small sites or projects with limited space allocated for utilities would be poor candidates. Additional costs include maintenance of the catchment area, conveyance systems, cistern, and treatment system. There is currently no U.S. guideline on rainwater harvesting. The local health code authority has jurisdiction, potentially mak-

ing a particular site infeasible due to requirements for backflow prevention, special separators, or additional treatment.

Ecosystems

Green design recognizes the important role ecosystems can play when integrated with the built environment. Such integration can synergistically provide services that supplant conventional manufactured systems and complex technologies such as controlling external building loads, processing waste, absorbing stormwater, growing food, and providing natural beauty, sometimes referred to as an *environmental amenity*. The need to dramatically reduce building and infrastructure energy consumption is a motivation for designers to better understand the ways in which ecosystems and buildings can mutually benefit each other. A detailed knowledge of ecology and ecological systems enables green designers to successfully weave nature into the built environment.

Natural systems can shade and cool buildings, yet allow sunlight through for heating during appropriate seasons. Natural systems can provide calories and nutrition and may be able to take up large quantities of stormwater, thus allowing the downsizing of conventional stormwater handling systems.

Natural or constructed wetlands can treat wastewater from buildings, which sustains both the wetlands and the human systems with which they cooperate. Many of the waste-processing functions, which have traditionally been performed at distant wastewater treatment plants to which building waste must be pumped—often through miles of piping using motive energy provided by a series of lift stations—can be decentralized.

Wetlands can break down organic waste, thus minimizing the need for complex infrastructure, while at the same time creating nutrients that benefit the species performing these services. Constructed wetlands can be characterized as passive systems for wastewater treatment. They mimic natural wetlands by using the same filtration processes to effectively remove contaminants from wastewater. Wetlands remove contaminants by several mechanisms including nutrient removal and recycling, sedimentation, biological oxygen demand, metals precipitation, pathogen removal, and toxic compound degradation. In addition to being able to remove organic nutrients, constructed wetlands also have the demonstrated ability to remove inorganic substances, and thus can be used to effectively treat industrial wastewater, landfill leachate, agricultural wastewater, acid mine drainage, and airport runoff.

In addition to treating wastewater, constructed wetlands can provide surge areas for stormwater and treat this often contaminated runoff. Constructed wetlands also provide the added benefit of environmental amenity and can blend into natural or rural landscapes.

A number of site-specific factors affect the consideration of using a constructed wetland for wastewater treatment. These include hydrology (groundwater, surface water, permeability of ground), native plant species, climate, seasonal temperature fluctuations, local soils, site topography, and available area.

Constructed wetlands are built for surface or subsurface flow. Surface flow systems consist of shallow basins containing wetland plants that are able to tolerate saturated soil and anaerobic conditions. The wastewater entering the surface system flows slowly through the basins and flows out at the end as cleaner water. In subsurface systems, the wastewater flows through a soil substrate. Subsurface systems have the advantages of higher rates of contaminant removal compared to surface flow systems and limited contact for humans and animals. They also work especially well in cold climates due to the earth's insulating properties. The high carbon (and/or sometimes iron) and saturated conditions of wetland soils provide effective subsurface contaminant removal.

Construction

The role of the construction team in executing a green building project is extremely important and should not be underestimated. The construction management firm or general contractor and its subcontractors must be oriented to the purpose of the green features in order to produce a successful outcome.

The construction team is responsible for construction waste management, erosion and sedimentation control, limiting the footprint of construction operations, and construction indoor air quality (IAQ). The construction team must be motivated to:

- Establish materials handling practices to reduce construction waste
- Extensively train subcontractors in construction waste management
- Recycle site materials such as topsoil, limerock, asphalt, and concrete into the new building project
- Install products and materials in ways that reduce the potential for IAQ problems
- Pay attention to moisture control in all aspects of construction to prevent future mold problems
- Minimize the environmental impact of construction operations, such as compaction and the unnecessary destruction of trees on the site

To minimize gaps between design intent and what is actually built, the green-conscious designer should provide a high level of construction administration. The design specifications should require that certain construction methods and procedures be followed to ensure a fully realized green project. These include reduced site disturbance, handling of construction waste, control of rainwater runoff, and IAQ management during construction. The contractor should particularly protect installed or stored absorptive materials, such as insulation or sheetrock, from water damage or other contamination. Water damage is especially insidious if materials get wet, are installed wet, and are then covered up. If air-handling equipment is run during construction, the contractor should be required to replace any filters before occupancy.

Site Protection Planning

A site protection plan spells out what is required of the builder to minimize ecological and other damage to the site. It should include:

- An erosion and sedimentation control plan, including wastewater runoff management.
- A protection plan for vegetation and trees.
- A site access plan, including a designated staging or "lay down" area designed to minimize damage to the environment. This plan should indicate storage areas for salvaged materials and easily accessible collection areas for recyclable materials, including day-to-day construction waste (paper, corrugated cardboard packaging, glass, plastics, metals, etc.). It should also designate site-sensitive areas where staging, stockpiling, and soil compaction are prohibited.
- Measures to salvage existing clean topsoil on site for reuse.
- Plans to mitigate dust, smoke, odors, and other impacts.
- Noise control measures, including schedules for particularly disruptive, high-decibel operations and procedures for compliance with state and local noise regulations.

Health and Safety Planning

A health and safety plan addresses the health of workers on the construction site and the health of the building's future occupants. The plan should take into account the building's air quality design and provide for:

- Protection of ducts and airways from accumulating dust, moisture, particulates, VOCs, and microbes during and resulting from construction or demolition activities.
- Adequate separation and protection of occupied areas from construction areas for building additions.

- Increased ventilation and exhaust air at the construction site.
- Scheduling of construction activities to minimize exposure of absorbant building materials to VOC emissions. "Wet" construction processes such as painting and sealing should occur before storing or installing "dry," absorbent materials such as carpets and ceiling tiles. These porous materials absorb and retain contaminants and release them during building occupancy.
- A flush-out period, beginning as soon as systems are operable and before or during the installation of furniture, fittings, and equipment. The building should be flushed with 100 percent outside air for a period of not less than 20 days.
- Control of vermin.
- Prevention of pest infestation once the building or renovated portion is occupied.

No matter how well designed a building may be, poor job-site construction practices can undermine the best design by allowing moisture and other contaminants to become potential long-term problems. Preventive job-site practices can eliminate future IAQ problems in the completed building and reduce health risks for workers.

Planning for Building Disposal/Recycling

Each designer should anticipate to the extent possible eventual demolition and removal requirements of the materials, equipment, and systems for which he or she is responsible. Green building designs should incorporate design for deconstruction as part of the overall building design strategy. Materials should be used that have future value for recycling.

Deconstruction is the whole or partial disassembly of buildings to facilitate component reuse and materials recycling. Deconstruction for disassembly is the deliberate effort during design to maximize the potential for disassembly—instead of demolishing the building totally or partially—to allow the recovery of components for reuse and materials for recycling and to reduce long-term waste generation.

The challenge in designing a green building is how to select materials and products that lower the overall impact of the building, including its site. Some products have desirable high-energy performance qualities but consume large amounts of energy to manufacture or transport, or they may be difficult or impossible to recycle.

It is important to note that buildings can have a very long life, which makes planning for deconstruction a challenge. Some materials and building systems have a long "resi-

dence" period, while others are typically updated or replaced at intervals during a building's lifetime (e.g., finishes at 5-year intervals, lighting at 10-year intervals, and HVAC equipment at 20-year intervals).

The following guidelines facilitate deconstruction:

- Use recycled and recyclable materials.
- Minimize the number of types of materials.
- Avoid toxic and hazardous materials.
- Avoid composite materials and make inseparable products from the same material.
- Avoid secondary finishes to materials.
- Provide standard and permanent identification of all components and material types.
- Minimize the number of types of components.
- Use mechanical rather than adhesive connections.
- Use an open building system with interchangeable parts.
- Use modular design.
- Use assembly technologies compatible with standard building practice.
- Separate the structure from the cladding.
- Provide access to all building components.
- Size components to suit handling at all stages.
- Provide for handling components during assembly and disassembly.
- Provide adequate tolerance to allow for disassembly.
- Minimize the types of fasteners and connectors.
- Design joints and connectors to withstand repeated assembly and disassembly.
- Allow for parallel disassembly.
- Use prefabricated subassemblies.
- Use lightweight materials and components.
- Permanently identify points of disassembly.
- Provide spare parts and storage for them.
- Retain information on the building and its assembly process.

CASE STUDY

An example of an environmentally friendly project is El Monte Sagrado Living Resort and Spa, in Taos, New Mexico, which was completed in 2003. Surrounded by mountains and high desert country, the 4-acre luxury resort is located in Taos's historic urban downtown yet sits amid a verdant collage of ponds and diverse plant life. The resort's exterior, a blend of new architecture and remodeled old casitas, pays homage to the area's Native American roots and incorporates local materials and green building techniques, while finely crafted interior spaces honor a variety of global cultures.

But what is particularly innovative is the resort's commitment to integrate cutting-edge technologies with sustainable practices. About 60 percent of the complex, including the 30,000-ft^2 main building, the grouping of 36 guest suites, and several of the 12 casitas, is geothermally heated and cooled; other areas rely on energy produced through an extensive photovoltaic installation and smaller solar panels. Fresh rainwater is collected, filtered, and used in the spa. Stormwater runoff is collected and managed for ponds and waterfalls.

Virtually all wastewater on the property is recycled through a purification system called the Living Machine (Figure 5.7), which collects, treats, and reuses it for irriga-

FIGURE 5.8. Solar tree at El Monte Sagrado. (Photo courtesy of El Monte Sagrado Living Resort and Spa.)

FIGURE 5.7. Waterfall in front of the Living Machine at El Monte Sagrado. (Photo courtesy of El Monte Sagrado Living Resort and Spa.)

tion. But unlike most water treatment plants—typically eyesores designed to be hidden from view—the Living Machine is the heart of the Biolarium, one of the resort's prominent features. A greenhouse and botanical garden, the Biolarium is an inviting architectural space on its own, and it also serves as a demonstration area for educating guests about responsible water-use issues.

The primary solar installations are positioned for optimal energy efficiency and minimal visual distraction. One series of photovoltaic panels is incorporated seamlessly into traditional shade structures that shelter the porte cochere and lobby entry areas; the other group is implanted in the skin of the Biolarium, where the panels that can be seen simply add to the sense of technology in the space. By contrast, in a sunny clearing near a cluster of casitas, individual panels adorn the branches of a sculptural "solar tree" (Figure 5.8) and power pumps for nearby ponds.

The payback period for the geothermal systems is estimated at 5 to 6 years. The photovoltaic system on the Biolarium roof produces enough power to run the Living Machine pumps.

SUMMARY

Most of the built environments designed and constructed today will still exist during the coming era of rising temperatures and sea levels. Global temperature increases must be considered in everything from the design criteria selected for cooling loads and passive design strategies to materials and equipment selection. Not only must contemporary designs cope with potentially significant climate changes, but they must be capable of operating without the availability of petroleum. Figure 5.9 illustrates a creative way of combining electricity generation from a renewable energy source and providing shelter from the heat.

The environmental performance of green design is measured in terms of a wide range of potential effects:

- Fossil fuel depletion
- Other nonrenewable resource use

FIGURE 5.9. This "Solar Grove" models the life process of natural trees by converting sunlight into energy without adding carbon dioxide or other greenhouse gases to the atmosphere—while providing structures that are both shade-producing and aesthetically pleasing. The array of 25 "solar trees" converts a 186-vehicle parking lot into a 235-kilowatt solar electric generating system. The carport thus formed over an employee parking lot at the Kyocera North American headquarters in San Diego, California, utilizes a total of 1,400 Kyocera solar photovoltaic (PV) modules tilted slightly to maximize solar exposure. This installation is capable of generating 421,000 kilowatt-hours per year. According to the U.S. EPA, generating this amount of electricity from fossil fuel resources would annually release nearly 340,000 pounds of carbon dioxide, a contributor to global warming; 420 pounds of nitrous oxide, which has been linked to the destruction of the earth's ozone layer; and 250 pounds of sulfur dioxide, the principal contributor to acid rain. (Photo courtesy of KYOCERA Solar, Inc.)

- Water use
- Global warming potential
- Stratospheric ozone depletion
- Ground-level ozone (smog) creation
- Nutrification/eutrophication of water bodies
- Acidification and acid deposition (dry and wet)
- Toxic releases to air, water, and land

Comparing these effects for a building takes careful analysis. For example, the total energy for a building's life cycle is comprised of the embodied energy invested in the extraction, manufacture, transport, and installation of its products and materials, plus the operational energy needed to run the building over its lifetime. For the average building, the operating energy is far greater than the embodied energy, perhaps 5 to 10 times higher. Consequently, the operational stage has far more energy impacts than those up through the construction stage. For other buildings, however, the effects of the stages up through construction can be far greater. Toxic releases during resource extraction and the manufacturing process can be far greater than those occurring during building operation. The net result is that the designer using these tools must keep in mind the entire life cycle of the building, not just the stages leading to construction.

An integrated design process is essential to successful green design. An integrated design is one in which each component is considered to be a critical part of a greater whole. The economic, environmental, and community interests of a building are all closely intertwined and interdependent, and must be implicitly considered part of the program for all green buildings. A green design also requires that the architectural design be carefully coordinated and fully integrated with the mechanical, electrical, and other systems.

KEY TERMS

Greenhouse effect	Microclimates	Reclaimed water
Direct effect	Light trespass	Irrigation
Green design	Light pollution	quality (IQ)
LEED	Heat island	water
Embodied energy	Green roof (eco-roof)	Rainwater harvesting
Greenfields	Vertical landscaping	Xeriscape
Brownfields	Blackwater	Environmental amenity
Grayfields	Graywater	
Blackfields		

STUDY QUESTIONS

1. In what way is ozone depletion harmful?
2. Explain the greenhouse effect.
3. Name three nonrenewable sources of energy.
4. Name three renewable sources of energy.
5. Name three methods for natural cooling.
6. What are the different water sources that can be provided for a building?
7. What is the difference between graywater and blackwater?
8. Should painting or carpet installation occur first?

BIBLIOGRAPHY

American Institute of Architects Committee on the Environment (COTE) and Architects, Designers and Planners for Social Responsibility (ADPSR Colorado). *The Sustainable Design Resource Guide,* 3rd ed., 1997.

American Society of Heating, Refrigerating, and Air-Conditioning Engineers, Inc. *ASHRAE Green Guide.*

American Society of Heating, Refrigerating, and Air-Conditioning Engineers, Inc. *Advanced Energy Design Guide for Small Office Buildings.*

Brand, S. *How Buildings Learn: What Happens After They're Built.* Viking Penguin, 1995.

Day, Christopher. *Places of the Soul: Architecture as a Healing Art.* Glasgow, Collins, 1990.

Demkin, J. (editor). *Environmental Resource Guide.* John Wiley & Sons, Inc., 1996.

Denver AIA Committee on the Environment. *The Sustainable Design Resource Guide: Colorado and the Western Mountain Region,* 2000.

Grumman, David L. (editor). *ASHRAE Green Guide.* American Society of Heating, Refrigerating and Air Conditioning Engineers, Inc., 2003.

Hawkin, Paul, Amory Lovins, and L. Hunter Lovins. *Natural Capitalism: Creating the Next Industrial Revolution.* Little, Brown and Company, 1999.

Hellmuth, Obata + Kassabaum, Inc. (HOK). *Sustainable Design Brochure.* St. Louis, 1998.

Hermannsson, John (editor). *Green Building Resource Guide.* Taunton Press, 1997.

Kibert, Charles J. *Sustainable Construction.* John Wiley & Sons, Inc., 2005. (An excellent detailed description of the LEED building assessment standard and a thorough survey of green design practices.)

Mendler, S.F., and W. Odell. *HOK Guidebook to Sustainable Design.* John Wiley & Sons, Inc., 2000.

Public Technology, Inc., U.S. Department of Energy, and the U.S. Green Building Council. *Sustainable Building Technical Manual—Green Building Design, Construction and Operations.* Public Technology, Inc., 1996.

U.S. Green Building Council, *LEED Reference Guide.*

Chapter 6

Active HVAC Systems

THE ARCHITECT'S ROLE
SYSTEMS OVERVIEW
FUELS AND ENERGY SOURCES

CENTRAL HEAT SOURCES
FURNACES
BOILERS
HOT WATER CONVERTERS
INCINERATORS
SOLAR ENERGY

CENTRAL COOLING SOURCES
VAPOR-COMPRESSION REFRIGERATION EQUIPMENT
CHILLERS
ABSORPTION CHILLERS
HEAT PUMPS
INTERNAL COMBUSTION ENGINE COMPRESSORS
CONDENSERS
EVAPORATIVE COOLING

DISTRIBUTION MEDIA
AIR SYSTEMS
WATER SYSTEMS
STEAM SYSTEMS

DELIVERY DEVICES
AIR INLETS AND OUTLETS
NATURAL CONVECTION HEATING UNITS

UNIT HEATERS
RADIANT HEATING/COOLING DEVICES
FAN COIL UNITS
UNIT VENTILATORS
INDUCTION UNITS
PACKAGED TERMINAL AIR CONDITIONERS (PTAC)

LOCALIZED EXHAUST SYSTEMS

HEAT RECOVERY
HEAT RECOVERY FROM EXHAUST AIR
CONDENSER HEAT RECOVERY
REDISTRIBUTION OF HEAT WITHIN A BUILDING
RECOVERY OF PROCESS WASTE HEAT

THERMAL STORAGE
HEAT STORAGE APPLICATIONS
COOLING STORAGE
ANNUAL CYCLE ENERGY SYSTEM (ACES)

AUTOMATIC CONTROLS
THERMOSTATS
DIRECT DIGITAL CONTROL (DDC)
CARBON DIOXIDE CONTROL

NOISE AND VIBRATION

DISTRICT HEATING AND COOLING

At almost any given moment, there is a transfer of heat into or out of a building. Heat enters as solar radiation, transmission due to a temperature difference, internal sources, outside air ventilation, or the effect of a heating system. Heat leaves by transmission through the envelope, exchange of air with the outside, or removal by a mechanical cooling system.

The prevailing temperature and humidity condition is a result of the balance between the heat entering and the heat leaving a building. Regulation of these heat transfers can provide a thermal environment that permits the human body to effortlessly and comfortably reject its metabolic heat.

The subject of this chapter is the mechanical heating and cooling systems that add heat to or remove it from buildings to offset the natural heat transfers, thereby maintaining a balance that meets the requirements of the people or process occupying the space. These systems are referred to as *HVAC* (*heating, ventilating, and air-conditioning*) *systems*.

Often, the term *air conditioning* is associated only with cooling. Its broader meaning, however, refers to the treatment of air so that its temperature, humidity, cleanliness, quality, and motion are maintained as appropriate for the occupants, a particular process, or some object in the space.

In addition to treating the air, HVAC systems sometimes alter the MRT of a space by direct radiant transfer devices. Radiant heat tends to be a more comfortable means of warming people than introducing heated air into the space. When air itself is warmed to transfer heat by convection, (1) it stratifies above people, (2) its motion can cause a cooling effect, and (3) it has a drying effect on the skin.

Radiant heating also has the advantage of being more energy efficient. Heat is transferred directly to occupants and objects in a room without heating up large volumes of air first. A warmer surrounding MRT means that more body heat can be lost by convection without altering the net heat balance (Figure 1.2). This allows the space temperature to be kept cooler, which in turn reduces the rate of heat loss through the building envelope.

For cooling purposes, cool air motion is more effective than radiant heat transfer. As air is circulated, it mixes with—and carries away—the warm air, humidity, and odors generated by the body, imparting a general feeling of freshness to all of the occupants' senses.

Many—and sometimes all—of the functions of air conditioning can be provided by utilizing the natural resources of a site without resorting to active HVAC systems. In general, it is best to build as much inherent thermal control into the building structure as is economically feasible. This is an important challenge for architects today and is the subject of Chapter 7. Designing a building for the economically optimum level of thermal self-sufficiency will not only minimize annual energy costs, but in most cases will also stabilize the interior surface temperature of the envelope, resulting in more uniform interior conditions and greater comfort.

The Architect's Role

The engineers who design HVAC systems make their selections on the basis of (1) initial and life-cycle costs, (2) suitability for the intended occupancy, (3) floor space required for equipment, (4) maintenance requirements and equipment reliability, and (5) simplicity of control.

It is not up to the architect to decide which system is to be used in a given building; that is the concern of the mechanical engineering consultant. However, it is important for the architect to be able to communicate and coordinate with the engineer and to ask intelligent questions.

Together, the architect and engineer must evaluate several issues related to optimizing the building's thermal regulation capability. The level of insulation, the extent of shading of the structure, and other qualities of the envelope greatly influence mechanical equipment size and fuel consumption. Architectural enhancements that reduce continuing operating expenses may increase the initial cost of the structure while decreasing the size and first cost of the HVAC equipment. A cost-effectiveness analysis, performed as a cooperative effort between the architect and engineer, can determine the economically optimum balance between passive and active approaches.

Another part of the architect's job with respect to HVAC systems is to allocate at the initial stage of a project an adequate amount of space for the mechanical equipment and the necessary pipes and ducts so that only minor adjustments to the floor plan will be needed to fit the final mechanical design. The architect must provide space to accommodate the size of equipment that will meet the building's loads and proper clearances required around the equipment for access during normal operation, inspection, routine maintenance, repair, and eventual replacement. A good rule of thumb is that mechanical rooms will need 5 to 10 percent of the gross building area. Space for a flue stack and access for fuel delivery and storage must be provided for as needed.

Not only is the amount of overall space significant, but the number, position, and size of HVAC central stations must also be considered. Air-handling equipment generally should be in a location central to the area served and have close access to outdoor air.

Other concerns of the architect are whether and where ductwork and piping will occur, along with their dimensions. Chases are required for ductwork and clusters of piping that

FIGURE 6.1. Rooftop equipment penthouses. (Photo by Vaughn Bradshaw.)

run vertically. Space must also be provided within the occupied areas for exposed terminal devices along outside walls. The form and position of these devices must be carefully coordinated to avoid conflicts between furniture arrangements and the location of grilles or wall-mounted units. The effect that grilles and openings in outside walls have on the exterior appearance of a building is also important.

The noise generated by mechanical equipment is a determining factor in locating it. In general, mechanical spaces should be located as far away as possible, both horizontally and vertically, from areas requiring especially quiet backgrounds (bedrooms, conference rooms, etc.). Frequently, mechanical rooms are located in equipment penthouses. Two examples are shown on the roof level of the office buildings in Figure 6.1.

It is imperative that the architect closely coordinate with all of the consultants from the beginning of a project so that realistic provisions are made for the various space requirements. The design of the mechanical systems must merge with the architectural and structural planning and should develop concurrently. Decisions such as whether to employ open office planning with modular furniture systems or private offices have a great impact on mechanical systems and should be shared early with the engineer.

The ensuing discussion of HVAC systems will not address the specific requirements of each type of building: single dwellings, hotels and motels, retail shops, offices, schools, hospitals, libraries, auditoriums, commercial kitchens, and so on. Working out the details and making system decisions is the job of experienced engineering consultants. Instead, this chapter provides an introductory survey of the subject, outlining the generic types of systems and their various components.

Systems Overview

In general, HVAC systems have the following elements in common, as shown in Figure 6.2:

- Equipment to generate heat or cooling
- A means of distributing heat, cooling, and/or filtered ventilation air where needed
- Devices that deliver the heat, cooling, and/or fresh air into the space

All of these functions are sometimes combined into a single unit that by itself converts an energy resource into heating or cooling and then delivers it to the space.

The first component is a central heat source, such as a furnace, boiler (steam or hot water), solar collector, geothermal well, or heating water converter; or a central cooling source, such as a chiller, DX (direct expansion) air conditioner, or evaporative cooler. This is the "head end" or "front end" of the system, where the electricity consumption, fuel consumption, or heat collection occurs. The equipment is

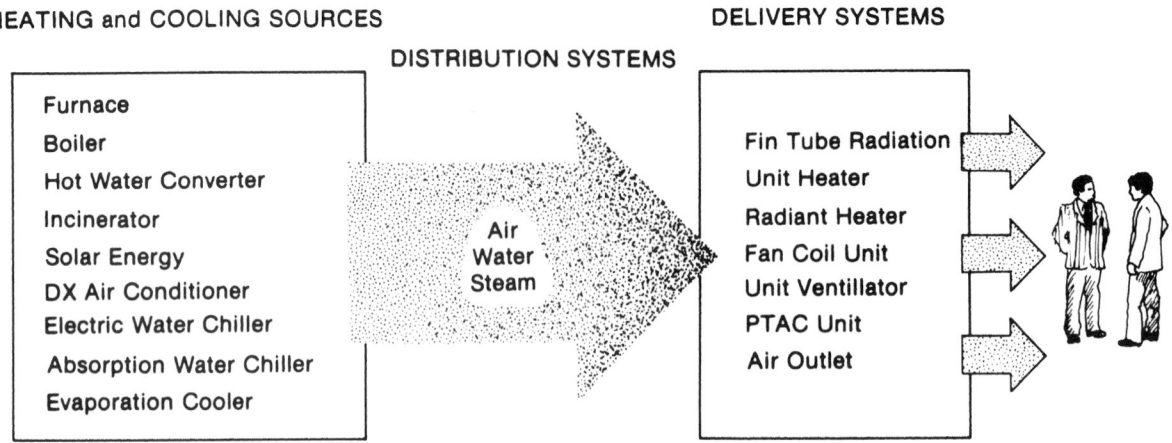

FIGURE 6.2. Basic elements of HVAC systems.

selected with a capacity to offset the peak load of the space or spaces served and to bring them from a setback temperature to their normal operating condition.

The heat or cooling is transferred to a conveying medium, which may be steam, water, or air. The medium is transported through a system of pipes or ducts leading to and from the various rooms or parts of the building. Except for steam, which travels under its own pressure, the mediums must be forced around the circuit by some motive power. For air this is a fan, and for water it is a pump.

Once the medium reaches the desired location, the heat or cooling is disseminated by some terminal delivery device. These devices are air registers and diffusers, hydronic radiators or convectors, and fan coil units. Self-contained package units that provide heating and/or cooling directly to a space are also in this category. Each terminal unit must have the capacity to offset its share of the load in that space.

Figures 6.3 and 6.4 illustrate an air-type and a water-type distribution system, respectively. Although some air systems are simpler than that shown in Figure 6.3, this example shows a general case where a heating and/or cooling medium is pumped to the air handler, where the heat or cooling is transferred to air that is blown through the ductwork to the destination space or spaces.

Figure 6.5 illustrates some common symbols and abbreviations used on mechanical building plans to show HVAC system components.

Fuels and Energy Sources

One of the first decisions to be made about the HVAC system for a particular building is which fuel or energy resource to employ. The selection should be socially responsible by considering the most beneficial use of our natural resources. The alternative with the least initial cost should be balanced with the one leading to the least cost over the life of the building and/or system under consideration. A life-cycle cost analysis must take into account and compare the likely increases in cost of the various fuels being considered over the life span of the building.

The other key factor in fuel selection is the need for storage and delivery facilities. The storage of combustible fuel not only requires space, but can also create safety hazards that may affect the owner's insurance costs. Listed below are the chief energy resources used in buildings today, and their forms of storage.

- Electricity (none)
- Natural gas (none)
- Fuel oil (liquid tank)
- Coal (solid)
- Propane and butane (pressurized liquid tank)
- Wood (solid)
- Solar (water tank or rock bed)
- Solid waste (solid)
- Geothermal (none)
- Ground temperature (none)

The fuels (natural gas, fuel oil, coal, propane, butane, wood, and solid waste) are converted to heat in some appropriate combustion apparatus that generally requires a flue extending up above the rooftop.

Electricity usually entails the least installation cost and is the easiest to distribute within a building. But in most regions its high billing rates more than offset those advantages. It also requires space for transformers, switchgear, or

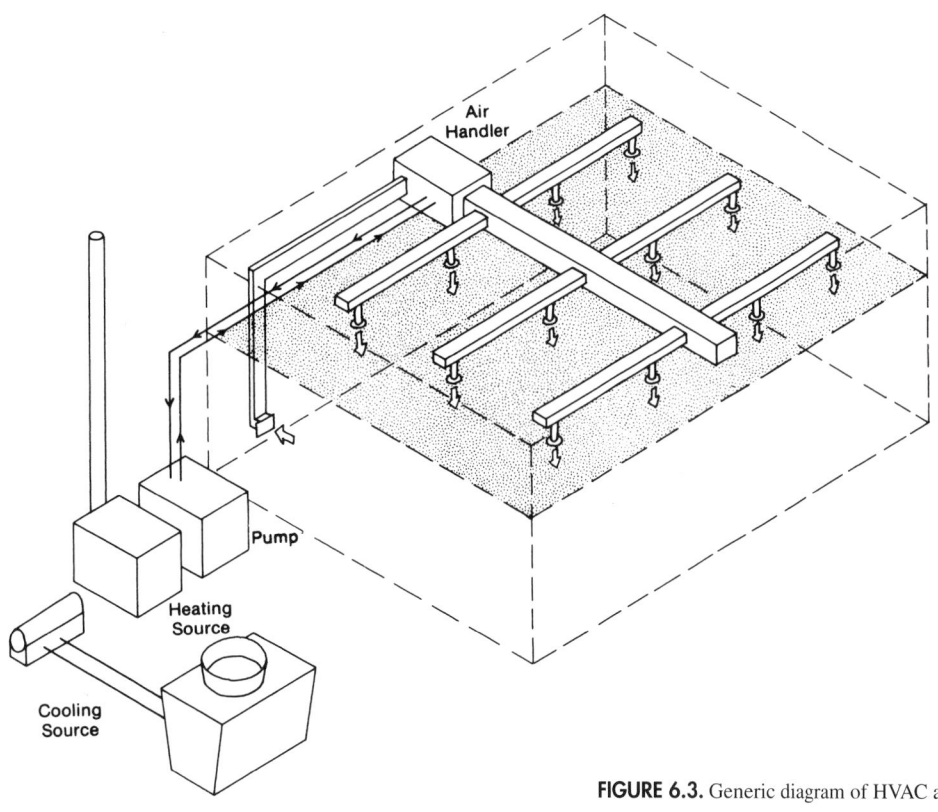

FIGURE 6.3. Generic diagram of HVAC air-distribution systems.

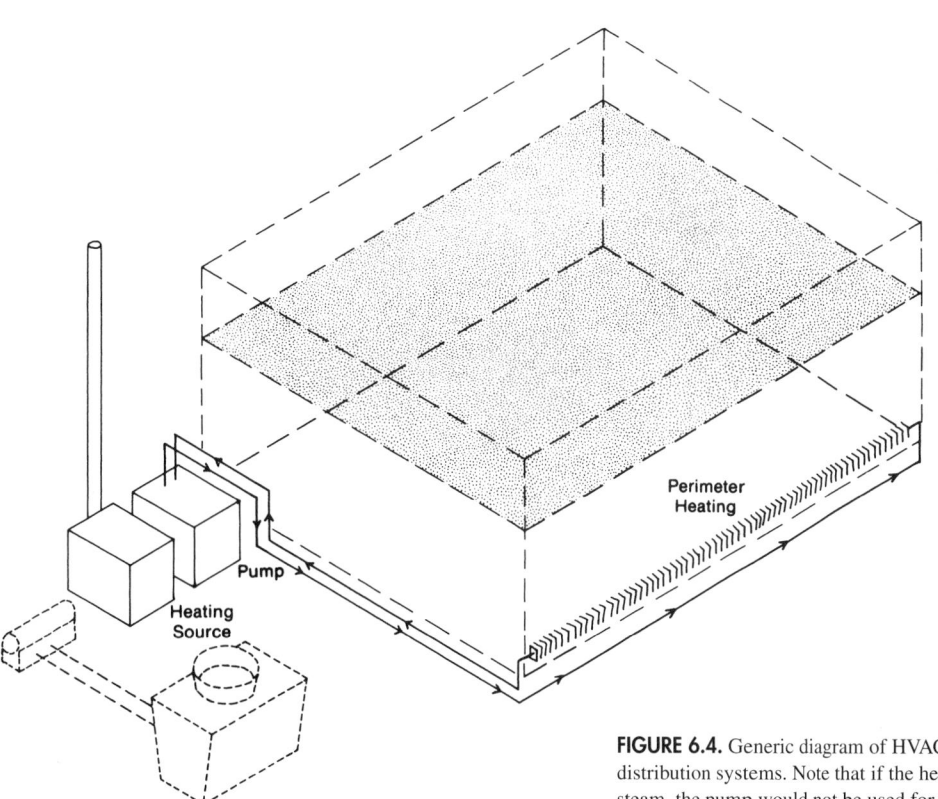

FIGURE 6.4. Generic diagram of HVAC water- or steam-distribution systems. Note that if the heating medium were steam, the pump would not be used for distribution.

143

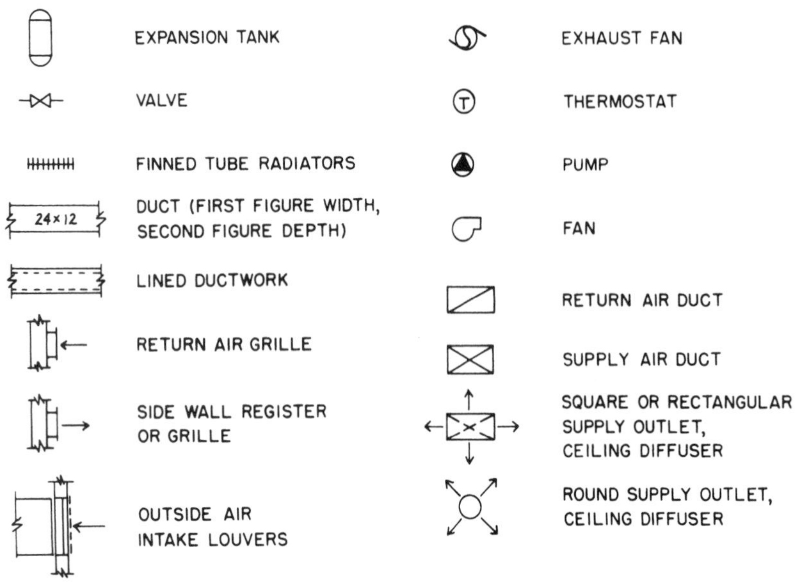

⬯	EXPANSION TANK	⟳	EXHAUST FAN
⟶⋈⟶	VALVE	Ⓣ	THERMOSTAT
⊢⊢⊢⊢⊢⊢	FINNED TUBE RADIATORS	⬤	PUMP
24×12	DUCT (FIRST FIGURE WIDTH, SECOND FIGURE DEPTH)	◗	FAN
	LINED DUCTWORK	▱	RETURN AIR DUCT
	RETURN AIR GRILLE	⊠	SUPPLY AIR DUCT
	SIDE WALL REGISTER OR GRILLE	⬚	SQUARE OR RECTANGULAR SUPPLY OUTLET, CEILING DIFFUSER
	OUTSIDE AIR INTAKE LOUVERS	⌖	ROUND SUPPLY OUTLET, CEILING DIFFUSER

HVAC Abbreviations

ACD	Air Conditioning Drain (Condensate)
AHU	Air Handling Unit
AP	Access Panel or Door
AS	Attenuator (Sound)
C	Condensate (Steam)
CD	Cold Duct
CDR	Condensing Water Return
CDS	Condensing Water Supply
CFM	Cubic Feet/Minute of Air
CHWR	Chilled and Heating Water Return
CHWS	Chilled and Heating Water Supply
CWR	Chilled Water Return
CWS	Chilled Water Supply
EA	Exhaust Air
EC	Electrical Contractor
EF	Exhaust Fan
EH	Entry Heater
FA	Fresh Air
FAD	Fresh Air Damper
FCU	Fan Coil Unit
FD	Fire Damper
FDC	Flexible Duct Connection
FPC	Flexible Pipe Connection
FTR	Fin Tube Radiation
GC	General Contractor
HD	Hot Duct
HP	Heat Pump
HPS	High-Pressure Steam
HWR	Heating Water Return
HWS	Heating Water Supply
LD	Louver Door
LPS	Low-Pressure Steam
MA	Mixed Air
MB	Mixing Box
MC	Mechanical Contractor
PC	Plumbing Contractor
RA	Return Air
RAD	Return Air Damper
SA	Supply Air
UH	Unit Heater
UV	Unit Ventilator
VD	Volume Damper

FIGURE 6.5. Symbols and abbreviations.

large substations. Natural gas and electricity installations must accommodate usage meters.

Fuel oil is derived from refining crude petroleum and is classified according to its viscosity. Number 2 oil is used in residential and light commercial burners, which require oil that flows readily and can be mechanically vaporized and mixed with air for combustion. The heavier grades (numbers 4, 5, and 6) are used in large commercial and industrial applications. Numbers 5 and 6 require preheating before they can be burned and for pumping, so they are better suited for continuously fired industrial processes than for intermittently fired space-heating systems.

Coal is classified according to its hardness. *Anthracite* coal is the hardest, *lignite* is the softest, and *bituminous* is inbetween. *Coke* is a derivative of bituminous coal. Fine coal is made into briquettes for ease of handling, and pulverized coal is used in power-generating plants. Heating values of coal range from about 7,000 to 14,000 Btu/lb (16.3 to 32.6 million joules/kg), with an average of about 13,000 Btu/lb (30.2 million joules/kg). The greatest problems with coal use are fuel storage (40 ft³ required per ton, or 1.25 m³ per metric ton) and the need for expensive emissions control equipment for air pollution abatement. It also requires ash disposal.

Propane and butane are petroleum gases that can be liquefied under moderate pressure. They are transported in pressurized cylinders and have a heating value of about 2,300 Btu/ft³ (85.7 million joules/m³).

One of the problems in burning wood as a fuel, in addition to storage requirements, is that it leaves a deposit of creosote on the inside walls of the chimney. The creosote is highly combustible, and once it accumulates and reaches a high enough temperature, it will spontaneously erupt into a chimney fire, which can be extremely hazardous. The best way to avoid chimney fires is to prevent creosote buildup by cleaning chimneys frequently.

Solid waste is increasingly being used as a fuel resource not only because it is a free replacement for costly conventional fuels, but also because it eliminates the cost and nuisance of disposing of the solid waste in some other manner. This latter cost avoidance, typically at least equal in value to the fuel cost savings, can help amortize the cost of the elaborate incinerator equipment required for clean burning.

Direct geothermal applications—primarily for space heating—have been in existence for several decades, and some even as far back as the nineteenth century. Geothermal heat can be used either in a central-district heating operation or through direct wells to individual buildings. It is appropriate for space and domestic water heating, industrial processing, and occasionally cooling. It is limited by the local availability of the resource.

Ground temperature is used as a source of heat or cooling in the form of groundwater, natural water bodies, or a heat exchanger in contact with the ground. Its use is based on the principle that the ground remains at a relatively uniform temperature, above the ambient air temperature in winter and below it in summer. Its use generally requires some form of heat pump, particularly in the winter, and its feasibility in any given application depends largely on the economics of the mechanical equipment required.

In general, energy resource selection is based on system economics. This relates to everything from fuel cost and equipment efficiency to the resulting air pollution. Where combustion emissions are a problem, stack scrubbers or other pollution-control devices will add to the installation cost. In the final analysis, the project budget will have a great effect on the type of HVAC system and fuel selected.

CENTRAL HEAT SOURCES

Heat-generating equipment is used in a central heating system to warm a fluid (air or water) or to generate a fluid (steam), which then conveys the heat to the various rooms and spaces throughout the building. Using one of the heat transfer processes, the heat source (electricity, fuel, solar radiation, etc.) transfers its thermal energy to the fluid,

thereby increasing the fluid's temperature or changing its state from a liquid to a gas.

The selection of central heating equipment in large buildings depends largely on economic factors once the total required capacity has been calculated. The choice of components depends on the types of fuels available and the availability of space and structural support, among other considerations. It is possible, but not always practical, to recover the internal heat gain from lights, people, and equipment in order to reduce both the size of the heating plant and its energy use.

Central heating systems are generally classified according to the heat-carrying medium (air, steam, or water) and the energy source. In combustion systems (gas, oil, coal, wood, solid waste), fuel is converted to heat in the central plant. The special requirements for combustion systems include a supply of air for combustion and for cooling the mechanical space, and a flue to remove combustion gases to a point where they won't be a nuisance or health hazard. While outside air can be ducted to interior or subterranean equipment rooms, it is better to locate these rooms on the perimeter or at the roof of multistory buildings to allow direct access to ventilation.

The natural movement of gases up a flue is caused by the buoyancy effect of the warmer combustion gases rising above the cooler outside air. The buoyant force is a function of the height of the chimney and the temperature difference between the flue gases and the outside air. Increases in either of these factors will increase the force, which is known as the *draft*. The forces are small and may be increased, decreased, or even reversed by wind currents.

Flues are normally extended up past the top of the highest part of a building to discharge the combustion products where they will not be drawn back into the building. In cases where it is not possible to provide an adequate natural draft because of insufficient chimney height, long horizontal runs of breeching between the stack or chimney and remote mechanical rooms, or where wind patterns around the building produce down drafts in the flue, a *forced draft* or *induced draft fan* can drive the flue gases upward. On large boilers, a forced or induced draft is necessary to control the pressure and quantity of combustion air for the optimum operating efficiency.

The American Gas Association (AGA) is a large organization that conducts research and disseminates information on the safe and efficient installation and operation of gas-fired equipment. It issues industry standards and proposes installation requirements which are usually incorporated into the major building codes.

The use of electrical energy for heating can substantially reduce the first cost of a system. Mechanical and architectural problems of fuel storage and flue stacks are eliminated, allowing greater flexibility in equipment location. Electric heating equipment generally requires less maintenance and offers the opportunity to eliminate the central heat-generating stage altogether. Electricity can easily be routed directly to the space to be heated with little impact on the architectural design, and can there be converted into heat by a wide variety of types and styles of heaters. A major drawback to using electrical heating is the high cost of electricity, which usually makes its operating expense much higher than that for fuels. This disadvantage often far outweighs the economy in first cost.

Other noncombustion sources that require no flues or access to outside air are solar energy, geothermal energy, and heating water converters. These must, however, be centrally situated and have a distribution system to deliver the heat where needed.

Systems using air as the primary distribution fluid have a heat-generating source known as a *furnace*. When the fluid is water—either as liquid or steam—the heating device is called a *boiler*. These terms generally refer to equipment using fuel or electricity as the energy source but are sometimes used in reference to alternatives, such as solar furnaces or heat recovery boilers.

The efficiency of a heat-generating system (Figure 6.6) is defined as:

$$\frac{\text{heat delivered}}{\text{energy input}} \qquad (6.1)$$

Fuel-burning equipment has a combustion efficiency of:

$$\frac{\text{energy input} - \text{flue stack loss}}{\text{energy input}} \qquad (6.2)$$

The overall system efficiency for fuel-burning equipment is the combustion efficiency multiplied by the distribution efficiency, which amounts to the same as Equation 6.1.

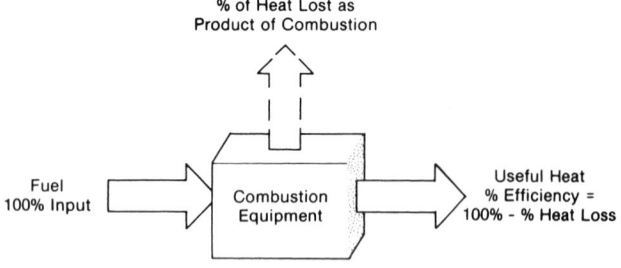

FIGURE 6.6. Diagram of combustion efficiency.

Conventional combustion technologies have relatively low efficiencies (75 to 85 percent). However, the newer *pulse-combustion* and *condensing combustion* processes developed and marketed in recent years achieve combustion efficiencies of 90 to 95 percent or more. This is accomplished by recovering much of the heat lost up the flue stack in conventional equipment. As a consequence, most of the combustion products are condensed before reaching the flue. Instead of the usual outside air intake and flue requirements, the only connections required are (1) a PVC plastic pipe 1½ to 2 inches (3.8 to 5.1 cm) in diameter to vent the 100°F (38°C) combustion exhaust gases, (2) a similar pipe to bring outside air to the combustion, and (3) a condensate drain pipe. The pulse-combustion process is available in the form of a furnace and as a residential-size hot water boiler.

Heat-generating equipment must have an output capacity greater than the space load calculated according to the method described in Chapter 4. This is necessary to account for distribution losses occurring between the central heating plant and the spaces to be heated and the warming up of the spaces from a setback temperature.

Furnaces

A typical furnace is shown in Figure 6.7. Cool return air from the occupied spaces at 60° to 70°F (16° to 21°C) passes through a filter, a fan or blower, and a heating chamber (combustion or electric), arriving at the supply air ductwork at 120° to 140°F (49° to 60°C). The unit may also include a humidification section that evaporates moisture into the air as it passes through. The typical expected life of a furnace is 15 to 20 years.

Boilers

A boiler is a closed vessel in which heat produced by combustion, electricity, or some other means is transferred to water. If the boiler produces steam, it must be able to withstand the resulting pressure.

Boilers are available in a wide variety of types and may be classified in numerous ways, including the categories represented in Table 6.1.

Fuels

Boilers may be designed to burn wood, coal, solid waste, various grades of fuel oil, or a variety of gas fuels. They can also utilize electricity as an energy source. Dual fuel boilers capable of burning gas and oil are common. Some boilers are set up to burn oil, coal, or gas interchangeably.

A boiler designed for one specific fuel may or may not be

(a)

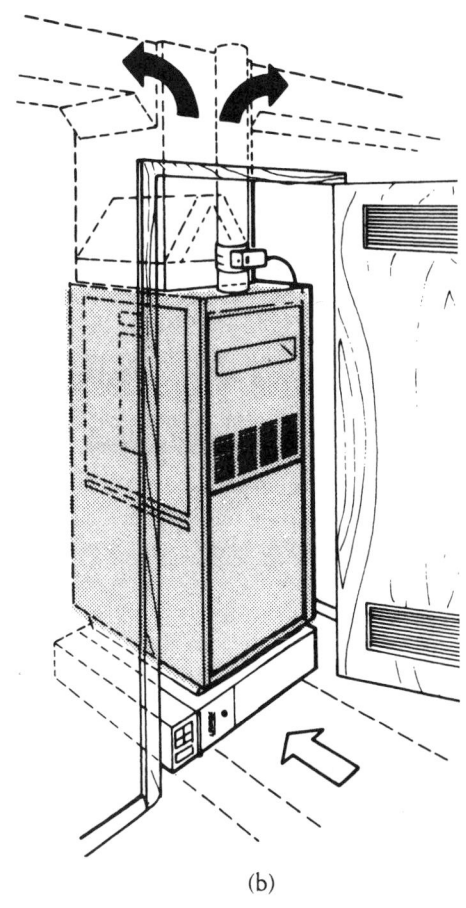
(b)

FIGURE 6.7. Typical Furnace. (a) An actual installation. (Photo by Vaughn Bradshaw.) (b) Diagram of a closet installation showing an A-coil added on top for cooling and an electronic air cleaner below. This airflow configuration is referred to as "up-flow." (Drawing courtesy of Lennox Industries Inc.)

convertible. For example, a boiler that fires a liquid fuel can be modified to accommodate another liquid fuel but not a solid fuel, whereas a boiler originally designed to burn coal can relatively easily add waste-firing capability or be adapted for gas, oil, or some other liquid fuel.

The decision to modify a boiler to burn wastes requires careful consideration of the proposed fuel's effect on combustion efficiency, the ash-handling system, and emissions. Some waste gases will burn in a boiler but will reduce efficiency and may not be worth using. Wastes that produce large quantities of ash or that corrode the boiler internals as they burn may not be economical fuels in the long run. Some industrial process waste gases condense inside boilers and

can destroy them. Others are explosive and require careful handling. A new fuel may completely change the boiler emissions, making them subject to different emissions standards. The cost of adding flue gas treatment may outweigh the benefits of burning the waste.

On the other hand, dramatic cost savings are possible by recovering industrial waste heat and using it to generate steam. Great versatility may be achieved with a combination boiler capable of oil- or gas-firing as well as waste heat consumption when available.

Coal, when used, must be conveyed to the boiler in usable form and at a steady rate. Manual shoveling of coal into the boiler's firebox and removal of ash has all but disappeared,

TABLE 6.1 BOILER CLASSIFICATIONS

Fuel	Number of Fuels	Fluid	Pressure	Material	Design	Flue	Application
Coal	Single	Steam	Low (below	Cast iron	Firetube	Natural	Space heating
Oil	Dual	Hot water	15 psi)	Steel	Watertube	draft	Process
Gas	Multiple		steam		Scotch	Induced	Power
Electricity			Medium		marine	draft	generation
Wood			(15–125 psi)		Sectional	Forced	Domestic hot
Solid waste			steam		Round	draft	water
			High (above		Special		
			125 psi)				
			steam				

replaced by a mechanized stoker and automated ash-removal system.

After the ash is removed from the boiler, it must be disposed of. In relatively small installations, it may be simply hauled away in trucks to the local landfill. Sometimes a buyer may be found for the ash, which has a number of uses, such as improving the qualities of concrete. Disposal of the huge amounts of ash produced by large coal-fired boilers is an important consideration in siting and space allocation. Some coal-fired power plants pulverize the ash, mix it with water, and pipe the slurry to an on-site ash pond. These ponds, the size of a small lake, are completely filled with ash over the 40-year design life of such plants.

In addition to the ash problem, coal combustion produces fly ash and flue gas air pollutants. Fly ash consists of particles ranging widely in size. The flue gases contain oxides of sulfur and nitrogen. State and federal regulations limit the quantities of pollutants that coal-fired boilers are allowed to release into the atmosphere.

Sulfur oxide emissions are responsible for "acid rain" by combining with water in the atmosphere to produce sulfuric acid. Acid- and moisture-laden clouds have traveled north from the United States to Canada, where they released the acid rain. This same problem has occurred in many other parts of the world as well. Even though the acid rain is very diluted, it has been sufficient to severely damage plant and animal life in some Canadian lakes. The rapid erosion of monuments and ruins that are thousands of years old, as well as of modern statues and buildings, has also been attributed to acid rain. The paper industry around the world has been seriously affected, since acid rain has polluted the water systems on which trees in various parts of the world depend.

The amount of sulfur oxide produced by a coal-fired boiler is related to the chemical composition of the coal burned. The so-called Western coal has a lower sulfur content and will produce fewer objectionable sulfur emissions, while the high-sulfur Midwestern and Eastern coals will produce more. Flue gas desulfurization (FGD) equipment removes sulfur from flue gases, fixing it in a sludge that must be disposed of. There is a trade-off between the higher cost of low-sulfur coal versus the cost of an FGD system and the need to dispose of the accompanying sludge when burning high-sulfur coal. It may be worth paying the premium in cost for low-sulfur coal if the boiler does not consume much coal. If the use of large quantities of coal is anticipated, desulfurization equipment enabling cheaper high-sulfur coal to be burned will be cost effective.

A relatively new technology that eliminates the sulfur and nitrogen oxide problems is called *fluidized bed combustion*. In this process, a low-grade (high-sulfur) coal is burned in a bed of granular limestone. The limestone reacts with the sulfur, capturing it in the form of calcium sulfate (gypsum). By controlling the bed temperature to 1,550°F (845°C), nitrogen oxide formation is inhibited.

Fluids

Steam and hot water boilers are very similar in appearance and operation. The differences are basically in the controls—which allow temperatures to reach the boiling point in the case of steam generators and maintain them below the boiling point in the case of hot water generators—and in the accessories attached.

Pressure

As the pressure of steam increases, so does its temperature and sensible heat content. By increasing the pressure of water, its boiling point can be raised above 212°F (100°C) so that it can be heated as a liquid to higher temperatures with a higher resultant sensible heat content. The advantage of high-pressure/temperature systems is that a given quantity of water (liquid or steam) can carry more heat, and consequently distribution piping can be smaller while still carrying the same amount of heat.

Steam boilers used for space or domestic hot water heating are usually of the low-pressure type, while those used for electrical power generation, hospitals, kitchens and many industrial processes are medium or high pressure. When steam boilers are used in high-rise buildings, they are normally medium or high pressure.

Low-pressure hot water boilers are usually limited to space- and domestic water-heating applications. When fitted with a pump to distribute the hot water, they can serve a fairly large area, including high-rise buildings.

High-pressure hot water and steam plants are more complex than low-pressure ones, and require added safety precautions and the presence of a qualified operating engineer whenever they are in use.

Materials and Design

Boilers are constructed of either cast iron or steel. The ASME (American Society of Mechanical Engineers) sets codes and standards in accordance with which all boilers must be constructed.

Cast-iron boilers are generally of the sectional type. They are restricted to low-pressure steam and low-temperature hot water systems. They are available in gross outputs ranging from the smallest residential load (50,000 Btuh, 14,650 W) to approximately 13 million Btuh (13,000 mbh, 3,800 kW).

Small steel boilers are assembled units of welded steel construction and are called *portable boilers.* They are prefabricated on a steel foundation that can be transported as a single package from the factory. Larger boilers are installed in refractory brick settings built at the site.

Above the combustion chamber, a group of tubes—usually horizontal, but sometimes vertical or on a slant—extend between two headers within a cylindrical shell. If flue gases pass through the tubes and water surrounds them, the boiler is termed a *firetube* type. If water flows through the tubes, it is called a *watertube* type. Since watertube boilers can withstand higher pressures than firetube units, they are capable of higher capacities. Both types can generate either steam or hot water.

Low-temperature hot water boilers and low-pressure steam boilers may be either cast iron, steel firetube, or steel watertube. High-temperature hot water boilers and high-pressure steam boilers are restricted to the steel watertube type. An example of a watertube boiler is shown in Figure 6.8.

Steel watertube boilers require a space on one end roughly the length of the boiler to allow the tubes to be pulled out and cleaned. For all boilers, space must be allowed as indicated by the manufacturer for cleaning the flue passage and for general operation and maintenance.

The water used in a boiler normally contains some impu-

FIGURE 6.8. Watertube boilers providing heat for a 500,000-ft² airport terminal. (Photo courtesy of Bryan Steam LLC / Bryan Boilers.)

rities in the form of suspended solids, dissolved solids, and gases. In a watertube boiler, the solids may be deposited in the tubes as scale, which impedes heat transfer and results in overheating and tube failure. Oxygen and carbon dioxide are corrosive agents that cause boiler parts to become brittle, lose their strength, and fail. Make-up water provided to a boiler to replace any that is lost is usually chemically treated to avoid these problems.

When steam gives up its heat, either at the intended delivery point or along the way, it condenses back into a liquid. Condensate return systems collect the liquid and transport it back to the boiler to be recycled. In the past, some systems were designed to discard their condensate down the drain to avoid the expense of installing condensate return equipment and of the energy required to operate it. Over the years, the cost of the water and of the water treatment has grown to the point that these once-through systems are now a liability. For new installations, they are no longer permitted, because they impose such a tremendous burden on the local water source and sewer systems.

There are, however, inevitable losses even in a closed-cycle system, so small amounts of water must be continually added. As boiler water circulates, evaporates, and is replaced, the impurities tend to become concentrated. For this reason, a portion of the water from the system is periodically discharged to prevent scale formation. Water removed in this way is called boiler *blowdown.*

Flues

A boiler stack in some cases may be terminated several feet above the top of the roof but in most cases must extend much higher. In general, local codes govern the height of the stack above the roof. Prefabricated metal stacks which come in

3-foot and 4-foot (0.9-m and 1.2-m) lengths are commonly used. The breeching is the portion of the flue between where it leaves the boiler and where the stack rises up. A rain cap or hood may be used at the top of the stack to keep out rain and snow.

The three means of maintaining a draft up the flue are:

1. *Natural draft.* The stack effect causes a natural draft up the chimney, which sucks the combustion air through the burner or fuel bed.
2. *Induced draft.* A power-driven fan in the flue draws the combustion air through the burner or fuel bed.
3. *Forced draft.* A blower in the burner section forces combustion air through the boiler and up the stack.

Capacity

Boiler capacities are typically rated in terms of:

1. *Boiler horsepower.* The heat required to produce 34.5 pounds of steam per hour. It is equivalent to 33,472 Btuh.
2. *Pounds of steam per hour (lb/hr).* There are about 1,000 Btuh/lb (2,300 joules/kg) of steam, depending on its temperature and pressure.
3. *Btuh.* Btu per hour. A Btu is the amount of heat required to raise the temperature of 1 pound of water 1 degree Fahrenheit.
4. *kW.* Electric boilers are rated in kilowatts. Steam boilers are available in capacities up to 3,000 kW and hot water boilers up to 2,000 kW. This also applies to nonelectric boilers in SI units.

Heating boilers are rated according to test procedures developed by the Hydronic Institute (formerly the Institute of Boiler and Radiator Manufacturers). The ratings, found in manufacturers' catalogs, are referred to as I = B = R ratings for cast-iron boilers and as SBI (from the Steel Boiler Industry Division of the Hydronic Institute) for steel boilers.

Boiler efficiencies depend on the type, unit, size, and, to some extent, age. Large (50,000 lb/hr or 6.3 kg/s) modern gas-, coal-, or oil-fired steam boilers in good condition should achieve peak efficiencies of 80 to 85 percent, while smaller boilers in the past were limited to efficiencies as low as 65 percent. A great deal of operating expense can be saved by raising the efficiency level of large boilers by even just a few percentage points.

In some cases, a higher average efficiency can be obtained by using multiple smaller boilers for heating water instead of a single large one. Multiple modular boilers are controlled to cycle on and off in sequence according to the demand. Since boilers are usually designed for maximum efficiency when operating at 100 percent of their rated out-put, the group efficiency is kept higher than that of a single large boiler with fluctuating or intermittent operation. Modular boilers also have the advantage of continuing to provide at least partial capacity when one or more of the units have to be taken out of service for maintenance or repair. Extra modules can even be installed to provide full backup capability. Their small individual size makes it easier to move them in and out for installation in existing facilities with limited access openings to mechanical spaces. These advantages must be weighed against the possible higher cost of multiple units and the increased number of piping connections, as well as the larger mechanical space requirement to determine whether multiple boilers are economical for a given installation.

Hot Water Converters

When the source of heat is a hot fluid (steam or hot water)—such as in the case of geothermal applications, district heating, or systems in which a central steam boiler is used for some process load—heating water or domestic hot water to serve the building needs can be generated in a *hot water converter.* This is simply a heat exchanger that transmits the heat from the steam or hot water source to the water to be heated.

These heat exchangers are a shell-and-tube type, which means that they consist of a cylindrical "shell" surrounding a bundle of tubes. Most often, the steam or liquid heat source is circulated through the shell part, while the water to be heated is circulated through the tubes. A shell-and-tube heat exchanger is shown in Figure 6.9.

When hot water is the source, the water leaving the shell is somewhat cooled down, and is normally returned to its source for recycling. When the shell contains steam, it condenses as it gives up its latent heat to the water in the tubes.

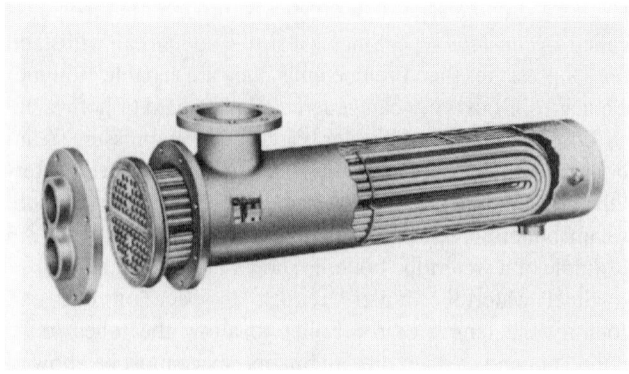

FIGURE 6.9. Shell-and-tube heat exchanger. (Photo courtesy of ITT Fluid Handling Division.)

The condensate collects at the bottom of the shell, drains off, and is returned to its source.

Sometimes converters are used to protect a boiler from scale and corrosion. The converter enables a relatively small fixed volume of water to be recirculated between the boiler and the converter. The water that flows through the converter tubes and out to the distant loads is subject to the inevitable losses and fresh water makeup that result in scale formation. The scale clings to the insides of the tubes, which can readily be cleaned or replaced.

Incinerators

In any building, solid waste must be disposed of in some manner. It can either be burned on site in an incinerator or be hauled away for disposal at a landfill. There are certain costs and handling provisions required in either case.

When local waste collection service is not available, adequate, or suitable for the type of waste involved, provisions for private incineration may be included in the building design. Due to air pollution concerns, however, private incineration is restricted in many areas by local or regional agencies. A review of such regulations should precede the consideration of on-site incineration.

Where quantities of solid waste are large, it may be cost effective to incinerate it and recover the heat for some useful purpose. Solid waste utilized in this manner is called *refuse-derived fuel* (RDF). Modern heat-recovery incinerators use a high-temperature process called *pyrolysis* which destroys the refuse and converts it into carbon dioxide and water vapor, flue gas and ash, and permits the system to meet EPA emissions standards even with pathological wastes and plastics.

A classification of the different types of solid waste, established by the Incinerator Institute of America and the Solid Waste Processing Division of ASME, is presented in Table 6.2. Types 5 and 6 waste cannot be covered by a blanket definition. In each individual case, the actual chemicals or solids must be analyzed to determine the ash content and the heating value.

Normally there is not enough solid waste to satisfy all of a building's heating and hot water loads, especially if the climate is cold and the heating loads are high. Thus, the heat-recovery incinerator cannot be used as the sole generator of heat, but only as a supplement to a conventional fuel-fired boiler.

The integration of piping and controls for adding an incinerator to a conventional boiler system is not a problem. The steam generated by the incinerator is introduced into the main steam header at the central heating plant. For this reason, the incinerator and its accessories should be located in or as close as possible to the central heating plant. Handling of solid waste requires additional area for separation, shredding, and the conveying system. The entire plant may require a ramp or elevator for transporting the waste.

Incinerators vary in size, burning rate (pounds per hour), type of waste consumed, and feeding method, that is, whether they are fed by a chute or hopper. Large-capacity equipment may require a flue 36 inches (1 m) in diameter and 50 feet (15 m) high. As much as 250 ft^2 (23 m^2) of floor space and an 11-foot (3.35-m) ceiling height might be needed to accommodate the unit and provide adequate space around it. Weights run about 600 lb/ft^2 (3,000 kg/m^2). Because of the high flue temperatures from commercial and industrial incinerators, chimneys must be 8-inch (20-cm) brick with 4½-inch (11.4-cm) refractory linings.

Small, preassembled, portable domestic incinerators are available with burning rates of up to 100 lb/hr (12.6 g/s). They require about 30 ft^2 of floor space, including the working space around them.

To determine the economics of on-site incineration or heat-recovery incineration, it is necessary to identify the total amount of solid waste generated daily, the percentage of each classification according to Table 6.2 together with their respective heating values, the availability of disposal services by the local municipality or a private contractor, and the cost of those services (pickup and transport of the waste and disposal at a landfill). The investment cost for installation of all necessary equipment (incinerator, shredders, storage, etc.) and the cost of its operation must then be weighed against the value of the recovered heat and the avoided cost of hauling and disposal. When the economics of RDF utilization seem marginal, the use of small-scale packaged heat-recovery units should be investigated.

Hospitals are prime candidates for RDF systems. They typically generate about 20 pounds (9.1 kg) of solid waste per patient bed per day under peak conditions. In addition, they have large year-round steam requirements for space heating, hot water, sterilization, kitchens, and laundries.

Solar Energy

It has been estimated that if the sunlight reaching the earth's surface in a single day could be converted to useful energy forms, it would satisfy the energy needs of the entire world for over 50 years. On a smaller scale, the sunlight falling on a given building in most cases carries enough energy to keep it comfortable year-round. Sunlight is an attractive energy source because it is widely distributed, environmentally harmless, inexhaustible, and, in its raw state, free.

Solar energy is useful in four different forms: (1) photo-

TABLE 6.2 CLASSIFICATIONS OF SOLID WASTE

Type		Composition	Incombustible Solids (%)	Heating Value	
				Btu/lb	kJ/kg
0	Trash	Highly combustible material such as paper, cardboard cartons, wood crates, sawdust, combustible floor sweepings from commercial and industrial activities, and up to 10% by weight of plastic bags, coated paper, laminated paper, treated corrugated cardboard, oily rags, and plastic or rubber scraps.	5	8,500	19,800
1	Rubbish	Combustible waste similar to Type 0 such as paper, cardboard cartons, wood scrap, foliage, and combustible floor sweepings from domestic, commercial, and industrial activities. The mixture can include up to 20% by weight of food wastes, but contains little or no treated papers, plastic, or rubber wastes.	10	6,500	15,100
2	Refuse	An approximately even mixture by weight of rubbish (Type 1) and garbage (Type 3). Common for residential occupancy.	7	4,300	10,000
3	Garbage	Animal and vegetable wastes from restaurants, cafeterias, hospitals, gardens, markets, and similar facilities.	5	2,500	5,800
4	Pathological waste	Human and animal remains consisting of carcasses, organs, and solid organic wastes from hospitals, laboratories, abattoirs, animal pounds, and similar sources.	5	1,000	2,300
5	Residues	By-product waste—liquid, semiliquid, gaseous, or condensed gases—such as tar, paints, solvents, sludge, fumes, etc., from industrial operations.	—	[a]	[a]
6	Industrial solid waste	Essentially inorganic matter with a high percentage of incombustible solids such as rubber, plastics, wood waste, etc.	—	[a]	[a]

[a] Heating values must be determined by the individual materials to be incinerated.

synthesis, which maintains life on this planet by producing food and converting carbon dioxide to oxygen, (2) photovoltaic cells, which convert sunlight directly into electricity (discussed in Chapter 10), (3) natural daylighting (discussed in Chapter 8), and (4) thermal energy, which can be used for space heating and cooling, domestic water heating, power generation, distillation, and process heating. It is this last form that is the subject of this section.

The equipment, systems, and overall technology for using solar energy for heating, cooling, humidification, and water heating are currently available in the form of off-the-shelf hardware. The only impediments to the use of solar energy are economic ones.

The motivations for using solar energy as a heat generation source are its special values:

Conservation. It displaces the use of, and thereby conserves, fossil fuels which are in limited supply and have a greater value as industrial chemicals.

Environmental. Compared with the combustion of the fuels that it displaces, solar energy use decreases the level of air pollution emissions. It also releases none of the harmful radioactivity into the environment that nuclear energy does.

Insurance. It is capable of substituting for conventional energy technologies which could suddenly become expensive, unavailable, or undesirable for a number of social, political, or physical reasons.

Decentralization. It is well suited for decentralized, on-site self sufficiency, which promotes social stability.

There are two principal categories of solar thermal energy systems: active and passive. *Active systems* are those using pumps, fans, heat pumps, or other mechanical equipment to transmit and distribute the thermal energy either by means of air or a liquid heat transfer fluid. Active systems must have continuous availability of electricity to operate the equipment.

Passive systems incorporate solar collection, storage, and distribution into the architectural design of the building structure itself, so that it takes maximum advantage of the solar energy by natural means, that is, without pumps and fans. Passive methods involve optimizing the location, type, and size of windows, overhangs, shading devices, and integral thermal storage mass, using little or no mechanical energy. When a building design incorporates passive solar features but still makes minor use of electrically driven fans or pumps, it is sometimes referred to as a *hybrid system.*

There are passive and active solar energy methods to accomplish space heating, space cooling, and domestic water heating, but the passive use of the sun's energy that directly penetrates into the building is the most efficient way. Passive thermal systems will be addressed in Chapter 7, and natural (passive) daylighting in Chapter 8. This section is limited to the active systems.

Solar Radiation

Solar energy approaches the earth as a spectrum of electromagnetic radiation with wavelengths ranging from X-rays (0.1 μm long) to radio waves (100 m long). By the time it reaches the earth's surface, however, 99 percent of the sun's radiant energy is contained in the range of wavelengths between 0.28 and 4.96 μm.

In solar HVAC applications in the United States, solar radiation is expressed in British thermal units per hour per square foot. In SI units and in the context of U.S. outer space activities, the standard units are watts per square meter. In the context of meteorology, the accepted unit is Langleys per minute (Ly/min). Table 6.3 shows the relationship between these three units.

For convenience, a quantity called the *solar constant* is

TABLE 6.3   SOLAR RADIATION UNITS

To Convert From	To		
	Btuh/ft^2	W/m^2	Ly/min
Btuh/ft^2	1	3.155	0.004521
W/m^2	0.3170	1	0.001435
Ly/min	221.2	696.8	1

To convert from a unit on the left side to a unit across the top, multiply by the intersecting number in the table.

defined as the amount of radiation that would be received at the outer limit of the earth's atmosphere by a surface perpendicular to the sun's rays when the earth is at the average distance from the sun. Since the sun is located at one focus of the earth's yearly elliptical orbit, the distance between the earth and the sun varies with the time of year. In late December and early January, we are only 91.4 million miles (147.2 million km) away from the sun, while on July 1 the earth–sun distance is about 94.4 million miles (152.0 million km). As a result, the normal incidence intensity on an extraterrestrial surface varies from 444.1 Btuh/ft^2 (1,399 W/m^2) on January 1 to 415.6 Btuh/ft^2 (1,309 W/m^2) on July 5. The currently accepted value of the solar constant is 429.5 Btuh/ft^2 (1,353 W/m^2).

Actually, the earth's spinning causes the sun's apparent motion. But to an observer standing at any given spot on the earth's surface, the sun appears to move across the sky in a regular pattern. The sun's position from a point on the earth can be defined in terms of its altitude, β, above the horizon and its azimuth, φ, as shown in Figure 4.4. At solar noon, the solar azimuth, by definition, is 0°.

As the sun's radiation passes through the earth's atmosphere, some is scattered by nitrogen, oxygen, and other molecules, and by aerosols, water droplets, dust, and other particles. This scattering of the sunlight is what causes the sky to appear blue on clear days. Some of the scattered rays reach the earth in the form of diffuse radiation.

Some solar energy is absorbed by the ozone in the outer atmosphere—which sharply cuts off the ultraviolet radiation—and by water vapor and carbon dioxide. The total amount of attenuation at any given location is determined by the length of the path through the atmosphere that the solar rays traverse and by the composition of the atmosphere. The path length is expressed in terms of *air mass,* m, which is illustrated in Figure 6.10 as the ratio of the amount of atmosphere in the actual earth–sun path to the amount that would exist at sea level with the sun directly over head (m = 1). Beyond the earth's atmosphere, m = 0. The effect of air mass

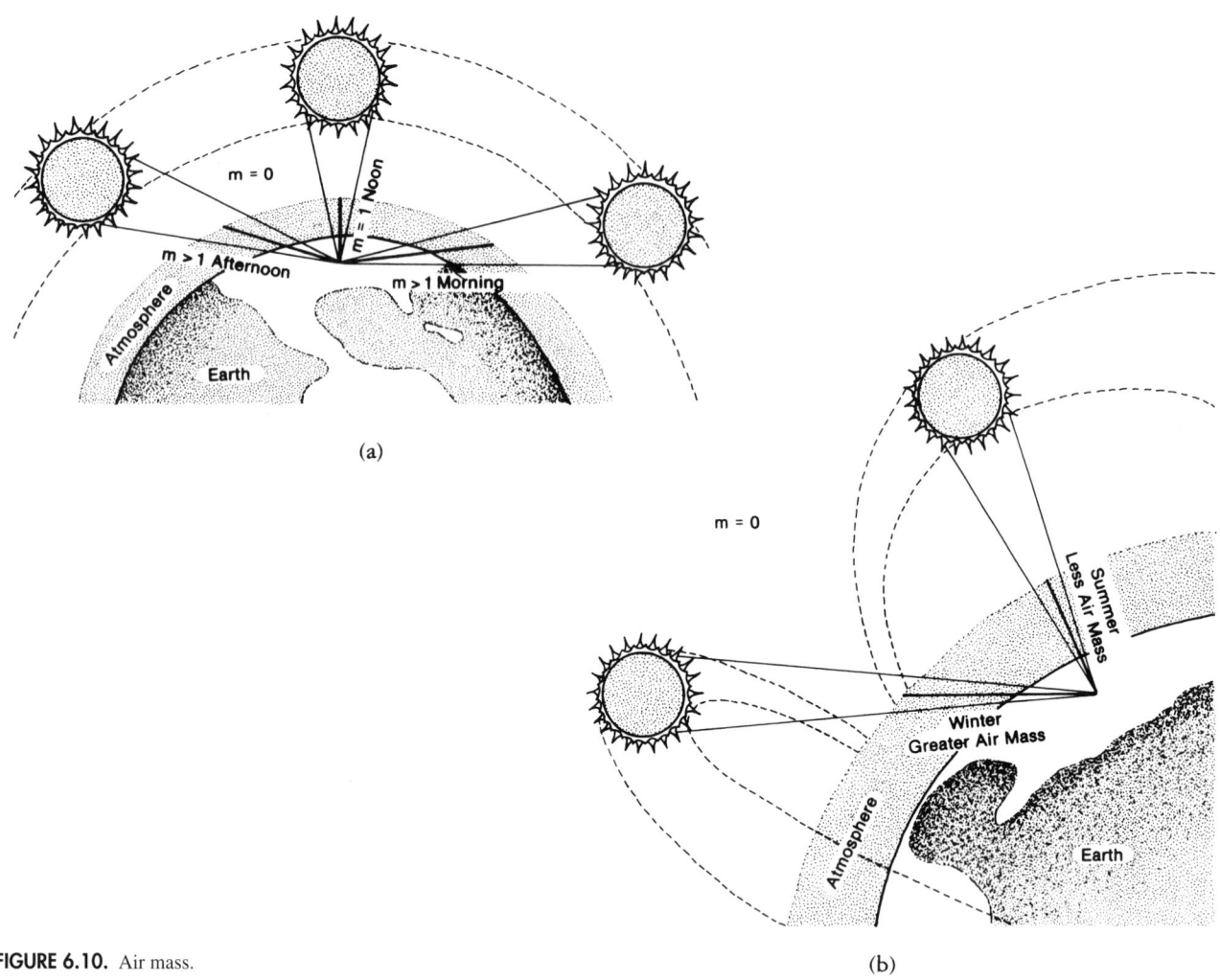

FIGURE 6.10. Air mass.

(a)

(b)

has a much greater impact on the difference in the intensity of solar radiation between summer and winter than the variation in earth–sun distance.

By the time the solar rays penetrate the earth's atmosphere and reach ground level in a clear air region, their intensity may be reduced to about two-thirds of the extraterrestrial value. In coastal or industrial areas where additional water vapor or smoke is in the air, the solar intensity is reduced even more. This reduction in solar intensity affects the level of sun light as well as heat.

The net effect of all these factors is that the available energy varies from zero on a very cloudy day or at night to approximately 350 Btuh/ft^2 (1,100 W/m^2) on a bright summer day. Clearly, in order to utilize solar energy, a system must have the ability to collect the energy, transfer it to the area where it's needed, and store it for use on cloudy days or at night.

Systems

Active solar energy systems gather solar radiation by absorbing it in collectors, from which the heat is removed by a heat transfer fluid and conveyed to storage. If the temperature in storage is adequate, the heat will be conveyed to the load upon demand. If the temperature in storage is inadequate, an auxiliary or standby backup heat source is called upon.

A complete solar energy system is composed of three interconnected subsystems: (1) collectors, (2) storage, and (3) delivery to load. The subsystems are interconnected by piping, ductwork, pumps, fans, and heat exchangers. Figure

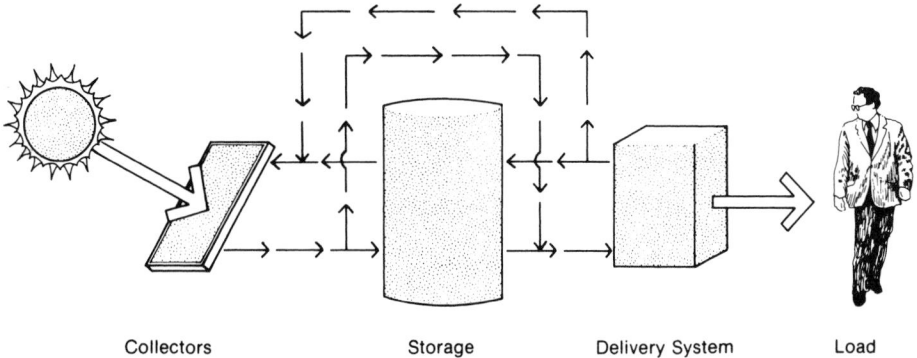

FIGURE 6.11. Schematic diagram of a solar energy system.

6.11 provides a simplified schematic diagram of a solar collection system.

Solar collectors operate much the way a parked car does on a sunny day with its windows rolled up. Sunlight passing through the glass strikes objects inside the car and is converted to heat. Since the rate of heat gain from the incoming sunlight is greater than the rate of heat loss by radiation and conduction-transmission through the shell of the car, the air temperature in the car rises well above the outdoor temperature. Solar collectors are specifically designed to take advantage of this same phenomenon and optimize the ratio of heat gain to heat loss.

Collectors are generally divided into flat-plate and concentrating types. *Flat-plate collectors* are less costly and much more common. They have been built in a wide variety of designs from many different materials.

The heart of the collector is a blackened metal surface called the *absorber plate*. Figure 6.12 shows a cross section of a typical flat-plate collector. When the sun's rays, passing through the cover plate(s), strike the absorber plate and are absorbed, the plate heats up. A fluid circulated through tubes or channels in the plate picks up the heat as it passes through and carries it away to an isolated storage unit. When the absorber's temperature exceeds the surrounding ambient temperature, heat is lost from the collector. The larger the difference between the collector and the outside temperatures, the larger the collector heat losses.

In order to reduce radiation and convection heat losses, the heat is trapped in by glazing over the absorber plate. A flat-plate collector generally has one, two, or no cover plates made of glass or plastic. Each cover plate blocks about 8 percent of the incident solar radiation from entering the collector but also slows down the heat loss back out to the surrounding environment. The glazing is usually glass, but fiberglass, plastics such as Tedlar, Plexiglas, polycarbonate,

or any other material that allows sunlight to pass through can be used. The heat loss out the back and sides of the collector is minimized by insulating the entire collector box.

Flat-plate collectors—referred to as *collector panels*—are typically about 4 × 8 feet (1.2 × 2.4 m). The materials used in their construction are important because they determine the collector's cost, performance, and life expectancy. The absorber surface can be a plain black paint or an expensive high-efficiency *selective surface coating*.

The number of transparent covers needed depends on the desired operating temperature of the collectors and on the ambient temperature. Two covers are used for collection temperatures above 150°F (65°C) or in cold climates, and one cover is used for lower collector temperatures or mild climates. In cases where only low collection temperatures are needed and solar collection is not important in the coldest part of the winter, such as with swimming pool heaters, the cover plate may be omitted altogether.

Flat-plate collectors are capable of absorbing both the direct and diffuse components of solar radiation.

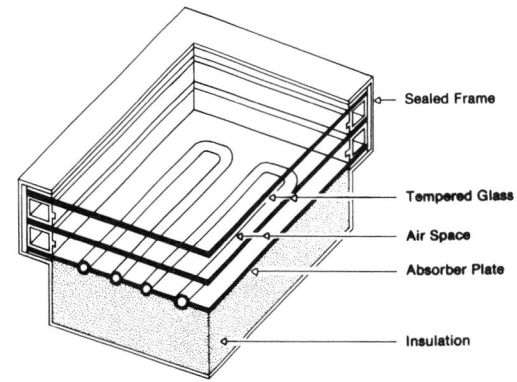

FIGURE 6.12. Cross section of a flat-plate collector.

Concentrating collectors, by contrast, can use only the direct rays of the sun, but can achieve much higher temperatures than flat-plate collectors. The moderate temperatures available from flat-plate collectors are suitable for space and domestic water heating. When high temperatures are needed for absorption or Rankine cycle cooling, or to produce steam to drive electricity-generating turbines or for other purposes, a concentrating collector is usually used. There are, however, some carefully designed flat-plate collectors that heat up to as high as 200°F (93°C) and can be used for absorption cooling.

In general, concentrating collectors focus the direct-beam solar radiation onto a point much smaller than their receiving aperture. By thus concentrating the energy, the higher temperatures can be achieved. A number of different methods are employed using configurations of optical lenses or reflectors. One method is to arrange simple flat reflectors in front of flat-plate collectors to increase the amount of radiation striking them. One popular device used to concentrate is the Fresnel lens, which is made out of an inexpensive plastic.

Because the angle of the sun's rays varies throughout the day, most concentrating collectors utilize some mechanism to track the sun during its daily path across the sky in order to receive the maximum amount of direct radiation. By being able to orient itself all the way to the horizon, a sun-tracking collector can begin to accept radiation directly from the sun long before a fixed, south-facing flat plate can receive anything other than diffuse radiation from the portion of the sky that it faces. Thus, over an entire day in a relatively cloudless area, a concentrating collector may capture more total radiation per unit of aperture area than a flat-plate collector can.

After the heat is absorbed in the collector, it is transported to the storage subsystem. Since heat loads generally continue when the sun does not shine (at night) or is insufficient (overcast days), excess energy, when available, must be stored for future use. The economically optimum storage capacity usually ends up as being from 1 to 3 days' output from the collector.

The predominant criterion for a thermal storage medium is that it hold a large amount of heat per unit volume. Since water has a high specific heat and a fairly high density, it is commonly used. Storage tanks are constructed of concrete, steel, or fiberglass.

There are some materials with an even greater storage capacity per unit volume and a melting point that occurs within the normal storage temperature range. The latent heat of fusion, which is the amount of heat required to melt a material and the same amount that is released when the material solidifies, is used to increase the storage capacity. One such material that has been employed with moderate success is Glauber's Salt, which melts at 88°F (31°C) and has a heat of fusion of 92 Btu/lb (214 kilojoules/kg). It can store about 7.3 times more heat than water in the same volume. Even though Glauber's Salt shows the most promise of all latent-heat materials to date, it still has a problem: its integrity is difficult to maintain over a large number of freeze–thaw cycles.

When the fluid circulating through the collectors freezes in cold weather, it usually causes costly damage to the collectors. To prevent such damage, some form of freeze protection must be employed.

Freeze protection for liquid systems can be accomplished either by draining the collector fluid into a tank within the building whenever solar energy is not being collected or by substituting a liquid that will not freeze, such as an ethylene glycol (antifreeze) mixture instead of water circulating through the collectors. For the latter method, a heat exchanger is usually placed between the fluid circulating through the collectors and the storage tank to avoid the high cost of filling the entire storage vessel with the antifreeze solution. The heat exchanger may be:

- A shell-and-tube type, as described previously
- A coil of tubing within the tank
- Tubing wrapped around the outside of the tank
- One fluid circulating within a jacket surrounding the tank containing the other fluid

Most building and plumbing codes require double-walled heat exchangers to provide absolute protection from contamination of any potable water supply with an antifreeze solution. This is accomplished by (1) using two shell-and-tube heat exchangers, (2) circulating the antifreeze collector fluid through one coil in the tank and circulating the potable water through another coil so that any leak in one coil will not automatically mix the collector and potable fluids, or (3) employing two concentric jackets around the storage tank. The option of wrapping tubing around the tank already complies with the requirement by having a double-walled separation between the fluids.

Freezing problems can also be avoided by using air as the collector fluid. Storage for air-type systems is typically by a large bed of river bed–quality rocks, which are smooth, nearly round, and have a radius of about 1 to 1½ inches (2.5 to 3.8 cm). The volume required by rock storage is larger than that required for water to provide the same storage capacity. On the other hand, rocks are usually not as expensive as large water-storage tanks.

Another means of freeze protection in warm or temperate climates, where freezing temperatures rarely occur, is the circulation of stored solar heat through the collectors as needed. This process wastes some of the stored solar heat, but may reduce special equipment costs and the inefficiencies of heat exchangers. This method is not suitable for cold northerly regions where subfreezing temperatures persist for long periods of time.

During the summertime, collectors may absorb much more heat than they can use or store. Liquid systems typically have upper temperature limits which must not be exceeded. When those temperatures are reached, the collectors may be drained to storage, or the liquid may be circulated through a purge unit which rejects heat to the outside air. Air-type collectors are sometimes purged with outside air if a limit temperature is exceeded. Other designs do not limit how hot an air-type collector may become.

Normally, a backup source of heat capable of supplying 100 percent of the load is included for times when there is insufficient stored heat due either to long periods of low sunlight or to the solar energy system being out of service. This full-sized backup source also routinely furnishes auxiliary heat when the storage temperature falls below that required by the load.

For applications with both a cooling and a heating load, a heat pump may be an appropriate backup. In the summer, a heat pump operates as a conventional air conditioner, and in the winter it operates in the heating mode to raise the temperature of the collected solar heat. A heat pump is able to extract heat from the solar transfer fluid at low temperatures and then reject it at a higher temperature to heat a building or hot water. Such systems enhance the solar energy system by (1) enabling the storage to be used even when its temperature drops below the level of direct usefulness and (2) using the solar heat at a lower temperature so that the collectors can operate at a lower temperature. This reduces losses to the ambient environment, resulting in a greater net amount of solar heat collected. Less collector area is thus required, which reduces the installation costs of the solar energy system. Typically, a solar-assisted heat pump system requires a collector area about half the size of that needed for a conventional direct-heating system.

Fuel-fired heaters may also be used as the auxiliary source. The least expensive backup system to install is electric resistance. However, if auxiliary heating is frequently required, the high cost of electricity may offset the low first cost. Both fuel-fired heaters and heat pumps are more expensive to install but have lower operating costs than an electric resistance system.

The use of solar energy for domestic or institutional hot water heating is typically one of the most cost-effective solar energy applications. In contrast to space heating and cooling, which are seasonal loads, water heating is usually relatively constant throughout the year except for variations in incoming cold water temperatures between summer and winter which alter the load slightly. Year-round use amortizes the initial equipment cost faster (that is, provides a faster return on the investment). These economic factors favor those areas where the alternative for water heating is high-priced electricity or propane. Solar water heaters are manufactured as packages that include collectors, a storage tank, and controls in Australia, Israel, Japan, Greece, the United States, and elsewhere.

Industrial-process heating loads, like service hot water, are typically year-round and thus also cost-effective applications. Very large amounts of low-temperature energy are used for such applications as drying lumber or food, cleaning as part of food processing, extraction operations in metallurgical or chemical processing, cooking, curing of masonry products, paint drying, and many others. Temperatures for these applications range from near-ambient temperature to that of low-pressure steam and can be provided by flat-plate collectors or concentrating collectors with low concentration ratios.

Industrial-process heat, the thermal energy used for manufacturing processes, needs to be supplied in the form of steam, hot air, or a hot liquid, depending on the application. If hot air is needed, an air-type collector should be considered. If steam is needed to operate some equipment, the solar energy system must be designed to produce steam, so concentrating collectors will probably be required.

Solar cooling can be used to provide either refrigeration for food or chemical preservation, or comfort cooling. Solar comfort cooling has the advantage that the highest cooling load occurs when the solar energy availability is greatest. For cooling applications, solar energy can be stored either in the form of hot water as it comes out of the collectors or as chilled water.

Solar air conditioning can be accomplished by three types of systems: absorption, Rankine cycle, and desiccant. The most common method is with absorption refrigeration equipment, which will be described in the next section on central cooling generation equipment. The *Rankine cycle* uses steam produced by solar energy to turn a turbine to create mechanical shaft power that is then used to operate a conventional air-conditioning device.

Solar desiccant cooling is based on a cycle of humidification and dehumidification. A solid or liquid desiccant first absorbs humidity from the air. The dry air is then cooled and humidified to the desired state by evaporative cooling from a different moisture source. The moisture absorbed by the desiccant is then evaporated away using solar heat, and the cycle repeats.

Equipment for solar cooling, like that for solar space heating, is expensive. Some economy is achieved by using the same solar collection/storage system for both heating in the winter and cooling in the summer. But reducing the cooling loads through careful building design and insulation is usually less expensive than providing the equivalent solar cooling. Solar energy systems should be used only for those cooling loads that cannot be avoided by thoughtful building design.

Economics

The primary objective of a solar energy system is to collect as much solar energy as possible at the lowest possible cost. At present, because of the high cost of solar equipment, solar systems take a fairly long time to pay back the investment in them. The energy is free, but the equipment investment is relatively high. The optimum collector and storage size is a trade-off between added investment and added savings.

The technology for solar energy utilization is not exceedingly complex, but the capital cost of the system installation is high. A system typically costs in the range of $60 to $140/ft² ($645 to $1500/m²) of collector. Pool heating systems range from $12 to $75/ft² ($110 to $800/m²).

Although it is technically possible to construct a solar heating and cooling system to supply 100 percent of the load, that system would usually be nowhere near economically competitive with the conventional alternatives. The cost for that last small percentage of capacity would be astronomical. Most solar installations are sized to handle 40 to 70 percent of the total load, depending on economic optimization.

The size—and thus the capacity—of a solar energy system is determined by a life-cycle cost analysis that balances the cost of energy saved over the expected life of the installation against the system cost. This analysis includes a thermal evaluation and optimization of the building. The evaluation of energy cost savings includes an estimate of utility costs now and in the future.

Although important in any building, minimizing heating and cooling loads is doubly important when considering solar energy. Particular attention to insulation, fenestration, ventilation, and energy management in order to reduce heating and cooling loads will improve the cost effectiveness of solar energy systems by reducing their size and cost. In order to minimize the initial investment cost, it is essential to reduce the load to a minimum before sizing the solar system.

Continued increases in the cost of fossil fuels will make solar energy use economically competitive for many applications.

Architectural Impacts

Active solar energy systems pose challenges to architectural design in that they must be incorporated into the building in such a way that thermal performance is satisfactory and the appearance is aesthetically satisfying. Active systems are characterized by the presence of solar collector panels facing the equator at an appropriate angle with the horizontal so that the incident rays of the solar radiation are approximately perpendicular to the collector surface. In the Northern Hemisphere, collectors oriented toward the equator are facing south, and in the Southern Hemisphere they face north. For simplicity, the *south* direction will be referred to here, but the reader should substitute *north* for Southern Hemisphere applications.

Collectors. Solar collectors can be located anywhere near or on a building as long as they are not shaded from exposure to the sun. Being the connecting link between the sun and the rest of the system, the collectors must be exposed to direct sunlight, which may make them highly visible features of the building.

Collectors may be located on the roof of a building, on the south wall, on the ground next to it, or on an adjacent accessory building oriented in the proper direction. Most active collectors are placed on the roof of the building they serve in order to obtain unobstructed sunlight, solid anchoring for the collector, and protection from vandalism and accidents, and to permit piping or ductwork to be run inside the structure, which reduces heat loss. One of the considerations in locating solar collectors is to avoid reflecting glare into neighboring windows or toward traffic during any time of the day.

Active collectors can be purchased ready-made, or they can be custom-designed for a particular building and fabricated at the construction site. Appearance is affected by the type of collector selected.

Most collectors have flat cover plates, but some use curved plates. Flat-plate collectors are usually framed in wood, plastic, or metal, and have an uncluttered appearance. Concentrating collectors vary so much in appearance that a generalization cannot be made.

The appearance of roof-mounted collectors depends on several factors; the most important is the size of the installation. In a residential application, the domestic water heating needs of a family of four can typically be satisfied by two or three collector panels, which would not impact very much on the dwelling's architecture. If, on the other hand, the panels are numerous, they can dominate the roof of a building. For large installations, such as those used for space heating, cooling, or industrial processes, collectors may cover the entire roof, making it appear to be surfaced with dark-colored glass.

An economic study needs to be made for each installation to determine the optimum size of the collector array. It will depend on the load, the desired storage performance, the type of application, and the cost of competing conventional energy sources in the locale. Considering all these factors, a rule of thumb for the optimum collector area is:

- Domestic water heating—about 15 ft² (5 m²) per residential user
- Space heating—in the range of 25 to 50 percent of the internal floor area served
- Space cooling—even larger than for space heating

Process loads must be considered on an individual basis. Most of the references cited in the bibliography on solar energy at the end of this chapter provide methods for estimating the appropriate size of a collector array.

Other factors that affect how roof-mounted active collectors look are the shape and orientation of the roof itself. Flat roofs present the fewest problems, since the collectors can be positioned almost anywhere on the roof and tilted up on a mounting frame to capture the sun's rays. If a pitched or sloping roof peaks on an east/west axis, a large area of the roof may be available for collectors, but complications arise if available roof areas are not facing south.

In general, collectors should face as close to solar south as possible. (Solar south differs from magnetic south to a varying degree, depending upon position on the earth.) If fog conditions, shading, or much colder temperatures prevail in the morning hours, it is desirable to face collectors slightly west of south, since it is more efficient to emphasize the afternoon collection period in those cases. However, under no condition should the collectors face more than 30° off of south.

Solar collectors are usually tilted from the horizontal plane toward the equator in order to collect the maximum amount of solar radiation. The best angle of tilt depends upon the latitude and the time of year. Whereas most concentrating collectors have elaborate tracking mechanisms to change their tilt angle with the time of year, flat-plate collectors are generally fixed in place at some angle. The optimum fixed tilt angle depends upon whether the dominant load occurs in the winter, summer, or evenly all year.

Figure 6.13 shows the monthly variation in solar radiation per clear day on surfaces of various slopes. Due to the combined effects of the varying length of days and changing solar altitudes, the incoming radiation on horizontal surfaces

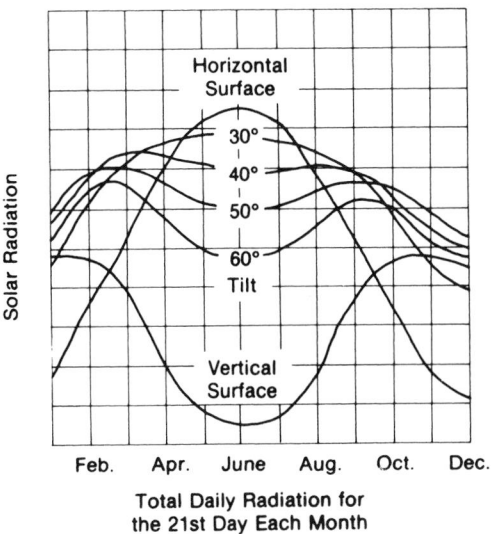

FIGURE 6.13. Solar variation with surface angle.

reaches its maximum in midsummer, while vertical south-facing surfaces experience their maximum during the winter months.

The optimum fixed collector tilt for a year-round application (process or domestic hot water) is usually considered to be 0.9 × latitude. Where maximum wintertime collection is desired (space heating), the optimum tilt angle is latitude + 10° or 15°. In the summertime, latitude − 10° or 15° gives the maximum collection. A few degrees of deviation from the optimum tilt or from due south does not seriously affect collector performance.

Solar collectors may be separate from the building or they may serve as part of the weatherproof enclosure, thus reducing the cost of roofing or siding. Such a reduction is a credit that reduces the cost of the collector. On the other hand, if solar collectors are part of the wall or roof architecture, the building's solar orientation becomes a rigid design constraint and, unlike separate collectors, limits the flexibility in creating more conventional (contemporary or traditional) designs.

If there is other mechanical equipment on the roof, or if the roof faces the wrong direction or is shaded, active collectors can be located on the south wall of a building. A south wall installation generally works best for space heating in northern climates (at high latitudes), where the winter sun is very low in the sky and strikes vertical walls directly. An added advantage is that vertical collectors are in a position to receive reflected sunlight from the foreground, which can be significant. They also avoid the problem of snow accumulating over the cover plate, although snow drifts can become a problem near the ground. Another solution is to mount collectors like awnings to function as solar shades that control heat gain through walls and windows in the summer. This sort of mounting can significantly increase the cost of the installation.

Designers generally have three broad alternatives for incorporating solar collectors into the design of a building. The solar features of a building can be (1) concealed, (2) integrated into the design of the structure, or (3) highlighted as a prominent design feature. The architect, or builder can choose one of these alternatives and adapt it to his or her own tastes, the client's tastes, and the design parameters of the community. The end product should be a building that is pleasing to look at, compatible with its neighbors, and energy-efficient.

Concealment. One way that a bank of collectors can be screened from view is by the use of parapets, provided that parapets are appropriate for the building's style. Parapets may be most useful in historic districts that have many flat-roofed buildings. It is likely that such an installation would be almost invisible from street level, but parapets will not conceal collectors from the view of people in taller adjacent buildings. The use of parapets as screening devices must be approached carefully so that solar access is not compromised.

On tall buildings with flat roofs, collectors may also be concealed by being located away from any visible edges of the roof so that they are out of the line of sight of people on the ground.

Landscaping can also be used to conceal a collector or to integrate a solar building into its surroundings. Ground-mounted active collectors can be completely concealed by properly placed low shrubs. When solar collectors are placed on the rooftops of low buildings surrounded by high-rises, the roof may be landscaped to create a more pleasing appearance. This is a traditional method of improving the appearance of rooftops covered with a great deal of mechanical equipment, and the technique is just as useful for integrating solar collectors. Although a good idea, rooftop landscaping is not always economically feasible.

Like collectors mounted on a flat roof, ground-mounted collectors must face south and should be tilted on a frame to best capture the sun's rays. They can also be concealed or integrated by being mounted on natural earth mounds (berms) that have been sloped at the proper angle facing south. Berming makes the collectors appear less obtrusive than a mounting frame does. Berms can also be used to obstruct the view of a ground-mounted or wall-mounted col-

lector. The use of berming is most appropriate for large sites with a diverse or naturally rolling terrain. The major disadvantage of ground-mounted collectors is that they must be located close to the building they serve in order to minimize the cost of connecting piping or ductwork.

Integration. There are three basic ways to integrate active collectors into buildings. First, if the collector array will cover only a part of a wall or roof, the collectors can be accented or emphasized as a distinct building feature. Second, when the collector array is large enough to cover an entire roof or wall, the array can be treated as if it were the roof or wall surface and used as an unconventional roofing or facing material. Finally, whether the collector array covers all or part of a roof or wall, it can highlight certain features (like the vertical or horizontal lines of windows), or those features can draw attention away from the collector installation.

Panels laid flush with a sloping roof line are usually the most compatible with the form of a building. If racks must be used to mount the collectors, the smaller the angle between the roof surface and the collector panels, the better the collectors will blend with the roof. When a thin steel mounting rack is not in balance with a solid roof, the triangular ends of the mounting frame may be covered with the same roofing material for a more integral appearance. When the color of the collector and its frame and piping matches the color of the roof, the installation is less obtrusive.

Highlighting. Collector mountings, which tilt collectors to their proper orientation, can be used as a design element and as a highlight for solar features. This strategy involves accentuating the collector mountings and emphasizing the contrast that results from changing the planes of the collectors and mounting surfaces. The accentuated mountings can be a dramatic addition to a building's design.

A collector array placed over the ridge line will usually make the installation dominate the roof. If spaces are left between collector panels, the installation looks busier and more noticeable than if the panels fit snugly together. Some collectors require a small space between them for piping connections. Part of the installation will be insulating piping or ducts leading from the collectors to the thermal storage area. Their visibility is usually minimized.

Figure 6.14 shows one solution to these architectural challenges in incorporating a bank of solar collectors.

Storage. In addition to space for the collectors themselves, space must be provided for storage units, piping, ducts, and all associated equipment. These are usually not

FIGURE 6.14. Hot water for this hotel is heated by the roof-mounted solar collectors. (Photo by Vaughn Bradshaw.)

major architectural problems as long as the appropriate space and access to it from within the structure are provided for. The volume of storage depends upon the collector area and the function intended. For preliminary planning purposes, allow space for a water tank with a capacity of 1½ to 2½ gallons per square foot of collector area (0.05 to 0.10 m³/m²). If rock-bed storage is selected instead, a volume of 0.5 to 1.15 ft³ should be allowed for every square foot of collector (0.15 to 0.35 m³/m²). The volume required for heat of fusion storage systems is less than that for water. The most common location for thermal storage systems is in a basement.

Provisions for Future Retrofit. In the event that solar energy is not cost effective at the time of building construction, the client would be well served if the building were designed to accommodate a solar retrofit at a later date. There is little doubt that the cost of fossil fuels will eventually increase to the point where solar energy will be a very economical alternative.

CENTRAL COOLING SOURCES

Strictly speaking, the expression "source of cooling" is not technically correct. The process of cooling is actually the removal of heat. However, for our purposes here, it will be useful to think of cooling as being added to a space as an analogy to adding heat to a space.

Unlike space heating, which can take any number of forms using radiation and convection of heat into the space, comfort cooling is almost always accomplished by cooling air and then distributing the air into the space, where it mixes with the room air and cools down the entire volume. The air can either be cooled centrally and then ducted throughout the building, or it can be cooled by incremental units in or near the space and then discharged directly into the space.

The four principal cooling processes are (1) vapor compression refrigeration, (2) absorption refrigeration, (3) evaporative cooling, and (4) natural ventilation. Another source of cooling will only be mentioned here in passing because it doesn't involve any complicated process: using groundwater or a direct heat exchange with ground that is at lower than room temperature. Ground source cooling will be included along with the other cooling sources in the next section, which deals with transporting the heat and cooling to their ultimate destination.

Natural ventilation is simply the use of cool breezes from outside when available. Some techniques for taking optimum advantage of natural ventilation will be discussed in Chapter 7. This section will deal only with the first three processes above, which are mechanical or active processes for generating a cooling effect.

The vapor compression and absorption processes are categorized below in terms of the type of equipment commercially available. Cooling processes almost always involve circulating air through a machine that cools the air down and blows it with a fan back into the space to be conditioned. The fluid that imparts the cooling effect to the air is either a refrigerant—which changes from a liquid to a gas—or water. When the fluid is a refrigerant, the central cooling generation equipment is referred to as DX, and when the fluid is water, it is referred to as a *chilled water system,* and the cooling generation machine is called a *water chiller.*

Vapor compression refrigeration
DX (refrigerant)
Water chiller
    Reciprocating (small)
    Centrifugal (large)
    Rotary

Absorption refrigeration
Water chiller (all sizes)

The capacity of a refrigeration machine can be measured in one of several units, including Btuh or kW. A common unit in the United States is the *ton of refrigeration,* which is equal to 12,000 Btu/hr of heat removal. The term is a carry-over from the days when ice was used for cooling, and represents the cooling effect of 1 ton of ice in a 24-hour period. It is derived as follows:

$$\text{Latent heat of fusion of ice} = 144\ \text{Btu/lb}$$
$$1\ \text{ton of ice} = 2{,}000\ \text{lb}$$
$$1\ \text{day} = 24\ \text{hr}$$

$$144\ \frac{\text{Btu}}{\text{lb}} \times \frac{2{,}000\ \text{lb}}{24\ \text{hr}} = 12{,}000\ \frac{\text{Btu}}{\text{hr}}$$

If the capacity of cooling needed is known in British thermal units per hour, the number of tons may be found by dividing the British thermal units per hour by 12,000.

Vapor-Compression Refrigeration Equipment

The *vapor-compression cycle* is a method for transferring heat from one location to another. This process can be thought of as a "heat pump" similar to a water pump which by the use of mechanical energy overcomes the natural gravitational tendency of water to flow downhill, forcing it up instead. By using a certain "trick" called a *compressor,* mechanical energy can similarly be used to force heat to flow from a lower temperature region to a higher temperature region against its natural tendency.

The concept of pumping heat is basic to the air-conditioning process, and the vapor-compression cycle used for comfort conditioning is also the refrigeration technique commonly used in the processing, packaging, and storing of food. When applied to cooling food and comfort conditioning, it is referred to as *refrigeration.* A machine that uses the same cycle for space heating, called a heat pump, will be discussed later.

The vapor-compression cycle heavily makes use of devices known as *heat exchangers.* These devices do exactly as their name implies, and they may take many forms. The heating water converter introduced earlier is an example of one such device that exchanges heat between steam and water. Heat exchangers in their various forms are the basis of nearly all HVAC systems.

The vapor-compression refrigeration cycle is illustrated diagramatically in Figure 6.15. Its four basic components are:

- Evap orator (heat exchanger to absorb heat from the space)
- Compressor
- Condenser (heat exchanger to reject heat to the outside)
- Expansion valve

A refrigerant fluid is circulated around the loop shown in Figure 6.15. The evaporator transfers heat from the space to the refrigerant, causing the liquid refrigerant to *evaporate* into a gaseous state.

The compressor *compresses* the refrigerant gas, causing it to become much warmer than the outside air. The compressor, which is the only component that uses energy (electricity), is the heart of the system. The refrigerant enters the compressor on the "suction side." After it leaves the compressor, the refrigerant is referred to as *hot gas.*

The condenser transfers the heat to the outdoors, thereby *condensing* the compressed refrigerant back into a liquid for use in the next cycle. The piping carrying the refrigerant away from the condenser is called the *liquid line.*

The final step is the *expansion* of the refrigerant in an

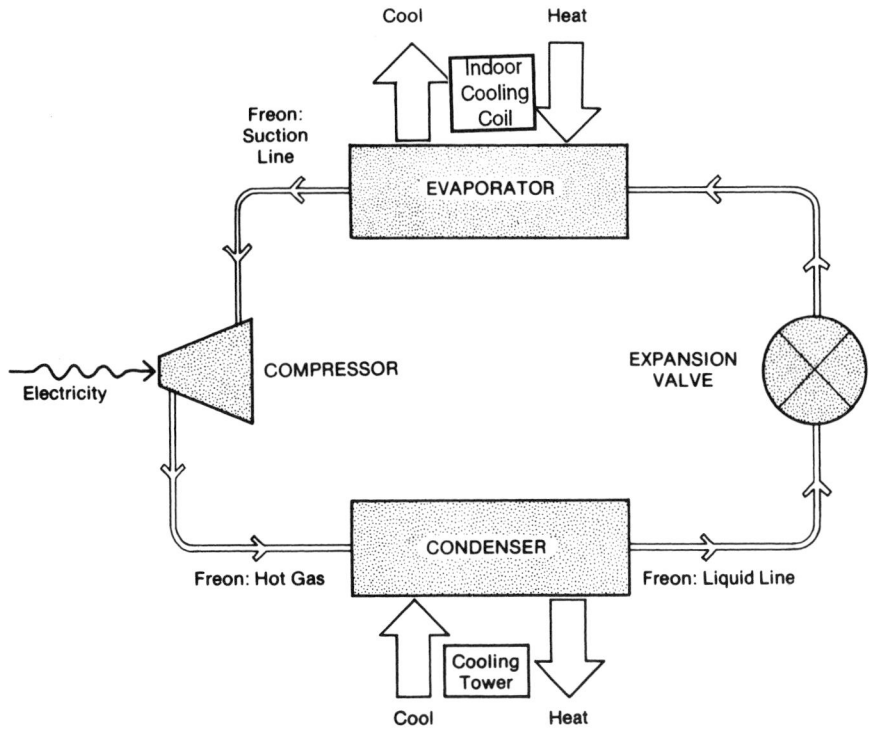

FIGURE 6.15. Vapor-compression refrigeration cycle.

expansion valve. This relieves the pressure built up by the compressor and, in so doing, reduces the temperature even further so that the refrigerant can absorb more heat from the interior space when it re-enters the evaporator.

The *refrigerant* is the working fluid of the refrigeration system. The suitability of any fluid for use as a refrigerant depends on a number of factors. A refrigerant should be non-flammable, not toxic to people, not irritating to eyes, nose, lungs, or skin, relatively inexpensive, and easy to detect in case of leaks. Many different fluids can be and have been used as refrigerants, including chloroform, ether, air, ammonia, methyl chloride, carbon dioxide, sulfur dioxide, butane, and propane. None of these fluids, however, have *all* of the characteristics desired. So a series of chemical compounds were developed especially for use in refrigeration.

One major architectural concern with the vapor-compression cycle is the considerable noise and vibration that a compressor creates when operating. For this reason, it is preferable to locate compressors outside the occupied space and as far from critical acoustical spaces as possible. The machine must also be securely anchored to the floor so that it cannot "walk" and place strains on the connecting piping by its vibration. To prevent the vibrations from being transmitted to the structure and causing damage and annoy-

ance, a spring or resilient rubber vibration isolator is always used.

When warm, moist room air is cooled by an evaporator, it often drops below its dew point, with the result that water—called *condensate*—is deposited on the heat transfer surface. This *dehumidifying* process requires that a drip pan be placed under the heat exchanger and that a drain line be installed to carry away the condensate.

When interior air is cooled directly by passing it over an evaporator in which the refrigerant is expanding from a liquid to a gas, the process is known as *direct expansion* (DX). One limitation of a DX system is that a remote condenser cannot be located farther than 200 feet (60 m) away from the evaporator. This limitation confines DX systems to small- and intermediate-tonnage air-conditioning equipment (½ to about 135 tons, or 1.75 to 475 kW) and requires proximity to a rooftop or an outside mounting location for the condenser. For higher tonnages and more spread-out loads, multiple DX units can be used.

An alternative is to employ a *chilled water system*. In this system, the entire refrigeration cycle occurs within a single piece of equipment known as a *chiller*. The chiller can be located just about anywhere, indoors or out, regardless of where the load is. Water is used to bring the heat from the space to the evaporator section of the chiller, and water is also often used to carry the rejected heat from the condenser to the outdoors. Another way to think about it is that water carries the cooling effect from the chiller (source) to the space to be cooled.

In large buildings, where it is not practical to connect ductwork to all the spaces needing cooling, it may be most economical to have one central refrigeration machine that can distribute its cooling effect through a network of piping rather than having a lot of small refrigeration machines distributed around the building. In theory, it is possible to connect several evaporator units to a single compressor-condenser unit, but in practice, complications arise if the distance between the evaporators is more than a few feet. In addition, there are limitations to the length refrigeration piping can be run. In such circumstances, the use of a chilled water system is most practical.

Chillers

An electrically driven water chiller uses the same vapor-compression refrigeration cycle as a DX system. It has five connecting points for inputs and outputs:

- *Electricity* input to operate the compressor
- *Chilled water return* from the space-cooling equipment
- *Chilled water supply* to the space-cooling equipment
- *Condenser water supply* from the outdoor heat-rejecting equipment
- *Condenser water return* to the outdoor heat-rejecting equipment

A water chiller is a factory-designed, prefabricated assembly (not necessarily shipped as a single package) containing one or more compressors, condenser, water cooler (evaporator), and interconnections and accessories. Cooling water carrying heat from the spaces returns to the evaporator at typically about 50° to 60°F (10° to 15°C). There, heat is removed and the chilled water is released at a lower temperature (usually 40° to 50°F, or 4° to 10°C). The chilled water is then supplied to the various cooling units distributed throughout the building.

The condenser meanwhile rejects the heat to the outside. The condenser may be either air-cooled, in which case it is remote from the rest of the chiller and located outside for direct contact with the outside air, or it may be water-cooled, in which case it is part of the chiller package and the heat is transferred to an outdoor cooling tower via a condenser water circuit. Small chillers with air-cooled condensers are sometimes packaged together for outdoor mounting.

Both indoor and outdoor versions of water chillers, depicted in Figure 6.16, are commercially available. The

evaporator on indoor units is usually a cylindrical shell-and-tube heat exchanger with the tubes within the cylinder carrying the refrigerant and water surrounding the tubes. An indoor water-cooled condenser is typically also a cylindrical shell-and-tube heat exchanger with the refrigerant flowing through the shell in which the water-carrying tubes are located.

Chillers are categorized as follows:

	Capacity
Reciprocating chillers	Up to 200 tons
	Up to 700 kW
Centrifugal chillers	Above 60 tons
	Above 210 kW
Rotary chillers	50 to 400 tons
	175 to 1400 kW
Absorption chillers	Above 60 tons
	Above 210 kW

In general, water-cooled chilled-water systems are available in larger capacities than air-cooled units.

There should always be a floor drain located near a water-cooled condenser in case of leaks and for convenient disposal of water when the system is drained for shutdown or repair. There should be sufficient area around a chiller to permit the installation, maintenance, and service personnel to work without restriction. A space of about the same length as the condenser shell must also be provided on one end of a water-cooled condenser for "rodding" (cleaning) or replacing the condenser tubes. If enough floor area cannot be provided, double doors, a louver, or a removable wall panel at the end of the chiller can allow for the occasional cleaning of the condenser tubes.

Absorption Chillers

The *absorption chiller* used for air conditioning is the lithium bromide–water cycle. Its units are factory designed, prefabricated assemblies (although not necessarily shipped as a single package). Its external connections are similar to those of the vapor-compression chillers discussed previously, but instead of electricity it is connected to a source of heat, and its internal functioning is very different. It does not use a compressor, but instead uses thermal energy—low-pressure steam, hot water, or other hot liquids—to produce the cooling effect. Direct gas-fired units are also available in smaller sizes. Although it uses thermal energy instead of electricity, it produces chilled water and discharges hot condensing water to an outside cooling tower just like the vapor compression unit. And though the cooling effect is accom-

(a)

(b)

FIGURE 6.16. (a) Indoor packaged centrifugal chiller. (b) Outdoor air-cooled electric chiller. (Images courtesy of Trane Commercial Systems.)

FIGURE 6.17. Indoor absorption chiller. (Image courtesy of Trane Commercial Systems.)

plished without electricity, electric power is still needed for pumping the chilled and condenser water. An absorption chiller is shown in Figure 6.17.

The advantages of an absorption chiller are that it is compact, has fewer moving parts, is more trouble-free, generates less vibration, and is quieter in operation. It can also make productive use of waste heat or other free or inexpensive sources of thermal energy.

Small absorption units have been used in conjunction with high-temperature solar collectors as the heat source to produce solar cooling. These installations have proven that the technology for solar cooling is already fully developed with commercially available equipment. For the most part, however, the cost for the equipment cannot be justified at the present prices of competing fossil fuel–derived electricity. As energy costs continue to rise, however, solar absorption cooling may become an economically competitive alternative.

Many installations of cogeneration systems (see Chapter 10) include absorption chillers to make use of steam exhaust from turbine generators in the summer. In the past, when fossil fuels were orders of magnitude less expensive than electricity, these systems were very economical.

Today a careful engineering analysis is required to determine whether such cogeneration/absorption chiller systems are cost effective. In most cases, they are theoretically advantageous, but a close look at actual equipment operating characteristics and maintenance costs determines whether in fact they are. Often, the thermal and electrical loads are not well enough matched, or controls are not precise enough, and the entire system is not as flexible as it needs to be in order to save money.

The ideal applications for absorption chillers are situations in which true waste steam or waste heat is available, such as with industrial processing or central electricity generating stations. Another potential application of absorption chillers is using geothermal energy. There are few such applications, but one very successful example is in the Rotorua International Hotel at Rotorua, New Zealand.

A variation of the absorption chiller is a unit that employs a compressive refrigeration machine with its compressor driven by a steam turbine. The exhaust steam is used as the heat source for an auxiliary absorption cycle. These two processes make an efficient combination. However, as with the absorption chiller alone, caution must be exercised not to apply such systems indiscriminately. An economic study comparing the relative costs of steam or other heat source versus purchased electricity must be made independently for each project.

Heat Pumps

Heat pumps are simply machines utilizing the vapor-compression refrigeration cycle the same way that a DX unit or chiller does. The difference is that a heat pump has valves that can reverse the direction of heat flow, turning the evaporator into a condenser and vice versa. In the cooling cycle, the heat pump operates like any piece of cooling equipment, absorbing heat from the inside air of the conditioned space and pumping it outdoors, where it is released into the atmosphere. It can thus supply heating or cooling to an interior space by either absorbing heat from or rejecting heat to the outside.

The heat pump derives its name from its ability to transfer heat against its natural direction. Heat normally flows from warmer to cooler areas just as water flows from higher to lower elevations. The driving force for heat flow may be regarded as analogous to the effect of gravity on water. A heat pump may be thought of then as analogous to a water pump.

While it may seem paradoxical that heat can be obtained from air when ambient temperatures are nearly freezing, remember that air always contains some thermal energy as long as it is above absolute zero. The higher the temperature, the more energy there is to be extracted from the air. That is why heat pump output drops as the outside temperature gets colder.

Depending on the outside temperature and the efficiency of a particular unit, a heat pump delivers from 1½ to 3⅓ units of heat for every unit of electricity it uses. It thus can save 30 to 60 percent on electric heating bills, depending on the geographic location and equipment used.

The measurement of efficiency of a heat pump is the *coefficient of performance* (COP). It is the ratio of heat output to electricity input using the same units (Btuh or kW). The annual average COP of a heat pump is called its *seasonal performance factor* (SPF). This is the total heat output of the heat pump divided by its total power consumption for one entire heating season.

The COP of a heat pump is generally greater than 1, while the COP of electric resistance heating is exactly 1, or even a little less if minor transmission line losses are included. For this reason, heat pumps are typically more economical to operate than electric resistance heating. But like electric resistance heating, heat pumps use only electricity, so no

provisions are necessary for outside combustion air and flue stacks.

Four types of heat pumps are commonly available:

- Air-to-air
- Water-to-air
- Water-to-water
- Air-to-water

The air-to-air heat pump is the most common type. It uses *air* as the outdoor source for heating or cooling and delivers the heat or cooling to *air* indoors.

Since the outside air is the source of heat for an air-source (air-to-air) heat pump, the heat output and COP decline as the weather gets colder. Due to equipment limitations, the heat pump cycle cannot operate below freezing temperatures. In order to allow heating service to continue in sub-freezing weather, air-source heat pumps are normally equipped with electric resistance heating elements. Thus, below freezing, the COP is exactly 1. There are some heat pumps available in Europe that have supplemental heating fired by fossil fuel instead of by electricity. These are referred to as *hybrid* or *bivalent*.

As an illustration, Table 6.4 contains sample manufacturers' data on heat pump heating capacities and efficiencies at various outside ambient temperatures.

During cold weather, the heat pump's capacity drops while the load increases. In most cases, supplementary heat will be required, especially at subfreezing temperatures. The expense of electric auxiliary heating makes air-source heat pumps undesirable in cold northern climates. They are most beneficial in southern climates with mild winters and a balance between heating and cooling loads that is compatible with the equipment capacities, or where electricity is the only option for heating.

A water-to-air heat pump uses *water* as the source for heating or cooling and delivers the heat or cooling to *air* in

TABLE 6.4 TYPICAL HEAT PUMP PERFORMANCE

Ambient Temperature		Heat Pump Capacity		
°F	°C	Btuh	kW	COP
65	18	60,000	18	2.5
55	13	35,000	10	2.1
45	7	20,000	6	1.6
35	2	5,000	1.5	1.1
25	−4	0	0	1.0

the conditioned space. Where groundwater of adequate quality is available, it is a particularly attractive source because of its relatively high and nearly constant temperature: 50°F (10°C) in northern areas and 60°F (16°C) or higher in the south. Even where well water or groundwater is abundant, it may cause corrosion in heat exchangers or induce scale formation, which makes it unsuitable unless expensive water treatment is employed. Other considerations are the cost of drilling, piping, and pumping the water. Industrial process wastewater, such as warm water from laundries, may be a source for heat pump operations if the water quality is such that it wouldn't cause mechanical problems.

Another application for water-source heat pumps is with a closed piping loop (as shown in Figure 6.18), in which the heat rejected by one heat pump in the cooling mode is circulated to another heat pump in the heating mode. In some cases and in some moderate seasons, the system can satisfy all the building's loads by pumping heat from one area to another without having to add any supplementary heat at all. However, more commonly, excess heat is rejected by an outdoor cooling tower, and supplementary heat is added to the entire loop by a central boiler as needed. For maximum heat pump efficiency (COP up to 3.3), the water in the loop is maintained in the 60° to 90°F (15° to 32°C) range, which

allows low-cost plastic piping to be used and does not require insulation of the piping.

An option for this type of system is to use the water loop for supplying water to sprinklers for fire protection. In the event of a fire, the circulating pump is stopped, and water is allowed to flow through the same pipes to the sprinkler outlets in a conventional manner. This double duty for the piping reduces the cost of installing both systems and eliminates the need for periodic testing of the sprinkler lines. This system has been accepted by many code-enforcement authorities, and in many cases, it is the only way that the high cost of closed-loop heat pump piping can be justified.

Water-to-water heat pumps are used in lieu of a combination boiler and chiller. Some dairy operations have used heat pumps to simultaneously chill the fresh milk and heat water for all necessary cleaning and sterilizing operations.

Air-to-water heat pumps are sometimes used to cool and dehumidify an interior space, pumping the heat into water for useful applications. Restaurants can cool the normally hot cooking area in this way while simultaneously producing hot water for food preparation and dish washing. Other cost-effective applications of this heat pump water heating technology in recent years have been indoor pools, athletic facilities, small-scale industrial processing operations, motels, hotels, and apartment buildings. In general, any facility with a high demand for hot water at the same time as a need for air cooling and dehumidification would be an appropriate application.

Because of the reversibility feature, heat pumps are slightly less efficient in the cooling mode than standard air conditioners. The purchase price of a heat pump is also slightly higher than that of a cooling-only unit with the same cooling capacity. Installation costs are about the same.

From the point of view of energy, heat pumps can sometimes be advantageous. For heating, a typical gas heating unit is between 60 and 80 percent efficient (20 to 40 percent of the energy contained within the fuel is lost up the flue). With electric resistance heating, every kilowatt-hour purchased is turned into useful heat. Heat pumps can put out 2½ or more units of heat for every unit of electricity input (250 percent efficiency).

In comparison with electric resistance heating, heat pumps are always more efficient to operate over the long run. Thus, energy costs are lower. On the other hand, heat pump equipment, being more elaborate and having moving parts, is more expensive to purchase, install, and maintain. All costs over the equipment life cycle should be carefully evaluated before selecting a heat pump over an electric resistance system.

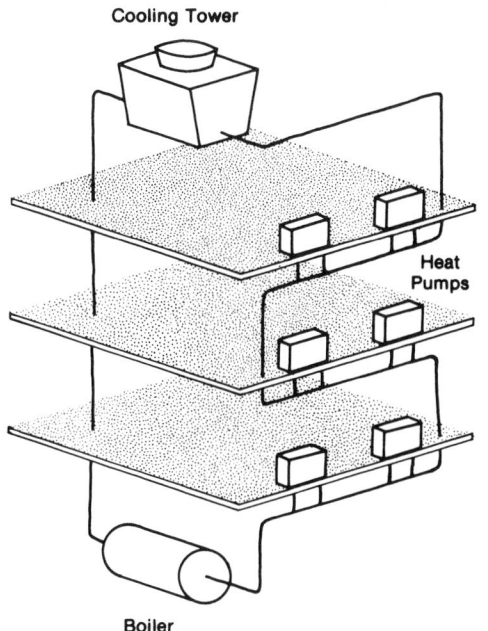

FIGURE 6.18. Closed-loop heat pumps.

Heat pumps should also be compared carefully with combustion heating equipment. Superficially, the heat pump appears to have a seasonal efficiency of well over 100 percent, while a fossil fuel heating system's seasonal efficiency might be 40 to 60 percent. A closer study of raw source energy, however, reveals that every unit of electricity generated from fossil fuels by a utility requires at least 3 units of fuel energy. So, on the basis of raw source energy, a heat pump may have the same efficiency as fuel-fired equipment or a little less. The raw source energy cost is usually reflected in utility rates, so a heat pump may or may not be more economical to run. Like electric resistance heating, heat pumps do not require exhaust flues or expensive fireproof enclosures, because the heating is achieved without combustion. A close scrutiny of energy, maintenance, and first cost factors is required to evaluate the cost effectiveness of heat pumps for any given installation.

Internal Combustion Engine Compressors

For flexibility, refrigeration compressors are available in a variety of drives: (1) electric motors, (2) gas and diesel engines, or (3) gas and steam turbines. These can be used in either water chillers or heat pumps. Engine-driven and turbine-driven compressors are particularly well suited for heat pump applications where the exhaust heat can be recovered and utilized for augmenting the heat pump output. Recovered exhaust heat can also be used for heating domestic hot water, operating an absorption chiller, or any other process requiring moderately high temperature heat at whatever rate is available.

Engine-driven compressors become less efficient at partial loads, and if operated below 20 percent of the full load for a substantial period of time (more than 3 or 4 hours at a stretch), maintenance problems on the engines markedly increase. Such engine drives for compressors are also larger and heavier than the standard electric motors.

Condensers

A condenser is the heat exchanger in a refrigeration cycle used to discharge heat to the outside environment. The three commonly used types of condensers may be classified as:

1. Water-cooled
2. Air-cooled
3. Evaporative

Some characteristics of each are compared in Table 6.5. Air-cooled condensers are widely used on smaller refrigeration systems, while medium-sized installations may use either air-cooled or water-cooled condensers, and most large-scale ones are restricted to water-cooled systems. Water-cooled systems generally yield higher efficiencies than air-cooled ones, but the additional initial cost of installation and greater maintenance requirements of water-cooled condensing systems are disadvantages on smaller installations.

Water-Cooled Condensers

In the early days of air conditioning, city water was passed through the condenser once to pick up the reject heat, and then it was discarded. Because of the burden this placed on city water supplies and on sewer systems, these once-through systems are now prohibited in most large municipalities. As a result, the condenser water is normally recirculated through a *cooling tower*, through which the heat is rejected into the atmosphere. Except in rare cases where once-through wastewater condensers are permitted and are cost effective, water-cooled condensers are always connected to a cooling tower.

In the cooling tower, the warm condenser water (95°F, 35°C) is sprayed into the air and allowed to trickle down over wood, steel, or plastic slats. Outdoor air is drawn through the tower over the trickling water and is then discharged into the atmosphere. The intimate mixing of air and water results in a portion of the condenser water evaporating. Each pound of water evaporated removes about 1,000 Btu of heat from the condenser water because of the latent heat of vaporization. The evaporative cooling effect decreases the water temperature about 10°F (5.5°C).

When outside ambient conditions are cold enough in the winter, an arrangement known as a *water-side economizer cycle* can use condenser water from a cooling tower to provide air conditioning without using the chiller. This can save a considerable amount of energy. Since cooling tower water is not clean enough to circulate directly through cooling coils, a flat-plate heat exchanger (Figure 6.19) transfers the heat absorbed from the cooling coils to the isolated cooling tower water loop.

Cooling towers are built in many different models, materials, and sizes. They are made to match the largest centrifugal or absorption machine built. In the larger sizes, redwood is the preferred construction material.

Cooling towers require considerable maintenance. The basin must be flushed out monthly to keep wind-borne dirt from clogging the condenser tubes.

Since some of the water is lost by evaporation, make-up water must be continually added to the condenser water circuit. Because of the large quantity of make-up water, chemical treatment must be added to inhibit corrosive action upon metals in contact with the water, as well as to prevent algae growth.

TABLE 6.5 CONDENSER TYPES

Type	Location	Capacity	Relative Costs		
			Initial	Energy	Maintenance
Water-cooled	Indoors as part of chiller package with cooling tower outdoors	All sizes	High	Medium	High
Evaporative-cooled	Outdoors	All sizes	Medium	Low	Medium
Air-cooled	Usually outdoors; can be indoors if intake and exhaust air are ducted to it	Up to 125 tons	Low	High	Low

One of the limitations of cooling towers is that when they are operated during the winter, the quantity of air must be carefully regulated to the point where the spray water will not freeze. An electric heater must sometimes be installed in the cooling tower to prevent freeze-ups in winter operation. If the cooling tower does not need to be operated during the winter, it must be drained of all the water.

The air flow through the cooling tower may be produced by fan power, stack effect, or natural wind currents. Cooling towers are classified as natural-draft, induced-draft, or forced-draft, according to the method of moving air through them.

The natural-draft cooling tower depends on the stack effect and natural breezes to move air at low and variable velocities through open spaces in the tower. Such towers are constructed of cyprus or redwood and are rarely used nowadays unless low initial cost and minimum power use are primary concerns. These cooling towers require a much greater amount of space, but are much quieter than mechanical draft towers.

An example of mechanical-draft (fan-powered) cooling towers is shown in Figure 6.20. Both the induced-draft and forced-draft cooling towers employ a fan—either centrifugal or propeller type—to bring outside air through the tower. The induced-draft tower has the fan mounted at the top and draws air up through the structure as the warm water falls down. The fan in the forced-draft tower pushes the air into and up through the structure, discharging it out the top.

FIGURE 6.19. Flat-plate heat exchanger. (Photo courtesy of Tranter PHE, Inc.)

FIGURE 6.20. Cooling towers. (Image courtesy of Trane Commercial Systems.)

Mechanical-draft cooling towers are also made for indoor installation, in which case they must have ducted air connections to the outdoors.

Water-cooled condensers can also be used with spray cooling ponds. Since they rely on natural breezes, spray ponds require large open spaces to catch the slightest air movement. Spray ponds are rarely used anymore, but occasionally they may be integrated into a large development. They generally require a flat-plate heat exchanger to avoid circulating pond water through the chiller tubes.

Air-Cooled Condensers

An air-cooled condenser is simply the condenser component of the refrigeration cycle placed outside in the form of a bundle of finned copper tubes (called a *coil*) mounted in a steel frame. One or more propeller or centrifugal fans blow large quantities of outdoor air across the coil, which contains the refrigerant. The propeller fan type (Figure 6.21) is generally used for outdoor applications, whereas the centrifugal fan type may be located indoors with ductwork to carry the outside air to and from the unit.

Air-cooled condensers are generally less expensive than water-cooled units of the same capacity. They also do not require draining in the winter and require a minimum of maintenance.

When the compressor is packaged together with the condenser, it is called a *condensing unit,* and the components are enclosed in the same cabinet.

Evaporative Condensers

The *evaporative condenser* is a cross between a cooling tower and an air-cooled condenser. The condenser does not have water circulating to it. Instead, hot gas discharged from the compressor is piped to the condensing coil within the casing of the outdoor evaporative condenser. In this respect, it is like an air-cooled condenser. However, it is similar to a cooling tower in that water is sprayed directly onto the coil, where it encounters an air stream, causing some of the water to evaporate. The water that does not evaporate drips to a sump where it is collected and recirculated by a pump.

The evaporative cooling principle makes the evaporative condenser, like the cooling tower, more effective at rejecting heat to the outdoors. Since the condenser water pump is eliminated, the evaporative condenser uses less energy than a water-cooled condenser with a cooling tower. It is also more efficient than the water-cooled condenser system because there is one less heat exchange step involved.

Evaporative condensers can be placed outdoors, or indoors if ductwork is provided to supply outdoor air and carry away the humid exhaust air from the units. As with a

FIGURE 6.21. Air-cooled condensers. (Image courtesy of Trane Commercial Systems.)

cooling tower, precautions need to be taken to prevent the freezing of spray water in cold weather if the unit is located outdoors. When evaporative condensers are mounted outdoors, they should be located where the direct rays of the sun can never strike them.

Evaporative condensers share with the air-cooled condenser the limitation that the condenser unit cannot be located too far from the compressor. The cost to install them is typically somewhere between the costs for air-cooled and water-cooled condenser systems.

Location

Any outdoor condenser equipment should be located so that there are no obstructions to air circulation above and around the unit. It should also be located so that no hot discharge air from the unit is directed into outside air inlets or toward walkways, shrubbery, building overhangs, or open windows.

It is generally desirable to place chillers and boiler units together in a central location to minimize piping. Refrigeration and boiler plants can be located in the basement, or along with air-handling equipment and cooling towers on the roof. If there is space available cooling towers may also be located on the site adjacent to the building.

Evaporative Cooling

Heat energy is required to transform a liquid into a vapor state. The amount of heat required for this transformation is called the *latent heat of vaporization.* As water is evaporated to vapor, sensible heat is drawn from the air, reducing the air temperature accordingly.

In an evaporative cooling system, a blower draws outdoor air in through grilles, passing it through pads kept moist by recirculated water. The cooled air is delivered directly to the indoor space. The gently moving cool air further cools by evaporating body moisture.

This method of cooling—simpler and less energy-consuming than the refrigeration processes—has been employed in hot, arid parts of the world for many years. Its chief drawback is that it consumes quite a bit of water, an increasingly scarce resource especially in dry, hot areas.

Under the right conditions, the evaporative cooler can create very comfortable interior conditions. For example, with outdoor air at 105°F (41°C) and 10 percent RH, the evaporative cooler can supply air to the indoors at 78°F (26°C) and 50 percent RH, which are normal design conditions for refrigerated cooling. This is accomplished only with power to operate the fan; no electricity is needed to drive a compressor, which is the primary power consumer in a conventional refrigeration cooling system.

Outdoor conditions in about half of the United States are well suited for use of the evaporative cooler. It is also occasionally used in more humid regions. It can provide some degree of temperature reduction in almost any climate, but the accompanying humidity increase may be unacceptable if the temperature is not reduced to comfortable levels.

Under some conditions, the units are more effectively used with the water pump shut off for ventilation only. This is useful for flushing a building with cool night air to pre-cool it for the next day's heat gain, or to set up some air motion to help evaporate body moisture when a building is occupied.

Evaporative coolers bring in 100 percent outside air and do not recycle any air from the space. This means that the air supplied to the space must have grilles or open windows through which to exit from the building. When windows must remain open for proper exhaust of the air, building security must be carefully considered.

DISTRIBUTION MEDIA

Complete air conditioning requires a source of heating, such as a boiler, furnace, or electric resistance coil; a method of cooling and dehumidification, such as the refrigeration cycle; equipment for introducing moisture if needed; filters or air washing devices to clean the air; and fans or blowers for proper air distribution. Fresh air may be introduced on the intake side of the conditioning equipment so that it can be treated before entering the space.

The four means of moving heat to a conditioned space are warm air, hot water, steam, and electricity. Likewise, cooling is delivered to a space by either chilled water or cool air. Electricity delivered to the space can produce cooling using the vapor-compression refrigeration cycle, but the heat still needs to be rejected to the outdoors in some way. When water or steam carries the heat or cooling to the conditioned space, it is referred to as a *piped system.* When warm or cool air is carried through ductwork, it is called a *ducted system.* In large buildings, *combination systems* are often used in which water or steam is piped to *air-handling units* (AHUs) where air is passed over a heat exchanger coil. The heat or cooling is transferred to the air, which in turn is ducted to various conditioned spaces. Distribution systems—whether piping or ductwork—contain a supply and return path.

The benefits of using ducted air for distribution to the spaces include: (1) necessary outside air may be brought into the interior for ventilation; (2) the air may be cleaned and filtered in the process; and (3) the air motion created causes rapid mixing of the air in the space, creating uniform conditions and preventing stagnation and stratification of the room air. When outside air is not essential, the decision to use a ducted system may depend on the clearance above the ceiling and whether the structural columns and beams would obstruct the necessary ductwork.

Ducts, piping, and other conduits carrying necessary resources to and from occupied spaces are normally located within a network of concealed spaces (above ceilings and in wall chases) seen only by the builders and repair people. Exposed piping and ductwork have been tried and can be employed for a particular architectural effect. But concealed services offer a number of advantages:

- Better isolation of noise and vibration from the equipment operation and from flowing air and water
- Less complicated surfaces to clean
- Less care needed in construction (leaks, not looks, are most important), which translates into lower construction costs
- Better architectural control over the appearance of ceiling and wall surfaces

Access to such hidden spaces for maintenance is more difficult, but access panels and doors are available for this purpose, and suspended lay-in-tile ceilings offer almost unlimited access. Contrary to popular belief, visual quality ductwork is more costly to install than a ceiling to hide standard ducts.

Within a building, a group of spaces with sufficiently

similar loads can be controlled by one common thermostat and still maintain comfort conditions. Each such grouping of spaces—separately controlled—is called a *zone*. The comfort conditions in a zone may be modulated with a dedicated circulator pump, fan, motorized valve, duct damper, or even a separate heating/cooling unit.

A single set of equipment with a single zone is the simplest and least expensive to install. This is common in most residential applications. One thermostat is located so that it senses the average building temperature and controls the single air-conditioning system according to that demand. If all the occupied areas do not deviate too far from that average condition, adequate comfort conditioning is accomplished. Generally, the larger the zone, the less likely it is that all occupants will simultaneously be comfortable.

Once the decision is made that a single-zone approach is inappropriate, the building must be carefully divided into appropriate zones. Some considerations that might require separate zones are different densities and schedules of occupancy, different solar exposures on the north and south sides, different solar exposures on the east and west sides at different times of the day, direction of prevailing winds, and internal load patterns. The smaller the zones and the more zones in a building, the more equipment and the more complicated the controls needed, and therefore the more expensive the system.

Normally, the more stable "core" or internal areas of large nonresidential buildings are separately zoned from spaces on the perimeter where loads through the envelope fluctuate with outdoor conditions. The core load is predominantly an internal cooling load with little or no heating required except for outside ventilation air brought in. The distribution of air conditioning to core areas is invariably ducted to enable ventilation with outside air at the same time. Where heating loads occur only at the perimeter, a separate perimeter heating system (baseboard fin tube radiation or fan coil units served by a central boiler, or incremental heating and cooling units) is often installed.

When building envelopes are sufficiently well insulated, the temperatures of interior surfaces do not fluctuate as much with outdoor temperatures, so inside air temperatures and mean radiant temperatures stay more uniform. As a result, the perimeter may be regarded as an extension of the core, and fewer separate zones are needed, so the separate perimeter heating system can be dispensed with.

Buildings housing a number of apartments or condominiums should be zoned with provisions for separately metering the air conditioning going to each residential unit. Making tenants responsible for their own air-conditioning costs encourages energy conservation and relieves a central owner or managing agency of the need to adjust billings for frequent energy cost increases. It is often most economical initially to install separate air-conditioning systems for each unit. This is most often what happens, even though a single central system would be more energy efficient.

Air Systems

The functions of an air-handling system are to heat, cool, humidify, dehumidify, filter, and transport air. It usually supplies conditioned air to a space, returns air from the space to the AHU for reconditioning, exhausts a portion of the return air and replaces it with fresh outside air, and supplies the conditioned mixture of return and fresh air back to the space. Thus, the heating or cooling load occurring in the space is offset, and at the same time the air volume in the room is continuously turned over so that the occupants are always in contact with fresh air.

An air-handling system may contain some or all of the following components:

- AHU
- Fresh air (F.A.) and exhaust air (E.A.) dampers at the building skin
- Ductwork
- Supply air (S.A.) diffusers or registers and return air (R.A.) grilles in the conditioned space
- Duct heaters, terminal boxes, and induction boxes
- Controls

The AHU (see Figure 6.22) is the heart of the system, where the temperature, humidity, and air quality control takes place. It is made up of fan(s) or blower(s), filter(s), heating coil(s), cooling coil(s), and a drain pan, all within an enclosure. An AHU either comes as a package or is constructed on site from the various components. To complete the system, the fan, coils, and accessory modules—such as a mixing box, dampers, filters, ducts, and humidifiers—can simply be bolted together in the desired arrangement.

Fans

Fans are available in two basic types: centrifugal and propeller types. The *centrifugal* fan is the most widely used in indoor air handling because it can move large or small quantities of air very efficiently. It contains a wheel with blades that turns in a scroll-shaped metal housing. Centrifugal force throws air outward from the blade tips, creating the air pressure and motion at the fan outlet.

A *propeller* fan has a propeller that rotates on a shaft at the center line of a ring or plate, moving air along the same axis as the shaft. It is applicable for moving air only against

(a)

(b)

FIGURE 6.22. (a) Typical small AHU. (b) Typical rooftop AHU. (Images courtesy of Trane Commercial Systems.)

very low resistance and is normally used without ductwork. Wide propeller blades run more slowly but more quietly than narrow blades. Propeller fans take up less space than centrifugal fans.

Tubeaxial fans have a propeller mounted in a cylinder for in-line insertion into ductwork. If vanes are added as air guides, the fan is classified as a *vaneaxial* type. Axial flow

fans are not able to push air against resistances as high as those of the centrifugal fan, and generally produce a higher noise level, but they can move large quantities of air against a lower resistance. Their main advantage is that they are light and compact.

Generally, sound levels are low when fans are operated at maximum efficiency.

When ductwork is used to distribute the air, circulation may rely on a single supply fan to both push the air to the conditioned space and draw the air from the space to return it back to the AHU. On large installations, a second fan, called a *return air fan,* is used in addition to the *supply air fan* to overcome the tremendous resistance to airflow through the ductwork.

Filters

Filters are important not only for providing a comfortable and healthful air supply to the occupants, but also for reducing dust deposits on room surfaces. If unfiltered air were delivered to a room, dirt smudges would quickly appear on walls and ceilings, particularly near the air outlets.

Filters and other air cleaning devices are available in four general types for four general purposes: (1) typical commercial filters to remove visible particles of dust, dirt, lint, and soot, (2) electrostatic filters to remove microscopic particles such as smoke and haze, (3) activated charcoal to destroy odors, and (4) ultraviolet lamps or chemicals to kill bacteria.

Both throwaway and cleanable filters are available. Throwaway filters are generally standard on smaller AHUs (less than 10,000 CFM or 4.7 m³/s capacity). Standard commercial grade filters remove about 75 to 85 percent of the particles in the air. High-efficiency filters are used where a high degree of cleanliness is called for, such as in hospitals and laboratories. Filters for large AHUs can cover an entire wall in a room-size air handler plenum.

Heating Coils

Heating and cooling coils are simply heat exchangers between a heating or cooling medium and the air stream. Heating mediums available for heating coils are steam, hot water, or electricity. Steam and hot water coils consist of banks of copper tubing surrounded by sheets of corrugated fins which guide the air toward the tubing to maximize the heat transfer surface in contact with the air.

Steam or hot water, when used, must be produced by a boiler and conveyed to the AHU by piping. Thus, they are economical only for medium- and large-size installations, and become more and more attractive as the number of air-handling units served increases. Direct-fired oil or gas heating sections are more desirable for systems whose size does not justify a boiler.

Electric coils are electric resistance elements. They have the convenience of electricity but also the high cost. They may be applicable in special cases where usage is small or intermittent or where alternatives are not available.

It is undesirable to have air entering an air handler at a temperature much below 55°F (13°C) because it inhibits proper blending of the F.A. and R.A. air streams. If entering outside air is lower than 32°F (0°C), water in the coils may freeze, expanding in the process and causing the tubes to rupture. For this reason a special heating coil, called a *preheat* coil, is sometimes located in the duct close to an outside air intake in order to raise the F.A. temperature to about 55°F (13°C) before it gets to the air handler.

Another special heating coil, known as a *reheat* coil, is sometimes placed downstream from the cooling coil for applications where humidity control is critical, such as in hospitals, laboratories, and some industries. The cooling coil dehumidifies the air to a precise point, and then the reheat coil warms it back up to the necessary temperature.

Cooling Coils

Cooling coils may carry either chilled water or refrigerant fluid. In some AHUs, the same water coil is used to carry both hot and chilled water at different times according to need. In others, separate banks of heating and cooling coils are used. Sometimes a continuous source of cool water, such as a stream or underground spring or well, is available for use in lieu of water from a chiller. This can result in significant energy savings.

A coil carrying refrigerant serves as the evaporator to a DX cycle. Liquid refrigerant from a condenser enters the coil through an expansion valve usually positioned right next to the coil. Upon absorbing heat from the air stream, the refrigerant expands and evaporates into a gas, then flows out to the compressor either in the same space or in a remote location.

Drain Pan and Enclosure

The drain pan that collects condensate from a cooling coil is constructed of heavy galvanized or stainless steel and insulated to prevent the enclosure from sweating. The enclosure is made of heavy-gauge zinc-coated steel panels insulated on the interior with fiberglass which functions as both thermal and acoustical insulation.

Other Accessories

Accessories for AHUs include mixing boxes, special air-cleaning devices, dampers, and humidifiers. A mixing box or mixing plenum combines fresh and recirculated air together before they enter the fan chamber. Humidification can be accomplished by direct injection of steam into the air stream, vaporizing water from a pan by heating it, passing air through a moist porous pad, or by spraying water from a nozzle into the air stream.

Location and Space Requirements

Air-handling equipment is normally located outside the conditioned space—in a basement, rooftop penthouse, or in fan rooms on the occupied floors. These rooms are normally accessible from the corridors.

Fan rooms must be wide enough to permit easy removal of the coils, or else a door to the room should be placed opposite the coil section. Another alternative is to provide a knockout panel in the wall large enough and situated to allow the coil to be removed.

For preliminary layouts, 12 feet (3.7 m) minimum clear height should be provided under the ceiling structure for AHUs in commercial buildings; 14 feet (4.3 m) for industrial buildings. The actual room height required may be 12 to 20 feet, depending on equipment sizes and the complexity of ductwork, piping, and conduit layout.

Openings in the building envelope (normally louvered) must be provided for the air systems to draw in fresh outside air and eliminate fouled return air. These openings may be located near the fan(s) or in some remote location, linked by ductwork. Outdoor air intakes must be sufficiently far from exhaust outlets, combustion equipment discharges, and plumbing vents to prevent contamination of the intake air stream. The separation should be as great as practical, but in no case less than 10 feet (3 m). The bottom of outdoor air intakes should be located no less than 8 feet (2.4 m) above ground level, and exhaust outlets should be at least 10 feet (3 m) above adjoining grade.

Rooftop AHUs with a weatherproof housing, as shown in Figure 6.22b, are also available for locating outside without the expense of a penthouse equipment room. However, exposure to the elements can shorten the life of the unit, and the lack of shelter from wind, rain, sun, snow, and cold temperatures makes maintenance difficult and therefore less likely to occur preventatively.

When serving multiple floors, the vertical duct shaft area is usually 1 to 2 ft² (1 to 2 m²) for every 1,000 ft² (1,000 m²) of floor area served. Fan rooms usually require about 2 to 4 percent of the total floor area served (6 to 8 percent for hospitals) and are located to serve specific zones or levels. One AHU can serve 8 to 20 floors. The fewer floors served by an AHU, the smaller the vertical duct shaft and fan room height requirements.

On the other hand, excessive numbers of expensive AHUs should be avoided. A good balance is required, and locations should allow for a natural flow of ductwork to the spaces served in order to avoid bulky crossovers which waste usable area and volume. Close and early coordination between the architect and engineer is extremely important.

Air-Handling System Applications

The various ways in which AHUs operate are classified as follows:

- Single zone
- Terminal reheat
- Multi-zone
- Dual duct
- Variable air volume (VAV)
- Induction
- Economizer cycle

Single Zone.   The simplest form of air-handling system is the *single-zone* system, indicated diagramatically in Figure 6.23. The AHU supplies conditioned air to the entire building or to a discrete portion of the building that can be controlled as one zone. Only one coil at a time operates—

either heating or cooling—so that all of the air supplied is either hot or cold as called for. The quantity of heating or cooling is controlled either by modulating the supply air temperature or by turning the system on and off.

Because they are limited to a single control zone, single-zone systems are generally used for relatively small areas (5,000 to 15,000 ft² or 450 to 1,400 m²) and small capacities. All other types of systems are capable of handling larger areas because they can supply multiple zones with different conditions simultaneously.

Terminal Reheat System.   A *reheat system* is a modification of a single-zone single-duct system. As illustrated in Figure 6.24, the AHU only contains a cooling coil (unless a preheat coil is needed). Air is ducted to all spaces at the same temperature, as in a single-zone system, but at a temperature low enough to satisfy the temperature and humidity requirements of the zone with the highest cooling load. A terminal box at each zone *reheats* the air as needed. The reheating can be accomplished by a steam or hot water coil, an electric resistance element, or a direct fuel-fired duct furnace if a flue can be accommodated.

Reheat systems are popular for multiple zone applications in which precise humidity control is necessary because

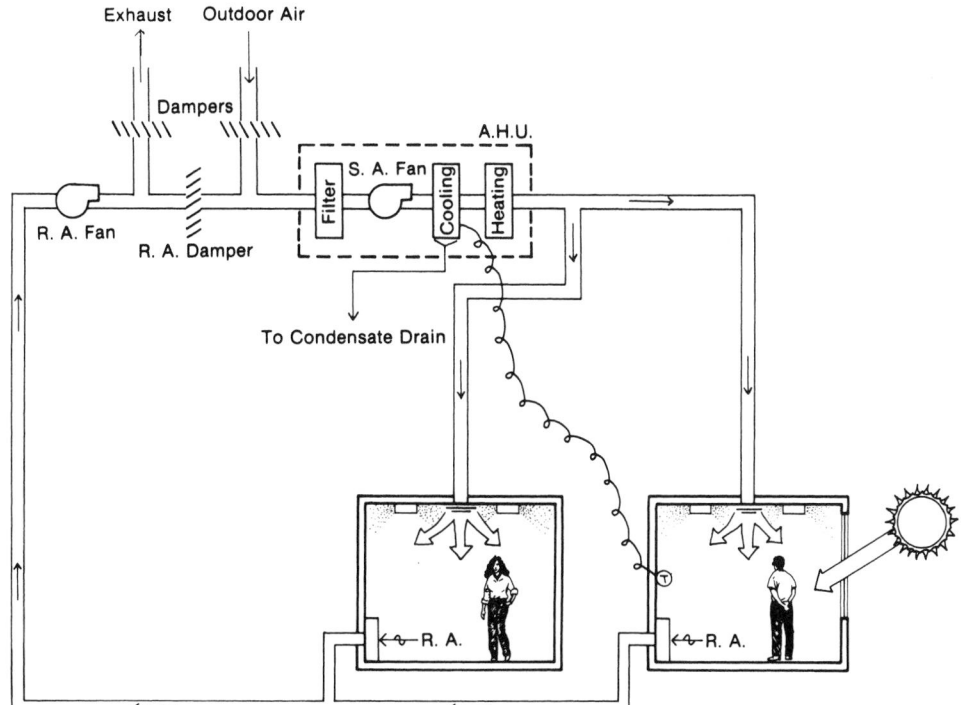

FIGURE 6.23. Single-zone system schematic diagram.

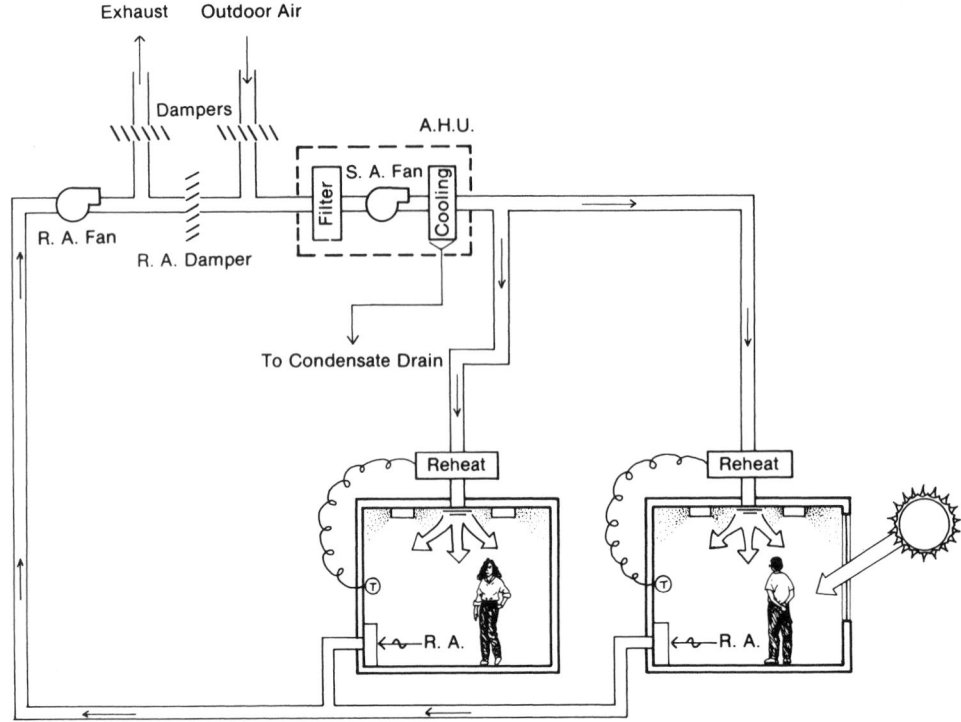

FIGURE 6.24. Terminal reheat system schematic diagram.

of their low initial cost. Unfortunately, they share with multi-zone and dual duct systems the inherent energy inefficiency of using both heating and cooling simultaneously for all spaces that have demands below that of the zone with the peak load. In fact, when reheat-system thermostat settings are raised during the summer with the intention of saving energy, more energy is actually used to reheat the supply air to a higher temperature.

Multi-zone System. Like single-zone and terminal reheat systems, a *multi-zone system* supplies a constant volume of air to each space it serves. Heating and cooling control is accomplished by separately varying the temperature of the air supplied to each zone.

As shown in Figure 6.25, one heating coil and one cooling coil separately heat air in a *hot deck* and cool air in a *cold deck*. The hot deck temperature must be sufficiently high to meet the heating demands of the coldest zone, and the cold deck temperature must be sufficiently low to meet the demands of the hottest zone. All intermediate zones are supplied with a mixture of air from the hot and cold decks.

A branch of the cold deck and a branch of the hot deck blend air together at the AHU in the correct proportion for each zone by modulating motorized dampers to satisfy the respective zone demands. The pairs of dampers move together in tandem—one closing while the other opens—to provide a constant volume of air to each zone. The maximum number of zones that can be accommodated in this way is usually 12.

Since the volume of air to each zone is constant and all the air passes through either the heating coil or the cooling coil, if the zone thermostat is satisfied and no heating or cooling is needed, the dampers position in a 50/50 mixing position to provide equal amounts of hot and cool air. Thus, energy is always being expended whether or not there is a load, which is extremely wasteful of energy. In some cases, the unneeded heating or cooling coil may be shut off in summer or winter, but during much of the time heating may be required in one zone while cooling is required in another, so both coils are needed.

With modern sophisticated controls, the waste can be reduced, but the system is inherently energy inefficient. One energy-conserving modification is the addition of a bypass deck from which air can be mixed with the hot or cold deck to provide temperature modulation. Another alternative is to install individual heating and cooling coils for each zone's supply duct to heat or cool the supply air to meet only that zone load. But damper leakage limits these gains. For these

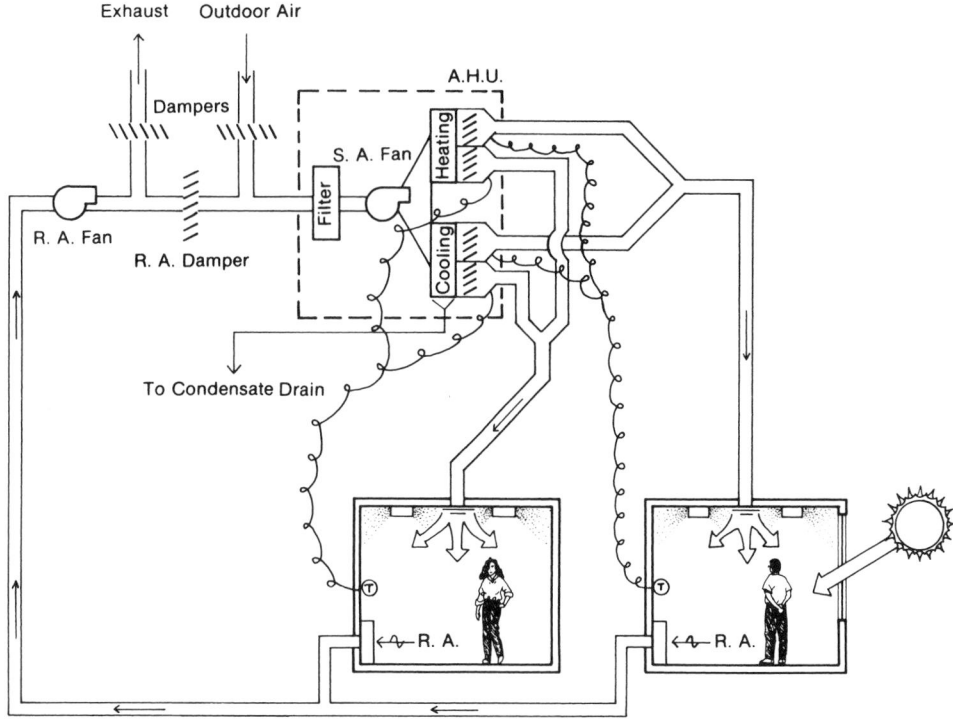

FIGURE 6.25. Multi-zone system schematic diagram.

reasons, multi-zone systems are installed nowadays only under special circumstances.

Dual Duct System. As illustrated in Figure 6.26, the *dual duct system* is similar to the multi-zone concept except that instead of mixing the hot and cold decks at the AHU, a separate hot duct and cold duct are paralleled throughout the distribution network and mixed at zone terminal mixing boxes. The box may include an outlet for releasing the mixed air directly into a conditioned space, or a duct may extend from the terminal to supply several outlets in a common zone with air at the same condition.

In most cases dual ducts are more costly than the multi-zone concept, but with some duct layouts they are more economical. Dual ducts also typically take up considerably more space than single-duct distribution. In order to reduce the size, and thus cost, of ductwork and to lower the floor-to-floor height, dual duct systems frequently employ high-pressure/high-velocity main ducts. The air velocity is reduced at the mixing boxes, and any ductwork downstream is the larger standard low-pressure/low-velocity type. Fan energy requirements are higher in order to create the high-pressure/high-velocity air flow.

Because of the energy inefficiency of using energy twice to both heat and cool air and then mixing the two air streams together, dual duct systems, like multi-zone ones, are not installed very often anymore.

VAV Systems. All of the above types of systems supply air to the conditioned space at a constant rate and modulate the heating or cooling by varying the air temperature. The *VAV system* instead provides air at a constant temperature and varies the air quantity supplied to each zone to match the variation in room load. Air flow can be modulated all the way down to zero when there is no conditioning required, although a minimum is usually set above zero to satisfy fresh air requirements.

Like the terminal reheat system, the VAV system uses a simple single-zone AHU. As shown in Figure 6.27, it cools or heats the air to accommodate the zone with the most extreme requirements and transports the air through a single duct to all zones. A terminal control device called a *VAV air valve* regulates with dampers the volume of air to the space.

The main advantages are low initial cost and lower energy costs than the other system options, since the redundant use of energy for both heating and cooling the same air is avoided. In addition, the system is virtually self-balancing. Although some heating may be done with a VAV system,

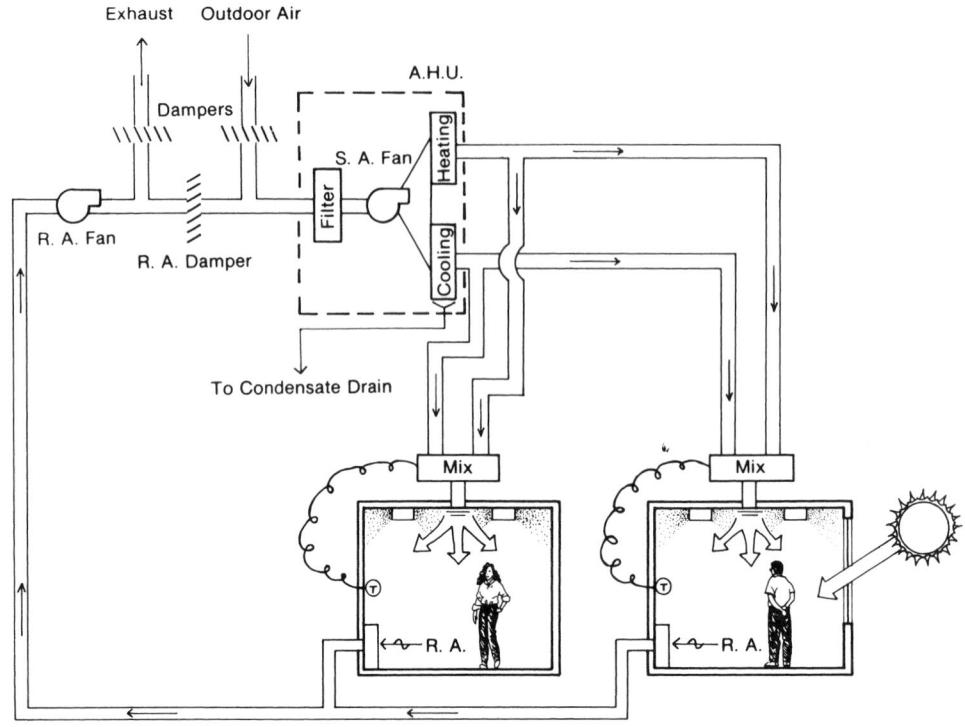

FIGURE 6.26. Dual duct system schematic diagram.

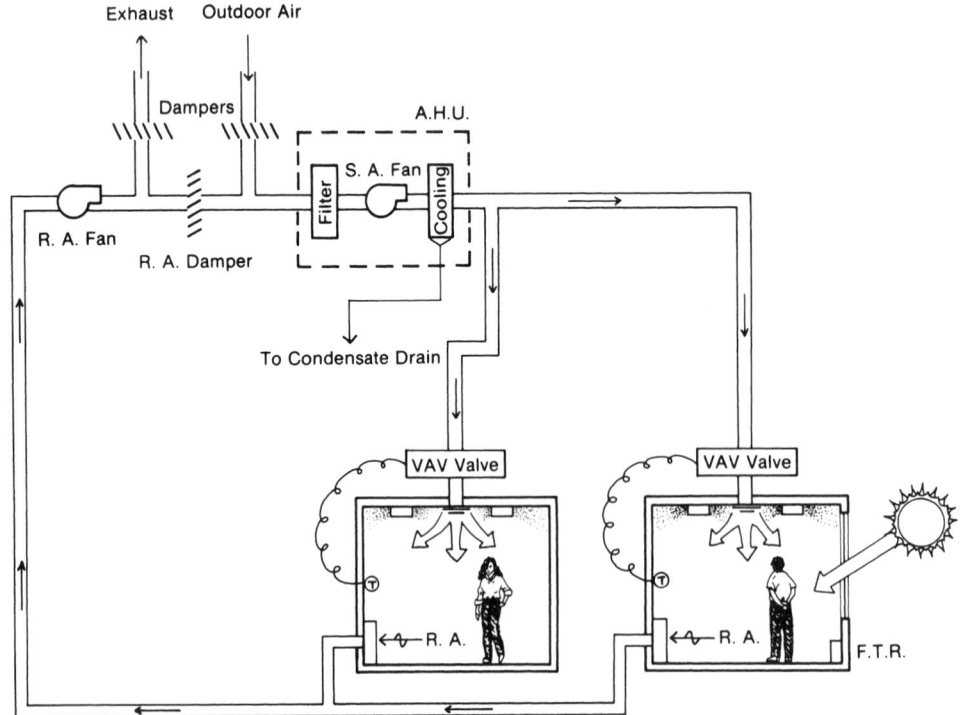

FIGURE 6.27. VAV system schematic diagram.

simultaneous heating and cooling is not possible. It is primarily a cooling-only system. Ideal applications are buildings with internal spaces that have large internal heat gain.

In general, VAV systems are not applicable in spaces with stringent ventilation- or humidity-control requirements, such as hospitals and laboratories. Another limitation of the VAV system is that much less air is discharged from the supply outlets at low-load conditions, causing impaired air distribution patterns within the rooms served. This drawback also makes it unacceptable for the high comfort standards in hospitals.

VAV systems are the most common systems for new institutional and office buildings where precise humidity control is not critical. Many older buildings with less efficient systems can be converted to VAV to reduce energy consumption.

Induction System. An *induction system* employs terminal units at the exterior perimeter of a building. A small amount of outdoor air, representing ventilation requirements, is cooled, heated, humidified, or dehumidified, depending on the season, and then delivered at high velocity through a small circular duct. The induction terminal device is in the form of either a floor-mounted cabinet or overhead outlet. At the terminal, the air velocity and sound level are reduced, and the air is released through nozzles or jets which *induce* room air into the flow pattern. The mixture of fresh and recirculated air then passes over a coil, which either heats or cools it, depending on the room requirements.

Induction systems are rarely installed anymore. The extremely high fan horsepower required is responsible for very high electricity consumption. Also, the expense of running both ductwork and piping as redundant sources of heating and cooling is usually prohibitive.

Economizer Cycle. Outside air can be used to cool a building actively or passively. Natural ventilation, discussed in Chapter 7, is the passive means, and the economizer cycle is its active counterpart.

Whenever the outside air is cooler than the return air from the space, it takes less cooling energy to supply 100 percent outside air. When the outside air temperature is at or below the normal supply-air temperature, no energy at all is required for cooling. This may occur during the fall, winter, and spring of the year, depending on the climate.

An *economizer cycle* is simply a control sequence that adjusts motorized dampers to draw in outside air when advantageous for reducing cooling energy. When it is not advantageous, the controls shut down the dampers to allow a minimum of outside air in—only enough for required ventilation. Any of the above types of air-handling systems except for the induction system can make use of an economizer cycle.

In spaces such as industrial buildings without air conditioning, the only way to remove heat is by "flushing" it out of the building with outside air. The air motion associated with the rapid air change contributes to the feeling of comfort. The necessary ventilation rates can be calculated, but they will typically vary with the seasons. Sometimes evaporative cooling is added for the hottest times of the summer.

Makeup-air units are AHUs designed expressly with the purpose of supplying 100 percent outside air to a space in order to replace air that is removed by exhaust fans within the space or that exfiltrates through the perimeter envelope. These are common in industrial spaces, laboratories, commercial kitchens, and other places where exhaust fans discharge large quantities of room air directly from the space; and usually in corridors of apartments, condominiums, and hotels. Since multiple-residence units are not permitted to be served by a common recirculation air-handling system, the makeup-air unit brings in all fresh air, filters it, tempers it with heat or cooling as needed, and delivers it to the corridor. The air then migrates into the residence units and pressurizes them to prevent infiltration. The air eventually exfiltrates through outside walls.

Classifications of AHUs

Air handlers are available in a wide range of sizes to handle everything from a small room to the simultaneous heating, cooling, and ventilation of several different zones in a large building. They come in a variety of forms suitable for different applications:

- Central system
- Unitary equipment
    Rooftop unit
    Unitary package unit
    Unitary split system
    Self-contained air conditioning unit
    Computer room unit
- Furnace (with or without a cooling coil)

Central System. A *central system* consists of more than one AHU or room delivery device served by the same central sources of heat and/or cooling. It is appropriate for large buildings with multiple zones of at least 5,000 ft^2 (450 m^2); tall multistory buildings; and applications of solar or geothermal utilization, heat recovery, or thermal storage. Hospitals, which require stringent air quality control, use central systems exclusively.

All equipment in central systems except air-cooled chillers, air-cooled condensers, evaporative condensers, and cooling towers are located indoors either within the building or in a rooftop penthouse equipment room. This makes servicing and maintenance more convenient than they would be for unitary equipment mounted outdoors, especially in inclement weather.

Routine daily checking and regularly scheduled maintenance are required for central systems. In some cases this can be handled by the building custodian, while in others a trained mechanic or full-time stationary engineer is required.

Central equipment normally has a life expectancy of at least 20 to 30 years, if properly maintained. Under certain circumstances, central systems are also more energy efficient. However, they are built on-site and typically take longer to install. They can be considerably more expensive if field labor costs are especially high.

Unitary Equipment.   Unitary equipment consists of a factory-assembled AHU and cooling compressor contained within a compact enclosure. It is distinguished from a room air conditioner by its capability of being connected to ductwork. In this category are:

Rooftop unit (RTU). A self-contained cooling unit within a single weatherproof housing; installed directly above the conditioned space. Available in capacities of 1 to 100 tons (3.5 to 350 kW). An example is shown in Figure 6.28.

Unitary package unit. An outdoor packaged unit similar to a rooftop unit except that the supply and return ductwork is intended to pass horizontally through an outside wall.

Split system. A system consisting of (1) an indoor module containing an evaporator coil, expansion valve, provision for heating, and an air handler, and (2) a separate outdoor module containing the relatively noisy compressor and an air-cooled condenser. The compressor/condenser combination is called a *condensing unit.* Available in capacities from 3 to 100 tons (10.5 to 350 kW).

Self-contained air conditioning unit. A packaged indoor unit containing a fan, cooling coil, and compressor. It may include a water-cooled condenser, in which case it needs to be connected to an external cooling tower or fluid cooler. It is also available without the condenser, for connection to a remote air-cooled condenser.

Computer room unit. A highly reliable AHU with extremely precise temperature, humidity, and dust control for sensitive electronic equipment. The high-quality components, reserve capacity, and redundancy of components all contribute to the high price for the necessary reliability. The units are located within or very near the space they serve. The cooling coil is either DX with a remote condenser or condensing unit, or chilled water. Heating can be electric resistance or hot water.

Unitary AHUs are available to handle single-zone, multi-zone, heat pump, makeup air, heating-only, cooling-only, or both heating and cooling functions. They are usually self-contained, with direct fuel-fired or electric heating, and cooling by a DX coil and an air-cooled condenser. The single package units are shipped completely assembled, piped, prewired, and charged with refrigerant—ready to install.

However, there are some unitary packages with coils for connection to a centralized heat or cooling source. Packaged reciprocating and centrifugal water chillers with air-cooled, evaporative, or water-cooled condensers are also available for connection to unitary blower/chilled water coil units. These are made for both indoor and outdoor mounting.

The fact that each unit is self-contained is a useful feature for individual metering of tenant utility services. In larger buildings, multiple units can be installed, each serving a separate portion of the building. Buildings with intermittent occupancy, or diversity of use are particularly well suited to this use of unitary equipment. Multiple single-zone RTUs can be spread out on the roof of a low-rise building for zone by zone control. Multi-zone RTUs can be used to regulate zones that are not too far apart.

FIGURE 6.28. Unitary RTUs. (Image courtesy of Trane Commercial Systems.)

The life expectancy of unitary equipment is only about 10 years, but within that time, little routine maintenance is required. Unitary equipment does not take up valuable floor space, which can amount to 5 to 7 percent of the total building area for central-system mechanical rooms.

Furnace.   Furnaces are a special form of AHU. They are a source of heat, from either a fuel combustion chamber or an electric resistance element, with a self-contained blower and filter. They are primarily used for supplying warm air to a space, but they often include a ventilation function when linked by ductwork to the outside, and they can fully air condition a space if a cooling coil and other accessories are added.

Gas-fired make-up air units, as shown in Figure 6.29, are AHUs that include an internal furnace section. They are typically used to provide large proportions of (or even 100 percent of) outside air to spaces that require it.

Ductwork

The purpose of the duct system is to transport air from the AHU to each outlet in the conditioned spaces—at the specified velocity—and then back to the AHU. The air distribution can be either high, medium, or low velocity. The range of air velocities for each category is:

- High velocity: above 2,500 fpm (above 12.7 m/s)
- Medium velocity: 2,200 to 2,500 fpm (11.2 to 12.7 m/s)
- Low velocity: 1,000 to 2,200 fpm (5.1 to 11.2 m/s)

High-velocity air can be distributed through much smaller ducts, which is a great advantage in tall or very spread-out buildings because it minimizes the bulkiness of ductwork carrying large air volumes. But high-velocity, high-pressure ductwork must be of heavier construction, which makes it more expensive.

Fans in high-velocity systems must be more powerful, and the fast-moving air causes loud noises. The higher-

FIGURE 6.29. Gas-fired make-up air unit. (Image courtesy of Trane Commercial Systems.)

powered fans consume considerably more energy than those in the standard low-velocity systems. This together with the higher installation cost is why high-pressure, high-velocity systems are not commonly installed unless duct space is an overriding concern. Even where air is supplied at high velocity, return-air ductwork is always the larger low-velocity type.

Noise levels are reduced by routing the air through a terminal box called a *sound trap* or *sound attenuator* which performs a function similar to that of an automobile muffler. These units, lined with acoustically absorbent material, reduce the sounds to acceptable levels before the air is discharged into the room. Their other function is to throttle the air from a high to a low velocity for introduction into the space. Terminal boxes are located near the spaces they serve and may distribute air to several outlets.

Return air may be drawn from the room either through ceiling grilles into a plenum between the ceiling and the floor structure above it or by systems of low-velocity ducts that return the air to the AHU. From there it is either recirculated or exhausted. Corridors are sometimes used as return-air plenums, but the need for undercut door thresholds or louvers in doors to permit air flow when doors are closed compromises acoustical privacy. Where this cannot be avoided, special transfer grilles that attenuate sound transmission can be used.

Ducts are constructed of galvanized sheet metal, fiberglass, or, in corrosive environments, aluminum or stainless steel. They are either round, rectangular, or flat oval in cross section. Round ducts are usually used in high-velocity systems, while rectangular ducts are typical for low-velocity systems. Ducts are attached to the building structure with "strap hangers" nailed to supports above.

Ductwork is usually insulated for the following purposes:

1. To reduce heat loss from ducted warm air
2. To reduce heat gain into ducted cold air
3. To prevent condensation on the exterior of cold air ducts when the surrounding air is warm and humid
4. To acoustically isolate air noises in the duct from the occupied space

Fiberglass duct liner is frequently used to accomplish all of the above functions, with the metal duct wall serving as the vapor barrier. Fiberglass ducts are self-insulating and do not need an additional duct liner.

The dimensions given for ducts, as shown on construction drawings, are usually the inside dimensions of the duct. An additional 2 inches should be allowed on each dimension for the duct wall and insulation.

Where coordination is considered early enough in the

design, ducts can be integrated within joist spaces and roof trusses and between bulky recessed lighting fixtures in order to minimize total space requirements. For preliminary allocations of duct space consider the following:

Supply air flow rate = 1 CFM/ft^2 (5 L/s·m^2) of building [based on 8–10 ft (2.43 m) ceilings]

Velocity of air through the duct = 1,000 fpm (5 m/s)

These relationships result in 1 ft^2 (1 m^2) of duct per 1,000 ft^2 (1,000 m^2) of building. The optimum ratio of duct width to duct height (*aspect ratio*) is 4:1. That means that 16,000 ft^2 (1,500 m^2) of space would be served by a supply air duct with a cross section of about 16 ft^2 (1.5 m^2), which would be 8 feet (2.5 m) wide and 2 feet (0.6 m) high.

Dampers are used to regulate air flow in ducts. They are categorized as:

1. *Balancing dampers.* Hand-operated dampers that are locked in position after adjustment to correctly proportion air flow to all outlets.
2. *Motorized control dampers.* Motor-operated dampers that, in response to remote signals, vary air flow as part of an automatic control system.
3. *Splitter dampers and turning vanes.* Air guides within ducts that prevent turbulence and consequent air flow resistance by making air turn smoothly. Splitter dampers are used where branch ducts leave larger trunk ducts. Turning vanes are used when a duct makes an abrupt 90° turn; they are required only where space is not available for sweeping large-radius turns.
4. *Fire dampers.* Dampers that are normally open but that automatically close in the event of a fire to prevent the possible spread of fire and smoke through a duct system. Large commercial structures have fire-resistant partitions, floors, and ceilings that confine a fire within an area for some specified time. When an air duct passes through one of these fire barriers, a fire damper is generally required by fire codes. Access doors are normally installed at fire dampers for servicing and inspection.

Figure 6.30 illustrates two of the many styles of ductwork. Figure 6.30a shows low-velocity round supply ductwork. Air noise is primarily controlled in this television studio by using large enough ducts to keep the velocity very low. The foam rubber insulation serves as a vapor barrier to prevent condensation and helps to further reduce air noise. Its smooth surface also conceals imperfections in the sheet metal installation and eliminates the need to paint the exposed ductwork.

Figure 6.30b is an example of fabric ductwork that is inflated by the internal pressure of the air it conveys. The

(a)

(b)

FIGURE 6.30. (a) Round exposed ductwork at the ceiling of a community access television studio. It is insulated with black, flexible, closed-cell, elastomeric foam rubber insulation and terminates in round diffusers. (Photo by Vaughn Bradshaw.) (b) Round fabric ductwork which evenly discharges air along its length was used in these renovated infield press rooms at the Daytona International Speedway. (Photo by permission of DuctSox Corp.)

material cost for fabric ductwork is normally higher than that of sheet metal, but it is very quick and therefore less costly to install. The net result is sometimes a less costly overall installation. The fabric is also undamaged by some chemical fumes that can corrode galvanized steel, and it can be readily removed and cleaned in a washing machine. Its disadvantages include the fact that it is not suitable for return or exhaust ductwork, for obvious reasons. It has the greatest value in exposed applications where it distributes supply air evenly along its entire length, but its appearance does not appeal to all designers.

The principal agency that sets standards for ductwork and its installation is SMACNA (Sheet Metal and Air Conditioning Contractors National Association).

Water Systems

Air is a bulky medium for conveying heat or cooling. One pound of air is about 13 to 14 ft^3 (0.368 to 0.396 m^3) and carries only 0.24 Btu/°F (0.14 joule/°C) temperature difference. Water, on the other hand, can carry the same amount of heat or cooling in much less volume, so a system of water piping fits much more easily into the building structure than ductwork for an equivalent amount of thermal distribution.

Water as a heat-carrying medium can be used for both heating and cooling. It is heated to 160° to 250°F (70° to 120°C) in a boiler, or cooled to 40° to 50°F (4° to 10°C) in a chiller, and then piped to terminal devices, which might be fin-tube radiation, convectors, or unit heaters for heating only, or to fan coil units or radiant panels for heating or cooling. The terminal device transfers the heat or cooling to the room air. The water may also be transported to AHUs, where the heat or cooling is transferred to an air stream for ducted delivery to the space.

The major advantages of a water system are its flexibility for zoning, its relatively low first cost, and the ease of fitting it into new or existing buildings. The disadvantage is that water cannot distribute fresh air along with the heat and cooling. Thus, the ventilation requirements of a space must be met by infiltration and open windows at the building perimeter, outside air vents at the terminal device, or a separate ventilation duct system. This last method is the most expensive and cancels out the advantages in cost and space savings by going to the water system in the first place.

When the water system is for heating only, it is referred to as a *hydronic system*. These are common in residences and at the perimeter of commercial buildings in which the interior zones are served by a separate cooling-only ducted air system.

Separate systems for handling (1) internal cooling loads and (2) perimeter heating loads result in the best comfort conditions. Cooling is most effective when introduced into a space from above, and heating is most effective when introduced at the base of perimeter walls, preferably below windows. The same outlet cannot perform both functions equally well. When a single combined outlet is provided in a room, it is an economical compromise.

Groundwater may occasionally be available at a low enough temperature for use as the cooling medium, but it usually requires expensive water treatment so that it doesn't corrode or clog the piping system. Normally a chiller is needed.

Distribution piping is classified according to the number of parallel pipelines connecting the system together:

1. *One-pipe systems* use less piping and are the least costly to install. The same water passes through each terminal in series and becomes progressively less effective as it does so. Thermal control is very poor, particularly between different zones.

2. *Two-pipe systems* consist of a separate supply and return pipe so that each terminal can draw from the supply pipe and return to the return pipe as needed. All terminals have access to water of approximately the same temperature. However, all the water must be hot or chilled, and all terminals on the system must be in either a heating or cooling mode at any given time. This is a disadvantage in some cases and at some times of the year when one zone needs heating and another needs cooling.

3. *Three-pipe systems* overcome this drawback by supplying one heating pipe and one cooling pipe to each terminal. The third pipe returns all the water to both the chiller and boiler. Unfortunately, mixing the cooling return water at, for example, 55°F (13°C) with the heating return water at, for example, 140°F (60°C) results in return water to both the boiler and chiller at somewhere around 95°F (35°C). A great deal of energy is wasted in heating the mixed water back up to 140°F (60°C) for the boiler or cooling it back down to 55°F (13°C) for the chiller. This costly energy waste makes the three-pipe system all but obsolete.

4. *Four-pipe systems* are the most expensive but provide year-round independent-zone comfort control without the energy waste of the three-pipe system. Each terminal is provided with a chilled water supply and return and a heating water supply and return.

Control of the flow of water to different zones can be achieved either by separate pumps or by motorized control valves for each zone. Zone control is particularly useful for two-pipe systems, since one zone may need chilled water while another needs heating water. But it can also be used to shut off water flow to unoccupied zones when others are still occupied. If thermal output cannot be controlled at the terminals, zone control can modulate the water temperature or flow rate to an entire zone in response to varying demand.

The piping used to carry chilled or hot water may be copper tubing up to about 4 inches (10 cm) in diameter, or steel or wrought iron for larger systems. For some chilled water lines, PVC plastic or fiberglass piping is permitted. Like ductwork, piping must be supported by hangers fastened to the building structure.

Pipes carrying chilled or hot water are insulated to minimize the transfer of heat to surrounding unconditioned air before it reaches its intended destination. Insulation can increase the diameter of a pipe by 2 to 8 inches (5 to 20 cm). Chilled-water piping, in particular, is insulated and covered with an airtight jacket to prevent moisture in the air from condensing on the pipe surface—which is usually at a temperature below the dew point of the air—and causing water damage where it drips.

Steam Systems

Because steam delivers its heat by releasing its latent heat of vaporization (1,000 Btu/lb, 2,326 joules/g), 1 pound (1 kg) of it can convey about 50 times as much heat as 1 pound (1 kg) of water. But when water is vaporized to steam, its volume increases about 1,600 times, which means that it can carry only one-thirtieth as much heat by volume. As a result, pipes for steam heating systems are larger than those for water systems, though still smaller than air ducts handling equal loads.

When water is boiled to steam, the pressure generated forces the steam through the distribution piping. Thus steam provides its own motive power for distribution, unlike water, which needs a pump.

Since steam travels under its own pressure, its flow rate cannot be controlled with a pump as precisely as water flow. Also, the temperature of water can be modulated, while steam temperatures are not as easily varied.

The inferior control over heat output by steam systems, together with noise problems arising from condensation of steam in the pipes, has led to a trend away from steam and toward hot water distribution in simple space-heating systems. If steam is available for some other purpose, as in hospitals and some industrial buildings, it is usually used for space-heating distribution to AHUs and perimeter terminal devices. It is also still practical for district-heating applications. Metering the condensate from steam can provide an easy way of estimating the amount of heat used, which is difficult in other systems.

When steam releases its latent heat and condenses, the liquid condensate must be returned to the boiler for recycling. The return piping for this condensate must slope downward, unless it is pumped from a receiving tank back to the boiler. Construction voids must be provided to accommodate the sloped pipes, and trenches or tunnels are often necessary under a slab-on-grade foundation to accommodate

steam and condensate piping below terminal devices at the base of perimeter walls. Steam piping is usually made of steel.

The destinations of steam in a space-heating system may be terminal devices similar to those served by hot water. Fin-tube radiation devices have taken the place of the old cast-iron radiators, which are not installed anymore. Other terminal devices suitable for steam service are fan coil units, unit heaters, and infrared heaters. Terminal devices that include a fan can modulate their heat output by controlling the fan. Unlike water systems, which have the option of two- or four-pipe distribution, steam terminal devices that include summer cooling must be served by a pair of steam pipes and a separate pair of chilled-water pipes.

Like hot water, steam can serve as the heating medium in an AHU heating coil. In addition, steam can heat water in a heating-water converter to take advantage of the better thermal control that water offers.

DELIVERY DEVICES

The HVAC terminal is a device that delivers heat or cooling to the conditioned space. A variety of terminal devices are available, and the selection depends on:

1. The functions required (heating only; heating, cooling, dehumidification, and air filtering; or conditioned ventilation)
2. Whether the room is on the building perimeter or not
3. Which energy source is selected
4. The type of distribution system, if any, to which the terminals will be connected

The only devices that are limited exclusively to heating are natural convection devices and unit heaters. Radiation devices are usually only for heating. Combination heating/cooling radiant panels are available but are useful in very limited applications, because they cannot dehumidify. Fan coil units can heat, cool, filter, and dehumidify. The unit ventilator, a variation of the fan coil unit, can also introduce outside air into a space. Packaged terminal air conditioners, including heat pumps, can perform all functions, as can air-handling systems that terminate at air-delivery devices.

Air Inlets and Outlets

The proper delivery of air for heating, cooling, humidification, dehumidification, or ventilation is a crucial part of the ducted air distribution system. Velocities are much greater in the supply ductwork than would be acceptable in the occupied zone of a room. Also, the temperature of the supply air is well above or below that acceptable in the occupied zone [floor level to 6 feet (1.8 m) above the floor, and greater than 1 foot (30 cm) away from the walls]. The delivery system provides a transition in which supply air mixes with room air outside the occupied zone. Suitable air delivery can also counteract natural convection currents, which cause drafts, and uneven radiant temperature effects within the room.

When conditioned air is supplied to a space by ductwork, the duct usually terminates in an outlet device capable of introducing the air into the space in an air pattern that maintains the desired air temperature, humidity, and motion with only minor horizontal or vertical variations. If the air is not properly introduced, the resulting conditions will be unsatisfactory. Any obstructions in the supply air stream can cause uneven, ineffective delivery of air within the conditioned space. Thermostats must not be located in the supply air stream because that would cause erratic operation of the system.

When a jet of air emerges from a supply outlet, surrounding room air is entrained or inducted into the air stream and is replaced by other room air. As a result, room air always moves toward the supply outlet.

This can cause smudge marks around outlets when dirt particles are suspended in the air from heavy pedestrian traffic (lobbies, stores, etc.) and heavy smoking. Beveled mounting frames and other special designs are available for minimizing smudging problems. Outlets on smooth ceilings, such as those made of plaster, show smudging as a narrow band of discoloration around the outlet. On textured ceiling surfaces, smudging occurs over a larger area.

If the supply air is a jet flowing parallel to and within a few inches of a surface, the room air moving toward the air stream pushes it toward the surface. This *surface effect,* illustrated in Figure 6.31, occurs when the angle between the discharge jet and the surface is less than about 40° and the jet is within about 1 foot of the surface. The jet from a floor outlet is drawn to the adjacent wall, while the jet from a ceiling outlet "hugs" the ceiling.

The horizontal distance that an air stream travels from an outlet before dropping is called its *throw.* The throw should be long enough to give the air stream time to mix sufficiently with room air so that its temperature and velocity reach comfortable levels before the air drops down into the occupied zone. Otherwise, uncomfortable drafts would result. If the air stream is warmer than the room air, it will tend not to drop at all.

There are essentially four types of supply outlets available as standard manufactured products. They are:

1. Ceiling diffusers
2. Grilles and registers
3. Perforated ceiling panels
4. Slotted diffusers

Representative examples of these air outlets are presented in Figure 6.32. Dampers that are fixed, adjustable, or automatically controlled are available for regulating the air flow through an outlet.

Ceiling outlets are usually diffusers that spread air out in a horizontal discharge pattern, creating a plane of conditioned air that blankets the ceiling. As the supply air mixes with entrained room air close to the ceiling, its velocity and temperature gradually decrease, until it finally drops to the occupied zone, where it comes into contact with the room occupants. *Ceiling diffusers* come in round, square, and rectangular shapes.

A *grille* is a rectangular opening with fixed vertical or horizontal vanes or louvers through which air passes. A *register* is a grille with a damper directly behind the louvered face to regulate the volume of air flow. Grilles and registers

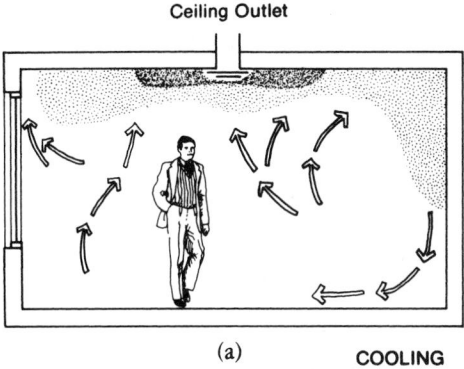

(a)   COOLING

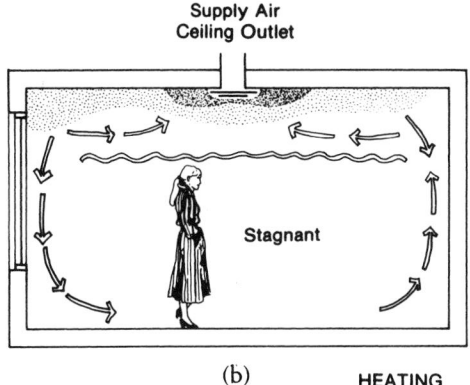

(b)   HEATING

FIGURE 6.31. Air diffusion patterns in a room: (a) summer cooling and (b) winter heating conditions.

(a)

(b)

(c)

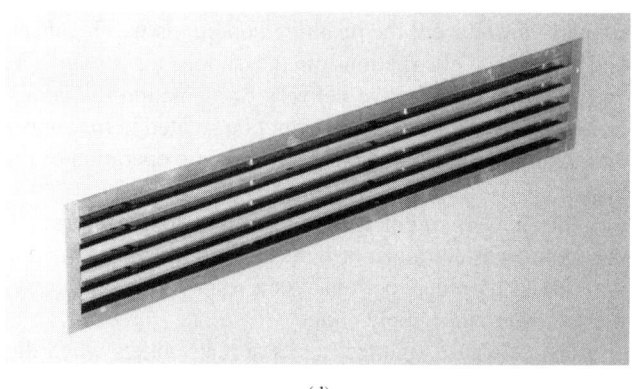

(d)

(e)

FIGURE 6.32. Air outlet styles: (a) square ceiling diffuser, (b) round ceiling diffuser, (c) perforated grille, and (d) linear bar wall grille. (e) Air is distributed in this airport terminal through exposed round ducts with slotted diffusers along the side. (Photos by Vaughn Bradshaw.)

are generally used for wall and floor outlets and discharge air nearly perpendicular to the surface in which they are mounted. Sidewall registers direct streams of supply air up above the conditioned space, parallel to the ceiling if possible. By the time the air falls into the occupied zone, it is mixed with the room air, so it doesn't cause uncomfortable drafts.

When a large proportion of the ceiling may be used as

an outlet, *perforated ceiling panels* provide a well-distributed air supply traveling straight down. Perforated metal face plates can also be placed over standard ceiling diffusers so that the individual air diffusers blend into a perforated ceiling.

Linear *slotted diffusers* are elongated outlets consisting of single or multiple slots usually installed in long continuous lengths. These are sometimes located in the floor at the base of glass doors or floor-length windows or along a sill under standard windows. In addition to collecting dirt, such outlets often become obstructed with books, papers, flower pots, or curtains hanging close to them. As a result, air flow deviates from its intended pattern, and the system does not operate as designed.

Slotted diffusers and other special designs are also available for special ceiling effects. The slotted diffusers are available in two basic flow patterns: perpendicular and parallel. The perpendicular flow type discharges air within 30° of perpendicular to the face of the diffuser. It creates long, narrow bands of conditioned air flowing into the room and is best suited for perimeter heating applications or high sidewall cooling. The parallel flow type discharges the air parallel to the surface of the diffuser and is suited for cooling applications.

No single type of outlet is best for both heating and cooling. The heating outlet location that most effectively provides comfort is near the floor at outside walls—especially below windows—so that a vertical stream of air flows upward to blanket the cold surfaces and counteract perimeter downdrafts of cold air. Adding heat at the area of highest heat loss prevents uneven air temperatures and an uneven radiant temperature distribution. The warm air also mixes with any cold air infiltration, preventing chilly drafts.

Proper cooling requires substantially larger quantities of air than are generally needed for heating. This air must be carefully and evenly diffused in order to provide general air motion in the room and to prevent drafts, cold spots, and warm spots. The best method is to deliver air through multiple outlets strategically located in the ceiling.

When a compromise is required for year-round operation of a single system, the optimum location is determined by the predominant application. An outlet placed above a window to blow air along the ceiling toward the interior induces some air to travel upward across the window to reduce downdrafts. This does not, however, eliminate downdrafts and is not advisable in cold climates unless supplemented by heated floors or a source of heat beneath the window. Ideally, a ducted air system provides cooling and ventilation while a separate perimeter system, such as fin-tube radiation, is used for heating. When this is not possible, it is particularly important that the air-diffusing equipment be carefully selected and located.

Another reason for careful design of the air delivery system is that the rapid movement of air through the outlets can cause considerable noise. The noise level from an outlet varies in proportion to the velocity of the air passing through it. The noise level is normally kept low by proper sizing, selection, and placement of the devices, but sometimes a constant low-level background air noise is desirable to mask other sounds in a space. An air outlet that is too quiet can actually create the impression of a noisier environment if it is not loud enough to drown out other sounds, such as the rustling of papers, soft voices, footsteps, and the moving of chairs. This is a particularly important consideration for libraries, reading rooms, sleeping quarters, and other quiet areas.

Whenever air is supplied to a room, provisions must be made either to return the room air to the air-handling equipment or to exhaust it to the outside. Air is typically drawn from the space through a return inlet or exhaust inlet located apart from the supply outlets so that the air has a chance to permeate the space.

When the return inlet is located too close to the supply outlet or at the same height, short circuiting can occur, which means that the supply air bypasses portions of the space without conditioning it. This can result in pockets of stagnation which are too warm in the summer, too cold in the winter, and stuffy from inadequate air change.

Any type of supply outlet can be used for a return or exhaust inlet. The most popular inlet styles tend to be either louvered (fixed bars or louvers set at angles to break the line of sight into the duct or space behind them), eggcrate (square grid pattern), or perforated. Inlets are normally referred to as grilles or registers.

Return air inlets may be connected to a duct; they may lead to an undivided plenum above the ceiling from which air is drawn back to the air handler; or they may simply transfer air from one area to another. Exhaust is almost always ducted.

Exhaust air inlets are normally located in ceilings or high on walls. Return inlets for cooling systems can be located in ceilings or high on walls too, provided that they do not cause short circuiting. Return air inlets for heating systems must be located near the floor to discourage stratification of warm air near the ceiling and across from the supply outlets to make sure that the air covers the entire room.

When it is necessary to locate heating outlets in the ceiling, the likelihood of poor mixing during the heating season is reduced by a low return, since the cool floor air is withdrawn first and is replaced by the warmer upper air strata. Return inlets actually in the floor are undesirable because they readily collect dirt, which clogs the filters and imposes an undue strain on the air-handling system.

Air flow patterns can be carefully designed for selective conditioning of small occupied areas within larger enclosures. This is especially useful in large, sparsely occupied industrial spaces or in large areas encompassing a wide range of activities.

The importance of proper air delivery cannot be overemphasized. An air-conditioning system may have the very best components, but may still be unsatisfactory if the delivery system is not providing comfort for the occupants. The architect and engineer should work closely together on the selection and placement of outlet and inlet devices so that both the desired architectural effect and proper comfort conditions can be achieved.

Natural Convection Heating Units

The original cast-iron radiators are included in this category even though they also transferred heat by radiation. They are rarely used anymore, replaced now by *finned-tube radiation* devices. Despite their name, finned-tube radiation devices actually transfer heat to the air by convection rather than radiation.

Finned-tube radiation is available in two basic forms: baseboard or convector. The former is shown in Figure 6.33. Enclosures are available in a variety of styles, but the heating element concealed inside is generally the same: aluminum or copper fins 2 to 4 inches (5 to 10 cm) square bonded to copper tubing. Steam or hot water from a central source is circulated within the tubing. Electric resistance units are also available; these have an electric element in place of the copper tubing and are autonomous from any central heat source.

In all cases, the element heats the air in direct contact with it. As the air warms up, it tends to rise, thereby creating natural convection air currents. This draws more air into the bottom of the enclosure to be heated and released through the outlet grille above.

FIGURE 6.33. Baseboard finned tube radiation (FTR). (Photo by Vaughn Bradshaw.)

Baseboard units are installed along the base of walls. The enclosure height varies, depending on whether one, two, or three tiers of elements are required to satisfy the load. Baseboard units are less conspicuous than convectors.

Convectors, on the other hand, have a larger output per length of wall occupied. The element(s) are enclosed within a cabinet at least 2 feet high by 3 feet wide (0.6 m high by 0.9 m wide). The cabinets may be freestanding, wall-hung, or recessed.

Enclosures for finned-tube units are available in a variety of architectural designs for finished spaces, as well as simple utilitarian covers for protection only.

Since cold air tends to fall at windows and outside walls, causing cold drafts along the floor, the most effective placement for finned-tube units is along the outside wall, and particularly beneath windows. The warm air rising from the units counteracts this cold air circulation pattern, and raises the MRT of the glass and wall surfaces at the same time.

If the device were placed along an interior wall, it would reinforce the circulation pattern of cold air, and occupants in the room would feel too cold on one side and too warm on the other. When a baseboard unit is used, its enclosure normally runs along the entire outside wall, even though the element within it may not extend the entire distance.

Natural convection devices are used for heating-only applications in residences and small light-commercial buildings. They are also used as a separate perimeter heating system in larger commercial and institutional buildings that have a ducted air-conditioning system for cooling throughout.

Unit Heaters

Unit heaters are essentially natural convection devices with a fan added for blowing forced air across the heating element and out into the room. But in addition to steam, hot water, and electricity, they can obtain their heat from direct combustion of gas or oil. For direct combustion, gas or oil is piped to the unit, and a flue vent to the outdoors is usually provided for the combustion products. Combustion products from natural gas are sometimes allowed to be released into the heated space if it is well ventilated. Models are available for through-the-wall application for venting the flue gases or for introducing outside air into the space.

As indicated by the term *unit,* the devices are factory-assembled components with a fan and heating mechanism enclosed within a casing. The casing provides an air inlet and vanes for directional heated-air outlet.

Unit heaters are not intended for use with ductwork, although a short length of duct can be used to bring outside air to the air inlet for ventilation purposes. The air outlet may be arranged for horizontal or vertical (up or down) delivery of air to the room. Figure 6.34 shows an example of each. Air velocities leaving the units may range from 300 to 2,500 fpm (1.5 to 12.5 m/s), and the warmed air may travel from 75 to 300 feet (23 to 91 m). The units may be suspended from the roof structure or floor-mounted, and are placed around the perimeter, where the major heat losses occur.

Applications for unit heaters are usually large, open areas, such as warehouses, storage spaces, industrial shops, garages, factories, and showrooms, where heating loads and the volume of heated space are too large to be handled by natural convection units. Smaller cabinet models are available for use in corridors, lobbies, vestibules, and similar auxiliary spaces.

Some advantages of unit heaters are: (1) they have large capacities and a wide effective radius of distribution, so few units are required; (2) space is saved if they are mounted overhead, out of the way; and (3) heating of cold spaces is rapid.

Radiant Heating/Cooling Devices

A heat- or cooling-delivery device that is exposed to view so that heat is transferred by radiation is classified as a radiant heating or cooling device. As mentioned at the beginning of this chapter, radiant heat tends to feel more comfortable than heated air and is a more energy efficient method of providing heat.

There are three general types of radiant heater: high temperature, medium temperature, and low temperature.

High-Temperature Radiation

High-temperature radiant devices, known as *infrared heaters,* can be electric, gas-fired, or oil-fired. Venting of combustion gases is necessary for oil-fired units and for gas-fired ones if adequate ventilation is not available. They rely on high temperatures of over 500°F (260°C) to radiate a large output of heat from a relatively small area. The units are typically mounted overhead and inclined toward the area to be warmed. Reflectors are used to direct the heat distribution pattern.

Infrared heaters are used to provide radiant warmth to people outdoors or in semi-enclosed buildings during cold weather. Applications include loading docks, grandstands, public waiting areas, garages, and hangers. They are also used to melt snow over limited areas. Figure 6.35 shows two examples of infrared heaters.

Medium-Temperature Radiation

Medium-temperature radiant panels are large, flat surfaces that operate at somewhat cooler temperatures but are still too hot to touch. Prefabricated metal panels in a variety of styles and sizes are commercially available. They are heated either by electric resistance elements or by tubes with hot water circulating within them bonded to the radiant surface. These panels can be mounted on the walls or ceiling.

Medium-temperature panels utilizing water circulation are the only radiant devices that can cool as well as heat. In the cooling mode, chilled water is circulated through the tub-

(a)

(b)

FIGURE 6.34. Unit heaters: (a) Horizontal. (Photo by Vaughn Bradshaw.) (b) Vertical. (Photo reprinted by permission of The Trane Company, La Crosse, WI.)

(a)

(b)

FIGURE 6.35. Infrared heaters. (a) These portable infrared heaters are used to extend the season for outdoor eating at restaurants beyond when it would otherwise be too cold. (b) The square radiant heaters interspersed with the round lights on the underside of this hospital entrance canopy provides a transition for people in a state of ill health entering and leaving in cold weather. The radiant heat also helps to melt snow to prevent people from slipping. (Photos by Vaughn Bradshaw.)

ing attached to the back of the panel. Operating much like a solar collector, the excess heat in the room is absorbed by the panel and carried away by the water. If the adjacent air is cooled to its dew point, condensation will form on the panel surface, so some form of dehumidification or air conditioning must generally be provided in addition to the cooling panel.

Low-Temperature Radiation

Radiant floor, wall, and ceiling systems operate near room temperature, as shown in Figure 6.36a. Besides providing a highly comfortable form of heat, radiant systems have the potential to reduce energy costs if designed carefully and in the right circumstances. Because of the higher MRT in the occupied space and the addition of a heat conduction component (in radiant floor systems), the air temperature needed to maintain comfort is much lower than with other forms of heating. Since the air itself is not being heated as much, the heat loss through walls and roof is greatly diminished. It is crucial to adequately insulate the exterior side of the radiant heat source to avoid losing too much of the heat generated.

The heat can come from electric elements, water piping, or ducts embedded in the floor or laid above the ceiling. In either case, the systems, constructed on site, should be well insulated from the environment outside the conditioned space. Underfloor heating can be used with the great majority of standard floor finishes, including carpeting.

Embedded radiant heating systems went out of favor in the United States in the 1970s due to certain drawbacks: (1) the large quantity of piping needed was relatively expensive, (2) the use of electrical elements became impractical because of high electrical costs, (3) malfunctions were difficult and inconvenient to correct, and (4) the thermal inertia of concrete slabs prevented embedded radiant systems from responding effectively to rapid changes in a room's thermal demands, such as warmup after a night setback.

In recent years, comfort heating applications have begun to reappear using new and improved materials, such as a flexible plastic piping capable of operating continuously at temperatures of 120°F (50°C) and higher, and with a life of 30 years. Figure 6.36b–d shows how this new tubing is used for floor, ceiling, and wall radiant heating systems.

Another innovative product now available is a manufactured gypsum board heating panel. It consists of an electric heating element embedded in a sheet of ⅝-inch (1.6-cm) fire-resistant gypsum wallboard. Standard sheet sizes are 4 feet (1.2 m) wide by 6, 8, 10, or 12 feet (1.8, 2.4, 3.0, or 3.7 m) long, with an output of approximately 15 W/ft^2. The panels

(a)

(b)

FIGURE 6.37. Snow melting system at a ski resort. (a) During installation. (b) Complete and in operation. (Photos courtesy of Uponor Wirsbo, Inc.)

plied with chilled water in the summer and heating water during the winter. Since the same pair of supply and return pipes are used to carry the chilled and heating water, this is called a *two-pipe system*, and all FCUs must be in either the heating or cooling mode at the same time. Typically, the whole distribution piping system is changed over between heating and cooling once in the fall and once in the spring. No matter when this changeover occurs, there is inevitably some discomfort as weather fluctuates, and as occupants on one side of a building need heating while those on the opposite side need cooling. This (hopefully brief) period of inconvenience is the trade-off for the economy of installing only one set of supply and return pipes.

One way to improve comfort is to provide an electric heating coil in addition to the chilled-water coil in each

FCU. This adds the expense of an additional electrical hookup to every FCU and limits the selection of heating energy sources to electrical power only. Nevertheless the construction cost is generally less than that of adding a whole set of separate heating pipes.

Three- and *four-pipe systems* have independent hot- and chilled-water coils in each FCU. This provides the flexibility for any unit to deliver either heating or cooling to a space, regardless of whether the other FCUs in the system are heating or cooling. In a three-pipe system, each FCU is connected to a separate heating water and chilled water supply pipe, but to only a single return pipe. Throughout the system, hot water and chilled water mix in the common return line, causing additional energy to be expended to heat the water back up or cool it down to the normal return temperature.

A four-pipe system eliminates this inefficiency by providing both supply and return pipes for the separate cooling and heating coils. Because of its autonomous supply and return lines, a four-pipe arrangement is the only one that can accommodate steam as the heat distribution media.

FCUs equipped with a cooling coil will also dehumidify. Built-in humidification is not practical for FCUs, so if necessary, a separate package humidifier must be provided in the room.

FCUs are commercially available in three basic forms:

- Wall
- Ceiling
- Vertical stacking

Figure 6.38 shows an example of wall and ceiling units.

(a)

(b)

FIGURE 6.38. Fan coil units. (a) Wall and (b) ceiling-mounted. (Images courtesy of Trane Commercial Systems.)

The wall-mounted type discharges air vertically upward. Various models are available which can be concealed in a custom enclosure, recessed into the wall, or installed exposed as a floor console (as shown) in a variety of cabinet styles. Concealed or console type units are usually located at the outside wall, typically below a window. The conditioned air is discharged through a grille at the top of the cabinet to allow mixing with entrained room air before impinging on the occupants. Console units can also be wall hung in areas where they would conflict with a cove or floor molding or semi-recessed for minimum projection of the cabinet. Recessed units are often used in corridors, and come with the manufacturer's standard wall plate, which includes return air grilles and access doors. Wall-mounted units can be operated either by controls inside an access door on the unit or by a remote thermostat on the wall.

Ceiling-mounted models discharge air horizontally. They may be enclosed within a cabinet for exposed mounting at the ceiling (as shown), or without a cabinet for concealed mounting. Ceiling-mounted FCUs can accommodate very short lengths of supply and return ducts to the conditioned space.

Mounting horizontal FCUs above a hard ceiling should be avoided because it limits the accessibility for maintenance. If not properly maintained, the condensate drain invariably becomes clogged over the years, which results in water overflow from the drain pan. If mounting above the ceiling cannot be avoided, the unit should be installed where it can be readily accessible for maintenance.

Since ceiling-mounted units are not usually located for air delivery at the perimeter windows, where the greatest heat loss is, they are best suited for warm climates. The advantages of these units are that they do not occupy any floor space, and that a single unit can be ducted to supply air to several small adjacent rooms.

Vertically stacking FCUs are designed for use in multiple-floor apartments, condominiums, office buildings, hotels, and other similar applications. Their chief advantage is the elimination of separate field-installed piping risers and horizontal run-outs. They are available in both individual room and ducted discharge models. The units can be provided with an integral front panel, or just grilles and thermostat exposed for encasement within drywall.

FCUs in general are best suited for conditioning a large number of small, individually controlled, and variably occupied rooms, such as those in hotels, motels, apartment buildings, nursing homes, and medical centers. There is no mixing of air from the conditioned spaces, and each space must have operable windows or some infiltration to provide ventilation air. Alternatively, a central ducted ventilation system may be used in conjunction with the fan coil system.

The disadvantages of FCUs are: (1) maintenance and service work must be performed inside the occupied areas; (2) as the units get older, the increased fan noise can bother occupants; (3) a condensate drain line is required for each unit that provides cooling; (4) it is difficult to prevent bacterial growth in the drain pans; and (5) filters are small and inefficient, and require frequent changing to maintain proper airflow.

Unit Ventilators

A *unit ventilator* (Figure 6.39) is an FCU with an opening through the exterior wall to admit outside air. The outside of the opening is covered with a louvered grille, and a damper on the inside controls the amount of fresh air taken in.

Unit ventilators are well suited to spaces having a high density of occupants and to perimeter rooms with no operable windows. Such spaces include school classrooms, meeting rooms, and patient rooms in hospitals and nursing homes. The units must be located at or very near an exterior wall or roof. Because of the provision for outside air, units fitted with automatic controls and a motorized damper can make use of an economizer cycle.

A drawback of such a unit is that the outside air vent may eventually leak in the closed position, allowing excessive infiltration into the room. Besides uncomfortable drafts and excessive energy use, this can also result in freezing and rupturing of coils. Also, it is sometimes a challenge to visually integrate the outdoor grilles into the building's exterior design.

(a)

(b)

FIGURE 6.39. Unit ventilator. (a) Interior. (Image courtesy of Trane Commercial Systems.) (b) Exterior. (Photo by Vaughn Bradshaw.)

Induction Units

An induction system, as described earlier in the section on distribution systems, is a hybrid between an air and a water system. Its terminal delivery device, called an *induction unit,* receives hot and chilled water like an FCU, as well as air.

Induction units are similar in appearance to FCUs, but they have no fan. The air circulation is provided by the high-pressure, high-velocity supply air as it passes through a nozzle. This relatively small amount of air at very high velocity induces additional room air into the air stream.

Room air enters at the lower front of the unit. The room air and supply air mix within the unit and then pass through a coil where reheating or cooling takes place. A drain pan is provided to collect any condensed moisture from the cooling coil.

Due to the excessive noise of the induction units along with the other reasons mentioned previously, these systems are rarely installed anymore.

Packaged Terminal Air Conditioners (PTAC)

These self-contained air-conditioning units, also known as *incremental units,* do not receive hot or cold water from a central plant. Instead, a compressor, condenser, expansion valve, and evaporator are contained within the unit, which operates in the DX mode like a miniature unitary packaged AHU. PTACs are also factory-assembled, but unlike their larger counterparts, they are located within the space they serve and are generally not designed for use with ductwork. Special units are available for use with a limited duct run to an adjacent room, but usually separate units are required.

The category of PTAC encompasses window air conditioners, through-the-wall room units, and heat pumps. Window units and through-the-wall units are functionally the same, but window units are designed for minimal installation cost and for portability. Their capacity is limited to between ½ and 2½ tons (1.75 and 8.8 kW), and they have no provision for heating. Through-the-wall units are permanently mounted and are generally available in sizes from ½ to about 1½ tons (1.75 to 5.25 kW). They may include heating by a steam coil, a hot water coil, or an electric element.

Both types are visible from the exterior of the building: the window unit is mounted in an open window so that the condenser side protrudes out, and the through-the-wall model is visible from the outside as a grille through which outside air passes to and from the condenser. A variety of alternative louver designs are available as manufacturers' accessories, or the architect may select one of his or her own design as long as it meets the airflow performance requirements. Since access to the outside is already available for the

condenser, ventilation air may also be brought in through the grille.

Through-the-Wall Units

Most through-the-wall units are mounted close to the floor, and in this position they are almost indistinguishable from the cabinet of a wall-mounted FCU. Both are located beneath a window sill for direct discharge of conditioned air upward across the glass to prevent drafts on the occupants. Return air is drawn from the room at the bottom of the cabinet. The on–off switch and thermostat settings are controlled either at the unit or by a remote wall-mounted thermostat. Figure 6.40 shows an interior and exterior view of a through-the-wall unit.

As an alternative, through-the-wall units can be installed high up in the wall near the ceiling when conditions prevent a lower installation. In these cases, the units must be remotely controlled from a wall thermostat, and conditioned air is discharged out the front of the cabinet.

During construction of a building exterior, a sleeve or *wall case* is inserted into the wall. The wall case is shipped with an external weather barrier to keep out rain and wind during the construction period. The outdoor louver is normally equipped with hand grips to allow the contractor to install the grille from the interior. This eliminates the need for scaffolding.

When the building has been cleaned and made ready for occupancy, the chassis module is set into the wall case. Sometimes the chassis is installed during construction to provide temporary heat, but this should be allowed only with the manufacturer's approval, and with an agreement with the contractor to clean the units prior to occupancy.

Finally, the cabinet front is installed. Cabinets may be painted or finished as desired. Some manufacturers include a line of decorative cabinets for their units.

The advantage of PTAC units are (1) simplicity of installation, (2) flexibility of direct thermal control by the occupant, making each room an individual zone, (3) suitability for individual metering of separate tenants, (4) unit modules can be easily removed for servicing or repair, and replaced with another, (5) failure of any one piece of equipment only affects one room and does not cut off service to the whole building, and (6) units do not need any ductwork, a central chiller, a cooling tower, pumps, or piping for chilled and condenser water, which saves space and installation costs. If heating is by self-contained electric elements, there is also no need for any mechanical equipment rooms. Since space normally occupied by the building wall is utilized, other spaces in the building are freed up for other purposes, and there may be some savings in wall construction costs.

(a)

(b)

FIGURE 6.40. Through-the-wall unit. (a) Interior and (b) exterior views. (Photos by Vaughn Bradshaw.)

The most popular applications of PTACs have been in apartment or condominium buildings, hotels and motels, and some office buildings. They are often used for renovations in general because they minimize HVAC space requirements and disruption of the rest of the building. They are also well suited for air conditioning only part of a building when a heating-only system would suffice for the building as a whole.

A control is available for most PTACs that does not allow room temperature to drop below about 40°F (4°C), even though the control on the unit is in the off position. This is a helpful feature for motels and some apartment and commercial applications because it provides enough heat to unattended spaces to prevent plumbing damage.

Although the initial costs for incremental units are less than those for multi-room conditioning by a central system, PTACs are usually less efficient than a central cooling system. This translates into higher energy costs over the life of the building. Another drawback is the effect of the numerous louvers on the appearance of the building exterior.

Heat Pumps

The third type of incremental unit is the room-size *heat pump.* Such units, smaller versions of the central heat pumps discussed earlier, are available as water-to-air or air-to-air units. The air source units are the same as through-the-wall units except that they have a reversing cycle to pump heat from the outside to the inside in the winter.

The water-source incremental heat pump, also known as a *hydronic heat pump,* is installed within each space in combination with a piping loop through which water circulates. The heat pump unit is enclosed within a console-style cabinet and does not need an outside wall opening unless ventilation through the unit is desired. A central boiler and cooling tower are usually required in the system. The entire system is referred to as a *closed-loop heat pump system.*

These systems are applicable for buildings where a number of interior areas as well as perimeter offices must be served. The heat pumps in the interior core of the building operate in the cooling mode most of the time. The absorbed heat is transferred to the water in the loop. Perimeter units in a heating mode can then extract heat from the water and supply it to the perimeter room. Heat is thus "pumped" from where it is not wanted to other locations where it is needed. By recovering heat in this way, large energy savings can result during the portions of the year when heating is needed at the perimeter. Even so, the energy savings alone may not be enough to justify the considerable installation cost for these systems.

A cooling tower is normally needed to reject heat from the loop when all the heat pumps are in the cooling mode or when more heat is rejected to the loop than is needed for perimeter heating. When the heat from the interior is insufficient to meet the perimeter heating load, a central boiler or solar energy system is required to make up the difference. The water in the loop must usually be maintained between 70° and 90°F (21° and 32°C). Because of its moderate operating temperature range, the loop piping does not usually need to be insulated.

Groundwater from a well, stream, or pond sometimes serves as a relatively constant heat source/sink for water-source heat pumps. This makes the heat pump capacity independent of the outside ambient temperature, thus reducing the supplementary energy requirements. This also improves the seasonal efficiency of the heat pump for both heating and cooling.

Factors that need to be considered before deciding on a groundwater system include (1) the adequacy of the water supply, (2) the water chemistry, and (3) local groundwater temperatures. It is possible to establish community wells to supply untreated water as a separate utility from the potable water service. The well water for use as a heat pump source could be individually metered the same way as the potable water is.

Limitations

One of the drawbacks of through-the-wall PTACs is that the outside air vent tends to leak, allowing excessive infiltration into the room. The hydronic heat pump eliminates that problem as long as it doesn't have an outside air intake for ventilation.

Incremental units are quite noisy compared to other options. They have a compressor, an indoor fan, and—in the case of through-the-wall units—a condenser fan all in or adjacent to the occupied space. In contrast, FCUs have only the indoor fan, and they are available in very quiet models.

Since the units are in the occupied space, disruption of the occupants for service and repair is inevitable. Each unit has its own compressor, which requires periodic maintenance or replacement, and the filter must be changed when dirty or reduced airflow reduces the capacity of a unit. Access panels on the cabinets must not be obstructed by the building structure. Maintenance costs are high to properly service the many units in a typical installation.

Through-the-wall units must be located at an exterior wall. The air distribution limits of PTACs in general confine their conditioning range to a throw of 20 to 30 feet (6 to 9 m). Thus, the applicability of through-the-wall systems may extend only that far in from the perimeter wall.

LOCALIZED EXHAUST SYSTEMS

Spaces such as industrial process areas, laboratories, critical medical care areas, kitchens, toilet rooms, smoking rooms, chemical storage rooms, and so on need an exhaust system to directly remove the offensive odors, toxic fumes, exces-

sive air contaminants, or excessive humidity which is generated in them. The exhaust system usually consists of one or more fans and, unless the space directly communicates with the outside, some ductwork to convey the air to a safe discharge location. Buildings containing a large proportion of such spaces have greater heating and cooling requirements because of the high rates of outside air intake needed to make up for the exhausted air.

Generally, exhaust fans moving smoke- and grease-laden kitchen air streams, as well as those serving hospitals and laboratories, are required to be at the discharge end of the exhaust duct system. This keeps the exhaust air stream under a negative pressure at all times, so that any duct leakage will be inward, and the exhaust is prevented from escaping to contaminate the building interior.

Roof- and wall-mounted exhaust fans such as those shown in Figures 6.41a and 6.41b are the last element in their respective exhaust system. The air discharges from them directly to the outside. In cases where an in-line fan may be located at the room air intake or elsewhere in the duct system, a wall louver such as the one shown in Figure 6.41c, or a rooftop opening is required.

In laboratories and some industrial process areas, a hood is placed over the point of origin of the contamination. Ducted exhaust hoods are also normally installed in kitchens—over ranges and steam tables—for localized removal of grease, moisture, and heat.

High rates of outdoor air to replace the exhausted room air can be introduced by the primary air handler or by a make-up air unit. This requires that all the air be conditioned for the comfort of the occupants before it is exhausted back outside.

A more energy-conserving approach is to employ special exhaust hoods that inject tempered air (partially heated in winter and not cooled in summer) directly to the area needing ventilation and exhaust only a small amount of open room air. In many cases, this system is not appropriate, such as when a whole room needs to be exhausted. In that event, some sort of air-to-air heat recovery device can conserve energy if large quantities of air are involved.

In some cases, air can first be used in rooms that do not contaminate it excessively or in corridors. The air is then allowed to flow through the area of contamination and is exhausted from there. In order for this to occur, the space to be exhausted must have a negative pressure (slight vacuum) compared to the other spaces. These kinds of spaces are normally at a negative pressure, so that any air movement between the rooms will be toward the contaminated space and not in the opposite direction.

Supplying fresh air to corridors in order to exhaust indi-

(a)

(b)

(c)

FIGURE 6.41. (a) Roof-mounted exhaust fan with round weatherproof enclosure. (b) Sidewall fan with its round weatherproof enclosure. (c) Where it is inconvenient to route ductwork all the way to the roof and a wall fan would be objectionable, an interior in-line fan can sometimes be used in conjunction with wall louvers such as these. (Photos by Vaughn Bradshaw.)

vidual residence units is common in apartment buildings, condominiums, hotels, motels, hospitals, and nursing homes. Keeping air flowing away from the corridor prevents odors from migrating into the corridors and neighboring spaces.

For some types of contaminants, odor-adsorbing devices such as activated charcoal filters can be used to decontaminate air for recirculation. This can save tremendous amounts

of energy by reducing the need for conditioning the outside air. As mentioned earlier, even when air is recirculated through activated charcoal, a minimal amount of outside air is still required for diluting carbon dioxide with fresh oxygen.

Another reason for air exhaust is moisture accumulation. Warmth and dampness together are conducive to bacteria growth. Thus, maintaining dry surfaces is important in minimizing surface bacteria. It is commonly thought that the aim of codes requiring ventilation of toilet rooms is the dilution of offensive odors. In fact, such ventilation is primarily intended to eject air-borne moisture in order to promote sanitary conditions.

In multistory buildings where spaces to be exhausted are on different floors, exhaust stacks need a chase to pass through the floors above. In buildings such as hotels, apartments, and hospitals, the vertical exhaust ducts from kitchens and bathrooms are often located in the same chases as the plumbing serving those same rooms. However, since kitchen exhaust ducts can accumulate grease coatings and are thus a fire and smoke hazard, they must be kept separate from bath and other exhaust systems.

In major laboratory buildings, considerable space is required for the numerous exhaust stacks. In order to prevent exhaust air from being drawn back into an occupied space, the stacks are discharged at a height somewhat above roof level.

HEAT RECOVERY

There are almost limitless applications of heat recovery in buildings. There are very few pieces of equipment specifically designed for heat recovery. For the most part, heat recovery systems are custom-designed by combining standard HVAC components such as heat exchangers, pumps, and fans.

Each building or industrial process should be analyzed to match up heat needs and excess or waste heat sources. The temperatures of heat source and heat needs, as well as the logistics of heat recovery, are important considerations. The value of the recovered heat must be carefully weighed against the cost of the heat recovery system. In all cases, piping and ductwork should be well insulated, and constructed of materials suited to the temperature and corrosiveness of the application.

Heat Recovery from Exhaust Air

Probably the most common application of heat recovery is the retaining of heat or cooling effect from exhaust air and the addition of it to incoming fresh air. This process is appropriate for buildings with very high outside air requirements as well as for numerous industrial processes requiring large quantities of fresh air. The systems in common use for such applications are categorized as (1) closed-loop run-around, (2) open run-around, (3) in-line duct-to-duct heat exchanger, (4) heat wheel, and (5) heat pipe.

Closed-Loop Run-Around

This system uses standard finned-tube coils installed in the incoming and exhaust air streams. A fluid, circulated between the coils by a pump, is alternately heated and cooled by the two air streams, transferring sensible heat in the process.

This type of system can operate in all seasons by preheating outside air when it is colder than the exhaust air and precooling warm outdoor air with the exhaust from an air-conditioned building. To avoid freezing the coil during the winter, the fluid usually contains some sort of antifreeze.

This system is well suited for retrofit applications since the air ducts need not be next to each other. Sensible heat recovery efficiencies of 40 to 60 percent are achievable with the closed run-around loop.

Open Run-Around

This system is able to transfer both sensible and latent heat between the air streams. The supply and exhaust air streams are alternately contacted with a hygroscopic fluid spray. The fluid, which is usually a salt solution such as lithium chloride in water, is pumped from one air stream to the other. It absorbs heat and moisture from the warmer air stream and releases it to the cooler one. As with the closed run-around loop, the two air ducts do not have to be adjacent.

Air-to-Air Heat Exchanger

These devices are especially developed for heat recovery applications. The intake and exhaust ducts must be brought together at the heat exchanger. The two air streams pass through the unit but are separated by a heat transfer surface through which sensible heat flows.

The heat transfer efficiency ranges from 40 to 60 percent. No auxiliary pumps or drives are required. Once the exchanger is in place, the heat transfer is automatic.

Heat Wheel

This is another device exclusively used for heat recovery. Again, both ducts must be connected to the same device. The revolving wheel has a large surface area exposed to the air streams and is packed with a dessicant-impregnated material that is permeable to air and can absorb moisture.

As the wheel turns, portions of it are alternately exposed to the incoming and exhaust air streams. The portion exposed to the warmer air stream picks up sensible heat and latent heat (moisture) and releases both to the cooler air stream when it rotates around to that point.

Special seals and carefully designed airflow reduce the likelihood of cross-contamination between supply and exhaust air as the wheel rotates. A further precaution is a purge process that uses outside air to blow away any objectionable residual effects of contaminated exhaust air on the wheel surfaces. Carryover is considered negligible.

Retrofit applications of the heat wheel are limited by their large size and by the requirement that ductwork be in close proximity. However, it can achieve a rather high—60 to 90 percent—efficiency of sensible and latent heat recovery.

Heat Pipe

The heat-pipe heat exchanger consists of a coil with a partition dividing the coil face into two sections. The coil contains a bundle of straight tubes sealed at both ends with radiating fins bonded to each tube. Each tube is a separate heat pipe that is a passive heat exchanger needing no pumps or fans. Since there are no moving parts, the units have an extremely long life and require no mechanical maintenance.

In order to operate properly, the heat tubes, and thus the whole coil, must slope down from the cold end to the hot end. The slope angle must be accurately set.

The heat pipe recovers only sensible heat. Since the air streams are separated by a partition, there is no cross-contamination. Like heat wheels and air-to-air heat exchangers, heat pipes are not usually applicable to retrofits of existing buildings because the ductwork usually would have to be modified to bring the intake and exhaust ducts together. The heat transfer efficiency of heat pipes is very high.

Condenser Heat Recovery

When large commercial buildings need simultaneous heating and cooling, a *double-bundle condenser* may be installed on the chiller. A double-bundle condenser is a water-cooled condenser with two separate condensing water circuits enclosed within the same shell or separate shells. One of the circuits goes to the conventional cooling tower and the other is the heat recovery circuit. The hot refrigerant gas from the compressor is passed through the condenser shell(s). When the recovered heat is not needed, the cooling tower circuit is activated and the water carries the condenser heat to the cooling tower for rejection to the outdoors. When the heat can be productively used, the heat recovery circuit is activated, and the water in that circuit carries the heat to wherever it is needed. When operated in this mode, the chiller functions as a heat pump.

The recovered heat is in the form of hot water typically in the range of 100° to 130°F (38° to 54°C). This can be used in a heating coil or for heating hot water for any number of purposes. If more heat is reclaimed than needed during the cool-

ing period, it may in some cases be appropriate to store the surplus in a storage tank for later use.

Another form of condenser heat recovery is the direct use of hot refrigerant gas. The refrigerant coming from the compressor can sometimes be diverted from the condenser and passed instead through a nearby heating coil. The refrigerant condenses as it gives up its heat to the air stream. Applications of this method include preheating outside air and making reheating more energy efficient.

Redistribution of Heat Within a Building

The closed-loop heat pump system described previously is a form of heat recovery since it reclaims heat in those parts of the building where there is excess heat and makes use of it in other parts of the building where it is needed.

There are standard fluorescent lighting fixtures available that directly recover the heat given off by the lights. The medium of recovery can be either air or water.

Recovery of Process Waste Heat

Any process that dumps hot gases into the atmosphere or hot liquids down the drain may be a good candidate for heat recovery. For example, hot drain lines from kitchens and laundries can be fitted with a heat exchanger for preheating incoming cold water that needs to be heated. A heat pump can also be used to extract heat at temperatures as low as 50°F (10°C) from cold drain lines, increasing the temperature and discharging the heat where needed.

Waste heat in the form of hot gases from a stationary engine (generator or engine-driven compressor) can be passed through a *waste heat boiler.* Waste heat recovery boilers are usually firetube boilers in which the hot gas is passed through the tubes that are surrounded by water. Some industrial processes exhaust gases at 800° to 3,500°F (425° to 1925°C). Reclaimed heat can be used for generating steam, heating water, or heating make-up air.

Other devices that recover heat from the flue gases of large industrial-scale boilers are variously called *heat traps, economizers,* and *air heaters.* The heat reclaimed by these units is used to make the boiler operate more efficiently rather than for some other purpose.

THERMAL STORAGE

Thermal energy storage (TES) systems can be used to store energy (either hot or cold), when available at no cost or at reduced rates, for later use when a heating or cooling load is present. By shifting those loads, energy and/or costs may be saved in several ways: Taking advantage of excess internal heat or solar heat gains when they are available can save heating energy and the cost of heat-generating equipment. Spreading out the demand for cooling over a 24-hour period can reduce the necessary investment in equipment and the peak electricity demand by as much as a third.

Similarly, by expanding the capacity of an existing central cooling or heating plant through the use of thermal storage, the addition of expensive chillers or boilers can sometimes be avoided. The avoided costs may include both the larger electrical power service and higher cooling tower capacities that would otherwise be needed. The cost of thermal storage may amount to half the cost of the equivalent added chiller or boiler capacity in some cases.

TES systems can be designed for either full or partial storage. A full storage system provides enough capacity to meet all of a building's thermal requirements with no other heating or cooling source. It is possible—but not economical in most cases—to heat or cool a building entirely from storage.

A partial storage system, on the other hand, does not shift the entire load to nonpeak periods. Instead it operates in conjunction with a conventional heating or cooling system. The combination is usually more cost effective than full storage because, typically, as the size of a storage system approaches 100 percent capacity, there is a diminishing return of energy cost savings.

The cost of thermal storage is determined by the cost of the storage medium, the container(s), and the space that must be provided.

The medium may be water, ice, or some other more exotic phase-change material. Water is normally the least expensive medium, while phase-change materials take up less space. Phase-change materials can store the same amount of heat or cooling as water in 15 percent of the volume by utilizing the latent heat of fusion as it melts and resolidifies. Ice is the usual phase-change material for cooling storage. Other chemicals with a suitable melting point have been tried for storing heat, but have so far proved to be too expensive or have tended to disintegrate into their constituent components after a limited number of freeze-thaw cycles. Some appropriate material may eventually be developed, but even so, if used for both heating and cooling, a different phase-change material would be required for each temperature range, so the same volume could not be used for both.

Steel tanks are relatively inexpensive containers that can be used for years if corrosion inhibitors are added to the water. For small potable-water applications, glass- or ceramic-lined steel tanks are customary. Resembling standard residential water heaters, they are available in capacities from 30 to 120 gallons (115 to 455 L). Fiberglass tanks do not corrode and weigh only one-fourth as much as steel tanks of the same capacity, but they are somewhat more expensive. Concrete tanks are frequently used for very large water storage capacities.

Storage units may be located wherever they can be fit into a building. That may be in a subbasement, in a penthouse, or beside the building. Outdoor concrete reservoirs, or buried tanks of steel or fiberglass, may be covered with a parking lot or tennis courts. Regardless of the location, it is critical that the storage container and the piping to and from it be well insulated to maintain the heating or cooling effect for the desired length of time.

The unique circumstances of each building determine the feasibility of TES and the optimum design solution. In general, it is recommended that thermal storage be considered for energy and peak load reductions in buildings larger than 200,000 ft^2 (20,000 m^2). As a rule of thumb, storage in buildings larger than 1,000,000 ft^2 (95,000 m^2) may reduce first costs as well as energy costs.

Heat Storage Applications

Heat Recovery

When occupied, most well-designed modern commercial buildings are self-heating because of their high internal gains. A small amount of supplemental heat may be required at the perimeter, but the bulk of heating energy is used for morning warmup. Most of the heat required can be provided by heat left over from the previous day's operation. When the building structure itself cannot be designed with sufficient thermal storage capability and insulation to retain the heat overnight, the excess heat from the day before can be mechanically stored in a water tank.

The stored heat can be obtained by the heat recovery techniques discussed in the previous section. These include a double-bundle condenser on a chiller, heat pumps, heat-of-lights recovery, or heat reclaimed from kitchen, laundry, or industrial processes.

Alternative Energy Sources

Thermal storage can also retain heat from alternative sources such as solar energy or solid-waste incineration. If such systems are not economical at present, but are expected to be sometime in the future, a thermal storage water tank can be incorporated during the initial construction or remodeling of a building. The alternative heat source can then be added later when it is determined to be cost effective, and the thermal storage can be put into use at that time. This sort of advance planning minimizes the disruption of retrofitting alternative energy systems and can make the difference between whether they are viable or not.

The performance and effectiveness of a solar energy system are strongly affected by the storage capacity and storage temperature.

Off-Peak Electric Heating

Another application of TES is the storage of heat from electric resistance heaters during less expensive off-peak periods for later use during peak-demand periods. Off-peak electricity can cost as little as half the normal rate.

Off-peak electric heating units, available as factory-assembled packages, have the same advantages as any other electric heater: simpler, less expensive installation with no chimney, fuel tank, or fuel piping; simpler zone control; cleaner operation; and minimal maintenance. The ability to take advantage of discount electricity rates reduces the high operating cost that is the usual disadvantage of electric heating. These heaters must be carefully placed because they are very heavy.

Standard single units are available from small-room sizes up to 10,000 gallons (38,000 L) storage, which could handle an entire small building. The smaller units have a special high-density core capable of storing a great deal of heat in a compact space and of blowing warm air out the bottom of the cabinet directly into the room. The larger ones resemble a firetube boiler and utilize water for storage.

Cooling Storage

Although cooling storage may not necessarily reduce a building's overall energy consumption, it can level out the electricity demand for cooling by spreading it over a greater period of time. The energy that would have been used during on-peak periods is deferred to off-peak periods when the storage medium is charged. This can significantly reduce on-peak demand charges, thereby decreasing overall energy expenses. In addition, when the refrigeration equipment operates at night to charge the storage for the next day, the cooler outdoor temperatures at night improve heat rejection and chiller efficiency.

In New York City, for example, the savings by operating an electric chiller at night when electricity rates are lower can exceed $20 per ton ($6/kW) each month. At this rate, it is not difficult to justify the cost of a chilled-water storage system.

The practice of cooling storage in the United States dates back more than a century to when businesses in the Northeast and Midwest cut ice from lakes during the winter and stored it to provide cooling the following summer. Beginning in the early 1920s, stored cooling systems became popular for retail stores, office buildings, hotels, and hospitals. Then, as low-cost electricity became more available, and mechanical cooling systems increased in reliability and efficiency, interest in stored cooling systems waned.

Today, the high cost of energy, combined with technological improvements in storage and control equipment, has made this time-tested technique quite cost-effective for virtually any type of building in some parts of the country.

Cooling storage systems are categorized by the storage medium used: (1) water, (2) ice, or (3) other phase-change materials. They function by chilling water or freezing phase-change materials (generally ice or Eutectic Salts) in an insulated tank during the evening or early morning hours. This stored coolness is then used for space conditioning during hot daytime hours, using only circulating pumps and fan energy. The result is a reduced demand for electricity during a period when rates are highest.

Ice requires as little as one-eighth the volume of chilled water for the equivalent storage capacity. The mechanical process of freezing water typically requires more energy than it takes to store the same capacity in a chilled-water reservoir. But ice is a lower-temperature cooling source, and thus needs less water or air to be circulated in the building for the same effect, which reduces distribution energy. Ice containers such as those in Figure 6.42 are available as factory-fabricated modules.

Ice storage is advantageous for small and retrofit applications. Water is typically preferred for large new systems where its bulk can be programmed into the building from the start.

One particular advantage of water is that chilled-water storage reservoirs are seldom used to capacity in the winter. If the total capacity is divided into two separate compartments, one part can be storing reclaimed heat for winter heating while the other part is still storing chilled water. Another benefit is that one module can be drained for repair while still maintaining partial capacity.

Annual Cycle Energy System (ACES)

The concept of an *annual-cycle energy system* (ACES) is that a huge thermal storage capacity is available to store excess heat during the summer for use as space heating in

FIGURE 6.42. Ice storage system. The outdoor packaged air-cooled chiller on the left freezes water in the ice storage modules on the right. (Photo courtesy of The Trane Company.)

the winter. As the heat is drawn off for winter heating, the storage medium is cooled. The following summer, the cool storage is available for summer cooling, and the cycle repeats.

The summer heat may be recovered from a chiller double-bundle condenser or from solar collectors. The winter cooling may be acquired as the chiller cools down the storage while it pumps the heat into the conditioned space or from direct heat transfer with cold outside air. Cooling may be stored in the form of chilled water or ice. Ice is generally used to achieve the needed capacity in a reasonable amount of space.

An ACES can be designed to independently provide the entire heating and cooling capacity or to serve as a supplement to instantaneous heating and cooling equipment, depending on the relative economics. In either case, the capacity of storage and other equipment must allow for a sufficient safety factor to accommodate weather uncertainty from year to year. The control strategy to optimize energy savings in light of that uncertainty is a major issue with ACES.

The benefits of an ACES are that a maximum amount of free heat and cooling are used, and the compressor for pumping the heat back and forth can be operated at off-peak times and at a greatly reduced capacity. When chiller heat is always rejected to a space, hot water loads, or storage, there is no need for a cooling tower. Some or all of the fuel-burning equipment may also be able to be dispensed with.

AUTOMATIC CONTROLS

The automatic control system of a building is analogous to the central nervous system of the human body. It senses the conditions outdoors or in the conditioned spaces, and transmits that information to control devices that adjust the HVAC equipment to produce the desired results. Since HVAC systems are sized to satisfy the heating and cooling needs during peak-load conditions, automatic controls are required to vary the output when loads are below the maximum. An automatic control system must also vary the heat or cooling delivered to each zone. The larger the building, the more complex the automatic controls required.

The specifications for the mechanical work on a building usually require an operating manual and an orientation session to instruct the building manager on how the systems work. These are important to enable the building's ultimate operators to maintain and operate the controls for proper comfort and maximum energy efficiency.

The control system must respond to the sometimes rapidly varying conditions at the perimeter and distinguish them from the relatively stable internal loads.It must automatically adjust the HVAC equipment operation to provide comfort and appropriate process conditions in a safe and energy-efficient manner. It must control fans in case of fire, stop a refrigeration condensing unit if there is no water in the system, and prevent steam coils from freezing when exposed to freezing air.

Automatic control systems are composed of (1) sensing devices, (2) transmitting networks, (3) actuating devices, and (4) indicator devices.

Sensing devices are the "eyes and ears" of any control system. They sense air or water temperature, humidity, and steam pressure or air pressure and convert the sensation into a transmittable signal. The thermostat is the most common example of a sensing device.

The signal is transmitted as electric impulses, pneumatic pressure, or a combination of the two. All-electric controls are usually most appropriate for smaller buildings. As the number of control devices and control points increases, a break-even point is reached beyond which pneumatic controls are more economical. Electric controls are usually less expensive for packaged air-conditioning equipment or small systems up to about 50 tons (175 kW). The economy of scale generally justifies the cost of an air compressor and pneumatic equipment for central systems above 50 tons (175 kW) with several zones. Compressed air is distributed in pneumatic systems through ½-inch (1.3-cm) copper tubing or sometimes low-cost plastic tubing.

The actuators, corresponding to the fingers and muscles of the system, are automatic dampers, valves, and switches. They may be motor-controlled, hydraulic or pneumatic cylinders, or air diaphragms. Through appropriate relays and linkages, they open and close in a preprogrammed sequence. Time-clock mechanisms can alter the sequence according to the time and date.

The indicators translate the environmental condition or actuator position into a visual image for the benefit of human monitors. These indicators include thermometers, gauges, and indicator lights.

Thermostats

The thermostat is a sensing device that signals the heating/cooling equipment when to start and stop and how much to modulate. The temperature difference between when the equipment is shut off and when it is called to restart is referred to as the *dead band*. Thermostats are available with night setback, multiple setbacks, and 7-day clock features.

Thermostats must be located where they can sense the average temperature of the space to be conditioned. They should not be located where they could be exposed to conditions that are not representative of the whole space, such as in direct sunlight, in a cold draft, on an outside wall, or near a heating or cooling surface.

If a thermostat must be concealed for aesthetic reasons or to avoid damage, a variation of the standard wall thermostat can be located in the return air duct. Thermostats are available that can be placed behind a decorative flush-mounted wall plate. Thermostats are also available with remote-sensing wires which may be wrapped inconspicuously around a picture, sculpture, molding, chandelier, or other feature.

Self-contained thermostatic control valves can be mounted directly on cast-iron radiators, finned-tube radiation units, FCUs, or unit heaters. These devices are an inexpensive way to obtain accurate room-by-room zone control for terminal devices served by water- or steam-distribution systems.

They are well suited for retrofit in dormitories, apartments, offices, and other buildings originally built with only a single thermostat per floor or for the entire building. As such they are a highly cost-effective energy conservation measure. Since these control valves provide individual room control of heat output, windows no longer need to be left open in the winter to maintain comfort in overheated rooms.

Self-contained microprocessor-based thermostats can be programmed for numerous daily setback cycles that can change for each day of the week. The program can be easily interrupted for holidays or unanticipated occupancy. These devices are simple to operate and can control a number of different zones with separate sensors.

Direct Digital Control (DDC)

Today HVAC controls are almost exclusively microprocessor based. They provide more accurate control, greater flexibility, and faster response than previous control technologies. They can monitor, control, and in some cases, diagnose equipment problems.

The compact stand-alone DDC module is capable of automatically starting and stopping equipment, controlling flows, detecting malfunctions, and signaling when service or repair is needed. It typically replaces numerous separate mechanically operated control components while providing solid state reliability.

DDC systems range from simple individual controllers, to sophisticated networks of distributed processor-controllers. In almost all cases they are field reprogrammable.

In building-wide systems, there is typically at least one operator console, which is nothing more than a PC computer loaded with software that monitors and communicates with all the intelligent DDC modules (Figure 6.43). Usually a peripheral printer stands by to provide a hard-copy log of operating data, to record alarms, and to print out periodic reports. Almost any number of remote operator stations can be added with a simple modem interface. The operator station(s) can be used to change set points, check the status of any connected piece of equipment, reprogram operations, or allow remote manual control. Different users can be allowed different levels of access through the use of passwords.

Carbon Dioxide Control

An electronic device is available for installation in the return air duct to measure the level of carbon dioxide in the air. If carbon dioxide is excessive, the device sends a signal to open the outdoor air damper wider to allow more fresh air into the building.

This device makes it possible to reduce the amount of ventilation to an absolute minimum and to take in only as much outside air as is needed for oxygen requirements. An economic analysis is required under given circumstances to determine whether the cost of the device is justified by the energy savings.

NOISE AND VIBRATION

Sound becomes noise when it becomes objectionable to the building occupants. HVAC systems are generally the principal sources of *background sound* within buildings. Without proper precautions, this sound can become irritating noise. Noticeable sounds can be critical problems in buildings such as theaters, and severe vibration can cause dangerous structural damage to a building.

Sound is a by-product of the energy supplied to the moving components of HVAC equipment. During the transformation of electrical or combustion energy into useful work, energy losses appear as heat, vibration, and sound. Those items contributing the most sound are pumps, fans, compressors, and room air diffusers.

The HVAC industry has established *noise criteria* (*NC*)

(a)

(b)

FIGURE 6.43. DDC system components. (a) Desktop computer used for operator interface. (b) Field control modules. (Images courtesy of Trane Commercial Systems.)

values for evaluating the acceptability of sound levels. The background sound level in a room due to mechanical equipment is measured by a sound meter. These measurements are then weighted according to the most intrusive frequencies.

The noise level and vibration transmission can be partially controlled by mechanical isolation, shields, baffles, and acoustical liners. Some equipment is quieter than others, and the quieter ones are usually specified for buildings in which noise control is especially critical. Larger equipment is usually mounted on support structures that distribute the vibration over a larger area. Massive concrete pads and vibration isolators are placed between the equipment and the building to avoid transmission of vibration and noise through the structure. Equipment rooms can be acoustically insulated if necessary.

Flexible fabric connections are used to join ductwork to equipment subject to vibration. Sound-absorbent interior duct lining is designed to attenuate the sound generated by the fan and the air whistling through the ducts. Rectangular ducts—particularly those with a high aspect ratio—can transmit excessive noise to occupied spaces if not properly supported. Rectangular ducts used in medium- and high-velocity designs must be properly supported, stiffened, and lined with sound-absorbing material. Round ducts are stronger and have better aerodynamic characteristics, and therefore generally experience fewer noise problems. VAV terminals, high-pressure duct system terminals, and sound attenuator boxes can reduce sound, but should be located at least 15 feet (4.6 m) away from the air diffuser for proper attenuation of residual effects. Grilles, registers, and diffusers can be selected to minimize noise output.

Noise control affects the building architecture in that mechanical spaces should be located carefully so that equipment sounds do not intrude into interior occupied spaces or neighboring structures. Equipment spaces should be separated as far as possible from spaces whose function demands low background noise levels. These include executive offices, conference rooms, sleeping areas, theaters, auditoriums, and worship halls. Spaces located farthest from the equipment and separated from it by the most barriers will usually be the quietest. Consideration should also be given to not locating equipment near outdoor assembly areas.

The acoustical effect of a room modifies the quality and offensiveness of sounds by the time they reach the listeners. A thorough treatment of room acoustics is beyond the scope of this text, but an overview of the subject is provided in Chapter 14. Basically, the ability of a room to absorb sound depends on room size and geometry, surface hardness, the frequency of the sound, and the listener's distance from the sound source. For critical spaces and special noise problems, it is common practice to engage the services of an acoustical consultant.

DISTRICT HEATING AND COOLING

The concept of *district heating and cooling* involves a central plant that provides heat and/or cooling to an associated group of buildings clustered together in fairly close proximity. Such installations represent the largest HVAC installations. Large hospitals, universities, airports, industrial facilities, and residential and multiuse complexes are good candidates, although metered district services are also available to serve downtown buildings under different ownership in some large cities. District heat distribution is a convenient way to utilize geothermal energy.

The advantages of the large scale are diversity of use, greater energy efficiency, centralized fuel handling, less maintenance, and lower operating costs than with separate plants providing the same service in the individual buildings. Maximum heat recovery and thermal storage potential are also available.

Boilers and water chillers can be located in single or multiple plants. The central equipment may be remote from the buildings served, but economics usually dictates a central location.

Underground piping systems are normally used for primary distribution from the central plant(s) to each building. Besides being buried in the ground, primary distribution piping can be run through building basements, tunnels, or a combination of all three. Other services may be carried along with the piping in the same passageways. Manholes are located periodically along tunnels to provide access for inspection and maintenance.

Heat may be distributed by high- or low-pressure steam or by high-temperature, high-pressure hot water (HTW) at temperatures ranging from 300° to 400°F (150° to 200°C) with corresponding pressures of 55 to 500 psi (380 to 3,450 kPa). Chilled water is used for cooling. High-pressure steam or hot water allows heating piping—and sometimes heat loss—to be smaller than that for low-pressure water or steam. High pressure may also be required for power generation or process applications.

Steam supply lines are larger than water lines of the same capacity, but steam condensate lines are smaller than water return mains. Thermal insulation is extremely important due to the large surface areas of the long runs of piping exposed to heat loss or gain. Piping may be steel or cast iron for steam or hot water; copper is sometimes used for hot water distribution. Fiberglass-reinforced plastic (FRP) or PVC plastic is typically used for chilled-water service. Piping generally comes in 20- to 40-foot (6- to 12-m) lengths that are assembled in the field. All field joints and the piping in general must be pressure-tested for leaks before they are buried.

At the building boundary, the heat or cooling may be converted to an air or water system or left as steam for distribution within the building. An equipment room is usually necessary to house pressure-reducing valves, heating water converters, pumps, AHUs, and other building services.

SUMMARY

HVAC systems can be classified by the following categories:

Central heat and cooling sources
   Furnaces
   Boilers
   Hot water converters
   Incinerators
   Solar energy
   DX air conditioners (with
     air-cooled condenser)
   Electric water chillers (with
     air-cooled condenser or
     cooling tower)
   Absorption water chillers
     (with cooling tower)
   Evaporative coolers

Distribution systems
   Air (Figure 6.44)
   Water (Figure 6.45)
   Steam (Figure 6.46)
Delivery devices
   Finned-tube radiation
   Unit heaters
   Radiant heaters
   FCUs
   Unit ventilators
   PTAC units
   Air outlets

There is usually no one absolute best system for a given building. The various benefits and drawbacks of each system must be considered in light of the architectural program. Determining factors include:

1. How spread out the building or complex is
2. Whether steam or chilled water must be available for other purposes
3. Available space for central boilers, chillers, cooling towers, and air-handling apparatus
4. Utility services available and fuel-handling capability
5. Capital investment, energy costs, and other operating costs
6. Possible obstructions to ductwork, such as stairwells, elevator shafts, structural members, or insufficient floor-to-floor height
7. Limitations on the location of outdoor air intakes, such as height above grade, prevailing wind direction, and dirt and other contaminants
8. Structural support available
9. Accessibility of equipment space for replacing equipment
10. How delivery devices can physically and aesthetically fit into the space
11. Whether through-the-wall grilles are acceptable for the external wall appearance
12. The number of zones or whether individual room control is necessary
13. Location of equipment space in relation to critical areas from the standpoint of noise and vibration control

There are many criteria for selecting a system. It is possible to design an air-conditioning system to achieve practically any set of conditions if cost is no object. The cheapest alternative is seldom the one that provides ideal comfort conditions.

Solar energy utilization is dependent upon certain characteristics of solar radiation: (1) it is relatively low in intensity, so that when large amounts of energy are needed, large collector areas must be used; (2) it is intermittent, varying from zero intensity at sunrise to a maximum at solar noon and back to zero at sunset; and (3) it is subject to unpredictable interruptions by clouds, dust, and other air types of pollution.

The flat-plate collector is used most commonly for solar space- and water-heating applications. It is essentially an insulated box with a transparent cover plate that traps heat inside. During daylight hours, when the available solar energy exceeds demand, heat is stored for use at night and during periods of cloudiness. The most important considerations are collector orientation, sizing, avoiding overheating in the summer, and freeze protection in cold climates. The size of the solar collector array and the storage capacity are usually determined by economics. The solar system is usually not capable of providing 100 percent of the energy required, so a full-sized backup heating system is required.

Heat recovery from exhaust air, process waste heat, or excess internal heat gain can reduce or eliminate the investment and operating costs for heat generation. Any application requires a careful engineering study to determine whether energy savings can justify the high cost of the additional equipment. If it is cost-effective, then the engineer and architect must decide whether space can be made available to accommodate the equipment.

Thermal storage can be very cost-effective if heat recovery is available. It is viable for retrofit applications as well as

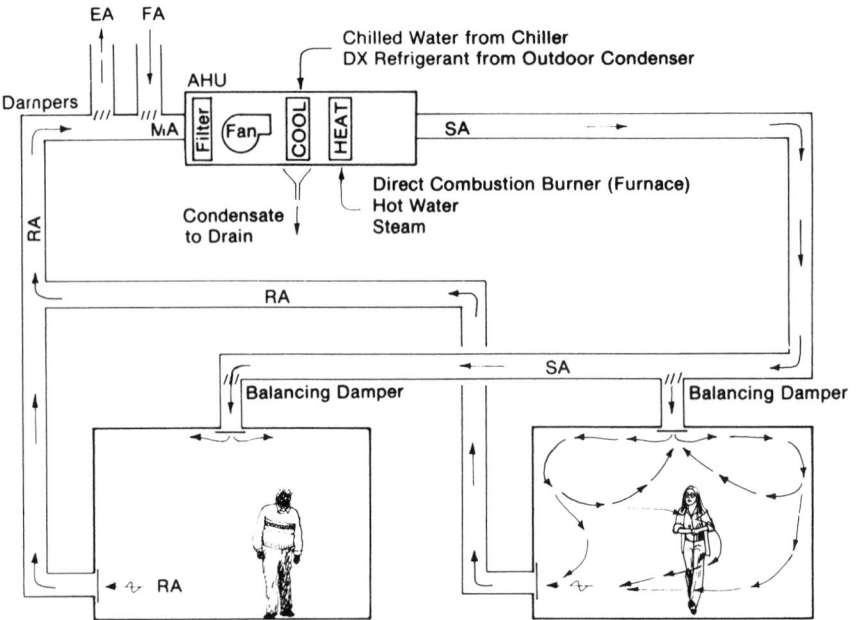

FIGURE 6.44. Air system schematic diagram.

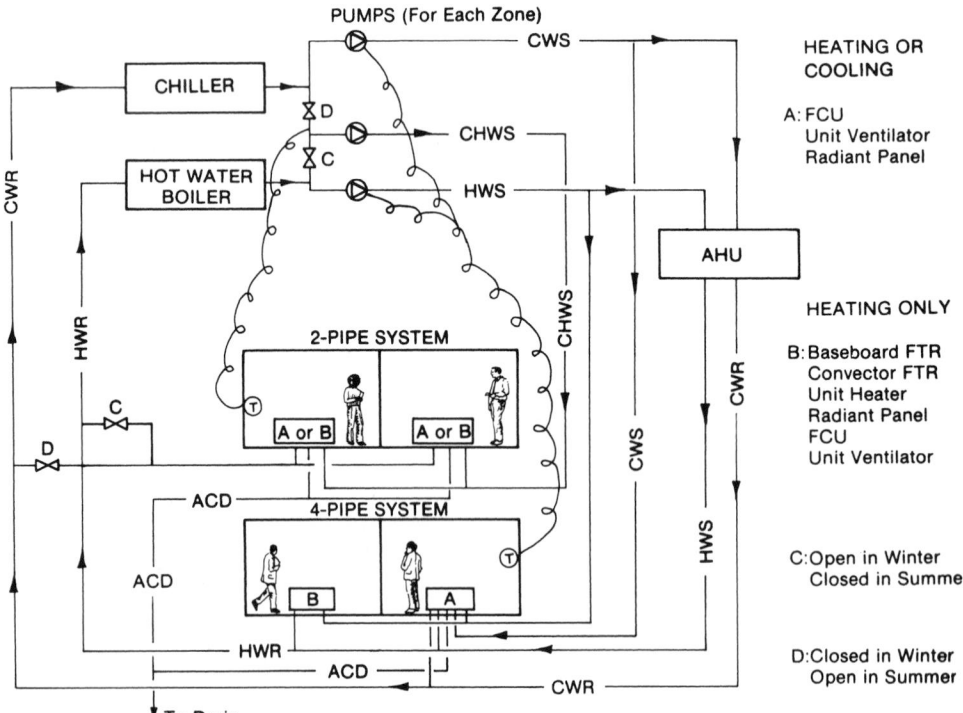

FIGURE 6.45. Water system schematic diagram.

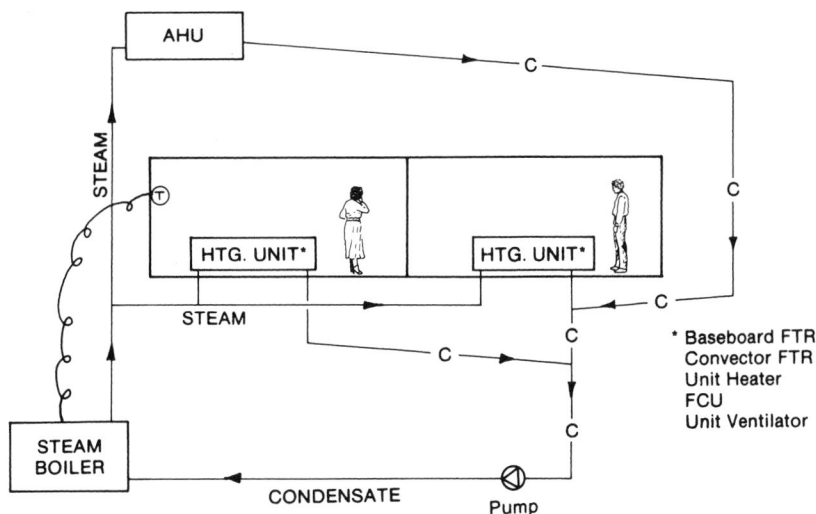

FIGURE 6.46. Steam system schematic diagram.

for new buildings. An economic feasibility study is required in each case to determine whether the energy or peak electrical demand cost savings can justify the expense.

Solar collectors can be added as a second stage to a thermal storage installation when they can contribute further economic benefits of their own.

KEY TERMS

Heating, ventilation, and air conditioning (HVAC)
Air conditioning
Evaporative cooler
Anthracite coal
Lignite coal
Bituminous coal
Coke
Draft
Forced draft
Induced draft fan
Furnace

Boiler
Pulse-combustion process
Condensing combustion process
Fluidized bed combustion
Portable boiler
Firetube boiler
Watertube boiler
Blowdown
Hot water converter
Refuse-derived fuel (RDF)
Pyrolysis

Active solar energy system
Passive solar energy system
Hybrid system
Solar constant
Solar collector
Flat-plate collector
Collector panel
Selective surface coating
Concentrating collector
Rankine cycle
Solar desiccant cooling

Chilled-water system
Vapor-compression cycle
Compressor
Refrigeration
Heat exchanger
Hot gas
Liquid line
Refrigerant
Condensate
Direct expansion (DX)
Chiller
Absorption chiller
Air-cooled condenser
Heat pump
Coefficient of performance (COP)
Seasonal performance factor (SPF)
Cooling tower
Water-side economizer cycle

Condensing unit
Evaporative condenser
Evaporative cooler
Latent heat of vaporization
Piped system
Ducted system
Combination system
Air-handling unit (AHU)
Zone
Coil
Condensate drain
Centrifugal fan
Propeller fan
Tubeaxial fan
Vaneaxial fan
Return air fan
Supply air fan
Preheat coil
Reheat coil
Single-zone system
Reheat system

Multi-zone system
Hot deck
Cold deck
Dual duct system
Terminal reheat system
Variable air volume (VAV) system
Induction system
Economizer cycle
Make-up air unit
Room exhaust fan
Central system
Rooftop unit
Unitary package unit
Unitary split system
Computer room unit
Two-pipe system

Four-pipe system
Perimeter heat
Sound trap (sound attenuator)
Aspect ratio
Hydronic system
Throw
Ceiling diffuser
Grille
Register
Perforated ceiling panel
Diffuser
Surface effect
Throw
Finned-tube radiation
Radiant panel
Infrared heater
Unit heater
Fan coil unit (FCU)
Unit ventilator
Induction unit

Packaged terminal air conditioner (PTAC) (incremental unit)
Through-the-wall unit
Wall case
Heat pump
Hydronic heat pump
Closed-loop heat pump system
Exhaust-air heat recovery
Double-bundle condenser
Hot refrigerant gas heat recovery
Waste heat boiler
Thermal energy storage (TES)
Time-of-use

metering
Annual Cycle Energy System (ACES)
Sensor
Electric control system
Pneumatic control system
Actuator
Indicator
Direct digital control (DDC) system
Noise criteria (NC)
Background mechanical noise
District heating and cooling

STUDY QUESTIONS

1. What factors influence the selection of a building's heat source? How do these concern the architectural program?

2. What are the options for building cooling sources? What are the characteristics that differentiate them?

3. What factors influence the selection of a building's HVAC distribution system? How do these concern the architectural program?

4. List the types of heating and cooling delivery devices. Indicate the various alternative locations for each. What are the factors that determine where each should be located in a given building?

5. What are the architectural implications of localized room exhaust fans?

6. List four types of heat that can potentially be recovered under cost-effective circumstances.

7. What are the five potential applications of thermal storage?

8. What are two purposes of automatic controls for an HVAC system?

9. What are the ways in which mechanical systems can be quieted? How can the architectural program minimize both the expense of sound and vibration control measures and the intrusion of noise into acoustically critical spaces?

10. Explain the entire process by which a single room within a building is heated and cooled by a district heating and cooling system.

BIBLIOGRAPHY

HVAC Systems, General

American Society of Heating, Refrigerating and Air-Conditioning Engineers, Inc. *ASHRAE Handbook of Systems and Equipment; ASHRAE Handbook of Applications; ASHRAE Handbook of Fundamentals.* (The four-volume standard of the industry.)

American Society of Heating, Refrigerating, and Air-Conditioning Engineers, Inc. *Standard 62.1—Ventilation for Acceptable Indoor Air Quality (ANSI Approved).*

American Society of Heating, Refrigerating, and Air-Conditioning Engineers, Inc. *Standard 62.2—Ventilation and Acceptable Indoor Air Quality in Low-Rise Residential Buildings (ANSI Approved).*

American Society of Heating, Refrigerating, and Air-Conditioning Engineers, Inc. *Standard 90.1—Energy Standard for Buildings Except Low-Rise Residential Buildings, SI Edition (ANSI Approved; IESNA Cosponsored).*

American Society of Heating, Refrigerating, and Air-Conditioning Engineers, Inc. *Standard 90.1—User's Manual.*

American Society of Heating, Refrigerating, and Air-Conditioning Engineers, Inc. *Standard 90.1—Energy Code for Commercial and High-Rise Residential Buildings.*

American Society of Heating, Refrigerating, and Air-Conditioning Engineers, Inc. *Standard 90.2—Energy-Efficient Design of Low-Rise Residential Buildings (ANSI Approved).*

American Society of Heating, Refrigerating, and Air-Conditioning Engineers, Inc. *Standard 90.2—Energy Code for New Low-Rise Residential Buildings Based on ASHRAE 90.2.*

American Society of Heating, Refrigerating, and Air-Conditioning Engineers, Inc. *Standard 100—Energy Conservation in Existing Buildings (ANSI Approved, IESNA Cosponsored).*

American Society of Heating, Refrigerating, and Air-Conditioning Engineers, Inc. *Standard 134—Graphic Symbols for Heating, Ventilating, Air-Conditioning, and Refrigerating Systems (ANSI Approved).*

American Society of Heating, Refrigerating, and Air-Conditioning Engineers, Inc. *Air-Conditioning System Design Manual.*

American Society of Heating, Refrigerating, and Air-Conditioning Engineers, Inc. *Designer's Guide to Ceiling-Based Air Diffusion.*

American Society of Heating, Refrigerating, and Air-Conditioning Engineers, Inc. *Ground-Source Heat Pumps—Design of Geothermal Systems for Commercial and Institutional Buildings.*

American Society of Heating, Refrigerating, and Air-Conditioning Engineers, Inc. *HVAC Design Guide for Tall Commercial Buildings.*

American Society of Heating, Refrigerating, and Air-Conditioning Engineers, Inc. *Humidity Control Design Guide for Commercial and Institutional Buildings.*

American Society of Heating, Refrigerating, and Air-Conditioning Engineers, Inc. *Underfloor Air Distribution Design Guide.*

Bobenhausen, William. *Simplified Design of HVAC Systems.* Wiley-Interscience, 1994.

Bovary, H. *Handbook of Mechanical and Electrical Systems for Buildings.* McGraw-Hill Book Company, 1981. (Detailed reference for making decisions on appropriate systems for new buildings and retrofits.)

Carrier Air Conditioning Co. *Handbook of Air-Conditioning System Design.* McGraw-Hill Book Company, 1966.

Haines, Roger W., and C. Lewis Wilson. *HVAC Systems Design Handbook,* 4th ed. McGraw-Hill Professional, 2003.

Johnson Controls. *Building Environments: HVAC Systems.* Delmar Learning, 2000.

Kreider, Jan F. (editor). *Handbook of Heating, Ventilation, and Air Conditioning.* CRC Press, 2000.

Lechner, Norbert. *Heating, Cooling, Lighting: Design Methods for Architects,* 2nd ed. John Wiley & Sons, Inc., 2001.

Levenhagen, John I. *HVAC Control System Design Diagrams.* McGraw-Hill Professional, 1998.

Love, Sydney F. *Planning and Creating Successful Engineered Designs,* rev. ed. Advanced Professional Development, Inc., 1986. (Oriented to the design process. Techniques for defining project objectives, establishing design requirements, determining the worth of any design feature, trading off one feature for another, and group decision making.)

McQuiston, Faye C., Jerald D. Parker, and Jeffrey D. Spitle. *Heating, Ventilating, and Air Conditioning: Analysis and Design,* 5th ed. John Wiley & Sons, Inc., 2000.

Porges, F. *Handbook of Heating, Ventilating and Air Conditioning,* 8th ed. Butterworth-Heinemann, 1982.

Porges, John. *Handbook of Heating, Ventilating and Air Conditioning: Ready-Reference Tables and Data,* 7th ed. Transatlantic Arts, 1976.

Shepherd, Keith. *Vav Air Conditioning Systems.* Blackwell Science, 1999.

Stamper, Eugene, and Richard L. Koral. *Handbook of Air Conditioning, Heating, and Ventilating,* 3rd ed. Industrial Press, Inc., 1979.

Wulfinghoff, Donald R., *Energy Efficiency Manual: For Everyone Who Uses Energy, Pays for Utilities, Designs and Builds, Is Interested in Energy Conservation and the Environment.* Energy Institute Press, 2000.

Automatic Controls

Panke, Richard A., C.E.M., *Energy Management Systems and Direct Digital Control.* Fairmont Press, 2002.

Edwards, Harry J., Jr. *Automatic Controls for Heating and Air Conditioning.* McGraw-Hill Book Company, 1980.

Haines, Roger W, and Douglas C. Hittle. *Control Systems for Heating, Ventilating, and Air Conditioning,* 6th ed. Springer, 2003.

Energy-Conscious Design

Clark, William H. *Retrofitting for Energy Conservation.* McGraw-Hill Professional, 1997.

Hunt, V. Daniel. *Handbook of Conservation and Solar Energy.* Van Nostrand Reinhold Company, 1982.

Shurcliff, William A. *Air-to-Air Heat Exchangers for Houses: How to Bring Fresh Air into Your Home and Expel Polluted Air, While Recovering Valuable Heat.* Brick House Publishing Company, 1982.

Shurcliff, William A. *Superinsulated Houses and Air-to-Air Heat Exchangers.* Brick House Publishing Company, 1988.

Stitt, Fred A. *The Ecological Design Handbook.* McGraw-Hill Professional, 1999.

Thumann, Albert. *Plant Engineers and Managers Guide to Energy Conservation,* 8th ed. Marcel Dekker, 2002.

Turner, Wayne C. (editor). *Energy Management Handbook,* 4th ed. Marcel Dekker, 2002.

U.S. Department of Commerce. *Energy Conservation in Buildings: An Economic Guidebook for Investment Decisions.* U.S. Government Printing Office, 1980.

Van Der Ryn, Sim, and Stuart Cowan. *Ecological Design.* Island Press, 1995.

Solar Energy

Anderson, Bruce. *Solar Energy: Fundamentals in Building Design.* McGraw-Hill Book Company, 1977.

Duffie, John A., and William A. Beckman. *Solar Engineering of Thermal Processes,* 2nd ed. John Wiley & Sons, Inc., 1991.

Klima, Jon. *The Solar Controls Book: Fundamentals of Domestic Hot Water and Space-Heating Solar Controls.* 4 volumes. Solar Training Publications, 1982.

Kreider, Jan F. *The Solar Heating Design Process: Active and Passive.* McGraw-Hill Book Company, 1988.

Kreith, Frank, and Jan Kreider. *Principles of Solar Engineering,* 2nd ed. Taylor & Francis Group, 2000.

Rideout, E., and O. Isacson (editors). *The Energy Systems Handbook,* 3rd ed. Technical Handbook Publications, 1982.

Schaeffer, John, and Doug Pratt (editors). *Real Goods Solar Living Source Book: The Complete Guide to Renewable Energy Technologies and Sustainable Living,* 11th ed. Gaiam Real Goods, 2001.

West, Ronald E., and Frank Kreith (editors). *Economic Analysis of Solar Thermal Energy Systems (Solar Heat Technologies).* MIT Press, 1988.

Chapter 7

Passive Methods

LOAD REDUCTIONS

SITE PLANNING

BUILDING FORM, ORIENTATION, AND COLOR

INSULATION AND THERMAL MASS

FENESTRATION

EARTH SHELTERING

INTERNAL GAIN REDUCTION

PASSIVE SOLAR HEATING

PASSIVE VS. ACTIVE

DISTINCTION BETWEEN RESIDENTIAL AND COMMERCIAL APPLICATIONS

PASSIVE SOLAR METHODS

PASSIVE COOLING

NIGHT SKY RADIATION

NATURAL VENTILATION

EVAPORATIVE COOLING

CASE STUDY

The comfort zone presented in Chapter 1 indicates conditions in which most people feel comfortable in the shade. When conditions of temperature and RH fall outside the comfort zone, corrective measures such as the addition of air motion, radiant heat, cooling, humidification, or dehumidification can produce comfortable conditions. In many cases, these may be achieved by natural means (landscaping, openings in the structure, building form, insulation, and other design arrangements), whereas mechanical air conditioning is more practical in others.

Active HVAC systems are intended to overcome the natural tendencies of heat flow. Passive systems, as an alternative, work with nature instead of against it. Generally they are low technology, are lower in cost, and should be designed to be aesthetically appropriate.

Passive methods can be divided into three categories:

- Load reduction
- Passive solar heating
- Passive cooling

If any of these methods reduce the loads on the active HVAC systems while still providing the necessary comfort conditions, a number of savings may be realized. If the peak heating or cooling loads are reduced, the HVAC equipment and distribution systems may be smaller and less costly. Generally, that will also result in lower annual energy usage, and thus lower heating and cooling costs. Since the systems and equipment would be smaller, they would consume less space within the building.

The most simple and inexpensive ways to provide comfort—and the ones to consider first—are those that reduce the heating and cooling loads. Next, consider ways of satisfying a portion or all of the reduced loads by passive solar heating or passive cooling methods. Only then, after the loads have been reduced or satisfied by passive methods as far as economically feasible, should active HVAC systems be designed. In some cases, it may be possible to dispense with a heating or cooling system entirely, or else the loads may be substantially reduced before the final selection and sizing of equipment.

It may sometimes cost little or nothing extra to incorporate passive methods in the initial stages of design, and they will result in lower mechanical costs. Adding passive methods as an afterthought is generally much more costly.

The optimum passive treatment depends on the type of building. Residential and some small commercial buildings with a low density of lighting, equipment, and occupants have relatively little internal heat gain and may benefit from passive solar heat gains. High-rise office buildings and other buildings in which internal loads dominate are only moderately influenced by the envelope. However, fenestration, when appropriately applied, can be beneficial for natural day-lighting or intentional heat dissipation. When carelessly included in conventional ways, fenestration can contribute significantly to heating and cooling loads. In any case, the local climate and site conditions affect the optimum building design.

There are a wide range of passive techniques available to accomplish the various objectives. These include proper siting of a building in relation to the sun, other buildings, vegetation, and other land forms; efficient design of the building form and envelope in response to climate; and internal load reductions. Normally, many of these methods should be integrated simultaneously on any given design. They are addressed separately here only for convenience in identifying them.

Many passive methods relate to the building envelope. These are concerned with wall and roof insulation, fenestration areas, and types of glazing.

The envelope plays a key role in the manipulation of energy flows. It should be thought of not so much as simply a barrier to heat and cold, but rather as an energy mediator between the continually fluctuating external environment and the desired stable internal environment maintained at comfortable conditions. It can be conceived of as a dynamic boundary that can be sensitively attuned to the local environment. Natural conditions such as sun, wind, and vegetation should be regarded as resources to be manipulated with the aid of passive techniques rather than as intruders.

A complete passive program for a building must consider the site as well. Since it may dictate certain design characteristics, the site is the natural place to begin.

LOAD REDUCTIONS

Site Planning

Cooperating with the site requires an overall strategy taking into account existing site conditions, climatic conditions, and the intended use of the building. An *internal-load-dominated* building requires a program for reducing heat gain and dissipating its surplus. Depending on the balance between the internal gains and the rate of winter heat loss through the *skin,* these buildings may be in a cooling mode during all or a large portion of the year. If passive cooling through the envelope is planned for these types of buildings, heat flow must be facilitated from the interior core, where the heat is generated, to the skin, where it can be dissipated; or else an imbalance may exist where it is so hot in the interior that mechanical cooling is needed and so cold at the perimeter that mechanical heating is needed.

Buildings without a high density of internal gains (such as residences and light commercial buildings) or assembly halls used only intermittently (such as places of worship occupied only a few times a week) are much more sensitive to climate. The annual temperature extremes and their duration dictate whether the building must be designed for heat retention and passive solar heating in the winter, heat dissipation in the summer, or both.

Climate

The principal climatic variables are temperature range and extremes on an annual and daily basis, humidity, solar radiation, winds, and precipitation. The microclimate at a particular site can be substantially different from the general regional climate.

The basic determinants of climate are the earth's heat-storage capacity, the atmosphere, and winds. Wind currents blow warm or cold air from other regions. During the daylight hours, the sunshine has a warming influence on the air and the ground. But the large heat-storage capacity prevents the earth from heating up too much during the day or cooling off too greatly at night. In addition, since moisture can hold a large amount of latent heat, humidity levels in the air are a major determinant of the magnitude of air temperature swings. The layer of atmosphere around the earth acts as a blanket to keep heat from escaping from the earth too rapidly. It selectively passes solar radiation emanating from a source at more than 5,000°F (2,700°C), but retards the passage of radiation from lower-temperature terrestrial sources. This is referred to as the *greenhouse effect.* Radiant heat escapes through our atmosphere much more quickly with a clear night sky. Cloud cover not only keeps sunshine out during the day, but also keeps radiant heat in at night.

Even though there is a symmetrical pattern of solar radiation around noon, afternoon temperatures are warmer than those in the morning because the earth's storage capacity creates a thermal lag. Likewise, the lowest daily temperature is usually just before sunrise because it takes that long for the previous day's heat to be drained away. As soon as the sun comes up, it begins to heat up the environment.

Because of the long-term effect of thermal storage, summer temperatures typically reach their peak in July or August, even though peak solar radiation occurs in June. The earth and its water bodies don't heat up to their maximum temperatures until about a month after the maximum solar input. It takes until January or February—a month or two past the winter solstice—to reach the coldest winter temperatures because the ground must first deplete its thermal storage.

Geographically speaking, climates are usually colder at higher latitudes (both north and south), since the days are shorter and there is less solar radiation. Microclimatic effects such as elevation, proximity to large water bodies, shading, and wind patterns can, however, alter this trend.

While climates can be described in numerous ways, they are usually classified into four types for which specific design responses have been established. These are:

Cold. Characterized by long, cold winters and short, very hot periods occasionally during the summer. Generally occurring above 45° latitude. The design responses emphasize heat retention, wind protection, passive solar utilization, and provisions for rain and snow.

Temperate. Characterized by cold winters, which require considerations for heating, and hot summers, which require some cooling efforts, especially if the summers are humid. Generally within a band between 35° and 45° latitude.

Hot arid. Characterized by long, hot summers, short, sunny winters, and a wide daily temperature range between dawn and the warmest part of the afternoon. Concerns are with heat and sun control, wind utilization, rain utilization, and increasing humidity. In the United States, generally the southwest portion.

Hot humid. Characterized by very long summers, only slight seasonal variation, and relatively constant temperatures. Heat and humidity are the chief comfort concerns. Solar shading, heat-gain reduction, and wind utilization are the focus of load reduction. Occurs in the southeast portion of the United States.

Victor Olgyay, an early pioneer on the subject, defined these climatic types and some responses to them in his book *Design with Climate.* This book is still considered a basic reference for climate-sensitive design.

Siting

An analysis of the heating and cooling requirements, the regional climate, and the local microclimate will determine if sunshine, shading, winds, or evaporative cooling is needed. The site should be studied to determine which are the prevailing seasonal winds so that the building can be placed to either take advantage of them or be sheltered from them. The topography—if anything other than flat—can influence the relative windiness of different locations on the site. Existing vegetation may be another determinant in building location on the site. Minor landscaping can augment the existing trees and shrubs. A careful analysis of shading and wind sheltering by both existing trees and proposed plantings at various times of the year is warranted.

Cold Climates.   The most important considerations in placing climate-sensitive buildings on a site in a cold climate are access to winter sun and shelter from winter winds. Exposure to morning sun is okay for most of the year, but shade should be provided to the west and northwest in the summer. Areas to be avoided include valley bottoms where cold air collects, north slopes with reduced sun exposure and greater exposure to cold winter winds, and hilltops where cold winter winds that carry off heat are the strongest. These buildings should ideally be placed on south-facing slopes. The buildings can even be set into the slope or buried entirely underground for greater isolation from the climate. In some cases, sites below hilltops are subject to cool air currents naturally flowing downhill.

Temperate Climates.   Climate sensitive-buildings here should be accessible to winter sun and summer breezes, yet be sheltered from winter storm winds. Summer shading is important to the east, west, and over the roof.

Hot Arid Climates.   Summer shading is most important here, especially to the west and over the roof. Some access to winter sun and sheltering from winds are desirable. Building sites in valleys near a water course can keep structures quite a bit cooler than poorly ventilated locations.

Hot Humid Climates.   Buildings in this type of climate should have their wall openings directed away from major noise sources so that the building can be opened up for nat-

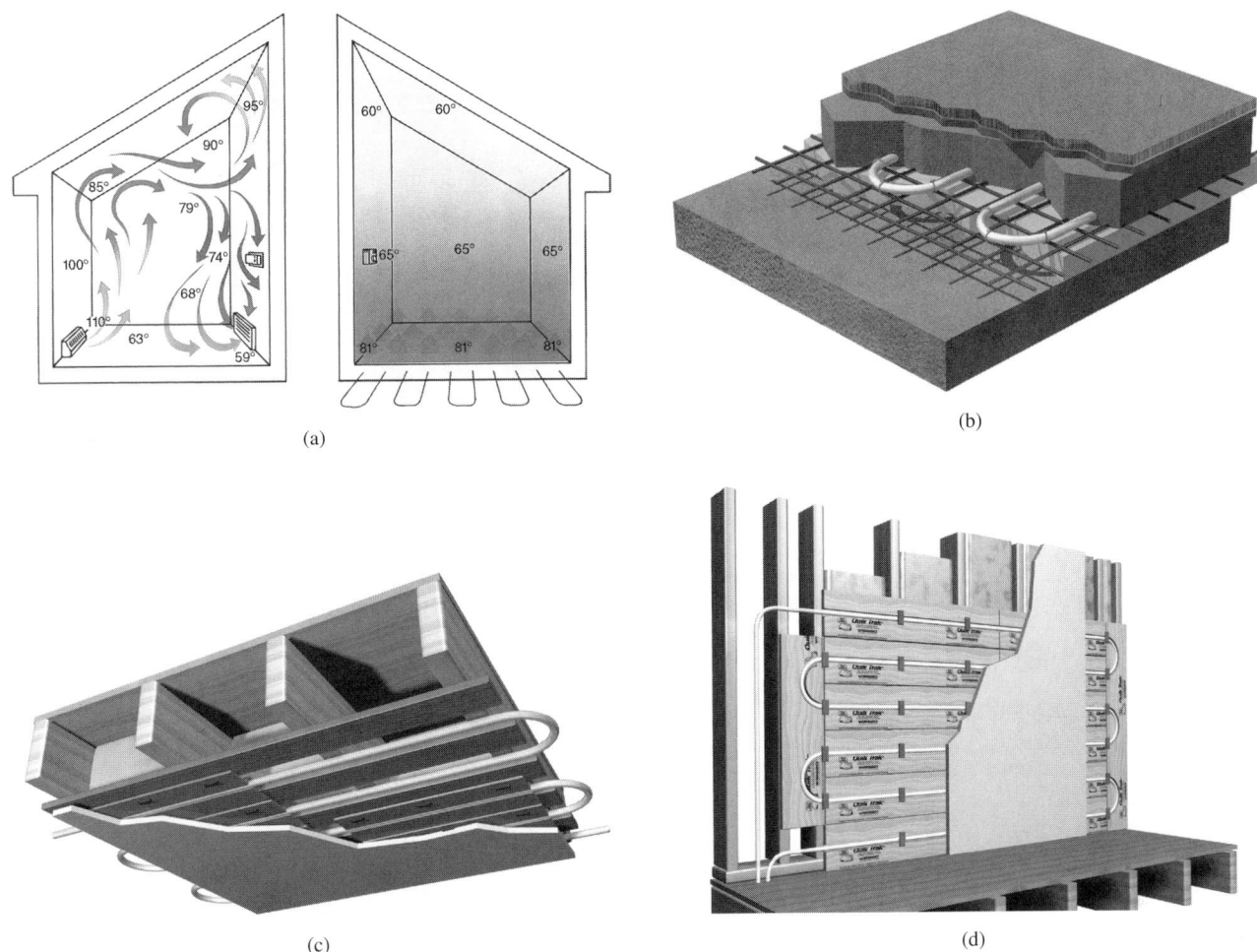

FIGURE 6.36. (a) Radiant floor heating—shown on the right-hand side—provides gentle low-temperature warmth and uses less energy to heat the air throughout the space compared to circulating warm air, shown on the left hand side. Radiant heating can be provided by circulating heated fluid through embedded tubing, as shown, or embedded electrical heating elements: (b) radiant floor, (c) radiant ceiling, (d) radiant wall. (Photos courtesy of Uponor Wirsbo, Inc.)

are installed using standard drywall methods and a simple wiring connection.

Once installed, all joints and fasteners are covered by conventional taping, and the radiant heating system—completely concealed in a normal sheetrock ceiling—is ready to receive paint or texture. The system is very conducive to room by room thermostat control.

One application that has remained in use is the melting of snow from driveways, walkways, and airport runways. A warm antifreeze solution is circulated through pipes embedded beneath the surface. Figure 6.37 shows flexible plastic tubing used for snow melting. As an alternative, electric heating cables are sometimes used.

Fan Coil Units

A *fan coil unit* (FCU) is a factory-assembled package consisting of a heating and/or cooling coil, fan, and filter. When it includes a cooling coil, it is equipped with a condensate drain pan which must be connected to a drain line. The fan draws air from the room and blows it through the coil. The heated or cooled air is then discharged into the room. An FCU is functionally similar to an AHU, but is classified separately because it serves only one room or a small group of rooms and is therefore a terminal device.

FCUs for heating and cooling service may contain either one or two coils. When only a single coil is present, it is sup-

ural ventilation. Ideally, floors should be raised up above the ground with crawl spaces underneath for good air circulation. Shading and access to breezes are important. Some access to winter sun may be desirable.

Solar Controls

Solar radiation potentially contributes heat, light, and glare to a site. The shadows cast by land forms, other structures, or vegetation should be analyzed for the critical times of year and periods of the day, and the building location selected on the basis of whether winter heating, summer shading, or both are desired.

Glare from nearby water bodies or a sea of parked cars can be controlled by proper siting or a strategic use of bushes, berms, and fences without obstructing desired solar heat.

To take advantage of the sun in the winter, the location selected on the site must be free of obstructions to winter sunshine or must receive the most sun during the hours of maximum solar radiation. Buildings blocked from exposure to the low winter sun between the hours of 9:00 a.m. and 3:00 p.m. cannot make direct use of the sun's energy for heating. During the winter months, approximately 90 percent of the sun's energy output occurs between those hours, so any surrounding objects, such as buildings or tall trees, that block the sun during that period will severely limit the use of solar energy as a heating source. Placing a building in the northern portion of a sunny area minimizes the possibility of shading by off-site developments in the future.

Summer shading of windows, walls, and roofs can be achieved by other structures (new or existing), shading elements that are part of the building, or surrounding vegetation (Figure 7.1). Shading devices such as roof overhangs, screens, and fins can be designed to allow winter sun to penetrate to buildings that need heating.

Other buildings can be useful in providing shade on a proposed structure, but such shade from off-site buildings cannot be entirely relied upon unless there is some certainty that they will remain in their existing condition. While other provisions can be retrofitted later in the event that a neighboring building is demolished, it is a good idea to formulate such backup plans in advance. Backup plans should similarly be prepared for shading effects of trees since they can die or be removed for other reasons. It may take many years for replacement trees to achieve an equal shading effectiveness.

Besides the uncertainty of trees being removed, they can also grow beyond their present proportions. Remember to anticipate where existing trees will eventually be, or else plan for periodic pruning and trimming.

Vegetation. Vegetation has many aesthetic advantages on a site in addition to its practical shading function. Other related benefits are decreased air pollution, noise, and glare. In order to permit access to summer breezes, aid drainage, and reduce problems from pests, forests or extensive vegetation should be discouraged near a building.

The types of vegetation that can be used for shading are trees, shrubs, and vines. Trees can be very tall and are capable of shading both the roof and walls of low rise buildings. They can be either evergreen or deciduous. Shrubs, bushes, or low trees may be used to screen the lower portions of walls. In addition to their value as ground covers, deciduous vines can be trained horizontally across a trellis or vertically along walls to provide both shading and insulation. Tree and bush sizes and shapes are shown in Ramsey and Sleeper's *Architectural Graphic Standards* and Lynch's *Site Planning.*

Shading can be provided by both deciduous and evergreen trees and vines if precisely placed. *Deciduous* plants have greater flexibility since their foliage is most dense in the summertime, but after losing their leaves in the fall, they allow the low winter sun to shine through between their bare branches. For this purpose, trees with high crowns, less dense branching structures, and large leaves are ideal. Good examples are members of the ash family, some of the maples, and eastern sycamores.

Another benefit of deciduous plants is that they grow most rapidly in response to higher temperatures rather than to more intense solar radiation. Thus, they provide the deepest shade during the hottest weather. In contrast, fixed structures provide shade only as a function of the sun's position. Deciduous plants do most of their shading from mid-June to early October (in the Northern Hemisphere)—providing south windows with access to solar radiation through much of the spring—while fixed sunscreens provide shading only in a symmetrical pattern around June 21, the summer solstice.

Not all deciduous trees are foliated at the same time, and some hold on to their leaves well into the heating season, especially when fertilized and irrigated. Moreover, some have a dense branch structure that blocks a relatively large percentage of the solar radiation even when bare. Agricultural extension services in most areas can provide information on foliation characteristics of local vegetation.

Trees are most effective in front of low-rise buildings at times of day when the sun is low in the sky. This makes them useful for intercepting early morning and late afternoon sun on the east-southeast and west-southwest sides of buildings in northern latitudes, or winter sunshine to the south of

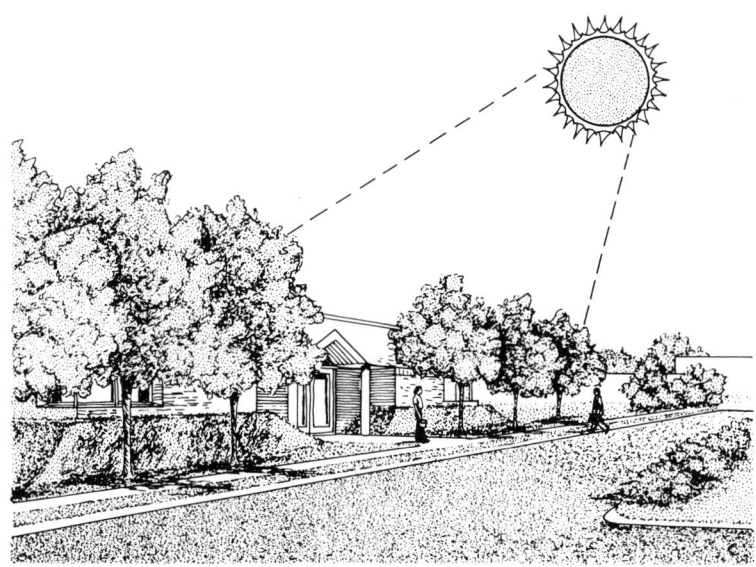

SUMMER

WINTER

FIGURE 7.1. Shading by deciduous vegetation.

buildings that are in a cooling mode all year long. At higher sun angles and for taller buildings, structural features such as louvers, screens, or overhangs are needed for shade.

The objective of shading is to minimize heat gain during the cooling season of any particular building. The air temperature in the shade of a tree may be 5° to 10°F (3° to 6°C) cooler than the surrounding ambient air temperature.

The effect on shaded walls is even more dramatic. Studies have shown that when shaded by a single large tree in direct sunlight, a wall may experience a drop in temperature of 20° to 25°F (11° to 14°C). Even when there is no direct sunlight on the walls, shading by a large tree can reduce the wall temperature by 5° to 10°F (3° to 6°C).

Wall temperatures are also significantly lowered by shrubs located immediately adjacent to the wall. Measurements have shown wall temperature reductions of 20° to 25°F (11° to 14°C) during periods of direct sunlight and 5° to 10°F (3° to 6°C) during periods of indirect sunlight for a moderate-size shrub (5 feet tall and 4 feet wide).

The effect is compounded when a wall shielded by a

hedge is also shaded by a tree. A wall temperature reduction of 10°F (6°C) has been found during periods of indirect sunshine, while temperature reductions approaching 30°F (17°C) have been observed when the wall is in direct sunlight.

These marked reductions in heat gain occur not only because vegetation shades the wall surface from the sun's rays, but also because the enormous surface areas of the leaves are engaged in *evapotranspiration,* which actually evaporatively cools the air. The shrubs adjacent to the wall also trap the cooled air, creating a blanket of cool, stagnant air next to the building surface. Another factor reducing loads is decreased air infiltration.

Shading west-southwest walls and the roof reduces envelope loads not only at the typical peak-load time of a building, but also at the time of day (late afternoon) that is usually the peak period of the electrical utility. If air-conditioning consumption can be decreased during the peak-load period, less utility capacity will be needed. Such reductions in both building peak load and utility peak load result in a positive environmental impact beyond the boundaries of the individual site. If a number of customers of the same utility all reduce their demand for cooling during the utility's peak period, it is possible that less electricity-generating capacity will need to be built.

In addition to shading, which reduces cooling needs in specific areas or for specific structures, trees cool whole neighborhoods by lowering air temperatures. During the day, for instance, the air temperature in a forest may be 25°F (14°C) cooler than the temperature at the top of the tree canopy. Similarly, neighborhoods with large trees may have maximum air temperatures up to 10°F (6°C) lower than similar neighborhoods without trees.

All sunlight is converted to heat, whether the sunlight is direct or reflected, and whether it strikes the outer surfaces of a building directly or streams through windows onto the floors or walls. Light striking nearby ground surfaces and objects heats these materials, which also warms the air immediately surrounding the building. As part of the overall landscape planning of a site, a lawn or other low ground cover, due to its evapotranspiration, can cool an area by as much as 1,500 Btuh/ft² (4,750 W/m²) during the summer. This can make a moist lawn 10° to 15°F (6° to 8°C) cooler than bare soil and 30°F (17°C) cooler than unshaded asphalt, even though it doesn't provide any shade of the building. Table 7.1 lists the relative cooling effects of various ground surfaces.

Excess glare from nearby unshaded ground, water bodies, or car windshields can also be minimized by the use of ground cover such as grass or ivy, which absorbs a fair amount of light. Reflectivities of some common surface materials are listed in Table B3.2 in the appendix.

Considering total energy use, lawn areas are relatively energy-intensive because of the gasoline consumed in lawn mowing. However, trees and shrubs may consume even more indirect fossil fuels per unit area because of pesticide and fertilizer requirements. Ideally, low-maintenance native or naturalized species should be utilized near a building on all ground areas that do not already contain trees and shrubs. Maximum local cooling occurs when grasses or ground covers are allowed to reach their maximum height.

A whole range of low-growing, low-maintenance, hardy ground covers are generally available for various appearances and textures. Porous paving or paving blocks that allow grass to grow up through them can be used to replace asphalt in order to create cool green surfaces that are sturdy enough to withstand moderate vehicular traffic.

In addition to serving as a useful ground cover, vines have the potential to cover a large portion of a building in a very short period. Vines are especially useful during the 3- to 5-year growth period required for trees and shrubs to become established in the landscape. They are also useful in situations where very limited ground space or tall buildings preclude the use of trees or shrubs. Vines, however, require a supporting trellis away from the wall to ensure adequate air circulation and minimize the potential for root damage to the wall.

Like trees and shrubs, vines growing on walls or trellises immediately adjacent to walls can significantly lower wall temperatures by shading and evapotranspiration. Wall temperature reductions of 5° to 10°F (3° to 6°C) have been measured for a moderately sparse vine (about 3 inches (7.6 cm) thick with a density of about 80 percent) in indirect light. When the direct solar radiation was at a maximum, the temperature of a vine-covered wall measured 10° to 15°F (6° to 8°C) cooler than an unshaded wall. Thicker vines provide larger temperature reductions, up to about 15°F (8°C). Thus vines can be an effective shading device, but not as effective as trees and shrubs.

Ponds, lakes, or other water bodies on the site can also cool summer breezes by evaporating water into them, as illustrated in Figure 3.4. Unshaded bodies of water on the sunny side of a building reflect sunlight and solar heat toward it. Thoughtful siting and use of strategic shading devices can block unwanted heat or glare. When locating a building in relation to large water bodies, it is best to build on high ground for both proper drainage and protection against flooding.

Outdoor Spaces. Outdoor spaces such as a plaza next to a building in a cold or temperate climate should be on a

TABLE 7.1   OUTDOOR SURFACE TEMPERATURES

Air Temperature Is 84°F (29°C)	Surface Temperature (°F)	Deviation from Air	Surface Temperature (°C)	Deviation from Air
Dark asphalt	124	+40	51	+22
Light asphalt	112	+28	44	+15
Concrete	108	+24	42	+13
Short grass	104	+20	40	+11
Bare ground	100	+16	38	+9
Tall grass	96	+12	36	+7

sunny side of the building. If placed on the north side, they will be in continual shade for most of the winter and will be wasted because people won't use them. By siting a building into a south-facing slope or by berming earth against the north wall, the shadow cast by the building will be reduced or eliminated. Besides allowing sunlight to reach the north side, covering a north wall with earth provides the additional benefit of reducing heat loss through the wall in winter and heat gain in summer, since ground temperatures are higher in winter and lower in summer than the outdoor air temperature.

In hot climates, outdoor spaces should be situated on a shaded side or else they, too, will be avoided. They should also be in the path of breezes as much as possible. In general, the suitable orientation of an outdoor space in a temperate or hot climate is dependent on the primary season of intended use.

Wind Control

In contrast to the sun, wind should be utilized during warm periods and blocked during cool periods to aid in the natural conditioning. In designing for wind protection and wind use, directions and velocities of the wind should be known in relation to cool and warm periods of the day and year. Of all climatic variables, wind is the most affected by individual site conditions; general climatic data will probably be insufficient.

Sheltering. Knowledge of the local prevailing winter wind direction and winter storm patterns will indicate where to locate windbreaks in order to block the frigid winds of winter. Like prevailing regional winds, both natural and artificial land forms and structures on or near a site can channel air movements into particular patterns. Site flexibility will sometimes allow a building to avoid the most extreme winds, but there may not be much choice on small sites.

Besides being able to provide shade, trees and shrubs are

fairly effective as windbreaks for low-rise buildings. Evergreens are best suited for this purpose because they maintain their dense growth during the winter. Placing fences, bushes, trees, and other objects in the direction of prevailing winter winds creates an area of relative calm on their leeward side and protects a building from cold winter winds. By providing this buffer, they reduce heat transmission loss and cold air infiltration (Figure 7.2).

Wind directly striking a solid windbreak (a wall, earth berm, or another building) is reduced in velocity from 100 percent at the break to about 50 percent at distances equivalent to about 10 or 15 times the height of the break. Open windbreaks, such as trees and bushes, offer a maximum reduction in wind speed of about 50 percent at a distance equivalent to about five times the height of the object. But the more open a windbreak, the greater the downwind distance of effectiveness because a larger downwind (leeward) side wake is formed.

Windbreaks may also be an integral part of a building structure. Such protrusions as parapets or fin walls on the

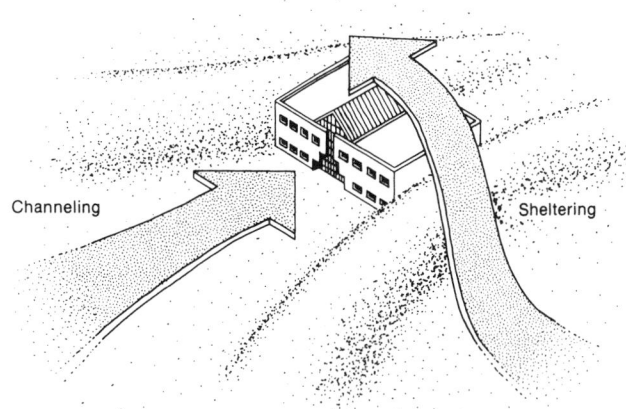

FIGURE 7.2. Windbreaks and channeling by site resources.

windward side of a building can divert air movement away from other wall and roof surfaces. If retaining winter heat is the objective, utilizing hills or below-grade construction can also shield buildings from cold winds. In a desert environment, windbreaks can also act as a desirable buffer to keep hot sandy winds from reaching a building.

When providing windbreaks, especially in urban areas, consideration should be given to the impact such redirection of wind will have on the surrounding environment. The "concrete canyons" of high-rise cities are an example of how funneling wind currents can produce blustery, eddying wind patterns. Buildings and their surrounding landscape should be designed so that wind speeds near pedestrian areas and openings do not exceed 10 mph (5 m/s). Large podium bases, wide canopies, and enclosed malls are some of the ways to shield pedestrian areas. In critical cases, scale models of tall buildings can be studied in wind tunnels to help predict wind speeds.

Channeling Wind. Site development, including landscaping and other structures, can be used to *channel* cooling breezes in order to carry away unwanted heat and moisture from a building. Wind convects heat away from roof and wall surfaces and can be funneled through the building for natural ventilation. Windbreaks adjacent to buildings can change the direction of airflow around and through the buildings while at the same time increasing or decreasing the wind speed (Figures 3.1, 3.2, and 7.2).

Ideally, careful placement of windbreaks allows summer breezes in while blocking out cold winter winds. While this is not always possible, in many localities the prevailing winds in summer and winter come from different directions, which simplifies the site strategy.

In areas where prevailing summer and winter winds come from the same direction, deciduous trees and hedges with a low, open branch structure may be used to direct summer winds toward the building. In winter, the bare branches allow cold winds to pass by, undeflected.

In areas where summer and winter winds come from different directions, ventilation openings and windows should be placed in the direction of summer winds. If, however, such a window arrangement conflicts with proper solar protection, summer winds can be directed into the building by additional windbreaks. In some cases, strategic plantings can be placed so as to combine shading and wind-directing benefits for optimum effect. In most cases, however, a compromise is necessary.

In hot humid climates, where maximum ventilation is required, the velocity of the wind, and hence its effectiveness as a cooling agent, can be increased by using windbreaks to constrict and accelerate wind flow in the vicinity of the building.

Building Form, Orientation, and Color
Indigenous Materials
Architects and the entire building team should work with local materials as much as possible. The least energy-intensive materials are those requiring the least energy for fabrication, transportation, erection, and maintenance.

Shape
A building may be thought of as forming a "footprint" on the land. The architecture should strive for a design that is both geographically and culturally appropriate by being within the regional stylistic vernacular. Other factors that influence building shape include the limitations of the site and the needs of the occupants. Beyond the demands of these factors, energy efficiency can be incorporated into the structure.

Again, bear in mind the primary distinction between an internal-load-dominated and a skin-load-dominated building in terms of their need for interaction with the environment. The former needs to dissipate the correct amount of heat throughout the year. The latter needs to lose a minimum of heat and gain the maximum amount of solar radiation in the winter while gaining a minimum of transmitted heat and solar heat in the summer.

The first basic concept in building shape is the ratio of building length to width to height. The ratio of length to width is given the name *aspect ratio*. The height is usually expressed in terms of the number of stories for a given floor area requirement.

According to Olgyay, the optimum shape for a skin-load-dominated building in any climate is a rectangular form elongated to some degree in the east-west direction. The longer southern facade of the structure provides a maximum of solar gain in the winter, when the bulk of the sunshine is low in the southern sky. The smaller eastern and western exposures minimize heat gain from the longer, low easterly and westerly sweeps of the sun in summer. The optimum amount of elongation depends on the climate and may be limited by the site.

Heat transfer between the interior and exterior is dependent on the surface area of materials separating the inside from the outside. A simple, compact form such as a sphere (dome), cylinder, cube, or other regular polygon exposes less surface area than an elongated or complex form with an equal floor area (see Figure 7.3). A form approaching one of these shapes minimizes heat transfer, thus conserving heat in cold weather and keeping heat out in hot weather. Likewise,

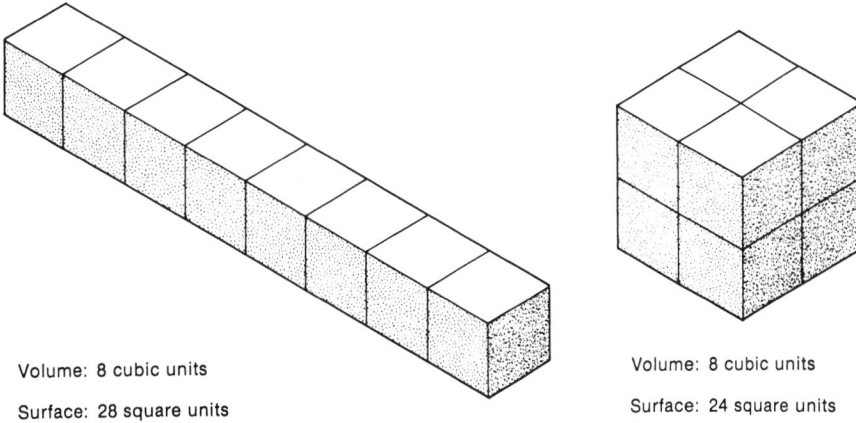

Volume: 8 cubic units

Surface: 28 square units

Volume: 8 cubic units

Surface: 24 square units

FIGURE 7.3. Surface/volume ratio.

window areas should be minimized to decrease unwanted heat loss and gain.

In areas where summer comfort is the most important consideration, other shapes may be called for. A thinner, more linear plan may take advantage of natural flow-through or cross ventilation. The building should be open and airy to facilitate the passage of breezes, and have overhangs and other shading devices to protect it from the sun.

Skin-load-dominated buildings in hot arid climates do well with thick, massive walls, which take advantage of the large daily temperature swings common wherever the air is dry. These walls absorb the solar heat during the day and reradiate it during cool nights. Where temperature and wind conditions do not lend themselves to natural ventilation, openings and fenestration should be turned away and shaded from the sun and made to face onto an interior courtyard. The courtyard shades, shelters, and keeps dust out of the interior. Vegetation planted within the courtyard can cool (by evapotranspiration) the air drawn into the building.

In climates with less severe weather, the designer has more flexibility in choosing building shape without any penalty of excessive heat gain or loss. Internal-load-dominated buildings in climates where the ambient outside temperature is generally cooler than inside temperatures may be low and spread out and have numerous extensions for a maximum surface/volume ratio. This will dissipate heat more effectively.

Roof areas should in general be minimized by being built up in multiple stories instead of spread out in a single story over a large site. In high-precipitation climates, a steep roof pitch more readily rejects snow and rain, thus avoiding structural and leak problems.

Figure 7.10 shows some shape options that accommodate passive solar utilization either for heating or for daylighting. If one of these designs results in a net gain, it is more effective than the compact design.

Clustering

The amount of heat gain from sunlight, like the heat gain and loss due to outside air temperature, is directly dependent on the exposed surface area of the building. As shown in Figure 7.4, *clustering* attached buildings both horizontally and vertically reduces the total exposed surface areas of walls and roof. Rows of skin-load-dominated buildings are most energy efficient when they share east and west walls, so that only the end units are exposed on the east or west face.

Minimum-surface structures include buried, semi-buried, or excavated buildings. These are only feasible, however, in reasonably dry soils.

Internal-load-dominated buildings located in cold climates where outside temperatures are below comfort levels most of the year benefit from exposed surface areas because these facilitate the dissipation of internally generated heat. The optimum amount of surface area may be calculated by determining the heat loss needed to offset the internal gain when the outdoor temperature is at its annual average. In hot climates, even internal-load-dominated buildings should have a minimum exposed surface area so as not to aggravate the cooling problem.

Orientation

Orientation of a building may be affected by function, site, or view. Energy requirements are affected by the solar and wind orientation of windows and other openings in the envelope as well as the dominant wall areas.

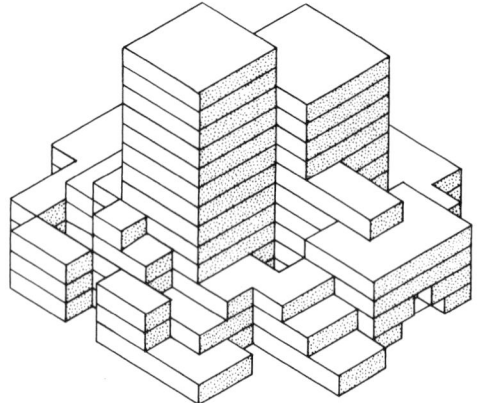

FIGURE 7.4. Clustering to reduce the exposed surface area.

Solar. Solar radiation strikes each surface orientation with a different intensity. At latitudes greater than 40°N, south walls receive nearly twice as much solar radiation in winter as in summer. East and west sides receive 2½ times more in summer than they do in winter. At latitudes below 35°N, solar gain is even greater on the south side in the winter than in the summer. Table 7.2 shows how solar radiation varies on the different exposures with different latitudes during different seasons.

Notice from Table 7.2 that a building with its long axis oriented east and west has a greater potential for winter solar heat gain than if it were oriented north-south. Windows oriented to the south provide the highest solar gain in winter but a low solar gain in summer. Orienting the shorter side of the building toward the east and west provides high solar gain in summer and low solar gain in winter and also minimizes summer cooling loads. This is especially important since the critical period for solar heat gain is usually in the late afternoon, when the sun is low in the west. For that reason, shading the west walls is very important. In lower latitudes, the east walls should be shaded as well.

For greatest passive solar heating benefit, the principal facade should face within 30° of due south (between south-southeast and south-southwest). Due south is generally preferred, although a slight variation may be beneficial, depending on local conditions at the site.

Since outdoor air temperatures tend to be lower in the morning than in the afternoon, an orientation slightly east of due south may not only take advantage of the early morning sun, when heat is most needed, but may also allow the building to face away from the west in order to avoid some of the summer afternoon solar heat gain. Where morning fog frequently occurs in the winter, morning sunshine is limited anyway, so an orientation slightly west of due south may be optimum. The optimum orientation varies from site to site and depends on the relative needs for heating and cooling, the time of day when the building is occupied, winds, and other factors.

Internal-load-dominated buildings, which are in a cooling mode all or most of the time, should minimize south, east, and west exposures as well as rooftop areas, which are subject to the highest solar gain.

Wind. Heat losses through building materials and through door and window cracks are directly dependent on exposure to and velocity of wind. Where cold winds are severe, skin-load-dominated buildings should be oriented away from the prevailing storm winds. Walls facing these winds should be windowless and well insulated. Large windows, openings, and outdoor areas should be located on the protected side.

If the size of the site or conflicting concerns do not allow for the proper configuration, or if it is not feasible to orient the building away from winds, the building should be protected by some sort of windbreak on the site. Orientation with respect to prevailing winds is especially important for tall buildings, many of which cannot be sheltered by trees and other structures. Devices can also be designed

TABLE 7.2   VARIATIONS IN SOLAR INTENSITY ON DIFFERENT ORIENTATIONS

Latitude	Season	Solar Heat Gain Relationships
Above 40°N	Winter	South more than 3 times the east and west
	Summer	East and west about 1.3 times the south
35°–40°N	Winter	South about 3 times the east and west
	Summer	East and west about 1.5 times the south
Below 35°N	Winter	South about 2 times the east and west
	Summer	East and west 2–3 times the north and south
		(at lower latitudes, north wall gains surpass south wall gains)

that offer solar shading when open and wind protection when closed.

Self-Shading

Because windows are usually the predominant path of entry for solar energy, economics dictates that efforts at sun control be concentrated on them. In hot climates, all fenestration exposed to direct sunshine should be shaded. Summer shading is vital if glazed areas are used for passive solar heating in the winter. Shading may also be used over exterior walls and roofs to reduce the amount of solar heat they absorb and to provide cooler outdoor spaces.

Exterior shading may be created by vegetation and other resources on the site, or it may be designed as a building element. The latter includes overhangs, vertical fins, louvers, screen panels, egg crate devices, and double roofs. Overhangs can serve other functions such as porches, balconies, verandas, and cantilevered upper floors.

Horizontal overhangs are particularly effective on southern exposures. They can be designed to provide shade in the summer while allowing the low-angle rays of the winter sun to shine through.

Because low early morning and late afternoon sun angles make it necessary that overhangs on east and west walls be excessively wide to be effective, vertical "fin" construction or some form of screening is best on those exposures. Since the west wall is exposed to direct sunshine in the late afternoon

when most buildings experience their peak cooling load, some form of shading there is very important (Figure 7.5).

Another type of shading device that can be used as a building component is live vines. Not only are vines able to shade out the sun, but they also provide some insulating capability and cool the air by evapotranspiration. In contrast to shades, which have a darkening effect, vine-covered windows have an airy quality of light and a pleasant atmosphere created by their greenery. This form of shading may reduce solar gain by 20 to 80 percent.

One method for supporting twining vines is to install a string, wire, or lath gridwork in front of the surface or area to be shaded or screened from view. Wire mesh, plastic netting, chain link fence, and fish nets are also suitable surfaces on which vines will twine and climb. For large surface areas, thin wire cable is appropriate. Turnbuckles can be installed and periodically tightened to keep the gridwork taut. Horizontal trellises and arbors can be constructed of a simple latticework of lathing lumber treated with a nontoxic preservative; larger versions can be made of heavy timbers. Detailed and elaborate designs can incorporate fancy millwork, columns, and arches for specific effects.

Color (Surface Reflectivity)

The exterior color of walls and roofs strongly affects the amount of heat that penetrates into the interior. The more sunlight is reflected off the building surfaces, the less is

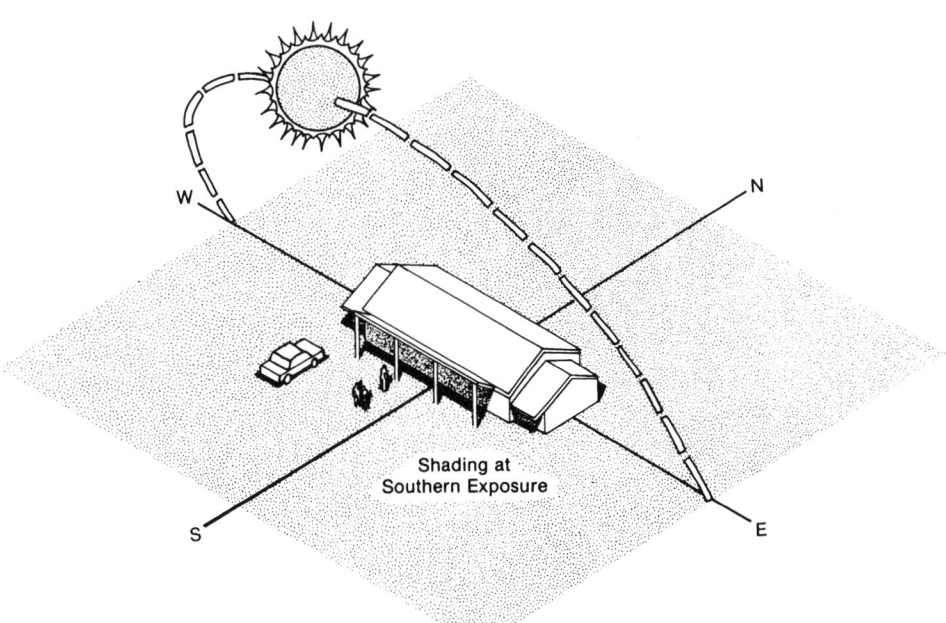

FIGURE 7.5. Overhangs are of most benefit on the south side (in the Northern Hemisphere).

absorbed into building materials. Since dark colors absorb much more sunlight than light colors do, white or metallic surfaces should be used in hot climates. Even in temperate areas, light colors should be used for protection during warm spells.

Color is particularly important when little or no insulation is used. It has less of an effect as the level of insulation is increased.

The ideal *dynamic envelope* of a skin-load-dominated building would be black in the winter and white in the summer. Since no such surface yet exists, the exterior surfaces on which the summer sun shines in warm climates should be light in color. In cold climates, dark surfaces on the south face can increase the absorption of solar heat.

Roof color is important in conjunction with pitch, climate, and the level of internal heat gains. If the need for heating is greater than the need for cooling, dark-colored, steeply sloped rooftops may absorb solar radiation and transmit some of it into the building.

If the roof is not sloped, the major heat gain will occur in the summer when the sun is high in the sky, and very little heat gain will occur in the winter when the sun is low on the horizon. Light-colored rooftops in those cases will reflect sunlight away from the building and reduce its cooling load.

When both heating and cooling are important, heavy snow conditions may negate the effect of dark surfaces anyway, so light-colored surfaces provide the best overall benefit. White gravel ballast or light-colored membranes, shingles, and other surface materials can suffice.

Interior Layout

While the building location on the site is being established, the rough shape defined, and shading and orientation selected, the interior spaces should be laid out at the same time for overall compatibility.

Spaces with the maximum heating and lighting requirements should be located along the south face of the building, while buffer areas (toilet rooms, kitchens, corridors, stairwells, storage, garage, and mechanical and utility spaces), which need less light and air conditioning, can be placed along the north or west wall. Rooms for occupations that require the greatest illumination levels, such as typing, accounting, reading, and drafting, should be located next to the windows and natural lighting, while rooms needing few or no windows for light or view, such as conference rooms, can be located away from window areas. High internal-gain spaces requiring extensive cooling should be located on north and east exposures.

In rooms with large windows, the windows should open to the south if winter heating is needed; summer sun pene-

tration can be controlled by overhangs or other shading devices. In climates where heat loss is not critical in the winter, windows can open both to the north and south. Windows in internal-load-dominated spaces should open to the north or east. In any case, windows on the western side, which are so troublesome for summer cooling, should be kept as small as possible.

Spaces that use passive solar heat or daylight should have a depth in from the south wall no greater than 2½ times the window height (from the floor to the top of the windows). In most cases, this amounts to a limitation of 15 to 20 feet. This ensures that the direct rays of the sun will penetrate the entire space. The limit is about the same even when the heating comes from an opaque radiant thermal storage wall, since this limit is considered the maximum distance for effective radiant heating from a solar wall.

Spaces that need to be deeper or in which large south-facing windows are undesirable can let sunlight in through south-facing clerestory windows or skylights. Note, however, that horizontal skylights gain the most solar heat in the summer, when the sun is directly overhead, and the least during the winter, when the sun angle is too low for solar heat to be effectively transmitted through. Admitting sunlight into a space through the roof has the advantage of allowing flexibility in distributing light and heat to interior parts of a building.

Building entries should be placed so that they are oriented away from and receive the greatest protection from prevailing cold winter winds. Another way to reduce infiltration is to design buffered entries. Providing an air lock, vestibule, or double entry decreases both the infiltration and conduction transmission through the entry area. Unheated areas such as garages, mud rooms, or sun spaces placed between doors leading to the conditioned space and the outside dramatically reduce the air change rate caused by people entering and leaving.

Conventional designs in recent years have made it necessary for fans to operate in order to supply ventilation and some cooling even during periods when outside temperatures don't require heating or cooling. The energy for fans and air conditioning in these cases can be avoided if the interior spaces are laid out in a linear plan to facilitate natural ventilation (Figure 7.6).

Energy consumption in both constructing and operating buildings can be minimized by grouping similar functions and needs in proximity to the resources required for those functions at the initial conceptual design stage. This consolidates and thereby minimizes the magnitude and complexity of the distribution networks providing the required mechanical services.

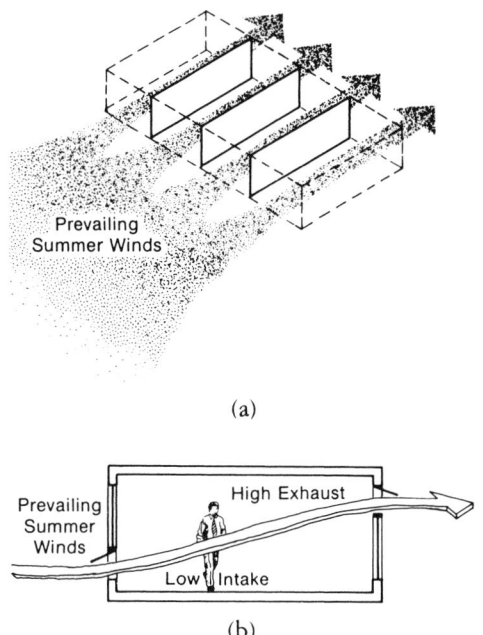

(a)

(b)

FIGURE 7.6. Interior layouts should be conducive to natural ventilation. (a) Partitions should be arranged so as to channel prevailing summer breezes straight through the building. (b) Openings on the leeward side should be higher than openings on the windward side.

For example, a lecture hall, auditorium, operating room, conference room, or any other room needing both an isolated and closely controlled environment is well suited for an interior or subsurface location. Areas with the most frequent public activity should be located on or near ground level. Closed offices or industrial activities with less frequent public contact can be removed to higher levels and more remote locations. Mechanical spaces should be located with consideration of the needs for acoustic isolation and restricting public access, in conjunction with outside air requirements, access for equipment replacement, and closeness to outside equipment such as condensers or cooling towers.

Creative Solutions

Ideally, a building envelope should moderate temperature extremes both daily and seasonally. On sunny winter days it should be able to be opened up to the sun's heat. At night, it should be able to close out the cold and keep the heat in. In the summer, it should be able to do just the opposite: close off the sun during the day, and open up at night in order to release heat into the cool night air.

A number of imaginative ideas have been suggested to allow a building to respond dynamically to its environment.

As illustrated in Figure 7.7, the entire structure can be built on a rotating turntable. In this way, the building can have one open exposure directed toward or away from the sun as desired. It can also track the sun as it moves across the sky in its daily cycle.

Another suggestion, also shown in Figure 7.7, is the use of barrier screens that can be moved around a building in response to the movement of the sun. This kind of shading device can also serve as a windbreak.

Space-age ideas such as these are generally impractical because of the tremendous construction cost involved. It is sometimes useful, however, for a designer to let his or her imagination run wild and then see how to adapt the results to more practical design solutions. There is certainly room for a great deal more creativity in the area of dynamic envelopes.

Insulation and Thermal Mass

The building envelope is a device through which heat exchange between the internal and external environments is controlled. The various modes of operation of an envelope are (1) admitting heat gain, (2) excluding heat gain, (3) containing internal heat, or (4) dissipating excess internal heat. The opaque portions of the envelope, once designed, are generally considered fixed controls. The dynamic elements of the envelope include operable window sashes, window shading devices, and insulating shutters.

The effect of insulation is to reduce heat gain and heat loss. The more insulation in a building's exterior envelope, the less heat transferred into or out of the building due to a temperature difference between the interior and exterior. Insulation also controls the interior MRT by isolating the interior surfaces from the influence of the exterior conditions, and also reduces drafts produced by temperature differences between walls and air.

Insulation along with infiltration control is important for reducing heating and cooling loads in skin-load-dominated buildings. Increased insulation levels in internal-load-dominated buildings, however, may cause an increase in energy usage for cooling when the outside air is cooler than the inside, unless natural ventilation or an economizer cycle on the HVAC system is available.

Types

Insulation is made from a variety of materials and in several forms. The forms generally fall into the following categories: (1) rigid or semirigid blocks or boards, (2) boards with impact- or weather-resistant surfaces, which are employed on building exteriors or below grade, (3) blankets,

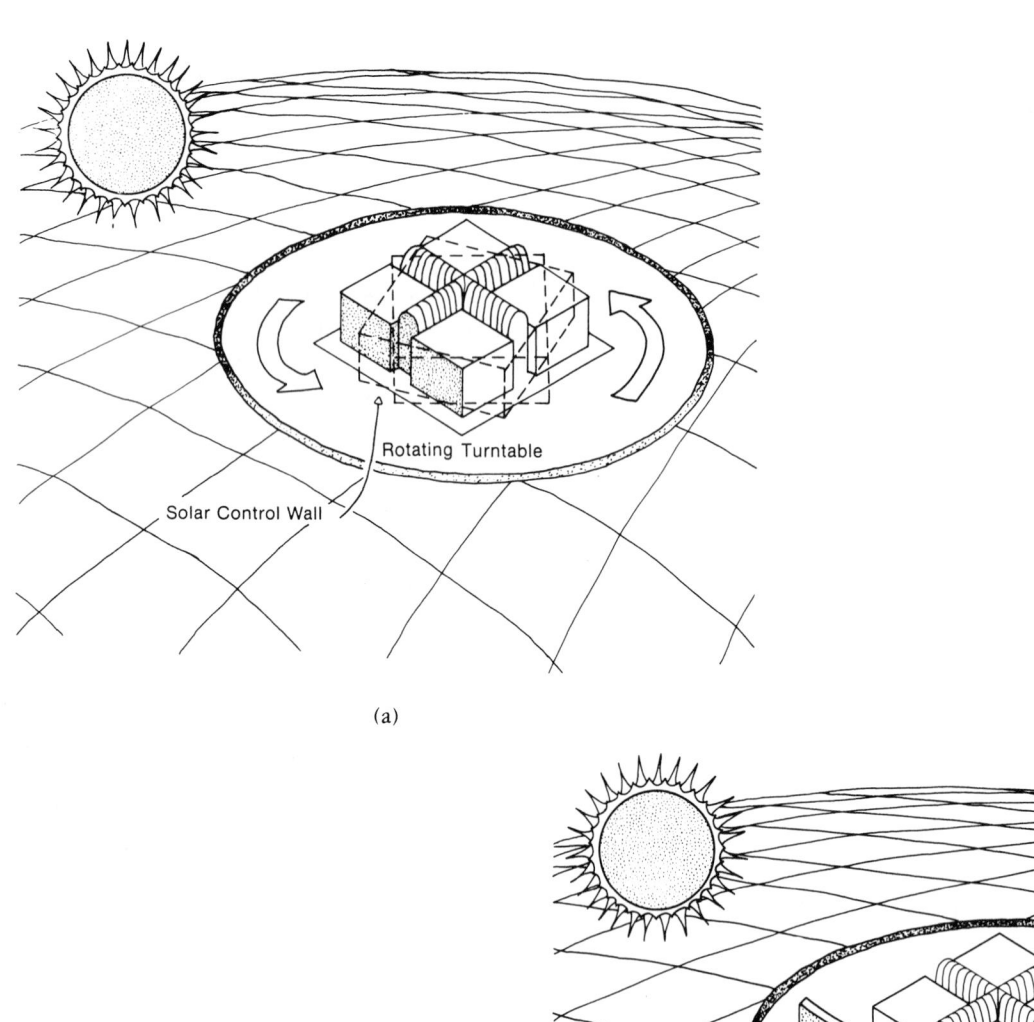

(a)

(b)

FIGURE 7.7. Dynamic envelope (high-tech solutions).

felts, or batts (sheets), which are either mechanically attached to vertical surfaces or laid flat on horizontal ones, (4) loose fill, which is poured or blown into cavities or onto flat surfaces such as above ceilings, and (5) foams and dry spray-on types, which can be pneumatically applied in a variety of ways. (The more rigid insulations may have some structural value.) When specifying insulation, both performance and any complications arising from the thickness required must be considered.

Rigid.   In the first category are polystyrene, polyurethane, and polyisocyanurate. Polystyrene comes in the form of *beadboard*—so named because it is manufactured from small styrofoam beads which are puffed up and fused together into slabs—and extruded polystyrene. The latter has the advantage of some compressive strength, which makes it suitable for insulating beneath heavy objects. High-density beadboard also has a high compressive strength.

Both burn readily and give off a dense, black smoke. Polyurethane and polyisocyanurate are harder to ignite but give off cyanide fumes in a fire. Since these fumes are a hazard for firefighters as well as for occupants attempting to evacuate a building, these materials are not permitted on any exposed interior surface and must be covered on the inside with a fireproof wall of plaster or sheet rock.

The most common way of insulating masonry walls is to affix rigid sheets of insulation to the wall surfaces and cover them with a protective material. For a given thickness, the most effective insulating material is polyurethane foam. One inch of urethane is equivalent to about 2 inches (5 cm) of fiberglass. The rigid sheets are available in ½- to 2-inch (1.3- to 5-cm) thicknesses.

Another type of rigid insulation is sheathing board made of rigid board foam with foil facing on both sides, or processed wood or vegetable products impregnated with an asphalt compound to provide water resistance. It is usually affixed to the exterior of a frame wall in addition to flexible blanket insulation between the studs. The sheathing board also serves as a significant structural component of the wall.

Blanket.   This type of insulation is most commonly used in standard cavity walls, where the depth of the stud determines the amount of insulation that can be placed in the wall. The material usually consists of glass fiber or mineral wool. It is manufactured in standard widths to match either 16-inch or 24-inch (40-cm or 61-cm) stud and joist spacing, is 3 to 7 inches (7.6 to 18 cm) thick, and comes in long rolls or in batts of a specific length. It is available with reflective foil or a vapor barrier on one side. One advantage of fiberglass is that it is highly fire-resistant. Its drawbacks are that it loses its effectiveness when wet and it is not self-supporting in its normal form.

Loose Fill.   Loose fill insulations that are commercially available include cellulose, vermiculite, and blown-in fiberglass. Sawdust, shavings, and shredded bark can also be used. These materials are principally used in existing walls that were not insulated during construction. They are also commonly added between ceiling joists in unheated attics. Vermiculite and perlite are mixed with concrete aggregates to reduce heat loss.

Foam-in-Place.   Foams such as polyurethane are also available in liquid form with a catalyst for on-the-job foaming. The liquid may be poured into forms or sprayed on with special equipment. In the hands of a skilled applicator, this material can be rendered into almost any sculptural form and will provide considerable structural support. It is applicable to odd-shaped structures but needs a weather-protective membrane.

Prefab Panels.   Premanufactured panels are available with various sidings combined with foam insulation. A semi-rigid blanket-type insulation faced with a vapor retarder is available for use on precast concrete and masonry walls. Another special product is a floor insulation system consisting of a PVC extrusion designed to be suspended below framing members in order to hold a foam board and blanket insulation above it, thereby completely isolating the framing members.

Some of these panels have been designed especially for retrofit markets as a way of avoiding the expense of opening up walls to install blanket insulation, as well as the health and flammability problems linked with some forms of blown insulation. Such products include drywall panels attached to foam for interior retrofit and exterior panels added directly to a building's face.

Natural Insulations.   The insulating property of vines and plant materials covering building walls is achieved by supporting them on a vertical trellis system set off from the building in order to protect the facade. A dense mat of ivy traps air in the space between the trellis and the building wall, creating an effective wind and air *buffer* zone. In the summer the same vines block the sun, cool the air next to the wall by evapotranspiration, and move the air by creating a thermal chimney effect between the vines and the building.

Additional benefits are the ability of the vegetation to catch dust and buffer noise.

Sinking a building into the earth insulates as well as buffers the building. About 30 inches (cm) of packed dry earth is equivalent to 1 inch (cm) of urethane. Blocks of sod and moss are commonly used in northern countries where insulation is particularly important.

As siding, wood is significantly better insulation than plaster. Sawdust is almost as effective as fiberglass, although it tends to settle in vertical wall cavities. Dried grass and straw matting work well as roof insulation although they are extremely flammable and decompose relatively quickly.

Roof Insulation.   Roof insulation should preferably be placed on the inside of a building for proper performance. Insulation, however, is relatively soft as a platform on which to build anything. If it is not initially well attached, or becomes delaminated or detached due to storm, construction, or service damage, it can compromise the integrity of the roof. Some new roofing systems place the insulation above the membrane.

The most common form of insulation for flat built-up roofs is rigid panels that are tapered for proper drainage. There are a variety of insulation systems to select from, and that selection should be made carefully to make sure that it is compatible with the total roofing system. If structural supports can accommodate it, snowfall can be retained by a parapet around the roof perimeter in order to provide additional insulation in winter.

Thermal Mass Effect

Conventionally constructed envelopes tend either to have controlled U-values through the use of insulating materials, or to be very dense and heavy in weight (Figure 7.8). Massive constructions result in thermal time lags (building temperature lags behind the outdoor temperature), which tend to produce more stable interior conditions (see Figure 3.9).

One illustration of the influence of thermal mass is the adobe structures indigenous to the southwestern United States. Due to the low thermal resistance of the masonry-like material, a calculation on the basis of U-value alone would predict that the structure should need a great deal more energy for heating and cooling than is actually the case. As heat penetrates the wall in the daytime, its heat capacity soaks it up like a sponge until the interior surface temperature of the wall rises above the interior air temperature. Only at that time will heat travel to the interior. As a result, there is a time lag between the outside temperature (and solar heat) and the heat gain inside the structure.

Similarly, since the dense, thick wall is holding so much heat at night, it takes a long time to dissipate the heat to the cold outdoors. Thus, as the outside temperature fluctuates, the indoor temperature remains more stable. The greater the mass and thermal capacity, the more stable the indoor temperature.

If the outside temperature fluctuates around a comfortable temperature and the building has enough mass, the structure may not need any active heating or cooling at all. Opening windows at appropriate times in the day/night cycle and collecting solar energy can temper the average temperature.

Massive west walls, east walls, and roofs can greatly minimize solar heat impacts in the summer. Lightweight constructions, by contrast, are more sensitive to short-term temperature and solar impacts. Table 7.3 gives approximate time lags for various materials.

The dynamic thermal performance of a masonry wall is even better when insulation is placed on the exterior side of the wall (Figure 7.8). Insulated thermal mass is the means of storing heat in passive solar applications. An insulated mass in cold climates can also absorb excess internal heat gain and release it to the space during unoccupied periods, thus minimizing overall heating requirements.

Earth-sheltered buildings employ a special kind of thermal mass. The thermal mass of earth against the exterior walls helps keep the interior at a uniform temperature day and night, throughout the year.

Superinsulation

Superinsulation is the application of greater than normal amounts of insulation in order to eliminate all need for mechanical space heating. Internal and solar heat gains become the primary source of heat. The savings on heating equipment and distribution systems may equal or outweigh the additional costs of extra insulation, extra thermal mass, and insulative window treatments.

The crucial concern in superinsulated buildings is preventing overheating both in summer and on moderate days in winter, when solar gains may be especially high. Designers of internal-load-dominated buildings must be particularly careful not to eliminate the positive aspects of heat dissipation in cool weather. The superinsulation level is generally much lower for these types of buildings.

Since insulation is needed at all exterior surfaces, it can add significantly to construction costs. The desirable amount of insulation must be carefully evaluated.

At some point, there is a diminishing return in energy savings on each increment of added insulation. The objective

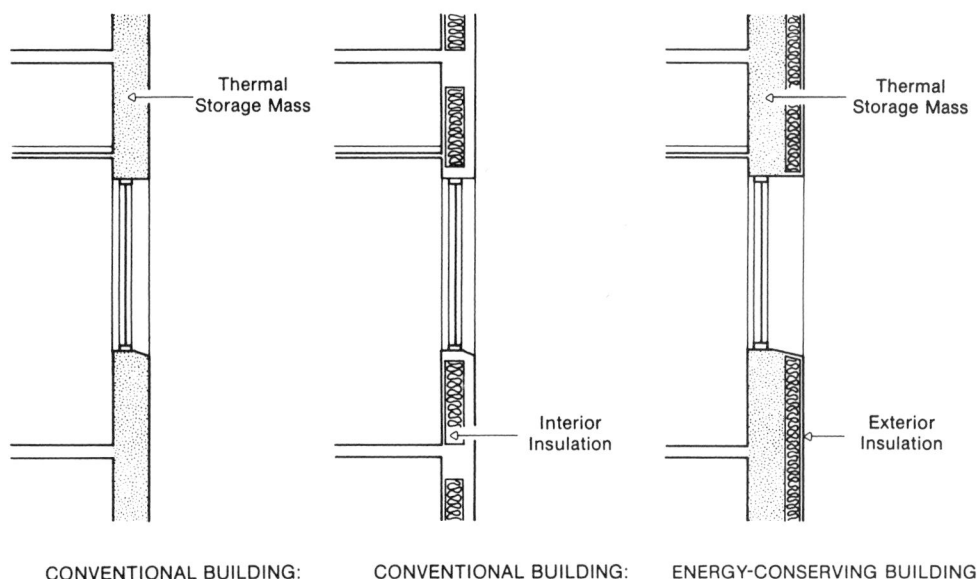

CONVENTIONAL BUILDING:
Thermal Storage Mass

CONVENTIONAL BUILDING:
Insulated

ENERGY-CONSERVING BUILDING

FIGURE 7.8. Insulation and thermal mass.

of superinsulation is to justify additional insulation on the basis of reduced mechanical equipment costs. Even so, the point of diminishing returns, at which more insulation ceases to provide sufficient economic benefits, may still be reached. This will be elaborated upon in Chapter 16.

Fenestration

A building's envelope forms a boundary between the interior environment and the variable external environment. *Fenes-*
tration refers to glazed openings in the envelope for the purposes of:

1. Illumination
2. View
3. Solar heat gain
4. Natural ventilation

Approximately half of the energy in the solar radiation that reaches the earth's surface is at wavelengths too long to be visible. While the visible light provides illumination, this

TABLE 7.3   TIME LAG FOR VARIOUS MATERIALS

Material	Thickness		Time Lag
	in.	cm	
Brick	4	10	2½ hr
	8	20	5½ hr
	12	30	8½ hr
Concrete (sand and gravel aggregate)	4	10	2½ hr
	8	20	5 hr
	12	30	8 hr
Insulating fiberboard	2	5	40 min
	4	10	3 hr
Wood	½	1	10 min
	1	2.5	25 min
	2	5	1 hr

infrared radiation is a useful source of heat. Ordinary clear glass transmits almost all wavelengths of solar radiation except the short ultraviolet component. It thus serves as an effective source of heat and light, as well as providing a feeling of spaciousness and freedom to the occupants.

Despite these potential benefits, fenestration can be an energy liability. Heat transmission through a window can be more than 12 times as much as through a well-insulated wall of equal area. Window frames also transmit heat to a varying extent, depending on materials and construction, and pass infiltration through cracks between the sash and the frame. Weather-stripping and gasketing are important components of any operable fenestration.

When the primary objective is minimizing heat transmission through the envelope, window areas should be reduced as far as the psychological need for visual contact with the outdoors will allow. But before automatically adopting that solution, consider the value of passive solar gain, light, and view through the fenestration.

Selection of Glazing Material

The choice of glazing material should be based on the desired overall performance. The pertinent factors include (1) light transmittance, (2) thermal performance, (3) life-cycle costs, (4) strength and safety, (5) sound reduction, and (6) aesthetic considerations.

Glazing materials commonly used are sheets (or panes) of glass or plastics that have good light transmittance. In some applications, such as artists' studios and showroom windows, clear glass may be desirable for maximum light transmittance for daylighting or merchandise visibility. Regular clear glass is also generally used for commercial building lobbies or any other area where visibility between the interior and the exterior is important.

The choice of colored glass depends on where and how it is used. A warm-toned bronze or gray glass may best complement an interior or exterior color scheme.

Tinted glazing, in general, can aid in controlling glare and year-round excess solar heat gain, or in modifying an unattractive or distracting view. Tinted or reflective glazing can also provide privacy to building occupants during daylight hours while maintaining a degree of outward view. Privacy through tinted glazing requires a substantially higher illumination level outside than inside. Reflective glass should be avoided on exposures that could reflect annoying glare onto nearby buildings or traffic.

Since reflective glass reduces the amount of solar gain all year, some form of dynamic shading may be more desirable when solar heat would be useful in the winter. An analysis of the energy impact over an entire yearly cycle of weather is required to ascertain any net energy benefit. Generally, reflective glass has a greater advantage in southern climates than in northern ones.

Heat-absorbing glass absorbs selected wavelengths of light. Since heat is absorbed into the glass rather than reflected, the glass itself becomes warm. This limits the illumination levels while retaining some of the solar heating qualities. Window glass manufacturers offer a range of combinations of reflective and absorptive properties for optimum selection.

Patterned, etched, or sandblasted glasses are available to provide diffuse lighting.

Even though the different types of glass—clear, heat-absorbing, reflecting, or light-diffusing—differ in their solar transmission characteristics, they all transfer heat by conduction at about the same rate.

The different sides of a building need not—and, with regard to thermal performance, generally should not—be identical in appearance. Each face may need a different response to solar, wind, and view exposure. For example, while reflecting and heat-absorbing glass is rarely needed on north, north-northeast, and north-northwest orientations (except for glare control), it may be useful on the west for limiting summer afternoon heat gain.

Although the primary factor in reducing sound transmission through a building envelope is its airtightness, heavier glazing also reduces noise intrusion from the outside.

In addition to its thermal, visual, and aesthetic functions, glazing must be selected for structural integrity. Glazing materials must be strong enough to resist breakage due to wind loads and thermal stress.

Local building codes specify glazing thicknesses for various conditions of height and wind speed. Manufacturers' recommendations should also be followed. It is suggested that wind tunnel tests be run for tall or unusually shaped buildings, and for structures whose surroundings may create unusual wind patterns.

Tinted or reflective glass is especially vulnerable to thermal stress. Warm air from floor registers directed at heat-absorbing glass in winter has been known to cause breakage from tension stresses on the glass edge. Manufacturers' recommendations should be followed regarding thermal stress capabilities.

For safety reasons, building codes may require glazing to be shatter-resistant in certain circumstances. Tempered, laminated, or wired glass as well as some plastics may be used to satisfy these requirements. Federal Standard 16 CRF 1201 sets breakage performance requirements.

Acrylic and other plastic glazing materials are available as an alternative to glass. Transmittance values of these plastics range from 10 to 92 percent and reflectance values from 4 to over 60 percent. Plastics are available in translucent as well as transparent forms. They may also be corrugated or tinted in a variety of colors.

The major drawbacks to plastic panels for glazing is that they are subject to deterioration from sunlight and damage by abrasion. Some expand and contract in response to temperature extremes to a greater extent than glass. Recently developed products have reduced these problems and are being used for special glazing applications.

Solar Control Devices

Window treatments are what enable fenestrations to be dynamic elements of the envelope. Some provide selective shade or open up to let sunshine in when desired.

Multiple glazing and operable thermal insulation reduce conduction transmission.

The shading effectiveness is expressed in terms of shading coefficient (SC), as discussed in Chapter 4. The lower the SC, the less the heat gain. The U-value describes the transmission quality of the device.

Retrofitting existing buildings with suitable solar control devices can achieve substantial reductions in energy consumption. It is important, however, to select the most appropriate method or combination of methods for a given building.

Window treatments generally fall into the categories of interior, exterior, and modifications of the glazing itself. Some examples of special window treatments commercially available are shown in Figure 7.9.

Interior.   Interior shading devices, such as roller shades, venetian blinds, and draperies, provide the least effective shading and thermal insulation, but they offer easy operation by room occupants. Traditionally, internal shading devices have been primarily intended to provide varying degrees of privacy from the outside, and to minimize discomfort from the MRT at the window.

Venetian blinds (either horizontal or vertical) afford full privacy when completely closed. When draperies are closed, the degree of privacy depends on their color and tightness of weave and on whether the principal illumination comes from the interior or exterior. Fully opaque materials obscure the view so completely that neither shadows nor silhouettes are detectable.

Curtains uniformly reduce sunlight, while blinds allow both a reduction in intensity and a redistribution of light. Like tinted glazing, draperies provide protection from glare, or can be used to modify an unattractive or distracting view. Open-weave dark-colored fabrics of uniform pattern permit maximum outward vision; unevenly patterned weaves and pale colors tend to reduce it. A semi-open weave modifies the view without completely blocking out the outdoors, whereas tightly woven fabrics cut off outward vision completely.

Indoor shading devices, especially draperies of closely woven fabrics, can effectively absorb sounds originating within a room and help reduce sound levels within a room, although they do not attenuate sound transmission through the fenestration.

When fitted with a foam or other insulative backing, conventional draperies or shades (Figure 7.9a) become *thermal barriers*. Operational insulative shutters, located on the interior or exterior, are rigid window insulations that may be in a hinged, sliding, folding, or bi-folding configuration. If on the inside, they may be *manually operated;* if on the outside, they need some form of *mechanical operation*. Interior insulating devices are limited by available wall storage area for them when they are not covering the window. It is important to provide an airtight seal around the edges of any operable insulation device to preserve its thermal performance and to prevent condensation.

Exterior.   The most effective way to reduce the solar heat gain through fenestration is to intercept the direct radiation from the sun before it reaches and penetrates the glass. Exterior shading devices dissipate their absorbed heat to the surrounding air, while inside drapes and blinds create a heat trap within the conditioned space. Fully shaded windows achieve a solar heat gain reduction of as much as 80 percent. The tables in Appendix B3 may be used to determine the effect of shading devices during the critical times when shading is needed.

The same techniques for shading exterior surfaces in general are also applicable for exterior shading of windows; however, a variety of other methods may also be employed. Fenestration can be shaded to some degree by roof overhangs, vertical and horizontal architectural projections, recessed glazing, awnings, trellises, exterior louvers, insect screening, mesh screening, and screens of miniature fixed metal louvers.

Since overhangs do not block the summer sun on east- or west-facing glass, vertical louvers or extensions may be used instead. These may be used instead of or in conjunction with reflecting or heat-absorbing glass. Vertical elements can be designed to vary their angle according to the sun's position. Such movable vertical louvers can provide SCs as low as 0.10 to 0.15.

Egg-crate shading devices are a combination of vertical and horizontal elements. They can be very effective shades,

(a)

(b)

FIGURE 7.9. Some window treatments commercially available. (a) Insulated window shade. When in the closed position, it isolates the interior space from the window by the magnetic strips on either side and a top and bottom seal. It conveniently rolls up to open like a standard window shade. (Photo courtesy of The Warm Company.) (b) Accordian-style operable exterior "hurricane" shutters such as these reduce heat loss and heat gain, reduce noise intrusion, protect patio doors from rain and wind damage, and can be locked from the inside or outside to provide security protection. The shutters shown are made from extruded aluminum. They slide from left to right or right to left and can be motor operated. When not needed, they retract completely for an unobstructed view. (Photo courtesy of Alutech United, Inc.) (c) Rolling shutters are available with curved guide rails mechanically bent to exactly fit over curved glass panels of any pattern. (Photo courtesy of Alutech United, Inc.) (d) Motorized blinds at the Eugene Public Library in Eugene, Oregon, automatically open and close to modulate solar heat gain. (Photo by Vaughn Bradshaw.) (e) Adjustable solar-collector louvers can be constructed to control heat gain and loss as venetian blinds would, while water circulating within the louvers collects solar heat and carries it to storage.

providing SCs below 0.10. The horizontal elements can also control glare from ground reflections.

East and west walls can be shaped in a "sawtooth" pattern to face glazing toward or away from the sun as the climate dictates.

Removable or retractable awnings, as well as operable exterior louvers or shutters, can be adjusted to the season. Such operable shading devices on the outsides of buildings are difficult to maintain and are subject to more rapid deterioration. Awnings may be the simplest and most reliable movable shading devices, but their aesthetic appeal is limited.

Mesh materials reduce the intensity of sunlight and provide some degree of diffusion. These materials typically consist of a loosely woven fiberglass fabric with a weave

(c)

(d)

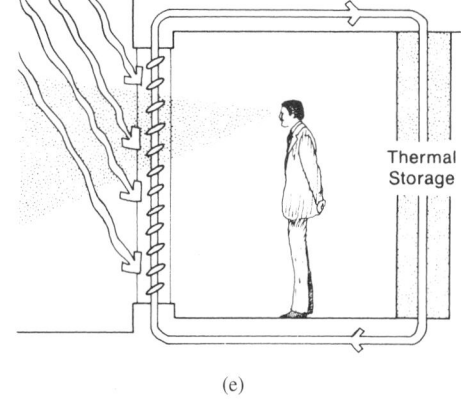

Thermal
Storage

(e)

FIGURE 7.9. *(Continued)*

especially designed for intercepting sunlight. They are mounted in frames over the windows. The reduction in solar intensity is due simply to the amount of opaque area of the fabric. Mesh materials have a fairly long life span.

Louver materials are an improvement over mesh materials since they have a vertical directionality. They are available as tiny fixed louvers forming a screen that allows an unobstructed view from inside to the outdoors. The louvers are usually arranged to admit much more sunlight when the sun is low. They thus transmit much of the solar energy in winter while blocking almost all direct radiation in the summer. The shading behavior may be varied in the manufacturing process by adjusting the dimensions and configuration of the louvers. There is less distortion of the view through the windows than with mesh materials, but louvers allow no upward view at all. The life expectancy of the metallic screen is fairly high, provided that it is protected from impact.

Exterior operable blinds have the operational advantages of interior blinds, together with the advantage of being able to intercept sunlight before it enters the building as external shades do. The tilt of the blinds may be controlled either by remote manual control, or automatically by a time clock or sun-tracking device. Like venetian blinds, these devices can be raised and lowered.

It is important that air be able to move freely around any exterior shading structure in order to carry away the heat absorbed by the shading and glazing materials. Manufacturers usually provide instructions for proper ventilation between the glazing and their shading devices.

Glazing Modifications. Operable shading such as miniature venetian blinds placed between two layers of glass is not as effective as an exterior device, but is more effective than an interior shading device.

Double glazing separated by a 1/2-inch (1-cm) air space cuts heat loss through a single-glazed window in half. Two separate windows (such as a single-glazed window with storm window) 1 to 4 inches (2.5 to 10 cm) apart produces the same reduction. Triple glazing generally cuts the heat loss to two-thirds of the loss through a double-pane window.

Double glazing is usually appropriate for climates with an average winter temperature above 30°F (0°C). Triple glazing is usually recommended when the average winter temperature is colder than 30°F (0°C) or if there are more than 4,500 degree days per year. When the appropriate number of glazing layers is applied for a given climate, the solar gain may turn the window from an energy liability into a source of net energy gain.

When heat-absorbing glass is used in multiple glazing, it should be installed in the outer panel so that the absorbed heat can more readily dissipate back outside.

Plastic films are available that are designed to be glued to the inside face of window glass to intercept the sun's energy before it enters. These offer the same properties as reflective and absorptive glass. They are also available in many combinations of reflectivity and absorptivity, and can be either reflective (converting windows into tinted one-way mirrors) or darkening (creating the effect of sunglasses). Silver- and gold-tinted reflective films block out slightly more total radiant energy than visible light. Bronze- and smoke-colored reflective films intercept more of the visible light. The silver-colored film is the most effective at reflecting solar radiation, and it can reject up to 80 percent of the incident solar energy.

During the winter, films reflect interior heat back inside, saving some of the heat that would otherwise be lost out the window and thus improving the MRT. Drafts due to cold glass surfaces are also reduced, and higher RHs can be maintained without condensation.

Films make window glass more shatter-resistant. In general, they should not be used on thermal pane or other self-contained insulated windows, since the film can change the coefficient of expansion of the glass, resulting in cracking under the pressure of thermal expansion and contraction. For the same reason, tinted film should never be applied to tinted glass, nor should it be applied to very large areas of glass. In the latter case, the differences in heat absorption between areas in full sun and other areas that are temporarily shaded can cause cracking. The principal drawback of films is the fragility of the plastic surface in comparison with treated glass and their limited service life.

An imaginative approach to a dynamic envelope is a radiant energy trap arranged to intercept the solar heat before it gets into the occupied space and to collect it for storage and later use. This concept is shown in Figure 7.9e. Large, hollow louvers either in the interior or between the two layers of double glass can adjust to let sunlight in or reflect it away. Water is circulated through the louvers to carry away the absorbed heat. When space heating is needed, heated water (preferably recovered heat) can be circulated through the louvers for improved occupant comfort.

Earth Sheltering

The temperature of the earth varies widely from region to region, from season to season, and even over the short distances between neighboring sites. Light, dry soil reaches a near-constant ground temperature at a shallower depth than heavy, damp soils. The average earth temperature at a given site is generally 2° to 3°F (1° to 1½°C) above the average annual air temperature, but the earth temperature fluctuates much less than the air temperature. Thus, *earth sheltering* provides a warmer environment in the winter and a cooler one in the summer.

Earth sheltering can minimize interior temperature swings during all seasons. But it should be considered within the broader context of the overall control strategy. If an earth-sheltered design compromises the effectiveness of other passive control techniques that are more effective, it is counterproductive.

Earth sheltering is appropriate for all climates except hot humid ones. In those regions the ground tends to be too damp, and the moist summer air condenses on interior walls as it is cooled by the ground.

In the southeastern United States, the most effective passive cooling strategy is natural ventilation. Earth sheltering is beneficial there only when it doesn't appreciably interfere with ventilation.

By contrast, ventilation is not a desirable passive control strategy in the southwestern United States. There, the high mass of the earth can be very beneficial. Buildings may be completely sunken with a recessed interior courtyard into which cool, dense night air tends to slide. Various passive evaporative cooling techniques may also be used. Growing plants on earthen roofs and courtyards helps cool by evapotranspiration.

In cold climates, earth sheltering can provide protective benefits from exposure to cold winds and the air temperature in general.

Underground construction practically eliminates excess infiltration, but it introduces a need for mechanical ventilation. Energy is conserved, however, since only the air required for adequate ventilation needs to be heated or cooled to room temperature.

Waterproofing, structural design, adequate fenestration area, and adequate ventilation are some of the major concerns of underground construction. A rule of thumb is that the air in underground spaces should be kept in motion by natural or mechanical ventilation to avoid condensation.

Concrete walls, if left unsealed, will absorb moisture from the air and drain it away into the earth. Where condensation is a potential problem, either the air must be dehumidified or the walls must be warmed.

Interior spaces should be oriented to make maximum use of daylighting, clerestory windows, and internal courtyards. However thermally effective earth sheltering may be, its acceptability depends on the designer's skill in introducing daylight, access, and views.

Under the right conditions, earth sheltering can result in substantial savings on exterior finish materials, as well as reducing transmission and infiltration through walls and openings. Other advantages of earth sheltering are that it is an effective sound insulator, so that the interior spaces are less intruded upon by ambient noise levels, and the construction does not interrupt any pre-existing land use (pedestrian or vehicular traffic, agriculture, etc.).

Internal Gain Reduction

Passive load reduction calls for minimizing all internal gains. Since the number of people in a building and their activity level are presumably a function of the building program, a portion of the internal gain cannot be reduced. Likewise, all equipment in use is presumed to be essential. Thus, the remaining potential for reducing internal gains lies with turning off unneeded equipment and minimizing energy for lighting.

Daylighting offers enormous opportunities for reducing cooling loads. The cooling load produced by incandescent lighting is 6 to 10 times greater than that due to daylighting for equal lighting levels.

Chapter 8 elaborates on daylighting. If daylighting is not feasible, fluorescent and H.I.D. lighting should be employed for maximum lighting efficiency.

Another area of great potential is the industrial environment. Process heating equipment can be better insulated and/or relocated where it won't contribute as much toward internal loads. Water tanks and drains can be covered. These measures not only reduce the internal heat gain, but may also save process energy.

Exhaust hoods can be used to capture heat rising from hot processes. If there are hot objects such as furnaces, molten material, or hot ingots, the major heat source is in the form of radiant heat. Radiant heat can be reduced by lowering the surface temperatures of the hot objects by insulation, water-cooling, or shielding.

PASSIVE SOLAR HEATING

Passive vs. Active

There are two distinct approaches to solar space heating: active and passive.

In general, active systems use outdoor collector panels to collect heat, and fans, pumps, and other mechanical equipment to transport it to and from an isolated storage unit. The medium for collecting and transporting the solar heat may be water or air. The solar-heated water or air is supplied from the storage unit to the spaces in the building in a completely regulated manner by a mechanical distribution system.

Passive systems, on the other hand, collect and transport heat by nonmechanical means. A passive solar heating or cooling system is defined as a system in which the thermal energy flows naturally by radiation, conduction, and natural convection, with no help from pumps or fans. The building is the system. There are no separate collectors, storage units, or mechanical elements. All of these functions are integral parts of the building structure and materials themselves. The cost of the system is therefore any additional cost of materials required for the passive design above that for an equivalent conventional building.

Since passive systems involve no mechanical equipment, they create little or no noise. Because of the absence of separate heat-delivery devices, passive systems can be essentially invisible from the interior.

Depending on the extent of the solar contribution and other design considerations, passive systems typically pay back this added investment with fuel cost savings in 1 to 13 years. A passive solar installation, with no moving parts, lasts as long as the building itself. The initial added cost of an active system normally takes more than 30 years to be amortized by fuel cost savings. An active system cannot be expected to last more than about 20 years before some of its major components need replacement.

Active systems are inherently able to more closely control the thermal environment within a building. Only as much heat is added to the space as is needed to satisfy a thermostat. Except for provisions for collector panels and storage space, active systems can be added to just about any building with no other design constraints. Since a passive system is the building itself, the entire building must be oriented toward the proper thermal performance characteristics. The form is highly dependent on local site and climate conditions.

Purely passive systems operate with the sun as the exclusive energy source. Active systems import energy—usually electricity—to power the fans and pumps that make the system work. Hybrid systems are primarily passive, but use a minor amount of mechanical aid for better control and to enhance their performance.

Design challenges resulting from the comprehensive impact of a passive solar system on building design include conflicts between solar access, space use, view, and ventilation. Special attention must be given to avoiding glare and overheating.

Every passive heating system (except roof ponds) includes two essential elements: south-facing glass (or transparent plastic) for solar collection and thermal mass for heat absorption, storage, and distribution. A third component included in many cases is some form of shade or covering for use when heating is not necessary. The passive features may be emphasized as an important element of a building's design, as in Figure 7.10, or they may be scaled to give the building a conventional appearance.

For example, glazing on the same plane as the building wall integrates the passive collection device with the building. Glazing sloped away from the wall plane increases its visibility and can dominate the form of the building. The impact of the glazing can be reduced by dividing a large expanse of glazing into smaller units with mullions or by modifying the appearance with trim details. Extensive window trim, however, can block a significant portion of the sunlight from the collector.

Since solar gain and storage are present in every building, nearly all buildings benefit from passive solar heating to some extent. It is only when a building is designed to optimize these effects and solar energy contributes substantially to the heating requirements that it is considered a solar-heated building.

Distinction Between Residential and Commercial Applications

Passive design principles are not limited to residential use, but it is important to distinguish between the small-scale externally loaded building—of which residences are a prime example—and internally loaded buildings. The form of skin-load-dominated buildings should be much more responsive to climate and site concerns. The form of internal-load-dominated buildings should also be designed in response to climate and site, but the appropriate responses have to do primarily with minimizing heat gain from the environment during the cooling season.

Large, internal-load-dominated buildings, such as office buildings with a well-insulated envelope, are generally more harmed than benefited by passive solar heating. Since internal gains satisfy most of the heating requirements, additional solar gain would overheat the space most of the time, creating a larger burden on the mechanical cooling system. The primary heating load on commercial buildings with well-insulated envelopes is the heating of ventilation air, which is not aided by passive solar systems. Thus, passive solar heat gain should generally be excluded from such buildings as much as possible at all times.

The *balance point temperature* (BPT) is defined as the exterior air temperature at which the internal heat generated exactly balances the heat loss through the skin or envelope

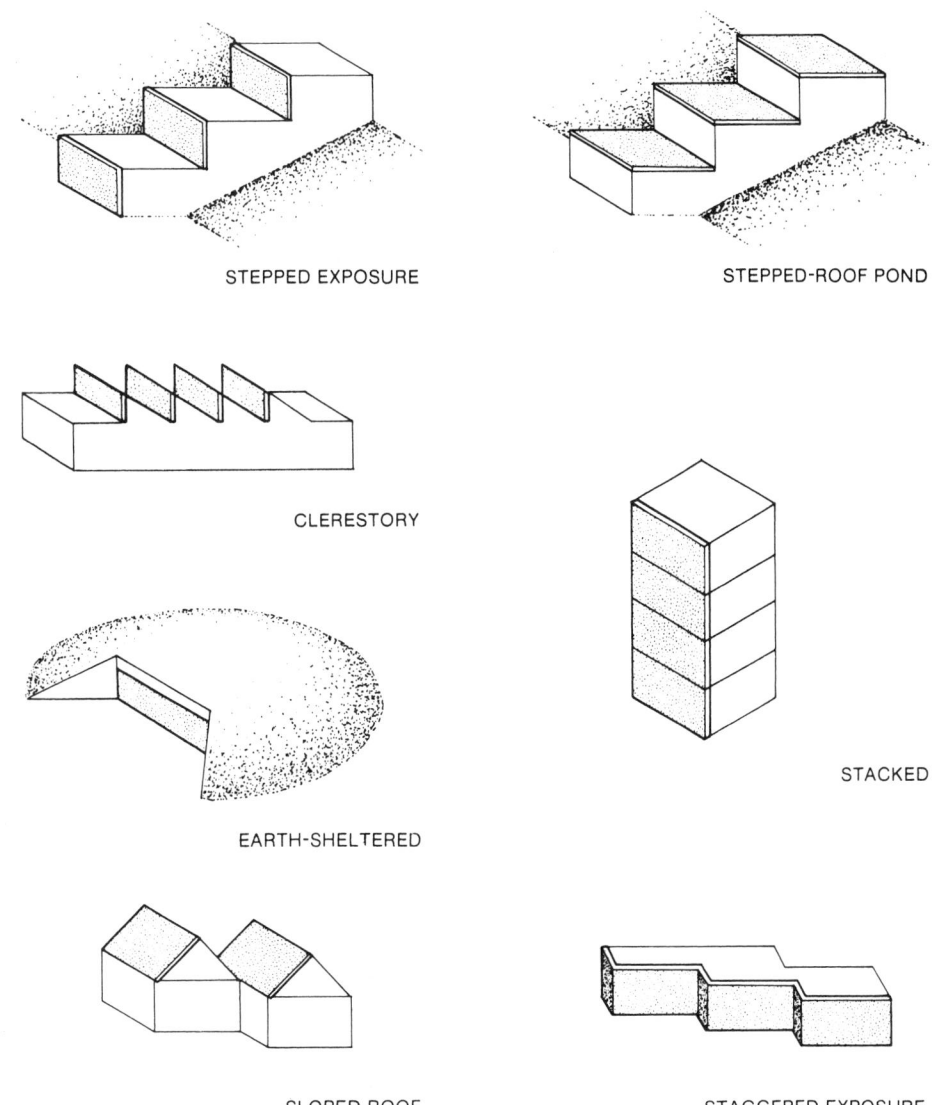

STEPPED EXPOSURE

STEPPED-ROOF POND

CLERESTORY

STACKED

EARTH-SHELTERED

SLOPED ROOF

STAGGERED EXPOSURE

FIGURE 7.10. Building shape options suggested by passive solar applications.

of the building. The building under full operation requires no heating energy input until the exterior air temperature drops below the balance point. Above the balance point, the internal heat gain must be rejected either passively or actively.

For example, consider a building with a heating load, as calculated in Chapter 4, of 1 million Btuh (294,000 W) at a design temperature difference of 50°F (28°C). Its heating load is 20,000 Btuh per °F (10,500 W per °C) of temperature difference between the inside and outside. If the building's internal heat gain (as calculated on the cooling load form in

Chapter 4) amounts to 600,000 Btuh (176,000 W), the balance point is 600,000/20,000, that is, 30°F (176,000/10,500, that is, 17°C) below the indoor design temperature. That means that no space heating is needed until the outside temperature drops to about that point (assuming that there is a free flow of heat between the areas where the internal gains are generated and the envelope).

If, on the other hand, the internal load is 1.2 million Btuh (352,000 W), the balance point would be 60°F (33°C) below the inside design temperature. In that case, no heating would be required at all as long as the outside temperature didn't drop below its design point, except for some possible heating during morning warm-up.

Passive cooling techniques and methods of reducing electrical lighting such as natural daylighting (discussed in the next chapter), task lighting, and the use of high-efficiency types of electrical lighting (fluorescent and H.I.D.) are the most important passive methods of reducing energy requirements in internal-load-dominated buildings.

Nonresidential buildings such as schools, airports, and other structures with low occupant densities can benefit from passive solar heating as well as daylighting.

The energy needs of a building must be defined at the earliest stage of design in order to integrate passive solutions into them.

Passive Solar Methods

There are three distinct categories into which passive systems can be divided: direct gain, indirect gain, and isolated gain. Each has advantages and limitations. Selection among them for any given application should be based on the specific design requirements of the space. More than one type may be combined on a project for optimum performance and cost.

The bibliography at the end of this chapter includes a number of books on passive solar heating. These present a great deal of design information, including various sizing methods. Edward Mazria's *The Passive Solar Energy Book* is especially recommended.

Direct Gain

The simplest and most efficient approach to passive solar heating is the concept of direct gain (Figure 7.11a). *Direct gain* refers to solar energy being introduced directly into the space through ordinary fenestration. The space itself is used as a solar collector, and it must also contain storage capacity to absorb excess daytime heat gain for later release at night or on overcast days. The occupied space thus serves as the collector, storage, and distribution system all together. The solar heating system is often indistinguishable from the space itself.

Such a system has an expanse of south-facing glazing—which can also serve other functions, such as providing ventilation, light, and views—and strategically located thermal mass within the space. Typically, the floor, walls, and/or ceiling are constructed of materials capable of storing heat. Otherwise, massive furnishings and other objects must be included in the design of the space.

Masonry and water are the most common materials used for heat storage. Typically, at least one-half to two-thirds of the total surface area enclosing the space—including interior partitions—is constructed of concrete, concrete block, brick, stone, adobe, or other thick, massive material. The ground around earth-sheltered buildings serves as thermal storage mass. When water is the thermal storage medium, it is usually limited to only one wall of stacked plastic or metal containers arranged so that direct sunlight strikes it for most of the day. It is important for the bulk of the mass to directly intercept the sunlight.

An additional benefit of the mass within the space is that it can passively keep room temperatures from rising too high during hot summers.

During the winter, the warm surfaces in direct-gain spaces provide an MRT higher than the air temperature, allowing comfortable conditions at air temperatures 5° to 10°F (3° to 6°C) below normal room temperature.

Site conditions and window treatments in direct-gain systems must be carefully designed to prevent the glare of bright winter sunshine from entering the space. The glazing material may be transparent or light-diffusing.

Operable window insulation can improve the thermal performance of the system by inhibiting loss of the solar heat back out through the glazing at night. Such insulation can improve system performance to such a degree that less glazed area is needed to supply adequate solar gain. Smaller window areas reduce summer heat gain as well.

Even though glass filters out much of the sun's ultraviolet radiation, enough is transmitted through to bleach paints, interior furnishings, and other building materials. Colors and materials that resist fading should be selected if they are to be exposed to direct sunlight for long periods of time.

Since the space is the solar energy system, retrofitting a building with a direct-gain system is not always feasible. Buildings constructed with masonry walls and floors and a southern orientation are more readily adaptable.

Relatively large daily space temperature fluctuations of 10° to 30°F (6° to 17°C) are characteristic of direct-gain systems. Uniform temperatures can be maintained by a conventional heating system, but some temperature swing is needed for the system to operate. Depending on circumstances, a properly designed direct-gain system can be 30 to 75 percent efficient in capturing and using incident solar energy.

In general, direct-gain system design demands a skillful and total integration of all the architectural elements—walls, floor, ceiling, and interior surface finishes—within each space.

Indirect Gain

An *indirect-gain* system places a thermal storage mass between the sun and the occupied space. The sunlight first

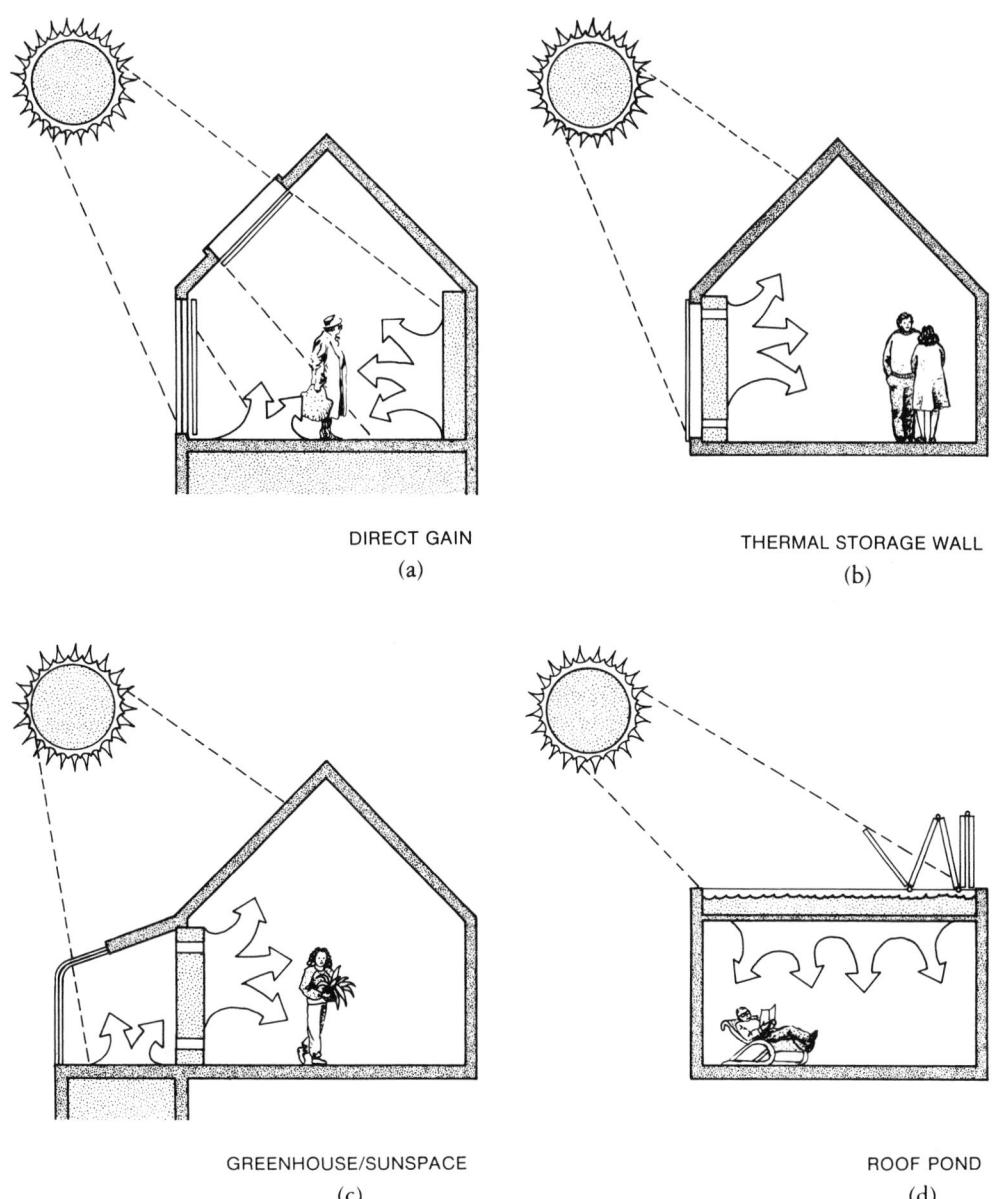

FIGURE 7.11. Passive solar heating methods. (a) Direct gain, (b) thermal storage wall, (c) sunspace/greenhouse, and (d) roof pond.

strikes the thermal mass, where it is absorbed, stored, and then slowly transferred into the occupied space.

There are three basic types of indirect gain systems: (1) thermal storage walls, (2) roof ponds, and (3) greenhouse/sunspaces. They all operate basically the same way, the only difference being the location of the storage mass. There are advantages and drawbacks associated with each.

Again, the thermal storage material may be either masonry or water. In addition to metal and plastic contain-

ers, a concrete vessel with a waterproof lining may be used to contain the water.

Thermal Storage Wall. The *thermal storage wall* (Figure 7.11b), usually constructed of masonry, absorbs sunlight and then conducts it through to the occupied space. The outside surface of the wall is usually painted black or some dark color and covered with a sheet of glazing. Vents at the top and bottom of the wall may or may not be added to circulate room air between the glazing and the wall for rapid warming when the sun is shining. Due to thermal lag, the heat doesn't conduct through the wall for hours.

The first and classic example of this type of system was the house built by Felix Trombe and Jacques Michel in Odeillo, France, in 1967. It features a 2-foot- (61-cm-) thick concrete wall and double glazing.

A thermal storage wall using water to collect and store the heat operates in essentially the same way as a masonry one. The only difference is that the water transfers the heat through the wall by convection instead of conduction, which means that the transfer occurs much more rapidly.

The classic example of a water wall system is the Steve Baer residence in Corrales, New Mexico. The wall consists of 55-gallon steel drums filled with water and stacked horizontally. The thermal storage of the water wall is supplemented by interior adobe walls and a concrete floor.

Some form of operable insulation over the wall glazing can significantly improve its performance. The Baers included a rigid insulation panel hinged at the bottom in their system. It has a reflective surface on the interior side so that when it is opened up flat on the ground in front of the collector wall, it reflects additional sunlight onto the system, thereby improving the performance.

Thermal storage walls are fairly versatile for retrofitting. They can simply be added to the south wall of an existing building with a clear southern exposure. The main challenge is to integrate it architecturally.

The overall efficiency of a thermal storage wall is about 30 to 45 percent. Water walls tend to be slightly more efficient than masonry walls.

Roof Pond.

The thermal mass in a *roof pond* system (Figure 7.11d) consists of water contained within large plastic bags located on the roof of the building. The water bags are supported by the roof structure, which is usually a metal deck. The metal decking, which serves as the finished ceiling of the room below, readily conducts the heat from storage and radiates it to the space. This system can also provide passive cooling in the summer.

In winter, the ponds are exposed to sunlight during the day and then covered with insulation at night. In summer, the ponds are covered during the day to insulate them from the sun and heat. The insulation is then removed at night to allow the ponds to be cooled by natural convection and night sky radiation. In this way, the storage mass stays warm in the winter and cool in the summer. The insulating panels are manipulated mechanically with electric motors.

Spraying or flooding the outside surface of the water bags can provide additional summer cooling by evaporation. This can increase the cooling capability by up to four times that provided by night sky radiation alone.

The original roof pond system was built by Harold Hay in Atascadero, California. This house has been 100 percent solar heated and passively cooled since it was occupied in 1973.

Roof ponds are normally 6 to 12 inches (15 to 30 cm) deep, so the roof structure must support the 32 to 65 lb/ft^2 (160 to 320 kg/m^2) dead load that the pond system adds to the roof. Interior masonry partitions can help moderate indoor temperature fluctuations and support the large roof weight. The system is limited to single-story structures.

Roof pond systems are characterized by very stable indoor temperatures due to the large radiative surface area (usually the entire ceiling). Daily space temperature fluctuations are typically only 9° to 14°F (5° to 8°C) when the building is constructed of all lightweight materials, and 5° to 8°F (3° to 4°C) when it is masonry.

Roof ponds are typically 30 to 45 percent efficient when double glazed with an inflated plastic air cell. The efficiency is somewhat dependent on how well the movable insulation is sealed.

Greenhouse/Sunspaces.

Attached greenhouses, solaria, and sunspaces (Figure 7.11c) actually combine direct- and indirect-gain systems. *Greenhouses* are heated directly by solar radiation. A thermal wall located between the greenhouse and the occupied space also receives direct sunlight and transmits the heat to the adjacent space. Heated air from the greenhouse may also be vented into the occupied space. The thermal wall releases some of its heat to the greenhouse at night to help maintain plant life. The mass also moderates the high temperatures during the day. The common wall can be made of masonry or water.

Sunspaces contain no thermal mass for nighttime heating, and they are not heated at night. They are used as an extension of the occupied space when weather permits. As heat builds up in the sunspace, it is transferred into the occupied space to satisfy immediate heating loads and to be stored in the mass of the space.

Fans are sometimes used to extract a greater amount of heated air from the greenhouse in order to heat adjoining spaces. This allows warm air ducted from the greenhouse to be stored in a rock bed located under the floor of the spaces to be heated. Heat is then delivered to the spaces from the radiant floor.

The efficiency of a solar greenhouse may be as high as 60 to 75 percent. Only 10 to 30 percent of the incident solar energy, however, is supplied to the occupied space, unless an active heat storage system is employed. The remainder is used for heating the greenhouse.

These systems can be readily retrofitted onto the south wall of an existing building. Care must be taken to provide adequate shading and venting in order to prevent serious overheating of these spaces in the summer.

Isolated Gain

Unlike direct and indirect passive solar heating systems, *isolated-gain* systems feature solar collection and storage units that are located outside the boundary of the occupied space. This allows the systems to function independently of the building so that heat is drawn from the system only when needed.

The most common application of this concept is the natural convective loop. This system includes a flat-plate collector and heat storage. Either water or an air/rock storage combination can be used as transfer and storage mediums.

The natural convection process requires that the storage unit be located at a higher elevation than the collector. As the water or air in the collector is heated by the sun, it rises and enters the top of the storage unit. In the meantime, cooler water or air is drawn from the bottom of the storage unit into the collector. The natural convection current continues as long as the sun is shining.

These systems are similar to active solar collection systems except that fluid transfer is accomplished by natural convection instead of by pumps or fans.

PASSIVE COOLING

Cooling accounts for 20 percent of the energy end use in the United States and 40 percent in just the southern part of the country. Internal-load-dominated buildings are the most concerned with cooling, but skin-load-dominated buildings in hot climates also need cooling in the summertime.

Internal-load-dominated buildings may sometimes benefit from maximizing their surface area and eliminating insulation in order to increase heat transfer to the outside, where the climate is cool enough. Optimizing the heat transfer characteristics of the envelope requires a thorough analysis of how many hours during the year the building would benefit from a certain envelope U-value. Reducing the R-value might increase heating requirements in cold weather and could increase the cooling load in the summer. An annual load analysis must be performed to determine how such moves would affect the overall energy requirements.

Greater insulation helps keep heat out in the summer and in during the winter. This may or may not be helpful. Similarly, greater or lesser quantities of fenestration will dissipate or conserve internal heat gain. The load calculation procedures in Chapter 4 can be used to determine whether internal heat gain exceeds winter heat requirements or whether the benefit of heat dissipation in the winter is worth the added heat gain in the summer. The more dynamic the envelope, the closer it can come to optimizing benefits for both summer and winter.

Outside air from an economizer cycle or natural ventilation can be used to dissipate excess internal heat in the winter as an alternative to increasing the envelope's U-value. Therefore, additional summer heat gain through the envelope can be avoided.

Daylighting can reduce the internal heat gain from artificial lights. This reduces not only the electricity consumed for lighting, but also the cooling load.

The cooling methods presented in this section affect only the sensible component of the cooling load. There are no passive techniques for dehumidifying. However, in high-humidity areas, passive sensible cooling may be accomplished, and dehumidification can be achieved by using an active desiccant moisture-absorption system. Desiccants are porous materials with a high affinity for water vapor. Solar energy can be used for drying out, or regenerating, the desiccant. Vapor-compression cooling can thus be avoided altogether.

Night Sky Radiation

The coldest heat sink in our environment is the night sky. Warm objects directly exposed to the sky radiate their heat out to it at night (Figure 7.12).

Clouds at night limit the amount of heat that can be radiated to outer space in this way, but on a clear night, the effective sky temperature can be 20° or 30°F (11° or 17°C) lower than the ambient air temperature. Because they have less cloud cover, areas with an arid climate can usually take advantage of this phenomenon more often than those with humid climates.

A roof pond system is perhaps the most effective method of radiant cooling. Spraying water onto the water bags to take advantage of evaporative cooling also substantially improves the system performance in dry climates, but helps only slightly in humid ones.

Another way of utilizing nocturnal radiation is with light-weight radiators in which a fluid is circulated. The cooled fluid can be used to cool a thermal storage system at night.

FIGURE 7.12. Night sky radiation.

The cold storage can be used the following day for space cooling.

Natural Ventilation

Natural ventilation is the movement of air into and out of a space through openings such as windows and doors intentionally provided for this purpose or through nonpowered ventilators. It can provide both fresh air ventilation and a cooling effect by replacing hot interior air with cool outside air, and by the air motion itself, if the temperature is not too warm.

Two natural circulation techniques can be used: wind-induced cross ventilation and gravity ventilation. The simplest way to ventilate is through windows.

The siting of buildings that rely on prevailing breezes for cooling must be selected according to local wind direction. Many sites experience winds from all directions. In those locations, natural ventilation designs must be nondirectional. If natural ventilation is a prime concern, locating a building on a hilltop will maximize cross ventilation from any direction.

Types of natural-ventilation openings include (1) windows, doors, monitor openings, and skylights, (2) standard air registers, (3) roof ventilators, and (4) specially designed inlet or outlet openings.

Operable Windows

Operable windows provide natural ventilation and take advantage of cooling breezes. Various types of operable sash are available to direct air into or above occupied areas. They may open by sliding vertically or horizontally, by swinging in or out on a vertical axis, or by tilting on horizontal pivots at the top, bottom, or middle (Figure 7.13).

Casement and pivot windows can open their entire area for airflow. Double-hung, single-hung, and sliding types have only half or less of their total area available for airflow. Awning and hopper windows are useful for directing air up or down, and can be left open during rainstorms.

For cooling purposes, the outdoor air stream should be directed toward where people are actually located. Air for ventilation only should be directed upward into the area above people's heads, whereas cooling air must be directed down to the level of the occupied zone.

Personal manual control of openings is necessary to allow each occupant to control wind velocity so that it is sufficient to enhance the evaporation of sweat from the skin but not so great that it is irritating.

Two openings are necessary for proper cross ventilation: one as an inlet—on the windward side of the building—and the other as an outlet on the leeward side. Rooms are most effectively ventilated by an inlet located near the bottom or middle of the windward wall and by an outlet in any position on the leeward wall. The inlets and outlets should be roughly the same size, or the outlet can be larger than the inlet.

Obstacles upstream from intake openings or downstream near outlets can significantly reduce wind velocity, and thereby the cooling effect. Roof overhangs and the judicious location of trees to increase wind velocity at ground level improve the air movement.

Air and sound are often inseparable. Natural ventilation sometimes cannot be used because of the noise that would accompany the breeze through an open window. Polluted air may be another deterrent to natural ventilation. Mechanical air conditioning may be needed if ventilation air must be cleaned. Almost any device that reduces sound or filters dirt particles will also reduce the velocity of the breeze.

For natural ventilation through operable windows, doors, louvers, or ventilating skylights opening to the outside, the total openable area should be at least one-twenty-fifth of the floor area served. Toilet rooms and bathrooms should have at least 3 ft^2 (0.3 m^2) of window or skylight opening area if it is to serve as an alternative to a mechanical exhaust fan.

Gravity Ventilation

Large open windows may not be desirable because of the lack of security and privacy, particularly at night. Although windows are needed for visual reasons and for egress in case of fire, they do not, by themselves, provide a well-insulated envelope when closed. Ideally, dynamic envelope elements should have high transmission when open and low transmission when closed. A gravity ventilation system eliminates

(a)

(b)

(c)

FIGURE 7.13. Operable sash options. (a) Casement, (b) double-hung (the single-hung sash looks similar, except that the upper window is fixed), (c) sliding, (d) awning, and (e) hopper. (Photos by Vaughn Bradshaw.)

(d)

(e)

FIGURE 7.13. *(Continued)*

some of the need for window ventilation and allows for more flexibility.

Gravity ventilation of spaces takes advantage of thermal buoyancy. Vents at different levels in interior spaces draw cool air in through the lower inlet while forcing warm air out through the higher outlet. The rate at which air circulates depends directly on the air temperature, the height difference between the two vents, and the sizes of the two apertures:

$$Q = 9.4 \, A \, \sqrt{H \, \Delta T} \qquad (7.1)$$

where

Q = airflow rate, CFM (L/sec)
A = free area of inlets or outlets, whichever is less, ft^2 (m^2)
H = height difference between inlets and outlets, ft (m)
ΔT = temperature difference between the incoming
and outgoing air °F (°C)
9.4 = constant of proportionality (= 116 for SI units)

Air inlets should be located as low as possible and in areas likely to have the lowest air temperature, such as the north side of a building. Drawing air through subterranean passageways before it enters a building will cool it.

These vents should be clear of obstruction. Outlets should be located as high as possible and preferably in areas where wind movement can be used to create a suction effect to aid in drawing air through the building.

Gravity ventilation can be used for summer cooling or for air exchange purposes. Inlets and outlets should close tightly in order to prevent infiltration when the entrance of outside air is undesirable.

An induced-draft system is a special kind of gravity ventilator. It uses a chimney that, when heated by sunlight, creates an additional updraft that pulls a breeze through the building. This is also called a *solar chimney* (Figure 7.14).

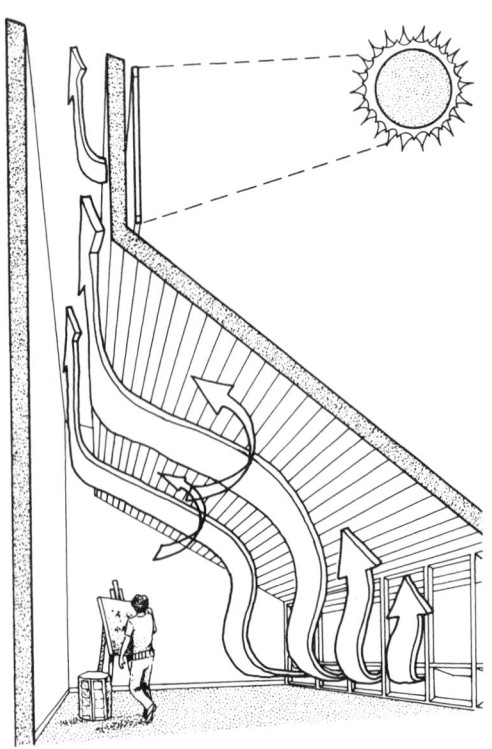

FIGURE 7.14. Solar chimney.

FIGURE 7.15. Wind-driven roof ventilators. (Photo by Vaughn Bradshaw.)

The rate of airflow is proportional to the intensity of the solar radiation striking the darkened upper part of the chimney.

Roof Ventilators

Roof ventilators (Figure 7.15) provide a rainproof outlet. They are designed to create a suction, drawing air from below, when wind blows over them. Their capacity depends on location on the roof, the resistance to airflow from duct-work, the model type, and the height of the unit above the outside air inlets.

A ventilator should be located on a part of the roof that receives the full wind without interference. If it is installed within the suction region created by the wind passing over the building, in a light well, or on a low roof between two higher buildings, its performance will be seriously impaired.

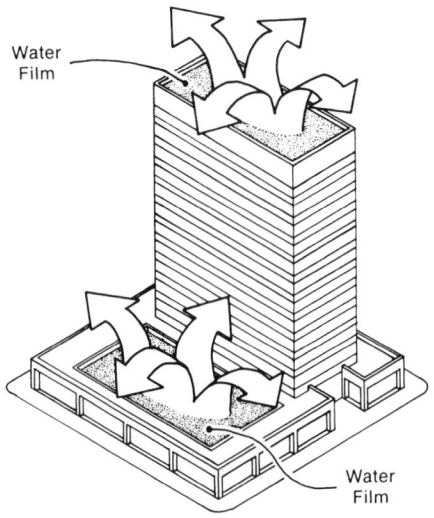

FIGURE 7.16. Passive evaporative cooling.

Ventilation stacks should be located so that wind can act on them from any direction. Even without wind, buoyancy effects alone will remove air from the room in which the inlets are located.

Flushing with Night Air

Nocturnal ventilation, whether naturally induced or fan-forced, can reduce cooling loads in climates with large daily temperature swings, that is, hot days and cool nights. Maximizing thermal mass and airflow rates through the building at night is important to achieve the greatest effect.

Evaporative Cooling

Evaporating water into hot, dry air can achieve substantial cooling. Fountains have traditionally been used to cool courtyards in hot arid climates.

Passive evaporative cooling systems, which are suitable primarily for hot arid climates, vary in appearance. One simple system is a sprinkler on the roof of a conventional building. During a hot day or night, the water evaporates off the roof surface and cools the building.

A variation uses a specially designed pond of water located on the roof to cool the building (Figure 7.16). A series of louvers can be installed to shade the roof pond in order to minimize heat gain to the pond, while circulating breezes that evaporate the water. This type of system is usually installed only on flat-roofed buildings.

If widely used, evaporative cooling systems could aggravate water shortages in dry regions. However, using them sparingly and in conjunction with other passive cooling methods can reduce the amount of water they use. The systems are ideal in hot dry climates where there is plenty of water.

CASE STUDY

The new headquarters for the Swiss Reinsurance Company is London's first green-built high-rise commercial office tower (Figure 7.17). Located in the heart of the City of London, England, the building at 30 St. Mary Axe is unquestionably innovative environmentally, socially, technologically, spatially, and architecturally.

The Swiss Re tower, designed by the London-based architecture firm of Foster and Partners and completed in 2004, is an instantly recognizable addition to the city's skyline. The building is nicknamed the "Gherkin" because of its distinctive tall, rounded, pickle-like shape. The 41-story building rises 590 feet (179.8 m) above the street and provides 822,000 ft^2 (76,400 m^2) of floor space. In addition to offices, the building provides a shopping arcade accessible from a newly created public plaza. Half of the tower's ground level is comprised of shops, and an adjacent outdoor

FIGURE 7.17. The Swiss Reinsurance Company's new headquarters— designed by Foster and Partners—is said to be London, England's, first environmentally sustainable skyscraper. It is estimated to use up to 50 percent less energy than a traditionally built office tower of similar size. (Photo courtesy of Nigel Young/Foster and Partners.)

café spills out onto the same plaza. The top floor of the building—London's highest occupied space—is a club room featuring a spectacular 360° panarama of the capital.

The highly progressive environmental strategy for the building maximizes the amount of natural lighting and ventilation cooling to significantly reduce its energy consumption. The building is expected to consume up to 50 percent less energy than a traditional air-conditioned prestigious office building.

The unique form was developed using a number of complex fluid dynamic studies of the local environmental conditions. The smooth aerodynamic shape directs air movement around the building, which reduces the amount of wind channeled to the ground-level surroundings compared to a rectilinear tower of similar size. This not only helps improve pedestrian comfort in the street-level plaza, it also creates external pressure differentials that drive a system of natural ventilation using the stack effect.

The plan view at each floor has a circular perimeter indented by six triangular light courts creating a radial effect. The indentations at each level remain a constant size, but the space between them varies with the size of the floor plan. The floor plan is then rotated at each successive floor, creating a series of spiraling five-story atria/light wells that link together, vertically stretching the full height of the building. These spiraling light wells function as the building's "lungs," drawing fresh air in through operable panels in the facade and up through the vertical wells. Windows into the light wells at each floor open automatically, distributing the fresh air to the office interiors to augment the air-conditioning system with natural ventilation. This minimizes reliance on artificial cooling and is anticipated to save energy for up to 40 percent of the year.

The light wells and the shape of the building maximize natural daylight, moderate the use of artificial lighting, and allow views out from deep within the building. Providing better views for more occupants improves the overall working environment. The balconies on the edge of each floor at the light wells also provide strong visual connections from floor to floor. These spaces provide a natural social focus, allowing places for refreshment and meeting areas.

The building's form is a departure from that of conventional box-like office buildings. It appears more slender than a rectangular block of the same size. The tower is tapered at the base to better integrate with its site and improve connections to the surrounding streets. The slimming of its profile near the base together with the circular shape maximizes the outdoor public space at ground level. This enables much of the site area to be used as a landscaped public plaza, including mature trees and low stone walls that subtly mark the site boundary and provide seating.

The building widens in profile as it rises and then tapers again toward its apex to allow the maximum amount of sunlight to the plaza level. Other advantages of this distinctive form are that annoying reflections are reduced and transparency of the tower is improved.

The exterior cladding consists of approximately 5,500 flat triangular and diamond-shaped glass panels, which vary at each level. The glazing to the office areas consists of a double-glazed outer layer and a single-glazed inner screen that sandwich a central ventilated cavity, which contains solar-control blinds. The cavities act as buffer zones to reduce the heat gains and losses, and are ventilated by exhaust air drawn from the offices. The glazing to the light wells that spiral up the tower consists of operable double-glazed panels including a gray-tinted layer and a high-performance coating that effectively reduces solar heat gain.

The diagonally braced steel structural envelope in place of conventional walls and roof is, by virtue of its triangulated geometry, inherently strong and light. This permits a flexible column-free interior floor space. It also allows a fully glazed facade, which opens up the building to light and views. The interior atria show up on the exterior as distinctive spiral bands of gray glazing.

This is more than just an ornate building for the client, Swiss Re. It's a manifestation of the company's increasing dedication to renewable energy and sustainable building design. Swiss Re has vigorously expressed their concerns about the effects that global warming could have on the earth in the form of catastrophic natural disasters. As a reinsurance company for insurance companies worldwide, they have a strong interest in the issue.

Together, architect Norman Foster of Foster and Partners, the City of London, and Swiss Re have pushed the state of the art of green building design. They have set an example and raised the standard for modern buildings worldwide. It is no surprise that this project won the 2004 RIBA Stirling Prize for Architecture. Its innovative form, plan, and envelope successfully integrate natural ventilation, daylighting, and view opportunities into an interesting package.

SUMMARY

Energy-conscious design calls for considering passive methods of thermal control as an alternative to active HVAC systems. The first step is to reduce the heating and cooling loads on the building by (1) understanding the microclimate and its impact on heating and cooling loads of the particular building type, (2) getting the lay of the land in order to utilize site resources of shading or wind sheltering to the fullest extent feasible, (3) arranging the form, orientation, color, and insulation level of the structure for minimum energy use, and (4) reducing internal gains. Then, if necessary, consider passive solar heating or passive cooling to establish the greatest degree of self-sufficiency. At that point, if adequate comfort conditions still cannot be provided, add the needed mechanical equipment.

Wind is significant as a potential ventilation aid and cooling agent and also as a potentially destructive force.

Outdoor spaces should be limited to the south and east sides of a building in cold climates, the south and north sides in temperate climates, and the north, south, and east sides in hot arid climates; they can be placed on any side of a building in hot humid climates, depending on the availability of breezes. The orientation of outdoor spaces should depend on which season intended activities take place.

Heat gains and losses through fenestration are considerably greater than those through a solid, insulated exterior wall because the U-value of glazing is generally higher. Thus, the area of fenestration has a large influence on the total heat gain or loss through the envelope.

Fenestration is a dynamic envelope element since it can open up to the sun's heat and light when available and close to keep heat in at night. Complete glazing systems are available, as are window insulation systems. Screens and shades for both internal and external installation prevent heat loss in the winter and excessive solar heat gain year-round. Internal-load-dominated buildings can modulate such dynamic elements in order to reject heat when it becomes excessive and to shield the building from excess solar heat during its cooling season.

Note that there are some major differences in the passive approaches appropriate to internal-load-dominated and skin-load-dominated buildings. If heat gain needs to be restricted, the appropriate solution is some combination of building form, thermal mass, insulation, shades, colors, and daylighting in response to the local microclimate.

The objective of passive solar heating is essentially to let the sunlight in through the windows when needed and keep it indoors.

In hot dry climates with wide daily temperature fluctuations, a massive envelope prevents the sun's heat from penetrating in during the day. At night, some of the heat stored all day gradually migrates into the interior, warming it as the outdoor temperature drops. The coolness of the night air is then stored in the walls and keeps them cool during the day.

The more mass, the more stable the interior temperature. The ultimate massive envelope is provided by earth sheltering—berming earth up against the outside walls, molding the structure into the side of a hill, or recessing the building completely underground.

Passive cooling methods include night sky radiation, natural ventilation, and evaporative cooling.

KEY TERMS

Internal-load-dominated
Skin-load-dominated
Deciduous vegetation
Evapo-transpiration
Wind sheltering
Wind channeling
Aspect ratio
Clustering
Dynamic envelope

Buffer space
Superinsulation
Fenestration
Operable thermal barrier
Manual operable shading
Mechanical operable shading
Earth sheltering
Balance point temperature (BPT)

Direct gain
Indirect gain
Thermal storage wall
Roof pond
Greenhouse
Sunspace
Isolated gain
Night sky radiation
Natural ventilation
Evaporative cooling

STUDY QUESTIONS

1. What are the purposes of fenestration?
2. Give three examples of internal-load-dominated buildings. Give three examples of skin-load-dominated buildings.
3. Why are overhangs suitable only for southern exposures?
4. Why is it preferable to place shading devices on the exterior rather than on the interior?
5. Define each type of passive solar heating system and list the efficiency of each.

BIBLIOGRAPHY

Energy-Conscious Architecture

American Society of Heating, Refrigerating, and Air-Conditioning Engineers, Inc. *Standard 90.1—Energy Standard for Buildings Except Low-Rise Residential Buildings (ANSI Approved; IESNA Cosponsored).*

American Society of Heating, Refrigerating, and Air-Conditioning Engineers, Inc. *Standard 90.1—User's Manual.*

American Society of Heating, Refrigerating, and Air-Conditioning Engineers, Inc. *Standard 90.1—Energy Code for Commercial and High-Rise Residential Buildings.*

American Society of Heating, Refrigerating, and Air-Conditioning Engineers, Inc. *Standard 90.2—Energy-Efficient Design of Low-Rise Residential Buildings (ANSI Approved).*

American Society of Heating, Refrigerating, and Air-Conditioning Engineers, Inc. *Standard 90.2—Energy Code for New Low-Rise Residential Buildings Based on ASHRAE 90.2.*

American Society of Heating, Refrigerating, and Air-Conditioning Engineers, Inc. *Standard 100—Energy Conservation in Existing Buildings (ANSI Approved; IESNA Cosponsored).*

American Society of Heating, Refrigerating, and Air-Conditioning Engineers, Inc. *Standard 105—Standard Methods of Measuring and Expressing Building Energy Performance (ANSI Approved).*

American Society of Heating, Refrigerating, and Air-Conditioning Engineers, Inc. *Standard 119—Air Leakage Performance for Detached Single-Family Residential Buildings (ANSI Approved).*

Banham, Reyner. *The Architecture of the Well-Tempered Environment.* University of Chicago Press, 1984.

Behling, Sophia, and Stefan Behling. *Solar Power: The Evolution of Sustainable Architecture.* Prestel Publishing, 2000.

Brown, G.Z., and Mark DeKay. *Sun, Wind and Light: Architectural Design Strategies,* 2nd ed. John Wiley & Sons, Inc., 2000.

Climatic Atlas of the United States CD-ROM version 2.0. NOAA National Climatic Data Center, Asheville, NC. US Government On-line store: http://nndc.noaa. gov/?http://ols.nndc.noaa.gov/plolstore/plsql/olstore.p rodspecific?prodnum=C00519-CDR-A0001. (This version, which includes Alaska and Hawaii, replaces the paper atlas last published in 1968 and supersedes version 1.0 of the *Climate Atlas of the Contiguous United States* published on CD in 2000. The new CD atlas contains 2023 color maps of climatic elements such as temperature, precipitation, snow, wind, pressure, etc. The period of record of the data for most of the maps is 1961–1990. There is an online version of the *Climate Atlas* at http://www.ncdc.noaa.gov/olca/ having restricted GIS capabilities, but viewable at no charge.)

Givoni, Baruch. *Man, Climate, and Architecture.* Van Nostrand Reinhold Company, 1981.

Hawkes, Dean, and Wayne Forster. *Energy Efficient Buildings: Architecture, Engineering, and Environment.* W.W. Norton & Company, 2002.

Knowles, R.L. *Energy and Form: An Ecological Approach to Urban Growth.* MIT Press, 1978. (A mathematical procedure for analyzing building form for optimum energy conservation through the architectural design.)

Lechner, Norbert. *Heating, Cooling, Lighting: Design Methods for Architects,* 2nd ed. John Wiley & Sons, Inc., 2001. (Good pictorial examples.)

Leckie, Jim, Gil Master, et al. *Other Homes and Garbage: Designs for Self-Sufficient Living.* Sierra Club Books, 1981.

Mendler, Sandra F., and William Odell. *The HOK Guidebook to Sustainable Design.* John Wiley & Sons, Inc., 2000.

Olgyay, Victor. *Design with Climate: Bioclimatic Approach to Architectural Regionalism.* John Wiley & Sons, Inc., 1992. (A classic.)

Stang, Alanna, and Christopher Hawthorne. *The Green House: New Directions in Sustainable Architecture.* Princeton Architectural Press, 2005.

Stitt, Fred A. *The Ecological Design Handbook.* McGraw-Hill Professional, 1999.

Van Der Ryn, Sim, and Stuart Cowan. *Ecological Design.* Island Press, 1995.

Passive Solar Heating

Anderson, Bruce (editor). *Solar Building Architecture (Solar Heat Technologies).* MIT Press, 1990.

Anderson, Bruce. *Passive Solar Design Handbook.* Van Nostrand Reinhold Company, 1998.

Anderson, Bruce, and Malcolm Wells. *Passive Solar Energy,* 2nd ed. Brick House Publishing Company, 1996.

Bennet, Robert. *Sun Angles for Design.* Melrose Plantation Press, 1978.

Chiras, Daniel D. *The Solar House: Passive Heating and Cooling.* Chelsea Green Publishing Company, 2002.

Crosbie, Michael J. (editor). *The Passive Solar Design and Construction Handbook,* 2nd ed. John Wiley & Sons, Inc., 1997.

Freeman, Mark. *The Solar Home: How to Design and Build a House You Heat with the Sun (How-to Guides).* Stackpole Books, 1994.

Johnson, Timothy E. *Solar Architecture: The Direct Gain Approach.* McGraw-Hill Book Company, 1982.

Kachadorian, James. *The Passive Solar House (Real Goods Independent Living Books).* Chelsea Green Publishing Company, 1997.

Kreider, Jan F. *The Solar Heating Design Process: Active and Passive.* McGraw-Hill Book Company, 1988.

Levy, M. Emanuel. *The Passive Solar Construction Handbook: Featuring Hundreds of Construction Details and Notes, Materials Specifications, and Design Rules of Thumb.* Rodale Press, 1983.

Mazria, Edward. *The Passive Solar Energy Book: A Complete Guide to Passive Solar Home, Greenhouse and Building Design.* Rodale Press, 1979.

Schwolsky, Rick, and James Williams. *The Builder's Guide to Solar Construction.* McGraw-Hill Book Company, 1982.

Van Dresser, Peter. *Passive Solar House Basics.* Ancient City Press, 1996.

Solar Control

Olgyay, Aladar. *Solar Control and Shading Devices.* Princeton University Press, 1977.

Watson, Donald. *Solar Control Workbook.* Association of Collegiate Schools of Architecture, 1981. (The effects of solar control on passive heating, cooling, and daylighting, and the energy trade-offs between the three. Analysis techniques for sun control, heat flow, and daylighting. Includes case studies, exercises, and studio applications.)

Ramsey, C.G., H.R. Sleeper, and J.R. Hoke, Jr. *Architectural Graphic Standards,* 9th ed. John Wiley & Sons, Inc., 2000.

Lynch, Kevin, and Gary Hack. *Site Planning.* MIT Press, 1984.

Wind Control

Melaragno, Mechiele G. *Wind in Architectural and Environmental Design.* Van Nostrand Reinhold Company, 1981. (Wind as an influence on building form, natural ventilation, evaporative cooling, wind sheltering, kinetic art, and the addition of life and movement to landscape features.)

Landscaping

Creasy, Rosalind. *The Complete Book of Edible Landscaping.* Sierra Club Books, 1982.

Robinette, Gary O., and Charles McClennon (editors). *Landscape Planning for Energy Conservation.* Van Nostrand Reinhold, 1983.

Earth Sheltering

Frenette, Ed, and T. Lanee Holthusen (editors). *Earth Sheltering: The Forms of Energy and the Energy of*

Form: Award Winning and Selected Entries from the 1981 American Underground Space Associations Design Competition. Pergamon Press, Inc., 1981.

Golany, Gideon. *Earth-Sheltered Habitat: History, Architecture and Urban Design.* Van Nostrand Reinhold Company, 1982.

Kern, Barbara, and Ken Kern. *Earth Sheltered: Owner-Built Home.* Owner-Builder Publication, 1982.

Sterling, Raymond. *Earth Sheltered Residential Design Manual.* Van Nostrand Reinhold Company, 1982.

University of Minnesota's Underground Space Center. *Earth Sheltered Homes: Plans and Designs.* Van Nostrand Reinhold Company, 1981.

University of Minnesota's Underground Space Center. *Earth Sheltered Housing Design,* 2nd ed. Van Nostrand Reinhold Company, 1985.

University of Minnesota's Underground Space Center. *Earth Sheltered Housing Design: Guidelines, Examples and References.* Van Nostrand Reinhold Company, 1997.

PART THREE

ELECTRICAL SYSTEMS

These four chapters are concerned with the electrical systems and equipment used for buildings. The first chapter, on lighting, presents the fundamental concepts of illumination. Some lighting systems are primarily functional, while others are primarily aesthetic. The challenge is to achieve both and at low energy-usage levels.

Daylighting, while not electrically powered itself, has a strong impact on the need for and control of electrical lighting. It is thus included as a special topic. This is not to underrate its importance. Daylighting is a major energy-conserving method in its own right and is the most important passive control option for many nonresidential buildings.

The normal power supply systems have a minimal impact on the architecture of a building, but space must be provided for them. The impact can become major when alternative on-site generation facilities are included. By contrast, signal and conveying systems require a great deal of thought on the part of the architect and are important to the safe and proper operation of the building.

Chapter 8

Lighting

PRINCIPLES OF ILLUMINATION

TASK LIGHTING

LIGHTING LEVELS

CONTRAST

COLOR OF LIGHT

SPACE GEOMETRY

ELECTRICAL LIGHTING

LIGHT SOURCE TYPES

LUMINAIRES

LIGHTING ENERGY CONSERVATION

DAYLIGHTING

RESOURCE AVAILABILITY

MODELING

LOCATING FENESTRATION

GLARE

ELECTRIC LIGHTING CONTROLS

EMERGENCY LIGHTING

OUTDOOR LIGHTING

Since the field of lighting has its own unique vocabulary, this chapter begins with some definitions to help familiarize the reader with the pertinent nomenclature. Figures 8.1, 8.2, and 8.3 should aid in understanding them.

Candlepower (cp). The unit of *luminous intensity* of a light source. In SI units, it is called *candela (cd)*. It is an index of the ability of a light source to produce illumination. It always has a directional connotation, such as horizontal candlepower, candlepower at 60°, or spherical candlepower. A typical candle has a luminous intensity in the horizontal direction of approximately 1 candlepower (candela), which is the source of the term. The candela and the candlepower have the same magnitude.

Lumen (lm). A quantitative unit for measuring the flow of light energy, which is referred to in the lighting field as *luminous flux*. In the English system of units, 1 lumen is equal to the luminous flux emanating from 1 ft^2 of a hypothetical surface all points of which are 1 foot from a uniform point source of 1 candlepower (see Figure 8.1). In SI units, 1 lumen is the luminous flux emanating from 1 m^2 of a surface all points of which are 1 m from the 1-candela source. Illumination is the density of luminous flux, expressed as lumens per unit area.

Footcandle (fc). One lumen of luminous flux spread uniformly over an area of 1 ft^2 produces an illumination of 1 footcandle. When an SI lumen is spread over 1 m^2, the illumination is expressed in *lux (lx)*.

$$\text{footcandles} = \frac{\text{lumens}}{\text{ft}^2} \tag{8.1}$$

$$\text{lux} = \frac{\text{lumens}}{\text{m}^2} \tag{8.2}$$

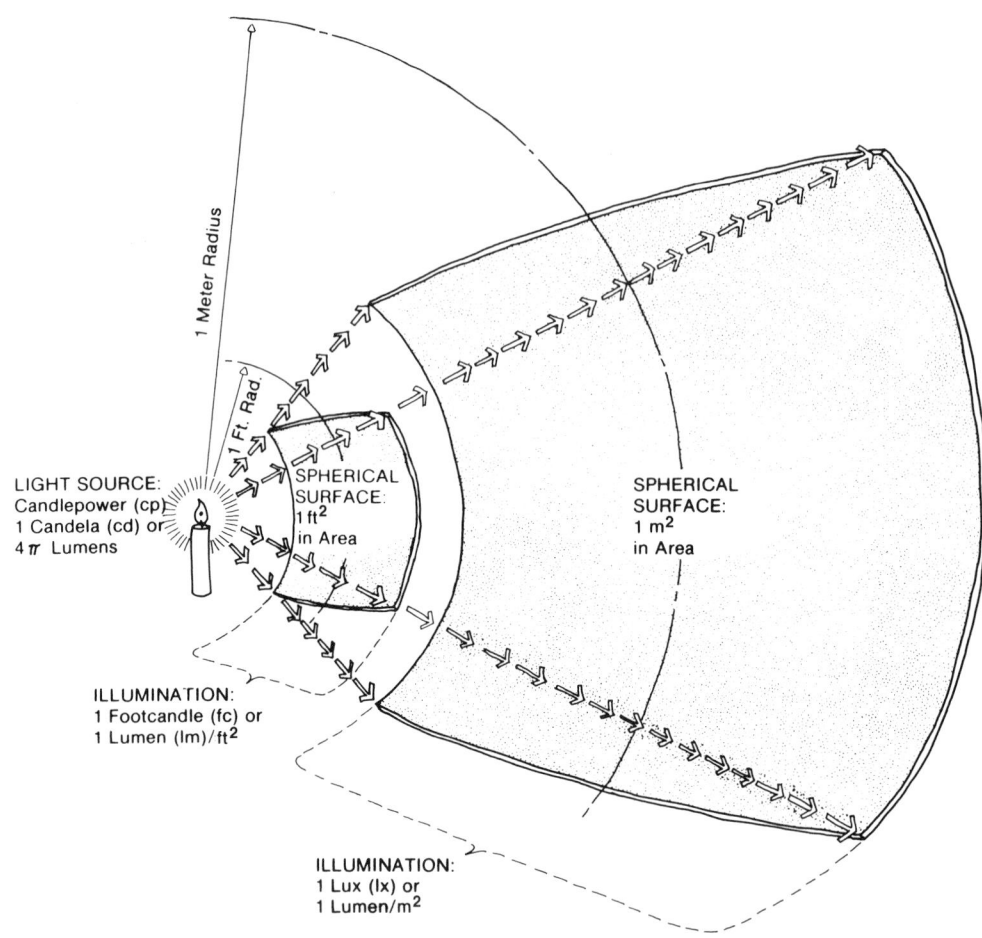

FIGURE 8.1. Definitions of luminosity.

A point source of 1 candlepower (candela) produces 4π lumens of light, which produces 1 footcandle (lux) of illumination over the entire surface of an imaginary globe each point of which is 1 foot (1 m) from the source. Since light continuously spreads out in ever-expanding spheres, the amount of illumination provided is inversely proportional to the square of the distance from the source (twice the distance = one-fourth the illumination). Measurements of illumination levels are commonly made with a *footcandle meter.*

It is useful to think of light as rays that travel in a straight line until they encounter some object. The rays are then either absorbed, reflected, or transmitted. In general, all of these phenomena occur at any given surface, but one or two of them may predominate. The proportion of light absorbed, transmitted, or reflected depends upon the type of material and the angle of incidence of the light rays as measured from the perpendicular to the surface. Light impinging upon a surface at "grazing" angles higher than 55° tends to be reflected rather than absorbed or transmitted. Even a clear piece of glass or plastic can become an effective mirror when the incident rays strike at a high enough angle.

Absorptance. The ratio of light absorbed by a material to the incident light falling on it. All materials absorb some light; darker objects absorb more than lighter-colored objects.

Reflectance, reflection factor, or reflectance coefficient. The ratio of light reflected by a surface to the incident light falling on it. Reflection of light is either specular, diffuse, or some combination of the two.

Specular reflection. Sometimes referred to as "regular reflection," this is the kind of reflection observed in a mirror. It is characterized by the angle of reflection being equal to the angle of incident light (see Figure 8.2). At some angle, the image projected by the light source can be seen reflected by the surface. All polished or shiny surfaces reflect light in this manner when the incident ray is directed to the surface at an angle between 45° and 55° from the perpendicular to the surface.

Diffuse reflection. The reflected light is scattered in all directions so that the reflecting surface appears equally bright from any angle of view (see Figure 8.2).

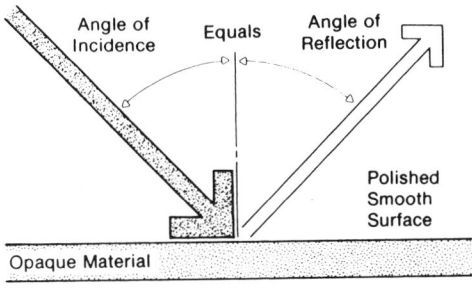

SPECULAR
REFLECTION

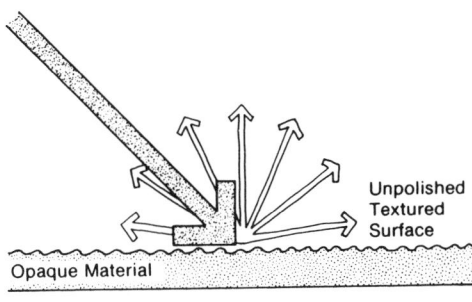

DIFFUSE
REFLECTION

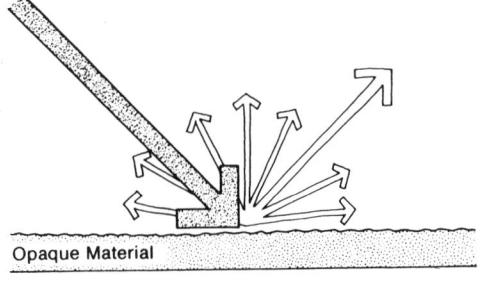

COMBINED SPECULAR
AND DIFFUSE REFLECTION

FIGURE 8.2. Specular and diffuse reflections.

Any surface with a sandpaper texture reflects light diffusely.

Transmittance, luminous transmittance, transmission factor, or coefficient of transmission. The ratio of the light transmitted through a material to the total incident light falling on it. Some materials selectively absorb certain colors more than others. Since those materials may have a different transmittance for each component of light, transmittances are generally used only in reference to materials that do not selectively absorb. Like reflection, light transmission may be either direct, diffuse, or a combination of the two.

Direct transmission. This occurs when light passes through clear, transparent materials (see Figure 8.3). The angle at which the light leaves is the same as that at which it enters.

Diffuse transmission. The transmitted light is scattered evenly in all directions, appearing equally bright from any angle of view (see Figure 8.3). Light, but no image, is transmitted. Materials that transmit light in a diffuse pattern are known as *translucent* materials. Translucent materials are widely used in light fixtures to diffuse or spread the light evenly in all directions.

Refraction. The bending of a ray of light as it passes obliquely through a material, as shown in Figure 8.3.

Footlambert (fL). A quantitative unit for measuring brightness. Brightness, or luminance, is an index of the intensity of light being emitted, transmitted, or reflected from a surface. The terms *brightness* and *luminance* are used almost synonymously. The distinction is that brightness is the perceived light, while luminance is a measured quantity. Everything visible has some brightness. The value of a footlambert is 1 lumen per square foot, which is expressed in the same units as the footcandle. When the illumination is *on* a surface, the lumens per square foot are measured as footcandles. When the brightness is *from* a surface, the lumens per square foot are measured as footlamberts. The footcandles falling on a surface multiplied by the reflection factor or transmission factor of the surface yield footlamberts.

$$\text{luminance} = \text{illumination} \times \text{reflectance} \quad (8.3)$$

$$\text{luminance} = \text{illumination} \times \text{transmittance} \quad (8.4)$$

$$\text{footlamberts} = \text{footcandles} \times \text{reflectance factor} \quad (8.5)$$

$$\text{footlamberts} = \text{footcandles} \times \text{transmission factor} \quad (8.6)$$

Brightness often depends on the direction of view, since many surfaces exhibit a different reflectance or transmittance at different angles. If a perfectly white,

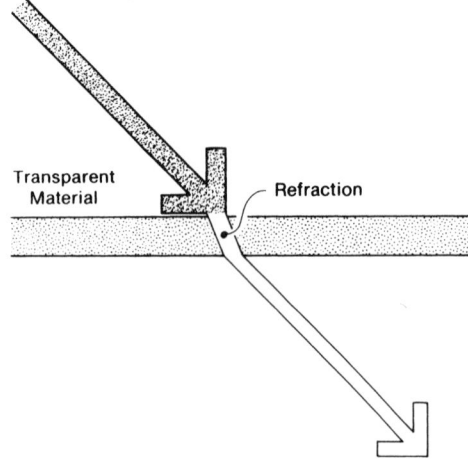

DIRECT TRANSMISSION

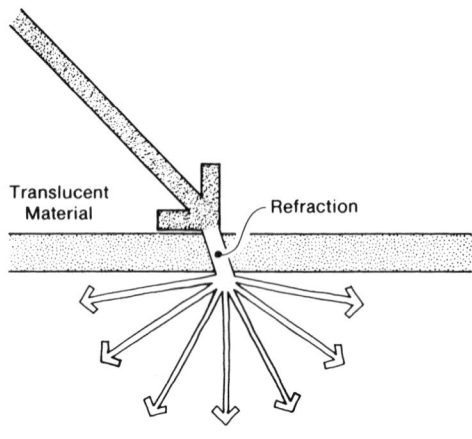

DIFFUSE TRANSMISSION

FIGURE 8.3. Direct and diffuse transmission and refraction.

diffuse surface is placed 1 foot from a 1-candlepower source, the brightness of that surface is approximately 1 lumen/ft², or 1 footlambert.

In SI units, the unit of brightness is the *lambert (L)*, which is defined as the luminance of a surface reflecting, transmitting, or emitting 1 lumen per square centimeter. Since 1 lambert is a much larger quantity than is normally encountered, the SI unit commonly used is the millilambert (mL).

$$1 \text{ lambert} = 1,076 \text{ footlamberts} \qquad (8.7)$$
$$1 \text{ millilambert} = 1.076 \text{ footlamberts} \qquad (8.8)$$

(Multiply footlamberts by 1.076 to obtain millilamberts.)

Glare. The effect of excessive brightness in the field of view, causing annoyance or discomfort and interfering with vision. It may be direct from a light source, or reflected from a shiny surface.

Work station. The immediate contiguous area in which a worker performs visual tasks. This may involve several surfaces, such as a desk, desk return, and table that function as one task location, or a conference table shared by several workers. Normally about 50 ft² (4.6 m²) is allotted per work station, but the area can exceed that.

Work plane. The surface on which the visual task is performed and at which the illumination is specified and measured. Unless otherwise indicated, it is normally assumed to be a horizontal plane 30 inches (0.76 m) above the floor.

Primary source. A luminous source where light energy is generated and transmitted directly to a task.

Secondary source. Surfaces that derive their brightness from reflected incident illumination. The sun is the ultimate primary source, whereas the moon is a secondary source of light.

Lamp. A generic term for an artificial light source. Incandescent filament lamps are commonly referred to as *bulbs,* fluorescent lamps are called *tubes,* and H.I.D. light sources are simply called *lamps.*

Reflector. A device for redirecting the radiant energy of a lamp by reflecting it in a desired direction.

Refractor. A device for redirecting the radiant energy of a lamp in the desired direction by refraction through a lense.

Luminaire. A complete lighting unit consisting of a lamp or lamps, together with parts designed to distribute the light, to position and protect the lamp(s), and to connect the lamp(s) to a power supply. A luminaire is commonly referred to as a *fixture.*

Efficiency. The efficiency of a luminaire is the ratio of light output (luminous flux) to the light produced by the lamp. The *efficacy* of a light source is the ratio of output of luminous flux, expressed in lumens, to the power input in watts and is expressed in lumens per watt.

Coefficient of utilization (CU). The ratio of the lumens from a luminaire received on a work plane to the lumens emitted by the luminaire's lamp(s). It depends on the particular space as well as on the fixture.

Coefficient of beam utilization (CBU). The ratio of the lumens reaching a specified area directly from a floodlight or spotlight to the total lumens emitted by the beam source.

Color rendition. The effect of a light source on the apparent color of objects in comparison (consciously or subconsciously) with their apparent color under a reference light source.

Ballast. A device used with fluorescent and H.I.D. lamps to provide the necessary circuit conditions for starting and operating.

IES. These initials stand for the *Illuminating Engineering Society of North America*, the only technical society on this continent devoted solely to illumination. It prepares and publishes standards, testing procedures, and nomenclature. The *IES Handbook* is the principal authority on these conventions. Membership in local chapters and sections is open to anyone interested in lighting.

PRINCIPLES OF ILLUMINATION

Task Lighting

The purpose of an indoor lighting system is to benefit the occupants. It should create for them a pleasant and suitable visual environment and enable them to satisfy their vision needs with efficiency, comfort, and a minimum energy consumption.

The object of *task lighting* is to provide task-related illumination while avoiding unnecessarily high levels of uniform lighting throughout the space. This is based on the understanding that all lighting costs energy, whether it is electric lighting or daylighting, so that an energy-conscious design dictates the use of only the amount of lighting necessary.

With that in mind, it is the lighting designer's responsibility, as a first step, to become familiar with the principal functions of a given facility. Information on commonly performed tasks should be obtained, including the types of tasks to be performed, the space allocated or necessary for each activity, the occupant density and traffic patterns, and the proportion of work space to circulation area.

Lighting should be planned according to the task performance needs. The tasks to be performed in each area of the building should be identified in order to coordinate the illu-

mination requirements, the orientation of the tasks, and the location of the lighting equipment. It is important to locate each task precisely, especially those that may require special treatment, so that the proper *level of illumination* can be provided at the task area with lower levels in the surrounding area.

It may be advantageous to group tasks that have the same lighting requirements into the same area, or into workstation clusters. In some cases, symmetrical, uniform lighting layouts may be appropriate if worker density is high throughout an entire room. If only a few workers are expected to occupy a building at certain periods of time (such as before or after hours), energy requirements are minimized by grouping their functions in one area or floor of the building. The most demanding visual tasks should be placed at the best daylight locations.

If changes in space use are anticipated, flexibility must be provided for the relocating of luminaires or the ability to redirect light. The bisecting of daylight monitors or luminaires by the addition of a stack of filing cabinets or by any other floor-to-ceiling divider added during remodeling compromises not only the design aesthetics, but also the lighting efficiency in the space.

If task lighting requirements vary from day to day, the lighting system should be flexible enough so that lighting levels can be easily reduced when task requirements do not call for full design levels. Individual luminaires or groups of luminaires may be controlled by separate switches so that selected units can be independently turned on or off, or dimmed as needed to vary the illumination in the area. Lights in unoccupied areas should be capable of being dimmed or shut off entirely. Daylight openings can be similarly adjusted by operable blinds, shutters, and reflectors.

When the same space is used for two or more totally distinct activities with widely varying lighting requirements, it is common practice to design separate lighting systems for each function. As much of the equipment should be used in common as possible.

The height and angle of the work plane should also be determined so that the proper level of illumination can be planned in the plane of the task. A work plane is usually horizontal, but some merchandise displays and library stacks are vertical, while some work surfaces may be at a variety of angles in between.

The design of a lighting system depends on the footcandle and lighting quality requirements of a space, the ceiling height, and the reflectance values of ceiling, walls, and floors. The decisions to be made include whether or not to incorporate daylighting, selecting the type of fixtures (appearance and fixture efficiency), fixture spacing versus mounting height ratio, and selecting the interior surface reflectances if they are not predetermined.

Lighting Levels

Once the lighting objectives for a space are set, it is necessary to establish the quantity of illumination. This is defined in terms of the amount of light falling on the working plane or the object of attention.

Recommended illumination levels for a wide range of applications are established by the IES and are listed in the *IES Lighting Handbook.* Table 8.1 provides guideline illumination levels for some tasks for initial planning purposes. Values are given in footcandles, which can be converted to lux by being multiplied by 10.76.

If the age of the worker, the need for speed and accuracy in performing the task, the contrast quality of the work, or the background reflectance are out of the ordinary, the values in Table 8.1 must be adjusted accordingly.

The values in Table 8.1 are also based on a typical task reflectance of 50 percent. For lower task reflectances, illumination levels must be increased by the inverse proportion. For example, if the task reflectance is 25 percent (one-half of 50 percent), the values in Table 8.1 must be multiplied by 2. If the task reflectance is 5 percent (one-tenth of 50 percent), multiply the illumination values by 10.

The *quantity* of illumination is only one aspect of task performance. Footcandles alone do not determine the visual effectiveness of a lighting system. This depends equally upon the *quality* of the light delivered on the task plane, including its brightness, brightness distribution, and color.

Contrast

With a system of task lighting, the background illumination levels are lower, but it may be desirable to provide some supplementary lighting for visual interest and balance in the space. For example, selective highlighting of paintings, murals, sculptures, plants, or wall segments, or general background illumination to define the boundary of a space can help create a pleasant environment, which is important for human well-being and efficient functioning.

In addition to providing adequate quantities of task light, a lighting system should modulate and distribute illumination in order to create a visually coherent environment. Inadequate luminance of the surroundings in relation to a task can result in a gloomy appearance which can be converted into a warm, cheerful environment by a slight increase in background illumination. Light can and should be used as an adjunct architectural material. Some background illumination is also necessary for control of glare.

When the amount of light in a space is increased, the *luminance* of the surfaces within the space also increases correspondingly. The human eye adapts and adjusts to varying light levels and becomes more sensitive to brightness differences as the brightness of surrounding surfaces increases. Therefore, to maintain visual comfort, brightness differences must be reduced as the general level of brightness is increased.

For example, it does not cause discomfort to look at a full moon against the night sky, even though the *contrast,* or *luminance ratio,* between the moon and sky is on the order of 5 million to 1. This is because the overall brightness is low. On the other hand, a bare lamp appears as a glare in a room lit to 10 footcandles (108 lux) when the contrast of lamp to background is only 400 to 1.

Contrast is just as important a concern as the level of illumination. In general, optimum work efficiency occurs when the brightness of the background is between 10 and 100 percent of that of the work itself.

As a general rule for initial planning purposes, areas surrounding task locations should be lit to an average of one-third the level of the task lighting, but in no case less than 20 footcandles (215 lux). In circulation and seating areas adjacent to task spaces, where no specific visual tasks actually occur, the average level of illumination can be as low as 10 footcandles (108 lux).

Visual efficiency falls off rapidly when the surrounding brightness exceeds that of the work. This condition is an example of *glare,* although glare can be more generally described as any condition in which brightness or contrast of brightnesses interferes with vision.

Glare

Glare can be divided into two categories: (1) *direct glare,* where the offending brightness is in the field of view, and (2) *reflected glare,* where the brightness is reflected from the work in such a manner as to reduce contrast or produce discomfort.

In general, direct glare is controlled by keeping the brightness of luminaires, ceiling areas, and walls below certain values. Any unavoidable glare sources should be located as far away from work stations as possible, or as far outside the field of vision as possible. Where glare is a potential problem, the light source(s) should be spread out while maintaining the necessary illumination levels.

Reflected glare, a function of brightness, can be specularly reflected from the work to the eye. These reflections are not only annoying, but can be visually fatiguing as well.

TABLE 8.1   RECOMMENDED ILLUMINATION LEVELS

Type of Space	Guideline (footcandles)[a]	Type of Space	Guideline (footcandles)[a]
Commercial and Institutional Interiors		Sorting, mailing	100
Art galleries	30–100	Restaurants	50
Auditoriums/assembly spaces	15–30	Schools	
Banks		Auditoriums	15–30
Lobby	50	Reading or writing, libraries	70
Customer areas	70	Lecture halls	70–150
Teller stations and accounting areas	150	Drafting labs, shops	100
Hospitals		Sewing rooms	150
Corridors, toilets, waiting rooms	20	Stores	
Patient rooms		Circulation areas, stock rooms	30
General	20	Merchandise areas	
Supplementary for reading	30	Serviced	100
Supplementary for examination	100	Self-service areas	200
Recovery rooms	30	Showcases and wall cases	
Lab, exam, treatment rooms		Serviced	200
General lighting	50	Self-service	500
Closework and examining table	100	Feature displays	
Autopsy		Serviced	500
General lighting	100	Self-service	1000
Supplementary lighting	1,000		
Emergency rooms		**Industrial Interiors (Manufacturing Areas)**	
General lighting	100	Storage areas	
Supplementary lighting	2,000	Inactive	5
Surgery		Active (rough, bulky)	10
General lighting	200	Active (medium)	20
Supplementary on table	2,500	Active (fine)	50
Hotels (rooms and lobbies)	10–50	Loading, stairways, washrooms	20
Labs	50–100	Ordinary tasks (rough bench and machine	
Libraries		work, light inspection, packing/wrapping)	50
Stacks	30	Difficult tasks (medium bench and machine	
Reading rooms, carrels, book repair and		work, medium inspection)	100
binding, checkout	70	Very difficult tasks (fine bench and	
Catalogs, card files	100	machine work, difficult inspection)	200–500
Offices		Extremely difficult tasks	500–1000
Corridors, stairways, washrooms	10–20[b]	Garages	
Filing cabinets, bookshelves,		Active traffic areas	20
conference tables	30	Service and repair	100
Secretarial desks (with task lighting			
as needed)	50–70	**Exteriors**	
Routine work (reading, transcribing,		Building security	1–5
filing, mail sorting, etc.)	100	Parking	
Accounting, auditing, bookkeeping	150	Self-parking	1
		Attendant parking	2
Post offices		Shopping centers (to attract customers)	5
Storage, corridors, stairways	20	Floodlighting	5–50
Lobby	30	Bulletins and poster panels	20–100

[a] To convert to lux, multiply the value by 10.76.
[b] In no case less than one-fifth of the level in adjacent work stations.

As an example, reflected glare is commonly experienced with tilted drafting tables, even when the paper or other medium has no sheen. The pencil or ink lines exhibit a high degree of specular reflectance, acting like tiny mirrors. At some angles, they can reflect as much or more brightness than the white paper and thus lose their contrast. The lines sometimes even disappear entirely. In this way, the illumination by an overly bright source can defeat its own purpose.

Reflectances

The reflectances of walls, ceilings, and floors have a marked effect on the utilization of light within a room. The inherent lighting qualities of a luminaire can be modified, accentuated, or nullified by the reflections from the surroundings. An extreme example would be attempting to use a high-quality indirect luminaire in a room with a dark ceiling. No matter how well the luminaire worked, the result would be insufficient for adequate working light.

For maximum lighting efficiency, major surface reflectances within a space should be high, that is, light colors should be used whenever possible, but excessively bright surfaces are to be avoided because they could cause glare or help create a "sterile" environment. The reflectance of paints and other surface finishes is usually readily available, or it may be determined by testing. As a guide, the IES recommends the following reflectances for office-type spaces:

- Ceilings: 80 to 92 percent
- Walls: 40 to 60 percent (average)
- Furniture, office machines, and equipment: 25 to 45 percent
- Floors: 20 to 40 percent

The reflectance of ceilings—when they are homogeneous—may be assumed to be the reflectance of the surface. This may not hold for perforated acoustical materials. Sometimes the reflectance of the finish between holes is given rather than that of the entire piece. When this is the case, the actual reflectance of the ceiling may be as much as 15 percent lower.

Ceiling reflectance values have little impact on totally direct lighting systems, but the luminance ratio between the luminaire and the ceiling is important for proper background brightness contrast. Indirect lighting systems, however, do require a high ceiling reflectance to ensure that the maximum quantity of light is reflected back into the room.

Luminance ratios are important considerations for walls, partitions, or other vertical elements in the space since visual comfort in the field of view is a factor affecting the comfort of the occupants of the area.

The reflectance of the walls is rarely that of the surface finish. Rather it is the average of the reflectances weighted by the area of all the reflecting surfaces: doors, window treatments, chalkboards, cabinets, and so on. Window brightness should be controlled or controllable by occupants to alleviate disabling glare.

Reflected glare from desktops can be prevented by using light-colored, diffusely reflecting surfaces with a 25 to 40 percent reflectance. The desktop surface most likely to produce glare is a dark, polished wood or glossy finish.

Floor reflectance has little importance for most tasks performed on surfaces such as a desktop at a height of about 2.5 feet (76 cm) above the floor. It does, however, make a useful contribution to the illumination of certain freestanding tasks, such as inspections of the sides or underside of machinery or other objects, for which additional lighting would be needed if floor reflectance was too low.

Light floors also contribute to general visual comfort because of reduced brightness contrasts. Even if they don't add much measurable light to the work plane, they do create the desirable impression of more and better distributed light.

Overall floor reflectances generally average about 10 to 30 percent, taking into account the chairs, desks, tables, files, and other usual furnishings.

Color of Light

The color of the light source and the objects to be illuminated within a space combined make up an important facet of the lighting quality. All visible light is electromagnetic radiation, as shown in Figure 2.8. Each color is composed of radiation at different wavelengths, and these various wavelengths are treated differently by any given material.

The combination of colored light spread equally over the entire spectrum of visible light, as is the case with the sun, produces white light. A source producing light energy over only a small portion of the spectrum characteristically produces light in only a single color. Mercury vapor lamps, for example, produce a blue-green-tinted light, and sodium lamps produce a yellowish light.

By definition, the color of an object is determined by that object's ability to modify the color of light incident upon it by selectively absorbing a portion of the light and reflecting or transmitting a spectrally modified light emphasizing a single hue.

For example, one material may reflect more blue than red, while another transmits more yellow than blue. The absorption, reflection, and transmission characteristics depend upon the spectral quality of the incident radiation in addition to the properties of the material.

The color reflected or transmitted is perceived by the eye

as the color of the object. An object is technically considered colorless if it does not exhibit selective absorption. Black objects absorb all light equally and reflect very little. White objects reflect all light equally and absorb very little. At near perpendicular angles of incidence, transparent materials transmit all colors equally while absorbing and reflecting very little.

These phenomena of selective absorption, reflection, and transmission are of tremendous importance for light control. For the greatest efficiency, incandescent light should be controlled (transmitted or reflected) with materials that absorb a minimum of red and yellow, while mercury light should be controlled with materials that absorb a minimum of blue and green. To make incandescent light appear whiter, absorb more red and yellow than green.

Note that these phenomena are only *selective.* Reflecting or transmitting materials cannot *add* wavelengths to radiation from a given source. They can only alter the proportions by selectively absorbing *out* certain wavelengths.

The color of light and reflected light in the environment also has psychological effects. Some colors help people relax and be cheerful; some assist in concentration. Others are stimulating and invigorating. A sense of spaciousness is enhanced by some colors.

Careful color selection can reduce eye fatigue, lift spirits, and improve production quality and quantity. Painting gray machinery light orange has been known to decrease accidents and improve the morale of factory workers.

Color therapy in hospitals has accelerated the recovery of patients and improved the effectiveness of the medical and nursing staff. Bright yellows and oranges (bright, warm colors) seem to make patients feel more cheerful and want to get well faster. Certain shades in hospital operating rooms can calm nerves and restore vitality. It should be possible in hospitals to flood the walls of rooms with courageous tones of color for those fearing impending operations, recuperative tones for the convalescent, sedative tones for the excitable, and stimulative tones for those who suffer from mental depression.

In general, red, orange, and yellow are considered warm, stimulating colors. Yellow is particularly cheerful and improves mental clarity. Blues and greens promote coolness and calmness. Green promotes healing, while blue is conducive to relaxation and tranquility. Blue light has been used to induce restfulness and sleep.

Light blue walls tend to impart a cool sensation, while orange makes people feel warmer at the same measured temperature. People tend to overestimate the passage of time in a red room and underestimate it in a green room. Cool colors, in general, seem to shorten the passage of time, and are

thus well suited to areas of dull, repetitive work. Red impels people to action, while green, the color of nature, seems to promote a feeling of well-being. Black can be depressing. Dark-colored objects seem heavier than light-colored ones.

Space Geometry

The geometry of a space—length, width, and ceiling height—affects the distribution of light within the area.

Open areas are more efficient, while the introduction of partitions reduces illumination. Large rooms utilize lighting energy better than do small areas of the same ceiling height. High ceilings are less efficient than lower ones because of the increase in distance of the light source from the work plane (this is more of a factor in electric lighting than in daylighting).

The relationship between the space proportions is illustrated by the following equation which defines the *room cavity ratio* (RCR). The smaller the RCR, the more efficient the space. Equation 8.9 is for the typical case of a task plane 2.5 ft (0.76 m) above the floor. Equation 8.9a is in English units and Equation 8.9b is in SI units.

$$\text{RCR} = \frac{2.5 \times \text{cavity height} \times \text{room perimeter}}{\text{floor area}} \quad (8.9a)$$

$$\text{RCR} = \frac{0.76 \times \text{cavity height} \times \text{room perimeter}}{\text{floor area}} \quad (8.9b)$$

For example, in a 10- × 10-foot (3.05- × 3.05-m) room with an 8-foot (2.44-m) ceiling, the cavity height is $8 - 2.5 = 5.5$ feet ($2.44 - 0.76 = 1.68$ m), and the RCR = 5.5(1.67). By increasing the room dimensions to 20 × 20 feet with the same ceiling height, the RCR is halved:

$$\frac{2.5 \times 5.5 \times 80}{400} = 2.75 \quad (8.10a)$$

$$\frac{0.76 \times 1.68 \times 24.4}{37.21} = 0.837 \quad (8.10b)$$

ELECTRICAL LIGHTING

Once the lighting objectives have been defined in terms of how much light is required to see by, how the light should look in order to allow proper use of the space, and what appearance the structure and space should have when rendered by light, provisions must be made for the lighting system(s).

A well-integrated lighting system makes the best use of

both natural and artificial light sources. As a supplement to whatever daylighting is available, electric lighting is used to ensure adequate task illumination, to provide architectural and emergency lighting, and for security purposes at night.

Natural daylighting is covered in the final section of this chapter. But first, since nearly all lighting systems include at least some artificial illumination, the basic concepts of electrical lighting will be presented here.

The development of a lighting design must take into consideration how to integrate the lighting with the architecture, acoustics, structural concerns, HVAC systems, electrical systems, and graphics and signage of the space, as well as the economic budget.

Lighting, for example, has thermal as well as visual aspects. Chapter 4, referring to the tables in Appendix B, showed that electrical lighting can contribute heavily to the internal gain of a space.

One watt of lighting produces about 3.4 Btuh (1 W) of heat gain in a space. Assuming a fairly large air-conditioning system, it takes about 1 kilowatt (1,000 W) of electricity to produce a ton (3.5 kW) of cooling (small systems use even more power per ton of cooling). Thus, every watt of lighting in an air-conditioned space requires the following amount of energy to cool the space:

$$1 \text{ W} \times \frac{3.4 \text{ Btuh}}{\text{watt}} \times \frac{\text{ton-hour}}{12,000 \text{ Btu}}$$

$$\times \frac{1,000 \text{ W}}{\text{ton}} = 0.28 \text{ W}$$

Every watt per square foot (m^2) reduction in lighting energy actually results in 1.28 W/ft^2 (W/m^2) savings in energy use. The implications for energy use are even further emphasized in Figure 8.4, which shows the flow of energy from the raw source to its final end use providing one unit of illumination.

As a note of caution, however, reducing electricity for lighting through the use of daylighting or high-efficiency luminaires may have a contradictory effect on the building's heating and cooling performance. The savings in lighting energy would be complemented in the summer by cooling savings, but the useful heat that would have been provided by the lighting systems in the winter might have to be made up by the heating system.

The total lighting cost of an installation should be studied from a life-cycle viewpoint, as discussed in Chapter 16. In addition to its beneficial or deleterious economic effect on other building systems, the lighting system itself has the following costs: (1) installation (including interest on the investment, taxes, and insurance), (2) energy, and (3) relamping, cleaning, and repair (including labor). Figure 8.5 illustrates approximately how these factors relate to the life-cycle

costs of the various light sources commonly used for indoor lighting.

Light Source Types

Tables 8.2a and 8.2b list the categories of available lamp types together with some of their distinguishing characteristics. The two types most commonly used are incandescent and fluorescent lamps.

Incandescent

Incandescent lamps are the least efficient of all lamps, having typical efficacies of less than 20 lumens per watt. Their efficiency increases with rated wattage, varying from 15 percent for a 10-W lamp to 25 percent for a 1,500-W lamp.

Their color is white with a large yellow-red component, which makes incandescent light flattering to skin tones. The exact color depends on the temperature, which in turn is determined by the wattage. High-wattage lamps are more blue, while low-wattage lamps are more yellow.

Dimmed lamps of a given wattage give off a yellow-red light.

The output, efficiency, and life of these bulbs are dramatically altered by even a small change in operating voltage. For example, burning a 120-volt lamp at 125 volts (104 percent of proper voltage) results in approximately:

16 percent more light (lumens)
7 percent more power consumption (watts)
8 percent higher efficacy (lumens per watt)
42 percent less life (hours)

Burning a 120-volt lamp at 115 volts (95.8 percent of proper voltage) provides approximately:

15 percent less light (lumens)
7 percent less power consumption (watts)
8 percent lower efficacy (lumens per watt)
72 percent more life (hours)

In general, a 1 percent change in voltage produces approximately a 3 percent change in light output and a 10 percent change in lamp life.

Reflector lamps are a specially designed type of incandescent lamp. Narrow-beam designs are called *spotlights,* and wide beam designs are called *floodlights.* They are available in soft glass for indoor use and hard glass for outdoor applications. Fixtures for these lamps act principally as a lamp holder, since beam control is built into the lamp. Special lamps are available with a "cool beam" or a variety of colored beams.

The advantages of incandescent lamps are low initial and

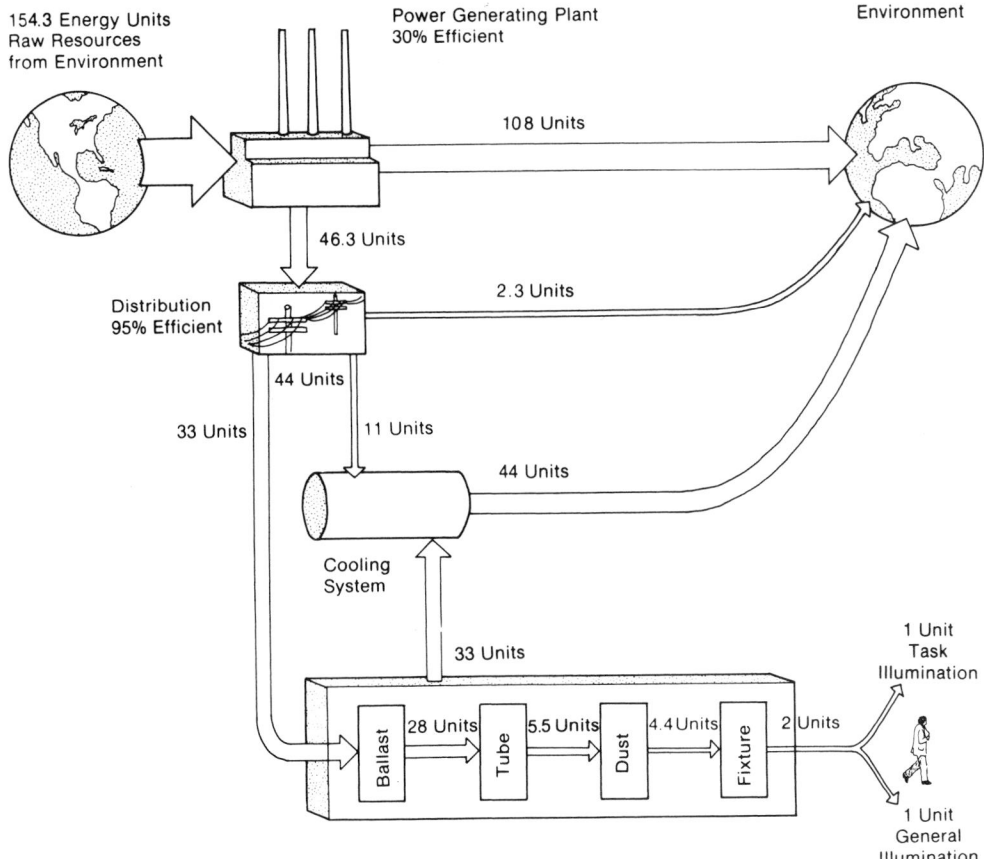

FIGURE 8.4. Energy use for illumination flow chart.

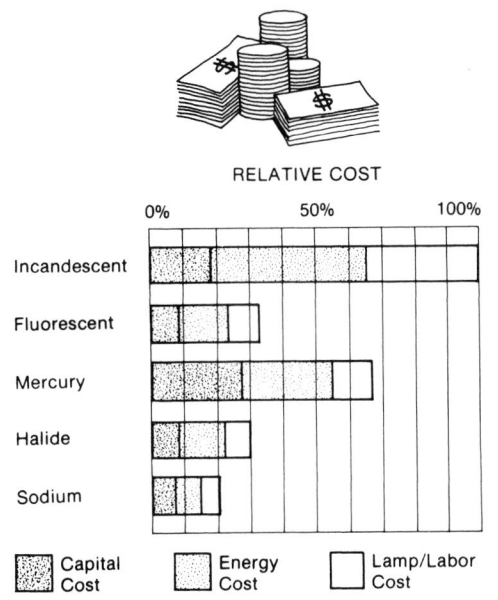

RELATIVE COST

0%   50%   100%

Incandescent

Fluorescent

Mercury

Halide

Sodium

▨ Capital
   Cost
⬚ Energy
   Cost
☐ Lamp/Labor
   Cost

FIGURE 8.5. Relative life-cycle economics of the various light sources.

replacement costs, inexpensive fixtures, compact size, simple installation, accurate color rendition, instant start, no ballast required, simple and inexpensive dimming, and high power factor; also, they are focusable as a point source, and their life is independent of the number of starts.

The disadvantages are low efficacy, large heat gain, short lamp life (resulting in high maintenance costs for the labor to replace them), and acute voltage sensitivity.

Because of their poor energy efficiency, incandescent lamps should be restricted to applications where (1) use is infrequent or of short duration, (2) low-cost dimming is required, (3) focusing fixtures are needed, or (4) minimum initial cost is crucial.

To overcome the disadvantage of short lamp life, extended service lamps have been designed for 2,500-hour life; these are useful in locations where maintenance is irregular or relamping is difficult. These lamps are really designed for slightly higher voltages than that at which they operate, and therefore efficacy is reduced.

So-called long-life lamps, which are guaranteed to burn

TABLE 8.2a   LAMPS AVAILABLE AND THEIR CHARACTERISTICS

Lamp	Power Consumption (Wattage Range)	Rated Initial Lumens/Watt (Incl. Ballast)	Average Rated Life (hours)[a]
Incandescent			
Standard	10–1,500	15–25	750–3,500
Tungsten halogen	100–1,500	15–25	2,000–12,000
Fluorescent			
Standard	15–100	40–95	9,000–20,000+
High-output	60–215	45–100	9,000–20,000+
High-intensity discharge (H.I.D.)			
Mercury vapor (standard)	40–1,000	24–60	12,000–24,000+
Mercury vapor (self-ballasted)	160–1,250	14–30	10,000–20,000
Metal-halide	175–1,500	69–115	7,500–20,000
High-pressure sodium	35–1,000	51–130	12,000–24,000+
Low-pressure sodium	18–180	62–150	12,000–18,000

Note: Data are based on commonly used lamps and are intended for comparison purposes only. Actual results depend on the specific application and products used. Consult manufacturers' literature for details.

[a] Representing the average life of a group of lamps. Approximately half of the lamps in any group are expected to burn out at this time.

for 2, 3, or 5 years, are designed for much higher voltages than that at which they operate. They are normally expensive and very inefficient, so their use is seldom advisable.

Krypton lamps, filled with krypton gas, have approximately 10 percent higher efficacy, longer life, and are smaller. Suitable applications are for increased efficacy as long-life lamps and increased output and life in exterior spotlights and floodlights. The cost premium for krypton lamps is approximately 50 percent.

Tungsten-halogen (quartz) lamps are another special type of incandescent lamp. The bulb is made of quartz in order to withstand high temperature. It is filled with iodine vapor that prevents the evaporation of the tungsten filament; this slows down the resulting deterioration of light output and the eventual burnout. Although it has approximately the same efficacy as an equivalent standard incandescent lamp, the tungsten-halogen lamp has the advantages of longer life, low lumen depreciation (90 percent output at 98 percent life), and a smaller size for a given wattage. Quartz lamps are not normally used for general lighting.

TABLE 8.2b   LAMPS AVAILABLE AND THEIR CHARACTERISTICS

Lamp	Lumen Maintenance @ 50% Life	Color Rendition	Start Time	Restart Time
Incandescent				
Standard	80–90%	Excellent	Instant	Immediate
Tungsten halogen	95%	Excellent	Instant	Immediate
Fluorescent				
Standard	85–90%	Very good	Instant	Immediate
High-output	85–90%	Very good	Instant	Immediate
High-intensity discharge (H.I.D.)				
Mercury vapor (standard)	70–80%	Poor	3–6 min	3–10 min
Mercury vapor (self-ballasted)	—	Fair–good	3–6 min	3–10 min
Metal-halide	75–85%	Very good	2–3 min	5–20 min
High-pressure sodium	80–90%	Fair–good	3–4 min	½–1½ min
Low-pressure sodium	—	Poor		Immediate

Fluorescent

Fluorescent lamps are much more efficient than incandescent lamps and have up to 20 times longer life than some incandescents. For these reasons they are widely used instead of incandescent lamps, except for specialty lighting and residential use. For example, more light is emitted by a 40-W fluorescent tube than by a 100-W incandescent bulb. The fluorescent luminaire is more than twice as efficient as the incandescent bulb, even including the electrical losses from the ballast needed to operate fluorescent lamps. The ballast typically consumes an added 15 to 20 percent of the rated wattage of the fluorescent lamp.

The efficacy of fluorescent lamps depends on their color-rendering capabilities. The most efficient are warm white, cool white, and white. These are best for industrial, institutional, and general office applications, where economical light production is of prime concern. Deluxe warm white, deluxe cool white, and other deluxe types should be used where color rendition is important and lower efficacies are acceptable. Prime-color lamps with lumen-per-watt ratings between those of deluxe and standard lamps offer a compromise for applications where excellent color rendition is required.

The lamp life of a fluorescent tube is dependent on the number of starts; the values listed in Table 8.2 are based on an average of 3 hours per start. If an area is not utilized for at least 15 or 20 minutes, the value of the energy saved by shutting off a fluorescent lamp will generally be greater than the value of the decreased lamp life caused by the on/off cycle. Otherwise, if the off period is less than 15 minutes, it does not pay to shut off the lamp.

The lumen output of a fluorescent lamp deteriorates rapidly during the first 100 hours of burning and thereafter much more slowly. The figures for initial lumens/watt in Table 8.2 represent output after 100 hours of burning. At 40 percent of a bulb's average rated life, its output typically drops to approximately 85 to 90 percent of the 100-hour initial value.

Fluorescent light output is also dependent on the operating temperature of the tube, which is affected by the ambient temperature. The output normally drops when the lamps are either above or below their rated temperature. Special "all-weather" lamps are available that maintain a fairly constant lumen output over a wide ambient temperature range.

The shape of fluorescent lamps is tubular, and most are straight tubes. The standard 4-foot (1.2-m) tube is rated at 40 watts. As an option, energy-conserving reduced-wattage lamps are available to decrease power consumption. A comparison between the wattages of standard and energy-conserving 4- and 8-foot (1.2- and 2.4-m) lamps is presented in Table 8.3.

TABLE 8.3 WATTAGE COMPARISON BETWEEN STANDARD AND ENERGY-CONSERVING FLUORESCENT LAMPS

Length	Standard Lamp (watts)	Energy-Conserving Lamp (watts)
4-foot	40	34
8-foot	75	60
8-foot, high-output	112	95
8-foot, very high output	215	185

Although the efficacy of these energy-saving lamps is higher than their standard counterparts, their light output is lower. In addition, the life of the 8-foot (2.4-m) lamps is shorter. Existing fluorescent lamps in overlit spaces can be replaced with these lamps to reduce their energy consumption, but with the loss of some illumination.

Reducing energy consumption in existing overlit installations can also be accomplished by removing some of the tubes from a group of fixtures. Caution must be exercised, however, that the remaining illumination is still uniform and has no dark spots. Also, removing one lamp from a multiple lamp fixture can have a deleterious effect on the ballast, deactivate the entire ballast, or not even affect the amount of energy used by the ballast. An electrical consultant should be called upon to check the luminaire before a lamp removal program is instituted. The ballast may need to be replaced or a "dummy" tube installed in order to ensure the success of such a program.

In addition to the common 4- and 8-foot (1.2- and 2.4-m) tube lengths, fluorescent lamps are available in a U shape for use in a 2-foot (60-cm) square fixture. The U lamp is essentially a standard 40-watt, 48-inch (1.2-m) fluorescent tube bent into a U shape. Such lamps are available with 3⅝-inch (9.2-cm) leg spacing, so that three can be accommodated in a 2-foot (60-cm) square fixture, or with 6-inch (15-cm) leg spacing, in which case two can be held in a 2-foot (60-cm) square fixture. They have slightly lower output than a straight 4-foot (1.2-m) tube, but are more desirable than straight 2-foot (60-cm) lamps for use in a 2-foot (60-cm) square fixture, as demonstrated by the comparison in Table 8.4.

Another novel fluorescent light on the market is a lamp that fits into an ordinary incandescent socket. General Electric's circular fluorescent lamp called the Circlite, for example, has its ballast contained within a screw-in adapter which fits into an incandescent socket. The lamp uses 44 W of

TABLE 8.4 STRAIGHT 2-FOOT VS. U-SHAPED FLUORESCENT LAMPS

Lamp Type	Watts[a]	Lumens	Lumens Watt	Hours of Life
Four 2-foot straight	110	5,200	47.3	9,000
Two U-shaped	100	5,800	58.0	12,000

[a] Including ballast.

power, but produces as much light as a 100-W incandescent bulb. Aimed primarily at the residential market, these units are also applicable in industry.

Fluorescent lighting may be dimmed, but high-quality SCR controls are necessary to avoid radio interference. The cost of such equipment is relatively high and is only justified when infinitely adjustable light levels are essential. If the lighting level simply needs to be reduced—such as in classrooms, lecture halls, or multipurpose areas—two- and three-level ballasts are available.

Ballasts should always be UL (Underwriters Laboratories) labeled to ensure intrinsic safety. The electrical energy referred to as *ballast loss* is converted to heat as part of the normal operation of a ballast.

Pendant-mounted luminaires located more than 6 inches (15 cm) below the ceiling allow this heat to easily dissipate from the fixture, as do fixtures recessed into ventilated-suspended ceilings. Whenever possible, a clearance of at least 1½ inches (3.8 cm) of air space should be allowed between a surface-mounted fixture and the ceiling in order to avoid overheating the ballast, which can seriously reduce ballast life.

Fixtures mounted on insulating surfaces, such as acoustic tile, or into unventilated or heated ceiling spaces inhibit the necessary heat dissipation, as do those fixtures boxed by a fire-rated enclosure and recessed into a fire-rated ceiling. More expensive cool-operation ballasts are available for such applications, as well as for high-ambient-temperature installations (normal ballasts are designed to start fluorescent lamps in an ambient temperature range of 50° to 105°F, 10° to 41°C).

Ballasts characteristically produce a hum or buzz when operating. Ballasts are sound-rated and classified by designations A, B, C, D, E, and F. A low noise rating is usually selected for offices and other quiet areas where the noise would be objectionable. The A rating, which is the quietest, is used in areas where the ambient noise level is below 24 decibels. The other ratings are recommended, in succession, as the ambient noise level increases by 6-decibel increments. The F rating, which is the noisiest, is for ambient noise levels of 48 decibels or above.

Even quiet ballasts, however, may be annoying because of amplification by resonant surfaces and conduction of sound through the metal of improperly mounted luminaires. If extremely low noise levels are required, ballast noise can be controlled by the use of soundproof cases or by locating the ballast outside of the room.

In addition to audible noise, fluorescent ballasts produce radio interference. This radio noise is negligible in most circumstances and can be abated by suppressors. It can be enough, however, to rule out the use of fluorescent lighting in some electronic laboratory installations.

The primary disadvantage of fluorescent lamps is their large size, which creates storage, handling, and relamping problems. The large lamps also necessitate a relatively bulky, expensive fixture both to hold the lamps and to control the light. Since the tubes emit light throughout their length, accurate beam control is not possible.

A special category of fluorescent light is neon-vapor lamps. They consist of glass tubes filled with ionized neon gas. The neon light has a pink to dark red color, depending on the gas pressure, and the efficiency of light production is not very high. The tubes are primarily used for electrical advertising signs. The tubing can be fashioned into any shape and operates satisfactorily in very cold weather. Different colors may be obtained by substituting helium gas for neon, by using various mixtures of the two, or by using colored glass tubing.

High-Intensity Discharge (H.I.D.)

These include mercury vapor, metal-halide, and high- and low-pressure sodium lamps. These lamps have inherently high efficacy and, with appropriate color correction, can be utilized in almost any application, indoor or outdoor, that does not have critical color criteria.

Mercury Vapor (M-V) Lamps. Clear M-V lamps produce a distinctive blue-green light. Color correction is possible by coating the outer bulb with phosphors that are excited by ultraviolet light and reradiate reds and oranges, which are entirely absent in the source color. The color of the emitted light can be corrected to make it acceptable for general indoor use. Lamps are available in clear, white, color-corrected, and white-deluxe. The deluxe lamp uses a stain on its envelope to filter out some of the blue-green, which reduces lamp output.

Efficacies of M-V lamps are higher than those of incandescents but lower than those of fluorescent lamps. Some self-ballasted units have even lower efficacies than incandescents. The ballast increases the wattage of the lamp alone by an additional 12 percent. Generally, the higher the wattage, the higher the efficacy. The efficacy also depends on the position of the lamp. Manufacturers' recommendations should be carefully followed.

Lamp life is extremely long. The values in Table 8.2 are based on 10 burning hours per start. Since the long life of these lamps is predicated on their being left on for long periods of time, they are not suitable for applications that are subject to frequent switching.

It takes 3 to 6 minutes for a M-V lamp to reach full output. Once extinguished, the lamp must cool before restart is possible, which means that a temporary power outage will leave an interior area completely in the dark for that length of time. Special fixtures incorporate small quartz lamps to supply light during such outages. Otherwise, if M-V lights are used, some incandescent lighting is advisable in order to provide minimum illumination for emergency egress.

Dimming is readily accomplished with 400-, 700-, and 1,000-watt units. Unlike the dimming of fluorescents, dimming of M-V lamps is a desirable and economical means of control. Since M-V lamps have such a high output, shutting off a unit often results in an imbalance in the lighting coverage—a problem readily solved by dimming. Such control is generally an economical means of saving energy, even when it includes the cost of sensing equipment that automatically maintains illumination at preset levels.

M-V lamps are applicable to indoor and outdoor use if proper attention is paid to color and fixture brightness. High mounting is necessary to avoid glare and to permit adequate area coverage. These lamps are commonly used in industrial spaces and stores. The smaller wattages with color correction are available with screw-in ballasts for direct replacement of incandescent bulbs in the same fixture. Their extremely long life make them attractive as a substitute for incandescents in locations where relamping is difficult and expensive.

In most instances, however, fluorescent lighting is less costly and has a longer life and higher efficacy. If fluorescent lighting is impossible, and frequent switching on and off is not planned, an M-V lamp with a separate ballast will be more energy-efficient and generally more economical than an incandescent source.

Metal-Halide Lamp. The metal-halide lamp is basically a mercury lamp with halides of metals, such as thallium, indium, or sodium, added. The addition of these salts causes light to be radiated at frequencies other than the basic mercury colors and increases efficacy, but reduces the life by about half and reduces lumen maintenance to 60 percent at two-thirds life.

The color is much warmer than that of the M-V light. Initial start-up time is faster than that for an M-V lamp, but the restart delay is much longer.

Metal-halide lamps are available with a base containing electronic circuitry that screws into an incandescent socket. They come in three sizes, which are the equivalent of 100-watt and 200-watt incandescents and a two-way 50/150-watt lamp. They are designed to use only about a third of the electricity that comparable incandescents use.

High-Pressure Sodium (HPS) Lamp. The light of these lamps is a yellow-tinted color due to the sodium contained within them under high pressure. The yellow color becomes less noticeable as the eye color-adapts, but complementary white sources used in conjunction can make the color of this light even more acceptable.

Special HPS units are available that can be substituted directly for existing M-V lamps. This improves illumination and reduces energy costs, since HPS lamps average double the efficacy of mercury lamps.

Low-Pressure Sodium (LPS) Lamp. LPS is the most efficient source available, but it is inappropriate for general lighting because of its distinctive deep yellow light output. Another desirable aspect of these lamps besides their very high efficacy is their 100 percent lumen maintenance. This, coupled with their long life, makes them a highly economical source in terms of cost per lumen.

LPS lamps are advantageous wherever color is not an important criterion and are widely used for highway lighting, where the discernment of objects is the primary objective. Their use for buildings is confined to applications where color rendition is of no concern, such as exterior applications, inside such spaces as warehouses, or as after-hours security lighting.

Luminaires

Once the suitable source(s) for a given installation have been determined, the fixtures available on the market can be reviewed in order to select those that are applicable. The essential function of a luminaire is to block the glare from the lamp(s) from direct view, and then to introduce the light into the space with the proper diffusion, modification, or redirection necessary to establish the desired lighting result.

Light may be concentrated so as to reach from a high mounting height to the working level while still providing uniformity of lighting. Alternatively, a low-mounted fixture may have a spread pattern so that its light covers a large area.

Luminaires are being continuously improved by manufacturers with the aid of computers to design shielding materials, reflector shapes, and light source positions for the best efficiency of their products. Such improvements are meant to allow as much of the light source output to reach the work plane as possible while maintaining comfortable vision conditions with minimum glare and contrast.

System Types

Lighting systems are conventionally divided into five categories according to how they control or distribute light: (1) indirect, (2) semi-indirect, (3) general-diffuse and direct-indirect, (4) semi-direct, and (5) direct. They differ principally in the proportion of light directed upward or downward. The technical definitions are listed in Table 8.5, and the categories are illustrated in Figure 8.6.

Indirect. Nearly all of the light in an indirect system arrives at the horizontal work plane indirectly. It is first reflected from the ceiling and upper walls. The walls and ceiling must therefore have a high-reflectance finish. The room illumination is diffuse, shadowless, uniform, and with low glare.

Luminaire mounting position, spacing between luminaires, and cove or valence dimensions must be carefully chosen to avoid spots of excessive brightness at the ceiling above the fixture. This generally limits the source to a position at least 18 inches (46 cm) below the ceiling. As a result, the ceiling must be at least 9 feet 6 inches (2.9 m) high.

The maximum attainable illumination on the horizontal work plane is approximately 75 footcandles. With higher levels of illumination, the ceiling may become so bright that it becomes a source of glare.

Indirect lighting is by nature inefficient, since all of the useful light reaches the working plane only after a double reflection—within the fixture and off the ceiling.

The shadowless light may be unsatisfactory for three-dimensional work.

Diffuse lighting may be desirable in offices, classrooms, drafting rooms, and machine shops, where shadows could be annoying, disruptive, or, in the case of a machine shop, even dangerous. Some directional lighting is desirable, however, if textures must be examined or if surface imperfections must be detected by shadows and reflections. Diffuse general lighting may be supplemented by some directional lighting to lend interest by creating shadows, brightness variations, sparkle, or a sense of depth.

Semi-Indirect. These systems are somewhat more efficient than indirect ones and allow higher levels of illumination without glare. Typically, a semi-indirect fixture employs a translucent diffusing element through which the downward component shines. The ceiling is still the principal radiating source, and the character of the room lighting is still diffuse.

General-Diffuse and Direct-Indirect. Fixtures for these systems distribute approximately an equal amount of light upward and downward. Since the ceiling is a major, although secondary, source of room illumination, there is some useful diffuseness, which provides satisfactory vertical-plane illumination. These fixtures also provide a bright ceiling and upper wall background for the luminaire. They should be suspended at least 12 inches (30 cm) below the ceiling in order to avoid excessive ceiling brightness.

General-diffuse fixtures, such as a diffusing globe, emit light equally in all directions. Direct-indirect fixtures may have an open top, luminous sides, and diffusing bottom. Their light output has very little horizontal component. A space lit with a general-diffuse source appears lighter than one lit with direct-indirect illumination because of the darker walls in the latter case.

Semi-Direct. The minor upward component serves to illuminate the ceiling. Shadowing in the space is generally not a problem if the upward component is at least 25 percent and the ceiling reflectance is at least 70 percent. With smaller upward components, the system is essentially direct lighting.

Direct. These systems are inherently efficient. Since basically all light is directed downward, illumination of the ceiling is entirely due to light reflected from the floor and room furnishings. These systems, more than any other, require a light, diffuse, high-reflectance floor, unless a dark ceiling is desired for aesthetic purposes. Since illumination is largely independent of wall reflectance, walls may be any color.

TABLE 8.5 LIGHTING SYSTEM TYPES

Type	Distribution of Light Emitted by Luminaire	
	Upward (%)	Downward (%)
Indirect	90–100	0–10
Semi-indirect	60–90	10–40
General-diffuse	40–60	40–60
Semi-direct	10–40	60–90
Direct	0–10	90–100

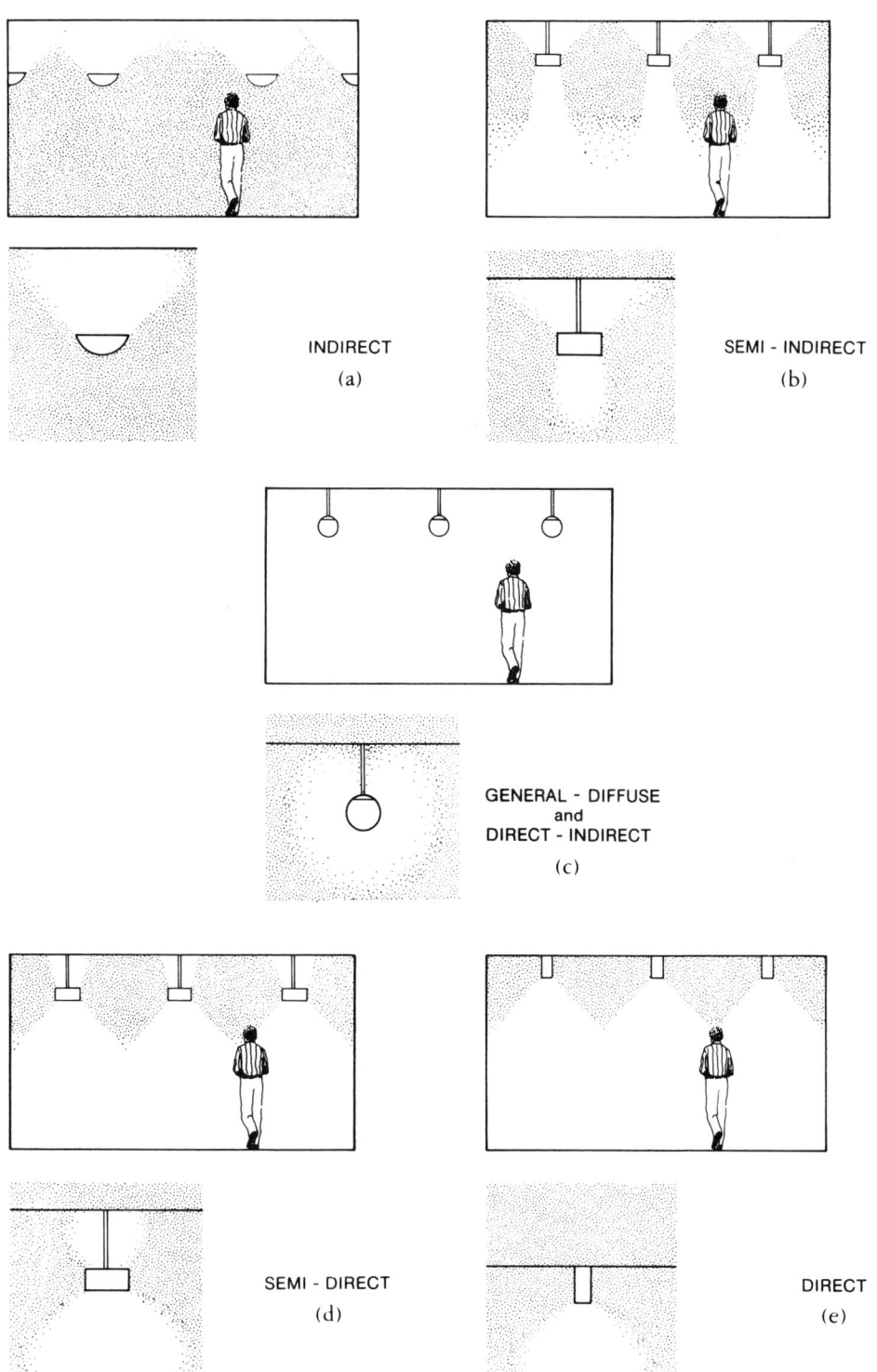

FIGURE 8.6. Lighting system types. (a) Indirect, (b) semi-indirect, (c) general-diffuse and direct-indirect, (d) semi-direct, and (e) direct.

Pendant-mounted direct fixtures may be used in conjunction with a ceiling deliberately painted a dark color in order to give the impression of lowering the ceiling of a poorly proportioned room, or to hide unsightly piping, ductwork, or other exposed objects.

Direct lighting provides very little vertical surface illumination and can produce shadows. Careful design is necessary to prevent disturbing reflected glare. Concentrated down-lights used alone are appropriate in restaurants and other areas where the visual privacy created by the limited horizontal illumination may be desirable.

A luminous ceiling, which is essentially one large fixture covering the entire ceiling, is a special form of direct lighting that greatly reduces glare. Such systems are sometimes known as *transilluminated* ceilings because light from sources above the ceiling surface is directed downward through cellular louvers or translucent diffusing material that forms the visual ceiling. These systems can be used to obtain higher illumination levels, but the brightness of the shielding or transmitting medium must be limited. They are appropriate only for even, uniform levels of illumination.

Fixture Selection

Because of its luminance, each luminaire or other luminous source is a point of visual attention. Numerous large or very bright luminaires, or those arranged in striking patterns, draw attention away from other surfaces. In some cases, accent color or accent lighting directed so as to emphasize some area or object may be desirable in order to avoid monotony.

Luminaire size should correlate with room size and ceiling height. In general, fluorescent fixtures larger than 2 × 4 feet (60 × 120 cm) should not be used in ceilings less than 10 feet (3 m) high unless their size is disguised by a smaller luminaire surface pattern.

Some spaces require overall, uniform illumination, while others can utilize either local lighting alone or local lighting in addition to general. Lighting can be mounted as part of a work station or permanently mounted in the ceiling above.

Lighting equipment should be unobtrusive, but not necessarily invisible. Fixtures can be chosen and arranged in various ways to complement the architecture or to create dominant or minor architectural features or patterns. Fixtures may also be decorative. Examples of some common fixture styles are shown in Figure 8.7.

Any fixture, properly applied, can provide a sufficient quantity of light. Other factors to consider in selecting a fixture are the quality of light, ease of installation, ease of maintenance, cost, and life.

Quantity and Arrangement

Once the required level of illumination is ascertained and a fixture is selected, it is necessary to determine the number of fixtures and the approximate spacing needed for the planned illumination level so that a reflected ceiling plan may be generated. Before outlining a couple of simplified procedures for finding the number of fixtures, the concept of maintenance factor should first be introduced.

Maintenance Factor. The light output from any fixture falls below the initial value over time due to a number of factors. The lamps themselves deteriorate with age and emit fewer lumens as they are used; dirt, dust, and foreign matter collect on the luminaires and reflecting surfaces of the room; and the reflecting and transmitting materials in some luminaires undergo gradual chemical changes in the presence of air and light. The output of some luminaires and lighting systems is more affected by dirt than that of others.

To compensate for this deterioration of illumination, a factor variously known as the *maintenance factor, depreciation factor,* or *light loss factor* is used in design calculations for lighting systems. The designer must exercise some judgment in the selection of a maintenance factor. A manufacturer's literature may provide some indication of the maintenance factor for its luminaire, but the rest of the system and the anticipated housekeeping practices must be evaluated on an individual basis. Maintenance factors vary from about 0.8 for clean areas with good maintenance to 0.5 for those with low maintenance.

All fixtures and reflecting surfaces in a room require regular cleaning in order to preserve efficiency. A mechanically air-conditioned space, due to the continuous filtering of the air, generally stays cleaner than a naturally ventilated space, which freely lets dirt and dust in from the outside. Smoking in a room accelerates the accumulation of residue, and urban locations are, in most cases, dirtier than suburban or rural areas.

In the design of a space, consideration should generally be given to the maintainability of fixtures and surfaces. Finish materials should be selected with an awareness of how easy they are to clean, as well as a knowledge of their surface reflectance.

The accessibility of fixtures for maintenance has a strong bearing on the likelihood that they'll be kept clean and that their lamps will be replaced. The maximum height that a custodian can safely reach from a conventional ladder is about 15 feet (4.6 m). Platform ladders or elevated towers can go higher but are considerably more expensive and cumbersome. A "lamp stick" can replace conventional filament

(a)

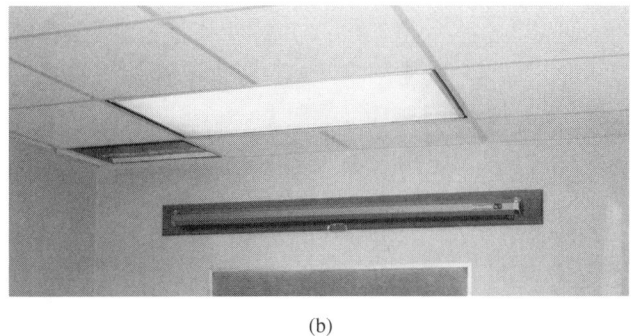

(b)

(c)

FIGURE 8.7. Common fixture styles. (a) Task lighting built into work station furniture to illuminate the desk surface. (Photo courtesy of Steelcase Inc.) (b) Ceiling fluorescent recessed, (c) Ceiling fluorescent surface-mounted, (d) Luminous ceiling in an art museum, (e) Ceiling incandescent recessed. (Photos b–e by Vaughn Bradshaw) (f) Ceiling incandescent surface-mounted, (g) ceiling eyeball, (h) track lighting, (i) incandescent integral ceiling, (j) wall-mounted fluorescent, (k) wall-mounted incandescent ambient lighting. (Photos f–k courtesy of Lightolier) (l) Emergency lighting, (m) H.I.D lights, (n) outdoor wall mounted light, (o) bollard light, (p) outdoor pole-mounted light. (Photos l–p by Vaughn Bradshaw)

or mercury lamps in open reflectors as high as approximately 25 feet (7.6 m). Alternative methods for luminaire maintenance are crawl spaces above the ceiling, catwalks, and disconnecting hangers allowing fixtures to be lowered to floor level.

Calculation Procedure. For this method, the following information must be obtained:

- Square feet to be illuminated
- Required footcandles
- Maintenance factor
- Number of lamps per fixture
- Number of lumens per lamp
- Coefficient of utilization (CU)

The CU is an index of the efficiency of the entire lighting system, taking into account both the lighting equipment and the room dimensions. The CU for a given fixture is readily available from the manufacturer. CUs for a variety of standard fixtures are summarized in such lighting manuals as the *IES Lighting Handbook.*

Each luminaire has a different CU for each type of space in which it is used. When looking up a CU for a given fixture, two standardized variables are required: (1) the room cavity ratio, as explained earlier, and (2) the reflection fac-

(d)

FIGURE 8.7. *(Continued)*

(e)

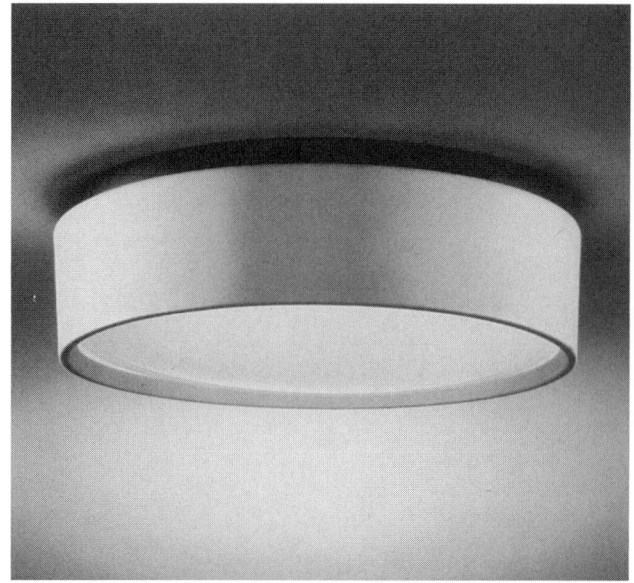

(f)

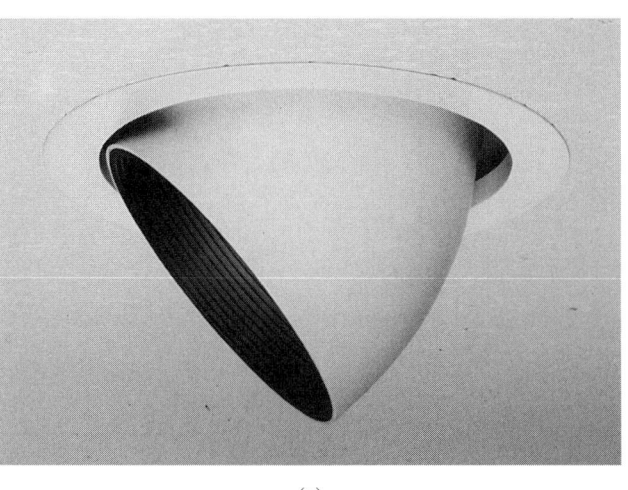

(g)

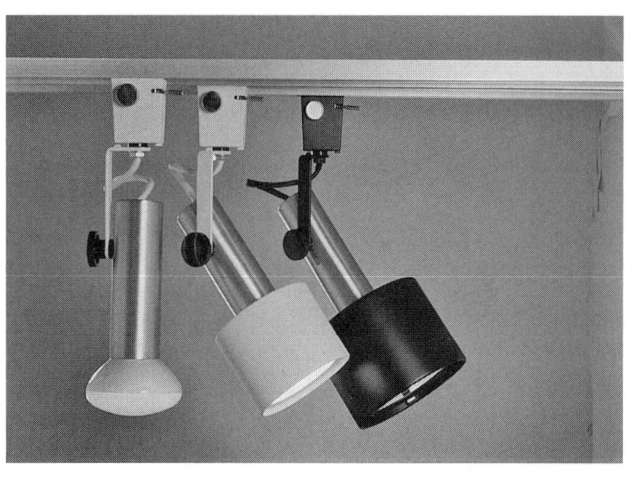

(h)

FIGURE 8.7. *(Continued)*

tors of walls and ceilings. Reflection factors generally vary from 0.27 for dark areas to 0.62 for light-colored areas.

Once the above information is gathered, the following formula may be applied:

number of fixtures =

$$\frac{\text{required footcandles} \times \text{ft}^2}{\substack{\text{maintenance} \times \text{lamps} \times \text{lumens} \times \text{CU} \\ \text{factor} \qquad \text{per} \qquad \text{per} \\ \text{fixture} \quad \text{lamp}}} \qquad (8.11)$$

Graphical Procedure

Step 1. Once the required footcandles and the maintenance factor are known, calculate the maintained footcandles as follows:

$$\text{maintained footcandles} = \frac{\text{required footcandles}}{\text{maintenance factor}} \qquad (8.12)$$

Step 2. From Figure 8.8, find the watts per square foot for the appropriate maintained footcandles and type of fixture.

(i)

(j)

FIGURE 8.7. *(Continued)*

Step 3. Calculate the number of fixtures as:

number of fixtures =

$$\frac{\text{watts per ft}^2 \times \text{ft}^2}{\text{lamps per fixture} \times \text{watts per lamp}^a} \quad (8.13)$$

Lighting Energy Conservation

The following methods are used to maximize the efficiency of electric lighting systems:

1. Provide light-colored ceiling, wall, and floor surfaces. Specify light-colored furniture.
2. Provide task lighting where higher illumination is necessary to allow lower lighting levels in surrounding areas.

a Including lamp and ballast loss.

3. Use primarily high-efficiency lamp types such as fluorescent and metal-halide lamps.
4. Use efficient luminaries and carefully arrange them to provide exactly the level and quality of lighting needed to meet the project's needs.
5. Provide manual and automatic switching and dimmers. People can usually be relied on to turn lights on when it is too dark, but they almost never turn lights off when they are no longer necessary. Occupancy (motion) sensors determine if anyone is present in a room and automatically turn the lights off a few minutes after the last person leaves. Timers can turn lights on and off according to a preset schedule.
6. Make use of daylighting, and complement with electric lighting automatically controlled by photo sensors and dimmers.

(k)

(l)

(m)

FIGURE 8.7. *(Continued)*

(n)

(o)

(p)

FIGURE 8.7. *(Continued)*

DAYLIGHTING

As mentioned in Chapter 7, the most practical method of passive solar energy utilization in commercial buildings is *daylighting*. It reduces not only lighting energy use, but the cooling load as well (see Figure 8.4).

Sunlight is a highly efficient source of illumination, making it a comparatively cool source. It provides approximately 90 to 120 lumens of illumination per watt of total energy, compared to 40 to 75 lumens per watt for common fluorescent and 15 to 25 lumens per watt for incandescent light. As electric light is created, it introduces about twice as much heat per unit of light into a space as does daylighting.

However, larger areas of glazing are needed for increased daylight, which can adversely affect heating and cooling loads. Proper shading can prevent much additional heat

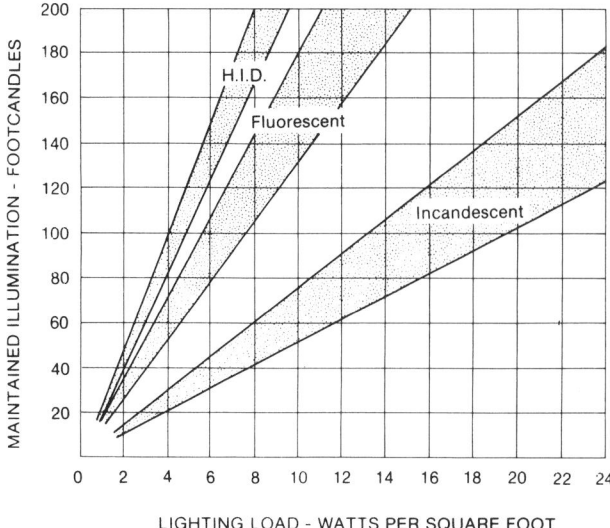

FIGURE 8.8. Quantity determination chart (watts/ft²).

gain in the summer, and additional heating loads can be minimized by double and triple glazing and operable thermal barriers for use during nondaylight hours. The same fenestration can even supply natural ventilation for added benefit.

Daylight has been ignored in most buildings because it is so variable. On a bright day, sunlight can produce levels of illumination 50 times as high as those recommended for artificial illumination. There is more of it in the summer than in the winter, and it varies each day from none before dawn to a maximum at noon and back to none after dusk. An overcast sky distributes daylight much differently than a clear sky, and sky conditions can change several times a day.

When energy was cheap, it was much easier to simply eliminate this troublesome variable by relying on all electric light sources except for some accent fenestration. But that luxury can rarely be afforded any longer.

Daylighting also has a much greater impact on the architecture of a building than does electric lighting, affecting not only the amount of fenestration and its appearance on the facade, but even the building orientation and shape. If the building is more than a single story, light wells may be needed to bring daylight into the interior core.

To be effective, daylighting must meet the same visual performance criteria as artificial lighting in providing adequate levels and quality of task illumination. In addition, daylighting presents other challenges because of its variability. Systems, however, need not and should not be overly complicated to achieve these goals.

In its favor, daylight provides pleasing color rendition and is generally available at the same times that light is needed in most commercial buildings. Although they can't use daylight all the time, even buildings with round-the-clock operation can use daylight for at least one shift and thus reduce their lighting energy consumption correspondingly.

Resource Availability

The sun can provide about 8,000 to 10,000 footcandles (86,000 to 108,000 lux) of light, yet only a fraction of that is needed for most tasks. Indirect sunlight provides illumination levels that are 10 to 20 percent as bright as direct sunlight. This is still more light than is needed, so a daylighting system can be somewhat inefficient and still be very useful.

The complication is that the quantity of daylight available varies widely with the season, latitude, time of day, and weather conditions. Using fixed fenestration that can't follow the sun, the variation within a building is usually even greater. Openings into a room from multiple directions can even out the discrepancies, but the inherent variations in daylight remain.

The composition of daylight, and hence the way it can be used, depend on sky conditions (see Figure 8.9). When the sky is overcast, the daylight is diffuse, coming from all directions, including a contribution from ground reflection (diffuse). When the sky is clear, daylight comes from direct sunlight, the clear blue sky, and ground reflection (both specular and diffuse). A partly cloudy sky has more diffuseness than a clear sky, but with strong directionality, and the lighting conditions can change by the minute.

In order to make use of the available daylight, it is important to know which of these conditions to design for. Overcast skies produce a relatively stable, uniform source of light over the entire sky. Clear skies provide a relatively stable blue background light in conjunction with a very intense sunlight that continuously varies in its direction.

Although both overcast and clear conditions occur almost everywhere, the latitude, time of year, air pollution levels, and general humidity levels of a given microclimate determine which condition prevails. The average number of clear days per year or the percentage of time that skies are clear during daylight hours can be researched through a local weather station. In some cases, a daylighting design may utilize both types of conditions when they occur. Local terrain, landscaping, and nearby buildings sometimes affect the type and quantity of light available.

Three design approaches can be taken in dealing with this resource variability:

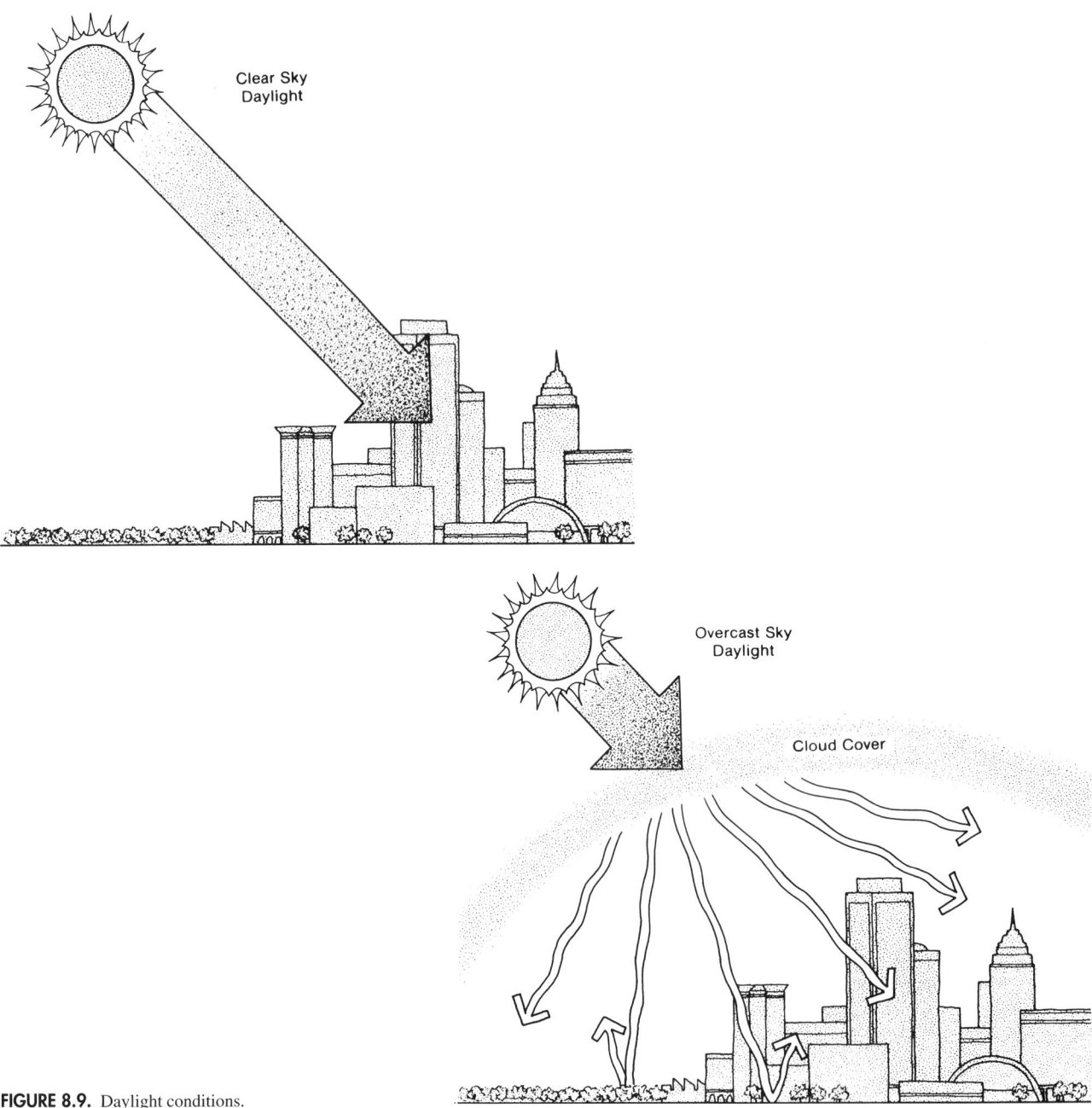

FIGURE 8.9. Daylight conditions.

1. Design for minimum acceptable natural illumination under minimum available daylight conditions (such as 9:00 a.m. in December) and screen out excess illumination at other times.
2. Design the system to provide adequate light under average sky conditions. Supplementary lighting would be necessary when less than average daylight is available.

3. Design for adequate natural illumination during the majority of typical working hours.

Modeling

There are methods for calculating daylight illumination levels, but because the effects of daylighting scale without distortion, physical scale models are recommended for conceptualization and simple evaluation of proposed design solutions. Even if crudely built and measured, scale models are considered more reliable tools for revealing the quantity, quality, and distribution of light in a space than theoretical mathematical calculations or drawings.

Once a model is constructed approximating the real conditions of the building, light meter readings can be taken in the model interior under real sky conditions. Some important points to remember when building a model are:

- The model should be built at a scale large enough to accommodate a standard light meter, or at least a test probe, and to make accurate construction easier. A scale of 1/2 inch = 1 ft (4 cm = 1 m) or 3/4 inch = 1 ft (6 cm = 1 m) is recommended.
- The opacity, surface texture, and reflectance of the surfaces in the model should match those of the real materials as closely as possible.
- Glazing is best simulated by using pieces of the actual material (i.e., 1/4-in. or 6-mm glass) over fenestration openings in the model. If that is not possible, the transmission characteristics should be matched as closely as possible, or the light meter measurements can be adjusted according to the ratio of the actual to the simulated transmissivity.

Heating, Cooling, Lighting: Design Methods for Architects (2nd ed.), by Norbert Lechner, provides instructions and helpful advice on modeling and measuring daylight in a model with a light meter. Flexibility and modular construction is advised to enable comparative modeling of design variations.

Locating Fenestration

The admission of daylight into structures is frequently accomplished by windows, which also provide views. Daylighting, however, needn't be confined to that method. Openings for illumination may also serve other functions, but they should be considered potentially different from openings for views, solar heat gain, or ventilation.

Fenestration for daylighting may be low windows at eye level, high windows such as transoms, horizontal *skylights,* or raised *clerestories,* or *monitors* extending above roof level, as shown in Figure 8.10. Openings through the roof are identified as *top-lighting* while vertical windows are referred to as *side-lighting.*

Since the quantity of light from the sky is generally greater than that reflected by the landscape, and since the light entering through fenestration is related to what is visible from it, high windows and top-lights are usually more effective than lower windows and less affected by surrounding obstructions. Lighting from above is most useful in areas where light is needed but visual contact with the outdoors is not. It affords more security, frees up the walls for other purposes, distributes illumination more uniformly to all the walls, and allows penetration of light into the interior of large, low buildings.

Horizontal skylights should be avoided in climates requiring artificial cooling, unless effective shading is used during the cooling season. They cannot effectively collect solar heat in the winter and can leak when it rains. They are, however, the most efficient means for daylighting under overcast sky conditions.

Clerestory monitors or sawtooth roofs, on the other hand, use standard weather-tight window construction and, when oriented south in the Northern Hemisphere (north in the Southern Hemisphere), can maximize solar heat gain during the winter. Light and heat can be increased with reflecting materials on the roof, such as white gravel or mirrors. Overhangs can minimize solar heat gain in the cooling season.

North-facing clerestories are appropriate if heat is never needed. They provide a comparatively steady light level, which can be enhanced with a white roof or an external reflector.

Figure 8.11 illustrates how window height affects light distribution in a room. High windows or top-lighting should be primarily used for horizontal tasks and low windows for vertical tasks. High windows and top-lights provide the best distribution of diffuse skylight and allow deeper penetration and better uniformity of daylight. Placing the top of windows as near the ceiling as possible allows the light to reflect off the ceiling for optimum penetration and indirect lighting effects. Tall, narrow windows provide deeper penetration than short windows of the same area but can create strong patterns of light and dark contrasts. Low windows best distribute ground-reflected light.

Using a system of mirrors, a periscope-like device, as shown in Figure 8.12, can not only provide daylighting, but also enable occupants completely underground to view above-ground scenery. Light and images can be reflected down into the space and through a vertical window if desired.

As an initial rule of thumb for standard lighting of offices, lobbies, or circulation areas, glazing should be approximately 5 to 10 percent of the floor area served. For more intensive lighting of spaces for such purposes as display, drafting, typing, or factory work, the area of glazing should be approximately 25 percent of the floor area served.

Glare

The quality of daylight, including glare, brightness contrasts, uniformity, and veiling reflections, is every bit as

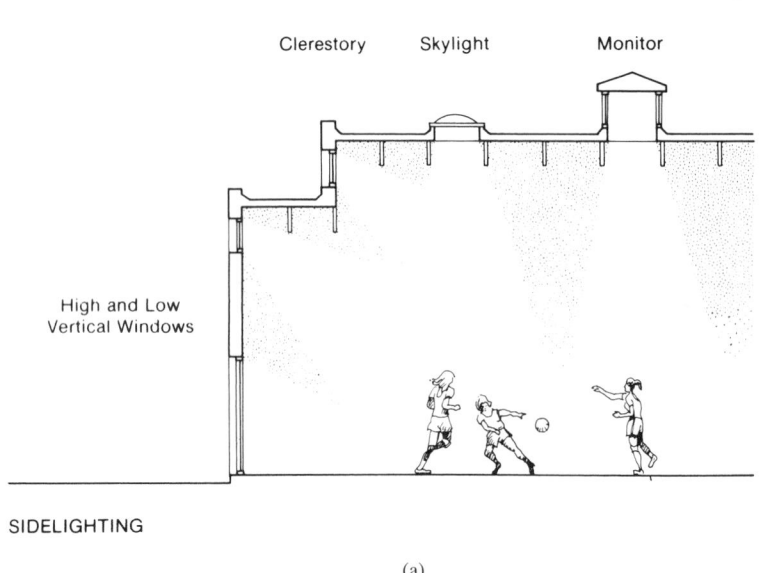

FIGURE 8.10. (a) Fenestration types. (b) Examples of vertical windows, skylights, and clerestories can be seen in this fitness center. (c) An example of a monitor. (Photos b and c by Vaughn Bradshaw.)

important as the quantity provided. Glare is one of the most important design considerations in both daylighting and direct-gain passive solar energy collection.

The contrast between a bright outside environment viewed through a window and the darkness of an interior space can cause discomfort and eye fatigue as the eye repeatedly readjusts from one lighting condition to the other. Direct sunlight or reflected sunlight from bright, shiny surfaces, which can be disturbing and even disabling, must never be permitted to enter the field of view of an occupant.

The methods for avoiding glare and reducing brightness contrasts fall into two categories: compatible interior design and light-control devices.

(c)

FIGURE 8.10. *(Continued)*

Compatible Interior Design

Surface brightness should change gradually from outside to inside. Contrast can be reduced by using light colors and high reflectances for window frames, walls, ceilings, and floors. Internal surface colors and reflectances should be selected according to the primary source of incoming light: Direct and reflected sky light hits the floor first; reflected sunlight hits the ceiling first. Light colors and high reflectances on surfaces far from the openings also help increase light in dim areas.

Windows should be placed adjacent to light-colored interior walls so that the light reflected from those walls produces a transition of light intensities rather than an extremely bright opening surrounded by unlit walls. Window openings should be beveled on the inside to form a transition surface between the window and the interior wall surface, and also to allow more light penetration.

The contrast between a window and its surroundings is reduced by placing fenestration in other walls or in the roof. Glare is avoided by positioning fenestration away from excessively bright views. Factories, for example, typically use north-facing clerestories to reduce the likelihood of machinery accidents caused by glare.

Supplementary artificial lighting should be used in dark areas, rather than oversupplying daylight in lighter areas in order to achieve adequate daylighting everywhere.

The illumination from windows diminishes rapidly with depth into the room, as shown in Figure 8.13. The uniformity of lighting within a room can be expressed as a ratio of illu-

mination from maximum to minimum. The maximum should be measured at a point 5 feet (1.5 m) in from a window, since work stations are usually not located right next to a window. This ratio should not exceed a value of 10. Another way to gauge uniformity in a room with windows on only one wall is to check that the average illumination level of the half of the room with the windows is not more

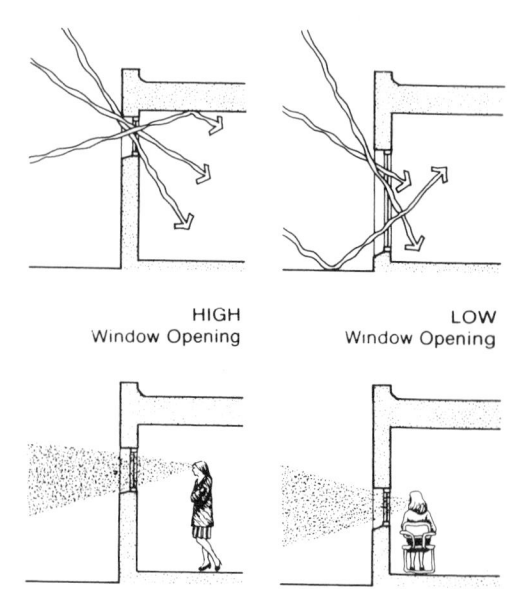

FIGURE 8.11. Window height and interior light distribution.

than three times greater than the average illumination level in the other half of the room.

Rooms with windows on only one wall should be proportioned and/or tasks located near windows in accordance with the limited depth that light can penetrate into a space. As a rule of thumb for avoiding glare problems and for proper daylight penetration and distribution in such a room, the height from the top of the window to the floor should equal about one-half the depth of the room, and whenever possible, additional daylight should be introduced at the back of the room. Since the lower part of the window contributes almost no light to the inner part of the room, sill height is not critical.

It is preferable to orient furniture so that daylight comes from the left side or the rear of the line of sight. A work station should never face a window unless the window has a northern exposure and there are no exterior glare sources in the line of sight.

Light-Control Devices

Light-control devices can be divided into three categories: exterior, interior, and integral with the glazing itself. Some are fixed and others are flexible. A design may combine more than one device for the same opening. In addition to controlling glare and limiting excessive brightness at the perimeter, control devices can be used to balance light patterns and distribute daylight throughout a room.

Exterior. Exterior light controls, as illustrated in Figure 8.14, include overhangs, outdoor light shelves, louvers, fins, egg crates, and site work. The advantage of an overhang is that it decreases light levels near the window, which has the effect of making light levels more uniform throughout the space. A louvered or translucent overhang performs the function of shading the direct sun to block glare while allowing more diffuse illumination in to the interior.

Light shelves reflect sun and sky light toward the ceiling for indirect distribution and penetration deeper into the space. At the same time, they shade any glazing below them from direct sunlight, while still allowing the use of ground-reflected light near the window. This increases the uniformity of illumination by increasing it farther toward the back of the room and decreasing it at the front.

Horizontal louvers and blinds are best used on south exposures and are good for redirecting light to high-reflectance ceilings. Vertical louvers are best suited for east and west exposures. Exterior louvers are usually fixed, although some mechanized ones that raise up out of the way when not needed are commercially available (see Chapter 7).

A light-colored wall to the north of north-facing fen-

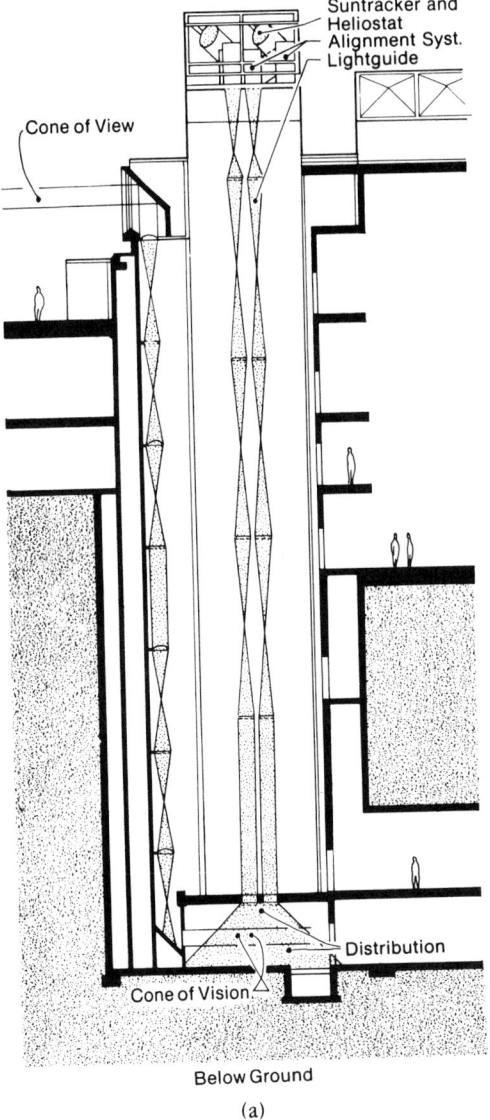

FIGURE 8.12. Underground daylighting for the Civil/Mineral Engineering Building at the University of Minnesota. (a) Diagram of the lighting and image transmission systems. (b) Close-up of solar optic components (photo courtesy of BRW Architects, Minneapolis.)

estration can increase the available light into a space by reflection. Thermal buffer zones, such as arbors, vestibules, courtyards, atria, and greenhouses, can control glare and heat gain. Snow is a highly effective reflector of light, but it can also cause glare. In climates with snow during part of the year, the seasonal changes in landscape reflectance must be taken into account.

(b)

FIGURE 8.12. *(Continued)*

Figure 8.14 also illustrates the concept of utilizing spill light. Transom-type glazing high up on walls of enclosed spaces can maintain privacy while borrowing enough spill lighting to sufficiently illuminate a corridor.

Interior. Interior controls are generally more flexible, offering the occupants a choice to adjust the amount of light and heat entering the space, but they do not keep out solar heat gain as well. In many buildings, these controls can be manual, but in larger buildings with many occupants, there is often less of a sense of individual responsibility and choice, so automatic control systems may be necessary.

The common types of devices are retractable shades (opaque or translucent), draperies, curtains, venetian blinds, or insulated panels. For maximum effectiveness, the outside-facing surface of any of these devices should be mirrored or highly reflective.

Roller shades are intended to diffuse direct sunlight, eliminate glare, and increase the uniformity of illumination. In fact, diffusing shades, as well as diffusing curtains, can become so bright when illuminated by direct sunshine that they themselves become a source of glare. Off-white fabric colors or an additional opaque drapery can reduce such brightness. If they are set up so that they can be pulled up from the bottom as needed, shades made of an opaque material can be used to eliminate glare, while the upper portion of the window still permits daylight into the room.

Venetian blinds permit considerable light to enter a room by the interreflection between their slats while still blocking glare. When mirrored on at least one side, they can be used to redirect daylight deeper into a room. Used only in the upper 2 feet (60 cm) of a window, they can beam daylight as deep as 30 to 40 feet (9 to 12 m), achieving 50 footcandles at the work plane. The optimum arrangement, however, is probably obtained by covering only the lower half of a window with these blinds.

Glazing Materials. The luminance of windows and top-lights can be reduced by using tinted or reflective glazing or glass blocks. However, since these low-transmission materials cut down the daylighting potential, clear glass with some form of shading is preferable in order to control glare and unwanted heat gain. Shading can maximize the utilization of light in all seasons and allow desirable solar gain. In addition, low-transmission glazing does not eliminate the need to shield occupants from the glare from direct sunlight.

FIGURE 8.13. How far daylight from vertical fenestration penetrates into a room can be clearly seen in this shopping mall. (Photo by Vaughn Bradshaw.)

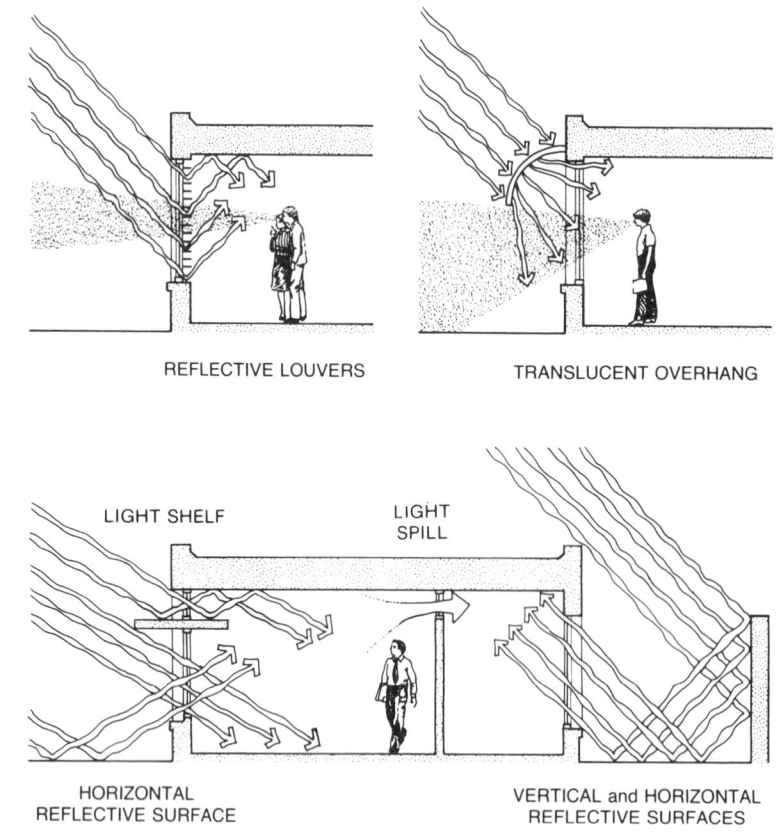

REFLECTIVE LOUVERS

TRANSLUCENT OVERHANG

LIGHT SHELF

LIGHT SPILL

HORIZONTAL REFLECTIVE SURFACE

VERTICAL and HORIZONTAL REFLECTIVE SURFACES

FIGURE 8.14. Light-control methods.

Diffusing, translucent materials are appropriate for use on skylights. The lack of contrast in illumination creates a restful long-term work environment. However, it is also as dull as an overcast sky and should not be used where the sparkle of sunlight is desirable, as in an entrance hall.

Electric Lighting Controls

Even with adequate daylight illumination, electric lighting must usually be provided for times when not enough natural light is available. Therefore, in order to reduce energy consumption, a control system is needed to switch off the artificial lighting when daylight is sufficient.

The controls may be either manual or automatic. Manual controls are simpler and less expensive, but not as reliable as automatic controls. In either case, the lighting in the daylight zone must be switched separately from that in other areas.

Automatic controls normally consist of photoelectric switches that automatically dim or turn off unnecessary electric lighting when daylight is sufficient and then turn them back on when needed. A time delay is typically built into the equipment in order to avoid frequent cycling of electric lighting when passing clouds briefly block the sunlight.

EMERGENCY LIGHTING

In accordance with the Life Safety Code and the National Electrical Code, emergency lighting is generally required for public buildings by most local codes. Its purpose is to automatically illuminate the means of egress when the normal lighting sources are not available due to a power failure or any other interruption of electrical service.

Certain critical functions require an entire emergency power system, such as standby generators or central battery systems. In those cases, selected portions of the normal lighting system are usually designated as emergency lighting and served by the emergency power. Otherwise, self-contained battery-powered lighting packages are strategically located to operate during power outages.

In either case, all exits must be illuminated so that traffic can flow smoothly. Most code authorities accept 1 footcandle of illumination as sufficient to avoid panic and permit orderly egress. An exit sign itself can be equipped with an integral battery in order to illuminate the sign and the area immediately below it.

OUTDOOR LIGHTING

Outdoor lighting can provide safety and security at night in just about any kind of building. It is sometimes employed at commercial, industrial, or institutional facilities also in order to support outdoor industrial production and to enhance appearance. Outdoor shopping malls and sporting events need lighting for after-sundown activities.

For safety and security, lighting should be provided at building access points, along pedestrian routes, and in parking lots. The same lighting that provides safety and security and that supports the function of the facility can add to the appearance of a building or surrounding area if carefully planned and executed.

Aesthetic and functional effectiveness depends on a well-designed program that goes beyond just decorative fixtures and a decorative use of light. However, the daytime appearance of the equipment must be considered.

Outdoor lighting fixtures can be classified as fixed or aimable. Luminaires that can be aimed include floodlights and spotlights. Both types of fixtures can be pole- or wall-mounted.

Many lighting manufacturers, with the aid of computer programs, can provide a detailed layout based on a designer's specifications. Where security is critical, lighting beams should overlap so that the area will still be covered if one lamp goes out.

SUMMARY

It is normally the job of the electrical or special lighting consultant to design the details of a lighting system. But it is up to the architect to develop the lighting program to meet the functional and aesthetic needs of the building. Also, a fundamental understanding of lighting is necessary if daylighting is to be used.

The following is a list of steps for a preliminary lighting design in accordance with the information presented in this chapter. The steps are not intended to be followed consecutively like a cookbook; in fact, several of them need to be considered concurrently. They are merely listed separately for the sake of clarity and identification.

1. Examine the architectural objectives. What are the functions or tasks in a space? What are the needs for general background lighting? How can the lighting scheme enhance the architectural design while accomplishing the specific lighting objectives for each area and maintaining

energy efficiency? All wall, ceiling, and floor surfaces must be taken into account as light reflectors and are just as important to the overall lighting system as the luminaire selections.

2. Explore the opportunities the building presents for integrating daylighting with artificial light sources. Study alternative solutions with the aid of diagrams, sketches, and especially physical models. Include appropriate fenestration in the building design.

3. Determine appropriate task-lighting levels and surrounding illumination levels.

4. Select an appropriate electric light source for proper color rendition at the highest efficacy. Consider the relative installation cost and the frequency of lamp replacement of the alternative sources.

5. Choose one of the five lighting system types and plan the general arrangement of fixtures in the space for proper distribution of light without glare or unwanted shadows. Incorporate any desired aesthetic effects into the system design.

6. Select luminaires that satisfy the needs of the overall system with proper light distribution and a suitable appearance.

7. Determine the required number of fixtures by the calculation or graphical method. Lay them out on a reflected ceiling plan according to the desired effects, being careful to avoid interference with structural, mechanical, or other architectural items.

KEY TERMS

Luminous intensity	luminance ratio	Specular reflection
Candlepower (candela)	Glare	Diffuse reflection
Lumen	Task lighting	Lamp
Luminous flux	Work station	Incandescent lamp
Illumination or lighting level	Work plane	
	Primary source	Fluorescent lamp
Footcandle (lux)	Secondary source	High-intensity discharge (H.I.D.) lamp
Refraction	Footcandle meter	
Brightness or luminance	Absorptance	Metal-halide lamp
	Reflectance, reflection factor, or reflectance coefficient	Mercury vapor (M-V) lamp
Footlambert (lambert or millilambert)		
Contrast or		

High-pressure sodium (HPS) lamp	tance, transmission factor, or coefficient of transmission	Room cavity ratio
Low-pressure sodium (LPS) lamp		Maintenance factor, depreciation factor, or light-loss factor
Ballast	Direct transmission	
Color rendition	Diffuse transmission	
Luminaire	Coefficient of utilization (CU)	Indirect or diffuse lighting
Reflector		Direct lighting
Refractor	Coefficient of beam utilization (CBU)	Daylighting
Lamp efficacy		Side-lighting
Luminaire efficiency		Top-lighting
Transmittance, luminous transmit-	IES	Skylights
	Luminance	Monitors
		Clerestories

STUDY QUESTIONS

1. Name three applications for which each of the following lamp types would be appropriate:
   a. Incandescent b. Fluorescent c. H.I.D. d. Low-pressure sodium

2. Determine the number of fluorescent fixtures needed to provide a uniform 70 footcandles (750 lux) on a standard work plane in a 20- × 20-foot (6- × 6-m) office with a 10-foot (3-m) ceiling. Based on the fixture selected and room reflectances, the CU equals the room cavity ratio times 0.1. The lamps are 40 watts and emit 3,150 initial lumens each. The maintenance factor is 0.75. The ballast loss is 8 watts per lamp. Use both the calculation and graphical procedures to determine how many fixtures would be needed first based on two-tube and then on four-tube fixtures.

3. Which types of buildings should utilize south-facing clerestories for daylighting? Which should use north-facing clerestories?

4. Under what conditions would horizontal skylights be appropriate?

BIBLIOGRAPHY

Brown, G.Z., and Mark DeKay. *Sun, Wind and Light: Architectural Design Strategies,* 2nd ed. John Wiley & Sons, Inc., 2000.

Flynn, John E., Jack A. Kremers, Arthur W. Segil, and Gary Steffy. *Architectural Interior Systems: Lighting, Acoustics & Air Conditioning,* 3rd ed. Van Nostrand Reinhold Company, 1992.

Kaufman, J.E. (editor). *IES Lighting Handbook Application Volume.* Illuminating Engineering Society of North America.

Lechner, Norbert. *Heating, Cooling, Lighting: Design Methods for Architects,* 2nd ed. John Wiley & Sons, Inc., 2001.

Stein, Benjamin, and John Reynolds. *Mechanical and Electrical Equipment for Buildings,* 9th ed. John Wiley & Sons, Inc., 1999.

Thumann, Albert. *Plant Engineers and Managers Guide to Energy Conservation,* 8th ed. Marcel Dekker, 2002.

Traister, John E. *Residential Electrical Design Revised.* Craftsman Book Company, 1994.

Turner, Wayne C. (editor). *Energy Management Handbook,* 4th ed. Marcel Dekker, 2002.

Chapter 9

Normal Electrical Service

ELECTRICAL THEORY

POWER AND ENERGY

ELECTRICAL CODES

COMPONENTS OF BUILDING ELECTRICAL SYSTEMS

LOADS

TYPES OF ELECTRICAL LOADS

MOTORS

SERVICE ENTRANCE

TRANSFORMERS

SERVICE DISCONNECT

METERING

INTERIOR DISTRIBUTION

CONDUCTORS

FEEDERS AND BRANCH CIRCUITS

GROUND FAULT CIRCUIT INTERRUPTER

MOTOR BRANCH CIRCUITS

BATTERY POWER SUPPLY

UTILITY DEMAND MANAGEMENT

UTILITY CHARGES

DEMAND CONTROL

POWER FACTOR CONTROL

Electricity constitutes a form of energy that occurs naturally only in uncontrolled forms such as lightning and other static electricity discharges, or in the natural galvanic reactions that cause corrosion.

No one knows exactly what electricity is or how it works. It does, however, behave in predictable ways; that is, when a light switch is thrown, the light consistently goes on, or if it doesn't go on, prescribed steps (such as replacing the lamp) can be taken to correct the problem. Because the experience is repeatable, observers have made up theories about what constitutes the electrical phenomenon. These theories have changed and evolved over time and undoubtedly will continue to be improved upon.

Electrical Theory

The currently accepted theory is that *electrical current* consists of the flow of electrons induced by an imbalance of positive and negative charges. Like charges repel and opposite charges attract, so electrons, being negatively charged, are repelled by a negative area and attracted by a positive area. When a positive area is connected to a negative area by a conducting material, the electrons flow from the negative to the positive side.

This phenomenon can be likened to the flow of water through a pipe. Just as water flow can be measured in gallons per minute, electron flow, referred to as current or *amperage,* is measured in *amperes,* which is usually abbreviated as *amps* or sometimes simply *A.* An ampere is technically 1 *coulomb* per second flowing past a given point, and 1 coulomb $= 6.28 \times 10^{18}$ electrons. When current is used in an equation or formula, it is commonly represented by the letter I.

Continuing the analogy with water flow, a difference in *electrical voltage* is considered the force driving the current, just as a difference in height or pressure drives the flow of water from one point to another. The higher the voltage in a given system, the more current flows. Voltage is sometimes referred to as *electromotive force,* or *emf,* and is expressed in the units of *volts* (*V* or *E*).

Electrical resistance is equivalent to friction or a constriction slowing the flow of water. The unit for resistance is *ohms* (*R*).

The relationship between voltage, current, and resistance, known as *Ohm's Law,* is represented in any one of the following ways:

$$V = IR; \quad R = V/I; \quad I = V/R \qquad (9.1)$$

By definition, 1 ohm of resistance allows 1 ampere of current to flow under the electromotive force of 1 volt. According to Ohm's Law, then, 2 ohms of resistance allow only 1/2 ampere under the same 1 volt.

An important characteristic of electric current is that it always flows through the path of least resistance. Materials that offer a low resistance to the flow of electricity are known as *conductors,* and those offering so much resistance that they virtually prevent the flow of electrical current are called *insulators.*

Among the materials that conduct electricity, some are better conductors than others. Silver and gold are the best known conductors, followed closely by copper. Aluminum has about 80 percent of the conductivity of copper. Since each material offers some degree of resistance (which is believed to be a form of friction on the atomic level), heat is generated by the electron flow through a conductor.

In order for electricity to flow through a conductor under a voltage difference (emf), it must have a continuous path forming a closed loop. Such a continuous closed path is called a *circuit.* A simple circuit including a voltage source and a load is shown in Figure 9.1.

When a circuit is interrupted—for example, by a switch in the off or "open" position—the condition is known as an *open circuit.* When the continuity of the path is completed, it is called a *closed circuit.* When a conducting material makes contact with two points in the loop in such a way that it forms a short cut around the load, the entire current will take the path of least resistance. This condition, called a *short circuit,* is extremely dangerous.

There are essentially two types of electrical current: *direct current (DC)* and *alternating current (AC).* DC has a constant flow rate induced by a constant voltage source, like a battery in which one terminal is always "+" (positive) and the other "−" (negative). Any current in which each wire is always the same polarity, one always positive and one always negative, is a DC.

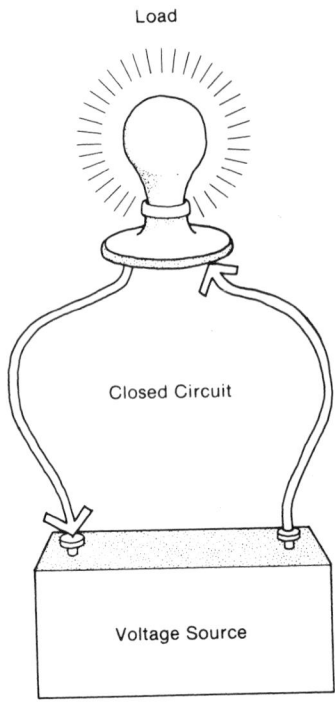

FIGURE 9.1. Simple electrical circuit.

In an AC, the voltage difference, and hence the current, reverses its direction at a fixed frequency. The change from positive to negative and back again to positive is considered a full cycle.

The commercial power from utility companies in the United States and Canada is AC, typically supplied at a frequency of 60 cycles per second, also known as 60 *hertz* (*Hz*). In many other countries, the frequency is 50 hertz. Equipment made for one frequency is not compatible with any other frequency. Motors will generally not perform as desired or will overheat, burn out, or at least have a shortened life if used at a different frequency from that intended.

Power and Energy

Power is defined as the ability to do work, or the rate at which energy is expended in performing work. The relationship between power and energy is expressed as

$$\text{Power} = \frac{\text{energy}}{\text{time}} \qquad (9.2)$$

which is the same as the relationship between rate of speed and distance traveled:

$$Speed = \frac{distance}{time} \left(e.g., mph = \frac{miles}{hour}\right) \qquad (9.3)$$

Just as *distance = speed × time, energy = power × time.* Therefore, the amount of energy used is directly proportional to the power of a system and to the length of time it is in operation.

Electrical power is expressed in *watts* (W) or *kilowatts* (kW), and time, in this case, is expressed in hours. Thus, the units of energy are *watt-hours* (Wh) or *kilowatt-hours* (kWh). The relationship between watts and kilowatts is 1,000 W = 1 kW. One watt-hour equals 1 watt of power in use for 1 hour, and 1 kilowatt-hour equals 1 kilowatt in use for 1 hour or 1 watt in use for 1,000 hours.

Watts and kilowatts represent the rate at which energy is being used at any given moment. Utility meters measure kilowatt-hours just as the odometer or mileage recorder of an automobile indicates distance traveled.

Power (P) is related to current, voltage, and resistance as follows:

$$P = VI = RI^2 = V^2/R \qquad (9.4)$$

In plain English, a watt is 1 ampere flowing under the electromotive force of 1 volt.

For example, a 60-watt incandescent light bulb requires 60 watts of power to operate. When the bulb is placed in an electrical circuit, as shown in Figure 9.1, with the voltage on one side 120 volts higher than the voltage on the other side, then 60/120 = 1/2 ampere of current will flow through the circuit. The current is said to be drawn through the circuit by the light bulb, which is called the *load*. If, however, the voltage difference were only 12 volts instead of 120 volts, the current draw would be 60/12 = 5 amperes.

As this example illustrates, more current is needed to get the same power at a lower voltage. This is significant because the size of wire required to conduct electricity in any given application is determined solely by the current that will need to flow through it rather than by the power.

Electrical Codes

Electricity is of tremendous use, but it must be treated with caution and respect. The potential for fire or shock (injurious or fatal) is ever-present when electricity is in use. The risks associated with these hazards can be minimized by proper wiring and adequate safety precautions.

The *National Electrical Code* (NEC) of the National Fire Protection Association (NFPA) specifies safety measures that must be followed in the selection, construction, and installation of all electrical equipment. It is the definitive authority and is referred to by all designers, engineers, contractors, and inspectors in the electrical discipline and by many operating personnel charged with the responsibility for safe operation. Having been incorporated into the Occupational Safety and Health Act (OSHA), this code effectively has the force of law.

Many large cities, including New York, Boston, and Washington, D.C., have their own local electrical codes that

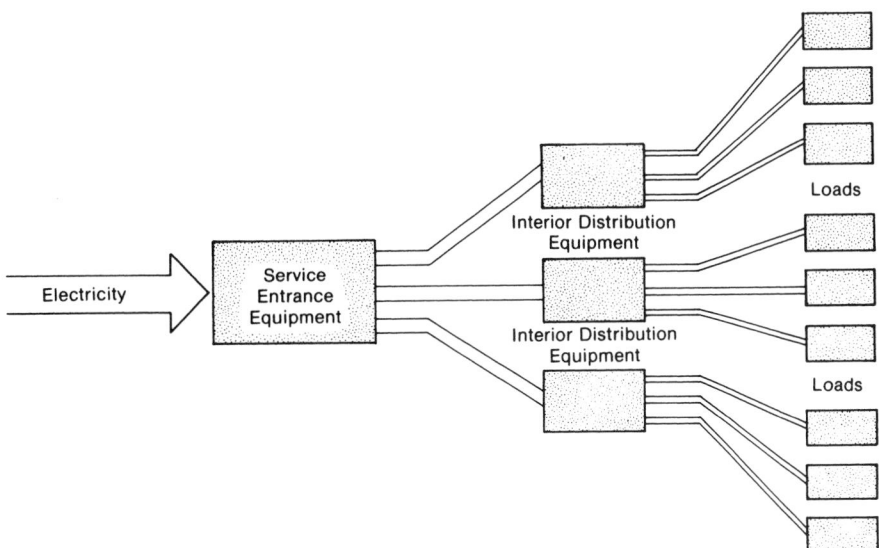

FIGURE 9.2. Major components of a building electrical system.

incorporate many of the provisions of the NEC but contain their own special requirements in addition.

Underwriters Laboratories (UL) Incorporated, establishes standards for, and actually tests and inspects, electrical equipment to ensure minimum standards of electrical safety. It publishes lists of inspected and approved electrical equipment. The validity of these listings is universally accepted, and many local codes accept only those electrical materials approved by, and bearing the label of, the UL. Electrical equipment or material that does not bear the UL label should not be permitted on any project.

Components of Building Electrical Systems

The major components of a building's electrical power system are illustrated in Figure 9.2. They can be classified into three major categories: (1) service entrance equipment, (2) interior distribution equipment, and (3) loads.

The first category includes transformers, service disconnect, fuses, circuit breakers, and meters. The second includes conductors, raceways, subpanels, and submeters. In the third is actual utilization equipment, such as lighting, motors, and miscellaneous outlet devices.

Referring to Figure 9.2, notice that the power system proceeds from the service entrance to the utilization points in a series of descending steps. At each branching point, the distribution capacity decreases.

LOADS

In an electrical context, a load is a device that consumes electrical energy in the process of performing work. Actually, as thermodynamics tells us, the electrical energy is not consumed, but is merely converted to another form of energy. However, the concept of consuming electricity as a resource is useful in understanding electrical power systems in buildings, so it is used for this purpose.

Types of Electrical Loads

Some common loads are:

1. Lighting
2. Motors
3. Convenience outlets
4. Heating elements
5. Electronic devices
6. Special process equipment (lab equipment, X-ray machines, etc.)

Lighting is the predominant user of electrical power in most buildings, since it typically extends all over a building and is on for a major portion of the workday. After lighting, motors are probably the next greatest user of electricity in nonresidential buildings.

Electric motors driving pumps and fans are essential components of HVAC and plumbing systems and most industrial processes. The power requirements of HVAC motors depend on the amount of ventilation and the amount of heat to be supplied to or removed from a building. Pump motors and air-handling fan motors often operate much of the time or even continuously, while the motors driving fans and compressors in cooling equipment may be used seasonally or year-round for those units serving the interior core of a large building. Elevator motor use is intermittent, but in large, tall buildings it can constitute a substantial load. Electric motors in industrial plants generally provide the basis for production and can be the largest load in such buildings.

Receptacles, also called *convenience outlets,* are devices into which portable electric appliances and equipment are plugged. Unlike mechanical equipment motors, which are usually concentrated in basements, penthouses, and fan or machine rooms, receptacles are placed at convenient locations distributed throughout a building along with lighting loads. Receptacle circuits are rated according to the total anticipated amperage requirements of all connected loads operating concurrently.

As pointed out above, the power consumed by an electric device is the product of voltage and current draw (watts = volts × amperes). Heavy loads such as large motors and heating elements require so much power that they would require an excessive current at the common 120 volts. By providing such loads with a higher voltage, the same power can be delivered at a substantially reduced amperage so that the wiring to the load can be much smaller and more economical.

Household electric ranges and clothes dryers operate at 240 volts. Electric heaters larger than 1,500 watts and window air conditioners larger than 1 1/4 tons (4.4 kW) normally require 240 volts. These must be either direct wired into a circuit or plugged into a special receptacle designed for that purpose.

HVAC equipment and other larger motors are available in a variety of voltages, depending on the availability in a given installation. Most manufacturers offer the option on most models of 208, 230, 277, or 480 volts. Extremely large equipment is available at 4,160 volts.

Loads can be divided into two broad categories: *resistive loads* and *inductive loads.* Nonballasted lights and electric resistance heating elements are examples of resistive loads. The current going to such loads rises quickly to its maximum value at the moment the load is energized and then stays there. Inductive loads, such as relay coils and electric motors, draw a peak current quickly and then drop back to a value one-third to one-fifth of the peak value. This peak is called *inrush current* or, in the case of motors, *locked rotor current.*

Motors

Electric motors differ widely in their operating characteristics and cost. Each motor is manufactured to operate under certain conditions and has its appropriate application.

Electric motors manufactured in the United States generally conform to the standards of the *National Electric Manufacturers' Association* (NEMA). The basic purpose of NEMA is to provide guidelines for the standardization of motor dimensions and performance characteristics by all motor manufacturers so that users can interchange motors without making adjustments. Conformance with NEMA standards is voluntary, but both domestic and foreign motor manufacturers nearly always follow the standards. Motor installation must conform to local utility regulations and local codes and ordinances.

Horsepower (HP)

Motors are rated in *horsepower,* and that rating is displayed on the nameplate of each motor. The horsepower rating of a motor is the maximum work output available from the shaft. Motor sizes are categorized as (1) *fractional* (less than 1 horsepower), (2) *integral* (above 1 horsepower), and (3) *medium to large.*

Like wattage, a horsepower is a measure of power. One horsepower is work done at the rate of 33,000 ft-lb/min. One foot-pound equals the work necessary to raise a 1-pound weight a vertical distance of 1 foot. No person can perform work at the rate of 1 horsepower. A person working hard can deliver no more than about 1/10 horsepower continuously over a period of several hours. Not even a horse can put out 1 horsepower, except for very short bursts.

Electrically, 1 horsepower is defined as equal to 746 watts. A motor that delivers 1 horsepower could just as easily be called a 746-watt motor, or a lamp that uses 746 watts could be called a 1-horsepower lamp. By convention, however, in the English system of units, horsepower is reserved for representing mechanical power, and watts are used in an electrical context.

Power Source

Another way motors are grouped is by the following three general categories: (1) DC, (2) *single-phase AC,* and (3) *three-phase AC.*

Direct Current. Large DC motors are used primarily when close control of speed over a wide range of loads and speeds is essential. They are used, for example, for elevators, cranes, certain types of fans, and some process machinery. In addition to ease of speed control, DC motors have the advantage of a comparatively high operating efficiency.

Since the power available from utilities is generally AC, it must be converted to DC in order to operate a DC motor. This requires added equipment and causes a DC motor drive to be more expensive than a standard AC motor of the same horsepower rating. Despite the added cost, however, DC motors are still desirable for many industrial applications.

Single-Phase AC. AC current supplied by a utility may be either single-phase or three-phase. Single-phase AC is graphically represented as a sine wave, as shown in Figure 9.3b, as compared to the straight line representation of DC current seen in Figure 9.3a. At 60 Hz frequency, the voltage in each wire is positive 60 times each second and negative 60 times each second. During every second, there are 120 times when there is no voltage at all in the wire.

The voltage is never constant, but is always changing between zero and its maximum in either the positive or negative direction. In a 120-volt service, the maximum voltage is about 170 volts, but the average amounts to 120 volts, and hence that is the effective voltage. Because the fluctuation occurs fast enough, there is no apparent interruption in the output of electrical equipment.

Three-Phase AC. In a three-phase service, the voltage reaches its maximum at a different time in each of the three wires, as graphically represented in Figure 9.3c. First, the voltage in one wire peaks, then that in the second, then that in the third, then again that in the first, and the cycle repeats.

Three-phase motors draw less current than single-phase motors of equal capacity, although the power required is generally about the same. Single-phase motors of 5 horsepower or more draw an excessively high amperage while starting, and it is generally uneconomical to wire them with the very heavy gauge wire necessary. A three-phase motor with the same power capacity draws less current per wire, and therefore can use a less expensive, lighter-gauge wire.

Single-phase motors are not usually available in sizes larger than 7½ horsepower, although a few larger ones are made. Single-phase air-conditioning compressors are avail-

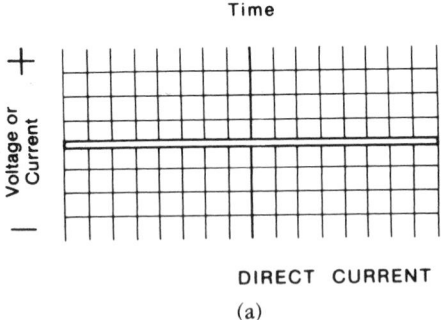

DIRECT CURRENT

(a)

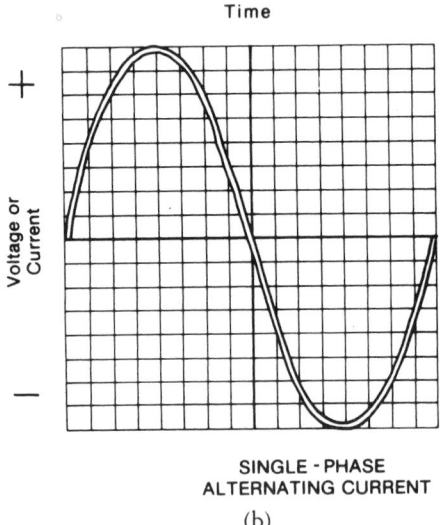

SINGLE - PHASE
ALTERNATING CURRENT

(b)

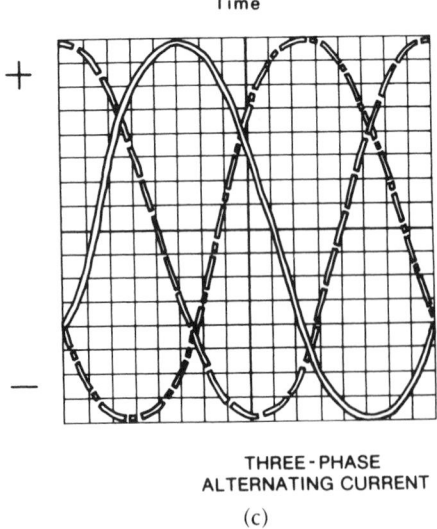

THREE - PHASE
ALTERNATING CURRENT

(c)

FIGURE 9.3. Graphical representation of current. (a) DC current, (b) single-phase AC current, and (c) three-phase AC current.

able only up to a 5-ton capacity. Beyond that, they are available only in three-phase.

Three-phase motors are simpler, lower in cost, more trouble-free, and more efficient over a wide range of loadings than their single-phase counterparts. On the other hand, a three-phase service requires three high-voltage transmission lines and three transformers, while a single-phase service requires only two high-voltage lines and one transformer.

If only a few small motors are in service, economics usually favors single-phase motors because of the lower service entrance costs. Three-phase power is used if large motors or many small ones are in operation. Certain designs of electric heaters also run on three-phase current. With a three-phase service, 120-volt single-phase is also available for lighting, control circuits, appliances, and other small loads.

If a three-phase motor is required to achieve a certain capacity and the available service is only single-phase, phase converters, which change the single-phase power into a sort of modified three-phase power, can be used. This modified three-phase power will operate ordinary three-phase motors and, at the same time, will greatly reduce the starting amperage.

Phase converters are relatively expensive, but their cost is partially offset by the lower cost of three-phase motors. Converters may be a necessity because they make possible the operation of larger motors than would be possible without them. Before applying a converter, the utility company serving a building must be consulted. Some do not favor or permit phase converters.

Efficiency

Every motor generates heat as well as power. Even though 1 horsepower equals 746 watts, a 1-horsepower motor will consume more than 746 watts because some power is wasted as heat in the process of overcoming resistance in the wires and friction in the mechanical motion. The amount of heat generated is a function of a motor's efficiency rating, or its *energy efficiency ratio* (EER). The less efficient the motor, the greater the amount of heat produced.

Thus, the EER indicates how well a motor converts electrical energy into mechanical energy. The larger the rating, the more efficient the process. It is the ratio of watts-out over watts-in:

$$EER = \frac{\text{shaft power (out)}}{\text{electrical power (in)}} = \frac{hp \times 746}{\text{watts}} \quad (9.5)$$

EERs also indicate the power required by a motor to perform a given amount of work. The electricity required by a motor is the amount of power being delivered at the shaft divided by the efficiency:

$$\text{Electrical power required} = \frac{\text{shaft power}}{\text{EER}} = \frac{\text{hp} \times 746}{\text{EER}} \quad (9.6)$$

The higher the EER, the lower the amount of power needed.

Motor efficiency generally falls within the range of 85 to 95 percent. The efficiency of small motors is generally less than that of larger motors. Normally, less expensive motors have lower EERs. High-efficiency motors cost a little more initially and are generally only 3 to 4 percent more efficient than standard motors. Even so, they often save a substantial amount of energy over their operating lives and more than make up for the higher initial expense. On a life-cycle basis, a motor's operating cost for electricity is usually far more significant than its first cost.

The amount of heat produced by a motor is dependent on both motor efficiency and load. Unless this heat is dissipated, the resultant temperature rise will cause serious damage to the motor. In order to dissipate the heat generated by motors in operation, equipment rooms where they are located should always have plenty of ventilation air for cooling.

Most motors are rated on the basis of how much their temperature will rise when they are running under full-load conditions, and this temperature rating is stated on the motor nameplate in degrees Celsius. The rated temperature rise is based on a maximum ambient temperature in the equipment room of 105°F (40°C). Higher-temperature environments require special (more expensive) motor selection.

Altitude also has an effect on the operation of a motor. Because the density of air decreases with increasing altitude, the cooling effect of air is less. Specially designed motors are available to remedy this problem for high-altitude applications.

If a motor can deliver its rated horsepower continuously without exceeding the temperature rise just explained, this is indicated on the nameplate by the term "cont." If the rated horsepower can only be delivered for brief periods of time without overheating, this is noted by a definite interval, such as "1/2 hour."

Life

The operational life of an electric motor is primarily a function of the hours of operation. The capability of routine cycling is built into most motors, but every stoppage and restart shortens operational life somewhat. Excessive duty cycling, called *short-cycling,* is usually very damaging to a motor and substantially shortens its useful life.

The amperage drawn from the power line depends on the horsepower delivered by a motor. When the motor is first turned on, it requires four to six times more power to accelerate up to speed than it uses under its normal load. Since the voltage remains constant or dips lower, the amperage drawn must rise proportionally to provide the electrical power. If a motor does not reach full speed because of too heavy a starting load, it will quickly overheat due to the higher than normal current pushing against the resistance in the wire.

Excessive duty cycling occurs when a motor is turned off and cycled back on again before it has had sufficient time to cool back down. This heat buildup damages a motor.

Duty cycling also has an effect on the life of contractors and relays that switch motors on and off. A contractor for a motor or any other electrical equipment can be turned on and off only a certain number of times before it fails.

Motors vary in the type of duty they are designed for. The *duty classification* indicates whether a motor is suitable for continuous duty at a nearly constant load for an indefinitely long period, periodic duty with alternating intervals of load and no load, or varying demands for operation at unpredictable intervals of time.

Selecting the proper motor capacity is just as important as selecting the correct duty classification. Undersize motors easily overheat and burn out, while oversize motors waste energy by operating at less than their optimum energy efficiency, and may wear out prematurely due to short-cycling.

Voltage

The *voltage rating* of a motor refers to the voltage of the circuit to which it should be connected. It is customary to assign a nameplate voltage slightly lower than the nominal system voltage in order to reflect the actual voltage level at the motor terminals, accounting for voltage drops occurring in the distribution system.

If two voltage values are listed, such as 115/230, the motor can be connected for use on either. Larger single-phase motors are usually designed for operation at either of these two voltages. At 230 volts, a motor consumes only half as many amperes as it does at 115 volts, thus allowing smaller wiring.

Motors are generally designed to operate satisfactorily at voltages 10 percent above or below that indicated on the nameplate. Overvoltage is better than undervoltage because it draws less current, resulting in less heating when delivering rated horsepower. A motor carrying heavy loads at voltages below the design point runs hot.

Unlike most other loads, motors perform at their rated output even when the voltage deviates from that intended. As discussed in Chapter 8, the performance of most light sources varies dramatically with over- or undervoltage. The output of an electric resistance heating element also varies with voltage. An element listed at 1,500 watts under 240 volts will consume and put out only 1,260 watts at 220 volts.

Service Factor

Almost all motors can develop 1 1/2 to 2 times their normal horsepower for short periods after coming to full speed. Continuous overloading, however, results in overheating and, in turn, a shorter life for the motor. All motors must be protected against overload by either heat-sensing or overcurrent sensing devices.

Some motors may operate at 15 to 25 percent above their nameplate horsepower without overheating. NEMA specifies how much over the nameplate rating a motor can be driven without overheating, and this amount is represented by what is called the *service factor*. Fractional horsepower motors have service factors in the 1.15 to 1.25 range, while most larger motors have a service factor of 1.15. Motors above 200 horsepower do not have any service factor, nor do motors manufactured outside of the United States.

The purposes of the service factor are:

1. To sustain an occasional but predetermined overload
2. To allow for a slight expansion in future capacity
3. To compensate for occasional excessive ambient temperature
4. To allow for uncertainty in estimating actual horsepower requirements
5. To extend the motor life by operation well below its normal temperature limit
6. To compensate for voltage irregularities

SERVICE ENTRANCE

Having discussed the end use of electricity in buildings—the loads—let us now look at the systems and equipment that bring the power for those loads into a building. First, consider the process of providing electricity to a building, as diagrammed in Figure 9.4.

Some form of fuel or energy source is converted to electricity by a generator at a utility's power plant. The electricity goes through a step-up transformer, where its voltage is increased so that less current will be needed to transport a given amount of power. This reduces transmission line power losses and allows smaller, more economical conductors to be used for transmitting the electricity to the many utility customers.

Power plants are sited for convenient access to fuel supplies and the dissipation of waste heat, which means that they are normally remote from the areas they serve. From the power station, electricity is distributed by a network of high-voltage transmission lines. The various standard transmission line voltages in the United States are 4,160, 12,500, 13,200, 13,800, 24,000, and 36,000 volts.

Area transformer stations convert the power to a lower voltage for overhead or underground distribution in a particular area. Some industrial users and large buildings with their own transformers can use this power directly. For most buildings, however, the supply voltage must be further

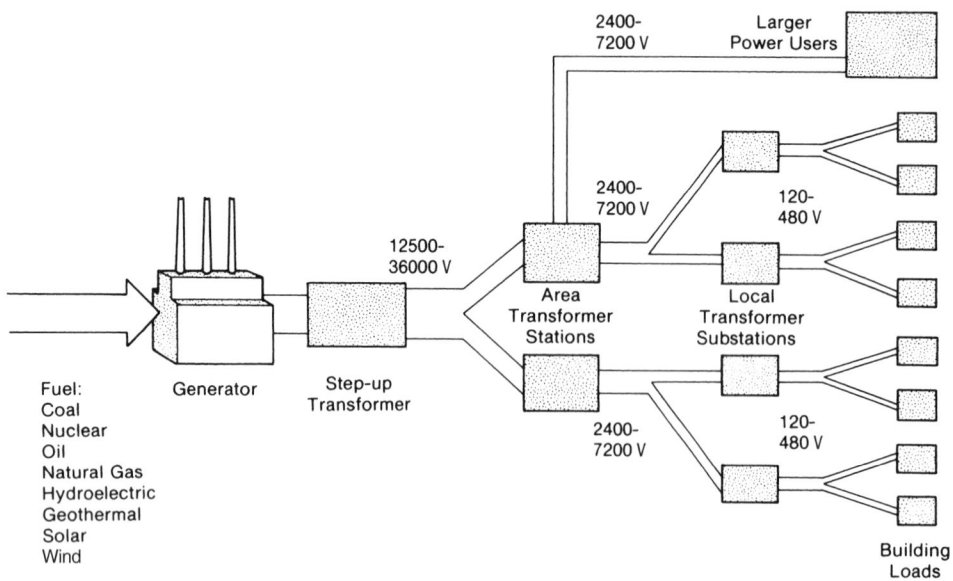

FIGURE 9.4. Supply of electricity to a building.

reduced by a local transformer substation serving the immediate locality.

Residential and small commercial buildings are normally supplied by such a substation with single-phase 120-volt or 120/240-volt power; the latter can be used for both large 240-volt appliances, such as ranges, electric clothes dryers, and electric heaters, and for smaller 120-volt loads.

Larger buildings may be supplied with single- or three-phase power at 120/208 volts, 120/240 volts, or 277/480 volts, depending upon their needs. The first two voltages are usable as-is for motors and general purpose power. A voltage of 277/480 is used for larger motors and may be reduced to 120 volts for general-purpose use by small transformers near the point of use.

The service equipment needed to connect a building to utility electrical service consists of a main disconnect switch and distribution switchgear, as shown diagrammatically in Figure 9.5. One or more step-down transformers are also needed by those buildings purchasing high-voltage power that does not come through a utility substation.

Transformers

A *transformer* is a device that changes or transforms AC from one voltage to another. The voltage on the incoming side of a transformer is known as the *primary voltage* and that on the outgoing side as the *secondary voltage*.

A *primary transformer* for a building is typically used to convert an incoming high-voltage service to some lower voltage, usually 480, 277, 240, 208, or 120 volts. If small, it may be mounted inexpensively on the nearest pole supporting the incoming transmission lines. Otherwise, it may be pad-mounted somewhere on the site outside the building, or installed in a room or vault inside the building.

In some cases, it may be more economical to distribute higher-voltage power, such as 480 volts, to remote areas within a large building and then step-down the voltage to 120 volts from floor to floor or as needed. These *secondary transformers* are located in electrical closets central to the various loads they serve. Secondary transformers are also used to convert AC power to DC for elevators or other DC motors.

Size

Transformers are identified by the amount of power they can handle, which is expressed in units of *kilovolt-amps (kva)*. That is the number of kilovolts (thousands of volts) times the rated maximum amperage. Table 9.1 provides approximate dimensions and weights for transformers of various capacities. Figure 9.6 depicts a representative example.

Transformers are available in single- or three-phase models. A three-phase transformer is simply three single-phase transformers in one case. For a given capacity, one three-phase transformer is less expensive and takes up less space than three individual single-phase transformers. However, the use of three single-phase transformers provides flexibility in use and some degree of backup protection, since all three are not likely to fail at the same time.

A preliminary assessment of transformer size may be made on the basis of present or probable future loads in order to get a rough idea of the space requirements. The types of loads are listed in the previous section.

For the purpose of sizing electrical systems, loads are divided into known and estimated loads. Lighting fixtures and other wired-in-place apparatus are called *known loads*. Because they are uncertain, the loads to be plugged into receptacles are called *estimated loads*. The sum of all known and estimated loads is called the *connected load*.

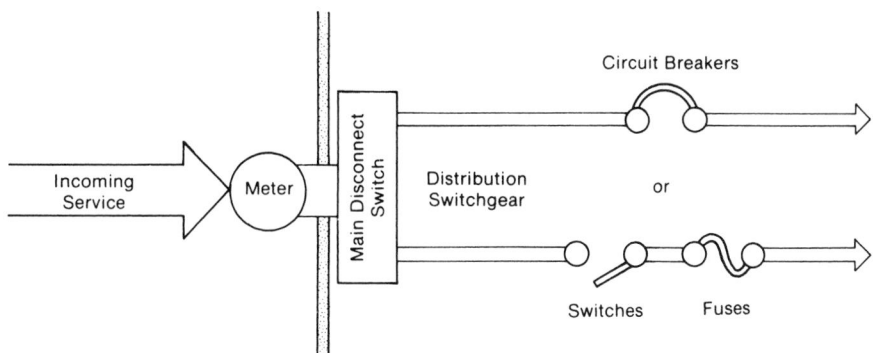

FIGURE 9.5. Service entrance components.

TABLE 9.1 TYPICAL TRANSFORMER DIMENSIONS AND WEIGHTS

kva	Approximate Dimensions (feet)			Approximate Weight (pounds)
	Height	Width	Depth	
3	1 1/2	1	1/2	100
5	1 1/2	1	3/4	150
10	1 1/2	1	3/4	250
15	2 1/2	2	1 1/2	325
25	3	2	1 1/2	375
50	3	3	2	600
75	3 1/2	3 1/2	2	900
100	3 1/2	4	2	1,100
150	3 1/2	4	2	1,500
250	4 1/2	5	3	2,500
300	5	5	3	3,100
500	6	6	3	4,100

FIGURE 9.6. Indoor 300-kva dry-type transformer. (Photo by Vaughn Bradshaw.)

Actual values of known loads should be used wherever possible, but if they have not been ascertained, the following rules of thumb may be used.

1. Lighting: watts per square foot (W/m^2) from Appendix Table B5.4.
2. Receptacles: 2 W/ft^2 (22 W/m^2) for the first 10 kilowatts and 1 W/ft^2 (11 W/m^2) for the remainder over 10 kilowatts for banks and offices only.
3. Air conditioning: Btuh/ft^2 (from Appendix Table B1) times kW/ton (from Appendix Table B7.5) divided by 12,000 Btuh/ton.
4. Elevator (if applicable): 1 W/ft^2 (11 W/m^2).
5. Miscellaneous motors and power: 1 W/ft^2 (11 W/m^2).
6. Commercial kitchen: manufacturer's input rating (watts) from Appendix Table B5.6A.

Once the wattage is established for each category of load, these wattages are added together, as in the following example of a five-story 50,000-ft^2 (4,600-m^2) office building.

___

Example

		Watts
1. Lighting:	3 W/SF × 50,000 SF	150,000
2. Receptacles:	2 W/SF × 5,000 SF	10,000
	1 W/SF × 45,000 SF	45,000
3. Air conditioning:		

$$\frac{43 \text{ Btuh/SF} \times 50{,}000 \text{ SF} \times .99 \text{ kW/ton}}{12{,}000 \text{ Btuh/ton}} \qquad 177{,}000$$

4. Elevator: 1 W/SF × 50,000 SF   50,000

5. Miscellaneous motors and power:

                  1 W/SF × 50,000 SF   <u>50,000</u>

                                               482,000

Kilowatts are approximately equal to kilovolt-amperes for this purpose, so a 500-kva transformer would be needed. Referring to Table 9.1, this transformer would be about 6 feet (1.8 m) high, 6 feet (1.8 m) wide, and 3 feet (0.9 m) deep, and would weigh about 4,100 pounds (6,900 kg).

The simultaneous use of all connected loads rarely occurs except when these loads consist almost entirely of everyday lighting. In most cases, the connected load is multiplied by a type of diversity factor (less than 1), called a *demand factor,* in order to reflect the actual demand on the system. For our purposes, however, a rough idea of size can be obtained on the basis of the approximate connected load.

Location

Transformers must be located so that they are accessible for periodic inspection. They give off about 1 to 1½ percent of their rated power as waste heat. The larger sizes are heavy and noisy, require considerable air circulation for cooling, and in some cases constitute a fire risk. Locating a transformer outdoors is desirable for these reasons and also because it doesn't take up valuable floor space within the building. It is also easier to maintain or replace an outdoor unit, and it is normally less costly.

However, it is sometimes hard to find a suitable outdoor location. Noise may be more disturbing to nearby courtyards and harder to contain outside than it would be in a basement. The heat generated by a transformer may be useful inside a building in the winter, although it is a liability during the summer.

An exterior transformer is subject to vandalism unless pole-mounted, and it may have an objectionable appearance. It can be screened for aesthetic and security reasons, and a massive construction such as a decorative brick wall may even be able to screen some of the sound. But any enclosure must be carefully designed to allow adequate air circulation for ventilation cooling. Furthermore, an outdoor transformer location must be relatively close to the building to minimize line power loss and wiring expense, and it must be shaded from direct sunlight so that increased temperatures will not hinder proper operation.

When, all things considered, an indoor installation is desirable, providing adequate ventilation for heat dissipation is the chief concern. This can be provided mechanically with fans or naturally with windows or louvers on an outside wall.

As a rule of thumb, 3 square inches (20 cm²) of ventilation opening should be provided for each kva of capacity, with a minimum of 1 ft² (80 cm²). A louvered opening, which usually has only 50 percent free area, should have double the opening size, or 6 in.² (40 cm²) per kva. For best natural convection currents, half the area of opening should be placed near the ceiling, and half near the floor.

Provisions for equipment removal must also be considered; ventilation openings in conjunction with removable panels may be used for that purpose. Bird screening must always be used with ventilation openings.

A transformer vault, which is simply a fire-rated enclosure, is necessary for those types of transformers that present a potential fire hazard in the event of damage or malfunction.

Customer-Owned Transformers

Power may be purchased at line or reduced voltage. If high-voltage service is purchased, all entrance equipment on the owner's side of the service connection is the responsibility of the owner. If reduced-voltage service is purchased, the step-down transformer and all other equipment necessary to provide the specified voltage must be furnished and maintained by the utility. That is why utilities charge higher rates for low-voltage service than for high-voltage service.

The cost savings in buying large quantities of power at the distribution voltage for reduced rates more than compensate large users for the cost of maintaining their own transformer. The small savings that would be accrued by small users, however, would not normally justify the high initial expense of installing a transformer, so they usually buy low-voltage power.

The cutoff point for cost effectiveness varies according to the utility rates, the transformer cost, and the amount of electricity used. For owners not equipped to maintain high-voltage equipment themselves, it is sometimes economically advantageous to purchase the high-voltage service and pay the utility company to provide and maintain transformers. This type of arrangement is not available from all utility companies.

Service Disconnect

A switch is a device to bridge or break the flow of electricity. When a switch is in the "off" position, it is said to be *open,* and its circuit is disconnected. When a switch is in the "on" position, it is said to be *closed,* and its circuit is completed, allowing current to flow.

The *main switch,* or *service disconnect,* is a means of disconnecting an entire building from the electrical service. Its principal purpose is to enable all electrical equipment and circuits to be quickly de-energized from one central location in order to make major repairs, or to make certain that there are no live wires inside the building in the event of a fire or other emergency.

These switches must be located near the point of service entry into the building. More than one, but no more than six, main switches may be used to connect all circuits in a building. Circuits for emergency and exit lights, fire alarm systems, and fire pumps may bypass the main switch and connect directly to the incoming electrical service.

Fuses and Circuit Breakers

Associated with each service disconnect is an *overcurrent protection device,* which may be a *fuse* or a *circuit breaker.* Either one is designed to disconnect a circuit automatically whenever the current reaches a predetermined value that would cause a dangerously excessive temperature in the conductor. This may be due to a short circuit, excess current draw from too much load, or a sudden surge from the power company. Excessive current can cause conductors to overheat, resulting in fires, explosions, or expensive equipment damage.

A fuse contains a short strip of metal with a low melting point. When installed in its socket or fuse holder, it becomes a link in the circuit. When the amperage flowing through the circuit exceeds the rating of the fuse, the metal strip melts and the fuse "blows," disconnecting the circuit just as if a switch were opened or the wire cut. The metal link is enclosed in a housing to prevent hot metal from spattering when the fuse blows, and to permit safe handling and replacement.

A fuse is a one-time device. When it blows, the circuit is broken, and the destroyed fuse must be replaced before power can be restored. A circuit breaker, on the other hand, contains a simple mechanism that "trips," opening a circuit in response to an overload condition, but it can be reset by a simple manual switch to restore power. A circuit breaker can also serve as a switch to turn a circuit on or off.

Circuit breakers are more expensive than fuses to install, but they are considerably more convenient and avoid the cost of additional fuses whenever they trip. Fuses, however, are more reliable and are still specified for that reason. Nonetheless, it is possible in some cases for a building owner or occupant to defeat the purpose of a fuse by replacing the proper fuse with a higher-rated one or a solid piece of conducting metal to avoid the "nuisance" of buying new fuses or correcting a minor overload. This can lead to extremely dangerous conditions.

Panelboards

A collection of service disconnect switches along with switches to major branch lines, called *feeders,* mounted together in a wall panel is known as an *electrical panel* or *panelboard.* Fuses or circuit breakers are generally included in the panel to protect the feeder circuits. An example is shown in Figure 9.7.

Panels may be either surface-mounted on solid walls or structural columns or recessed flush with the finish wall surface. They are often labeled to identify the circuit or equipment they serve. By convention, LP generally stands for lighting panel, KP for kitchen panel, AHU for air-handling unit, and AC for air conditioner. Panels may also be labeled according to the area served, such as the floor number or some abbreviation for a zone.

FIGURE 9.7. Panelboard. (Photo by Vaughn Bradshaw.)

Switchgear and Switchboards

Large buildings often have their primary (disconnect) switches together with their secondary (feeder) switches, fuses, or circuit breakers all encased in a freestanding assembly known as a *switchboard* (Figure 9.8). All switches, fuses, circuit breakers, and electrically live components are completely enclosed in an insulated metal structure, and all devices are controlled with insulated handles on the front of the panel.

The term *switchgear* is used for such a package designed for high-voltage (above 600-volt) service. Switchgear also describes switching equipment in multiple individual units not consolidated in a single panel.

Switchboards and switchgear are generally located in the basement within a separate, well-ventilated "electrical switch-gear" room designated for that purpose. When naturally ventilated, an outside air opening of 1 square inch per kva capacity should be allowed for in addition to the opening area provided for any transformer located in the same room.

Exact space requirements should be obtained from the electrical consultant when planning for such a room. In general, switchboards and switchgear range from 1½ to 4½ feet (½ to 1½ m) deep, with the height and width dependent on the size and number of circuits contained within it. Clearance requirements are typically 3½ to 4 feet (1 to 1¼ m) behind and 5 feet (1½ m) in front of the panel.

Switchgear in weatherproof housings are available for outdoor use. In some cases, a weatherproof enclosure is built around standard indoor equipment.

Unit Substations

A *unit substation,* as shown in Figure 9.9, is a single package combining a step-down transformer with a complete switchboard and meter(s). Such an assembly, also known as a *load center substation,* takes advantage of the economies of prefabricated construction and factory-matched components, and performs all the functions required between the incoming high-voltage service and the interior distribution network.

Unit substations are available in indoor or outdoor models. Indoor units are typically located in a basement electrical equipment room with adequate ventilation, as previously described.

Metering

A meter is normally provided at the service entrance by the utility company to measure how much electricity a facility

FIGURE 9.8. Switchboard. (Photo reprinted by permission of Schneider Electric.)

FIGURE 9.9. Unit substation. (Photo reprinted by permission of Schneider Electric.)

FIGURE 9.10. Meters. (Photo by Vaughn Bradshaw.)

uses. The meter may be either inside or outside, often at the owner's discretion. The meter is the property of the electric utility, whose rules govern its location and installation. Utility personnel must have access to inside meters. Regardless of the physical location, meters are always inserted ahead of the main service switch to prevent them from being disconnected.

The utility company may install a single master meter for an entire facility, or multiple meters for different types of service or different customer/tenants. Multiple- or submetering is desirable in apartment houses, and in rental office buildings where individual accountability for energy usage encourages conservation efforts. An example of multiple meters is shown in Figure 9.10.

INTERIOR DISTRIBUTION

Conductors

All electrical current within a building travels between the service entrance and the loads by means of conductors. These conductors are analogous to piping that carries water. The generic category of conductor encompasses wires, cables,

and busbars. A system of conductors connecting one or more loads to a source of power is referred to as *wiring*.

Conductor Material

Because it has one of the highest electrical conductivities, copper is commonly used for conductors, although aluminum and copper-clad aluminum are also used because of their lower cost. Since aluminum has a higher resistivity to current than copper does, an aluminum- or copper-clad aluminum conductor must be larger than an all-copper one in order to conduct the same amount of current.

Aluminum wire has other disadvantages. Joints between two aluminum wires or terminal connections between an aluminum wire and another device tend to loosen over time. Also, a film of aluminum oxide forms very quickly on any exposed aluminum surface, lowering conductivity and causing the joint to heat up enough to potentially cause a fire. For a proper and safe connection, this film must be removed and prevented from reforming.

These size and safety advantages favor copper for small conductors, especially for those in residential construction. However, aluminum's lighter weight and lower cost make it desirable for large conductors and high-voltage transmission lines in installations where proper equipment, techniques, and skilled workers can overcome the safety concerns.

Size

The maximum current-carrying capacity of conductors is called *ampacity,* a combination of "ampere" and "capacity." It is a function of the conductor material, cross-sectional area, and the insulation around it. It is standard practice to install a larger ampacity in the overall electrical system than is immediately anticipated in order to allow for future additions of more loads. Once in place, concealed behind walls, ceilings, or floors, a wiring system can be altered only with considerable trouble and therefore expense.

Another reason for initially oversizing conductors is that the amount of power lost by wiring as heat is proportional to I^2R, that is, the product of the square of the current and the resistance. A larger wire with less resistance, R, may pay for its added cost many times over with energy cost savings during the building life.

In the United States, the sizes of conductors are indicated by a rather eccentric numbering system. For the smaller conductors of round cross section, the American Wire Gauge (AWG) system is used. Under this system, decreasing numbers indicate larger diameters, from the smallest, No. 18, to the largest, No. 0000 (also written as No. 4/0 and pronounced "four-ought"). No. 18 is used mostly in flexible cords and for low-voltage (less than 30-volt) wiring for signal equipment and thermostats. No. 12 and 14 AWG, commonly used for lighting and general receptacle wiring, are about the width of a pencil lead, and No. 4/0 is almost 1/2 inch (13 mm) in diameter.

Larger conductors are designated by MCMs (thousand circular mils), which increase in number with increasing diameter. A circular mil is the square of the cable diameter in mils, which are thousandths of an inch. A 1/2-inch (13-mm) diameter conductor is 500 mils in diameter; $500^2 = 250,000$ circular mils (250 MCM). Common sizes are 250, 300, 400, and 500 MCM. The NEC prescribes the type and size of wires allowable for any particular application.

Countries outside the United States that use metric units express wire sizes as the diameter in millimeters.

Insulation

The risks of shock, electrocution, and fire from exposure to "hot" (live) conductors must always be guarded against. Wires are shielded from one another and from contact with the building and its inhabitants by insulating materials that do not conduct current.

Wires are insulated with a variety of insulations which are suitable for different applications. The most common insulations are thermoplastics, such as PVC (polyvinyl chloride). Insulation on wire smaller than No. 4 AWG is usually color-coded. White or gray insulation indicates neutral wires, green is for grounding wires, and all other colors (red, black, blue, etc.) are used to identify hot.

Types of Conductors

Busbars are large flat conductors, usually solid copper, used for carrying very high currents. The term *cable* conventionally refers to two or more separately insulated conductors grouped together within a common sheathing. It also refers to a single insulated conductor size No. 6 AWG (0.162-inch diameter) or larger. The term *wire* is reserved only for single conductors size No. 8 AWG (0.1285-inch diameter) or smaller sheathed with insulation.

Flexible cord (lamp cord) consists of many strands of fine wire, which allows for more flexibility. It is used only to connect portable lamps, appliances, equipment, and tools to plug-in receptacles. It is never permitted as permanent wiring.

Cable is available as nonmetallic sheathed cable or as armored cable. Armored cable is also called *Type AC* or *BX,* and may be sheathed with steel, aluminum, copper, or bronze. In some environments, plastic jackets are applied over the metal jacket to protect against rust or other corrosion. Cables often include a separate bare ground wire within the same outer sheath.

A ground line is used as an emergency path for current. In the event of a short circuit in a load or a fault in the wiring system, it conducts the current to the ground outside, where it completes a circuit with an overcurrent protection device and blows a fuse or trips a circuit breaker.

Cables, by themselves, are almost universally used for residential wiring because of the relative ease of installation and lower cost. In CII (commercial, industrial, and institutional) buildings, where electrical requirements change frequently, running the conductors in a raceway where they can be replaced more easily provides the needed flexibility at a minimal overall cost.

Raceways

A *raceway* is any channel for supporting and protecting conductors. Included in that category are (1) conduits, (2) wireways, (3) surface metal raceways, (4) cable trays, (5) floor raceways, (6) ceiling raceways, (7) busways, and (8) cablebus.

Raceways are constructed of a variety of metal or insulating materials. The insulating materials, usually cement-asbestos or impregnated fiber, are generally used in direct-buried installations because of resistance to corrosion.

Metal raceways, which may be run exposed or concealed, usually serve as the ground line. Raceways are typically installed first, and the wire or cable is then pulled through or laid in place later.

Conduits. A *conduit* is a protective sleeve usually used for individual cables. There are five types of conduit currently in use: (1) rigid metal conduit, (2) intermediate metal conduit (IMC), (3) thin-wall conduit, sometimes called EMT (electrical metallic tubing), (4) flexible conduit of interlocked spiral-wound steel tape, usually galvanized (similar to armored cable), and (5) rigid nonmetallic conduit of PVC and other materials that provide moisture and corrosion resistance, light weight, and ease of installation. Examples of conduits are illustrated in Figure 9.11.

The NEC limits the use of flexible conduit, but it is useful for making equipment connections and getting around obstructions. A liquid-tight version is also available for wet environments.

Conduit may be concealed behind a wall or ceiling, buried in a floor slab, exposed on a wall surface, or hung from a ceiling. Exposed conduit is primarily used for renovations and electrical additions where the budget does not permit the cutting and patching required to conceal conduit behind finished surfaces.

Wireways. *Wireways,* sheet-metal troughs with hinged or removable covers, are used to carry up to 30 cables. Open-

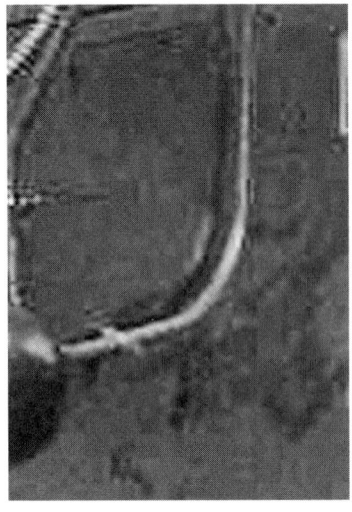

(a)

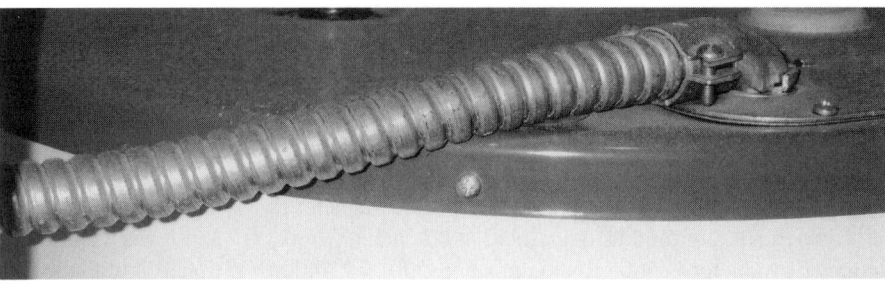

(b)

FIGURE 9.11. Conduits. (a) Electrical metallic tubing (EMT). (Image courtesy of Trane Commercial Systems.) (b) Flex. (Photo by Vaughn Bradshaw.)

FIGURE 9.12. Surface wireway. For block and brick buildings, a surface-mounted raceway such as this is usually necessary for distribution of power wiring. It can also accommodate signal wiring. (Photo by Vaughn Bradshaw.)

ing the cover permits full access for inspection, addition, or replacement of wiring. An example is shown in Figure 9.12.

They are normally permitted only for exposed applications in dry locations, where they are not subject to physical abuse. Special gasketed versions are available for outdoor use.

Surface Metal Raceways. Surface raceways are similar to wireways. They generally consist of a backplate fastened to a wall surface and a snap-on cover in a variety of shapes and styles. An example is shown in Figure 9.13.

Surface metal raceways are commonly used for electrical additions where concealed raceway would be too expensive or difficult to install. They are sometimes used as part of the baseboard of metal partitions. Prefabricated, prewired assemblies are also available with convenience outlets at regular intervals.

Cable Trays. *Cable trays* are open track supports for heavily insulated cables. Examples are shown in Figure 9.14.

Since the raceway does not completely enclose the cables, it cannot provide continuous protection or support, so the cables must be sturdily jacketed. This type of raceway is primarily used in industrial applications.

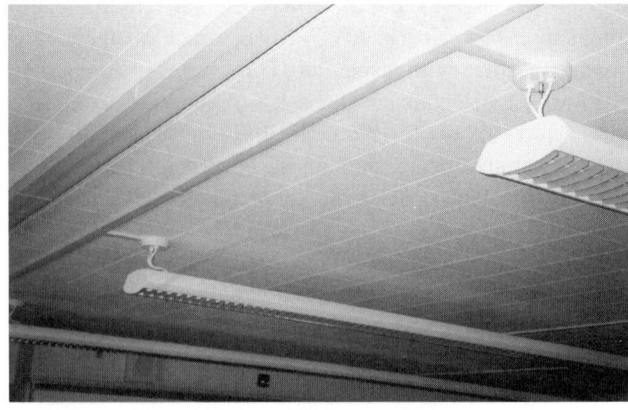

FIGURE 9.13. Surface raceways connecting light fixtures such as these are very helpful in remodel projects. (Photo by Vaughn Bradshaw.)

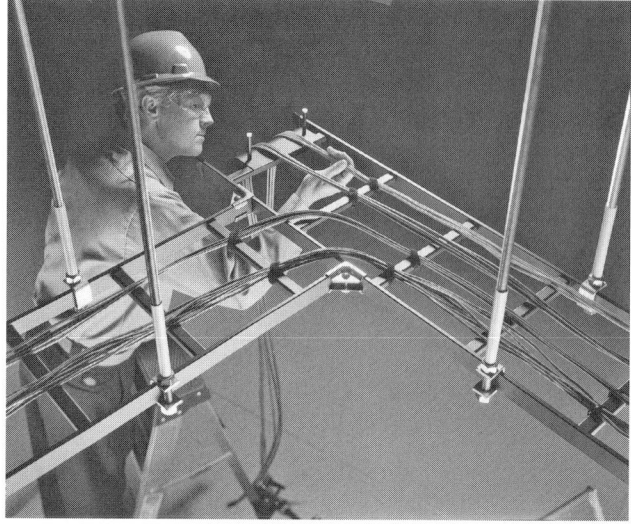

(a)

(b)

FIGURE 9.14. Cable tray. (a) Wire basket trays are available in zinc-plated finish and a variety of colors. (b) Solid-bottom cable tray hides the cables and accommodates an internal divider to maintain complete separation between electrical and data wiring. A smooth transition shown here from the horizontal cable tray to a connecting vertical raceway that feeds into a manufactured architectural column avoids the unfinished appearance of signal cables out in the open. (Photos courtesy of Wiremold/Legrand.)

Floor Raceways. Floor raceways may be either a system of underfloor distribution ducts, called *under-floor raceways* (Figure 9.15a), or integral with the floor itself, and then referred to as *cellular floor raceways* (metal or precast concrete) (Figure 9.15b). Their purpose is to allow versatility in the layout of partitions or furniture in open office plans, exhibit halls, merchandising areas, industrial facilities, or any other open space by providing electrical power and communications connections virtually anywhere on the floor regardless of proximity to walls.

Under-floor raceways may be installed beneath the floor, within the floor, or in a floor trench with a removable cover flush with the finished floor. The latter type is similar to wireway.

A cellular floor raceway simply channels wiring through cells or longitudinal voids serving as part of the structural system of a floor. Header ducts for feeding the cells run perpendicular to the cells and are encased either in the floor resting on top of the cells or in the ceiling plenum below. This type of raceway tends to be much more flexible than the under-floor type because wiring can run almost anywhere in the floor, not just where distribution ducts are run. This allows for rearrangement of layout at a lower expense, without regard for existing wiring limitations.

In either case, communications and low-voltage systems must be isolated from electrical power circuits in separate channels. The under- or in-floor duct systems must use two or three separate ducts, while the trench ducts employ metal partitions to separate the various systems within the same duct. In a cellular floor, electrical wiring may not be run in any cell containing communications, steam, water, gas, air, or any other service.

The principal drawback to these systems is their high initial cost. Aesthetically, they require either exposed metal cover plates on outlets not in use or square carpet tiles, which limit the floor finish options. An alternative for fixed layouts is to run individual conduits to specific floor outlets.

A more economical and increasingly popular alternative is the modular raised-floor system (Figure 9.15c). It creates an interstitial space below the floor for the distribution of building services including power, voice, and data wiring. At the same time, the interstitial space can serve as a plenum for the distribution of HVAC airflow. In some cases, these raised-floor systems can be as little as 2½ inches (6.4 cm) above the floor slab.

These systems offer the advantage of complete flexibility to change floor plans, layouts, and work station arrangements quickly and with a minimum of disruption. If a new computer or phone needs to be added, a couple of floor tiles

(a)

(c)

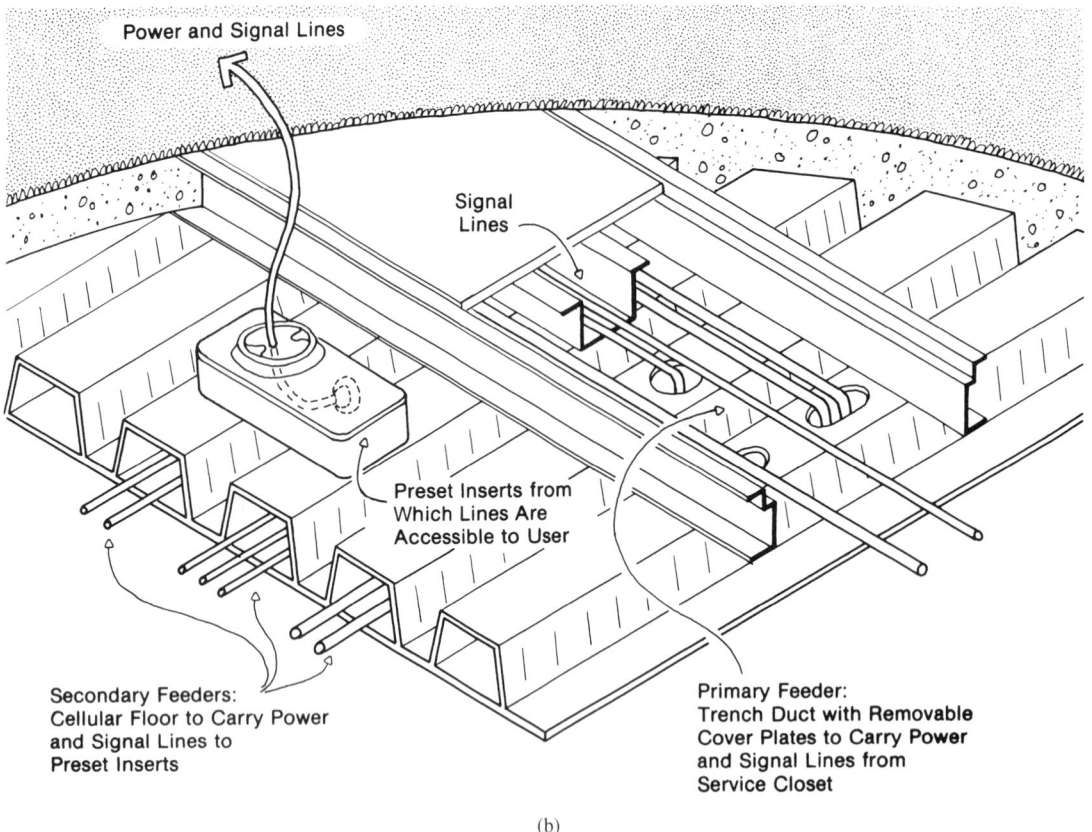

(b)

FIGURE 9.15. Under-floor wiring distribution. (a) Under-floor raceway can be cast into a concrete slab or mounted below wood flooring. All that is visible from above are floor boxes such as this. (Photo courtesy of Wiremold/Legrand.) (b) Cellular floor raceway. As with an under-floor raceway system, the termination points are fixed in place. (c) Raised flooring provides the advantage of greater flexibility. Power and signal wiring can be laid below the floor and then rerouted as room layouts change. HVAC air may also be supplied or returned below a raised floor. (Photo courtesy of Steelcase Inc.)

are removed, the hardware is connected, and the tiles pop back into place. People can then go right back to work.

Ceiling Raceways. Another lower-cost alternative to floor raceways is a ceiling raceway system. It consists of a network of wireways run above a suspended lay-in tile ceiling, accessed with relocatable floor-to-ceiling *power poles.* Connections are made to the power poles with simple plugs and receptacles. In addition to serving convenience outlets and communications systems, the ceiling raceway can provide power to ceiling-mounted lighting fixtures.

The lower cost of these systems derives from their not having to integrate with the structural floor and less disruptive electrical alterations. The more convenient access to the wiring also facilitates more rapid changes in layout. The ability to accommodate rewiring and electrical capability so readily makes a ceiling raceway system flexible in function as well as in the arrangement of a space.

The principal disadvantage of these systems is the difficulty of visually integrating the floor-to-ceiling poles into the interior design. Placing them in conjunction with acoustical partitions may help, or they can be highlighted as signposts.

Busways. When more power is needed than a single large cable can accommodate, the capacity is sometimes increased by using multiple parallel cables. At some point, it becomes more economical to use copper or aluminum busbars to conduct the very large current. A *busway,* also known as a *bus duct,* is a rigid assembly consisting of one or more

FIGURE 9.16. Busway or bus duct. (Photo by Vaughn Bradshaw.)

busbars bolted securely to insulating supports within a vented steel housing. This assembly is illustrated in Figure 9.16.

Since the NEC does not permit busways to be concealed, they are usually found in electrical equipment rooms or as feeders in industrial applications. When lighting is fed directly from a busway, it is known as *lighting track.* Prefabricated factory-assembled busduct and enclosure combinations are available for economical installations. For outdoor applications, special designs with weatherproof housings are available.

Cablebus. A *cablebus* resembles a busway from the outside, but instead of busbars, it uses insulated cables rigidly mounted in an open space frame. Since the enclosure is ventilated, the air-cooled cables can be undersize without overheating. Like the busway, a cablebus consumes more power and gives off more heat than the other wiring systems.

Feeders and Branch Circuits

Feeders are the first stage of distribution from the service entrance. They typically lead to local distribution points from which smaller capacity circuits branch out. *Branch circuits,* those portions of the wiring system that supply the ultimate loads, are the final connection to an electric device. A branch circuit may supply one large load, such as a motor or heating element, or it may serve a group of smaller devices, such as lights, receptacles, and motors.

Simple systems for single-family dwellings normally consist of branch circuits emanating directly from the service entrance. Buildings with a larger capacity and larger number of loads distribute feeders from the service entrance to subpanels located nearer to the loads.

The feeders are generally extended as near as possible to the center of the area served in order to minimize the length of the branch circuits. The resistance in a long conductor causes a voltage drop between one end and the other, which can cause serious problems for some loads. The larger the wire, the smaller the voltage drop. Since it is normally more economical to oversize one feeder to compensate for resistance than to oversize a multitude of branch circuits, *home runs* are kept short. A home run is the distance from the first outlet on a circuit to the panelboard and is preferably kept below 100 feet (30 m).

Sometimes these branch circuit distribution panels are shallow enough to be installed flush in stud partitions, and can thus be located almost anywhere out of the way. In large-scale operations, however, electrical closets are often needed to contain branch circuit panelboards, submetering equipment, or small transformers. In multistory buildings, it is desirable to place these one above the other for economical feeder runs and optimum accessibility.

The space requirements for electrical closets depend on circumstances and design, so no generalizations can be made. However, electrical closets containing a panelboard should allow at least 4 feet of clearance in front of the panel. It is wise to check with the electrical consultant on a project at the earliest opportunity to ascertain how much space to allow.

Ground Fault Circuit Interrupter

A *ground fault circuit interrupter* (GFCI), as shown in Figure 9.17, is a device that is substituted for regular receptacles in kitchens, bathrooms, garages, other wet areas, and outdoors to protect against electric shocks. A GFCI supplies power as any other receptacle does, but at the same time it monitors the amount of incoming and outgoing current.

Whenever the entering current does not equal the leaving current—indicating current leakage (a "ground fault")—the GFCI instantly opens the circuit. It provides a quicker and more sensitive response than the fuse or circuit breaker on a circuit, offering added protection for those critical areas subject to an increased likelihood of severe shock.

Motor Branch Circuits

Direct-wired motor loads are typically larger than other lighting and general power loads. For that reason and because of the surge of current when motors are started, they need their own branch circuit. The characteristics of motors and the work they perform make it necessary to pay careful attention to protective devices and conductor capacity.

Like any electrical circuit, motors must have overcurrent protection in the form of a fuse or circuit breaker at the point of supply. In case of a short circuit or prolonged overload, the overcurrent protection device will open the circuit.

Since motors need periodic maintenance and repair, there

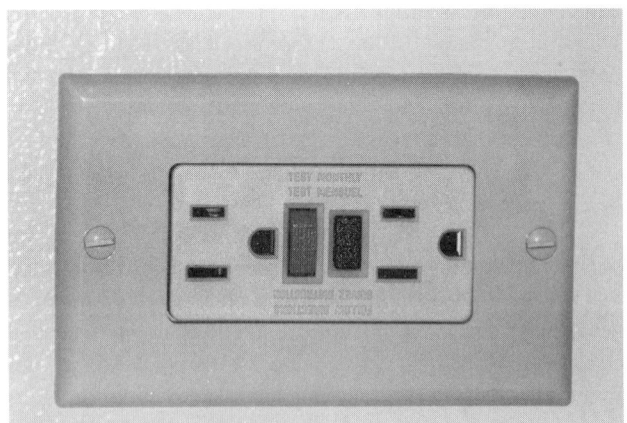

FIGURE 9.17. GFCI receptacle. (Photo by Vaughn Bradshaw.)

must be some means of opening the circuit in order to carry out these activities in safety. The *motor disconnect* may be incorporated with the branch circuit overcurrent protective device as either a fused safety switch or a circuit breaker. However, the disconnect must be within sight of the motor to prevent the circuit from being closed while someone is working on the motor. If the branch circuit panelboard is out of sight of the motor and its controller, another disconnect must be located at the motor or controller. A *motor controller* is any device that may be used to start or stop a motor.

Motor starters are essentially switches for connecting and disconnecting motors to and from their power source. They may be manual or automatic. The process of starting a motor at its full-line voltage level is referred to as *across-the-line* starting. Across-the-line starters are the simplest and least expensive means of starting a three-phase motor.

The starting surge of current sometimes affects other loads adversely, and in this case, the surge must be reduced by reducing the starting voltage. Using electronic circuits, a *closed transition* starter makes the smoothest transition between the reduced starting voltage and the full running voltage. It is also the most expensive method.

The *part winding* starter reduces the inrush current to approximately 65 percent and provides only about 65 percent of the starting horsepower. After a time delay of 1/2 or 2 seconds, the full voltage is applied, which causes a slight current surge.

The *autotransformer* starter allows a selection of 50, 65, or 80 percent of full-line voltage for starting.

BATTERY POWER SUPPLY

Although the same or similar equipment is used for both cases, the NEC distinguishes between emergency power and standby power systems. *Emergency systems* provide legally required power and lighting for human safety and property protection. This includes lighting for emergency egress and power for elevators, fire alarm systems, and fire pumps.

Standby systems are intended to provide power, in the event of a disruption in normal utility service, for economic protection, security, or convenience beyond those legally required safety measures. They are installed, for example, to protect critical industrial processes, research projects, and mainframe computer equipment when a power interruption would cause serious property damage or financial loss.

Emergency electrical power is also required by law for certain functions that are not as critical to life and safety as the emergency systems. Examples of these are water and sewage treatment and some communication systems. All three of these categories may utilize either on-site power generation or battery equipment. The latter, used for uninterruptible needs, supplies power until an orderly shutdown can be implemented, thus "bridging the gap." On-site power generation is used for longer-term outages or larger critical loads.

H.I.D. lighting, which cannot restart immediately, cannot tolerate the delay that occurs before an emergency generator is brought on-line. When used for emergency lighting, H.I.D. lamps must use a battery for emergency power.

On-site power generation for emergency, standby, or continuous service is a separate subject that is covered in the next chapter. Automatically charging battery equipment, being more of an appurtenance to the normal electrical system, is discussed below.

Due to their limited storage capacity, batteries are most often used only for a small amount of emergency lighting. Larger installations of batteries are, however, used as part of an *uninterruptible power system* (UPS). These systems provide the necessary reliability for computer facilities, microprocessor-based demand controllers, or wherever even a momentary power interruption could have disastrous conse-quences. A bank of batteries such as that used in a moderate-sized UPS is shown in Figure 9.18.

Fluctuations, spikes, or discontinuities in voltage can result in the loss of valuable data storage or damage to sensitive computer hardware. When a UPS is placed between the load and the utility source in the circuit, the load is always supplied by the battery. Under normal operation, the UPS filters out any aberrations in the power supply and keeps the batteries continuously charged. When the utility source is interrupted, the UPS continues to supply power to the critical load. UPSs are usually designed for specific applications, are very expensive and take up a considerable amount of space.

Batteries are rated in ampere-hours, which expresses the length of time a current output in amperes can be sustained. For small-scale applications, cabinet-mounted battery packs may suffice, while larger loads may require long racks of batteries. Small battery packs may be mounted at various locations within a building as needed, or battery power may be distributed throughout a building from one central location. Since batteries are quiet and highly reliable, with low maintenance requirements, they can be placed almost anywhere their heavy weight can be accommodated.

FIGURE 9.18. Room-size battery installation. (Photo courtesy of GNB Batteries Inc.)

Batteries need not be installed in a room dedicated specifically to house them, although that is customary for very large installations. In any case, a space containing batteries must be well ventilated in order to dilute the explosive hydrogen and oxygen gases given off during charging. It is important that smoking not be permitted in the same room as the batteries.

UTILITY DEMAND MANAGEMENT

Most electricity used in the United States is generated from some heat source used to drive turbine generators. The most common heat sources are coal, oil, natural gas, and nuclear reaction. Other heat sources used to a minor degree are geothermal energy and concentrated high-temperature solar energy.

It is helpful to remember that the efficiency of heat-to-electricity conversion on a commercial scale is limited to 30 to 40 percent. In terms of natural resources, electricity is an expensive form of energy. In particular, using heat to create electricity in order to re-create heat is generally wasteful of resources.

A small but significant portion of all electricity used is generated hydroelectrically directly from the motion of water. In some parts of the United States with an abundance of flowing water, such as the Northwest and the Tennessee Valley, the bulk of the electricity consumed is hydroelectric. Most of the country, however, does not have that resource locally available.

Research and development are currently underway to use low-temperature thermal energy in the oceans as a renewable power source. These methods, referred to as *ocean thermal energy conversion* (OTEC) systems, may play a significant role in the future. Wind-power electric generating systems are increasingly being used.

The cost of electricity to the user is dependent to some extent on which of the above processes are used to generate the electricity. Some means of generation are more expensive initially, but the fuel used is inexpensive or free. With others, the plant construction cost is lower, but the fuel cost is high. Nuclear power plants are generally the most expensive to construct and hydroelectric plants the least.

Utility Charges

Electric utilities bill their customers in a variety of ways. Residential customers generally receive a bill based on the amount of electricity in kilowatt-hours (kWh) they used in a given month. Except for very small users, CII customers typically receive a more complicated bill. The four potential components of a large user's bill are:

1. Usage (kWh)
2. Demand (kW)
3. Load factor
4. Power factor

The kilowatt-hour *usage* component of the bill is based on the utility's cost of providing the electrical energy actually consumed. It represents the amount of fuel the utility had to burn to provide that much electrical energy.

The kilowatt demand component of the bill, referred to as the *demand charge,* is based on the customer's highest *rate* of electrical consumption and can account for one-third to more than one-half of the total bill. Its purpose is to compensate the utility for the capital investment needed to provide 100 percent of the required generation, transmission, and distribution capacity, regardless of whether its customers collectively use it all the time or only a small fraction of the time. It is a means of allocating the fixed cost of the plant construction among the users according to their share of the maximum demand on the utility.

Some utility companies levy an additional penalty against users who consume very few kilowatt-hours but occasionally have a high kilowatt demand. The penalty is based on the customer's *load factor.* The load factor is the monthly kilowatt-hour usage divided by the product of the kilowatt demand and the number of hours in the month. It represents the ratio of the average demand to the peak demand.

Demand Control

The effect of demand charges is that it encourages users to reduce their peak demand. This may be accomplished by disconnecting some loads that are not really needed, in which case energy use is also reduced. If the loads are simply rescheduled, however, the same demand is spread out over a longer period of time, and the energy use is the same or possibly increased slightly if the rescheduling results in some waste.

It is most profitable for the utility and also the most efficient use of resources if the overall demand on the generators from all users is as constant as possible, with a minimum of extreme peaks and valleys. This would make use of the generating capacity most effectively. Otherwise, the utility company must provide expensive generating capacity for a small amount of time, while it stands idle most of the time.

Figure 9.19 illustrates how the demand on a utility for power varies with time. The utility must provide enough equipment to satisfy the system peak. It behooves the utility to shift the use of electricity from the peak periods to the

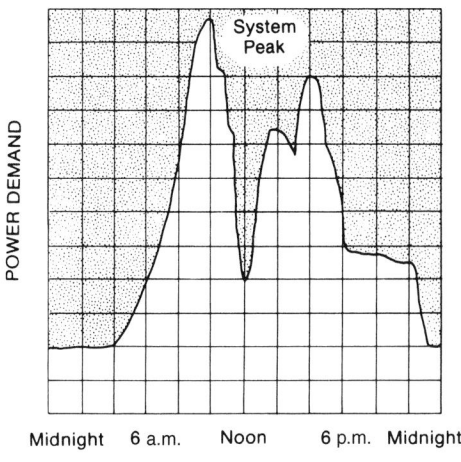

FIGURE 9.19. Utility demand fluctuations.

periods when excess capacity is available. Since the peak demand determines the need for additional electric generating plants, new construction can be delayed or even eliminated by such *demand leveling.*

A strategy some utility companies use to encourage customers to shift their usage to off-peak times is to offer lower rates during those times and charge a premium during periods of high demand. Customers subject to this type of rate schedule can often save a great deal of money by shifting appropriate loads either manually or automatically to the off-peak times, as shown in Figure 9.20.

Strategies for users to manage their loads in order to limit peak demand are (1) load shedding, (2) thermal storage, and (3) utilizing alternate sources of power. *Load shedding*

involves measuring the demand and, when it is too high, turning equipment off according to a predetermined order of priority. Some loads are eliminated altogether, while others are merely deferred to the off-peak time. Thermal storage equipment, as discussed in Chapter 6, allows continuous electric heating or cooling service while using only off-peak electricity. On-site generation of electricity, as discussed in Chapter 10, can pick up some loads in order to reduce or eliminate utility power usage during peak times. This is especially appropriate for facilities such as hospitals that already have their own emergency generating equipment.

Microprocessor-based controllers known as *energy management control systems* (EMCS) can be preprogrammed to automatically reschedule or disconnect loads to achieve the desired demand management. Customers not subject to "time-of-day" rate schedules can sometimes still benefit from load-management techniques to reduce their peak building load. Whether or not to use expensive equipment to shift loads automatically should be determined independently for each building by a cost-effectiveness analysis.

Power Factor Control

Motors and other inductive loads create a magnetic field which results in what can be thought of as momentum in an electric circuit. As the voltage alternates in an AC circuit, the "momentum" of the current tends to keep it constant. This causes a delay, or *lag,* in the current as it follows the voltage, as depicted in Figure 9.21.

This condition affects the utility company because the load uses kilowatt power, which is the instantaneous product of voltage times current. But, in fact, the utility must supply

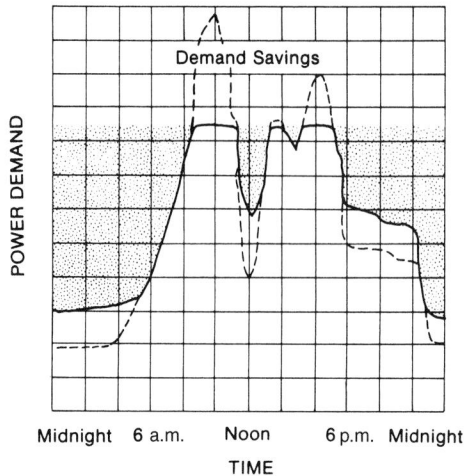

FIGURE 9.20. Demand leveling.

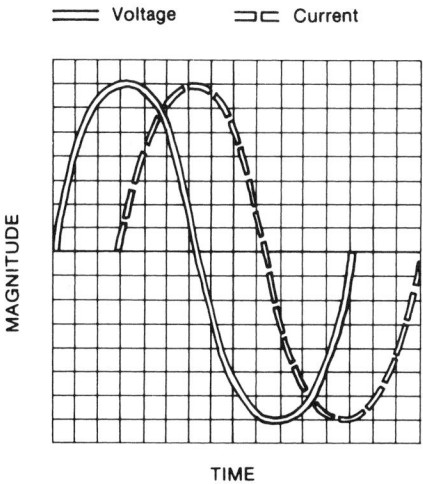

FIGURE 9.21. Voltage and current out-of-phase.

the full current at the full voltage. When many such loads are served by the utility, it must supply much more current than is actually consumed by the load and paid for by the customer.

The *actual power* used by the load is expressed in kilowatts. The power that the utility perceives as drawn by the load is called the *apparent power* and is denoted by the number of volt-amperes or, more conveniently, kilovolt-amperes (kva).

The difference between the actual power and the apparent power, in the form of the fraction kW/kva, is known as the *power factor.* Each load has a power factor, and there is an aggregate power factor for all loads in a facility. Some utilities assess a power factor surcharge to account for the difference. The power factor can be easily and relatively economically improved (made closer to 1) by the addition of *capacitors,* also called *condensers,* to the inductive loads.

These devices have exactly the opposite effect on a circuit, causing the current to lead the voltage. By matching the degree of current lag caused by an inductance load with the proper size of capacitor, the power factor can be adjusted to whatever value is desired. The objective is to balance the amount spent for power-factor-correcting capacitors against the savings on the utility bill.

Since a power factor closer to 1 can reduce the current flow while achieving the same actual power, capacitors can also reduce the size of the electrical conductors and service equipment required for a given load. This can save on installation costs and distribution losses within a building, further helping to justify the cost.

SUMMARY

Electricity is a concept which represents a form of energy. It can no more be held in the hand and examined than fire, the wind, or sunshine. Its behavior is predictable and repeatable,

which has allowed scientists to formulate theories about it in order to be better able to control it.

In accordance with those theories, electrical current (amperes) is analogous to the flow of water, and voltage has the same effect on electrical current as pressure or a difference in height has on the flow of water. A conductor, such as a wire, cable, or busbar, channels electrical current in the same way that a pipe channels the flow of water. And just as pipes cause a certain amount of friction, conductors put up some resistance to current flow.

The electrical resistance results in heat generation when current is forced through a conductor under the electromotive force of a voltage difference. The amount of current that flows depends on how much voltage is applied and how much resistance there is. The amount of heat generated depends on the amount of current flow and the amount of resistance.

Electrical power is the rate at which electrical energy flows. It is equal to the product of volts times amps and is measured in watts. Conversely, electrical energy represents an amount of power over some length of time. Nonresidential utility bills reflect both the amount of energy used (in kilowatt-hours) and the rate (in kilowatts) at which it is used.

Like kilowatts, horsepower is a rate of energy flow. However, while kilowatts refer to electrical power, horsepower refers to the mechanical power delivered by a motor.

The lights, motors, heating elements, and other end-use devices that connect to the electrical current in order to operate are called loads. The power supplying the loads may be at a variety of voltages and may be single-phase AC, three-phase AC, or DC. The equipment used to safely control electricity and distribute it to the many loads within a building can be divided into service entrance equipment and interior distribution equipment.

The service entrance equipment consists of a main disconnect switch and either fuses or circuit breakers for overcurrent protection. This equipment can be packaged in a variety of forms, and it may include transformers to step a higher voltage down to a level usable within the building.

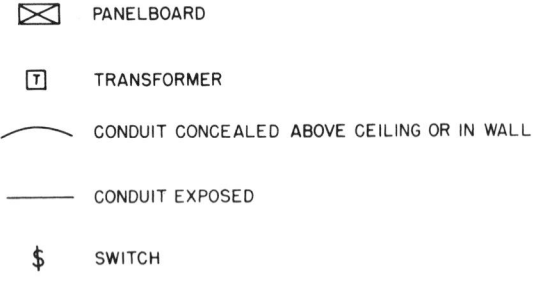

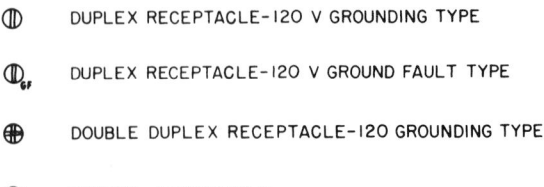

FIGURE 9.22. Common electrical symbols.

Electricity is transmitted from a generation plant and distributed to customers at a high voltage in order to reduce the amount of current necessary to supply a given amount of power. This allows for smaller, more economical conductors and reduces the amount of waste heat generated by the resistance in the conductor.

Some common symbols used on electrical engineering drawings are defined in Figure 9.22.

KEY TERMS

Electrical current (amperes)	Inductive load	Closed switch
Amperage	Inrush current	Main switch or service disconnect
Voltage	Locked rotor current	
Electrical resistance	National Electrical Manufacturers' Association (NEMA)	Overcurrent protection device
Conductor		Fuse
Insulator		Circuit breaker
Circuit	Horsepower	Electrical panel or panelboard
Open circuit	Single-phase AC	
Closed circuit	Three-phase AC	Switchboard
Short circuit		Switchgear
Direct current (DC)	Energy efficient ratio (EER)	Unit substation or load center substation
Alternating current (AC)	Short-cycling	
Frequency (hertz)	Duty classification	Wiring
Power	Voltage rating	Ampacity
Electrical power (watts or kilowatts) kW	Service factor	Busbar
	Service entrance	Cable
		Flexible cord
	Transformer (primary and secondary)	Raceway
Electrical energy (kWh)		Conduit
Load		Wireway
National Electrical Code (NEC)	Primary voltage	Cable tray
	Secondary voltage	Power pole
Underwriters Laboratories (UL)	Known load	Busway or bus duct
	Estimated load	Cablebus
Receptacle or convenience outlet	Connected load	Feeder
		Branch circuit
	Demand factor	Home run
Resistive load	Open switch	Ground fault circuit interrupter (GFCI)

Emergency system	control	Apparent power
Standby system	Demand leveling	Power factor
Uninterruptible power system (UPS)	Load shedding	Capacitor or condenser
	Energy management control system (EMCS)	Subpanel, branch circuit panelboard
Usage charge		
Demand charge		
Load factor	Lag	
Demand	Actual power	

STUDY QUESTIONS

1. What is the difference between electrical power and electrical energy? Give the units of each.
2. Under what circumstances would a primary transformer be required in a building? A secondary transformer?
3. What types of equipment require voltages higher than 120 volts? Which loads require three-phase power?
4. Under what circumstances are electrical closets required?
5. What are the options for providing flexible electrical power and communications connections to an open space where the layout is subject to change? What are the advantages and drawbacks to each method?
6. What are the characteristics of batteries that might influence their location in a building?

BIBLIOGRAPHY

Clark, William H. *Electrical Design Guide for Commercial Buildings.* McGraw-Hill/TAB Electronics, 1998.

Gottschalk, Charles M. (compiler). *Industrial Energy Conservation.* John Wiley & Sons Inc., 1996.

National Electrical Code (NFPA-70). National Fire Protection Association, Batterymarch Park, Quincy, MA 02269.

Stein, Benjamin, and John Reynolds. *Mechanical and Electrical Equipment for Buildings,* 9th ed. John Wiley & Sons, Inc., 1999.

Thumann, Albert. *Plant Engineers and Managers Guide to Energy Conservation,* 8th ed. Marcel Dekker, 2002.

Turner, Wayne C. (editor). *Energy Management Handbook,* 4th ed. Marcel Dekker, 2002.

Various publications may be obtained from the following sources (complete listings are available upon request):

The Electrification Council, 90 Park Avenue, New York, NY 10016.

IEEE (Institute of Electrical and Electronics Engineers), 345 East 47th Street, New York, NY 10017.

National Electrical Contractors' Association (NECA), 7315 Wisconsin Avenue, N.W., Washington, DC 20014.

National Electrical Manufacturers' Association (NEMA), 2101 L Street, N.W., Washington, DC 20037.

National Fire Protection Association, Batterymarch Park, Quincy, MA 02269.

Underwriters' Laboratories, Inc., 333 Pfingsten Road, Northbrook, IL 60062.

Chapter 10

On-Site Power Generation

OPTIONS

CONSERVATION

STORAGE

PURPA

THERMAL-SOURCE GENERATION

PRIME MOVERS

HISTORICAL PERSPECTIVE

TOTAL ENERGY AND COGENERATION SYSTEMS

ARCHITECTURAL IMPACT

ECONOMICS

OTHER ADVANTAGES

MIUS

WIND TURBINES

WIND VARIATION WITH LOCATION

TYPES OF WIND TURBINES

ARCHITECTURAL INTEGRATION

MATCHING SUPPLY WITH DEMAND

SOLAR PHOTOVOLTAIC CELLS

HYDROPOWER

WATER TURBINES

WATER WHEELS

AVAILABLE POWER

ECONOMICS

ENVIRONMENTAL IMPACT

On-site generation of power is desirable when utility-supplied power is:

1. Too expensive
2. Too unreliable
3. Unavailable

On-site generation can also be used to provide peak demand shedding in order to obtain lower electricity rates where applicable.

It is usually clear whether or not utility service is available to a given building or area. Reliability, however, can be a relative term, which may be sensitive to economics. Some critical functions, such as those in a hospital, cannot do without an emergency power supply regardless of its cost, while other building programs must weigh the cost of standby power generators against the value of having a more reliable supply.

The question of whether to generate power on-site for normal electricity requirements is not so easily answered. The initial cost of power generation equipment is usually very high, while the operating and maintenance costs are much less than the cost of utility-supplied power. In order to be cost effective, the total life-cycle cost of the on-site generation systems must be less than the expense of purchasing the necessary power from the utility over the same life-cycle period.

The life-cycle cost (discussed in Chapter 16) cannot be known in advance with complete certainty, so a reasonable estimate, taking into account all known factors, has to be made. Even though utility rates seem high, a decision based on economics can be made only after all these life-cycle costs are considered and after the best estimate is made of how fast utility costs will escalate.

Rates for purchased power are greatly influenced by the fuel mix used by the local utility. In many cases, a combination of some purchased power and some on-site generation is the most economical.

Options

The methods for generating electricity on-site are:

1. Thermal source
2. Wind turbine
3. Solar photovoltaic cells
4. Hydropower

Conservation

Whichever method of on-site power generation is considered, the importance of conservation cannot be emphasized strongly enough. The first step in carrying out any of the above options should always be to reduce building electrical loads to a minimum. It is almost always less expensive to reduce energy needs than to supply that energy from non-utility sources. Furthermore, reducing a building's needs will reduce its environmental impact.

Storage

One drawback to wind and solar generating systems is a consequence of the nature of electricity: Unlike fuels or heat, electricity is not readily stored, and therefore it is more convenient to generate and utilize it at the same time.

The inconsistency of weather and day/night cycles makes the use of some form of storage unavoidable to interface between the time of generation and the consumer use patterns. Batteries can store electricity, but some is inevitably lost as heat in the process. Also, a large volume of batteries is required to store an appreciable amount.

Other potential means for storing electro-mechanical energy are:

• Flywheel momentum
• Pumped water reservoirs
• Compressed air
• Hydrogen fuel cells

The first three are suited to wind power or hydropower, which generate electricity by mechanical means. The energy is stored before it is converted to electricity. The last method, hydrogen fuel cells, is applicable to any form of generation because it involves storing electricity after it has been generated.

By passing electrical current through water, it can be broken up into its constituent components: hydrogen and oxy-gen. This is known as *electrolysis*. The hydrogen can be stored and then later recombined with oxygen as needed in order to recover the energy. Hydrogen fuel cells are being developed that will enable a controlled release of the stored energy.

The drawback to storage, in general, is that each storage option involves the conversion from one form of energy to another. Therefore, as thermodynamics tells us, some energy is "lost" (made unavailable) in the process. Economics is normally the deciding factor in which, if any, storage option is to be used.

PURPA

An alternative to storing surplus on-site generated power is to feed it back into the utility's power grid. The *Public Utility Regulatory Policies Act* (PURPA) of 1978 requires utility companies in the United States to buy on-site generated electricity from small (under 30 megawatts capacity) private power producers. Furthermore, they must purchase this power at a rate reflecting the utility's *avoided costs*. These are the costs that a utility would have incurred if it had to generate the same amount of electricity itself or buy it from another utility on the nationwide power grid. In some cases, the rate includes costs avoided by delaying or not having to build additional generating stations.

This avoided cost is not necessarily the same as the utility's existing sales rate. Sometimes, a utility must buy power from a customer at a higher price than that at which it sells power to the same customer.

The federal government has provided this support because it has perceived on-site power generation as a method of reducing national energy consumption and dependence on imported fuel supplies. It has provided additional financial incentives in the form of an investment tax credit for a portion of the installation cost to further encourage private electricity generation.

THERMAL-SOURCE GENERATION

Prime Movers

The basic components of an on-site generation system, illustrated in Figure 10.1, are a *prime mover* and an electricity generator. The prime mover is a device that transforms the energy in fuel or steam into rotating shaft energy. A generator coupled to the shaft then converts the energy of the rotating shaft into electricity.

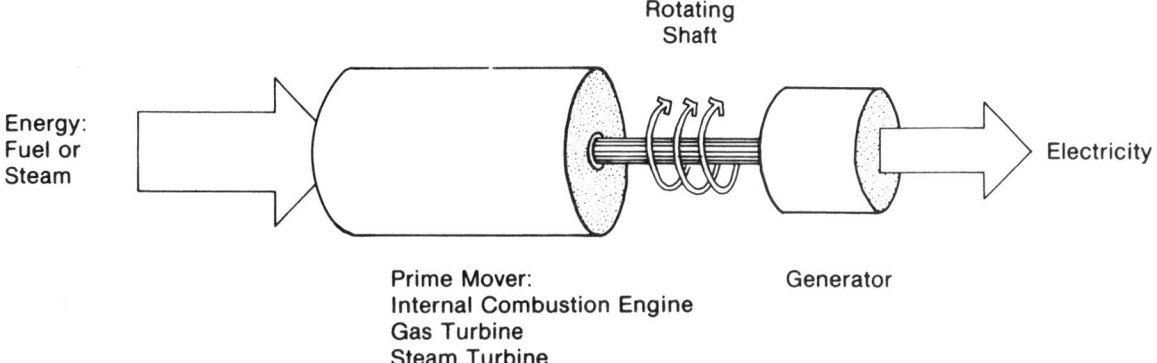

FIGURE 10.1. Prime mover and generator system schematic diagram.

The types of prime movers used for on-site generation are:

1. Internal combustion engines
2. Gas turbines
3. Steam turbines

In addition to these three, there is another type of on-site power generation equipment: *fuel cells.* They operate quietly on natural gas and other fuels, can achieve an almost 80 percent efficiency in converting fuel directly to electricity with almost no air emissions, and promise low maintenance. Although still in the development process, fuel cells are expected to be commercially available in the near future.

Internal Combustion Engines

Internal combustion engines burn a variety of liquid or gaseous fuels, such as diesel, oil, fuel alcohol, natural gas, and methane gas. They are available in a range of sizes from 7.5 to 25,000 kilowatts and are generally 20 to 40 percent efficient in converting fuel to shaft energy.

Their disadvantages, compared to the other types of prime movers, are higher noise and vibration levels and the need for more frequent maintenance. Remote location or sound isolation is necessary, as are large amounts of ventilation for cooling and combustion air. The latter requires extremely large louvers for indoor generators.

Factory-prefabricated packages combining a generator and prime mover as one unit on the same skid, called *engine-generator sets,* are available. Typical indoor and outdoor engine-generators are shown in Figure 10.2.

Gas Turbines

Gas turbines, which also burn a variety of liquid and gaseous fuels directly, are available in sizes from 37 to 75,000 kilowatts. They are relatively free of vibration and are lighter in

(a)

(b)

FIGURE 10.2. Engine generators. (a) Indoor. (Photo courtesy of NC Power Systems.) (b) Outdoors in a weatherproof enclosure. (Photo by Vaughn Bradshaw.)

weight but have a lower efficiency of 12 to 20 percent. An example of a gas turbine is shown in Figure 10.3.

Steam Turbines

Steam turbines require some source of high-pressure steam. The turbine converts the thermal energy in the steam into shaft energy.

The steam can come from a conventional high-pressure boiler burning liquid fuel, gas fuels, coal, or solid waste. It can also come from high-temperature solar collectors, geothermal resources, or the discharge from some industrial process.

The conversion of solar heat to shaft rotation in a turbine was demonstrated more than a century ago. Experimental systems have operated successfully, producing shaft power for water pumping, air conditioning, industrial machines, and electricity generation. The turbine converts the thermal energy in the steam into shaft energy.

Because they operate at higher pressures than those allowed for unattended operation, steam turbines are applicable only for large-scale operations where a stationary engineer is already on duty or in applications where the additional personnel to operate turbines will not signifi-

cantly affect the overall economics. For this reason, they are available only in sizes above 4,000 kW.

Historical Perspective

Prior to 1900, and in the early years of the twentieth century, nearly all large buildings and groups of buildings were supplied with power generated on or near the premises. Generated electricity was supplied for elevators, ventilators, call bells, fire alarms, and lighting.

Many homes of the very rich had their own private electric plants. George J. Gould's estate in Lakewood, New Jersey, had generator dynamos driven by two gasoline engines. They lit his house and grounds, operated a refrigerating plant, ran a laundry, and powered electric cigar lighters as well as Mrs. Gould's hair curler. The Vanderbilt mansion in the heart of New York was powered for a time by its own generating plant designed by Thomas Edison himself. It was removed because of its noise, which points up one of the problems with local power production: it can be very noisy.

The motive power in those early installations was usually a steam-driven reciprocating engine with belt connections to a direct current generator. The reason for local generation

FIGURE 10.3. Gas turbine generator. (Photo by Vaughn Bradshaw.)

was that the technology existed only for DC electricity. DC cannot be transformed from one voltage to another; thus, it must be generated and distributed at the voltage used in the building. The power loss is too high at such low voltages to permit distribution over very great distances.

In 1882, Edison's Pearl Street Station became the first centralized electric utility and began providing electricity to shops and homes within a 1-square-mile area. Its success stimulated the building of other utilities in cities all over the country. Purchased power, with its advantages of lower cost, less noise, and smaller space requirements, rapidly displaced private on-site generation in the vicinity of a utility station.

With the development and use of AC machines, electricity could be transmitted at high voltage over long distances and throughout extensive distribution systems. At the point of use, the electricity could be transformed down to usable voltages. High-voltage transmission substantially reduced power losses and soon became the universal method.

In the 1920s, utility companies established larger central power stations with lower construction and operating costs per kilowatt. They also bought fuel under large, long-term fuel contracts at a lower cost than what private individual or CII consumers had to pay. Utility power became more reliable and less expensive.

The decreasing costs of purchased power meant savings in operating expenses for industrial users. So, during the 1920s and 1930s, most owners, losing interest in on-site generation, removed their private generating equipment and accepted utility service. A few, however, did continue to generate electricity on site instead of, or in addition to, purchasing power, since utility service tended to be unreliable and expensive in some areas.

Today, fuel-fired generators are used extensively for emergency and standby power in hospitals and in other buildings housing critical processes. In some situations, they may also provide primary electrical power more economically than central utility service can. In many of those cases, the cost effectiveness is contingent upon using the by-product heat in some productive manner.

Total Energy and Cogeneration Systems

The coproduction of electricity (or power) and heat is one way to utilize energy input more fully in order to obtain the maximum total energy output. This can be accomplished by either generating electricity using heat recovered from some other process or generating electricity first and reusing the leftover heat for some productive purpose.

The term *total energy* (TE) was coined in the early 1960s to describe an independent system that produces electricity and heat simultaneously from the same energy source. Since that time, TE systems have been installed in shopping centers, industrial plants, commercial buildings, motels, hospitals, schools, and multifamily residential projects.

As an independent system, TE plants did not interconnect with their local electric utility, and often had to meet a facility's electrical demand regardless of the heat demand. Any excess heat was wasted by rejecting it through a cooling tower, river, or cooling pond or by simply releasing steam into the atmosphere. Any shortfall in heat had to be made up by auxiliary boiler equipment.

In addition, without a utility tie-in, users had to provide their own peak power and standby or backup provisions. This often meant redundant generating equipment (1) to supply the minimum (base) load, (2) to supply the peak load, and (3) for standby.

While a TE system implies isolation from utility electricity, the advent of PURPA has made it much more convenient to interface a combined electricity and heat production system with the utility. Such TE systems, which are linked to an electrical utility network in order to buy supplementary power and sell surplus power, are known as *cogeneration* or *combined heat and power* (CHP) systems (Figure 10.4).

FIGURE 10.4. CHP natural gas microturbine power plant at Harbec Plastics, Inc., in Ontario, New York. Heat recovered from the on-site generators is used to completely heat and air-condition the plastic injection molding production facility. The remaining power requirements are provided by a wind turbine and the electrical utility grid. The utility power is purchased from a retail green power provider. (Photo courtesy of Harbec Plastics, Inc., www.harbec.com.)

Topping and Bottoming Cycles

Cogeneration systems are classified into two categories: topping and bottoming cycles, as illustrated in Figure 10.5. The difference between them refers to where in the process the electricity is generated.

Topping Cycle.   When fuel is used to generate electricity first, and the leftover fuel energy not converted by the prime mover into shaft energy is then recovered as heat and used for some productive purpose, the system is known as a *topping cycle.* Counting that thermal use, the TE efficiency of the entire operation can be improved to a maximum of 85 percent.

The recovered steam or hot water can be used for space heating, space cooling (by means of an absorption chiller), or hot water requirements. Normally, the recovered heat is used as a supplement to a standard boiler system that provides for a facility's full thermal needs.

The recovered heat, which would otherwise be wasted, represents a cost savings since it takes the place of fuel that would have had to be paid for separately and burned to produce the necessary steam or heat. The total bill for electricity and heat is thus reduced.

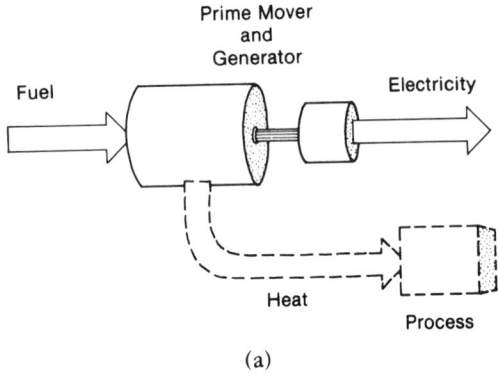

(a)

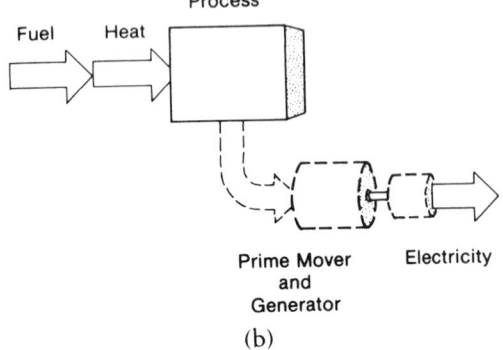

(b)

FIGURE 10.5. (a) Topping and (b) bottoming cycles.

All three types of prime movers—internal combustion engines, gas turbines, and steam turbines—are appropriate for this application.

Heat can be reclaimed from internal combustion engines by extracting it from the internal cooling system and from the exhaust. Engines normally require extensive cooling to remove the heat of combustion and the heat resulting from friction. Coolant fluids and lubricating oil are generally circulated through radiators (or cooling towers on larger systems) in order to remove the engine heat in much the same way that automobile engines are cooled.

The fluids can instead be circulated through heat exchangers to recover the heat. Jackets can also be placed around the engine manifolds to recover more of the waste heat. Finally, the hot engine exhaust can be passed through a heat-recovery boiler. Using any of these three methods, the engine heat can be converted into steam or hot water for some useful purpose. The overall effectiveness of energy use is thus increased to about 55 to 60 percent.

Most of the energy not converted to shaft power by a gas turbine is exhausted as heat. The heat is then recovered by passing the hot exhaust gases through a waste-heat boiler to produce steam, which in turn is used productively. In this way, the overall efficiency of fuel use for gas turbines can be improved to 75 to 80 percent.

The high-pressure steam that drives a turbine and generator to produce electricity becomes either low-pressure steam or hot water, which is available for productive use. The overall fuel efficiency using steam turbines can be as high as 85 percent.

In a topping cycle, the generation of electricity is the primary product, and useful heat is the by-product. A bottoming cycle is just the opposite.

Bottoming Cycle.   In a *bottoming cycle,* fuel is first burned to provide heat energy for some primary use. The still usable residual heat in the form of steam is extracted as it is exhausted from the process so that it can be used to drive a turbine that generates electricity or mechanical energy. Electricity is the secondary product.

If the heat comes from process discharge that would otherwise be wasted, the electric power is essentially free, and the only cost is for the equipment that must be purchased and installed.

Bottoming cycles are applicable where there is a primary need for high-pressure steam. Facilities for energy-intensive industries—such as chemicals, food processing, metal, mining, petrochemical, pulp and paper, refining, and textiles—fall into this category. Other good candidates are hospitals, shopping centers, and large university campuses.

Combined Cycle.

A variation of these two types is known as a *combined-cycle topping system*. It produces electricity first, uses the exhausted heat to fuel a heat-recovery boiler, and then applies the steam from the boiler to drive a steam turbine in order to produce more electricity.

Environmental Impact

Because TE and cogeneration systems use the same energy source for both electricity generation and thermal purposes, they consume one-half to one-third as much fossil fuel as that consumed by a central utility providing the equivalent amount of energy. Also, on-site generation eliminates transmission line losses resulting from large distribution networks.

By burning less fuel, such systems emit correspondingly less air pollution. Furthermore, on-site power generation shifts the location of air and thermal pollution sources. Central utilities usually have no handy use for exhaust heat and must reject it into streams, oceans, or the atmosphere. The more diffuse dispersion of air pollution and waste heat generally allows them to be more readily accepted by the environment.

On the other hand, in dense urban areas, local heat and air pollution sources can contribute to already existing pollution problems, while a central utility can consolidate the control and disposal of contaminants.

Architectural Impact

The differences in the equipment needed for a TE or cogeneration system and one without electricity generation capability are generally limited to the equipment room. No change is needed in chilled and hot water piping, fan coil units, absorption machines, or AHUs.

The ideal location for an equipment room housing generating equipment is in the center of a facility, since this minimizes the length of thermal and electrical distribution runs. Considerations of aesthetics, noise control, prime space requirements, and accessibility for service, however, may preclude a central location.

Generally, the same principles that apply to the design and location of central heating and cooling plants also apply to TE and cogeneration plants, since much the same equipment is used. However, TE and cogeneration plants typically require about 10 to 15 percent more space than a conventional plant.

The central plant houses all the prime movers, generators, electric switchgear, waste-heat recovery equipment, boilers, chillers, pumps, and auxiliaries. Cooling towers to dissipate excess heat or other outside components should be nearby, if possible, in order to keep the piping runs short. Air intake louvers and exhaust openings are required for combustion engines. If any fuel is used other than natural gas piped in directly from a central utility, fuel storage must be accommodated.

Sound and vibration are controlled in the same manner as for a conventional heating and cooling plant, although there is more noise from a generating plant. Obviously, areas sensitive to noise should not be immediately adjacent to the equipment space.

Economics

TE and cogeneration systems typically raise construction costs 2 to 5 percent compared to conventional heating and cooling sources. Under the proper conditions, the added first cost can be quickly amortized through operating cost savings. The system should not only pay for itself in a short period, but thereafter should represent continuous financial gain.

As an alternative, a recent innovation promises to practically eliminate the first-cost obstacle to the installation of cogeneration equipment. Package cogeneration systems are now available in 40- to 500-kilowatt modules that include prime mover, generator, and heat recovery equipment. This has enabled cogeneration systems to be offered on a lease or third-party ownership basis.

Instead of buying the expensive equipment outright, a building operator contracts to purchase electricity and heat from an on-site cogeneration system that is owned, operated, and maintained by a third party. The building operator can enjoy substantial energy cost savings without the responsibility and worry about the system's operation.

Balance Between Heat and Electricity Demands

Analysis of the magnitude, duration, and coincidence of electrical and thermal loads, and the selection of prime movers and waste-heat recovery equipment, determine both the project feasibility and the design. Full utilization of the excess heat determines the overall system efficiency and is one of the critical factors of economic feasibility.

A typical internal combustion engine may produce electricity and hot water in a ratio of 1:1 or 1.5:1. A small gas turbine, being an even less efficient producer of electricity, has a ratio closer to 0.7:1.

In comparison, a commercial office building is more likely to have a 6:1 or 8:1 ratio of electricity to thermal energy requirements. Unless a facility needs extremely large quantities of hot water—for example, for an on-site laundry—its thermal load in relation to the system's thermal output is the limiting factor.

Fuel cells, when they become available, will offer a more promising solution with their 3.5:1 ratio of electricity to thermal output. This ratio more closely matches the load balance of most buildings than the three principal prime movers.

The development of absorption refrigeration equipment that can productively use reclaimed heat during the summer was an important factor stimulating renewed interest in on-site power production and TE systems. Cooling, formerly something of a luxury, is now considered essential in many buildings. If air conditioning is not planned for climatic, economic, or other reasons, then cogeneration is unlikely to be economical unless there are substantial hot water or process loads to take advantage of waste heat in the summer.

Determining Criteria

The practicality of cogeneration is heavily dependent on specific local conditions, building and system design, how the plant will be operated, and local fuel and electricity rates. High and fairly constant electric power demand over a large portion of the day and over most of the year is most desirable. Generating equipment maintains its highest efficiency when operated at constant demand levels over long periods of time. The amount of energy saved is proportional to the amount of energy used, and a large savings is needed to justify the high initial installation expense.

Economic feasibility requires that the combustion fuels be inexpensive enough to compete with the prevailing electric rates. This refers to a comparison based on dollars and not Btus. A rule of thumb is that when the first digit of the fuel cost (in dollars per million Btu) is one-half of the first digit of the utility electricity rate (in cents per kilowatt-hour), cogeneration should be considered.

The estimating of electrical and thermal loads is very important for identifying the scale of the project, and the size of the equipment can have a great impact on integration with project design. It can change the whole aesthetic nature of the building or determine location and space requirements for equipment.

The following conditions can be used as prerequisites to judge whether a cogeneration system *may* prove feasible:

- Building floor area greater than 50,000 ft² (15,000 m²).
- Occupancy greater than 40 hours per week (additional hours will enhance the economic feasibility).
- Building program that includes both heating and cooling.
- Gas or oil rates low enough to compete with electric rates.

- Skilled operating personnel available in-house or by contract.
- Owner's fiscal policy that permits payback periods in excess of 8 years.
- Adequate space available to accommodate necessary equipment.
- Electrical and thermal loads in appropriate balance (must have use for at least 50 percent of the waste heat).
- Where a number of buildings with different functions can be served together, the diversity of loads may improve the load balance and thereby enhance the economic feasibility.

If the architect or owner determines that a building has good potential on the basis of the above guidelines, the next step is a thorough cost-effectiveness analysis by an experienced engineer. If a project seems at all a likely candidate, the cost-effectiveness analysis is very worthwhile. The cost of fossil fuels continues to rise, but as fuel costs escalate, most utility electric rates will most likely go up at least as much. However, the proper evaluation of conditions and the analysis of feasibility require the attention of qualified engineering advisors.

Other Advantages

One great advantage of cogeneration is that it can be considerably more reliable than utility sources in locations where blackouts or brownouts occur. On the other hand, most central utilities have a diversity of generating sources. This can be an advantage when certain fuels are in short supply.

Another potential advantage of on-site generation is that it provides the opportunity for producing power at frequencies other than the standard 60 Hz. High-frequency lighting powered by high-frequency electricity is more efficient than the standard arrangement. High-frequency (420 or 840 Hz) power for lighting may be produced by one generator, while standard-frequency (60 Hz) power is produced by another generator on the same drive shaft. DC power can also be produced by itself or in tandem with AC power.

MIUS

Cogeneration can be a component of what is called a *modular integrated utility System* (MIUS), illustrated in Figure 10.6. A MIUS, also called an *integrated utility system,* is designed to link together the five basic utilities—electricity, heating and cooling, solid waste, sewage, and potable water—so that the waste generated by one can be used as input to another.

Normally, when large building complexes or communities are built, power, fuel, trash collection, water treatment,

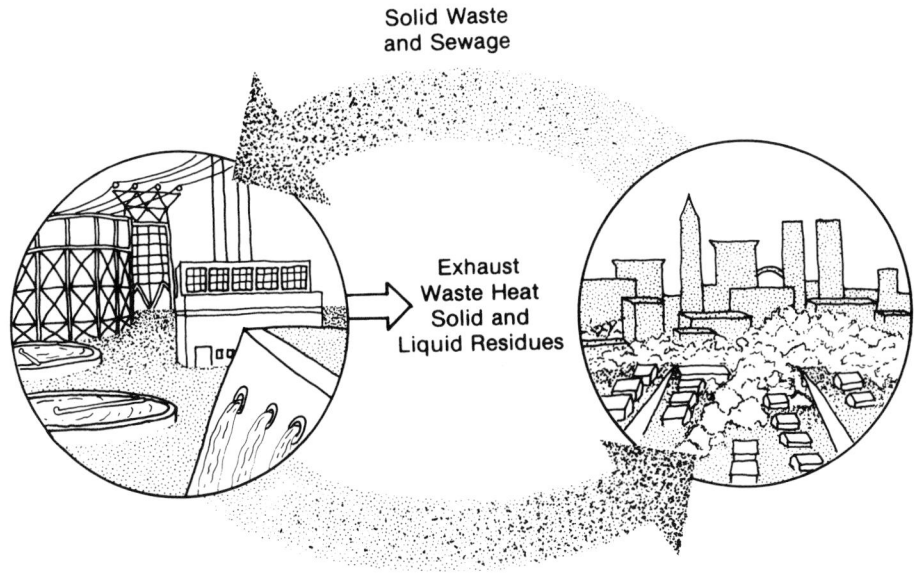

FIGURE 10.6. MIUS.

and other services are provided externally, each with its own energy requirements. The MIUS concept, in contrast, provides all these utility services onsite in an integrated system.

The word *modular* indicates that the system is located near the appropriate users and installed in coordination with the actual needs of the facility or community. The *integrated* approach means that the requirements of one service are met by using the effluent of another process.

For example, heat from a solid waste incinerator might replace or augment a boiler that produces steam for a turbine used to generate electricity. The residual heat could be recovered and used for hot water, space heating and cooling, water treatment, etc. Treated waste water might be recycled for cooling tower use, subsurface irrigation, and other secondary purposes, or, if the public would accept it, for potable use. Sewage could be treated in a biological treatment plant, in which case bioenergy recovery might be an added bonus, or in a physical/chemical system. Facilities could utilize thermal storage of hot and/or cold fluids to assist in load leveling.

All this, of course, requires careful management and coordination. MIUS systems require constant monitoring, valves need to be opened and closed fairly frequently, and adjustments are necessary as conditions change. Due to the complexity involved, a MIUS is typically operated by a minicomputer with process control capabilities.

The objectives for a MIUS are to provide utility services

at a lower installation and operating cost, while reducing the use of natural resources and minimizing the environmental impact. But the concept has other benefits, too. It enables utility services to be brought on-line in about half the normal time. A self-contained utility system, independent of the existing infrastructure, allows more flexibility in site selection and can expand the land-development options. If public services are in short supply, a MIUS can make a project more acceptable to local governing bodies. And finally, a MIUS, with its lower consumer utility costs and more reliable service, can make a development more attractive to its ultimate occupants.

MIUS applications are appropriate for moderate-size communities or large building complexes. They are best suited for high-density developments or redevelopments. The types of utilities to be integrated and the technologies employed depend on factors such as climate, population density, total population, consumption patterns, local geology, space availability, and access to water.

WIND TURBINES

Devices that convert the power in the wind to electrical power are referred to by many names. The term *windmill* originally referred to wind-powered mills used for grinding grain into flour or for sawing wood. A *wind generator* implies a machine that produces wind, that is, a fan. More accurate terms in use are:

- Wind turbines
- Wind-driven generators
- Wind-driven power plants
- Wind-driven electric power systems
- Wind energy conversion system (WECS)

The last one was established by the U.S. government as the official term for government use. All of these terms are clumsy, so most people—including some avid wind-power proponents—at times slip into using *windmill*. Despite the insistence of some people that one term or another is the only accurate one, there is a universal understanding of the simple word *windmill*.

Wind energy has been utilized for thousands of years to propel sailboats, grind grain, and pump water. It has been used intermittently to generate electricity for over 75 years. During the 1930s and 1940s, hundreds of thousands of small-capacity wind-electric systems were successfully used on remote farms and homesteads until the spread of the rural electrification program in the early 1950s underpriced them.

Once the capital investment for the machinery is made, the operating cost for wind power is essentially free. However, the cost of the machinery is so high that, even at today's electric utility rates, the investment would not be amortized by utility cost savings for a long time. In industrialized areas, wind power can only be economically justified on the basis of anticipating rapidly rising utility electric rates.

Aside from anticipating a rapid escalation of utility rates in the future, the motivations for installing wind turbines are (1) to minimize the environmental impact, (2) to achieve energy independence, and (3) being a wind-power enthusiast.

If the site is remote, without present electricity service, wind power is likely to be more economical than either (1) installing and buying oil or natural gas for an engine generator set or (2) extending utility power at $5,000 to $10,000/mile ($3,000 to $6,000/km). A rule of thumb is that if wind speeds average greater than 10 mph (5 m/s), and if at least 1/2 mile (1 km) of new power lines need to be paid for, then consider wind power. The more expensive electricity rates are, the more economical wind power is in comparison. Under almost any circumstances, wind speeds below 8 mph (3½ m/s) are not worth trying to use.

Wind Variation with Location

Topography determines the local wind speed. Constricting the wind increases its velocity in certain areas while shelter-

ing other areas from the wind (see Figures 3.1 and 3.2). Since the same volume of air must pass, except for slight air compression, the velocity increases as the cross section decreases. Wind speed also increases with height above the ground. The behavior of wind around obstructions, illustrated in Figure 10.7, can become quite complex for complicated structures.

The power contained within the wind is proportional to the wind speed cubed. But a windmill cannot extract all of the energy from the wind, since that would require that the air downstream come to a complete standstill. A windmill derives the maximum amount of power when the wind is slowed to one-third of its initial velocity, in which case the power output is about 60 percent of the power contained in the wind.

If the wind velocity is known, the maximum power, P, theoretically extractable from the wind with a perfectly efficient machine can be calculated by the following formula:

$$P = 0.0024 \, D^2 V^3 \qquad (10.1)$$

where P is in watts, D is the propeller diameter expressed in feet, and V is the wind speed in miles per hour. Any real wind turbine will produce less power than this because of friction and other limitations.

The structure supporting a windmill should be as open as possible while still having the structural strength to withstand the wind loading, since otherwise it can interfere with the wind and reduce its potential.

Types of Wind Turbines

Machines for tapping wind power are classified as either vertical axis or horizontal axis. A horizontal axis type is illustrated in Figure 10.8. The more vanes or blades included, the slower the rotation and the higher the torque (turning force). Multivaned windmills are good for pumping, while a two- or three-blade machine can produce the high rpm (revolutions per minute) suitable for electricity generation.

Architectural Integration

Wind machines are commonly mounted on high towers located some distance from buildings and not downwind of structures that would interfere with free wind patterns. They can be mounted atop a building, and in some cases the increase in wind speed as it moves around the building can be used to advantage.

Air patterns around projections into the wind stream are complicated and difficult to predict. It is best to measure the

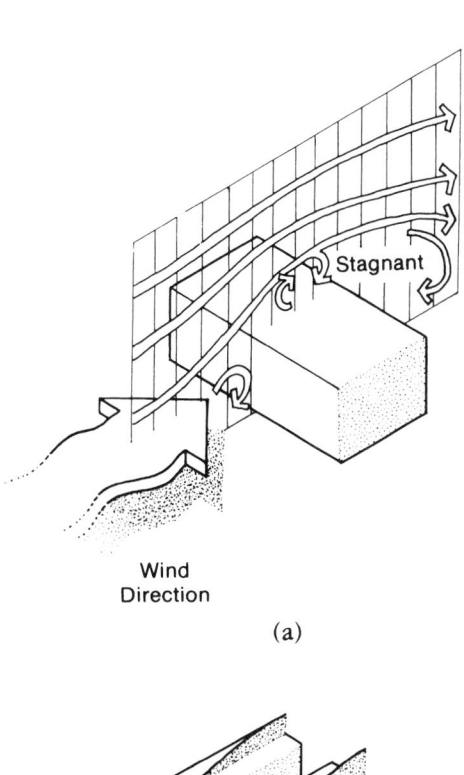

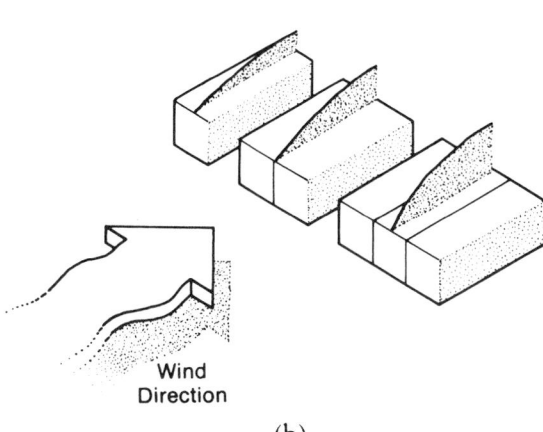

(b)

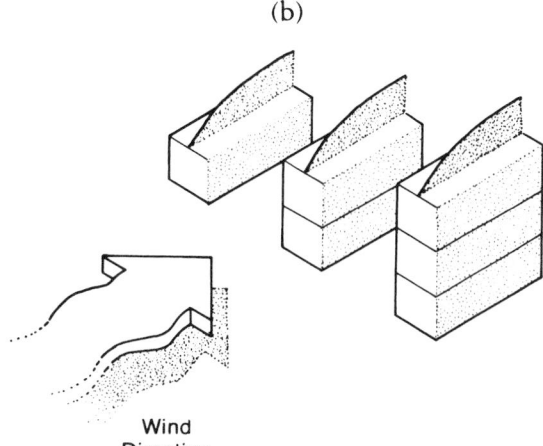

FIGURE 10.7. Wind patterns

FIGURE 10.8. A 250-kilowatt Fuhrlaender wind turbine at the Harbec Plastics plant in Ontario, New York. The wind power system provides 20 to 25 percent of the company's electricity requirements and is estimated to pay back its capital investment in 6 to 8 years. (Photo courtesy of Harbec Plastics, Inc., www.harbec.com.)

wind speeds and direction at the exact proposed location and height for as long a period as possible.

Matching Supply with Demand

One underlying problem with utilizing wind power, regardless of the cost of the windmill itself, is that the demand for electricity may not necessarily occur when the wind is blowing, since the wind is uncontrollable and unpredictable. The noncoincidence of supply with demand necessitates a storage system for wind power systems. A basic system including battery storage is shown schematically in Figure 10.9.

The implications of battery storage, however, involve more than just the additional cost of the battery. The process of storing and unstoring electricity in a battery represents a conversion of energy from electric to chemical and back to electric. As with any energy conversion, this process results in a loss of energy.

The inverter, which converts DC to AC, adds further losses. It is possible to generate AC power initially, but it cannot be stored. In order to have storage capability, DC power must be produced.

If the need is for heated water, the storage problem can be circumvented by electrically heating water concurrently as the electricity is being generated. This avoids the battery storage losses, and the hot water can then readily be stored. This concept is referred to as a *wind furnace.*

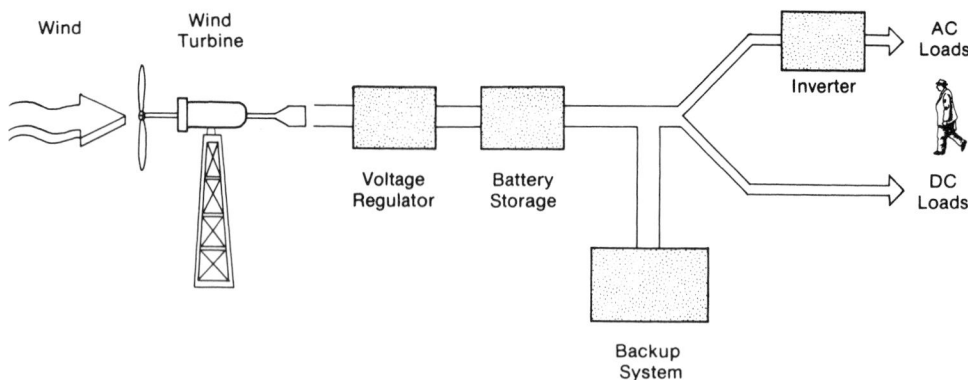

FIGURE 10.9. Diagram of a wind power system.

Pumping, heat storage, cold storage, or any other function that can accommodate an interruptible or variable power supply are the most efficient uses of wind power because they avoid both the cost and inefficiency of storage. When steady high winds occur in the winter, space heating is an especially practical application.

Another storage possibility involves the use of a flywheel. It conserves rotational energy for later electricity generation. The only losses are due to friction as the flywheel turns, and these can be minimal. The reason flywheels are not used is that a large mass is required in order to store a significant amount of energy, and the high centrifugal force due to the large mass ends up pulling the flywheel apart. Materials able to withstand such high centrifugal forces are under development.

Hydrogen production through wind-driven water electrolysis is currently being researched and experimented with. Hydrogen can be stored and then be used for fuel in a fuel cell as needed.

A final potential storage means is compressed air storage. A wind-powered air compressor can pressurize air. Later, as needed, the air can be released through a turbine generator to produce electricity.

SOLAR PHOTOVOLTAIC CELLS

Photovoltaic solar cells have been in use since they were developed for the space program in 1954.

All photovoltaics are DC and operate in a circuit just like a battery. One terminal is always positive and the other negative. Sufficient light or heat shining on the cell creates a voltage difference, which in turn causes current to flow when the cell is connected into a circuit. There is no voltage difference in the dark or at low temperatures.

All silicon (Si) cells are 0.45 volts per cell regardless of size. The maximum current varies according to the manufacturing technique and cell surface area. The maximum current of silicon cells is typically 500 milliamperes to 1.2 amperes or more; the actual current output is directly proportional to the illumination level, ranging from zero in the dark to its maximum in direct sunlight.

If connected to both a battery and a load, the cell will power the load and charge the battery when the sun is shining, and let the battery power the load when the sun is not shining. Electronic controls are needed to keep the voltage and current to the load uniform and to prevent overcharging the battery.

The photovoltaic process can only convert part of the sunlight into electricity; the rest is left over as heat. The more energy converted to electricity and the less heat produced, the more efficient the cell.

The theoretical maximum efficiency for this process is 22 to 23 percent. The current production-line cells are 10 to 12 percent efficient, and laboratory models are reaching 16 to 18 percent efficiencies.

The actual voltage of a solar cell is dependent upon the operating temperature; it drops slightly with higher temperatures. For this reason, photovoltaic cells are best assembled in an arrangement that cools the cells and uses the heat for useful purposes, similar to a TE plant concept. Besides recovering waste heat, this makes the cell more efficient and powerful.

In the future, solar energy may be a practical source for large-scale TE plants. Photovoltaic cells may be used in the manner just described, or solar heat may be collected to drive a steam turbine with recovery of the leftover heat.

Silicon solar cells are made of a semiconductor crystal. They are currently produced by carefully growing large single crystals of silicon for maximum conductive uniformity and purity. The crystals are then sliced into thin wafers which serve as the base for the cell production.

The manufacturing methods generally produce circular cells. A typical silicon cell is 3 inches (7.6 cm) in diameter. A number of cells are commonly grouped together into panels, as shown in Figure 10.10.

This production process is time-consuming and delicate, which accounts for the high cost. People in the industry believe that larger-scale production can result in more economical production methods and lower product cost. To stimulate quantity production, the U.S. government has bought large amounts of cells for demonstration projects.

Other materials are used as photovoltaics besides silicon. Silicon happened to produce a high efficiency at an affordable cost for the space program, so most work has been done with it. Another promising product, though, is the gallium-arsenide cell. It is less efficient but much less costly than silicon. Thus, it is actually more cost effective.

The number of expensive photovoltaic cells, and hence the cost of an installation, can be substantially reduced by utilizing some means of concentrating a large area of sunlight onto a smaller receiving area. This can be accomplished by either the reflection or focusing of the sunlight onto small areas of cells. The *concentration ratio,* as illustrated in Figure 10.11, is the total aperture area open to incoming sunlight divided by the area of the cells finally receiving the energy. Concentration ratios of up to 1,000:1 are practical.

As with wind power, some means of storage is usually required in order to couple loads with the available resource. This means is usually batteries, but could be hydrogen production. A typical system including battery storage is diagrammed in Figure 10.12.

In order to minimize storage costs, it is important to match the available supply with the demand for power. Electrically powered air conditioning is a good match because in general, the hotter the weather, the greater the cooling load and the more sunshine.

Interruptible demands are an efficient application because they eliminate the need for storage with its inevitable losses. The maximum solar utilization occurs when the energy can be used whenever it is available but is not required otherwise.

Some technical problems with photovoltaics that still need to be resolved are:

1. Improving the efficiency (the goal is 15 to 20 percent).
2. Reducing the cost of storage ($30/kWh is needed).
3. Fulfilling the need for some method of encapsulation that is low in cost, impervious to atmospheric contaminants, highly transparent, and nondegrading. Glass will be used as long as no less expensive material is developed.
4. Assurance of reliability for a minimum system lifetime greater than 20 years.

FIGURE 10.10. An array of 144 photovoltaic panels mounted on the state capitol building in Olympia, Washington. The panels generate 20 kilowatts of electricity, enough to light the dome and lantern of the Legislative Building from dusk to midnight every day. To avoid the expense of power storage, the solar-generated electricity is fed into the utility grid to offset the electricity purchased to run the lights in the dome and lantern. The panels—each measuring 63 inches (1.6 m) long and 31 inches (0.79 m) wide—were integrated into this historic building by placing them on the southern side of the fifth-floor roof and positioning them nearly parallel to the horizontal roof surface. Since the array is only about 1 foot (30.5 cm) high, it cannot be seen from the ground or nearby buildings. Concealing the futuristic-looking panels allowed the project to meet federal standards for historic preservation. (Photo courtesy of Michael L. Dean.)

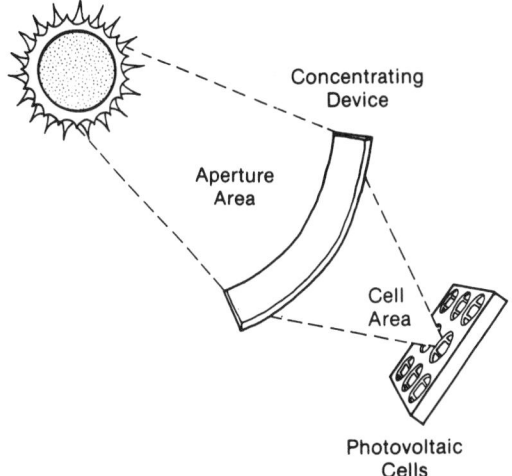

FIGURE 10.11. Concentration of sunlight on a photovoltaic cell.

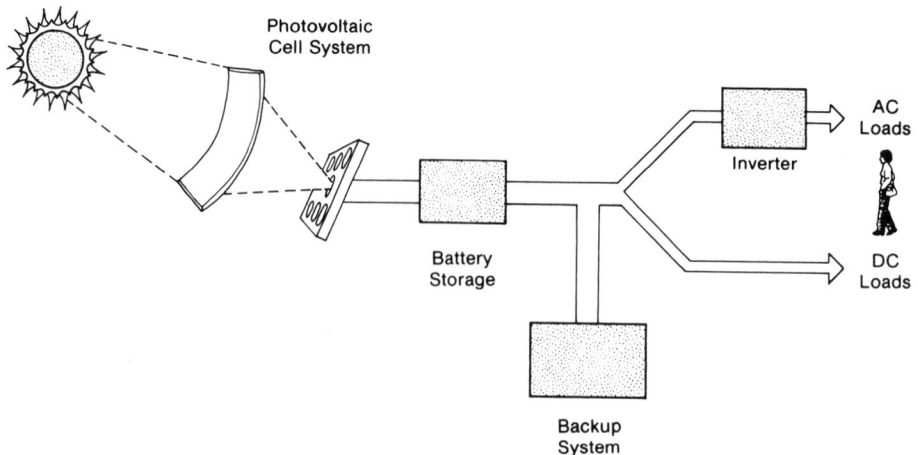

FIGURE 10.12. Schematic diagram of a photovoltaic power system.

HYDROPOWER

Hydroelectric power generation refers to the generation of power from flowing and falling water. *Small-scale hydro* is defined as a generation facility with a capacity under 15 megawatts. *Micro-hydro* is a hydropower facility with a capacity under 100 kilowatts.

Hydropower can provide great potential for generating electricity, and the technology is relatively simple. But a ready source of nearby flowing water is needed. Even though the ideal conditions necessary are pretty rare, there is a growing number of installations in the United States. Potential applications include industrial facilities, universities, and small communities.

The prime movers for utilizing hydropower are divided into two categories: turbines and water wheels. Both water wheels and turbines deliver their power as torque on a shaft. Pulleys, belts, chains, or gear boxes are connected to the shaft to deliver power to such things as grinding wheels, compressors, pumps, or electric generators.

Water Turbines

Water turbines are preferable for producing electricity from rivers and streams. Generally small in diameter and high in rpm, these devices are driven by water flowing under pressure through a pipe or nozzle. Such a turbine is typically coupled with an electricity generator.

A water turbine is basically a device that converts the energy in falling water into rotating mechanical energy. This energy, available through a rotating shaft, may be either used directly to operate mill and grinding equipment or linked to a generator to produce electricity.

Water turbines are used for producing either DC or AC electricity. A typical hydroelectric installation using a water turbine is pictured in Figure 10.13.

Water Wheels

Water wheels are the old-fashioned, large-diameter, slow-turning devices that are driven by the force of flowing water. Because they only turn at 2 to 12 rpm, they are better suited for producing mechanical power for grinding and pumping than for producing electrical power. The mechanical power can also be connected by pulleys and belts to directly drive lightweight process machinery or tools, such as lathes, drill presses, and saws. Water wheels can also provide a limited amount of electrical power by "gearing up" to the necessary generator speeds, but this is not a very efficient process.

Water wheels offer some advantages over higher-speed turbines. First, the technology is simpler and less expensive, and they can be site built. Water wheels can offer high torque and are thus capable of driving heavy, slow-turning mechanical equipment. They also require minimal maintenance and repair.

The disadvantages of water wheels are that they are significantly less efficient than higher-speed turbines, they operate at slow speed, and they are bulky. They also need to be housed in fairly large structures or otherwise protected against freezing in cold climates if year-round operation is required.

The two principal types of water wheels are the *overshot* and the *undershot* models, as illustrated in Figure 10.14. The open channel that transports water from behind the dam to

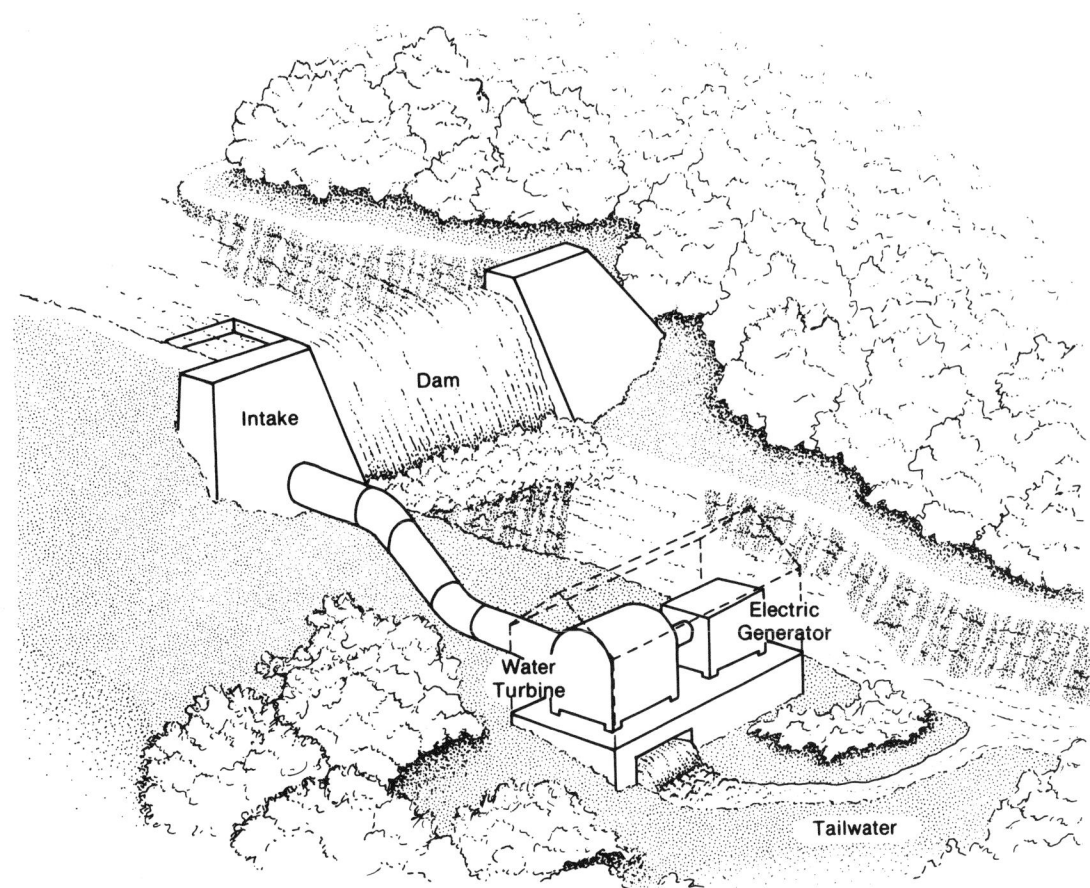

FIGURE 10.13. Dam and water turbine.

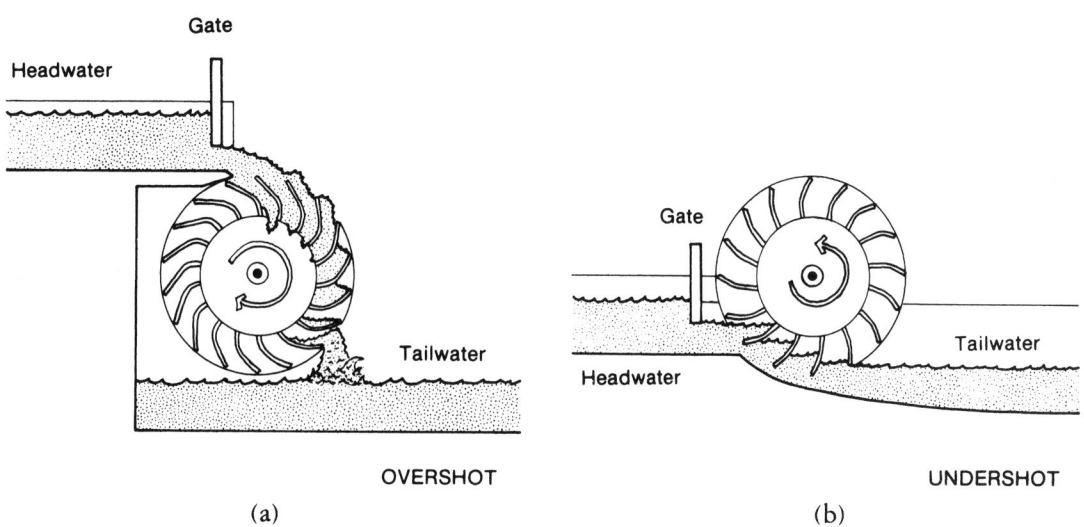

OVERSHOT

UNDERSHOT

(a)

(b)

FIGURE 10.14. Water wheels. (a) Overshot and (b) undershot.

the water wheel is known as a *sluice*. A larger channel is called a *canal*. A turbine requires a closed pipe for transporting the water under pressure.

Available Power

When a potential hydropower site is discovered, the first step is to assess the amount of power available from a river or stream. The amount of hydropower available is determined by (1) the quantity of water (flow rate) and (2) the drop or change in elevation (head) along the water course.

The *flow rate* is the quantity of water flowing past a point in a given period of time. Flow rates are typically measured in gallons per minute (gpm), cubic feet per minute (cfm), cubic meters per hour (m^3/hr), or liters per second (L/s). The *head* is defined as the vertical height (in feet or meters) that the water drops. These two concepts are illustrated in Figure 10.15. The power that can be developed at a site is a function of the flow rate multiplied by the head.

All water courses have a variation in flow. There may be daily as well as seasonal differences. The normal flow and the minimum flow must be determined either by checking documentation or by physical measurement in order to adequately assess the minimum continuous power output.

It is also important to determine what portion of the flow can be used for power generation. The percentage of the minimum flow that can be diverted for power generation must take into consideration the needs of fisheries (fish movement up and down the stream) and questions of aes-

thetics. Some people suggest that only 25 percent of the dry season flow be used for power generation. This depends upon the actual circumstances.

The *theoretical power* available, P_{th}, in horsepower and kilowatts is given by the following equations:

$$\text{horsepower: } P_{th} = \frac{Q \times h}{529} \qquad (10.2)$$

$$\text{kilowatts: } P_{th} = \frac{Q \times h}{709} \qquad (10.3)$$

where

$$Q = \text{usable flow, in cfm}$$
$$h = \text{head, in feet}$$

The theoretical power available represents more power than can be obtained from any real equipment. Any machinery or other equipment used to convert the power available in the flowing water into mechanical shaft power or electricity is less than 100 percent efficient. Water wheels range from 25 to 75 percent efficient, while turbines are 60 to 85 percent efficient. Water wheels are sometimes used for micro-hydro projects, but due to the high capital involved in small-scale hydro projects, the more efficient turbines are normally used to create the highest possible return for the investment.

The water wheel or turbine is called the prime mover. In addition to the efficiency of the prime mover, the efficiency of the belt drive or gear box that transmits the power from the water wheel or turbine to the generator also needs to be taken into account. Transmission efficiency can be approxi-

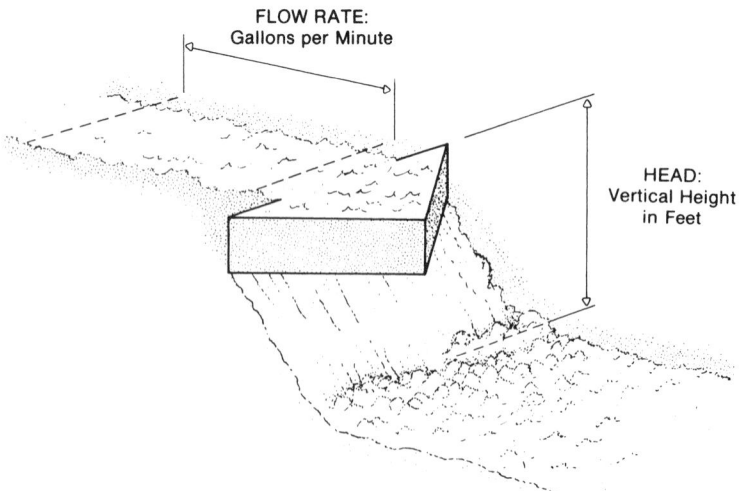

FIGURE 10.15. Concepts of hydropower head and flow.

mated at 95 percent. The efficiency of the generators themselves is about 80 percent.

The overall efficiency, then, is the product of all these efficiencies: prime mover multiplied by transmission and generator efficiencies. Typical overall efficiencies for electrical generation systems can vary from 70 percent for a high-head, high-speed turbine to 50 percent for more economical, lower-efficiency turbines. Overall efficiencies of systems using water wheels are usually well under 50 percent.

For example, diverting 25 percent of a river with a flow rate of 10,000 cfm and a net head of 100 feet would result in a theoretical power available of

$$P_{th} = \frac{0.25 \times 10,000 \times 100}{709} = 353 \text{ kW}$$

The actual power available, however, would be

353 kW × 0.8 (turbine) × 0.95 (transmission)
× 0.8 (generator) = 215 kW

The energy potential available at a hydro site is a function of the head and flow. Sites with a net head of more than 60 feet (18 m) are generally considered high-head sites. It is generally agreed that a head of 3 feet (1 m) is the absolute minimum needed to develop hydro power, but a site with a head of less than 10 feet (3 m) is probably not economical to develop.

Because low-head projects require large amounts of water and thus physically large equipment, the resulting unit cost of power generated will be higher than for high-head sites with similar power potential. For this reason, high-head sites are preferred. However, under the proper circumstances, low-head systems can also be practical.

Economics

The cost of the turbine or water wheel is only a portion of the overall cost of installing a hydropower system. Most hydroelectric installations require the construction of a dam, which also costs money.

The purposes of a dam are to (1) divert the river or stream flow to the turbine or water wheel, (2) store the energy of the flowing water, and (3) increase the reliability and power available from a stream or river. It can provide a means by which to regulate the flow of water and can raise the elevation and head of the water by making it deeper. In addition to helping to develop power, a dam may be useful for fire protection or irrigation needs.

The expense of a dam may be decreased by using indigenous materials, but more expensive concrete and steel rein-forcing methods are usually needed. Many hydroelectric projects involve refurbishing old dams below the 5-megawatt size, which originally supplied power to the nation's communities and industries. This saves a tremendous amount of money.

Other costs are for pipes or sluices to carry the water to the wheel or turbine. Aside from the dam and wheel or turbine, the largest expense is the gears or pulleys, shafting, and electrical generator. The pulleys, drive belts, and chains or gears are needed to "gear up" the power of the turbine or water wheel. As it transfers the power, the transmission converts the low rpm produced at the turbine or water wheel shaft to the appropriate rpm for the intended purpose.

One characteristic difference between hydropower and the other methods of on-site generation presented is that, except for seasonal variations, hydropower is available in a fairly consistent quantity. Engine generators and steam turbines using fossil fuels are usually run only as needed because the energy is stored in the fuel. Wind and solar photovoltaic power are intermittent and unpredictable resources.

Since hydropower is so consistent, it is available round the clock every day, whereas a facility's load is likely to vary or stop entirely some of the time. Thus, excess power is generally available to sell to the local utility. In most cases, however, the excess power is available during the utility's off-peak period, so it is of less value. Consequently, utilities pay lower rates for this off-peak power.

Environmental Impact

Turbines and water wheels in themselves have a negligible effect on the environment. However, the damming of a river or stream is necessary for most installations, and this has an important and sometimes irrevocable effect upon the long-term ecological balance of a particular environment.

Dams can create a better environment for some animals and plants. But by building a dam, a pond or lake is created where a stream or river used to exist. An already existing river ecosystem is flooded, encouraging the accumulation of silt and perhaps providing a breeding ground for mosquitoes. Any alterations made to water flow could harm the wildlife or fish in an area and may interfere with someone else's use of that same water downstream.

The resulting lake or pond also usually raises the water table behind the dam as a result of seepage and lowers it below the dam. Innumerable other changes occur as a result of the construction of a dam, and in general, the larger the dam, the greater the changes. Therefore, a specific environmental impact study should be undertaken before a final decision is made to build a dam.

Hydroelectric development is very site-specific. Each town, municipality, and state has different regulations and licensing and permit procedures. Access rights to the water and right-of-way for any needed pipeline across adjacent properties are crucial to a project.

Once it is established that a potential exists for a hydropower installation, the next step is to have a full feasibility study made by a qualified engineer. The study must consider the technological and economic feasibility, water rights, and any other legal or regulatory barriers that need to be overcome.

SUMMARY

Electricity is generated on site for emergency or standby power or as an alternative to expensive utility-produced power. It can also be used for peak-demand shaving in order to limit the peak kilowatt demand.

The methods used to generate electricity on site are thermally driven prime movers, wind turbines, solar photovoltaics, and hydropower. The first category includes internal combustion engines, gas turbines, and steam turbines. The steam turbine is the most versatile in terms of energy sources. It can use steam produced from liquid, solid, and gaseous fuels as well as from geothermal resources and high-temperature solar collection.

Wind turbines can be fast-moving two- or three-blade machines designed for generating electricity or slower multivane ones meant for direct connected pumping. Hydropower is represented by either water turbines for generating electricity or water wheels for pumping or for operating direct-connected machinery.

TE systems simultaneously produce electricity and heat as an integrated process. When the system is interconnected with an electric utility in order to enable a facility to buy and sell electricity, it is called a cogeneration system. This was made possible by the federal PURPA legislation.

TE or cogeneration systems can be either a topping or bottoming cycle type. The topping cycle generates electricity first and then makes the residual heat available for another use. A bottoming cycle is just the opposite, creating heat first for some process and then using the waste heat to generate electricity.

TE or cogeneration systems can be incorporated into a MIUS. A MIUS provides the five basic utilities—electricity, heating and cooling, potable water, sewage treatment, and solid-waste disposal—in one integrated on-site package in which the requirements of each process are met by the by-products of another.

Whichever method is used, a facility should reduce its loads as much as is economically feasible before adding on-site generation.

KEY TERMS

Electrolysis
Public Utility Regulatory Policies Act (PURPA)
Avoided costs
Prime mover
Fuel cell
Internal combustion engine
Engine-generator set
Gas turbine
Steam turbine
Total energy (TE)
Cogeneration or combined heat and power (CHP) system

Topping cycle
Bottoming cycle
Modular integrated utility system (MIUS)
Windmill, wind generator, wind turbine, wind-driven generator, wind-driven power plant, wind-driven electric power system, wind energy conservation system (WECS)
Wind furnace

Photovoltaic solar cell
Silicon solar cell
Concentration ratio
Hydroelectric power generation
Small-scale hydro
Micro-hydro
Water turbine
Overshot water wheel
Undershot water wheel
Canal
Flow rate
Head
Theoretical power

STUDY QUESTIONS

1. Describe a set of circumstances (including economic factors, site conditions, and type of facility) under which each of the following would be good prospects for on-site generation:
   a. Cogeneration
   b. MIUS
   c. Wind power
   d. Solar photovoltaic cells
   e. Hydropower
2. What is the maximum power obtainable from a wind turbine with a 50-foot-diameter propeller in a 10-mph wind?
3. How many 1-amp silicon cells would be needed to produce a maximum of 1 kilowatt? How much area would that many cells take up?

4. Approximately how much electrical power could be expected from a hydroelectric installation on a river flowing at the rate of 20,000 cfm and having a net head of 50 feet? About 25 percent of the river can be diverted for this purpose, and the overall system efficiency is 60 percent.

BIBLIOGRAPHY

Total Energy and Cogeneration

American Society of Heating, Refrigerating, and Air-Conditioning Engineers, Inc. *Cogeneration Design Guide.*

ASHRAE Handbook: Systems and Equipment. American Society of Heating, Refrigerating and Air-Conditioning Engineers, Inc.,

Wind Turbines

Boyle, Godfrey. *Renewable Energy,* 2nd ed. Oxford University Press, 2004.

Burton, Tony, David Sharpe, Nick Jenkins, and Ervin Bossanyi. *Wind Energy Handbook.* John Wiley & Sons, Inc., 2001.

Calvert, N.G. *Windpower Principles: Their Application on the Small Scale.* Butterworth-Heinemann, 1979.

Ewing, Rex. *Power with Nature: Solar and Wind Energy Demystified.* Pixyjack Press, 2003.

Gipe, Paul. *Wind Energy Basics: A Guide to Small and Micro Wind Systems (Real Goods Solar Living Book).* Chelsea Green Publishing Company, 1999.

Gipe, Paul. *Wind Power: Renewable Energy for Home, Farm, and Business,* revised and expanded ed. Chelsea Green Publishing Company, 2004.

Manwell, J.F., J.G. McGowan, and A.L. Rogers. *Wind Energy Explained.* John Wiley & Sons, Inc., 2002.

Spera, David A. (editor). *Wind Turbine Technology: Fundamental Concepts of Wind Turbine Engineering.* American Society of Mechanical Engineers, 1994.

Solar Photovoltaic Cells

Komp, Richard J. *Practical Photovoltaics: Electricity from Solar Cells,* 3.1 ed. Aatec Publications, 1995.

Markvart, Tomas (editor). *Solar Electricity,* 2nd ed. John Wiley & Sons, Inc., 2000.

Messenger, Roger A., and Jerry Ventre. *Photovoltaic Systems Engineering,* 2nd ed. CRC Press, 2003.

Patel, Mukund R. *Wind and Solar Power Systems.* CRC Press, 1999.

Photovoltaic System Design. SolarVision Publications, 1982.

Pulfrey, David L. *Photovoltaic Power Generation.* Krieger Publishing Company, 1978.

Sklar, Scott. *Consumer Guide to Solar Energy,* 3rd ed. Bonus Books, 2002.

Solar Energy International. *Photovoltaics: Design and Installation Manual.* New Society Publishers, 2004.

Thumann, Albert. *Plant Engineers and Managers Guide to Energy Conservation,* 8th ed. Marcel Dekker, 2002.

Turner, Wayne C. (editor). *Energy Management Handbook,* 4th ed. Marcel Dekker, 2002.

Watts, R.L., S.A. Smith, and R.P. Mazzuchi. *Photovoltaic Product Directory and Buyers Guide,* 2nd ed. Van Nostrand Reinhold Company, 1984.

Hydropower

The best resources in the field of hydropower were books published prior to 1930, which can only be found in libraries.

Bureau of Reclamation. *Water Measurement Manual.* Superintendent of Documents, U.S. Government Printing Office, 1987. (Techniques for measuring flow rates with weirs, flumes, gates, pipes, and orifices.)

Department of the Interior. *Design of Small Dams.* U.S. Government Printing Office, 1987.

Hamm, H.W. *Low-Cost Development of Small Water-Power Sites.* Volunteers in Technical Assistance (VITA), Inc., 1982.

Marks, Vic (editor). *Cloudburst: A Handbook of Rural Skills and Technology.* Cloudburst Press, Ltd., 1973. (Contains standard techniques of measuring head and flow rate, techniques of micro-dam building, and water wheel design.)

Owens, W.G. *A Design Manual for Water Wheels.* Volunteers in Technical Assistance (VITA), Inc., 1988. (Design and construction details of a water wheel for mechanical power.)

Warnick, Calvin C. *Hydropower Engineering.* Prentice-Hall, Inc., 1984. (Very technical. Relates to small-scale hydro.)

Special Systems

SIGNAL SYSTEMS

FIRE ALARMS

TELEPHONE SYSTEMS

SECURITY SYSTEMS

CENTRAL MONITORING AND CONTROL SYSTEMS

CONVEYING SYSTEMS

PASSENGER ELEVATORS

FREIGHT ELEVATORS

ESCALATORS

MOVING WALKS AND RAMPS

MATERIALS-HANDLING SYSTEMS

SIGNAL SYSTEMS

The subject of signal systems encompasses all signal, communication, and control systems. What they all have in common is the sending and receiving of electronically coded information.

The types of equipment include fire detection and alarm systems, telephones and intercoms, broadcast television with very high frequency/ultra high frequency (VHF/UHF) reception, closed-circuit television (CCTV) for security or educational purposes, paging and sound systems with AM/FM tuners, master clock systems (interconnected clocks and bells), data transmission, and HVAC controls ranging from simple thermostats to computerized energy management systems. These systems are designed by the electrical consultant, or special fire protection, audiovisual, or acoustical consultants.

Public buildings such as hotels, motels, hospitals, civic buildings, schools, and museums typically have many special service requirements beyond those of office buildings. In addition to normal data processing and telephone services, these buildings may require public address, piped music, CCTV, and telephone booth facilities. Passenger terminals require highly specialized heavy-density communication provisions.

All signal systems consist of:

1. A signal *source*
2. A means of conveying the signal
3. *Indicating* equipment at the destination

Figure 11.1 shows the relationship between these components for some representative devices. Figure 11.2 illustrates the common symbols for these devices as used on construction drawings.

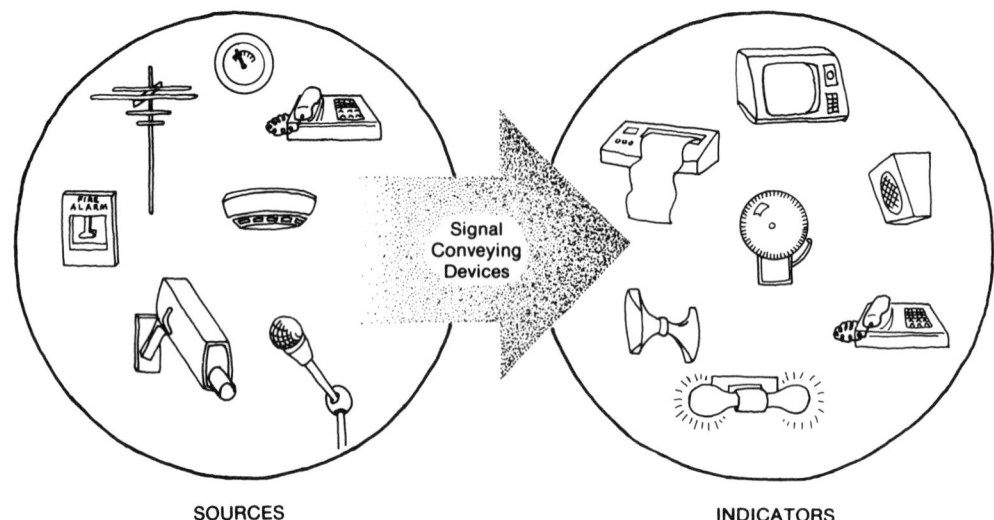

SOURCES

INDICATORS

FIGURE 11.1. Generic diagram of signal systems.

The source requires a *sensor* to pick up the information and some sort of device to process and transmit it. This could be a fire or smoke detector, manual fire alarm pull station, CCTV camera, telephone or intercom device, TV or radio signal antenna, or any other device *initiating* a signal or alarm.

The means of conveying a signal is usually small, low-voltage wiring, although signals can also be carried on radio airwaves. Signal transmission lines cannot be placed in the same raceway with electrical power wiring. TV antenna cables and closed-circuit connections must be shielded and generally should not be grouped with telephone lines because of possible signal interference.

At its destination, the signal may be indicated audibly, visually, or printed out as permanent hard copy. Signal indicators include loudspeakers, computer monitors, bells, horns, sirens, and flashing lights.

Fire Alarms

The signal source that senses a fire or smoke is called an *alarm initiating device.* It can be automatic, such as a *fire detector, smoke detector,* or *water flow switch,* or it may be manually operated, in which case it is called a *pull station.*

Audio indicators may be bells or horns. It is possible to have fire alarm initiating devices directly connected to indi-

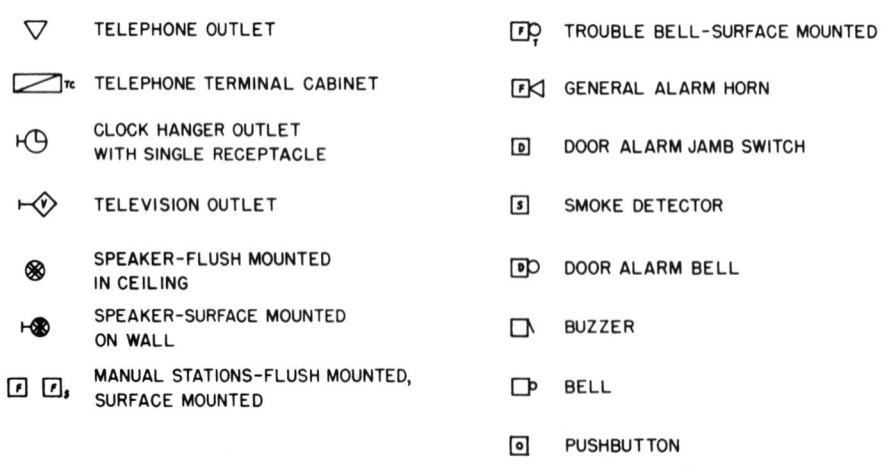

▽	TELEPHONE OUTLET
▱тс	TELEPHONE TERMINAL CABINET
⊦⊕	CLOCK HANGER OUTLET WITH SINGLE RECEPTACLE
⊦◇	TELEVISION OUTLET
⊗	SPEAKER-FLUSH MOUNTED IN CEILING
⊦⊗	SPEAKER-SURFACE MOUNTED ON WALL
▣ ▣,	MANUAL STATIONS-FLUSH MOUNTED, SURFACE MOUNTED

⊡	TROUBLE BELL-SURFACE MOUNTED
⊠	GENERAL ALARM HORN
▣	DOOR ALARM JAMB SWITCH
▣	SMOKE DETECTOR
⊡	DOOR ALARM BELL
◻	BUZZER
◻	BELL
▣	PUSHBUTTON

FIGURE 11.2. Common symbols for signal devices.

cating devices in a remote fire station or police headquarters. In that event, additional alarm indicators are usually required on site to signal the building occupants to evacuate.

The location of automatic detecting devices, pull stations, and audio and *visual* indicators is generally dictated by code. Examples of typical audio and visual indicators are presented in Figure 11.3. A pull station and automatic detecting devices are shown in Figure 13.2 in Chapter 13.

Manual pull stations are placed along the normal path of egress from a building so that the person who has detected a fire may activate an alarm as he or she exits. In order to function as intended, these stations must be well marked and easily noticeable. Concealing or camouflaging them because they would "spoil the decor" defeats their purpose and could jeopardize the safety of the building's occupants.

Similarly, alarm indicators are designed to attract attention when needed. Concealing bells or horns above hung ceilings because they are unattractive makes them harder to hear and could result in the loss of property or even lives. In most cases, it is possible to work with the consultant to select a location that satisfies the alarm function while not violating the aesthetic quality of the space.

Fire and smoke detectors monitor automatically, and are therefore more appropriate in buildings with sleeping residents or in unoccupied spaces. Many local building codes require automatic detectors, and the Federal Housing Authority (FHA) requires at least one in each residence it finances.

Since fires generally smolder before bursting into flame, and most fire fatalities are caused by smoke poisoning, smoke detectors, particularly the photoelectric units, are more effective than heat-sensing types.

Many buildings are protected against fire by automatic sprinklers, as discussed in Chapter 13. Flow switches are commonly installed in the sprinkler piping. When water sprays out of the sprinkler heads, the switch senses the flow and initiates an alarm.

Another type of automatic alarm is interconnected with some doors that are normally locked for security, but needed for an emergency exit. These doors are fitted with special panic hardware which automatically sounds an alarm when opened for emergency exit.

Telephone Systems

A schematic diagram of a telephone system is presented in Figure 11.4. The equipment may be either privately owned or provided as part of the local telephone service, and a variety of styles and capabilities are available.

Like electrical power, telephone connection involves a service entrance. The incoming cable terminates at a wall

(a)

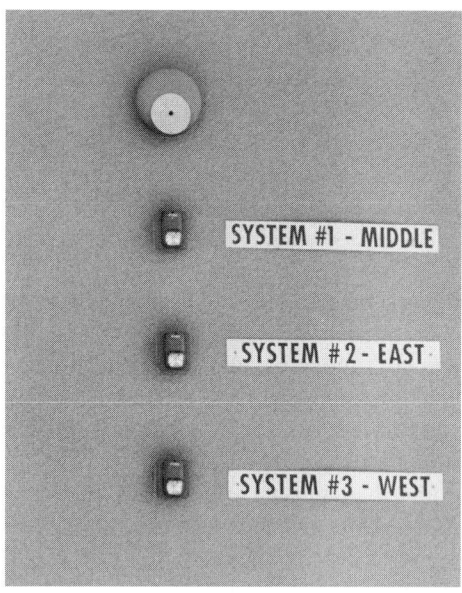

(b)

FIGURE 11.3. (a) Typical interior fire alarm horn and strobe light for the hearing impaired. (b) Typical exterior fire alarm bell and flashing lights. (Photos by Vaughn Bradshaw.)

box. A sheet of 3/4-inch (1.9-cm) marine plywood must be provided for mounting the necessary equipment.

The space required for a telephone service entrance varies with the size and type of building. In very large buildings, a telephone terminal room is required, whereas a small apartment building (three stories or less) merely needs a clear

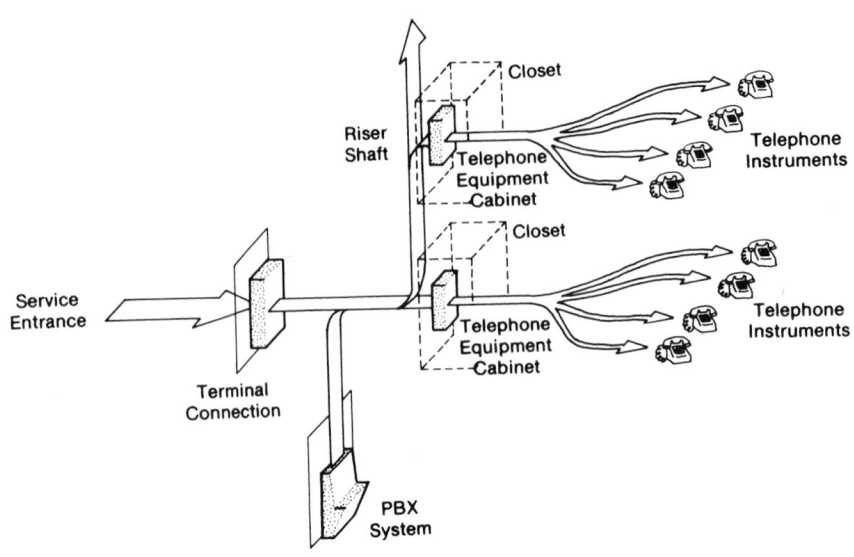

FIGURE 11.4. Schematic diagram of a telephone system.

wall space of 2 to 4 linear feet (60 to 120 linear cm). If located within a finished space, the entrance equipment can be enclosed within a recessed cabinet with a 3/4-inch (1.9-cm) plywood back. Generally, commercial buildings require a much greater telephone capacity and thus need more space than residential structures do.

The transmission wiring within the building is run within either conduit or sleeves to provide protection and facilitate changes. Under-floor and ceiling raceway systems, as described in Chapter 9, are often used.

Depending on the number of lines included, riser spaces or shafts may be needed for vertical wiring. If the floor plans of all floors are similar, the arrangement of wiring risers is relatively simple, since the risers can extend through vertically aligned closets. A sleeve is provided through the floor between the closets to accommodate these cables.

Large office buildings typically require apparatus closets to contain telephone switching equipment for each area or floor. The closet interior should be surfaced with 3/4-inch (1.9-cm) plywood on which the necessary cabinets, raceway, and fittings can be mounted. As an initial approximation, 1 linear foot (1 linear meter) of wall space is needed for every 800 square feet (240 m²) of floor area. Adequate lighting and electrical service are also required. Exact space requirements must be obtained by the electrical or special communications consultant for each application.

Security Systems

The purpose of an electronic *security system* is to extend the surveillance abilities of a limited security force in order to remove or reduce the incidence of crime (theft, assault, or vandalism). It does not eliminate the need for people to monitor the detection devices and apprehend criminals. Detection devices can automatically alert public police, or both monitoring and apprehending functions can be performed by private security guards.

The first need in designing for security is to identify crime risk areas in the given building, complex, or shopping mall. Detecting devices can be CCTV, *motion detectors, intrusion detectors,* or smoke and fire detectors. An integrated electronic security system can include all or some of these, or they can be separate systems.

Use of CCTV cameras is very common in banks, retail stores, high-rise apartment buildings, and industrial complexes. These combined with visual and audible alarms can alert a security officer on duty at the system's consolidated panel of a breach in the building's security. A typical remote-controlled camera is shown in Figure 11.5.

Potential surveillance locations include parking lots, parking garages, elevators, and all possible means of access from doors and windows to exterior openings for ventilation fans and ducts. In addition to CCTV cameras, sensors for a *voice-activated alarm system* are recommended for high-risk parking garages.

Adequate lighting, as recommended by the CCTV system manufacturer, is important for optimum camera resolution, and uniform lighting eliminates dark spots. Exterior lighting and cameras both need to be positioned and protected with impact-resistant lenses to resist vandalism.

Security systems consist of everything from a simple burglar alarm to a full central alarm system. The latter can pro-

FIGURE 11.5. CCTV security camera. (Photo by Vaughn Bradshaw.)

vide remote visual surveillance of critical areas through CCTV cameras. The cameras can pan, tilt, and zoom automatically or by remote control from the central monitoring console (Figure 11.6) to provide flexible coverage of critical areas. The surveillance of more than one area from the security control center can be either simultaneous, employing a monitor screen for each camera, or sequential, using one TV monitor to scan the signal from each camera one-by-one. A single TV monitor can also track security guards on their rounds.

Intrusion detectors at doors and windows, and motion detectors within the building, trigger an alarm if someone enters an unauthorized area. The system responds to any intrusion or other emergency alarm by instantly informing the central operator of the type, location, and time of each alarm and then printing out the same information for later reference.

Scream alarms can be installed in elevators to signal the central system operator about an assault. The system can perform two-way operation to indicate the activity at the remote location and also allow the operator to inform perpetrators that they are under surveillance and that security personnel have been dispatched. The elevator can then be controlled by the operator and held at the ground floor until the security force arrives.

One-way and two-way emergency voice communications systems can be added to a security system to enable fire fighters throughout a facility to contact each other and the command station. Special portable phones can be plugged into phone jacks located in stairwells, elevator lobbies, and other key locations.

Apartment buildings commonly combine their security

FIGURE 11.6. Central security system monitoring station. (Photo courtesy of GE Security.)

and doorbell functions. Restricted access is provided by a two-way intercom between the building entrance and each apartment (Figure 11.7). This provides the tenants with a remote means of screening callers before pushing their release button to open the lobby entrance door. CCTV can be added, at a much greater expense, to enable tenants to see callers as well as hear them.

Highly classified or high-risk areas in CII buildings can use key-carded access control systems, in which each authorized person carries a card coded with their identification. Inserting the card into a slot by the entrance, as shown in Figure 11.8, unlocks the door. These mechanisms can be centrally reprogrammed to allow deletions, changes, and modifications of employee accessibility at any time.

Alternative alarm systems are sometimes provided for critical applications in the event of failure of the primary system. Some systems may need an automatic standby

FIGURE 11.7. Apartment building lobby access control intercom. (Photo courtesy of Nutone Inc.).

FIGURE 11.8. Access control system ID card reader. (Photo by Vaughn Bradshaw).

power source (usually batteries) in the event of a power failure.

Equipment manufacturers can provide quite a bit of helpful information for planning a security system and the general security aspects of a building design. Consulting firms specializing in security system design are also available.

Proper security considerations should begin in the initial stages of design. It is much more difficult and expensive to retrofit electronic monitoring and detection systems after a facility is built than to integrate it from the beginning. Even more important is anticipating potential crime vulnerabilities and preventing them through thoughtful planning.

Central Monitoring and Control Systems

Large high-rise buildings are often equipped with central control centers that can control the HVAC and electrical services and monitor fire alarms from one point. Some of these are computerized for automatic control in response to a present sequence of instructions.

The more sophisticated of these act as Energy Management Control Systems (EMCS) to turn off equipment not in use, optimize HVAC operation, and cycle electrical loads to limit the overall demand. These systems need little space but must be accommodated where desired. The systems are usually tailored to the specific requirements of a building, so no guidelines can be given here for space requirements, but lighting, ventilation, and space for raceways are needed. The security monitoring system is sometimes combined with an EMCS.

CONVEYING SYSTEMS

Passenger Elevators

Elevators have made possible the existence of commercial buildings more than two stories in height. Mechanical devices for the vertical movement of people and goods are now essential equipment in many types of buildings, and the quality of elevator service has a strong bearing on the desirability of lease space in a building.

Vertical transportation represents a substantial portion of a building's total construction costs and can account for as much as 11 percent of the cost of a high-rise office building. Their spatial requirements and traffic patterns are a determining factor in interior planning.

Although the final decision for vertical transportation equipment rests with the architect, the factors affecting it are so numerous that consultation with an elevator authority is generally necessary. Such consultation service is readily available from independent consultants in the field and, to some extent, from the major elevator and escalator manufacturers.

Cable vs. Hydraulic

The two types of elevators are *cable* (also known as *electric* or *traction*) and *hydraulic*. Hydraulic elevators are limited to only a few stories, while the more common cable type can travel well beyond 30 stories.

Cable. The basic components of a cable-type elevator are the car, *hoistway shaft*, penthouse, and pit, as illustrated in Figure 11.9. The elevator car is suspended from cables,

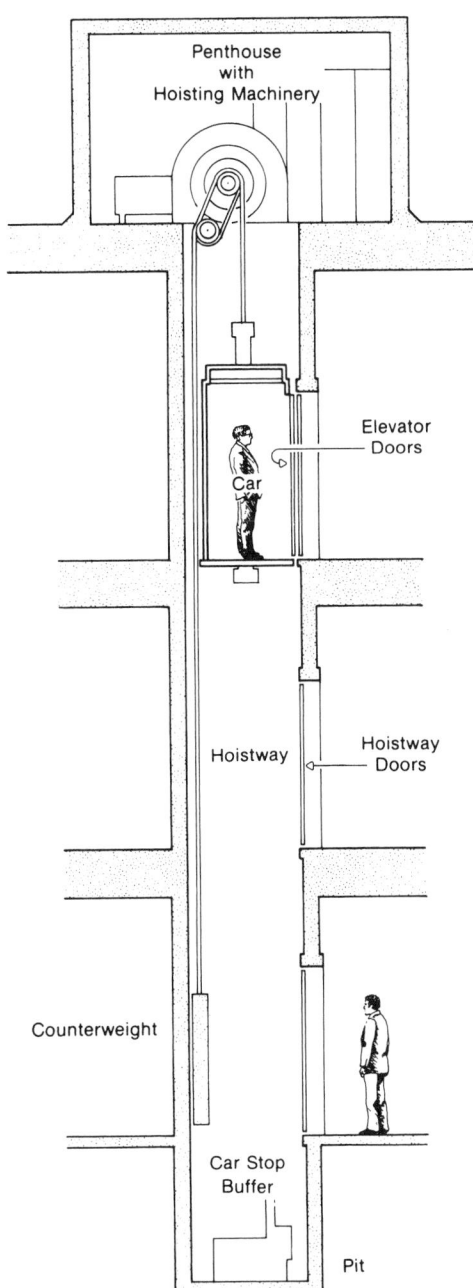

FIGURE 11.9. Section through a typical cable-type hoistway.

Labels in figure:
Penthouse with Hoisting Machinery
Elevator Doors
Car
Hoistway
Hoistway Doors
Counterweight
Car Stop Buffer
Pit

which are supported by the hoisting machinery located in the penthouse. The hoisting machinery raises and lowers the car as directed. The car is guided in its vertical travel up and down the shaft by a system of rollers and guide rails.

When the hoisting machinery is placed at the top of the hoistway, the building structure must be able to carry the weight of this machinery plus the fully loaded elevator and its counterweights. Counterweights are normally equal to the car weight plus 40 percent of the carrying capacity. If top mounting is impractical, and the application is low-rise and slow speed, the hoisting machinery can be placed in the basement adjacent to the hoistway.

Hydraulic. A hydraulic elevator contains the same basic elements as a cable elevator, except that instead of being suspended from a penthouse, it is supported by a hydraulic mechanism, called a *plunger,* extending down into a deep well, as shown in Figure 11.10. The car is raised and lowered by the plunger, or *jack,* attached to the bottom of it. Oil from a reservoir is pumped under the plunger, thereby pushing the car up. Once the car is raised to the desired level, the pump is stopped. The elevator is then lowered by slowly releasing the oil back into the reservoir so that the car descends by gravity.

The operating machinery consists of a motor-driven pump and a control valve. They are linked to the plunger by hydraulic piping, and can therefore be placed anywhere in a building, although preferably in a basement machinery room. Sound isolation is not critical since hydraulic elevator operation is extremely smooth and practically noiseless, but it is still good practice to treat the machine room acoustically in case of vibration in the hydraulic piping.

Although electric elevators are more commonly used, hydraulic elevators have certain advantages for low-rise applications. The absence of the cables, elaborate traction equipment, safety devices, and penthouse makes hydraulic elevators less expensive. Since the full weight of the car and load is carried on the plunger—and ultimately on the pit floor—structural requirements are decreased and the elevator has a larger carrying capacity. Furthermore, the shaft can be smaller for a given cab size because no space is needed for counterweights.

The hydraulic plunger and its casing are installed as a unit, similar to a well. Since drilling in rock is very expensive, soil conditions can be a prohibiting factor. The rigid plunger limits hydraulic elevator service to applications with a maximum of 75 feet (25 m) of car travel. Because of the hydraulic system, speed cannot exceed 200 fpm (100 cm/s).

A significant disadvantage of hydraulic elevators is their inherently larger energy cost. Since they are not counterweighted, they require a larger motor and more power to drive the oil pump and raise the car. Even accounting for the fact that the motor operates only in the up direction, the energy use can be nearly double that of a cable elevator providing the same service. When circumstances permit a hydraulic elevator, a life-cycle cost analysis is warranted to

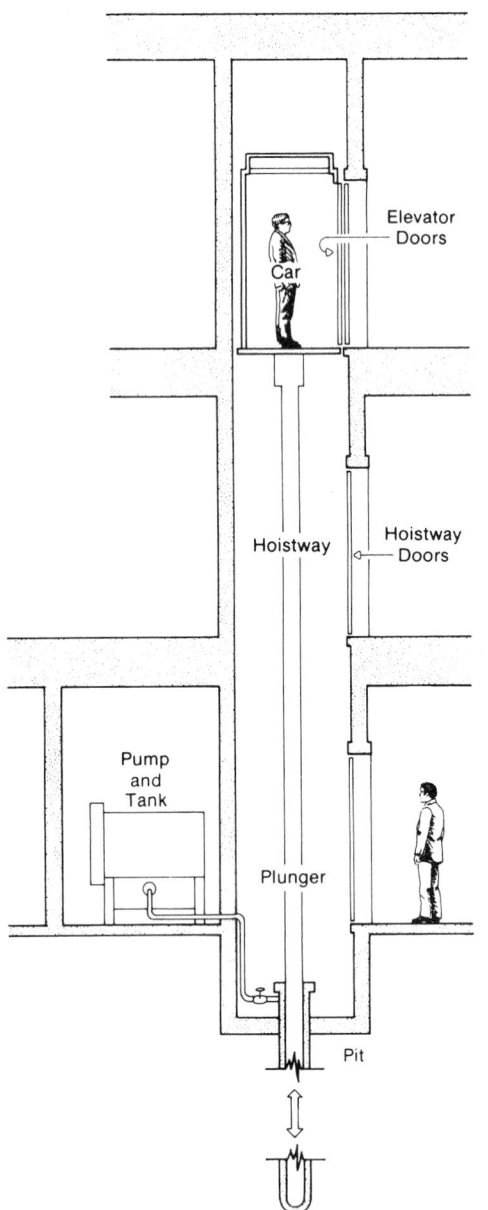

FIGURE 11.10. Section through a typical hydraulic hoistway.

determine the most economical installation. The machinery room for either type must be well ventilated to dissipate the heat generated by the motor and friction.

Safety

The most important factor in elevator design is safety. The primary reason for the great expense of cable elevator systems is that they are overdesigned far beyond what is cus-

tomary in most other fields, with safety factors of between 7 and 12 applied to the critical equipment and materials.

The hoisting motor for an electric elevator is provided with a speed governor which cuts off power to the motor and applies a brake to the motor drum if it exceeds the proper speed. Another brake operates directly on the cables, and if the car itself speeds out of control, a braking device is activated between the car and the guide rail.

Hydraulic elevators are held by a locked column of hydraulic fluid when the car is at rest. The standard safety devices required by law for cable-type elevators are not generally required for hydraulic elevators but may be part of local codes. But hydraulic elevators are inherently safer, since even if a leak develops in the hydraulic system, the elevator descends slowly, without danger to the passengers.

Elevator installations are governed by strict codes. The principal one is the American National Standards Institute (ANSI) code A 17.1, *Safety Code for Elevators, Dumbwaiters, Escalators and Moving Walks*. Architects working on buildings with elevators should keep an up-to-date copy on hand.

This code has the force of law in most parts of the United States, and its requirements are strictly enforced due to the safety risk at stake. Some states and municipalities (notably Massachusetts, Pennsylvania, Wisconsin, New York City, Boston, and Seattle) have their own elevator codes, which are based on, but more stringent than, the ANSI code.

In addition to the elevator code, the following National Fire Protection Association (NFPA) codes relate to elevator installations:

- NFPA No. 101, *Life Safety Code* (fire safety requirements)
- NFPA No. 70, *National Electric Code* (electrical aspects)

State and local laws add further requirements and restrictions for fire safety, emergency power, security, and special accommodation for handicapped persons. Provisions for the handicapped are required by law and are covered by the ANSI code, a special elevator industry code, and local ordinances.

The elevator industry is self-regulated and standardized. The *National Elevator Industry, Inc.* (NEII) publishes standard elevator layouts and, as mentioned above, its own elevator standard for the handicapped.

Local elevator consultants and companies are usually knowledgeable about all codes and standards in effect, but this does not relieve the architect/engineer of legal responsibility for the design. Therefore, it is wise and strongly recommended that the architect/engineer assemble all pertinent regulations concerning vertical transportation in the preliminary planning stage.

Elevator Selection

The ideal features of elevator service include:

- Minimum waiting time for a car at any floor level
- Rapid transportation
- Comfortable acceleration
- Rapid, smooth deceleration and stopping
- Rapid loading and unloading at all stops
- Clear visual floor indication in the cars and at landings
- Comfortable lighting and a generally pleasant car atmosphere

The overall objective in selecting elevators is to provide adequate service in terms of the above list for the least cost. The variables to determine are:

- Number of elevators
- Carrying capacity of each car
- Elevator speed

Each of these variables has an effect on the other two, and often several alternative combinations will provide the desired result. In that event, total cost of the installation or total life-cycle cost will determine the preferable solution.

Number of Elevators. The first consideration in determining the number of elevators required is the *population in the building*. Code occupancy restrictions in terms of square feet per person can be used if the exact design population is not known. Otherwise, the guidelines on cooling loads due to people given in Chapter 4 and the estimates in Appendix Table B6.3 can be used to form an estimate.

Rules of thumb can be used for a gross estimate. For example, in most buildings, the number of elevators can be based on 250 to 300 people per elevator. The FHA requires a minimum of two cars in all apartment buildings more than seven stories high and, beyond the minimum, one car per every 120 bedrooms.

Carrying Capacity. Establishing the carrying capacity of the elevator system and of each car results in a more precise estimate of the number of elevator cars required.

The desired passenger-carrying capacity, or *handling capacity* (HC), of an elevator system is expressed as the percentage of the building population to be carried one way in 5 minutes (1 hour for department stores). This percentage varies with the type of building and ranges from 12 to 18 percent for offices, 5 to 8 percent for apartments, 10 to 11 percent for dormitories, and 10 to 15 percent for hotels.

The actual number of passengers that the overall elevator system can carry in 5 minutes (population times HC) is the carrying capacity of each elevator times the number of elevators, N. The carrying capacity of each elevator is a function of its capacity and the time required for a round trip.

The *car capacity* is the maximum number of people that can fit into a single car. Elevators may be loaded to full capacity during peak hours. At most other times, they carry less than their capacity, and the calculations given below are based on a number called the *passenger load, car loading,* or *number of passengers per trip* (NPT), which is the average number of people that will actually be carried per load. Statistically, it is about 80 percent of the car capacity.

The *round-trip time* (RTT) is the average time required for a car to complete a round trip, starting from the ground floor and returning to it. It consists of the sum of the following four components:

1. Running time (distance traveled divided by the car speed)
2. Acceleration and deceleration time (typically about 2 seconds per stop)
3. Door opening and closing time (typically about 8 seconds per stop)
4. Loading and unloading time (depends on the number of people and the width of the doors)

The most significant of the above components is the running time, which depends on the speed of the elevator. The cost of an elevator is directly related to its speed; the higher the speed, the higher the cost. However, if one elevator capable of traveling 300 feet per minute (fpm) (1.5 m/s) is able to do the work of two at 150 fpm (0.75 m/s), the faster elevator would result in a lower cost for the whole installation.

Cable and hydraulic elevators with the same nominal speed may have different round trip times. The down speed of hydraulic elevators can be somewhat greater than the up speed, which raises the average round-trip speed and reduces the running time. The acceleration and deceleration rates, however, are slower than those of cable-type elevators.

There is a relationship between speed and the maximum distance a car can travel in order to keep the running time within reasonable limits. That relationship, along with a range of appropriate car loadings for different building types, is given in Table 11.1. These values can be used for preliminary elevator selection.

TABLE 11.1 PRELIMINARY ELEVATOR SELECTION DATA

Building Type	Normal Passenger Load per Car	Maximum Car Travel (feet)	Maximum Car Travel (meters)	Recommended Car Speed (fpm)	Recommended Car Speed (m/s)
Apartment	5–13	75	23	100	0.5
		100	30	200	1.0
		125	38	250	1.25
		150	46	350	1.75
		175	53	400	2.0
		250	76	500	2.5
		350	107	700	3.5
Department store	5–28	100	30	200	1.0
		125	38	350	1.75
		175	53	400	2.0
		250	76	500	2.5
		350	107	700	3.5
Hospital	19–22	60	18	150	0.75
		100	30	200	1.0
		125	38	250	1.25
		150	46	350	1.75
		175	53	400	2.0
		250	76	500	2.5
		350	107	700	3.5
Hotel	13–16	125	38	350	1.75
		200	61	500	2.5
		250	76	700	3.5
		350	107	800	4.0
		over 350	over 107	1,000	5.0
Office	10–22	100	30	200	1.0
		125	38	250	1.25
		150	46	350	1.75
		175	53	500	2.5
		250	76	700	3.5
		350	107	800	4.0
		over 350	over 107	1,000	5.0

The number of people that can be carried by one elevator in 5 minutes (300 seconds), NP_5, is related to NPT and RTT by the following equation:

$$NP_5 = \frac{300 \times NPT}{RTT} \tag{11.1}$$

By definition:

$$\text{building population} \times HC = NP_5 \times N \tag{11.2}$$

Combining Equations 11.1 and 11.2 yields a formula for calculating the number of cars, N, needed:

$$N = \frac{\text{population} \times HC \times RTT}{300 \times NPT} \tag{11.3}$$

Elevator Speed. In the final design of an elevator installation, the exact specification for elevator speed is determined on the basis of HC, interval, and travel time.

The *interval,* or *lobby dispatch time,* is the average waiting time between the departure of cars from the lobby. It can also be thought of as the frequency with which any car appears at the lobby. Since a given passenger arrives in the lobby sometime after the last car has left, the average wait in the lobby before boarding a car is less than this interval. An interval of 25 to 30 seconds is considered excellent for office buildings, 35 is borderline, and 40 seconds is the upper limit of acceptability, only suitable for moderate traffic conditions. Multiunit residential buildings should have intervals of 40 to 80 seconds, with only low-cost apartments reaching as high as 120 seconds. Intervals longer than these ranges create an awareness of delay.

The *travel time,* or *average trip time,* is the average amount of time a passenger spends from the moment of arrival in the lobby until departure from the car at an upper floor. In commercial buildings, travel time is preferably limited to 1 minute. A 75-second trip is considered acceptable, a 90-second trip annoying, and 2 minutes is the outside limit. Residential elevator trips can be longer.

Once the interval is established, the remainder of the travel time is determined by the car speed, making that a very important variable in elevator selection. Elevator speeds are classified as low speed and high speed as follows:

- Low speed: 100 to 800 fpm
- High speed: 600 to 1,200 fpm

Architectural Considerations

Elevators take up space and must therefore be architecturally integrated into a building. The spaces requiring attention are the elevator lobbies, shafts, and machine rooms.

Lobbies. The elevator lobby is usually the first place people see on each floor. Corridors then branch out from that point. Elevator lobbies also serve as waiting areas, so they should be clear of other circulation for the benefit of both waiting passengers and the general circulation. They must be large enough to allow the peak number of passengers to gather and wait comfortably. Each elevator lobby should provide approximately 4 ft^2 (0.4 m^2) of floor area per person waiting there at the peak time.

The elevator lobbies serving a given elevator or bank of elevators are generally located one above the other. The main lower lobby, usually at street level, should be located for convenient access from main entrances. When different main entrances are at different levels, basement or mezzanine entrances can be linked with a primary ground floor lobby by escalator. This is an economical way to transport large numbers of people rapidly, which breaks down the distinction between floors.

The main lobby should include the building directory, elevator indicators, and public telephones either within the lobby or adjacent to it. A locked elevator control panel is usually located in the main lobby within view of the elevators. Public telephones may also be located at upper floor lobbies as needed.

Elevators should generally be grouped together so that a person entering the building may immediately use any car that is available. If elevators are scattered, it is possible to have both idle elevators and waiting passengers.

Ideally, elevators should be placed at the center of the population served, but this is not always convenient. On the entrance floor, this could make the elevator lobby too remote from the entrance. It could also create planning problems for the entrance floor, which may serve a different purpose from the floors above. If elevators are placed too close to the entrance, general circulation may be impeded and waiting made unpleasant by people passing.

Therefore, the ground floor lobby should usually be established first. It should be large enough to accommodate the number of people likely to be waiting and located so that it will neither affect nor be affected by the general traffic through the entrance. It should, however, be readily visible and accessible from the entrance. Users are usually better prepared to walk distances on upper floors than on the entrance level.

In very large buildings, splitting up the elevator service into separate vertical circulation towers may be justified in order to provide more convenient access by widely dispersed passengers. This is called *zoning*. A *zone* is a grouping of floors or portions of a building served by a common elevator or bank of elevators.

It is most economical and efficient to divide tall buildings into vertical zones—an upper half and a lower half—for elevator operation. One lobby and group of elevators serves the ground floor to midheight, and the other (called an *express*) passes from the ground floor directly to midheight, stopping in the normal way from midheight to the top. Very tall buildings may even be divided into more than two vertical zones. Since all cars don't have to stop at all floors, upper-floor passengers get to their destination faster and the waiting time for all passengers is reduced. In most high-rise buildings, vertical zoning is necessary to meet elevator design criteria. And, not only does this arrangement provide more convenient service, it also frees up usable floor area on the floors above the lower zone(s).

It is preferable to have a separate ground floor lobby for each vertical zone. If that is not feasible, express elevators should be clearly differentiated from low-rise ones.

The way in which elevators are grouped is important. Two or three may be placed side by side, but if more than that are placed in a row, passengers may not notice the arrival of an elevator. Even if the warning-of-arrival system is good, it may be difficult for passengers to squeeze through a crowded lobby to get in before the doors close. Grouping elevators to face each other provides better visibility and quicker access than arranging them in one long bank. Elevators should never be separated by corridors or stairs, because this not only worsens the difficulties of view and access just described, but also introduces conflicting circulation arrangements.

The selection of cars and hoistway doors should be consistent with the overall architecture of a building. Car doors can be single, double, or four panel, and may open in the center or to the side, as shown in Figure 11.11.

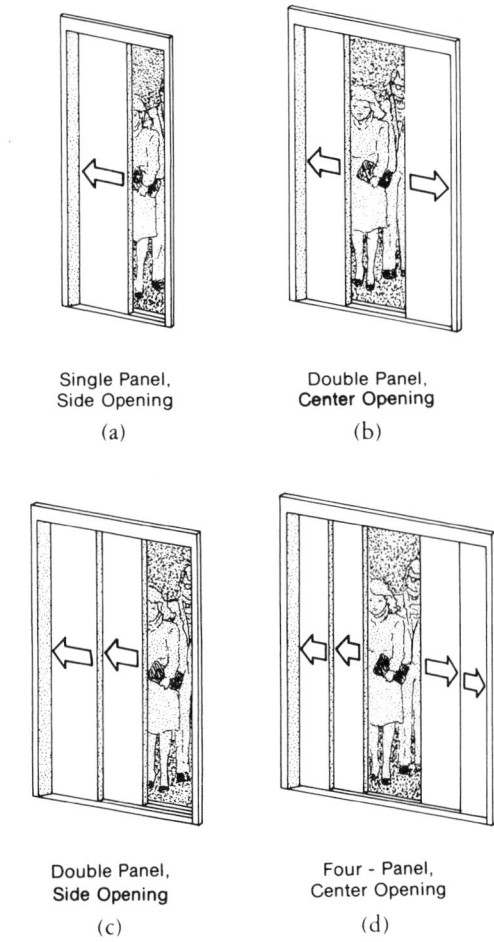

Single Panel,
Side Opening
(a)

Double Panel,
Center Opening
(b)

Double Panel,
Side Opening
(c)

Four - Panel,
Center Opening
(d)

FIGURE 11.11. Elevator door variations.

Cab size, door dimensions, and other elevator characteristics have been standardized by the NEII. Car and hoistway doorways must be the same width, which is generally in the range of 3 to 4 feet (0.9 to 1.2 m). Local fire and safety codes may specify further limits. Openings smaller than 3 feet 6 inches (1.1 m) make it difficult for more than one passenger to pass through at one time (Figure 11.12), which increases loading time and, consequently, RTT.

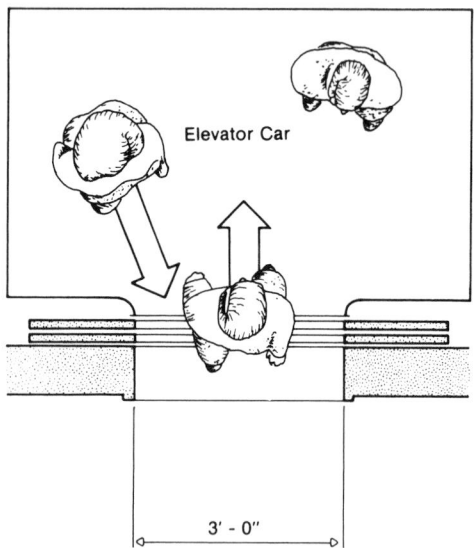

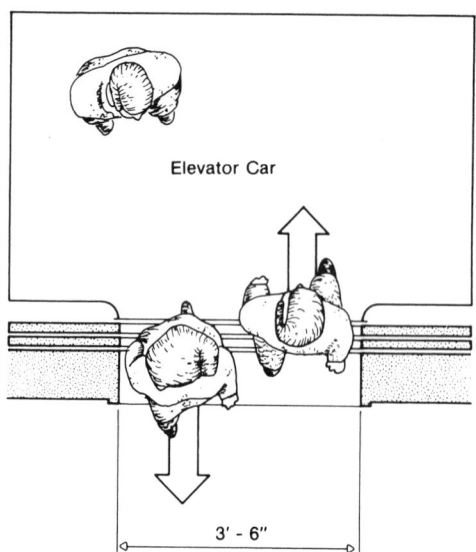

FIGURE 11.12. Passenger loading and unloading with door openings of different widths.

Door operation is generally automatic and provided with a protective mechanism to reopen the doors if they meet an obstruction while closing. Hoistway doors are similar to car doors, and all doors are interlocked so that the car cannot be operated unless all shaft doors are closed and locked and the car doors are shut. The speed of closing is regulated by code, but the amount of time the doors are open can be adjusted to meet building requirements.

Shafts. The shafts themselves are major space elements that must be integrated into a building. Table 11.2 lists typical dimensions of elevator shafts as a function of capacity and elevator type. Hospital elevators are normally much deeper than the normal passenger-type cars in order to accommodate stretcher carts, wheelchairs, beds, linen carts, laundry trucks, and other bulky vehicles. Shaft depth is typically about 9 feet 6 inches (2.9 m). The final design must be based on exact dimensions furnished by the manufacturer.

Machine Rooms. Cable-type elevators usually require a penthouse located directly above the shaft of each elevator. This penthouse needs approximately two stories of height above the top of the elevator when it is standing at the top floor. A preliminary planning figure for the required floor area is roughly twice the area of the shaft itself. The gearless traction drive, designed for speeds greater than 350 fpm (1.8 m/s), takes up more space than the geared traction type, which is meant for lower speeds.

A basement traction unit, also referred to as an *under-*

TABLE 11.2 TYPICAL ELEVATOR SHAFT SIZES

Elevator Type	Car Capacity		Shaft	
	Pounds	Passengers[a]	Width	Depth
Cable	1,200	8	6'-6"	5'-0"
	2,000	13	7'-6"	6'-0"
	2,500	16	8'-6"	7'-6"
	3,000	20	9'-0"	8'-0"
	3,500	23	9'-6"	8'-6"
	4,000	26	10'-0"	8'-6"
Hydraulic	1,500	10	6'-6"	4'-6"
	2,000	13	7'-6"	5'-0"
	2,500	16	8'-6"	5'-6"
	3,000	20	8'-6"	6'-0"
	3,500	23	8'-6"	7'-0"
	4,000	26	9'-0"	7'-0"

[a] Note that the NPT is about 80 percent of the car capacity.

slung arrangement, does not need a penthouse. It is suitable only for low-speed, limited-rise, and light- to medium-traffic applications.

The space required for the power unit of hydraulic elevators is approximately 3 × 6 feet (1 × 2 m) at the base and 5 feet 8 inches (1.7 m) high. Additional space must be provided around the unit in order to allow for maintenance.

Structural Requirements. The amount of structural support needed for a given elevator must be obtained from the manufacturer. The foundation, structural columns extending all the way up to the penthouse, and main beams supporting the penthouse floor must support the dead weight of the equipment when the elevator is not in motion, plus the added weight caused by the momentum of all moving parts and passengers when the elevator is at top speed and is suddenly caused to stop by the safety devices.

Freight Elevators

Freight elevators are characterized by slower speed and greater carrying capacity. Standard models are available with rated capacities of up to 10 tons, but electric elevators can be specially ordered with capacities of up to 15 tons and hydraulic elevators of up to 50 tons.

A careful materials-flow study should be made before selecting freight elevators. The factors to be considered are:

- Tonnage movement per hour
- Size of load
- Method of loading
- Distance of vertical travel
- Type of load

Once the materials-handling requirements are established, the number of elevators needed, their capacity (pounds), speed, door width, and location in the building can be decided. The car and door dimensions must be determined by analyzing the types of freight to be handled, as well as the pallets, skids, or hand trucks to be used. The requirements should be based on the combined size and weight of all items expected to be on the elevator at one time. Elevator manufacturers can provide help with the analysis.

For preliminary planning purposes in the absence of any more specific information, one freight (service) elevator should be allowed for every 10 passenger cars, or one for every 300,000 ft² (28,000 m²) of building floor space, whichever results in the greater number of cars. In some cases, freight elevators can serve as secondary passenger elevators during peak periods.

The rated load of freight elevators is based on 30 to 50 pounds per square foot (psf) (150 to 240 kg/m²) of net inside platform area. This relationship can be used as a guideline for preliminary space allocation for shafts. Door widths are normally 4 to 4½ feet (1.2 to 1.4 m).

Car speeds are generally 50 to 200 fpm (25 to 100 cm/s). Hydraulic types are commonly used in low-rise applications. They are limited to 60 feet (18 m) in height and a speed of 125 fpm (64 cm/s).

Escalators

Escalators, also known as *moving stairs* or *electric stairs,* provide an efficient and economical means of vertical transportation where a number of people need to be carried to a second floor or beyond. They can carry up to 10,000 people per hour. While elevators are more convenient for trips of five stories or more, escalators are superior for large passenger loads for two or three stories.

Escalators, placed directly in the line of circulation, operate continuously without floor stops or the opening and closing of doors. They are thus able to provide approximately the same service during peak periods and slack times.

Previously limited to department stores and passenger terminals, escalators are now becoming popular in office buildings, banks, and hotel atria for traffic to second and third floors, which reduces elevator requirements. Escalators, however, cannot carry people in wheelchairs or freight, so at least one elevator must still be provided for those purposes.

Carrying Capacity

The safety code allows escalator speeds of up to 125 fpm (64 cm/s), but the standard speeds available are 90 and 120 fpm (46 and 61 cm/s) (other speeds can be specially ordered). Escalators can be two-speed, so that the slower speed can be used normally and the higher speed used during peak traffic periods. For applications with no rush hour, 90 fpm (46 cm/s) is preferable since less agile riders may find the higher speed difficult to negotiate.

Three widths, measured between the balustrades, are available as standard models: 32, 40, and 48 inches (81, 102, and 122 cm). The 32-inch (81-cm) size has a step width of 24 inches (61 cm) and will accommodate one adult and one child (or 1¼ persons) side by side per step. The 40-inch (102-cm) model has a step width of 32 inches (81 cm), and the 48-inch (122-cm) size has a step width of 40 inches (102 cm); both are rated for two adults side by side per step. All treads have a 16-inch (41-cm) depth and an 8-inch (20-cm) rise.

The passenger-carrying capacity of an escalator is a function of the speed and width. The maximum rated capacities based on the passenger densities per step indicated above are as follows:

32-inch width	90 fpm	5,062 passengers per hour
	120 fpm	6,750 passengers per hour
40-inch or		
48-inch width	90 fpm	8,100 passengers per hour
	120 fpm	10,800 passengers per hour

For planning purposes, the number of passengers per hour is decreased by 25 percent to take into account luggage, briefcases, and packages being carried.

Spatial Requirements

Escalators must be placed so that riders can easily locate them from the building entrance, recognize their destination, and move toward them. Adequate lobby space is important at the embarking points to accommodate occasional congestion, and especially at the discharge points, where traffic backup could be very dangerous. This is crucial in theaters, stadiums, schools, and wherever else large traffic peaks occur.

At the top or bottom landing, escalators should discharge into an open area with no turns or choice of direction. Where confronting the riders with a decision is unavoidable, large signs clearly identifying the options should be used to make hesitation unnecessary. The landing space in front of an escalator should be at least 6 to 8 feet (2 m) deep if the speed is 90 fpm (46 cm/s), and 10 to 12 feet (3.5 m) deep for 120 fpm (61 cm/s).

Due to its ability to continuously carry large numbers of riders, an escalator takes up considerably less floor space than a bank of elevators providing the same passenger capacity. The entire stairway mechanism is built around a bridge-type truss which is supported at the top and bottom by structural beams (see Figure 11.13). The incline angle of all escalators in the United States is 30°.

Conventional escalators with a 10- to 25-foot (3- to 7.5-m)

rise, which are standard production models, contain an electric drive motor within the escalator assembly. A rise higher than 25 feet (7.5 m) requires a separate machine room below to accommodate the large motor. Modular escalators made by Westinghouse contain multiple modular drive motors within the escalator assembly to power the higher rises. A machine room is not necessary for these units.

Multiple escalators may be positioned either *crisscross* or *parallel*. Parallel arrangements normally utilize reversible escalators so that all but one in a bank can be operated in the direction of heaviest traffic at any given time. As an alternative for short rises, light traffic in the reverse direction can be accommodated by a staircase adjacent to the escalator.

The specific arrangement should take into account the views of passengers. The plan for multistory retail stores, for example, exposes patrons to the most expansive view of the merchandise displayed on each floor.

Once a preliminary layout is established, an escalator manufacturer should be consulted for a full specification of what is necessary for a complete installation.

Safety

Electric stairways are extremely safe. All side surfaces are smooth; handrails are designed so that fingers cannot be caught under them; and the top and bottom comb plates are trip-proof, providing safe transfer to and from the moving steps. The comb plates prevent jamming.

A brake mounted on the motor shaft can be applied manually by pushing a stop button, or automatically when an overload or overspeed condition exists. For additional protection, an emergency brake is applied if the main drive chain breaks. Either brake can hold a fully loaded escalator motionless. People may then walk up or down as on any stairway. Since passengers can't be trapped on an escalator, emergency power is usually not required, as it is for elevators.

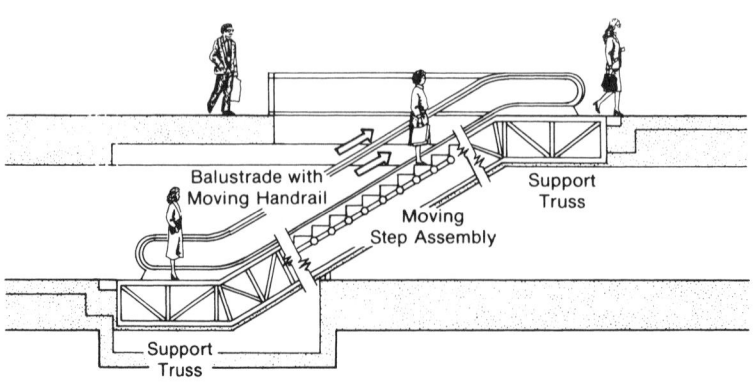

FIGURE 11.13. Escalator.

Moving Walks and Ramps

Moving walks and ramps are also available for transporting people either horizontally or at a slight incline. These are special conveyor belts that come in widths ranging from 2 to 9 feet (60 cm to 2.75 m), accommodating 3,600 to 18,000 people per hour at a maximum speed of 140 fpm (71 cm/s). The incline angle can be from 0° to 15°.

By definition, *moving walks,* also known as *moving sidewalks,* have an incline no greater than 5°. The moving walk's function is to transport people rapidly and effortlessly in the horizontal direction, and to easily transport large and bulky objects over long distances. It can also reduce congestion by moving people quickly away from a potentially congested area. Passenger terminals of any kind are prime candidates for moving walks. They are also useful for transporting people to and from remote parking lots.

A *moving ramp,* as distinguished from a moving walk, has an incline that is limited by code to no more than 15°. Moving ramps can rapidly transport large numbers of people up an incline like an escalator, but they are better suited for transporting baggage, large parcels, wheelchairs, and other wheeled vehicles. Another advantage of ramps is that they can be faster (up to 140 fpm) (71 cm/s) than escalators if necessary. Ramps can be used in multilevel stores if the use of shopping carts makes escalators awkward, if not impossible, to use. Moving ramps are also more useful than escalators in multilevel airports because they are more convenient for wheeled luggage carriers.

Speed and physical dimensions are not standardized for moving walks and ramps, and equipment is usually custom-designed for a given application. Most installations, however, are either 26 inches (66 cm) (single file) or 40 inches (102 cm) wide (two passengers abreast). Each job must be referred to a manufacturer for the development of construction details.

Materials-Handling Systems

Equipment that satisfies materials-handling needs includes *dumbwaiters, conveyors,* and *pneumatic systems.* Chutes are an economical passive alternative for conveying materials solely downwards. Mechanical systems can be grouped into four categories:

1. Dumbwaiters (vertical lift systems using a car similar to a miniature elevator)
2. Continuous conveyors (horizontal, vertical, or inclined)
3. Pneumatic systems (pneumatic tube systems and pneumatic trash and linen systems)

4. Other systems (self-propelled messenger carts and track-type automatic delivery systems, for the most part proprietary)

Dumbwaiters are used in department stores to transport merchandise from stock areas to selling or pickup counters; in hospitals to transport food, drugs, linens, and other small items; and in multilevel restaurants to deliver food from the kitchen and return soiled dishes.

The platform area of dumbwaiter cars is limited to 9 ft² (0.8 m²) and the height to 4 feet (1.2 m). Car speeds range from 45 to 150 fpm (75 cm/s) and a capacity of up to 500 pounds (230 kg). Automated delivery, or ejector-type, dumbwaiters, which automatically push carts or baskets out at the delivery terminal, are available with capacities of up to 1,000 pounds (450 kg) and car speeds of up to 300 fpm (150 cm/s).

Horizontal conveyors are commonly used to transport check-in baggage in airports, soiled dishes in cafeterias, and, occasionally, parcels in post offices or commercial buildings such as mail-order houses. Vertical conveyors provide a continuous-loop delivery system as a faster alternative to a dumbwaiter.

Pneumatic tube (PT) systems are used for delivering paper communications between two points. They are fast, efficient, reliable, and relatively inexpensive, but the compressors to operate these systems require a machine room and are somewhat noisy.

Pneumatic trash and linen systems are generally more efficient and less expensive than dumbwaiters or conveyors for rapidly moving bagged or packaged trash and linen to a central collecting point. The linen systems are primarily used in hospitals. Trash systems, together with automatic compactors, have numerous applications. The transporting pipes are 16, 18, or 20 inches (40, 46, or 50 cm) in diameter. The compressors are large and noisy, requiring a considerable amount of acoustically isolated space.

The above-mentioned systems represent only generic types, and Sweets catalogue and manufacturers' representatives should be consulted for the most up-to-date technology for material handling. First, however, a study of the material transfer requirements for a given facility should be undertaken. Specialized consultants in the field of materials handling are available, and some preliminary information can be obtained from manufacturers' representatives.

SUMMARY

Chapter 11 covers two categories of special systems: signal systems and conveying systems. Signal systems are those that transmit electronically coded information within a building. They include fire alarms, telecommunications systems, security systems, and controls for HVAC systems and electrical loads.

Signal systems have three physical components: initiating devices, indicating devices, and small low-voltage wiring linking them together. Modern highly automated buildings contain a substantial amount of signal wiring running throughout. The exposed devices on either end of the wiring systems include:

- Fire and smoke detectors
- Fire alarm pull stations
- Fire alarm lights, bells, horns, and sirens
- Telephone and switchboard equipment
- Computer terminals
- CCTV cameras
- Television monitor screens
- Intrusion and motion detectors
- Intercoms
- Public address speakers
- Thermostats
- Devices to monitor and control HVAC equipment and electrical loads

Conveying systems consist of:

1. Vertical transportation
   Passenger elevators
   Freight elevators
2. Inclined transportation
   Escalators
   Moving walks and ramps
3. Materials-handling systems

All conveying systems are noisy and should be physically or acoustically isolated from spaces requiring low noise levels. Elevators can be either the cable or hydraulic type. Hydraulic elevators are less expensive and quieter, but also slower and limited to about 75 feet (23 m) of car travel. They also use more energy. The number, speed, and capacity of elevators are determined by the amount of traffic (number of people served) and the building height.

For transporting large numbers of people only two or three stories, escalators are less expensive, take up less space, and provide more continuous service with little or no delay. Escalators, however, are not well suited for wheelchairs, wheeled luggage carriers, shopping carts, or large, bulky parcels, so moving ramps may be more useful for some applications. Moving walks can transport people and bulky objects horizontally, or at a slight incline, rapidly over relatively long distances.

Materials-handling systems include dumbwaiters; horizontal, inclined, or vertical conveyors; pneumatic tube systems; pneumatic trash and linen systems; and other specialized vehicles for carrying objects and paper communications. For some applications, more than one type of system may be suitable, in which case service and cost should be compared. Wherever mechanized material handling is considered, a study should be made of the quantities of materials involved, how quickly and how far they need to be moved, and other factors.

KEY TERMS

Initiating device	Hydraulic-type	Running time
Indicating	elevator	Interval, lobby
equipment	Hoistway, shaft	dispatch time
Fire and smoke	Plunger, jack	Travel time,
detectors	National	average trip
Pull station	Elevator	time
Audio	Industry, Inc.	Elevator zone
indicators	(NEII)	Underslung
Visual	Building	arrangement
indicators	population	Escalator,
Security	Handling	moving
system	capacity (HC)	stairs, electric
Motion	Car capacity	stairs
detector	Passenger load,	Moving walk,
Intrusion	car loading,	moving
detector	number of	sidewalk
Voice-activated	passengers	Moving ramp
alarm system	per trip	Dumbwaiters
Cable- (electric	(NPT)	Conveyors
or traction)	Round trip time	Pneumatic
type elevator	(RTT)	systems

STUDY QUESTIONS

1. List the signal systems that might be required in each of the following types of buildings:
   a. Office
   b. Primary or secondary school
   c. College or university
   d. Auditorium
   e. House of worship
   f. Restaurant
   g. High-rise apartment building

h. High-rise hotel

i. Industrial facility

j. Hospital

k. Library or museum

l. Department store

2. How many elevators, and of what capacity, would be required in the 20-story office building described below?

- 12-foot (3. 7-m) floor-to-floor height
- Square overall building plan 100×100 feet (30×30 m)
- 100 ft^2 (9.3 m^2) per person
- Necessary handling capacity of 13 percent
- Average loading and unloading time per floor of 3 seconds

3. How much area would each elevator shaft in Question 2 take up on each floor?

4. Also referring to Question 2, how many freight elevators should be provided, and how much space must be allocated for them?

BIBLIOGRAPHY

Life Safety Code. National Fire Protection Association, Batterymarch Park, Quincy, MA 02269.

Stein, Benjamin, and John S. Reynolds. *Mechanical and Electrical Equipment for Buildings,* 9th ed. John Wiley & Sons, Inc., 1999.

Strakosch, G.R. *Vertical Transportation, Elevators and Escalators,* 2nd ed. John Wiley & Sons, Inc., 1983.

PART FOUR

OTHER PIPED SYSTEMS

Plumbing systems are primarily designed by mechanical or plumbing consultants. However, architects are usually involved in specifying fixtures and in locating stormwater drains. Architects also typically size gutters and down spouts and coordinate the overall drainage plan. Otherwise, plumbing design is handled by consultants.

Fire protection is an integral aspect of the entire building. It includes specific alarm and extinguishment systems, but it also is concerned with the structural fire resistance of the building for containing a fire and the layout for safe emergency egress. Mechanical ventilation systems are also a key component of modern fire protection.

Chapter 12

Plumbing Systems

WATER SUPPLY

WATER SOURCES

DISTRIBUTION SYSTEMS

PLUMBING FIXTURES

TRAPS

CROSS-CONTAMINATION

ALTERNATIVES TO THE STANDARD FLUSH TOILET

SANITARY PIPING

PIPING

VENTS

INTERIOR SPACE PLANNING

FLOOR DRAINS

CLEAN-OUTS

INTERCEPTORS

BACKWATER VALVES

SEWAGE EJECTOR PUMPS

PRIVATE SEWAGE DISPOSAL

STORMWATER DRAINAGE

ROOF DRAINAGE

WATER DISPOSAL

Plumbing systems in buildings serve two basic functions: (1) providing water and drainage for sanitation and potable water needs and (2) disposal of precipitation falling on the building. The latter system is referred to as the *stormwater system.*

Only a century ago, urban dwellers disposed of their sewage and garbage by dumping it into the gutter. In rural areas, people dumped their waste into nearby lakes and rivers. Where there was no natural water depository, an artificial lake was created, consisting of a hole in the ground filled with rain and spring water. It was, and still is, called a *cesspool.*

All of these practices created a foul-smelling health hazard. Today, the presence of a potable water supply allows wastes to be made water-borne and disposed of through a piping system. Both cesspools and the dumping of waste into the street are now illegal. The former has been replaced by safe, fully enclosed septic tanks, and in cities, sanitary sewers carry waste to treatment plants which neutralize the toxicity before releasing the water into a waterway.

Buildings are generally supplied with running potable water which is distributed to plumbing fixtures throughout the building. The waste materials generated from human occupancy such as human excreta, food wastes, and other undesirable matter are introduced into appropriate plumbing fixtures where they are carried away by water and discharged into a drainage system which transports them safely to their ultimate disposal point. Our large multistory buildings could not be inhabited without a modern water-distributing and waste disposal system.

There are three components to the typical sanitation system:

1. A distribution system of piping that supplies water to the fixtures

2. The fixtures at which the water is consumed or used
3. Another piping system that drains the used water away

The fixtures are generally the only portion that is visible.

The typical utility connections at the building boundary for these plumbing systems are:

- Potable water supply (and meter)
- Sanitary drain
- Storm drain

Natural gas piping, where required, is often under the jurisdiction of the plumbing trade as well. The connection to a gas company pipeline is at the building boundary.

Almost every aspect of plumbing design and the materials used are governed by the local building code, which has the force of law. There are three national standard codes which are listed in the bibliography at the end of this chapter, and most local building authorities have adopted one of them. The primary purpose of these plumbing codes is to protect the health and safety of building occupants against inadvertent contamination of the water supply, contamination of the air and objectionable odor due to the escape of sewage gas, and chronically blocked drains due to improper size or pitch.

WATER SUPPLY

Water Sources

The source of water for most developed areas is either a lake or river. Since most of the waterways in industrialized areas are polluted, the water must be treated before it is supplied to buildings. A small proportion of the public water systems in the United States use groundwater from deep wells.

When buildings are planned for locations remote from public water services, a private water supply is necessary. Water coming straight from the ground from a well or spring is usually more pure than water from a stream or pond, which may have been exposed to pollution.

Special consultants are available to ensure both the proper quantity and quality of *supply water,* and the local health department requires certain specifications to be satisfied. The purity of a water supply and its suitability for drinking purposes, steam generation, or other industrial process can be assessed by a water-quality analyst.

Containers for collecting water samples and instructions for taking the sample can be supplied by the analyst. The analyst's report provides numerical values for the mineral content, pH, contamination, turbidity (cloudiness), total solids, and biological purity, together with an opinion as to the water's suitability for the intended purpose.

The pH is a measure of the acidity or alkalinity of the water. Acidic water has a pH of less than 7.0 and may accelerate corrosion of pipes and tanks unless neutralized.

Water may also contain bacteria that grow at ordinary temperatures and normally don't cause disease. Their presence in very large numbers, however, is undesirable. Bacteria known as *coliform,* originating from animal pollution, do produce diseases.

If necessary, the water may be treated with limestone to neutralize acidity or be filtered, sterilized, or softened. "Hard" water, which contains an excessive percentage of minerals, may cause scale to be deposited in hot water pipes, prevent soap from lathering well, and adversely affect some industrial processes.

If a private water supply source is needed, it is usually necessary to make private arrangements to dispose of sewage, and care must be taken not to pollute the water source. On-site treatment can be provided, if necessary.

Distribution Systems

Water Pressure

Potable water is distributed to buildings in developed areas through large pipes, called *street mains,* which usually run underground down the length of streets. The water company supplies clean water under pressure, which normally amounts to between 50 and 70 psi (pounds per square inch) (340 to 480 kPa) by the time it gets to a building. The exact pressure depends on the elevation of the building in relation to the source and on the fluctuation in demand on the system. The service entrance pipe to the building must be below the frost line in cold climates in order to prevent freezing. The ground usually freezes to a depth of somewhere between 2 and 7 feet (0.6 and 2.1 m), depending on the climate.

If a municipal water supply is not available, water must be obtained from a private well or some other source, and pumps must be used to create enough pressure for the fixtures to operate properly. The pressure of the water supply to the building must be great enough to overcome both the resistance in the distribution piping system and the *static pressure* of water standing in the vertical piping, and still be able to provide the necessary pressure at the fixtures.

The concept of static pressure is illustrated in Figure 12.1. One cubic foot of water weighs 62.4 lb. A 1-ft³ cube of water, therefore, exerts by gravity the following pressure on whatever surface it rests:

$$\frac{62.4 \text{ lb}}{\text{ft}^2} = \frac{62.4 \text{ lb}}{144 \text{ in.}^2} = 0.433 \text{ psi} \qquad (12.1)$$

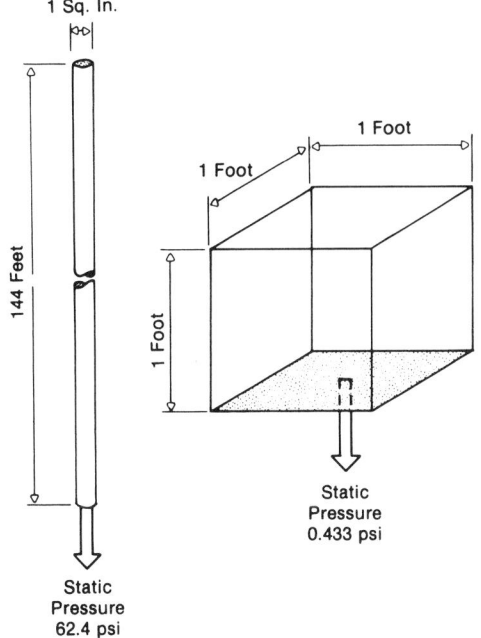

FIGURE 12.1. Static pressure.

If that same volume of water were placed in a column with a 1 in.^{2} cross section, it would extend up vertically 144 feet. The pressure exerted downward would increase to:

$$\frac{62.4 \text{ lb}}{1 \text{ in.}^2} = 62.4 \text{ psi} \qquad (12.2)$$

That 1-in.^{2} column of water exerts a downward pressure of 5.2 psi for every 12 feet of height. The supply water must be under 5.2 psi of pressure just to hold that 12-foot column of water in place. When the valve is opened on a second-floor fixture in a building with 12 feet from floor to floor, a pressure of more than 5.2 psi is needed just to get the water moving. A fixture on the tenth floor needs more than 52 psi in addition to the pressure necessary to operate the fixture.

In tall buildings, the pressure in the street main may not be great enough to operate fixtures on the upper floors, and in very tall buildings, there may not be enough pressure to even get water to the upper floors at all. In that event, additional pumps may be installed in the building to add the required pressure. This arrangement is called an *upfeed distribution.*

Another option is to feed water up to an elevated rooftop storage tank, from which water is supplied to the floors below by gravity. A rooftop storage tank can also be used to supply a fire hose system under emergency conditions. In cold climates, the water in the tank must be heated in order to prevent freezing.

Some very tall buildings may have to be divided into distribution zones in order to keep the water pressure within proper limits. If all the water were supplied from a rooftop tank, the static pressure from the water column leading all the way up to the roof might be excessive. Therefore, the bottom zone may be upfed, an intermediate tank halfway up may supply water to the lower half of the building, and a rooftop tank may supply the top half.

An additional tank in the basement, called a *suction tank,* is sometimes used to provide an additional buffer between the street main and the pumps that replenish the depleted water in the storage tanks. This keeps neighboring water users from being adversely affected by sudden demands within adjacent large buildings.

Rooftop water tanks are commonly enclosed by a surrounding screen two or more stories high, which can also conceal other equipment and irregularly shaped objects such as two-story elevator penthouses, cooling towers, exhaust fans, numerous plumbing vents, and exterior window-washing rigging.

Disadvantages of rooftop tanks are the greater structural support required and the need for periodic cleaning. Their main advantage is that they provide reserve storage in case of an electrical power shortage. An upfeed distribution system requires a standby power source in order to provide that same benefit. In most cases, the decision on whether or not to have a rooftop tank is based on economic considerations.

Domestic Hot Water

Although sometimes referred to as *building service hot water* (BSHW) in nonresidential applications, *domestic hot water* (DHW) is still the most common term. Water may be heated by a wide variety of devices that burn fuel or use electricity, the sun, or some other thermal energy source. Solar water heaters rarely provide 100 percent of DHW requirements. Thus, they are usually placed in-line with a conventional fuel-fired or electric heater and serve to preheat the water. Some parts of the United States now require solar water heaters on most new construction.

Hot water heaters fall into two general categories: *storage* (*tank-type*) and *instantaneous* (*tankless or in-line type*). The storage-type water heater (Figure 12.2a) consists of an insulated water-storage tank and a heating means, which heats the fixed quantity of water in the tank and stores it until it is needed. Moderate heat input can be used over a long period of time to accumulate hot water for a sudden heavy draw-off. Electrically powered storage-type water heaters are capable of taking advantage of off-peak electric power rates. For example, they can heat water during the night for use the next day.

The heaters are rated by their tank capacity in gallons and

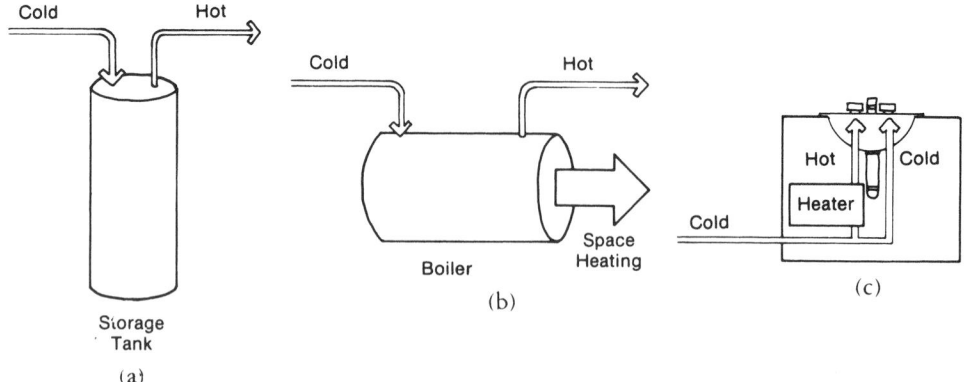

FIGURE 12.2. Water heater types. (a) Tank type, (b) with space heating boiler, and (c) point-of-use type.

by the recovery time, which is the time it takes for the tank to come up to the desired temperature when it is filled with cold water. The tank size must accommodate the quantity of hot water needed, and the capacity of the heating means must provide a sufficient *recovery rate* in accordance with the frequency of use.

Instantaneous water heaters, as the name implies, instantly heat water as it passes through on the way to the hot water fixture. Cold water enters at one end and is instantly converted into hot water, and the process continues indefinitely until hot water is no longer drawn. Instantaneous water heaters eliminate standby losses from hot water tanks.

The terms *direct/immersed* and *indirect/takeoff* refer to instantaneous water heaters that obtain their heat from a space-heating boiler (see Figure 12.2b), which must operate whenever hot water is needed, winter or summer. Usually, summertime use of a large space heating boiler is very inefficient for hot water heating.

Another type of instantaneous heater, illustrated in Figure 12.2c, is referred to as *point-of-use* or *local* water heaters. These are self-contained units that are located close to the fixture or fixtures requiring hot water. They are essentially small heat exchangers, either directly gas-fired or electrically powered. They can be installed easily in cabinets, vanities, or closets near the point of use. An example is shown in Figure 12.3.

Point-of-use water heaters, like booster heaters, eliminate standby heat losses from both hot water storage tanks and long piping runs. By avoiding the continuous temperature difference, instantaneous in-line heaters help to reduce energy consumption.

Because of the tremendous amount of heat required to instantaneously heat water up from ground temperature to

FIGURE 12.3. Typical point-of-use electric water heater mounted on the wall below a pedestal-type lavatory. (Photo courtesy of Stiebel Eltron Inc.)

the required use temperature, local water heaters are usually very limited in the capacity they can handle. If electric, these devices cause a large peak demand. For this reason, *electric in-line heaters are usually not economical on a life-cycle basis* where demand charges are levied.

Any water heater, regardless of the type, should bear the label of the National Board of Fire Underwriters, indicating approval for safety.

Temperatures. For energy conservation purposes, hot-water supply temperature should be kept as low as possible. The required hot-water supply temperatures for various uses are as follows:

180°F (82°C): Minimum needed for the sanitizing rinse on commercial dishwashers and laundries. Some types of equipment have built-in booster heaters to bring water up to this temperature so that the main distribution can be at a lower temperature.

140°F (60°C): For general-purpose cleaning and food preparation. General-purpose sinks should generally not have water above 140°F (60°C) because higher temperatures can cause serious burns.

105° to 120°F (41° to 49°C): For showers. Often provided by blending 140°F (60°C) general-purpose water with cold water in a mixing valve.

Distribution. Hot water is distributed by either (1) direct make-up to faucets under municipal pressure or (2) a *recirculating* hot water pump. The recirculating hot water pump delivers hot water instantly at the faucets. This is more convenient and eliminates the wasting of cold water that would otherwise need to be drawn off before hot water begins flowing. This can be a large quantity in large buildings.

The disadvantages of a recirculating system are that the pump is continuously consuming electrical power whenever operating, and the entire piping system is always filled with hot water, causing a greater heat loss than if the pipes were allowed to cool down for long periods of time between use. When hot water is drawn frequently from a number of taps throughout the system, this latter form of loss is equally great for either type of system. Operating recirculating pumps on a timer switch that shuts the system off during nighttime hours when there is no usage can reduce these aspects of energy consumption. In smaller systems in which all faucets are close to the source (such as single-family residences), the energy wasted in drawing off cold water first is less than the energy used by continuously recirculating hot water.

Chilled Water

Chilled drinking water is provided in most public buildings. Drinking fountains with their own self-contained refrigeration equipment for chilling are called *water coolers*. When a large number of drinking fountains are needed in a building, a central chilling unit may be installed, and the chilled water circulated to the various outlets.

Water is normally circulated continuously to the drinking fountains from the refrigeration device in order to overcome heat gain through the piping and have chilled water instantly available at any outlet. Like circulating hot-water systems, these systems use a pump to generate the flow. The pipes must be insulated so that they will not sweat and the water will not be warmed any more than necessary. The insulation must be covered with a vapor-tight material to prevent the surrounding warm, moist air from entering the insulation and condensing.

Piping Materials

Piping consists of pipe or tubing, and fittings used to join pipe to valves, fixtures, water heaters, or other pipe.

Copper Piping. Piping used for water supply within buildings is usually made of copper because of its strong resistance to corrosion. It is called *copper tubing* rather than *pipe* because it has such a thin wall. The walls are too thin for screw-type joints, so the fittings must be soldered to the copper tubing in what are called "sweat joints."

Copper piping for supply water purposes is denoted as type K, type L, or type M, in decreasing order of wall thickness. Types K and L are available in either a rigid or flexible tube known as hard or soft temper tube. The soft temper tube permits changes in direction without the use of fittings, and can be fed into existing construction voids. Type M is made in hard temper only.

Plastic Piping. Plastic pipe and fittings are produced from synthetic resins that do not appear in nature, but are derived from fossil fuels such as coal and petroleum. Plastic piping comes in a variety of materials known as PE, ABS, PVC, and PVDC. The PVDC pipe is the only one suitable for carrying hot water, and it is limited to temperatures no greater than 180°F (82°C).

Like copper, plastic piping does not present corrosion problems. It is lighter in weight and, at the present time, cheaper than copper. The National Sanitation Foundation (NSF) tests and certifies plastic pipe, and all pipes that are to carry potable water must bear the NSF seal. Local codes vary in their acceptance of plastic piping. Some allow it only for waste piping, some allow it also for cold water, and some even allow PVDC for hot water supply.

Pipe Insulation

Supply water piping is insulated for the same reasons that chilled and heating water pipes in HVAC systems are insulated: to prevent condensation and to reduce heat transfer between the air and the contents of the pipe. For example, when ground water is 50°F (10°C), the supply piping carrying that water has a surface temperature of about 60°F (16°C). According to the psychrometric chart, air at a typical summer temperature of 85°F (29°C) will condense on that surface whenever the relative humidity is above 40 percent.

All cold water piping should be covered with insulation and a tight vapor barrier. All hot water pipes in a recirculating hot water system should be insulated in order to conserve energy.

PLUMBING FIXTURES

All plumbing fixtures are supplied with clean, potable water and discharge contaminated fluids. Each one is designed to perform a specific function easily and safely. The types of fixtures include the following, which are illustrated in Figure 12.4:

- Water closets
- Urinals
- Bidets
- Lavatories
- Sinks
- Service sinks
- Wash fountains
- Drinking fountains (water coolers)
- Showers
- Bath tubs

Water closets (toilets) are the most common type of soil fixture. Closet bowls are available for floor- or wall-mounting. The wall-hung unit facilitates floor cleaning but requires a substantial fixture carrier. Bowl contours are available with round or elongated fronts. Water closets for use by handicapped persons are wall-hung and have an elongated front.

Urinals are desirable in men's public toilet rooms in order to reduce the contamination of water closet seats that occurs when closets are the only soil fixture. Also, each unit requires only about 18 inches (46 cm) of width. Wall-hung units help keep adjacent surfaces cleaner than the stall type does, but their standard height is too high for convenient use by young boys. One or more urinals in a group are sometimes positioned lower than the rest for this reason.

Water closets and urinals may be operated from either a flush tank or a flush valve, the latter being most common. A water-closet flush tank may be separate from the closet bowl, in which case it is hung on the wall, or an integral part of it. Flush tanks for urinals, always separate, are mounted above the urinals on the wall. Flush valves supply water at a high rate to the fixture for proper flushing, so they require a higher water pressure. In any case, water closets and urinals should be of the water-conserving type and should use no more than 4 gallons (14.1 L) of water per flushing cycle.

Lavatories in public facilities should have self-closing faucets in order to minimize water wastage. Self-closing faucets also minimize energy used for heating hot water. The flow rate from the faucets should be limited to a maximum of 0.5 gpm (1.9 L/min). Water flow through shower heads and kitchen faucets should be restricted by water-saving devices to a maximum of 3 gpm (11.4 L/min).

Service sinks, sometimes called *slop sinks,* are located in janitors' rooms for filling buckets, cleaning mops, and for general cleaning purposes.

Drinking fountains are made in a variety of forms for either indoor or outdoor use. Indoor units may be free-standing or wall-mounted. If projection into corridors is undesirable, recessed units are available. Fountains accessible to handicapped people are surface-mounted and have a push valve on the front. Outdoor fountains are normally pedestal-mounted and may have a frost-proof, foot-operated valve.

Sometimes a drinking fountain is provided by installing a bubbler in a lavatory or sink. Most codes, however, prohibit drinking fountains within toilet rooms. In addition, the bubbler arrangement does not provide chilled water, which is almost universally desired.

Plumbing fixtures must be constructed of a nonporous material. The materials commonly used are enameled cast iron, vitreous china, stainless steel, copper, and brass. Commercially available fixtures have rounded interior corners and are smooth, strong, hard, and capable of withstanding years of rugged use.

The selection of fixtures for a particular facility is usually a joint decision of the client, architect, and mechanical engineer. However, the minimum number of fixtures required is dictated by code. Different codes and local jurisdictions may vary concerning the number of each fixture type required for various categories of buildings. As an example, the fixture requirements of the *Uniform Plumbing Code* are presented in Table 12.1.

(a)	(b)
(c)	(d)

FIGURE 12.4. Examples of fixtures. (a) Wall-mounted water closet with flush valve, (b) floor-mounted tank-type water closet, (c) bidet, (d) wall-hung urinal, (e) wheelchair-accessible lavatory, (f) counter-mounted lavatory, (g) wall-mounted lavatory, (h) pedestal-mounted lavatory, (i) sink, (j) mop sink, (k) wash fountain with infrared control that starts water flow when hands are placed in the bowl and shuts off when hands are removed (photo courtesy of Bradley Corporation), (l) barrier-free wall-hung water cooler, (m) recessed water cooler, (n) freestanding water cooler, (o) institutional showers, (p) bathtub, and (q) waterless urinal using a chemical-type drain trap. (Photo a courtesy of American Standard, Inc. Photos b, d, e, f, g, h copyright 2005 American Standard. Photos j and o compliments of Acorn Engineering Company. Photos l–n courtesy of Halsey Taylor. Photos c, i, p, q by Vaughn Bradshaw.)

(e)

(f)

(g)

(h)

FIGURE 12.4. *(Continued)*

362

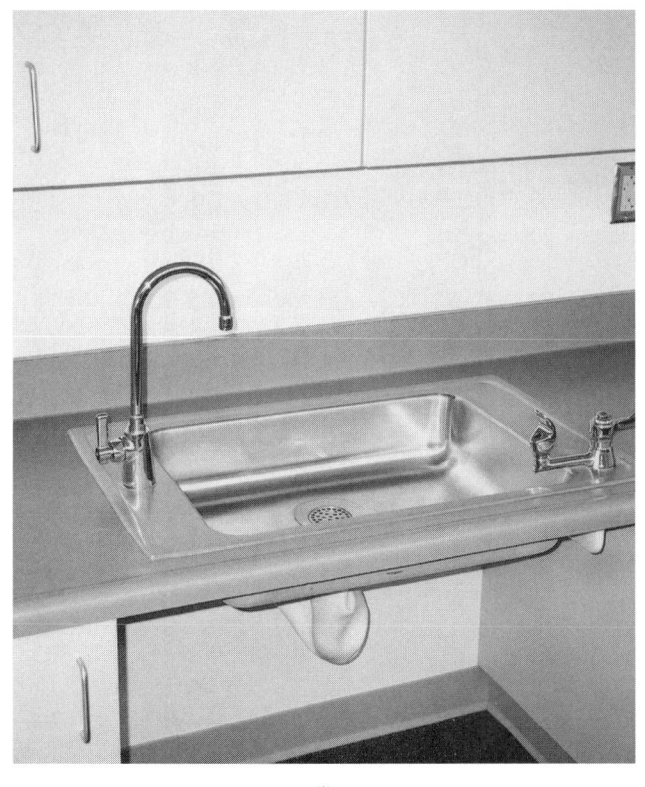

(i)

(j)

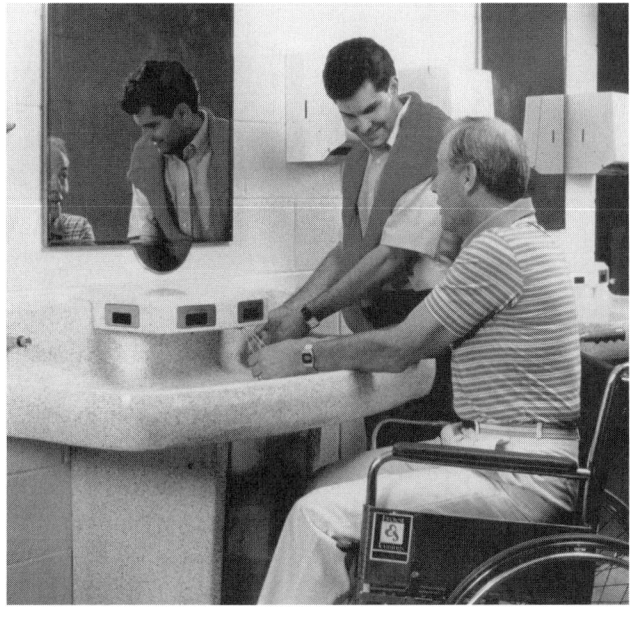

(k)

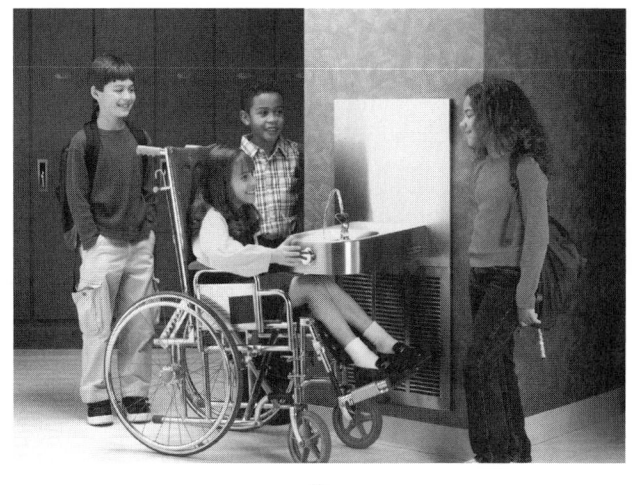

(l)

FIGURE 12.4. *(Continued)*

363

(m)

(n)

(o)

(p)

FIGURE 12.4. *(Continued)*

(q)

FIGURE 12.4. *(Continued)*

Some fixture mounting heights are dictated by the Americans with Disabilities Act or other codes. Fixture heights are usually dimensioned on the architectural drawings.

Traps

The putrefaction (decomposition) of the waste materials discharged from fixtures generates various gases. All of them are odorous, some are toxic, and some, such as methane, are explosive. The waste disposal piping within a building contains these gases, called *sewer gases,* which not only have an unpleasant odor but are also unhealthy to breathe. In order to isolate these gases from the interior room air, there must be a *trap* downstream from the discharge point of each fixture. A plug of water caught in the trap allows the next discharge of water to flow through, but prevents gases from passing back into the room.

When water is released down the drain, some of it replaces the previous water. A sufficient amount of water must flow through the trap after each discharge from the fixture to completely flush out the trap so that the residual water is clean. This makes the trap self-cleaning, leaving no sediment inside to decompose and emit noxious gases.

Water closets have a self-contained trap cast integrally

with the fixture (see Figure 12.5a). Some urinals have an integral trap similar to the water closet bowl, in which case they flush in the same way, while others require a separate trap. The wall-hung type generally has an integral trap with the outlet to the wall. Stall types discharge through the floor, and generally require a separate trap. All other fixtures require a separate trap in the drain piping within 2 feet (60 cm) of the fixture. These may be letter-shaped traps (Figure 12.5b) or drum traps (Figure 12.5c). P-, S-, and U-traps are named for their general shape.

When fixtures are not used frequently, the water in their traps tends to evaporate into the air. The trap then no longer seals out the sewer gases. This is usually a problem only with floor drains. Since floor drains tend not to be used very often, their traps may often dry out, creating a potentially very hazardous condition. For this reason, some codes prohibit connecting floor drains to sewer lines, making it necessary to arrange for some other provision. A hose bibb is sometimes provided above a floor drain to enable the trap to be periodically refilled.

Traps are desirable and necessary in the plumbing system, even though they tend to collect all sorts of debris and are the first place that stoppages occur. In public facilities, an endless variety of nonsoluble objects end up in drains, and when stoppages do occur, the traps must be accessible for clearing out. P-traps located beneath lavatories are readily accessible, but access panels may be needed in order to access traps to other fixtures concealed behind walls or beneath the floor.

Cross-Contamination

Because of the proximity of sewage to potable water in fixtures, there is a chance that sewage can accidentally be back-siphoned into the pipe carrying the potable water. This could happen, for example, if the outlet from a faucet is placed below the rim of a fixture and the fixture's bowl is full. The supply water could easily be contaminated by the contents of the bowl. And if the bowl is drained while the faucet is open, it is possible that the supply water will become contaminated by sewer water.

It is therefore important that an air gap exist between the supply outlet and the highest possible level of water in the bowl of all fixtures. In some fixtures and direct-connected equipment, a cross connection is unavoidable. Special devices known as *vacuum breakers, backflow preventers,* and *check valves* are used in such cases to prevent the backflow of polluted water into pipes carrying potable (hot or cold) water.

TABLE 12.1 MINIMUM NUMBER OF FIXTURES REQUIRED

Building Type	Water Closets		Urinals	Lavatories	Bathtub or Showers	Drinking Fountains
			Fixtures per Person			
Assembly Places Auditoriums, theaters, convention halls (employee use)	Male 1: 1–15 2: 16–35 3: 36–55 Add 1 fixture for each additional 40 persons	Female 1:1–15 3:16–35 4:36–55	0: 1–9 males 1: 10–50 males Add 1 fixture for each additional 50 males	One for every 40 people		
Assembly Places Auditoriums, theaters, convention halls (public use)	Male 1: 1–100 2: 101–200 3: 201–400 Add 1 fixture for additional 500 males, and 1 for each additional 300 females	Female 3: 1–50 4: 51–100 8: 101–200 11: 201–400	1: 1–100 2: 101–200 3: 201–400 4: 401–600 Add 1 fixture for each additional 125 males	1: 1–200 2: 201–400 3: 401–750 Add 1 fixture for each additional 500 people		One for every 75 people
Dormitories (residents' use)	Male 1: 1–10 Add 1 fixture for each additional 25 males, and 1 fixture for each additional 20 females	Female 1: 1–8	One for every 25 males Over 150, add 1 fixture for each additional 50 males	1: 1–12 Add 1 fixture for each additional 20 males, and 1 for each 15 additional females	1 shower: 1–8 Add 1 bathtub for every 30 females Over 150, add 1 bathtub for every 20 females	One for every 75 people
Dormitories (staff use)	Male 1: 1–15 2: 16–35 3: 36–55 Add 1 fixture for each additional 40 people	Female 1: 1–15 3: 16–35 4: 36–55	One for every 50 males	One for every 40 people	One for every 8 people	
Dwellings Single-family Apartments	One per dwelling One per dwelling or apartment unit			One per dwelling One per dwelling or apartment unit	One per dwelling One per dwelling or apartment unit	
Hospitals Patient rooms Wards Waiting rooms Employee use	One per room One for every 8 patients One per room Male 1: 1–15 2: 16–35 3: 36–55 Add 1 fixture for each additional 40 people	Female 1: 1–15 3: 16–35 4: 36–55	0: 1–9 males 1: 10–50 males Add 1 fixture for each additional 50 males	One per room One for every 10 patients One per room One for every 40 people	One per room One for every 20 patients	One for every 75 people One for every 75people
Industrial Mfg. plants Warehouses	1: 1–10 2: 11–25 3: 26–50 4: 51–75 5: 76–100 Add 1 fixture for each additional 30 people			One for every 10 people up to 100 people One for every 15 people over 100 people	One shower for each 15 people exposed to excessive heat or to skin contamination with poisonous, infectious, or irritating material	One for every 75 people

TABLE 12.1   MINIMUM NUMBER OF FIXTURES REQUIRED *(Continued)*

Building Type	Water Closets		Urinals	Lavatories	Bathtub or Showers	Drinking Fountains
			Fixtures per Person			
Offices Employee use	Male 1: 1–15 2: 16–35 3: 36–55 Add 1 fixture for each additional 40 people	Female 1: 1–15 3: 16–35 4: 36–55	0: 1–9 males 1: 10–50 males Add 1 fixture for each additional 50 males	One for every 40 people		
Public Buildings	Male 1: 1–100 2: 101–200 3: 201–400 Add 1 fixture for each additional 500 males, and 1 for each additional 150 females	Female 3: 1–50 4: 51–100 8: 101–200 11: 201–400	1: 1–100 2: 101–200 3: 201–400 4: 401–600 Add 1 fixture for each additional 300 males	1: 1–200 2: 201–400 3: 401–750 Add 1 fixture for each additional 500 people		One for every 75 people

Penal Institutions
  Inmate use
    Cell
	One per cell			One per cell		One per cell block floor

Exercise Room	One per exercise room		One per male exercise room	One per exercise room		One per exercise room
Employee use	Male 1: 1–15 2: 16–35 3: 36–55 Add 1 fixture for each additional 40 people	Female 1: 1–15 3: 16–35 4: 36–55	0: 1–9 males 1: 10–50 males Add 1 fixture for each additional 50 males	One for every 40 people		One for every 75 people
Restaurants, Pubs and Lounges	1: 1–50 2: 51–150 3: 151–300 Add 1 fixture for each additional 200 people		1: 1–150 males Add 1 fixture for each additional 150 males	1: 1–150 2: 151–200 3: 201–400 Add 1 fixture for each additional 400 people		

Schools
  Student use
    Nursery
	1: 1–20
	2: 21–50
	Add 1 fixture for each additional 50 students
				1: 1–25
				2: 26–50
				Add 1 fixture for each additional 50 students
						One for every 75 people

Elementary	One for every 30 males One for every 25 females		One for every 75 males	One for every 35 students		One for every 75 people
Secondary	One for every 40 males One for every 30 females		One for every 35 males	One for every 40 students		One for every 75 people
Colleges, universities, adult centers, etc.	One for every 40 males One for every 30 females		One for every 35 males	One for every 40 students		One for every 75 people
Staff use	1: 1–15 2: 16–35 3: 36–55 Add 1 fixture for each additional 40 people		One for every 50 males	One for every 40 people		

TABLE 12.1 MINIMUM NUMBER OF FIXTURES REQUIRED *(Continued)*

Building Type	Water Closets		Urinals	Lavatories	Bathtub or Showers	Drinking Fountains
			Fixtures per Person			
Worship Places						
Principal assembly place	Male 1: 1–150 2: 151–300	Female 1: 1–75 2: 76–150 3: 151–300	One for every 150 males	One for every 2 water closets		One for every 75 people
Educational and activity areas	Male 1: 1–125 2: 126–250	Female 1: 1–75 2: 76–125 3: 126–250	One for every 125 males	One for every 2 water closets		One for every 75 people

1. The number of water closets may be reduced by the number of urinals provided down to a minimum of ⅔ of the minimum number of water closets specified.

2. The total number of water closets for females should at least equal the total number of water closets and urinals for males.

3. Schools should have toilet facilities on each floor.

4. There should be a minimum of one drinking fountain per occupied floor in schools, theaters, auditoriums, dormitories, offices, or public buildings.

5. Refer to the local plumbing code having jurisdiction over a given project for other requirements.

6. All public places should have at least one toilet room for each sex, with a minimum of one water closet and one lavatory in each.

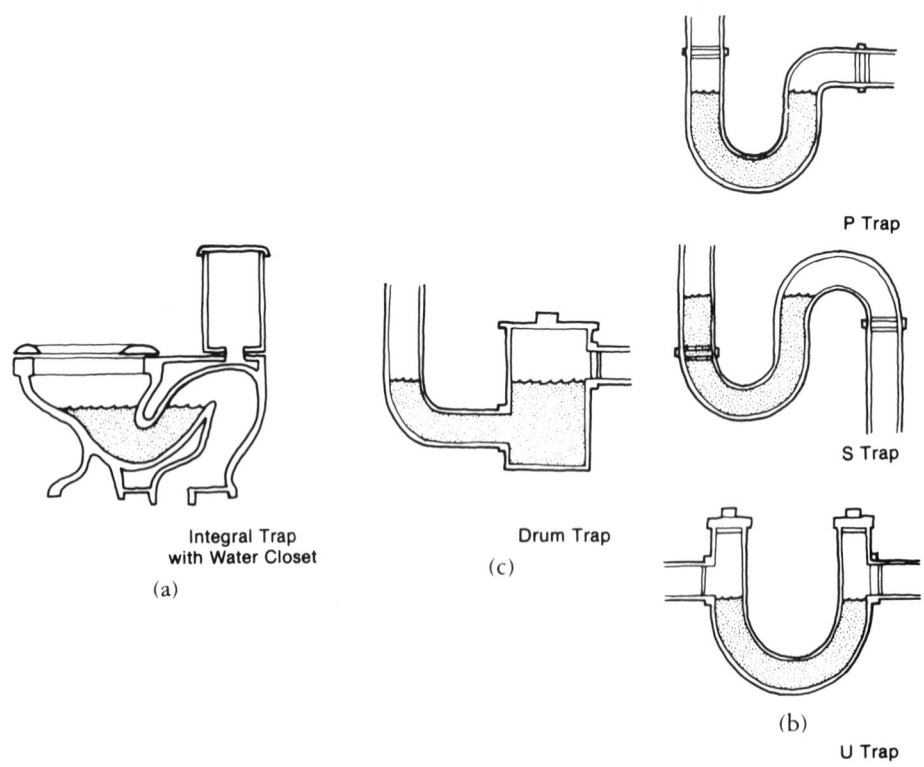

FIGURE 12.5. Traps. (a) Integral with water closet, (b) letter-shaped traps, and (c) drum trap.

Alternatives to the Standard Flush Toilet

There are essentially three types of toilets that reduce water requirements while still performing the function of a standard water closet. As illustrated in Figure 12.6, the types are (1) *pressurized air with water,* (2) *incineration,* and (3) *composting.*

The first type goes by the tradename of Microflush and is made by Microphor, Inc. It is an air-assisted toilet that uses only two quarts of water per flush. An air compressor, which can be remote from the fixture, supplies the compressed air that assists the water in the flushing process.

The incineration-type toilet is represented by the Incinolet from the Research Products Blankenship Corporation. It comes in either a toilet or urinal model and has no plumbing connections. It requires only the space occupied by the standard-size unit above the floor line. The waste deposited in the unit is completely reduced to a small volume of ash. The only connections that need to be made are to an electric power source—either 120/208 or 240 volts—and a 4-inch-diameter (10-cm-diameter) vent to outside air.

The third type of alternative toilet is called a *composting* or *humus* type. Examples are the Clivus Multrum, Mullbänk,

Carousel, and Humus 80. Except for the last one, these all consist of a toilet above the floor and a larger chamber below the floor, which holds the waste in ideal conditions, allowing it to decompose by aerobic digestion. The Humus 80 is a single unit that fits entirely above the floor. In all cases, room air is continuously drawn into the composting chamber and vented out through the roof. The one-way airflow prevents odors from passing back into the room and provides the composting chamber with plenty of oxygen for the digestion process.

The above-mentioned toilets reduce or eliminate water consumption for flushing at the expense of increased energy consumption. The pressurized-air type requires an air compressor that uses a modest amount of energy. The incineration type uses a large electrical current over a brief period of time to rapidly incinerate the waste. Most of the composting types use electricity for keeping the compost warm and for operating the ventilation fan. These uses probably would not amount to a substantial increase in energy use for most applications, but the inherent trade-off should be kept in mind. The requirement for electricity also makes these systems dependent on electrical service and inoperative in the event of power failure. However, this is no more of a prob-

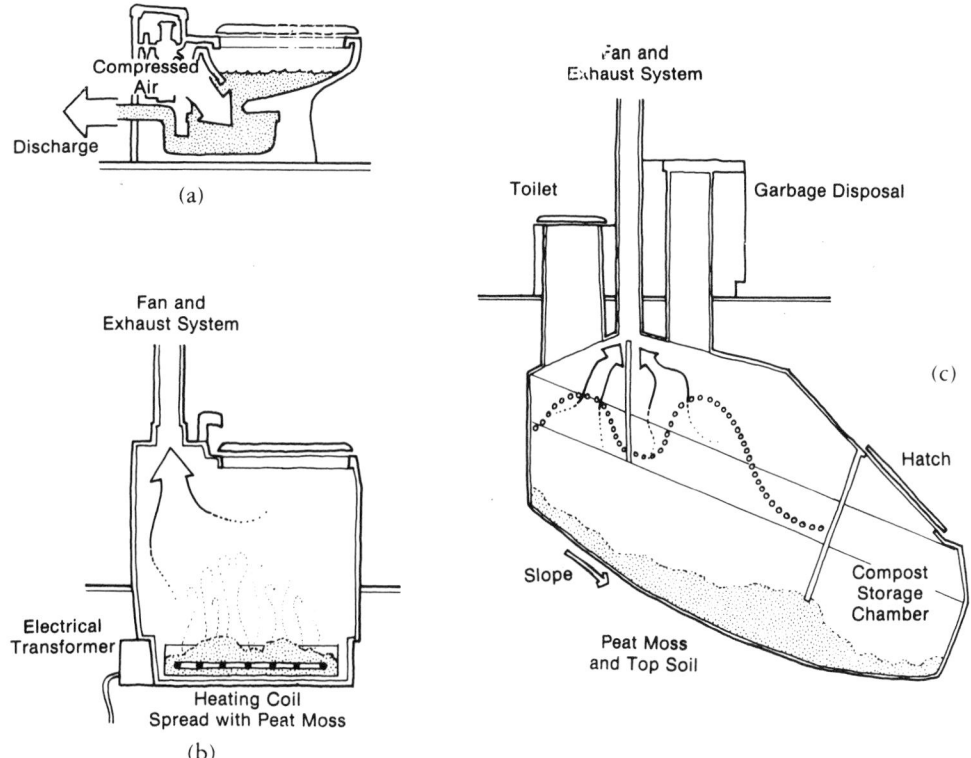

FIGURE 12.6. Alternative toilet types. (a) Pressurized air with water, (b) incineration, and (c) composting.

lem than the electrical requirements for pumping water to water closets on upper floors of high-rise buildings.

SANITARY PIPING

After water is used at the fixtures, it runs down a series of pipes and is led out of the building to a municipal sewer line or to a septic tank. The piping within the building, known as the sanitary system, is composed of the following elements:

- Waste stacks
- Soil stacks
- Vent stacks
- Branch waste lines
- Branch soil lines
- Branch vents
- Floor drains
- Clean-outs
- Fresh-air inlet

Drainpipes carrying nothing but dirty water from sinks and the like are referred to as *waste pipes*. Those carrying human excreta along with the water are referred to as *soil pipes*. A stack is simply a pipe stood on end. Vertical drainpipes are thus either *waste stacks* or *soil stacks*. A drain stack is a convenient, central drain that can serve a number of fixtures one above the other on different floors. *Branch drains* are drainpipes that are slightly pitched from the horizontal and connect to a stack or to the main building drain. A system of vertical air vent pipes are extended through the roof in order to admit air to, and discharge gases from, the sanitary piping system. These are called *vent stacks*, as illustrated in Figure 12.7.

Each fixture discharges through a trap into branch lines which are collected by vertical drain stacks. The bases of the stacks are connected together by the sloping building drain. The main building drain delivers the sewage to the building sewer, which by definition begins outside the building wall. The building sewer empties into the public sewer or a private disposal system.

Piping

The materials currently used for soil and waste piping are cast iron, type DWV (drainage, waste, and vent) copper, and plastic. While supply water is distributed under pressure and can change direction through 90° fittings, sewage flow travels entirely by gravity, and its piping requires "easy" downward bends of 45°. Drainage pipes, as shown in Figure 12.8,

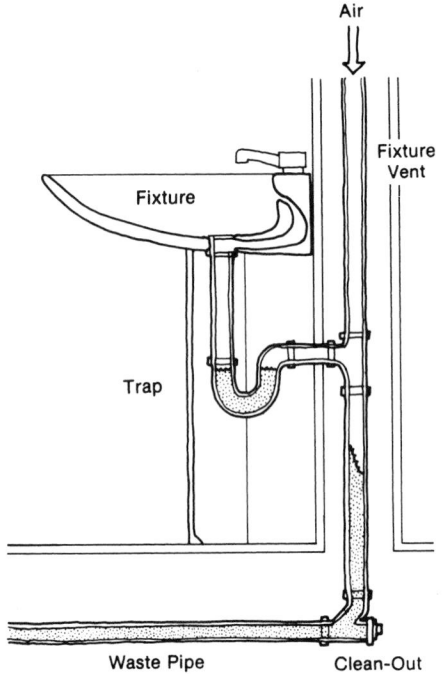

FIGURE 12.7. Vent stack.

are also much larger and must be carefully pitched to ensure an even, steady flow.

All drainage pipes are sloped downward toward the sewer pipe. Soil flows best in a pipe pitched at ¼ inch per foot (2 cm/m), but the pitch may be as little as ⅛ inch per foot (1 cm/m) or as much as ½ inch per foot (4 cm/m). If this range of pitches is impossible or impractical, the pipe must be pitched at 45° or more.

A vertical drop greater than ½ inch per foot (4 cm/m) causes the liquids to flow too rapidly, leaving the solids behind to plug up the drain over time. Pipes pitched at less than ⅛ inch per foot (1 cm/m) do not provide sufficient water velocity, and solids tend to settle and clog. There is also insufficient scouring action. An ⅛-inch (1-cm) pitch is so close to horizontal that it requires extreme care to prevent "valleys" in the pipe run. Pipes pitched at 45° or more have no problem, and the solids slide right down.

Vents

Air vents serve a number of purposes, among which are the following:

1. They relieve any potential vacuum that might suck all the water out of a trap. If the entire cross section of a drain pipe fills with water due to a heavy flow condition, a

siphon effect may readily develop which might empty one or more traps. With the sewer side of the traps open to outside air, this cannot happen.

2. They lead sewer gas pressure safely out of the building. Should a strong gas pressure develop in the sewer, the gas could conceivably drive the water in the trap into the building, and the gas could follow.

3. Air from the fresh-air inlet (or from the sewer if there is no fresh-air inlet) rises through the system and out of the ventilating stacks, carrying away offensive gases. This venting releases whatever gases and smells may form within the building's drainage system to the open air, so that in the event of a failure of the trap seal at a fixture, less or no gas and odor will enter the building.

4. The presence of fresh air circulating through the drain and sewer lines reduces corrosion and the growth of slime.

5. Venting aids the flow of large volumes of liquid through the drainpipes by relieving air pressure that could build up ahead of the water as it moves.

A *fresh-air vent,* or *fresh-air inlet,* is a short vent pipe connected to the main building drain just before it leaves the building. It is usually brought directly outside the building.

The pipe end on the outside is protected by a screen. Although not required by all codes, fresh-air vents provide a lower opening to the ventilating system. Air entering the low vent can rise up through all the drainpipes and pass out through the roof, providing much more ventilation than that provided by the openings at the top of the vent stacks alone. In addition, the fresh-air vent provides an overflow release. Should the main building trap plug up, the excess can flow out the fresh-air vent rather than backing up into the lowest elevation fixtures.

More than one fixture can be served by the same vent stack, but the maximum length of drain pipe between the trap and an air vent for proper functioning is limited to 48 times the pipe diameter. This length is called the *critical distance.* This requirement may be satisfied by running a vent pipe from within the critical distance over to the nearest vent stack. Any number of vent pipes leading from their respective fixtures may connect to one central vent stack. The choice between running a vent pipe horizontally to a central vent or installing a separate one for an isolated fixture is a design decision based on the distances and economics involved.

Vent stacks emit noxious odors and potentially hazardous gases. According to code, the end of a vent stack must be left

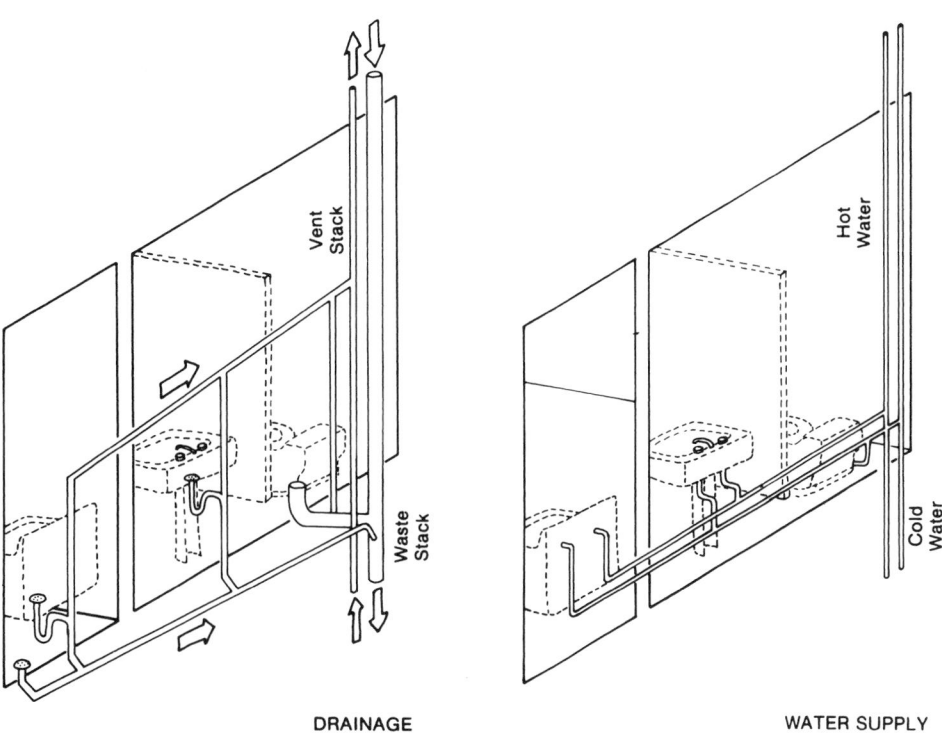

DRAINAGE WATER SUPPLY

FIGURE 12.8. Typical water supply and drainage configurations.

open and must extend at least 1 foot (30 cm) above the roof surface, as well as 3 feet (90 cm) or more above any nearby window or similar opening. It should be at least 10 feet (3 m) from any windows and 5 feet (1.5 m) above any occupied roof deck.

Small vent pipes are susceptible to frost closure, which occurs when warm, moist sewer gases reach the cold top of a vent and condense on the inner surface of the pipe. In freezing weather, the condensation becomes frost, which can build up and reduce or seal the opening. The diameters of smaller pipes are sometimes enlarged to a minimum of 3 or 4 inches (8 or 10 cm) before the pipes pass through the roof in areas where freezing temperatures are common.

Interior Space Planning

Plumbing should never be installed in exterior walls in regions where there is the remotest chance of subfreezing weather. Consequently, in these regions, fixtures should never be situated on outside walls. Small-scale piping assemblies normally fit into a 6-inch (15-cm) interior partition, although wall-hung fixtures require wall chases that are 18 to 24 inches (46 to 61 cm) thick. If there are more than two or three fixtures, a plumbing chase is typically required to accommodate all the piping.

Locating plumbing fixtures back to back or one above the other where possible, as shown in Figure 12.9, minimizes the piping runs and thus provides the most economical installation. It also conserves space by consolidating the plumbing in a given area into a single "wet" wall, and incidentally allows greater flexibility in changing partitions during later remodeling. Similarly, all fixtures in a given room should be placed along the same wall, if possible.

High-rise buildings are often organized around *service cores,* which include plumbing risers along with ductwork and conveying systems. If fixtures are grouped together, vertical piping can form plumbing cores in order to free the surrounding areas and allow for the flexible planning of partitions. The pipe risers for smaller groupings or isolated fixtures can be integrated with structural columns as *wet columns* (Figure 12.10). These arrangements must be well coordinated with the structural requirements early in the planning stages.

Toilet rooms are probably the most common plumbing concern. They are necessary in almost every building. In addition to providing a sufficient number of fixtures (see Table 12.1), the layout should take into account how the plumbing will get to the fixtures. Entrances and/or partitions need to be arranged so that the view from the corridor to the soil fixtures is blocked.

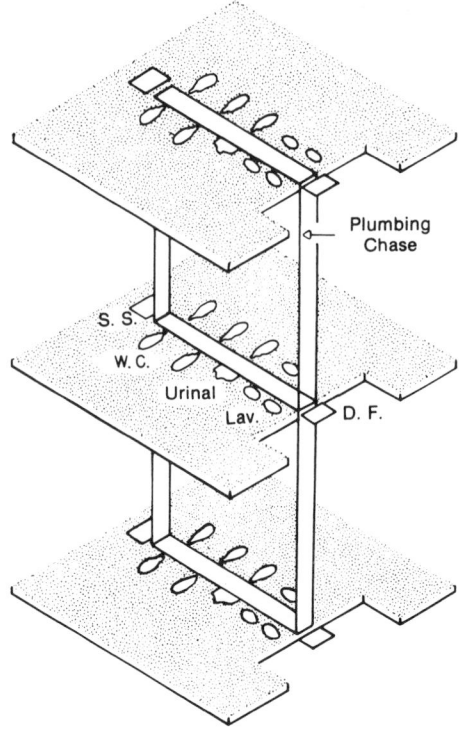

FIGURE 12.9. Stacked and back-to-back arrangements.

Another whole area of concern is the accommodations for handicapped persons. Among these are the provision of at least one handicapped-type water closet and lavatory in each public toilet room, wider water closet stalls for easier maneuvering, grab bars at water closets and bathing facilities, extra-wide entrance and stall doors for wheelchair passage, and a minimum 5-foot- (1.5-m-) diameter clear space within the room for wheelchair turning. These and other requirements are now incorporated into most codes.

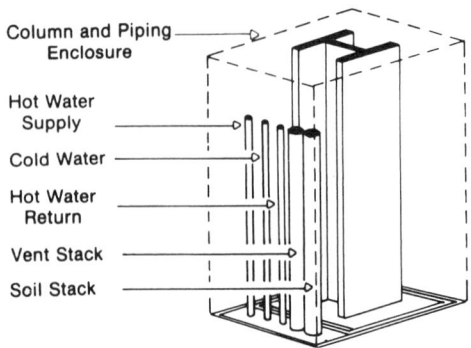

FIGURE 12.10. Wet column.

Floor Drains

Floor drains are employed to carry away water used in washing floors or drained from heating equipment. They are necessary in areas where floors are washed down after food preparation and cooking, mechanical rooms, and toilet rooms. They are also needed in storage closets and any other spaces that are sprinklered for fire protection.

Clean-Outs

Drainage lines stop up from time to time no matter how well designed, so access must be provided for cleaning them out. At fixtures, opening the trap sometimes provides access to the beginning of a branch. However, additional openings are necessary for full access to all parts of the drainage system.

Distributed throughout sanitary systems between the fixtures and the outside sewer connection are *clean-outs* which can be opened to unclog stopped-up piping. Clean-outs must be placed no more than 50 feet (15 m) apart in branch lines and building drains up to 4 inches (10 cm) in diameter. Lines larger than 4 inches (10 cm) in diameter must have clean-outs not more than 100 feet (30 m) apart. They are also required at the base of each stack, at every change in direction greater than 45°, and at the point where the building drain leaves the building. These clean-outs must be accessible for opening and have a working clearance around them.

Interceptors

Certain fixtures discharge waste containing materials that can clog drainage lines or cause problems at the public or private sewage treatment plant. Whenever materials must be separated from waste water, a device known as an *interceptor* or *separator* is used for the fixture discharge instead of a trap.

Interceptors are available in models designed to catch hair, grease, or a variety of other troublesome waste materials from industrial processes. They require periodic servicing and so must be readily accessible.

Grease from food processing and cooking, for example, can congeal inside cold piping, causing a blockage or, if it makes it to the waste-treatment plant, retarding the sewage-digestion process. Grease interceptors are generally installed in food-processing plants and in commercial, institutional, and sometimes residential kitchens. All waste from the kitchen sink passes through a circuitous route within the interceptor, where the grease floats to the top and is trapped between baffles as the remaining fluid waste continues on through. Grease is removed through the removable top cover or it can be drawn off with a valve. The interceptor can be located on the floor adjacent to a sink, on a floor below, or recessed into a pit. Capacities range from 14 pounds (6.4 kg) of grease to over 1 ton (900 kg) of grease for packing-house use.

Other interceptors are similar to grease interceptors for such applications as separating out oil, which represents a hazardous waste from industrial plants and service stations, and hair and lint, which can clog lines in barber and beauty shops. Filings of precious metals are worth recovering even if they are not detrimental to the drainage system. Hospitals and clinics discharge a variety of wastes that tend to foul lines and interfere with the processing at a sewage-treatment plant.

Backwater Valves

Backwater valves are used on lower-level soil or waste branches at or near the bottom of a stack connected to upper-story drains. The device prevents the backflow of sewage from the upper stories into the lower sanitary branch and out through the connected fixtures due to the static pressure that develops.

A backwater valve may also be used wherever a backup of drainage might be expected because of abnormal conditions. Overtaxed or clogged public sewers can otherwise back up into drainage lines within the building and flow up through floor drains and fixtures.

Sewage Ejector Pumps

Normally, the flow of sewage from the building sanitary system into the public sewer is caused by gravity. This requires that the building drain be higher than the municipal sewer or septic tank. This critical elevation should serve as a starting point for the designing and positioning of the building.

When some or all of the fixtures or equipment must be lower in elevation than the sewer connection, gravity drainage is not possible. In that case, a *sewage ejector pump* must be installed. The drainage from the low fixtures flows by gravity into a sump pit or receptacle and is then lifted up into the sewer by the sewage ejector.

While sewage ejectors can enable the building drain to be lower than the public sewer, this is not an ideal arrangement and should be avoided where possible, since power outages or equipment failures can render all sanitary drains inoperative. Ejector pumps should be used only on fixtures that need them, leaving the higher fixtures to drain by gravity. This option should be considered only as a last resort and only if it makes possible some important and otherwise unattainable planning requirement.

Private Sewage Disposal

Most new construction occurs in developed areas where sewage disposal from the site can be accomplished by means of discharge into a municipal sewer, as shown in Figure 12.11a. If a public sewer is not available, the building must be provided with a safe private means of disposal.

Septic systems, as illustrated in Figure 12.11b., are safe, practical, and accepted by all communities that do not have municipal sewer systems. The building drain is connected to the *septic tank* via a sewer line, just as it would be connected to the municipal sewer main. Waste and soil flow by gravity from the building into the tank.

The tank itself is a watertight container made of asphalt-protected metal, vitrified clay, poured concrete, concrete block, or fiberglass and buried underground. The inflow con-

nection is positioned opposite and slightly higher than the outflow connection. Both are located near the top of the tank. Baffles are placed in the path of flow to prevent sewage from passing directly from the inflow to the outflow. The tank capacity is sized so as to allow the waste to remain in the tank for at least 16 hours before being displaced by incoming sewage.

While inside the tank, in the absence of air, the soil and waste are acted upon by anaerobic bacteria, which dissolve most of the solids. Any matter not dissolved sinks to the bottom of the tank. The remaining liquid effluent is led by closed piping to a *drain field* or *disposal field,* which consists of short, separated lengths of perforated pipe embedded in a trench of loose, coarse gravel close to the surface of the ground. The effluent flows out of the pipe into the gravel and soil, where it becomes oxidized. There, aerobic bacteria

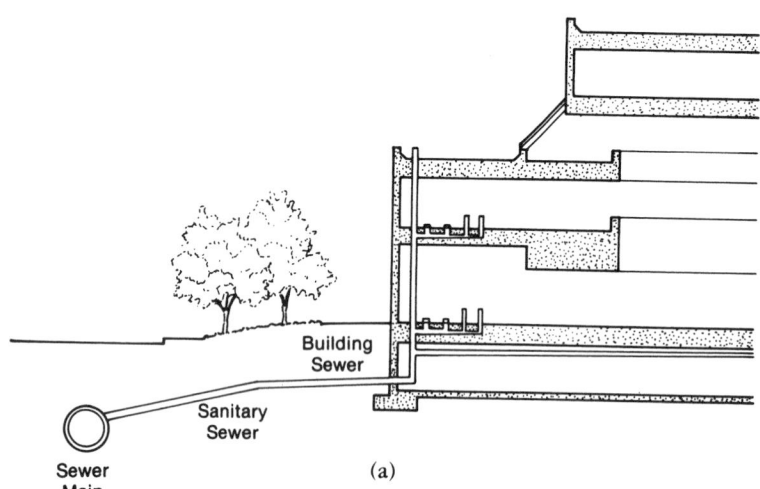

Building Sewer

Sanitary Sewer

Sewer Main

(a)

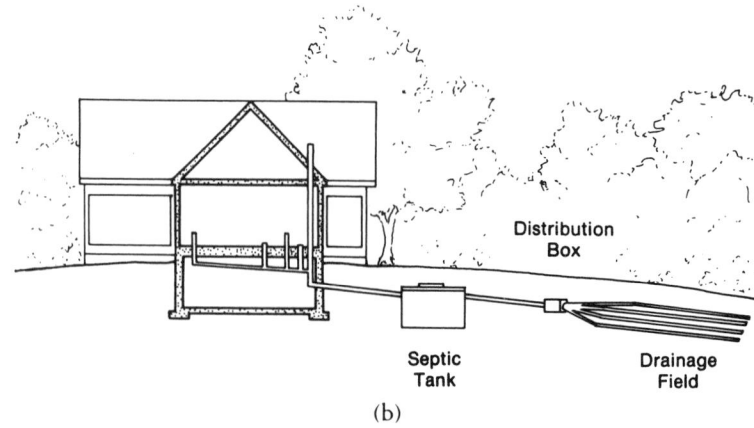

Distribution Box

Septic Tank

Drainage Field

(b)

FIGURE 12.11. Sewage disposal. (a) Municipal sewer and (b) septic system.

complete the decomposition of the effluent and render it harmless.

Operation of the field requires that the soil be reasonably porous. In tightly packed clay soils, the field must be laid over a sand filter bed. At the bottom of the bed, a collection piping system, which functions like a footing subdrain (described below with storm drains), must collect the treated effluent for discharge into a nearby stream. The sand filter bed is rarely used due to its high cost and the need for a suitable disposal point nearby.

Septic tanks should be located sufficiently lower than the building sewer so that the ¼-inch-per-foot rule can be followed and in an area that slopes gently away from the building. The disposal field must be located below a clear, grassy, sunny surface. Any trees would tend to clog the field with their roots. The entire system must be separated from water supply pipes and wells in order to avoid contamination of the potable water supply. The necessary size of a field and the distance required from potable water sources are dependent on the soil absorptivity, which is measured by a percolation test.

Seepage pits are sometimes used in lieu of drain fields. They are shallow holes lined with masonry with a reinforced concrete cover. Their placement requires the same precautions that are necessary when locating drain fields, and the pits must not extend into the water table level.

Eventually, septic tanks fill up with the solid sediment, which must be removed. Removable reinforced concrete covers or manholes are used to provide access to the inside of the tank.

It is sometimes desirable to provide on-site treatment of sewage and industrial wastes in order to make them suitable for discharge into open water bodies or even for recycling on-site. For example, a MIUS system, introduced in Chapter 10, would utilize such a process. Small-scale treatment facilities are available as package units and are merely inserted into the building sewer. For larger projects, a special system must be designed to satisfy the particular needs.

STORMWATER DRAINAGE

The control and disposal of stormwater is important for building integrity. As precipitation collects on rooftops, it must be diverted so that it will not run down the walls and cause leaks at windows or other penetrations in the wall membrane, drip on people at entrances, cause the building to settle by washing soil away from foundations, subject basement walls to groundwater pressure and possible leakage, or

erode away the surrounding ground. Flat roofs must be well drained or else water accumulating in dished areas could add enough weight to overload the structure.

Paved areas prevent water from soaking into the earth and draining away; instead, they catch and hold the water, creating puddles. "Ponding" of paved areas surrounding buildings is avoided by draining the water away in the same way as on a flat roof. Drainage from street curbs is normally collected in a catch basin.

Areas that need to be drained of water include roofs, pavements, balconies, and terraces. These areas are served by a storm drainage system that must be laid out separately from the sanitary system. Occasionally, storm drainage is permitted into the sanitary sewer, and the connection is made at one trapped point. Preferably, however, stormwater should be drained into a separate storm sewer. If that is not available, drainage may sometimes be disposed of in street gutters or in a nearby gravel bed, stream, or dry well.

An example of an arrangement of storm drain piping for draining the roof and balconies of a typical multistory, flat-roofed building as well as the surrounding paved areas is shown in Figure 12.12. Most nominally flat areas should have a slight slope, so that storm water is drained away at the low points.

Roof Drainage

The roof form has traditionally been sensitive to precipitation. Steeply pitched roofs have been designed in response to heavy rain and snowfall. Extending roof overhangs out far enough on low-rise buildings can provide some protection for walls exposed to storm winds. In areas with heavy rainfall, greater attention is also given to gutter and downspout details.

Rain falling on a sloped roof flows down to *gutters* along the lower edges. At one end of a gutter, a downspout channels the water either to the ground for surface runoff or into an underground storm sewer. If it is located in a low-rainfall region, a one-story structure without a basement may be built without gutters if its roof has a wide overhang and if a gravel-filled trench skirts the perimeter directly below the edge of the eaves to catch the water flowing off the roof.

Flat roofs actually have a slight pitch to guide stormwater to interior roof drains. *Roof drains* are similar to *areaway drains* serving pedestrian areas, except that the strainers for the former are usually dome-shaped while those for the latter must be flat. A wide variety of drains are available for different applications, and special drains are designed to fit against parapets and drain cornices.

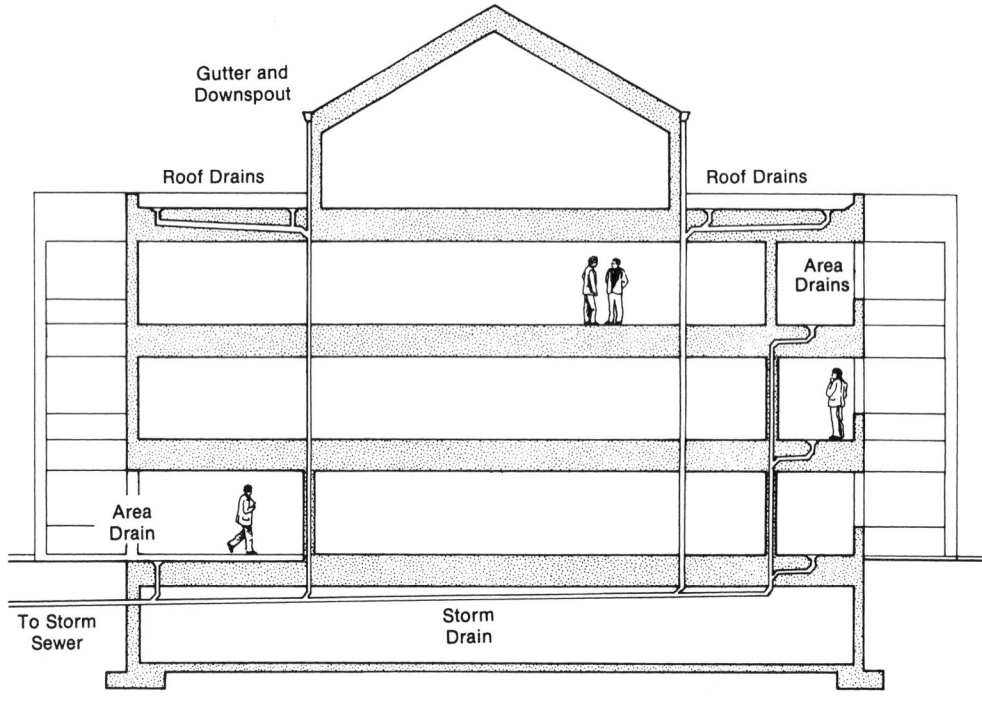

FIGURE 12.12. Stormwater drainage.

Leaders concealed within partitions carry the water from the roof drain down through the structure. These interior drainage systems using sturdy piping are generally more expensive than outside sheet-metal gutters and leaders.

The sizing of gutters and leaders is based on the area to be drained and the maximum recorded rate of rainfall at the particular location. The *National Standard Plumbing Code* lists the maximum rate of rainfall in about 200 cities in the United States, or rainfall rates may be obtained from local weather stations. The rates are typically listed in terms of the maximum number of inches/hour gauged during a 5-minute period. If the maximum rainfall rate is not known, the average value of 4 inches per hour may be used for preliminary purposes.

Once the rainfall rate, R, is obtained, it can be used along with the area to be drained, A, to determine the number of gallons per minute (gpm) needing to be drained away from the roof in the following equation:

$$gpm = \frac{A\ (ft^2)}{96} \times R\ (in./hr) \qquad (12.3)$$

The gpm value is used in Table 12.2 to find the appropriate size of roof gutters, as applicable, and in Table 12.3 to find the size of a vertical leader needed to serve a gutter or a roof drain.

Water Disposal

Stormwater can ultimately be discharged into municipal storm sewers; nearby rivers, streams or lakes; the ground; or a cistern. The easiest way to dispose of stormwater is to discharge it into a *storm sewer,* if available. If access to a storm sewer is not available, stormwater may be channeled into a nearby river, stream, or lake, subject to any regulatory limitations concerning flood control and water pollution control. A *catch basin* may need to be provided as a buffer to moderate the flow rate of water into storm sewers or surface water bodies that cannot accommodate the peak drainage.

Recharging Groundwater

If the first two options—a municipal storm sewer or surface water bodies—aren't available, water may be allowed to soak into the ground. This is possible only if the soil is permeable enough to absorb water at a sufficient rate. Sand and gravel are ideal soil constituents. The ground must also be

TABLE 12.2 ROOF GUTTER SIZE

Diameter of Gutter (inches)	Maximum (gpm)
3	7
4	15
5	26
6	40
7	57
8	83
10	150

These figures are based on *National Plumbing Code* requirements. Gutters that are not semicircular must have an equivalent cross-sectional area.

dry enough to accept the water. This solution, generally applicable only in sparsely built areas, could create problems for neighboring buildings if it is tried in a crowded city.

When buildings are built, they divert the normal course of stormwater dispersion, either into the ground or as runoff, and if the water is to be redirected, its total effect on the surrounding area must be carefully studied. Redirecting stormwater that is blocked by impermeable building surfaces back into the ground is referred to as recharging groundwater.

Surface Runoff. In some rare cases, a building can simply rely on *surface runoff,* but in most cases, special provisions must be made to prevent a concentration of runoff at one spot, which can cause erosion or flooding of neighboring lower areas.

On a small scale, a splash pad can lead the water from a gutter to an adjacent area of very permeable ground a few feet from the building. This method can accommodate only a low rate of flow.

Dry Well. If the surrounding soil is not very permeable—perhaps, for example because it is made of clay—or if water flow is very heavy, the stormwater can be piped farther from the structure to a *dry well.* The dry well has an extended area and many perforations through which the water can soak into the ground.

Footing Drain. A perimeter *footing drain* (Figure 12.13) is sometimes necessary if a building is surrounded by permeable ground at a higher elevation. As a result, groundwater flows downhill toward underground walls and accumulates against them.

The footing drain reduces the likelihood of basement

wall leakage by collecting the groundwater accumulating around the foundation in open-joint clay tile or perforated PVC pipe and leading it away. This water, possibly combined with roof drainage, is carried in tight-joint clay tile or bituminous fiber pipe to a surface absorption area of rock and gravel, where it flows through the stone and into the earth. At the absorption area, the storm drain outcrops from a protective head wall.

Obviously, this method is applicable only if the site is sufficiently large and sloped, but it has the advantage of being easily maintained. Whether or not the drain is functioning properly can also be readily observed.

Recharge Basin. An alternative method involves the use of a *recharge basin.* The collected stormwater is piped to an open, unpaved pit, where it gradually sinks into the earth. For safety, a fence is required around the basin to deter unauthorized access. A recharge basin cannot be used in areas of dense, impervious clay soil.

Cistern

Its relative purity makes rainwater desirable in areas where public supplies have a mineral content or contain other undesirable elements. In arid regions, where water is scarce and/or expensive, it may be worthwhile to collect rainwater for general building use. In either case, the stormwater collected from roof surfaces may be filtered and retained in a *cistern* for later use.

If the cistern is elevated above the end use, it may be supplied simply by the flow of gravity. Otherwise, a pump is needed. The volume of water able to be captured by a cistern (V) can be calculated by multiplying the rainfall rate (R) by

TABLE 12.3 VERTICAL LEADER SIZE

Diameter of Leader (inches)	Maximum (gpm)
2	23
2½	41
3	67
4	144
5	261
6	424
8	913

These figures are based on *National Plumbing Code* requirements. The cross-sectional area of a rectangular leader must be equivalent to that of the circular leader required. The ratio of width to depth of rectangular leaders may not exceed 3:1.

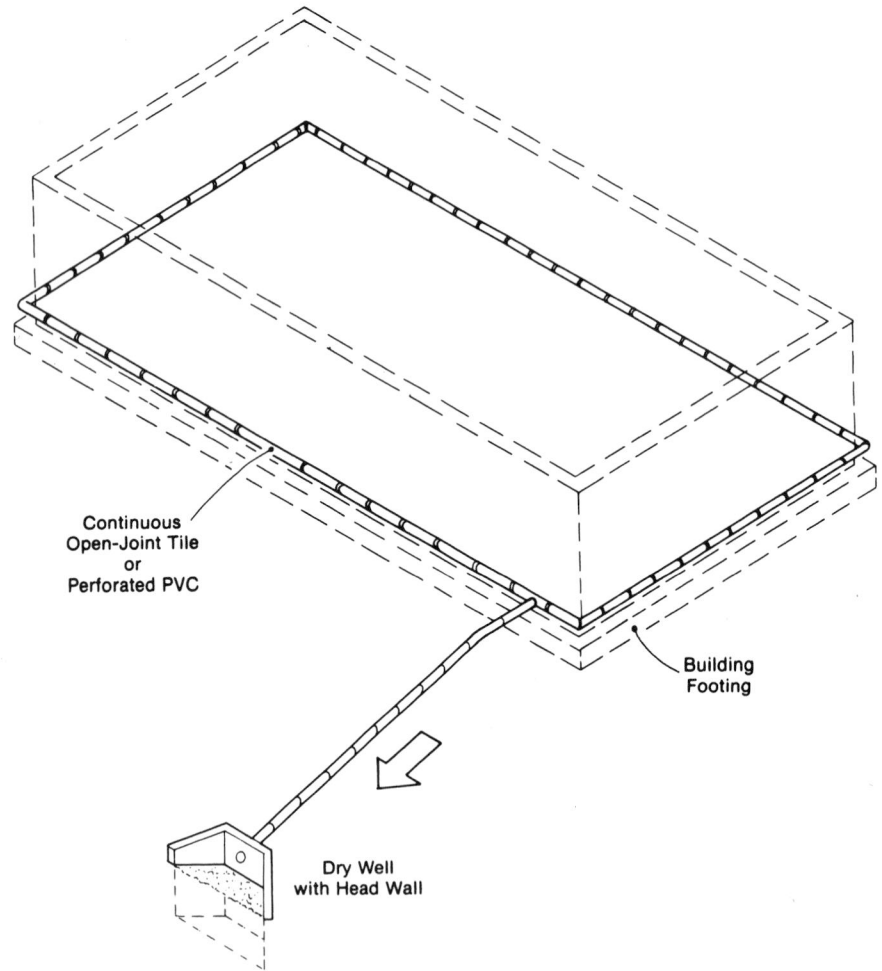

FIGURE 12.13. Footing drain.

the roof area contributing to the cistern (A) and by the time duration (t):

$$V(ft^3) = \frac{R(in./hr) \times A(ft^2) \times t(hr)}{12\ in./ft} \quad (12.2a)$$

$$V(m^3) = \frac{R(cm/hr) \times A(m^2) \times t(hr)}{100\ cm/m} \quad (12.2b)$$

One problem with using collected rainwater is that the water picks up air contaminants as it falls, as well as other impurities caused by leaves, birds, and dust as it runs off building surfaces. These can render the water unsafe to drink, so if the water is intended for potable purposes, a filtration system is necessary. Moreover, a flushing cycle that allows the first and dirtiest water of each rainfall to bypass

the cistern should be set up. Gutters and other catchment surfaces should also be accessible for occasional cleaning.

Using rainwater for irrigating surrounding vegetation does not require filtering. Rainwater is especially well suited for this purpose because it lacks the additives that most public supplies contain for potability. Irrigating with rainwater saves the potable water that would otherwise have been used for that purpose, thereby reducing the demand on the public water supply.

Disposal of Condenser Water

The use of municipal water in a once-through air-conditioning condenser is very wasteful and is thus prohibited in most areas. If it is permitted, however, it must be properly disposed of. Most localities also prohibit the use of the sanitary sewer for the disposal of waste from condensing equipment, since large volumes of clean water upset the bacterial action at the sewage-treatment plants. Some localities permit the use of sanitary sewers but charge a special sewer tax to help offset the cost of enlarged facilities.

Where such a once-through arrangement is unavoidable, it is preferable to discharge condenser water into separate storm sewers. In some cases, the used condenser water can be sprayed on the building roof to reduce the solar heat gain.

SUMMARY

The plumbing connections made to a building are:

- Cold water supply (with meter)
- Sewer
- Storm drain
- Sometimes natural gas (with meter)

Inside the building, a piping system distributes the cold water to fixtures and water-heating equipment. Hot water is then distributed to those fixtures needing it. If the pressure from the water supply is insufficient to push the water up to the upper stories and provide the necessary pressure for the fixtures to operate properly, a pumping system may be required to boost the pressure. This arrangement is known as upfeed distribution. As an alternative, water can be gradually pumped up to an elevated storage tank to downfeed water to the fixtures below.

At the fixtures, water is used and then discharged down the drain. The sanitary piping system carries it to the municipal sewer or septic tank. Public sewers are usually located below roads.

The sewage leaves a building at a point below the lowest fixture but generally higher than the sewer pipe so that the discharge will flow by gravity. A feasible and economical drain layout may influence the siting and design of a new building, and should therefore be considered at the initial planning stage. Vertical drainage piping within a large building can be a significant spatial element to be contended with. Consolidating fixtures along one or more plumbing cores is both an efficient use of space and the most economical installation.

The drainage system within a building is more complex than the supply system and must be carefully arranged to protect the health and safety of the building occupants. Without the proper precautions, contamination of the potable water supply or the room air may occur. The system consists of three essential parts:

1. *Traps.* A trap, generally at each fixture, contains a water seal to prevent undesirable sewer gases and odors from entering the occupied space. Since these traps occasionally become clogged, they must be accessible under a lavatory or through an access panel. Traps are built into toilet bowls and some urinals. Commercial kitchen drains and certain other equipment must have interceptors, and these must be frequently cleaned for proper drainage.
2. *Drainage lines.* These vertical stacks and horizontal branches carry the discharge from the fixtures to the disposal point. These are usually concealed in walls or floors. Drainpipes carrying the discharge from water closets or urinals are known as soil pipes, while those carrying the discharge from other fixtures are called waste pipes.
3. *Vent lines.* These pipes extend upward from the fixtures through the roof. They allow air to flow into or out of the drainage pipes, thus equalizing air pressure in the drainage system and preventing the water seal in the traps from being siphoned away.

Drainage installations for buildings must deal not only with flows from sanitary fixtures, but also with rainwater from roofs and paving. Although storm drainage does not contain putrescible solids that pose a danger to health, such drainage from roofs, walkway areas, and subdrains for foundations must be carried away in order to avoid pools of standing water, soil erosion, and rapid degradation of the structure.

Abbreviations

A.D.	Area Drain	R.D.	Roof Drain
A.V.	Air Vent	S.	Sink
C.B.	Catch Basin	S.O.V.	Shutoff Valve
C.I.	Cast Iron	S.S.	Service Sink
C.O.	Clean-Out	UR	Urinal
D.F.	Drinking Fountain	V.	Vent
D.S.	Downspout	V.C.P.	Vitrified Clay Pipe
D.T.	Drain Tile	V.C.T.	Vitrified Clay Tile
E.W.C.	Electric Water Cooler	V.S.	Vent Stack
F.C.O.	Floor Clean-Out	V.T.R.	Vent Through Roof
F.D.	Floor Drain	W.	Waste
F.L.	Flow Line	W.C.	Water Closet
H.B.	Hose Bibb	W.C.O.	Wall Clean-Out
K.S.	Kitchen Sink	W.H.	Water Heater
LAV	Lavatory	W.S.	Waste Stack
M.H.	Manhole	Y.D.	Yard Drain
R.C.P.	Reinforced Concrete Pipe		

FIGURE 12.14. Common plumbing symbols and abbreviations.

Small quantities of storm drainage from pitched roofs can be controlled by roof gutters and outside downspouts discharging onto splash blocks or into curb gutters. Large, flat roof areas, enclosed courts, and buildings in urban areas must have sloped surfaces leading from roof or areaway drains to interior downspouts, or leaders.

Stormwater can be ultimately disposed of into a municipal storm sewer, a nearby surface water body, or the ground. If rainwater is desirable for supply water, the storm drainage can be collected in a cistern.

Figure 12.14 illustrates standard plumbing symbols and abbreviations used on construction drawings.

KEY TERMS

Stormwater system
Cesspool
Supply water
Street main
Sanitary piping
Storm water drainage
Static pressure
Upfeed distribution
Downfeed distribution
Suction tank
Domestic hot water (DHW)
Storage (tank-type) water heater
Storage tank capacity
Instantaneous (tankless or in-line) water heater
Recovery rate
Direct (immersed) water heater
Indirect (takeoff) water heater

Point-of-use (local) water heater
Recirculating hot water distribution
Water cooler
Water closet
Urinal
Bidet
Lavatory
Sink
Service sink, slop sink
Wash fountain
Drinking fountain
Sewer gases
Trap
Cross-contamination
Vacuum breaker
Backflow preventer
Check valve
Pressurized-air toilet
Incinerator toilet
Composting toilet

Waste pipe
Soil pipe
Waste stack, soil stack
Vent stack
Vent pipe
Branch drain, branch line
Fresh air vent, fresh air inlet
Critical distance
Plumbing core
Wet column
Floor drain
Clean-out
Interceptor, separator
Backwater valve
Sewage ejector pump
Building drain
Building sewer
Public sewer
Septic system
Septic tank
Drain field
Disposal field
Seepage pit
Gutter
Roof drain
Areaway drain

Leader, down-spout
Storm sewer
Catch basin
Surface runoff
Dry well
Footing drain
Recharge basin
Cistern

STUDY QUESTIONS

1. Name four possible sources of potable water for a building water supply.
2. Given: (a) A water pressure of 15 psi (103 kPa) is needed to operate flush valves on the top floor of a building, (b) 10 psi (69 kPa) of pressure is lost per 100 feet (30 m) of pipe, and (c) the street main can provide 50 psi (345 kPa) of pressure.
   Find: How high the top-floor fixture can be above the street main without requiring upfeed pumping or a rooftop water tank.
3. List how many of each type of plumbing fixture are required for an office building occupied by 55 women and 50 men.
4. If a water closet is located 20 feet (6 m) from a waste stack, how much vertical space is required below the floor for the branch soil line?
5. If a building is 30 feet (9 m) horizontally from the nearest public sewer connection, what is the minimum vertical distance between the sewer connection and the lowest point of the building drain in order to have gravity flow?

BIBLIOGRAPHY

Alth, Max, Charlotte Alth, and S. Blackwell Duncan. *Wells and Septic Systems,* 2nd ed. McGraw-Hill Professional, 1991.

American Society of Heating, Refrigerating, and Air-Conditioning Engineers, Inc. *Guideline 12—Minimizing the Risk of Legionellosis Associated with Building Water Systems.*

Burns, Max. *Cottage Water Systems: An Out-of-the-City Guide to Pumps, Plumbing, Water Purification, and Privies.* Cottage Life, 1993.

Clean Water Act with Amendments. Water Pollution Control Federation,

Del Porto, David, and Carol Steinfeld. *Composting Toilet System Book: A Practical Guide to Choosing, Planning and Maintaining Composting Toilet Systems.* Center for Ecological Pollution Prevention, 2000.

Design and Construction of Sanitary and Storm Sewers. Water Pollution Control Federation,

Gravity Sanitary Sewer Design and Construction. Water Pollution Control Federation,

Hazeltine, Barrett, and Christopher Bull (editors). *Field Guide of Appropriate Technology.* Academic Press, 2002.

Ludwig, Art. *Builder's Greywater Guide: Installation of Greywater Systems in New Construction and Remodeling: A Supplement to the Book "Create an Oasis with Greywater."* Oasis Design, 1999.

Ludwig, Art. *Create an Oasis with Greywater: Your Complete Guide to Choosing, Building and Using Greywater Systems,* 4th ed. Oasis Design, 2000.

Ludwig, Art. *Branched Drain Greywater Systems.* Oasis Design, 2000.

Mendler, Sandra F., and William Odell. *The HOK Guidebook to Sustainable Design.* John Wiley & Sons, Inc., 2000.

Petersen, Erik Nissen. *Rainwater Catchment Systems for Domestic Supply: Design, Construction and Implementation.* ITDG Publishing, 2000.

Reed, Richard J., J. Pickford, and R. Franceys. *A Guide to the Development of On-Site Sanitation.* World Health Organization, 1992.

Woodson, R. Dodge. *Builder's Guide to Wells and Septic Systems.* McGraw-Hill Professional, 1997.

Woodson, R. Dodge. *Water Wells and Septic Systems Handbook.* McGraw-Hill Professional, 2003.

Codes

International Plumbing Code. International Code Council.

National Standard Plumbing Code. National Association of Plumbing, Heating, and Cooling Contractors.

Southern Standard Plumbing Code. Southern Building Code Congress.

Uniform Plumbing Code. International Association of Plumbing and Mechanical Officials (IAPMO).

Chapter 13

Fire Protection

STANDARDS

LEVEL OF PROTECTION

DETECTION AND ALARM SYSTEMS

DETECTORS

ALARM SYSTEMS

EXITS: MEANS OF EGRESS

EXIT AVAILABILITY

ELEVATORS

EXIT SIGNS

CONFLICTS WITH SECURITY

COMPARTMENTATION

CONFINEMENT

REFUGE AREAS

SMOKE CONTROL

HAZARDS OF SMOKE

SMOKE FROM PLASTICS

LIMITING FLAMMABILITY AND TOXICITY OF
BUILDING CONTENTS

METHODS OF SMOKE CONTROL

FIRE SUPPRESSION

EXTINGUISHMENT

EXTINGUISHING SYSTEMS

The extreme danger that fire represents to a building and its occupants requires a holistic approach to the architectural, mechanical, plumbing, and signal systems. Because an integrated approach includes features of all of these systems, this chapter discusses them together in the context of an overall building fire-protection system. It is important to remember that the technology for fire protection is still being developed, as is information about the effectiveness of each method.

The objectives of a fire protection program are to maintain conditions conducive to (1) life safety, (2) property protection, and (3) minimizing business interruption. The first and primary objective is to reduce the probability of loss of life by fire to an acceptably low level.

The key elements of fire protection are:

- Early detection and alarm system
- Means of egress
- Compartmentation
- Smoke control
- Fire-suppression systems
- Emergency power

Each method has its advocates, and the relative benefits of each are still being debated. However, no single feature or combination of features is both technically and economically correct for all buildings. There is no appropriate "cookbook" approach. Every building needs to be assessed on the basis of its specific conditions. Instead of looking at individual systems, the overall concept of fire safety must be considered. Safety is determined by the synergistic effect of the whole building.

There is no such thing as a completely fireproof building in the real world. But a combination of many life-safety systems can be used to provide the safest possible environment for inhabitants at a reasonable price.

The chief hazards of fire can be represented by a triangle of smoke, heat, and time, as shown in Figure 13.1. The first two factors are the cause of fatalities and property loss, while time is critical to the evacuation of people and the suppression of fire. The first 5 minutes of a fire are more important than the next 5 hours. Smoke can spread and overcome people in a matter of moments, and fire can spread at the rate of 15 feet per second (4.6 m/s). A phenomenon known as *radiant plate* enables fire to spread by radiation across a clear, open space in a fraction of a second.

Standards

The National Fire Protection Association (NFPA) generates standards covering all aspects of fire control. Once adopted by the NFPA membership, the standards are published and made available for voluntary adoption. Local building departments typically require that systems installed in their jurisdiction meet all or part of the standards. The standards are also used as the basis for national legislation and regulations.

The NFPA was organized in the 1800s. Its mission is to safeguard people and their environment from destructive fires. NFPA standards currently number about 260. Each NFPA document is published as a booklet. All of the documents also appear in the annual multivolume set of the *National Fire Codes* and code supplement. The most widely used of the official documents include the *National Electrical Code* (NFPA Standard No. 70) and the *Life Safety Code* (NFPA Standard No. 101). NFPA also compiles and publishes annual listings of fire statistics.

Local building departments review plans and make sure that the requirements of the building code are met. They also employ inspectors to inspect each building as it is erected and, after it is completed, to ensure that the plans have been fol-

lowed and that the building meets code requirements. When a building is approved by the inspector, the department issues a Certificate of Occupancy (CO), and the building may be used.

Level of Protection

The various model building codes and nationally accepted standards establish minimum levels of protection for public safety, but in actuality, the level of protection achieved when applying these codes and standards varies widely. The codes are general in nature, since their provisions must cover all buildings and all occupancies. Specific design solutions cannot be codified, and the codes and standards cannot specify exact methods of protection for each unique building design.

The guidelines promoted by insurance companies tend toward the ultimate in protection in order to minimize their exposure to loss. It is also in the interest of manufacturers of protective devices and equipment to advise a high level of protection in order to encourage greater expenditures for their products and services.

Many of the fire codes themselves are controversial. The ultimate responsibility is left up to the designer. Compliance with minimum fire codes does not shelter a specifier from exposure to liability. The level of protection that should be achieved should therefore be determined by a knowledgeable, objective analysis of the overall exposure to loss.

Proper fire protection design is a matter of conscience and professional ethics. People living and working in buildings have a right to expect personal safety as well as the accompanying peace of mind resulting from an expectation of survival in the event of fire or similar building emergencies. Moreover, proper life-safety design is cost effective. If stricter regulations later force the retrofit of an existing structure built with inadequate systems, it may cost three to three and a half times as much as the same installation in a new building. There are liability costs and downtime costs. Insurance rates are often dependent on life-safety and fire-protective measures. Cutting corners on life-safety systems is false economy, and much greater costs are often paid later. If the issue of fire safety is underemphasized by the building team until disaster strikes, making it a priority after the fact is too late.

DETECTION AND ALARM SYSTEMS

Immediate and reliable detection of fire and smoke is essential. Early detection of fire is the only way to prevent loss of lives and property. Fires generate large volumes of smoke

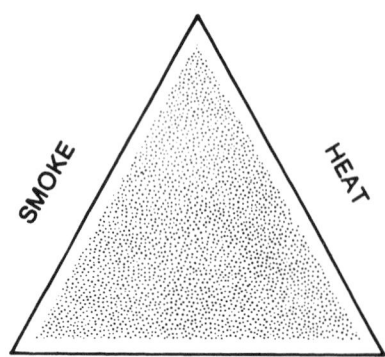

FIGURE 13.1. Fire hazards triangle.

and toxic gases in addition to heat, and if detection is delayed, paths of egress can become blocked and the occupants may be trapped in portions of the building. Detection of a fire in its incipient stage is also desirable in order to limit damage and to initiate fire-control activities while the fire is still small.

If all areas of a building were occupied at all times, the occupants could be relied upon to detect fires. However, since most buildings and parts of buildings are vacant at some time, automatic fire detection is usually required for reliability.

Automatic fire-detection systems can be used to initiate an audible and/or visible alarm locally, remotely, or by both means. The purpose of a local alarm is to warn occupants to evacuate or take some action to extinguish the fire. Remote alarms generally serve to notify a police or fire department or a building control station so that people there may take the appropriate action. At the same time, detection systems can be arranged to actuate automatic fire-suppression systems.

Detectors

As mentioned in Chapter 11, signal initiators in a fire-detection and alarm system may be automatic detectors or manual pull stations. Automatic detectors fall primarily into the following general categories: (1) heat detectors, (2) ionization detectors, (3) smoke detectors, (4) flame detectors, and (5) waterflow detectors in sprinkler systems. Examples of these devices are pictured in Figure 13.2. Other available devices sense concentrations of combustible and toxic gases.

Heat Detectors

Heat (thermal) detectors are the oldest, least expensive, and most widely used. They are the simplest and most reliable fire-detection device. However, they are also the slowest to respond to a fire, since it takes a period of time for the detector to heat up enough to respond. By that time, life safety could already be jeopardized and the fire seriously out of control.

Because they sense only heat, these detectors are subject to fewer false alarms than other categories of detectors. The general shortcoming of heat detectors is that they do not respond to smoke or other products of combustion.

Ionization (Products of Combustion) Detectors

Fires often develop gradually, first emitting invisible products of combustion, then visible smoke, and finally visible flames. A considerable amount of time can pass and damage can occur before there is sufficient heat to actuate a thermal detector. *Ionization detectors* are designed to sense the products of combustion emitted during the incipient stage of a fire, making these devices more sensitive than heat detectors and ordinary smoke detectors.

Ionization detectors are extremely sensitive to accumulations of dirt or insulating films. Contamination can prevent an alarm signal or create false alarms. For these reasons, NFPA, Standard 72 requires that smoke detectors be checked and cleaned periodically, so they should be situated in an accessible location. They are not suitable for dusty environments.

The disadvantages of ionization detectors are that they are expensive and that more false alarms are possible due to their sensitivity. They are most effective in detecting fires arising from electrical overload, such as in control and switchgear rooms, and on open-flame fires from loose sheets of paper. In general, ionization detectors are used in applications where fast-burning open combustion is anticipated.

Smoke Detectors

During the second stage of a fire, before the flames actually surface, smoke is generated. *Smoke (photoelectric) detectors* can detect this visible smoke. Although they sense fires in the smoldering stage, photoelectric detectors do not detect them as quickly as ionization detectors.

Smoke detectors are best suited for areas where the type of fire anticipated would produce large volumes of smoke before temperature increases would be sufficient to operate a heat detector. They are recommended for use in areas where it is not practical to employ the ionization type due to high ambient concentrations of products of combustion, or where the material expected to burn produces more visible smoke than flame. Office and apartment buildings, for example, commonly employ photoelectric detectors because of the possibility of an undetected smoldering fire in a vacant area.

Since photoelectric detectors are actuated by the presence of particulates in the air, they are prone to false-alarm conditions created by insects, dust, or other airborne particulate matter. Thus, they should not be used in dusty areas.

Smoke detectors in apartments are particularly susceptible to false alarms from kitchen smoke and excessive dust. Self-contained detectors are sometimes used to sound a local alarm only in the apartment for evacuation purposes. A separate centralized heat detector system sounds a remote alarm. This trade-off reduces the problem of false alarms but increases the likelihood of a fire reaching major proportions before the automatic actuation of fire-suppression systems or the dispatch of firefighting crews.

Flame Detectors

In cases where fires are likely not to generate smoke first (such as with gasoline), *flame detectors* may be used.

FIGURE 13.2. Detectors. (a) Manual pull station, (b) heat detector, (c) smoke detector, (d) ionization detector, (e) gas detector, and (f) waterflow switch. The waterflow indicator bolts onto a sprinkler pipe with the paddle inside the pipe. Any water motion deflects the paddle, causing a signal to be transmitted from the microswitch mounted in the box on top of the pipe. (Photos b, d, e, f courtesy of Ademco. Photos a and c by Vaughn Bradshaw.)

Flame detectors respond to the presence of either infrared or ultraviolet radiation, which are characteristic of flames. They are the most rapid fire-detection devices available, and due to their sensitivity, they are subject to frequent false alarms caused by nonfire light sources. In addition, they do not detect smoldering fires, which produce little or no flame.

Waterflow Detectors

Buildings with complete or partial sprinkler coverage commonly include a *waterflow detector*. In smaller buildings and in some industrial facilities that are completely sprinklered, this may be the only alarm-initiating device. In many applications, however, redundant arrangements with additional automatic detectors are used to speed up the detection and to increase the likelihood of successfully containing a fire and rescuing the occupants.

Manual Pull Stations

Pull stations are available in a variety of forms. The most common type contains a glass rod or window that must be broken in order to move the handle and thereby actuate the switch. The glass must then be replaced before the station can be reset.

Other designs locate the handle behind a cover that must be opened or have a directly accessible handle that is restrained by a spring requiring about the same force as breaking a glass rod. The advantage of these types is that a stock of replacement rods or glass panels is not needed. Most stations need to be reset by a key so that the initiating station can be identified before being reset.

Once the detectors have done their job, it is up to the control system to take over and ensure the safety of lives and the preservation of property.

Alarm Systems

Life safety usually depends on being able to safely evacuate the occupants of a building in the event of a fire emergency. This requires a reliable means of alerting all occupants to the existence of an emergency so that evacuation can proceed if possible.

Building Fire Alarms

A fire alarm control panel for a commercial or light industrial application is shown in Figure 13.3. Complete systems can detect a fire, sound an alarm, and actuate extinguishing functions. They can even be programmed to perform such duties as closing fire doors, closing fire dampers in air-conditioning and heating ducts, providing auxiliary power

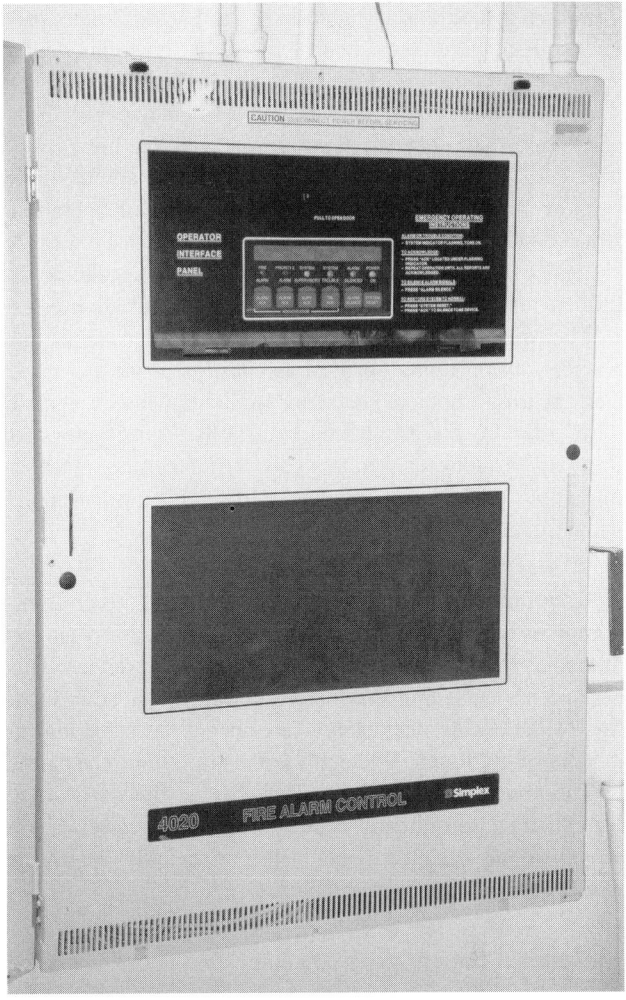

FIGURE 13.3. Fire alarm control panel. (Photo by Vaughn Bradshaw).

for exhaust fans to remove smoke, shutting down the building fan system, and turning off other machinery.

Normally, an alarm is automatically sounded upon activation of a manual fire alarm station or upon the operation of an automatic fire detector. In some cases, however, it is desirable for smoke or fire detection not to automatically initiate building evacuation. In public assembly buildings, for example, where a fire alarm could cause panic, extreme disruption, and possibly injuries, false alarms must be avoided. In such cases, occupants can be verbally notified and given clear instructions by trained employees after the alarm signal has been investigated. This method is intended to provide reasonable and prompt notification without causing undue alarm. Most sprinkler waterflow detectors directly initiate an automatic alarm.

Alarm Indicators. Signal-indicating devices must produce a distinctive sound and be located so as to be clearly audible in all portions of the building.

The two most common alarm-indicating devices are bells and horns. Since coded systems operate with intermittent signals, audible devices in such systems must be single-stroke bells or chimes that are capable of that type of operation. Noncoded signals operate continuously, so devices in those systems can be continuous bells or horns. For special applications, louder signals such as sirens or air horns are used.

The Americans with Disabilities Act (ADA) requires that flashing strobe lights be used as additional fire alarms to alert hearing-impaired persons. Lighted signs that flash "fire" may be placed above alarm bells for this purpose. Rotating beacons or strobe lights are sometimes used to make sure that hearing-impaired persons become aware of the alarm condition.

Annunciation. After a fire is detected and the firefighting forces have arrived, it is essential that they receive information as to the location of the alarm-initiating device. This is done through *annunciation.* To delineate the exact location of the occurrence, each control panel has a red alarm-indicator light for each detection zone covered by the panel.

The control panel may be located in a fire-department command station directly accessible from the street. Upon arrival at the site, the firefighters can quickly determine which devices have been actuated.

Electrical Supervision. Building fire alarms are often electrically supervised, which means that the absence of a current due to an open or short circuit in the system wiring will automatically sound a trouble signal. Supervisory trouble signals can annunciate according to device and location.

Emergency Power. NFPA Standard 72 requires alternate power sources for an alarm system in order to provide for continuity of protection. Most detection and alarm systems operate at 24 volts DC in order to make use of electronic circuitry, which makes battery backup especially convenient. Even if the project is equipped with an emergency generator, a battery may be supplied having a sufficient capacity to maintain the system for the transition period before the generator is in full operation (a maximum of about 10 minutes).

Combining Functions

Integrated systems can control (1) HVAC systems for energy management, (2) security and intercom functions, and (3) fire and life-safety functions. A central control panel serves as the building's control center and provides annunciation. The system can automatically coordinate operation of the HVAC system for dynamic smoke control.

Such systems have become common in larger projects in recent years. Combining the different functions results in a considerable savings over having three separate systems for fire, security, and automatic HVAC control. Most of these newer systems are sophisticated computers that can be programmed to account for variations in building uses and conditions.

Voice Fire Alarms

The upper floors of high-rises are beyond the reach of most fire-department rescue apparatus. This fact, together with the long travel distances up stair towers and the large numbers of occupants, often makes evacuation of high-rise buildings during a fire emergency impractical or undesirable. A *voice alarm* system allows specific instructions to be issued to occupants of each part of the building regarding safe areas of refuge and rescue efforts in progress. Such vital information cannot be transmitted to building occupants by a conventional bell or horn alarm system.

Some type of voice fire-alarm system is almost universally required by the fire codes of most major cities, primarily for high-rise construction. These systems are, however, equally beneficial in any large building, whether a high-rise structure or not, that contains many transient occupants who may not be familiar with the building, its evacuation procedures, and its alarm systems. In buildings such as hotels and convention centers, visitors tend to ignore or misunderstand bells, horns, or other signaling devices that do not transmit specific verbal instructions.

A voice fire-alarm system can be as simple as a standard unsupervised public address system operating totally independently of an otherwise conventional fire alarm system. At the other end of the spectrum, it can be electrically supervised and an integral part of the overall fire-alarm system. In this latter system, the same loudspeakers can be used for both sounding fire-warning tones (either manually or automatically initiated) and passing along voice instructions. Verbal communications may be in the form of prerecorded messages or live instructions issued from a fire command center, usually located at the ground-floor level.

A voice fire-alarm system incorporates the same actuating devices as those found in an automatic fire-alarm system—pull stations, smoke detectors, and waterflow switches—but utilizes special loudspeakers in place of bells or horns. These loudspeakers must be UL listed, and, in order to meet the requirements, they must employ a metal diaphragm rather than a paper cone. This provides a suitable frequency response for voice communications and tone signaling but is far from high-fidelity. Thus, even though combination voice fire-alarm/background music systems are permitted by many codes, the limited music reproduction quality makes such systems impractical in most cases.

Firefighters' Communications Systems

A *firefighters' communications system* is also generally required in most high-rise construction. It is used to provide communications between a fire command center and firefighters who may be located in any part of the building. Such a system is required in large structures because the signal from the portable radios normally carried by firefighters may be unable to penetrate throughout all portions of a tall or massive building to provide reliable communications.

Most firefighters' communications systems consist of a simple intercom system for use between firefighters' communications stations and the fire command center. Such communications stations, with self-contained handsets or intercom speakers, are typically located at all stair tower doors and at each elevator lobby. As an alternative, firefighters' communications stations may take the form of a phone jack into which a portable firefighters' telephone handset may be plugged.

Public Emergency Reporting Systems

A *public emergency reporting system* (PERS) consists of several emergency reporting stations (Figure 13.4) located in

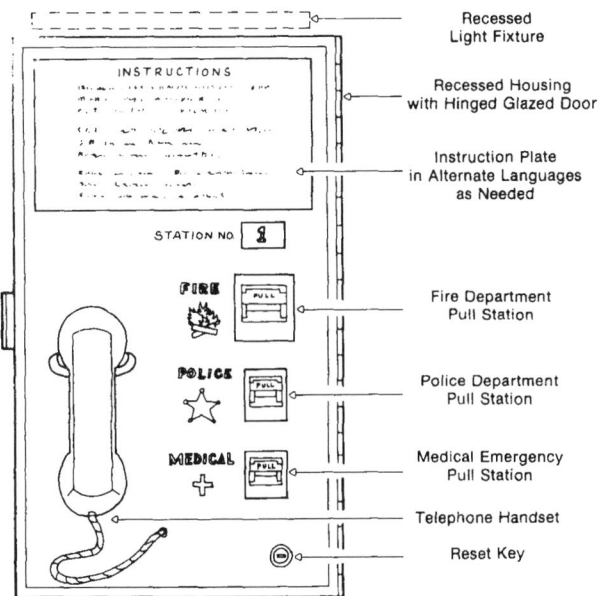

FIGURE 13.4. Public emergency reporting station.

key egress and public gathering areas. From each station, fire, police, or medical emergencies can be reported to qualified operators within the facility, who may then initiate the necessary actions.

When a fire condition is reported by pulling the fire handle, the building control room receives an audible alarm and a visual indication of the alarm location. When the handset is lifted from the telephone cradle, a flashing light and an audible alarm are activated in the central control room. By lifting the console handset, the operator can be in direct voice communication with the PERS station.

The municipal fire department can be directly notified of any fire alarm condition through central station monitoring. When the PERS station is used to report a police or security condition, the signals are received at the central control room and/or a staffed security room. Upon activation of the medical pull handle, a medical or paramedical service can be automatically summoned.

EXITS: MEANS OF EGRESS

Exit Availability

Once an alarm is sounded for building evacuation, a safe means of leaving the building must be available. A *means of egress* is a continuous path of travel from any point in a building to the ground level outside. Such paths or passageways to the outside must be readily available at all times and must lead outside as directly as possible. Doorways, corridors, stairways, and exterior doors must be arranged and designed so that the anticipated number of people in the building can exit safely. Straight-run stairs are capable of discharging people more rapidly than switchback stairs.

All areas and spaces within a building should have at least one exit door or exit way that cannot be locked against egress. In all rooms, spaces, and areas that can be occupied by more than 50 people, at least two unlockable exits must be provided.

Provisions for egress should accommodate handicapped or infirm occupants who may be unable to use stairs or escalators. The narrowest corridor should have no less than 8-foot 6-inch (2.6-m) clearance to allow room for two passing wheelchairs. At least one horizontal or ramped exit route should be provided, if possible, for wheelchair egress.

Elevators

It has long been standard practice to restrict evacuation by elevators in a fire emergency. Codes generally require eleva-

FIGURE 13.5. Panic hardware alarm. (Photo by Vaughn Bradshaw.)

tor shafts to be provided with smoke vents at the top. This transforms the hoistway into a smoke shaft carrying smoke upward by the stack effect and out of the vent opening. If the fire is on a lower floor, the shaft usually fills up completely with smoke. While this smoke venting is helpful in clearing away smoke from the floor of the fire, "smoke-logging" the elevator shaft makes it unusable for evacuating occupants or for the movement of firefighters within the building.

People have died while riding elevators down through smoke-logged shafts or while being trapped in cars that stopped in midshaft. To prevent this, elevator codes require an automatic fire *recall (capture) system.* When a fire alarm is activated, all elevators return immediately to a predetermined lower floor (usually the ground floor lobby) and park with the doors open. This enables firefighters to determine that all cars are secured and that no one is trapped in an elevator.

If elevators are located far apart, one or another may be far enough from the fire to be usable in an emergency. The feasibility of designing fire-safe elevators for emergency evacuation is being investigated.

Exit Signs

Exit routes from all areas of the building should be carefully thought out and marked appropriately. Even though means of egress are adequately designed and have sufficient capacity to allow safe evacuation, they may not be obvious to all occupants, particularly in public buildings where transient visitors are unfamiliar with all the paths and passageways to exits. Exit signs should be provided to delineate exits from other areas, rooms, and spaces. In order to indicate reliably the proper path by which to evacuate from a given point, signage must be clear, logical, and immediately identifiable.

Conflicts with Security

For the purpose of fire safety, it is desirable to have as many exits as are economically possible. For security reasons, however, it is better to have as few exits as possible. Life and property losses can result from crime as well as fire. In general, a compromise is necessary, taking into account the needs of the occupants and the method of operation of the building.

The location and type of building as well as the value of the contents are among the factors that determine the risk of criminal attack. The subject of building security is beyond the scope of this text, since every building requires a different treatment. In general, careful building layout and selec-

tion of hardware can minimize the security risks while still providing the necessary means of egress.

How occupants are expected to use entrances and exits and their attitudes toward them help to define the requirements. Frequent use of an exit door and failure to ensure that it is properly closed can allow an unauthorized intruder to enter. Front or main entrances are locked or unlocked according to the amount of supervision available, the general level of security, and sometimes the time of day; they should preferably be observable from a reception desk or from patrol cars on a main road. These doors are often not as vulnerable as exit stairs that are completely enclosed and exit directly to the outside.

Enclosed exits may require special security treatment. The problems with such exits are that (1) they provide an easy escape for thieves, (2) occupants can use the stairs to travel to prohibited floors or to avoid being questioned by receptionists, and (3) they can be inconspicuously used to gain unauthorized entry. The last problem is aggravated if the exit is used on a day-to-day basis, thereby wearing out the hardware.

Exit doors must not be lockable against the direction of exit travel, but they may be locked to entry. Standard panic hardware, as shown in Figure 13.5, may be used for this purpose. Remote-release latch mechanisms can control entry while allowing free exit. Emergency exit doors may also be equipped with a local and/or remote alarm that sounds when the door opens.

COMPARTMENTATION

Compartmentation means dividing a building or large space into two or more separate enclosures, each totally enclosed within a *fire-barrier* envelope consisting of floor/ceiling assemblies and walls. The purpose of compartmentalizing is to protect building occupants and property by (1) confining the fire, heat, smoke, and toxic gases to the area of origin until the fire is extinguished or completely burns itself out and (2) providing areas of refuge for the occupants and protecting firefighters.

Floors, walls, and ceiling systems serving as the fire barriers must have adequate fire endurance to withstand the stresses and strains imposed by developed fires. Noncombustible materials such as concrete can lend such fire resistance to the building structure.

The level of endurance is indicated by the barrier's *fire-resistance rating.* The fire rating required by code is a function of the likely severity of a fully developed fire within the compartment. The fire severity depends on the amount, distribution, and burning rates of the combustible contents, the ventilation characteristics, and the size and configuration of the compartment. These vary with the type of occupancy, and can vary between compartments within the same classification. It is usually required that stairways be enclosed with 1-hour fire-resistive construction.

The choice of structural and finish materials used for barrier construction also influences the fire severity. Heat-reflecting materials add to the intensity of a fire and cause quicker flashover, which is the temperature at which every combustible item in the room ignites. Under equivalent conditions, fire intensity will generally be less severe in compartments constructed of heat-absorbing materials.

Compartmentation is very effective, and properly rated and constructed fire barriers rarely fail because of structural damage. Since fire usually spreads into adjacent spaces because the compartment is not totally firetight, openings for penetrations of pipes, conduits, ducts, and the like must be carefully sealed.

Considering the possible variations in construction quality, the limitations in inspection capabilities, and the tendency of owners and contractors to cut costs, close supervision of the construction of fire barriers is critical to providing a firetight barrier. Dynamic smoke-control systems, such as those discussed below, create air-pressure differences in order to provide further protection across fire barriers.

Doors penetrating a fire barrier should be equipped with self-closing or automatic-closing hardware to ensure that the integrity of the compartment is maintained throughout the course of a fire. Doors may be normally held open with electromagnetic devices arranged to release upon activation of the building fire alarm system, allowing the doors to close.

If penetration through a fire barrier is needed for access to utilities, flush-mounted *access panel* doors, as shown in Figure 13.6, may be installed. They are designed to maintain the fire barrier ratings of the walls of stairwells, corridors, shafts, and anywhere else fire resistance ratings are required. Fire-rated access panel doors must be UL listed. They are available with a variety of features and in sizes ranging from 8 × 8 inches (20 × 20 cm) to 36 × 48 inches (91 × 122 cm).

Confinement

If the effects of a fire are to be minimized and occupants allowed an opportunity to evacuate, fires must be contained within their immediate areas of origin. Walls and partitions with fire resistance ratings should be provided in order to prevent the spread of fire to adjacent areas.

FIGURE 13.6. Fire-rated access panel. (Courtesy of INRYCO, Inc., Deerfield, Illinois.)

Normally, the vertical movement of smoke should be limited by fire-rated barriers between floors. Since atria cannot conform to this concept, they usually contain smoke vents at the top and are protected by sprinklers in order to keep large fires from developing.

The more compartments within an area, the better the chances for effecting fire control, minimizing damage, and preventing the fire from spreading and destroying other parts of the building. The suggested size of each compartment is no greater than 5,000 to 7,500 ft^2 (460 to 700 m^2) in buildings with a central air system. If each floor has its own air-handling system, compartments can be one to a floor.

Refuge Areas

Once smoke and fire take over the core of a building, anyone not able to evacuate or reach a point on the perimeter and be rescued usually dies. If everyone will not be able to get out of a building during a fire, safe *refuge areas* must be provided. This is usually the case in high-rise buildings and for wheelchair-bound persons in any multistory building. During multistory hospital fires, confined patients are normally moved horizontally to safe areas because it would be virtually impossible to move them downstairs. Refuge areas should be created for these people in the core of large buildings.

Ideally, even if a total burnout occurred, the refuge areas would remain safe. They would remain free of smoke, gases, heat, and fire throughout the duration of the incident and until rescue is possible. They would be capable of structurally withstanding whatever fire-related damages were inflicted upon it while sustaining essential services. This performance is impossible to provide in modern buildings short of a bomb shelter in the basement.

Limited refuge areas can be built out of noncombustible materials around double-vestibule, pressurized stairwells. They can normally serve as libraries, conference rooms, restrooms, or similar spaces. They can resemble all the other rooms in the building in appearance, but should have direct access to the stairs, self-closing fire doors, and a voice-activated intercom system connected to the master fire-control center on the ground floor.

Smoke inhalation is the greatest threat to life when a fire breaks out. If patients in a hospital or nursing room cannot be evacuated, or if time elapses before they are evacuated, self-closing doors on patient rooms can inhibit the spread of smoke into the room.

SMOKE CONTROL

Hazards of Smoke

Present-day buildings, furnishings, and trim tend to be highly combustible and give off toxic smoke when they burn. Smoke rather than the actual flames of a fire accounts for about 80 percent of all casualties from fires. Only 20 percent of fire-related deaths result from heat exposure. People die either directly from inhaling smoke and gas, or from becoming overwhelmed by smoke and then killed by flames. Even when people don't die, loss of memory and lingering physical effects can result from smoke inhalation. Smoke also causes a great deal of property damage during fires.

Before any flame is even visible, smoke may move faster than 50 fpm (25 cm/s) from the point of origin and may kill people before they are even aware that a fire is developing. Once a flame becomes visible, smoke may move well over 100 fpm (50 cm/s) before the flame has even started to spread. Fires in modern buildings usually last less than 30 minutes, while the smoke problem is present for hours.

Smoke may be acrid, biting, or choking. It can tear at the eyes and temporarily blind a person. At the same time, it can obscure exits. Stairwells and elevator shafts can become filled with smoke (smoke-logged), thereby blocking people's efforts to escape and impeding firefighters.

The toxicity of some smoke affects the nervous system immediately. Smoke may contain gases that dull a person's senses and cause the loss of all feeling. Victims may mentally float away, or they may comprehend what they need to do but be unable to coordinate their muscles. Peo-

ple will claw at a door knob, for example, instead of turning the handle.

The stress of hot, tearing smoke washing over them can cause normally rational people to panic and lose their composure. Many survivors report that their logical senses vanished and that fear overwhelmed them.

Smoke is actually composed of gases and particulate matter representing hundreds of different chemical materials. Very little is known about what is in the particulate matter, especially when it comes to smoke from plastics. The various chemical constituents seem to have a very short life span, lasting for 3 to 5 minutes, then reacting with another chemical and changing into something else.

Although many components of smoke are still a mystery, some are clearly understood. Carbon monoxide interacts with hemoglobin in the blood and prevents the hemoglobin from carrying oxygen. Death results from an oxygen deficiency.

Hydrogen cyanide, which is released from virtually any nitrogen-containing polymer, also enters the body by means of smoke in the lungs. It is carried by the blood to the muscles and then to the central nervous system, where the cyanide does not prevent the transport of oxygen, but rather the use of oxygen by the cells that need it.

The lungs themselves are extremely vulnerable to damage. These spongy organs readily absorb any substances in smoke. The effect smoke can have on a person varies with the level of the smoke's concentration, the person's respiratory rate and volume, and the length of the exposure.

Smoke from Plastics

The smoke problem is further compounded by the increased toxicity and volumes of smoke and other fire gases generated by many modern polymers. Some contemporary materials, such as plasticized polyvinyl chloride (PVC), produce up to 500 times as much smoke per ft^2 (m^2) of exposed surface area as red oak does during the critical first 3 minutes of escape time.

While natural materials emit toxic gases when burned, synthetic materials add exponentially to the poisonous brew, so their increased use in modern buildings complicates the smoke problem. Smoke containing toxic gases is given off by some of these substances through pyrolysis (chemical decomposition by heat) even before any visible flame or combustion occurs. About 10,000 new chemicals are introduced every year, and there are not nearly enough funds to research all their potential hazards.

Firefighting services have recognized this situation, and most firefighters won't even go near a fire today without self-contained breathing apparatus. They know they can't get in, make rescues, and safely get out without such protection.

Most plastics are synthesized petroleum distillates. For this reason, they often burn faster and hotter than other materials and can be considered to be molded gasoline. When the plastic materials found in furniture, carpeting, draperies, wall coverings, plumbing systems, electrical wiring, and other products and equipment are heated up, they break down into their component parts.

Fires involving such a decomposition can cause lung damage due to a disease called *metabolic acidosis,* a bodily reaction that abnormally reduces the alkalinity in the blood and tissue. Some chemicals are thought to cause disorientation and a loss of the sense of smell, lead to pulmonary edema (fluid in the lungs), and bring about respiratory failure. These can be classified as short-term effects.

Deadly in themselves, these components can also have a synergistic effect, that is, the total effect can be greater than the sum of its parts. Toxic materials at sublethal levels can combine to become deadly.

PVC, in particular, is capable of having this effect. A rubber substitute, it is commonly found in electrical wiring insulation, electrical fixtures, interior plastic plumbing, and office copying machines, to name only a few of its uses. PVC decomposes to form over 60 different gases, including hydrogen chloride, phosgene, and a host of other products, some known to be carcinogens.

For some time now, the burning of PVC has been known to produce hydrogen chloride, a sensory and pulmonary irritant. Hydrogen chloride incapacitates a person, and the immobilized victim succumbs to lethal carbon monoxide poisoning. This is the most widely credited cause of death from smoke inhalation.

Phosgene (carbonyl chloride) gas is produced in substantial quantities when PVC wiring insulation is subjected to electrical arcing. This could occur in electrical equipment, such as transformers or control panels of commercial buildings, involved in a fire.

Since it has a relatively long smoke life, phosgene remains toxic not only during the actual arcing period, but also for some time thereafter. Moreover, hydrogen chloride from the combustion lingers even longer, presenting a special danger to firefighters during the cleanup phase of a fire, when they often remove their breathing apparatus.

Urea formaldehyde (UF) foam does not support combustion readily unless it is exposed to the intense heat of an extensive fire. UF foam has a high nitrogen content and releases substantial quantities of toxic hydrogen cyanide gas when exposed to the elevated temperatures of a well-developed fire. The large quantity of hydrogen cyanide produced presents a danger not only to firefighters, but also to nearby spectators downwind of a major fire in which UF was ignited.

While smoke is a serious enough problem for fire victims, firefighters must face hundreds of fires during the course of a career. The rise in death and injury among firefighters is correlated to the rapid proliferation of new toxic substances in the marketplace, so much so that smoke poisoning is now considered an occupational disease.

Limiting Flammability and Toxicity of Building Contents

One solution to this problem is to minimize the use of furnishings and construction materials that produce such large quantities of toxic smoke. The fumes given off during combustion should be considered when selecting materials. All parties in the building project should be urged to keep smoke and fire safety in mind and to use fire- and smoke-resistant materials in order to make the building safer and to reduce everyone's potential liability.

Designers should ask suppliers how their materials have performed in real fires. Modern polymer materials often have a complex response to fire and may have good fire retardancy but high toxicity and smoke production. If the suppliers will not respond or if they claim they do not know, the designer should be cautious.

The architect should ask if building carpeting has passed the NBS Radiant Panel Test, ASTM E-6.648. He or she should find out if the local fire department has tested the carpeting for its potential behavior in an arson attempt. Good carpet won't spread fire. Office furniture should be as noncombustible as possible; wall coverings should be noncombustible or low hazard; and file cabinets should be steel.

Everything in plenums, whether part of an HVAC system or not, must be noncombustible or encased in noncombustible materials. In many major fires, combustible materials in hidden spaces have played a major role in injuries, deaths, and property losses. Shielded materials decomposing before they ignite may be far more hazardous than the same materials burning openly.

A risk assessment may be made before a material or system of materials is specified. The following nine factors should be considered when judging the relative hazards:

1. Ease of ignition
2. Rate of flame spread
3. Rate of heat release
4. Rate of smoke release

5. Toxicity of combustion products
   a. Rate of toxic gas release under:
      Flaming condition
      Nonflaming condition
   b. Flammability of gas released
   c. Physiological effects
   d. Damage to other materials
6. Rate of carbon monoxide production
7. Ability to be extinguished
8. Basic integrity of product
9. Nature of burning

These considerations apply to service materials, building contents, and, in some cases, structural construction materials and systems. The geometry of a material or system in relation to its environment, along with the nature of the occupancy, also affects the hazard risk in the event of a fire.

Methods of Smoke Control

Smoke Management

Traditionally, *smoke management* has consisted of passive methods of modifying smoke movement in order to protect building occupants and firefighters and to reduce property damage. These methods included (1) compartmentation using fire barriers, (2) smoke vents, and (3) smoke shafts. They were used singly or in combination to reduce the undesirable spread of smoke.

A barrier's effectiveness in limiting smoke movement depends on the leakage paths within it and on the pressure differences across it. Air tends to flow from spaces at a high pressure to those at a lower pressure. Pressure differences depend on stack effect, buoyancy, wind, and the HVAC system.

The principal flow paths are doors and windows, which may be either opened or closed. Smoke control doors can be used to maintain the barrier. They are normally held open magnetically and released when the fire alarm system is actuated. Door releases can be arranged by zone in order to respond only to smoke or fire detection in a specified area.

Airflow can also occur through cracks in partitions, floors, exterior walls, and roofs. Possible leakage paths include holes where pipes penetrate walls or floors, cracks where walls meet floors, and cracks around doors.

A typical automatic smoke vent is pictured in Figure 13.7. The effectiveness of passive smoke vents and smoke shafts depends on their proximity to the fire, the buoyancy of the smoke, and the presence of other driving forces. When sprinklers cool smoke, the effectiveness of *smoke vents* and *smoke shafts* is greatly reduced.

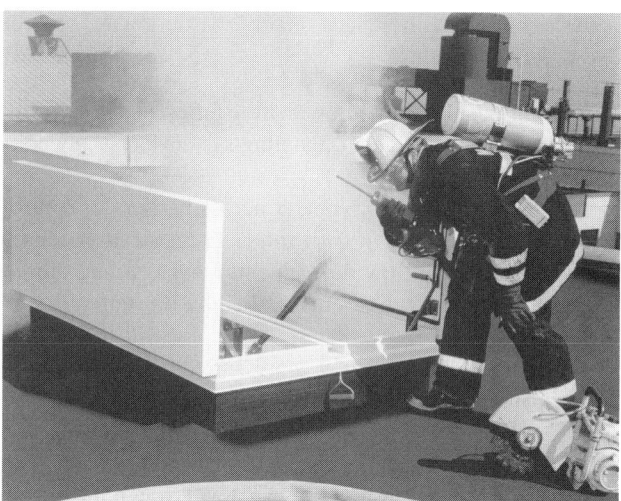

FIGURE 13.7. Automatic fire and smoke vent. (Photo courtesy of the Bilco Company.)

Elevator shafts in buildings have been employed as smoke shafts. Unfortunately, this renders them useless for evacuation, and these shafts often carry smoke far from the fire. Independent smoke shafts that do not leak to other floors can also be specially designed.

Part of the passive approach to smoke management involves shutting down the HVAC system in the event of a fire so that it doesn't contribute to the spread of smoke. There have been instances where centralized HVAC systems have pumped lethal carbon monoxide throughout a building. In recent years, the preferred approach has shifted to a more dynamic system that makes use of the fans to control the movement of smoke.

Smoke Control Systems

Systems using fans, called *smoke control systems,* are now employed to overcome the limitations of these traditional systems. They rely on fan-produced airflow and pressure differences to augment compartmentation. This active approach makes use of fans and special HVAC system controls that reposition dampers so as to dump all the building air outside and bring 100 percent outdoor air into the building. In fire situations, smoke control systems can limit smoke movement, or move it in a desired direction in order to help the passive systems control and manage the spread of smoke.

Principles. The two basic principles of smoke control are:

1. Airflow by itself can control smoke movement if the average air velocity is sufficiently large.

2. Air pressure differences across barriers can control smoke movement.

Cold air pushes warm air, with little difference in pressure necessary. Therefore, an air-handling system can easily control smoke.

Using *air pressure differences* across barriers to control smoke is often called *pressurization.* Pressurization causes high-velocity air to flow through small gaps around closed doors and in construction cracks, thereby preventing backflow.

Figure 13.8 illustrates how a pressure difference across a barrier can act to control smoke movement. Within the barrier is a door. The high-pressure side of the door can be either a refuge area or an escape route. The low-pressure side is exposed to smoke from a fire. Airflow through the cracks around the door and through other construction cracks prevents smoke from infiltrating to the high-pressure side.

When the door in the barrier is opened, air flows through it. If the air velocity is low, smoke by virtue of its buoyancy can flow up and over the airflow into the refuge area or escape routes, as shown in Figure 13.9. This smoke backflow can be prevented if the air velocity is sufficiently high, as shown in Figure 13.10. The velocity needed to prevent smoke backflow depends on the heat release rate of the fire.

Pressurizing Stairwells. In order for the occupants of any building to evacuate safely or reach an area of refuge, the paths of travel must be protected from fire, smoke, and hot gases. The quicker the smoke is controlled, the more likely the occupants are to be able to evacuate the building or get to a safe refuge area away from the fire zone. Since the evacuation routes are mostly stairwells, the primary smoke containment element is the *pressurization of the stairwells.*

Stairways, corridors, and exit passageways of a building are fire-separated from the rest of the building by enclosing the exitways with fire-resistive construction. All doorways and other openings in these enclosure walls must be protected with doors, fire shutters, or dampers having equivalent fire-resistive ratings. This construction is intended to fully protect the exitways until safe evacuation is accomplished. During evacuation and firefighting, however, stairwell doors are intermittently opened, and doors are sometimes accidentally left or propped open throughout fires.

Thus, to protect against smoke infiltration in case of fire, each of the stairwells is pressurized relative to the rest of the building by supplying it with outside air. As long as a positive pressure difference is maintained across the stairwell doorway on the fire floor, smoke cannot infiltrate into the

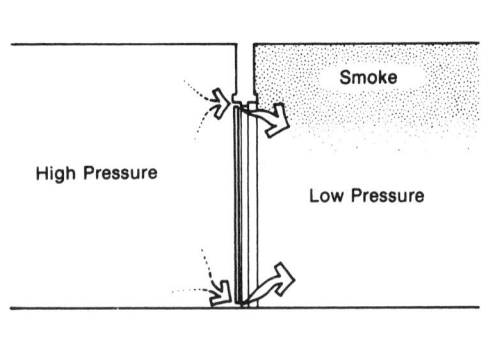

FIGURE 13.8. Pressure difference across a barrier.

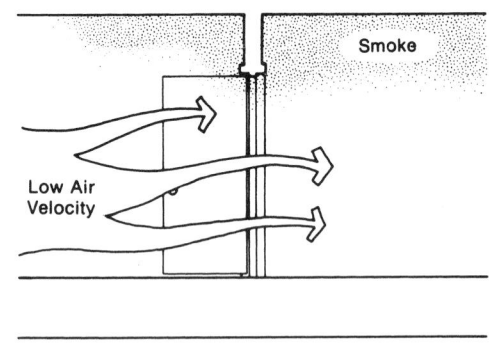

FIGURE 13.9. Smoke backflow.

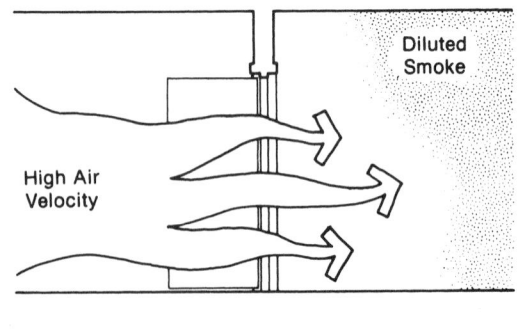

FIGURE 13.10. High air velocity through an open doorway.

stairwells and they will remain clear, enhancing the chances for escape.

Stairwell pressurization systems are divided into two categories: single- and multiple-injection systems. A single-injection system is one that supplies pressurization air to the stairwell at only one location.

The pressure in the stairwell, however, fluctuates as the doors are opened and closed by people evacuating. Too low a pressure allows smoke to contaminate the evacuation routes, but pressure that is higher than necessary may hinder people from opening doors.

Tall stairwells with single-injection systems can lose all their pressurization air when a few doors are open near the supply air injection point. The system can then fail to maintain positive pressure across doors further from the injection point. For tall stairwells, outside air should be supplied at various levels throughout the height of the stair shaft. Pressure-differential sensors are located at various levels in the stair shaft. These are linked to modulating dampers which automatically control the supply air quantity in order to maintain the pressure differential.

The stairwell system, typically independent of the main HVAC system, is activated only when smoke is detected.

Stairwell Compartmentation. An alternative to multiple injection is compartmentation of the stairwell into a number of sections, as shown in Figure 13.11. When the doors between compartments are open, the effect of compartmentation is lost. For this reason, compartmentation is not appropriate for densely populated buildings that are to be evacuated solely by the stairwell in the event of fire. However, if staged evacuation is planned, and if the system is designed to operate even when the maximum number of doors between compartments are open, compartmentation can be an effective means of providing pressurization for tall stairwells.

Zone Smoke Control. Pressurized stairwells are intended to prevent smoke infiltration into stairwells. However, in a building with just stairwell pressurization, smoke can flow through cracks in floors and partitions and through shafts to damage property and threaten life at locations remote from the fire. Except in summer, the stack effect causes a general upward movement of air, so that in the event of a fire on a lower floor, smoke will migrate to the upper floors through vertical shafts and openings in the floor construction. The technique of *zone smoke control*, illustrated in Figure 13.11, is intended to limit such a spread of smoke by evacuating smoke from the zone in which the fire initiated while pressurizing the surrounding zones.

The basic principle of zone smoke control is to provide a positive air pressure in the areas of safe refuge and a negative pressure in the areas of smoke contamination or incidence of fire. The pressure differences and airflow thus prevent the passage of smoke to the area of refuge.

First, the building must be divided into a number of smoke control zones, each separated from the others by partitions, floors, and doors that can be closed to inhibit the smoke movement. Frequently, each floor of a building is chosen to be a separate smoke control zone. However, a smoke control zone can consist of more than one floor, or a floor can consist of more than one smoke control zone.

Some arrangements of smoke control zones are illustrated in Figure 13.11. The term *pressure sandwich* is used to describe the venting or exhausting of the fire zone and the pressurizing of the surrounding zones.

The building's normal HVAC system is frequently used for smoke control. It is arranged so that it can either pressurize a zone or evacuate smoke from it as needed.

In the event of a fire on a particular floor, the supply air to the fire floor is cut off by a fire damper in the supply duct. Exhaust flow is maintained either through the return air, or by closing dampers in both supply and return ducts and activating an independent exhaust system. These

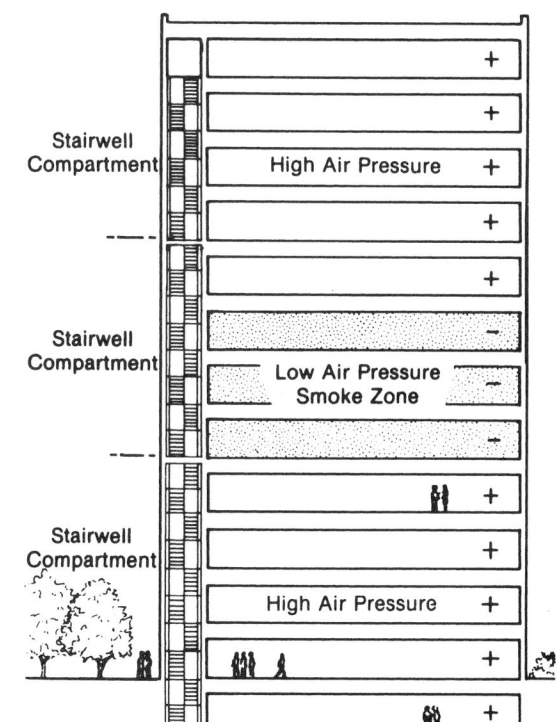

FIGURE 13.11. Air pressure sandwich.

actions produce a negative pressure on the fire floor relative to the floors above and below. To further ensure this pressure relationship, air can be supplied to the floors above and below the fire floor at a higher rate. The effect is to contain the smoke within the fire floor. Since the smoke is concentrated in the smoke zone, the building's occupants should evacuate the smoke zone as soon as possible after fire detection.

Venting of smoke from a smoke zone is important because significant overpressure can occur due to the thermal expansion of gases resulting from the fire. In addition, venting results in some reduction of smoke concentration in the smoke zone. If a mechanical exhaust system is not used, venting should be accomplished by passive smoke management methods, such as exterior wall vents or smoke shafts.

Fire Dampers. All ducts that pass through fire-rated walls, which are designed to contain the spread of the fire, must be equipped with fire dampers. These dampers must be located wherever a duct passes through a fire-rated wall, floor, or ceiling. Fire dampers are typically spring-loaded and have fusible links. When the temperature in the duct exceeds a safe level, the link melts, causing the damper to close and preventing flames and smoke from spreading throughout the building by means of the ductwork.

Fire and Smoke Detectors. Another commonly used control device is a thermostat, referred to as a *firestat,* placed in the return air ductwork. It is set to shut down the air-conditioning system or to initiate smoke control when the temperature rise in the duct exceeds a set value. At that point, it is apparent that a fire is present and, if the system continues under normal operation, that it will spread fire and smoke throughout the building.

Some air-handling systems may be provided with an optical sensor to detect the presence of smoke, since smoke may very well be present before the duct temperature rises enough to set off the firestat. Ionization-type detectors can be located in the return air ducts along with the smoke detectors for early detection of smoke.

Smoke Removal. Smoke evacuation can be part of a smoke control system. However, simply supplying and exhausting large quantities of air from the space or zone in which a fire is located cannot by itself control smoke movement, since it does not provide the needed airflow at open doorways and the pressure differences across barriers. This process, sometimes called *purging* the smoke, merely dilutes the smoke. Because of the large quantities of smoke produced during a fire, purging cannot ensure that air will be breathable in the fire space. But in spaces separated from the fire area by smoke barriers, purging can significantly limit the level of smoke.

Advantages

The stack effect can draw so much smoke up through a stairwell that having people exit through an unpressurized stairway would be equivalent to having them exit through a chimney.

Active smoke control has the following advantages over the traditional passive methods of smoke management:

1. Since smoke control is less dependent on tight barriers, leakage through barriers is less of a concern during design and construction.
2. The stack effect, buoyancy, and wind are less likely to overcome smoke control than passive smoke management. In the absence of smoke control, these natural driving forces cause smoke to spread to the extent that leakage paths allow.
3. Smoke control can prevent smoke flow through open doorways in barriers. In the absence of smoke control, smoke flows through these doorways.
4. Where smoke flows, fire often follows. Thus, smoke control can also help to control fires.
5. Smoke control systems aren't just for life safety; they also reduce property loss. In recognition of this fact, some insurance companies give credits for installed and tested smoke control systems.

Furthermore, while fires are developing or if they are shielded, they do not usually give off enough heat to activate an automatic sprinkler system until well after dangerous quantities of smoke have begun to move through a structure. Smoke control systems are often included as part of the fire safety systems of contemporary sprinklered buildings in order to handle the smoke generated prior to activation of the sprinkler heads and, subsequently, the smoke from the sprinklered fire.

Smoke control systems are required in all apartment buildings with seven or more stories. The exit access corridors are required to be continuously pressurized.

The Life Safety Code (NFPA 101) requires that every patient sleeping room in a hospital have an outside window that can be opened from the inside to permit venting of products of combustion. As an exception, buildings designed with engineered smoke control systems need not comply with this requirement.

FIRE SUPPRESSION

Smoke control systems can maintain tolerable conditions along critical exit routes and in refuge areas, but they cannot extinguish the fire itself.

Extinguishment

If a fire were totally confined to its area of origin, it could be allowed to burn itself out. It would consume all the available fuel in that area, however, which translates into property loss. To avoid unnecessary property loss, the fire must be actively extinguished. The most efficient fire-control technique is the use of automatic fire-extinguishing systems. These systems usually utilize a minimum amount of extinguishing agent and apply it directly to the fire.

High-Rise Buildings

Tall buildings and those extending over large areas have portions that cannot be reached from the outside by ladders and hoses. Large buildings therefore require their own firefighting systems.

Although most fire deaths occur in smaller buildings, more attention is usually paid to fire prevention in large commercial, institutional, and industrial (CII) buildings, particularly high-rises, because of the potential for multiple deaths or injuries.

For fire safety purposes, a high-rise building is one that (1) is too high for fire departments to reach all floors from the ground, (2) requires an inordinate length of time to evacuate, and (3) is capable of creating a stack effect internally.

Due to these characteristics, every high-rise building must have its own internal fire-suppression system, which usually means sprinklering. The exact definition of what height constitutes a high-rise building varies. All the model codes define a high-rise as being at least 75 feet (23 m) tall, but some local jurisdictions have set a lower definition. The state of Nevada, for example, defines a high-rise as 55 feet (17 m) and higher. West Virginia sets the limit at 40 feet (12 m).

Extinguishing Process

The three essential requirements for sustaining a fire are shown in Figure 13.12. If one of these is missing, the fire may not continue.

The most common means of fire suppression is the use of water supplied by automatic sprinkler systems or standpipe systems. Water extinguishes fires by removing heat and cutting off the oxygen supply from the source of combustion. It absorbs heat energy as the water is converted into steam. If

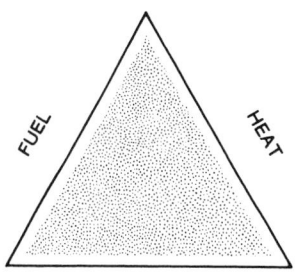

FIGURE 13.12. Necessary fire ingredients.

enough water is applied, the temperature of the burning materials can be reduced below their ignition point.

Besides water, carbon dioxide, other gases, foaming agents, and dry chemicals are also used for special-purpose suppression systems. These agents extinguish flames by chemically inhibiting flame propagation, suffocating flames by excluding oxygen, interrupting the chemical action of oxygen uniting with fuel, or sealing and cooling the combustion center.

Extinguishing Systems

The major fire-extinguishing systems for the protection of life and property against fire will be discussed in the remaining pages of this chapter.

Automatic Sprinklers

Sprinkler installations are reliable and very effective at extinguishing fires or at holding them back until firefighters arrive. Their success lies in the fact that automatic actuation occurs while the fire is still small and while control can still be initiated with moderate amounts of water. Once installed, sprinklers remain a continuous and reliable means of controlling losses due to fire.

Sprinkler systems consist of an array of water pipes fitted with automatic delivery heads. The sprinkler heads break up the water into fine droplets for wide dispersion.

Sprinkler Heads. Some representative examples of *sprinkler heads* are shown in Figure 13.13. Quick-response sprinklers were developed mainly for use in residential occupancies. A number of heads are arranged in a grid pattern close to the ceiling in such a way that areas may be fully covered with water spraying from them.

Each head is fitted with either a *fusible link* or a glass or quartzoid bulb containing a liquid to prevent the water from emerging. The fusible links are available for fusing at vari-

FIGURE 13.13. Sprinkler heads. (a) Sidewall, (b) concealed, (c) upright, and (d) pendant. (Photos courtesy of the Viking Corporation.)

ous temperatures. When heat melts the solder in the link, the cap held by the link and covering the opening is released and thrown from the head by water pressure. Similarly, with the bulb type, the rise in temperature causes the liquid to expand and break the bulb, allowing the water to pass. Bulbs are available that break at temperatures ranging from 155° to 355°F (68° to 180°C).

When the sprinklers are activated, the water stream strikes a shaped deflector plate which spreads it out over the area it protects. The flow of water helps to minimize and contain the fire, and at the same time may sound the alarm.

Sprinklers are available for installation in standard *upright, pendant,* and *sidewall* positions. The first two are illustrated in Figure 13.14. Upright sprinkler heads are installed above the supply pipe with the orifice facing upward and the deflector situated on top. Pendant heads hang down from the pipe with the orifice facing downward

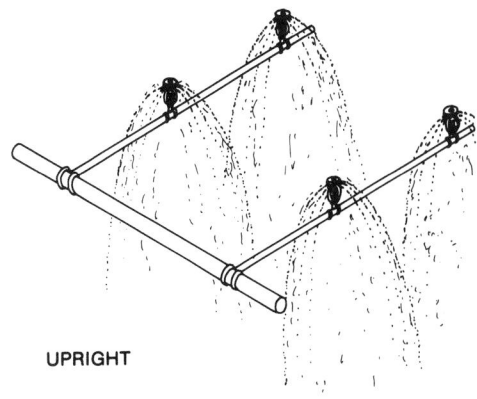

UPRIGHT

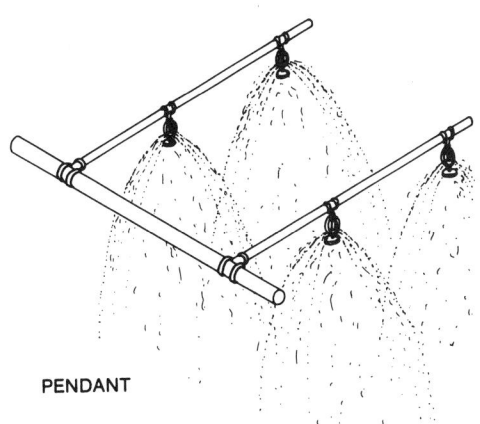

PENDANT

FIGURE 13.14. Upright and pendant sprinkler positions.

and the deflector located below them. They are not interchangeable because of differences in water patterns.

All types of sprinklers are available in various finishes, including plain or polished brass, satin, polished chrome, stainless steel, and even gold. The cover plate on a concealed sprinkler can be painted, as can the escutcheons on the semirecessed and flush types.

Water Supply. The sprinkler heads are served by a piping grid under or in the various ceilings which in turn is supplied from large risers connected to house tanks or public water mains.

A reliable water supply is critical for sprinklers. A good public water system is the best and most economical source for private fire protection. It is continuous supply and presents few or no maintenance problems.

The fire protection engineer compares existing water supplies to the demand by the building's sprinklers and hose

streams. If the volume of water available from the public water supply is inadequate for full fire protection, additional separate water supplies must be considered. Among these are wells, reservoirs, ponds, elevated tanks, or booster fire pumps with a stored water supply.

Since these sources may become inoperable during a conflagration, at least one connection on the outside of the building should be provided for fire department hoses. The firefighters can then hook up their engine pumps to supply additional water from hydrants to the sprinkler systems through these connections. Since the connection usually provides for the attachment of two hoses, it is called a *Siamese connection,* as shown in Figure 13.15. Such auxiliary outside connections are desirable, but not always required.

Occupancy Hazard Classification. The fundamental objective of sprinkler systems is the automatic discharge of water in sufficient density to control or extinguish a fire at its very beginning—before significant damage occurs, before significant quantities of water are required, and before firefighters are able to reach the scene. The density of water discharge is expressed in terms of area per head. The maximum area coverage permitted is 200 ft^2 (18 m^2) per head.

Sprinkler system requirements vary, depending on the classification of occupancy as follows:

Light Hazard. Included in this category are apartments, auditoriums, churches, hospitals, hotels, libraries, museums, nursing homes, office buildings, restaurants, schools, theaters, and similar structures in which effective fire protection can be provided relatively easily. The quantity and/or combustibility of the building contents is considered low, as is the rate of heat release from possible fires.

FIGURE 13.15. Siamese outdoor fire department connection. (Photo by Vaughn Bradshaw.)

The protection area allotted per sprinkler may not exceed 200 ft² (18 m²), with the maximum allowable distance of 15 feet (4.6 m) between supply lines and between sprinklers on each line. The sprinklers do not need to be staggered.

Ordinary Hazard. These occupancies include automotive garages, bakeries, boiler houses, electric-generating stations, feed mills, grain elevators, laundries, machine shops, manufacturing facilities, paper mills, printing and publishing establishments, warehouses, and other industrial properties. Combustibility ranges from relatively low to high, and quantity of combustibles ranges from moderate to high. Fires with a moderate to high rate of heat release are expected, and materials present may cause rapidly developing fires.

The protection area allotted per sprinkler may not exceed 130 ft² (12 m²) for a noncombustible ceiling and 120 ft² (11 m²) for a combustible ceiling. The maximum allowable distance between sprinkler lines and between sprinklers on a line is 15 feet (4.6 m). Sprinklers must be staggered if the distance between heads exceeds 12 feet (3.7 m).

Extra Hazard. This group includes buildings, or portions of buildings, in which the fire hazard is considered severe. Some examples are aircraft hangars, chemical works, explosives plants, linoleum manufacturing, linseed oil mills, oil refineries, paint shops, shade cloth manufacturing, solvent extracting, varnish works, and other operations involving the processing, mixing, storing, or dispensing of volatile flammable materials. Quantity and combustibility of content are very high, and rapidly developing fires with high heat-release rates may occur.

The protection area allotted per sprinkler may not exceed 90 ft² (8.4 m²) of noncombustible ceiling and 80 ft² (7.4 m²) of combustible ceiling. The maximum allowable distance between lines and between sprinklers on a line is 12 feet (3.7 m). Sprinklers on alternate lines must be staggered if the distance between sprinklers on lines exceeds 8 feet (2.4 m).

It may sometimes be necessary to separate high-hazard areas. Such high-hazard areas include kitchens and laundries of hotels, as compared to the lower-hazard guest rooms.

The basic types of automatic water sprinkler systems are: (1) wet pipe, (2) dry pipe, (3) pre-action, and (4) deluge. Secondary ones are water fog and liquid foam systems.

Wet Pipe Systems. The *wet pipe system,* the one most commonly in use, employs a piping system that contains water and is connected to a water supply under pressure at all times. When a fire occurs, the individual sprinklers are opened by the increased temperature, and water discharges immediately on the fire. This system is generally used when there is no danger of the water in the pipes freezing and wherever no special conditions require one of the other systems.

Since only those heads over the fire open, water flow and damage are limited. Wet pipe systems are fast-acting and are the most reliable type of automatic sprinkler system. Only one condition (the opening of the individual sprinkler) is necessary in order to start fire extinguishment.

Closed-head systems are the lowest in cost and require the least maintenance. The two major disadvantages of this type of system are: (1) accidental water discharge is possible through mechanical damage to a sprinkler head and (2) activation of the system can be somewhat delayed if the fire does not immediately liberate sufficient quantities of heat (the sprinkler head and surrounding area must reach the specific release temperature).

A variation of the wet pipe system is the *antifreeze system,* which is usually connected to a standard wet pipe system but protects small, unheated areas. Consisting of 20 sprinklers or less, the system is filled with an antifreeze-and-water solution.

Dry Pipe Systems. When unheated buildings are to be protected by sprinklers, there is a danger of freezing in the winter. To overcome this problem, a *dry pipe system* is used. Special valve arrangements enable the delivery pipes to be filled with air under pressure. Whenever a sprinkler head is opened by heat from a fire, air escapes and water is admitted to fill the piping and flow from the open sprinkler.

A dry pipe system operates more slowly than a wet pipe type and is more expensive to install and maintain. For these reasons, it is normally installed only where necessary, generally where freezing could be a problem.

Dry pipe sprinkler heads are normally installed upright. If the pendant type is required, they must be the dry model so that they drain fully to prevent freezing of trapped water.

Pre-Action Systems. The *pre-action system* is designed primarily to counteract the operational delay of a conventional dry pipe system and at the same time to eliminate the danger of water damage resulting from accidental discharge of automatic sprinklers or piping. A water-supply valve (deluge valve) is opened independently of the opening of the automatic sprinklers.

A pre-action system employs automatic sprinklers attached to a piping system containing air. The deluge valve holds the water back until it is opened by an automatic fire-detection system, which is generally more sensitive than the automatic sprinklers themselves. The detectors are installed in the same areas as the sprinklers and may sense heat, products of combustion, or ultraviolet or infrared radiation from a fire. Actuation of the detection system opens the valve, allowing water to enter the piping system, but water is not discharged until an individual sprinkler opens from the heat of a fire. The deluge valve can also be opened manually.

Pre-action systems are designed primarily to protect areas, such as rooms containing computers or other sensitive electronic equipment, where there is danger of serious water damage as a result of damaged automatic sprinklers or broken piping.

The pre-action system has several advantages over wet pipe systems. Automatic sprinklers don't operate until an area is likely to be beyond human tolerance. The fusible links in a sprinkler head are tested by UL to separate at $165°F \pm 5°F$ ($74°C \pm 3°C$) underwater and within 6 minutes of exposure to air at $290°F$ ($143°C$). Breathing air at this temperature would kill a person even if it were smoke-free.

The detection system can automatically sound an alarm and summon firefighters to the problem area with the possibility of effecting fire control before an automatic sprinkler head opens. Life safety is thereby enhanced and postfire water damage is minimized. Accidental discharge is almost impossible because two events must occur at the same time.

The principal disadvantage of pre-action systems is a degree of unreliability that is introduced by the necessary operation of electrical and mechanical devices before water can be introduced into the sprinkler system piping. This disadvantage is minimized by using especially reliable equipment and by a proper program of inspection, maintenance, and testing.

Deluge Systems. The purpose of a *deluge system* is to deliver the most water in the least time. It wets down an entire fire area by admitting water to sprinklers or spray nozzles that are open at all times. The water is supplied through a valve that is opened by the operation of an automatic detection system.

By using either automatic fire-detection devices of the type used in pre-action systems or controls designed for individual hazards, water is applied to a fire more quickly than with systems in which operation depends on opening the sprinklers only as the fire spreads. Deluge systems are suitable for extra-hazard occupancies in which flammable liquids are handled or stored, and in which fire may flash ahead of the operation of ordinary automatic sprinklers.

Deluge and pre-action systems are similar in design. The primary difference is that the sprinkler heads of a deluge system are normally *open,* while those in a pre-action system are normally *closed.*

Water Fogs. Water fogs can be dispensed from standard sprinkler or standpipe systems fitted with spray heads or nozzles. Fogs are effective for control of fires involving highly flammable solids or liquids, such as petroleum oils and gases and fast-burning explosive powders. They cool vapors that might otherwise reach ignition temperature.

Fog sprinkler heads do not contain a fire-detection device, as regular sprinkler heads do. A separate detection unit is installed which can be activated by heat, vapor, or smoke. The detector causes the supply valve to the fog system to open.

Liquid Foams. In order to provide a particularly effective smothering action and a minimum use of water, a liquid foaming agent can be introduced into the water in a standard sprinkler system. An air/water mixture in the form of a stabilized foam is produced. When the foam supply is exhausted, the system can operate as a deluge sprinkler system.

Damage Control. It is important to limit the destructive effects of the fire-extinguishing process on property. Hose streams and sprinkler discharge cannot be directed exclusively on the fire, and large volumes of water may be necessary for fire control and extinguishment. Also, water typically flows downward to any floors below. However, floor drains can and should be provided to drain away all excess water and to prevent its penetration into nonfire areas. Salvage covers or other means should be provided to cover sensitive objects and to direct water toward drainage points.

Water damage can be held to a minimum by cutting off the system as soon as its operation is no longer needed. For this purpose, all sprinkler systems should have a readily accessible outside valve that controls all of the normal sources of water supply to the system.

Combined Systems. Combining sprinkler piping with other nonfire protection functions is gaining acceptance by most code authorities. Systems that can be combined with sprinkler distribution piping include various types of hydronic heating/cooling systems, heat-recovery systems, hot and chilled water thermal-storage systems, and solar energy systems.

If a sprinkler system is required by code or insurance, combining it with another function offers a way to hold construction costs down. In cases where sprinklers are not required by law, combining them with another system can be a way to greatly upgrade the life and fire-safety aspects of a building at a lesser cost.

Insurance Savings. Sprinklering is primarily concerned with building and property protection. The cost of sprinkler systems can sometimes be offset by insurance savings because the insurance industry recognizes the value and effectiveness of sprinklers in protecting property. Sometimes, however, sprinklers can increase insurance costs because of concern over water damage.

Special Systems

Sprinklers can detect a fire, sound an alarm, suppress or extinguish a fire, and protect structural integrity. But they also agitate the air and redistribute the smoke.

Automatic fire-protection (suppression) systems other than water sprinklers may be needed for special environments, high-risk areas, isolated locations, or unusual hazards. Some examples include electronic data storage, paint dip tanks and spray booths, petroleum storage, securities vaults, and transformer rooms. Sprinklers may not be located in electrical rooms because they would run the risk of conducting electricity dangerously or could permanently ruin the equipment. Water can also cause irreparable damage to sensitive electronic equipment such as computers. In such cases, alternative suppression systems using carbon dioxide, FM-200, high-expansion foam, or a dry chemical as a suppression agent may be selected for special requirements.

Automatic Gas Systems. Various gas systems are available for automatic fire extinguishment. The most common gases are carbon dioxide, which is stored as a liquid and smothers flames when released into the atmosphere, and FM-200, a chemical that interferes with and stops the combustion process.

While the extinguishing mechanism differs for various gases, they all depend upon first detecting the presence of a fire by an automatic fire detection system and then discharging the gas into the fire area. In all cases, a specific concentration of gas must be attained within a specific period of time and then must be maintained. A bank of compressed gas cylinders is provided. The area to be protected must be sealed off in order to prevent the gas from escaping to surrounding areas and to maintain the designed concentration long enough to totally extinguish the fire. In some cases, there must be a time delay between fire detection and agent release so that personnel can safely evacuate the area.

Carbon Dioxide Systems. Carbon dioxide extinguishes or prevents fire solely by excluding oxygen from the fuel. It displaces the oxygen that would otherwise be present to support combustion. Systems to protect paper and other carbonaceous materials must attain a concentration of 30 to 50 percent carbon dioxide in air and maintain that concentration for half an hour to 1 hour. This concentration cannot support life, so a predischarge alarm signaling total evacuation of personnel from the area is essential.

Carbon dioxide has the advantage of causing no damage to unburned materials and leaving no residue. No cleanup is required, and the area needs only to be purged of the gas before operations can be resumed.

The principal disadvantages of carbon dioxide systems are the life-safety hazard and high cost. In addition, the length of time between detection and system discharge that is required for safe evacuation could allow the fire to gain considerable headway before extinguishment begins.

The reliability of carbon dioxide systems depends upon how well the area is enclosed. An open doorway would render the system ineffective.

Carbon dioxide is particularly suitable for fire prevention in areas where dangerous concentrations of flammable vapors may arise. Vapor detection devices are required to operate these systems.

Automatic Extinguishing Systems. Gaseous systems using FM-200 or other agents are used as an alternative to carbon dioxide systems.

High-Expansion Foam Systems. High-expansion foam for controlling and extinguishing fires is particularly well suited for confined areas. The foam is generated by blowing air through a screen sprayed with a detergent. To control a fire effectively, it is necessary to completely cover the area and the materials involved in the fire. To accomplish this, it is necessary to generate foam at a fast rate and possibly at several locations.

High-expansion foam systems have the advantage of utilizing only small quantities of water to extinguish a fire, thereby minimizing water damage and soaking of materials. However, a soapy film or residue remains when the foam dissipates, and some cleanup is still required. The foam system is automatically actuated by any of the detection systems.

The foam system's other disadvantage is that areas must be compartmentalized. While the compartments need not be airtight, as in the gaseous systems, wire mesh must be provided to contain the foam in the fire compartment, introducing a life-safety hazard. While the air in the foam bubbles will support life, vision is totally obscured and panic is probable. The installed cost is considerably less than that of either the carbon dioxide or FM-200 systems but more than that of a wet pipe sprinkler system.

protect these areas, dry chemical agent nozzles are located over the cooking areas and in the ducts.

Manual Fire Fighting

Manual fire suppression systems include *portable fire extinguishers* and *fire hose cabinets* (F.H.C.), as shown in Figure 13.16. Portable fire extinguishers may be charged with a variety of dry chemical agents. They are available in a wide variety of styles and capacities to meet any application.

If fire extinguishers are required, more and lighter units are recommended for applications involving the likely use by wheelchair-bound persons. If placed within easy reach, the 2½- and 5-pound (1.1- and 2.3-kg) units can even be operated by children in wheelchairs.

Hose standpipe (S.P.) systems distribute water to hose stations. The hoses can be operated by building occupants or fire department personnel. Outdoor Siamese connections separate from those serving automatic sprinkler systems can supply auxiliary water to standpipe systems.

SUMMARY

Fire protection systems are designed to detect the existence of a fire as early as possible, to sound an alarm, and to extinguish the fire or at least contain the fire and its effects until firefighters can bring it under control.

Two factors must be considered: the destructive capability of the intense heat generated and the harmful effects of the smoke upon people. Smoke is often the greater life hazard.

Smoke can spread far beyond the site of the fire to remote building areas, sometimes even before flaming occurs; in many instances, smoke victims have been found far from where the fire actually occurred. Smoke inhibits evacuation of a building by obscuring safe exit routes. It can temporarily blind and choke occupants, causing normally rational people to become disoriented and to panic. It can then overcome them and cause injury or death. Smoke from a fire rather than the flames themselves accounts for up to 80 percent of fire deaths. Moreover, while today's buildings are reasonably fire-resistant and are seldom structurally weakened by fire, smoke can cause a great deal of property damage.

Protection of life from both fire and smoke is predicated on the ability of the occupants to evacuate the building safely in the event of an emergency. Although evacuation is safest, if that is not possible, an alternative is to provide safe places of refuge within the structure. A reliable means must be pro-

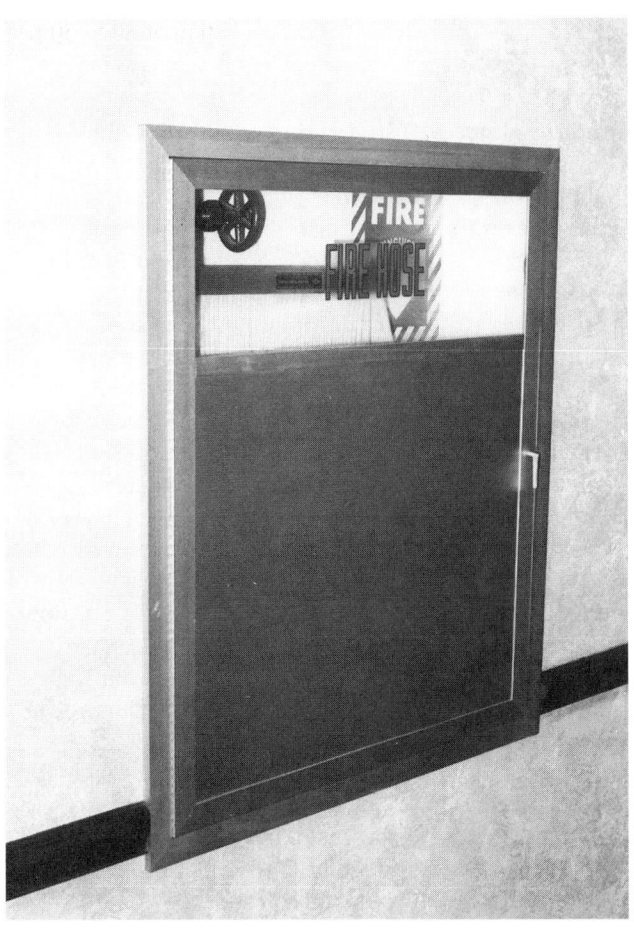

FIGURE 13.16. F.H.C. with glass door. (Photo by Vaughn Bradshaw.)

Dry Chemicals. Dry chemical systems may be used for fire suppression in commercial kitchens. The exhaust plenum and duct system typically become coated with grease, which, combined with high operating temperatures, produces a hazard area. Ignition of grease in a plenum or duct is often caused by a flash fire in one of the cooking appliances, and even though these surface fires could be extinguished by adequate portable extinguishers, this is impractical, considering that temperatures in excess of 2,000°F (1,093°C) are easily reached within a few minutes after ignition.

The automatic systems introduce a dry chemical with a sodium bicarbonate base into the plenum chamber and ducts in sufficient concentration to extinguish the fire in a matter of seconds. When the system discharges, it automatically cuts off the supply of heat to the stove or appliance, whether electrical or gas.

Grease fires can distort a duct, allowing flaming grease to spill over into the concealed spaces, walls, and ceilings. To

vided to alert all occupants about the existence of a fire emergency so that evacuation can proceed.

Since buildings are becoming more accessible to handicapped people, provisions must be made for their safe egress. As part of the design process, some designers spend a period of time in a wheelchair in order to sensitize themselves to the potential barriers. In addition to consideration of people in wheelchairs, signal indication and clear traffic flow should be provided for people with impaired vision or hearing.

The furnishings and finishes in modern buildings produce so much toxic smoke that it is physically impossible for occupants to leave a typical high-rise building before being overwhelmed by these fumes. Even if a building has enough exits for normal traffic, they become death traps when smoke completely obscures them and makes breathing impossible. One solution to this problem is to minimize the use of materials that produce these large quantities of toxic smoke.

Another solution is to control the flow of smoke by creating air-pressure differentials (positive air-pressure sandwiches) around the fire area so that the surrounding areas are at a higher pressure than the fire zone. Smoke is thus confined to the fire area, providing pressurized refuge areas for occupants and firefighters. Exhaust fans can also remove smoke from the fire zone in a controlled manner. By special design, these smoke-control systems may utilize the building HVAC system's supply and exhaust fans and air distribution system.

Special outside air fans are also added to pressurize corridors, exit stairwells, and, sometimes, elevator shafts. These keep the stairs free of smoke, permitting some degree of refuge, a safe escape route for occupants, and easier access to the fire areas by firefighters.

To facilitate active smoke control or improve passive smoke management techniques, vertical penetrations and open shafts should be minimized, reducing the opportunity for smoke migration and the stack effect.

Smoke control should be used in conjunction with water sprinklers. While they are important for suppressing a fire, sprinklers require substantial heat from a well-developed fire before they activate. They can also aggravate the smoke problem by cooling the smoke and thus causing larger particles to develop.

In any event, early detection is crucial in order to alert occupants and to initiate smoke-control and fire-suppression efforts. Detecting and locating the fire as soon as possible results in faster extinguishment, reduced hazard time to occupants and firefighters, and reduced property damage.

KEY TERMS

Radiant plate
Automatic fire detection
Heat (thermal) detector
Ionization (products of combustion) detector
Smoke (photo-electric) detector
Flame detector
Waterflow detector
Pull station
Annunciation
Electrical supervision
Voice fire alarm
Firefighters' communication system
Public emergency reporting system (PERS)
Means of egress
Elevator recall (capture) system
Compartmentation

Fire barrier
Fire-resistance rating
Fire door
Fire-rated access panel
Fire damper
Refuge area
Smoke management
Smoke vents
Smoke shafts
Smoke-control systems
Pressure differential
Pressurization
Pressurized stairwell
Zone smoke control
Firestat
Automatic sprinkler system
Purging
Sprinkler head
Upright sprinkler head
Pendant sprinkler head

Sidewall sprinkler head
Siamese connection
Wet pipe system
Antifreeze system
Dry pipe system
Pre-action system
Deluge system
Carbon dioxide fire-suppression system
Automatic fire-suppression system
High-expansion foam fire-suppression system
Dry chemical fire-suppression system
Portable fire extinguisher
Fire hose cabinet (F.H.C.)
Hose standpipe (S.P.)

STUDY QUESTIONS

1. What are the architectural concerns regarding:
   a. Fire detection and alarm systems?
   b. Means of egress?
   c. Compartmentation?
   d. Smoke control?
   e. Fire-suppression systems?
   f. Emergency power supply?

2. What are three ways to reduce the security risk of emergency exits?
3. Under what conditions would a refuge area be required?
4. What architectural means can be employed to minimize the hazards from toxic smoke?

BIBLIOGRAPHY

American Society of Heating, Refrigerating, and Air-Conditioning Engineers, Inc. *Guideline 5—Commissioning Smoke Management Systems.*

Boring, Delbert F. *Fire Protection Through Modern Building Codes.* American Iron and Steel Institute (AISI), Codes and Standards Division, 1982. (Covers allowable building areas and heights, requisites for suitable fire resistance of a building's structural components, and egress requirements.)

Catalog of National Fire Protection Association Codes, Standards, Manuals, and Recommended Practices. National Fire Protection Association (NFPA), Batterymarch Park, Quincy, MA.

International Conference of Building Officials. *Uniform Building Code.* Pasadena, CA.

NFPA. *Fire Protection Handbook.* Quincy, MA.

NFPA. *National Fire Codes.* Quincy, MA.

Various Publications. Smoke Control Association, Box 1057 Lee Branch Finance, Cleveland, OH 44120.

PART FIVE

RELATED SUPPORT SERVICES

The final three chapters address the subjects of acoustics, building commissioning, and economics for making design decisions. These are issues that relate to, and support all of, the other systems covered in this book.

Noise control in critical spaces normally is handled by consultants, but it is important for architects to be aware of the basic design principles in order to achieve proper sound control.

Building commissioning is a formal start-up and testing process that identifies and corrects operational deficiencies, saves energy, and helps to ensure that the building owner receives the intended performance from all systems.

Every architectural decision has an economic impact. While the best use of energy is important, a building owner is usually more interested in the best use of money. Both interests may result in the same decision, but not necessarily. However, by learning the "language of investment," a designer is more likely to develop cost-effective energy conservation proposals and be able to present them in a persuasive manner to the client.

Chapter 14

Architectural Acoustics

SOUND THEORY

SOUND GENERATION

PROPAGATION AND SOUND INTENSITY

FREQUENCY AND WAVELENGTH

HUMAN HEARING

SOUND REFLECTION, ABSORPTION, AND TRANSMISSION

DECIBELS

ROOM ACOUSTICS

BACKGROUND NOISE AND SOUND MASKING

GEOMETRIC ACOUSTICS

REVERBERATION AND ROOM RESONANCE

PUBLIC ASSEMBLY SPACES

SOUND ISOLATION

NOISE SOURCES

DEGREE OF NOISE REDUCTION

METHODS

AIRBORNE AND STRUCTURE-BORNE SOUND

OPEN OFFICE PLANS

MECHANICAL EQUIPMENT NOISE CONTROL

Acoustics is the branch of physics concerned with sound. Any sound that is objectionable or not desired is considered *noise*. What is considered desirable music by one listener may be regarded as noise if it intrudes into a space where, for example, people are trying to sleep or are playing different music.

The effect of building design on the control of sound in buildings is the subject of architectural acoustics. Good architectural acoustics does not happen by accident, and some appreciation of the nature of sound and the principles of acoustics is important for creating a satisfactory acoustical environment.

Conversations of congregating people can create unacceptably high noise levels nearby. Reverberant school corridors can create annoying sound intrusions in adjacent classrooms. Mechanical and electrical equipment can also be a source of noise. Designing machinery for quiet operation is the province of the manufacturer. It is an engineer's responsibility to keep the sound from mechanical equipment at an acceptable level.

Generally, it is the architect's role to recognize a potential noise problem in a proposed building and to take steps to solve it. An acoustical defect that appears in the completed building cannot be readily corrected, resulting in inadequate acoustical quality.

The acoustical design of buildings generally has three aspects: (1) planning to keep noise sources as far as possible from quiet areas, (2) the internal acoustics of rooms, and (3) structural precautions to reduce noise penetration. The first step in considering the acoustical aspects of a design is a detailed analysis of the design purpose of the structure. The planned usage of all individual spaces should be investigated before an acoustical plan is established.

The architect must visualize the overall building design and function in terms of acoustical qualities. For special concerns, the architect should seek the advice of an acoustical consultant in the early stages of design in order to plan for the acoustical aspects of the design from the beginning. The architect is also in a position to compare costs and coordinate the work of all trades in order to come up with the most cost-effective combination of modified sound sources and acoustical treatments.

Any decision affecting both acoustics and other architectural requirements must be made by the architect. Acoustical consultants can point out possible solutions that will provide satisfactory noise control. Utilizing this advice, it is the architect's role as coordinator to consider which acoustical solution can be most successfully integrated with the solutions to the other demands made on the buildings.

SOUND THEORY

Sound Generation

Physically, *sound* is a rapid fluctuation of air pressure. Vibrations from a source of sound set the surrounding air molecules into a similar physical motion. The result is a series of pressure pulses moving outward from the source (Figure 14.1).

These vibrations can be transmitted through air or any other "elastic" medium, including most building construction materials. In a vacuum, where there is no medium, sound cannot be transmitted.

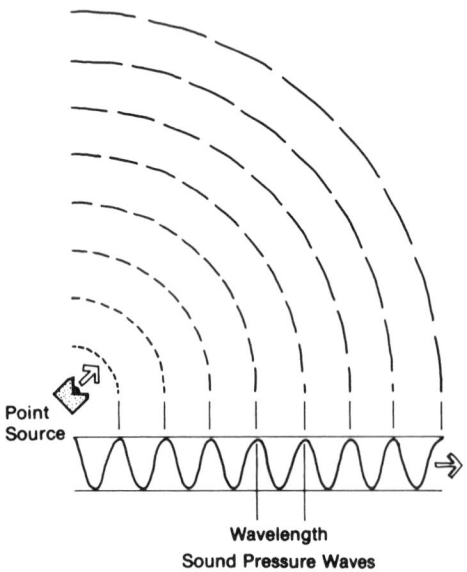

FIGURE 14.1. Wavefront propagation from a point source.

Propagation and Sound Intensity

When sound is generated by a point source in a free field (free from reflective surfaces or other interference), it moves outward as a wave front in all directions in an ever-increasing sphere. The energy increasingly spreads out as the sound waves move outward from the source, and its intensity diminishes in proportion to the square of the distance from the source (Figure 14.2). For instance, as the distance a sound wave travels from its source doubles, the spherical front of the sound quadruples in area, and the sound energy is consequently distributed over four times the area.

This dispersal of sound energy explains why it becomes increasingly difficult to hear a sound in the open air as the distance from the source is increased. In rooms, the sound waves impinge upon reflecting surfaces, which help to maintain the sound intensity and the audibility further from the source.

Frequency and Wavelength

The rate of oscillation of molecules by sound is known as the *frequency of vibration,* which is measured in cycles per second, otherwise known as hertz (Hz). The frequency of sound vibration is referred to as *pitch.* Normal sounds are made up of a combination of many frequencies. A sound of only a single frequency is known as a *tone.*

Sound travels a certain distance during a cycle of vibration. The distance is called its *wavelength.* The product of frequency times wavelength is the velocity of sound.

The lowest frequency of sound humans can hear is about 16 cycles per second (cps or Hz). Since the velocity of sound in air under normal atmospheric conditions is a constant 1,130 feet per second (345 m/s), sound at 16 Hz has a wavelength in air of 1,130/16 (345/16), or 70.625 feet (21.5 m). At a high frequency of 20,000 Hz, the wavelength is 0.0565 foot, or 0.6780 inch (1.72 cm). This wide range in the dimensions of wavelengths is significant in room acoustics.

Human Hearing

Sound vibrations impinging upon the ear create similar vibrations of the ear's receiving mechanism. If the wave motion is within the audible frequency range, it is perceived as sound. The ear is generally able to discern frequencies of vibration of 16 to 20,000 Hz. The ability to respond to the higher frequencies diminishes with age.

The ultimate criterion for evaluating an acoustical environment is its effect on the occupants. While it is relatively

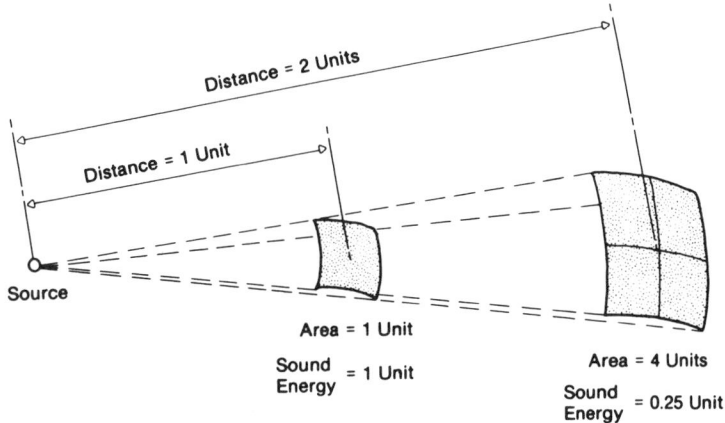

FIGURE 14.2. Sound energy spreading out.

easy to define the physical characteristics of sound, such as frequency or sound pressure level, how people hear and respond to sounds is a very complex subject. The human response to sound involves physiological and psychological reactions that are complicated by the conditioning of persons as a result of various experiences and by personal prejudices.

Sound Reflection, Absorption, and Transmission

Architectural acoustics rarely deals with sound in a free field. When sound strikes a boundary or any sizable surface, part of its energy is absorbed, part is reflected, and part is transmitted through the construction, as shown in Figure 14.3. The sum of all three components is equal to the total incident sound energy.

The proportion falling into each category depends on the physical properties of the material and of the impinging sound. Hard surfaces, such as concrete and dense plaster, reflect nearly all of the incident sound energy, while soft, porous materials, such as wood, fabrics, furnishings, and people, absorb a large part of the energy striking them.

When sound is absorbed, part of its energy is dissipated as friction by the air movement in the pores of the material, but a large portion is also usually transmitted on through the material. This is why fiberboard and acoustical tile are not good sound insulators. They may reduce the sound energy level within a space, resulting from noise generated in that space, but they don't prevent sound transmission between spaces. In other words, sound absorption and sound insulation are two very different phenomena.

The ratio of sound energy absorbed to the sound energy impinging upon a surface is called the *absorption coefficient* of the material. Denoted by the symbol α (alpha), it is a function of the frequency and angle of incidence of the sound. Since sound waves in an enclosed space generally travel in many directions at once, published values of α are averages over all possible angles of incidence. An absorption coefficient of 1 represents total absorption, such as would occur from an open window. When the absorption coefficients for 250, 500, 1,000, and 2,000 Hz are averaged together, the result is called the *noise reduction coefficient* (NRC).

Sound transmission from space to space depends on the

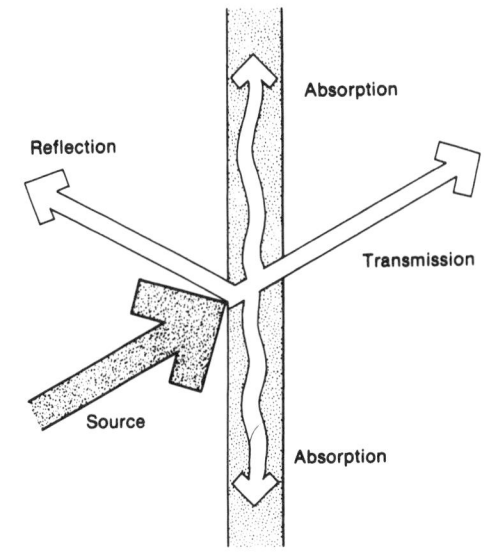

FIGURE 14.3. Reflection, absorption, and transmission.

sound-insulative qualities of the construction in between. This property of the construction materials is known as *transmission loss* (TL), which represents the difference in sound pressure level between the incident side and the opposite side of the construction. The TL is generally greater for more dense, heavy construction. Since it is more difficult for sound energy to set in motion a heavy partition than a light one, the heavy one is a better insulator. For the same construction, the TL is greater at higher frequencies. The values of TL are measured in decibels.

Decibels

Sound is normally measured in terms of sound pressure, which is the air pressure created by the sound. The unit of measure of sound pressure is the Pascal (Pa). The range of sound pressures of importance to us is so wide that a logarithmic scale, called the *decibel scale,* is used to compress it to a manageable size. A *decibel* (dB) is the ratio of sound

pressure to a base level chosen at the threshold of human hearing. Examples of decibel levels of some common sounds are listed in Table 14.1, along with the corresponding sound pressure level in Pascals.

Generally, a 3-dB change in sound pressure level is barely perceptible to the human ear. An increase of 10 dB is perceived as a doubling of loudness, and a decrease of 10 dB is perceived as half as loud.

For convenience, the decibel scale is also used to represent the energy of sound. The acoustical power generated by some pulsating source can be expressed in watts, but the full range of human hearing from the threshold of perception to the threshold of pain spans 10^{14} watts. So, sound power is condensed into the logarithmic decibel scale. Unlike sound pressure levels, sound power cannot be directly measured. It must be calculated from a sound pressure measurement.

Because the decibel scale is logarithmic, decibels cannot be summed up by simple addition. The decibel values of two or more simultaneously occurring sounds must be added

TABLE 14.1 TYPICAL SOUND PRESSURE LEVELS

Source	Sound Pressure (Pa)	Sound Pressure Level dB re 20μPa	Subjective Reaction
Military jet takeoff at 100 ft	200	140	Extreme
Artillery fire at 10 ft	63.2	130	danger
Passenger's ramp at jet airliner (peak)	20	120	Threshold of pain
Loud rock band[a]	6.3	110	Threshold of discomfort
Platform of subway station (steel wheels)	2	100	
Unmuffled large diesel engine at 130 ft	0.6	90	Very loud
Computer printout room[a]	0.2	80	
Freight train at 100 ft	0.06	70	
Conversational speech at 3 ft	0.02	60	
Window air conditioner[a]	0.006	50	Moderate
Quiet residential area[a]	0.002	40	
Whispered conversation at 6 ft	0.0006	30	
Buzzing insect at 3 ft	0.0002	20	
Threshold of good hearing	0.00006	10	Faint
Threshold of excellent youthful hearing	0.00002	0	Threshold of hearing

[a] Ambient.

Source: © 1989 Reprinted from ASHRAE *Handbook of Fundamentals 1989* by permission of the American Society of Heating, Refrigerating and Air-Conditioning Engineers, Inc.

TABLE 14.2 CORRELATION OF SONES, dBA, AND NC

Sones	dBA	NC
1	28	22
2	38	32
3	44	38
4	48	42
5	51	45
6	54	48
7	56	50
8	58	52
9	60	54
10	61	55
11	63	57
12	64	58
13	65	59
14	66	60
15	67	61
16	68	62
17	69	63
18	70	64
19–20	71	65
21	72	66
22–23	73	67
24–25	74	68
26	75	69
27–28	76	70
29–30	77	71
31–33	78	72
34–35	79	73
36–38	80	74
39–40	81	75
41–43	82	76
44–46	83	77
47–50	84	78
51–53	85	79
54–57	86	80
58–61	87	81
62–66	88	82
67–71	89	83
72–76	90	84

NC level = dBA level − 6 ± 2

dBA = $33.2 \log_{10}$ (sones) + 28 ± 2

logarithmically. Thus, when two sounds of the same pressure level combine, the resultant dB is only 3 higher than the dB of one of the sounds alone. If one of the sounds is 10 or more dB higher than the other, the lower pressure sound has no impact on the combined sound pressure.

In addition to defining sound levels, the decibel scale is used to describe the sound reduction ratio achieved by an insulative construction element. Unlike thermal insulation, the acoustical insulating value of a wall made of more than one type of construction is not the weighted average of decibel values in proportion to their areas. Instead, the overall sound reduction value is usually little more than the value of the least effective part of the wall, unless the area of poor insulation is extremely small.

For most purposes, decibel values are used to represent the total sound energy or pressure. However, sounds at different frequencies, while having the same pressure or energy, may be perceived subjectively as having different loudnesses. Several scales have been developed that group sounds into octave frequency bands so as to enable significant frequencies to be indicated. *Noise criteria* (NC), *speech interference level* (SIL), and A-weighted sound level (dBA) are examples of such scales. The dBA scale correlates well with subjective judgments of loudness. Most noise data, recommended standards, and sound insulation performances of various constructions are presented in decibel values averaged over 100 to 3,200 Hz.

In some cases, loudness levels are given in units of phons or sones. Sound loudness can be measured in terms of *phons*. The loudness level of a sound in phons is defined to be the decibel sound pressure level of a standard 1,000-Hz tone that is judged to be equally loud by a statistically representative sampling of people. However, because the human perception of sound does not follow the phon scale proportionally, another loudness scale was developed. A *sone* is equal to the loudness level of 40 phons. Table 14.2 shows the correspondence between the sone, dBA, and NC loudness scales.

ROOM ACOUSTICS

Architectural acoustics is concerned primarily with the behavior of sound in enclosed spaces, where sound is acted upon by the room boundaries and any obstructions present within the room. The shape of a room and the placement of reflective and absorptive surfaces determine the distribution of sound in the space.

The subject of room acoustics is concerned with the acoustical environment of an interior space and how this environment affects the functions of or activities in a space. It includes, for example, the control of excessive reverberant noise in a factory or the enhancement of music in a church.

Background Noise and Sound Masking

Any noise other than those sounds that an occupant wants to hear is called background noise. A continuously maintained level of background noise is called the *background noise level*. The presence of high background noise levels makes it difficult to hear desired sounds, or, to put it another way, the background noise raises the threshold of *audibility* for any sound that an occupant may want to hear.

The acoustical objective for an auditorium is a low background noise level so that spoken lines can be understood and musical notes heard clearly. In offices, the objective is to reduce the noise from business machines and peripheral conversations to a level that is not annoying and that allows for comfortable conversation among occupants. But in offices, the existence of some background noise can assist in acoustical privacy by *masking* lower levels of sound, as illustrated in Figure 14.4.

Acoustical privacy in an office depends on the *speech intelligibility* of what is being said, that is, the ability to understand it. The ability to make out a spoken sentence depends largely on the sound level of background noises. In a very quiet environment, private conversations may be easily overheard from some distance or sometimes even in an adjacent room. A sufficient background noise, however, can mask the conversation so that no eavesdropping can take place.

Electronic sound-masking systems are sometimes used to create a windlike "white noise" in order to unobtrusively raise the background noise level. This provides acoustical privacy and makes undesirable sounds, such as a neighbor's conversations, less comprehensible and therefore less distracting.

A sound-masking system is adjustable and can be carefully tailored to produce a combination of low-pitched and high-pitched sounds that effectively accomplishes its objectives while being generally unnoticeable to the worker who walks in off the street. Sound masking can also be effectively achieved with background noise or intentional noise from room air outlets.

Geometric Acoustics

If a wave front strikes a large reflective surface, a reflected wave front is generated. A line from the sound source to the reflective surface can be visualized as a ray similar to a beam of light. When the incident ray is reflected from the surface, the angle of reflection is equal to the angle of incidence. In this and other respects, the behavior of sound can be considered analogous to the behavior of light.

The analogy to light, however, is not always valid. For example, sound has much longer wavelengths and a much smaller velocity than light does, which affects some considerations. When acoustics is analyzed in a similar way to optics, the approach is called *geometric acoustics*. The more rigorous approach that recognizes the differences between sound and light is called *physical acoustics*.

High-frequency sounds have a short wavelength and can be thought of as a beam or ray similar to light. They are reflected by hard surfaces and proceed unaltered through large openings.

Low-frequency sounds, however, have wavelengths as long as 70 feet (21.4 m). A small surface does not reflect these sounds, and they may be diffracted or bent when they pass through openings in walls such as doors and windows. Recesses, surface protrusions, and smaller combinations of reflective and absorptive surfaces also diffract low-frequency sound. Geometric acoustics is not applicable to frequencies below about 250 Hz. Despite its limitations, the concept of geometric acoustics is commonly used to determine the proper room shaping conducive to satisfactory distribution of sound in auditoriums.

Figure 14.5a illustrates sound reflection. A point source of sound, S_1, generates a ray, R_1. If R_1 strikes a wall, the wall will reflect the sound along ray R_2 at a reflected angle equal to the incident angle. At any point along R_2, if only the sound from it can be heard, the sound will seem to be coming from

FIGURE 14.4. Sound masking.

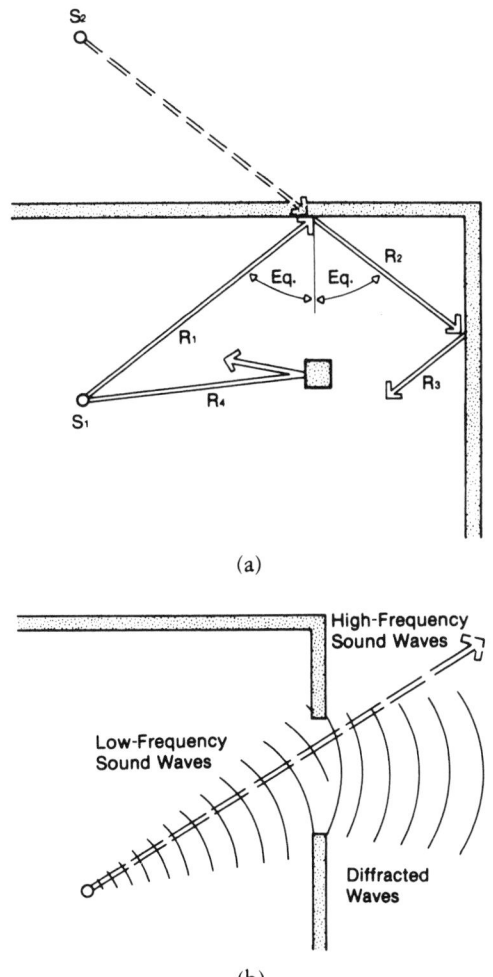

(a)

(b)

FIGURE 14.5. (a) Reflection of sound waves and (b) altered sound waves.

a source at S_2. Location S_2 is said to be the source of a first-order image. If R_2 is reflected from another surface, the reflected ray, R_3, can give rise to a second-order image.

Consider another ray, R_4, striking a column. High-frequency sounds (those having a short wavelength) will be reflected. Low-frequency sounds, however, go around it and are not reflected.

Sound passing through an opening may or may not be altered, depending on its frequency. A low-frequency sound may form a circular wave front, as shown in Figure 14.5b, at the opening such that the opening seems to be the source. The resulting divergence spreads out the sound from that point. A high-pitched sound, as a directional beam, may pass through the same opening with little or no change if the aperature is not too small.

Reverberation and Room Resonance

Sound Absorption

A space with an abundance of absorptive surfaces does not support sound and is therefore called *dead*. When surfaces are reflective, sounds can be sustained through several reflections before sound attenuation makes them inaudible, and the space is said to be *live*. This makes speech easier and gives life to music. If a room is too live, sound becomes distorted.

Reflective surfaces can be covered with *absorptive materials* in order to increase the amount of absorption. This causes reflected sound present in the room to die out quickly, resulting in a degree of quietness and a lower sound level.

If internally generated sound is excessive, acoustical treatment may be applied to the ceiling and carpets may be added. These *acoustical materials* are effective in sound control, but they affect only the two horizontal surfaces of a room, leaving the opposing vertical surfaces of glass and bare wall to reflect sound.

Various window treatments can be included in the sound absorption program. The NRC for venetian blinds, for example, is about 0.10 compared to 0.02 for bare glass and 0.03 for plaster. Drapery fabrics at 100 percent fullness have NRC values ranging from 0.10 to 0.65, depending on the tightness of the weave.

Note that these NRC values refer to reductions in sound levels *within* a space and not between spaces. Absorptive materials are poor insulators and do not keep out intruding sound. Absorptive treatments also have no effect on the pressure level of the source. They reduce only reflected sound.

The acoustical behavior of regular-shaped rooms can be predicted by the Sabin equation. Named after the American physicist W. C. Sabine, it provides a simple way to compare the sound absorption effectiveness of various materials of a given area. The equation is as follows:

$$A = \alpha \times S \qquad (14.1)$$

where

A = panel absorption in Sabins
α = absorption coefficient
S = surface area in square feet of panel

The absorption coefficient, α, is a dimensionless unit that must be determined experimentally by the manufacturer. Typical values for various building materials are shown in Table 14.3, but actual values should be obtained from the manufacturer. If several materials are used in a space, they must all be included, and the total value is the sum of the individual calculations.

TABLE 14.3   TYPICAL ABSORPTION COEFFICIENTS (α)

Material	Frequency (Hz)			
	250	**500**	**1,000**	**2,000**
Acoustical ceiling tile	0.15–0.95	0.35–0.95	0.45–0.99	0.45–0.99
Brick				
Unglazed	0.03	0.03	0.04	0.05
Unglazed, painted	0.01	0.02	0.02	0.02
Carpet				
Heavy, on concrete	0.06	0.14	0.37	0.60
Heavy, with pad	0.26	0.48	0.52	0.60
Light	0.05	0.10	0.20	0.45
Concrete block				
Course	0.44	0.31	0.29	0.39
Painted	0.05	0.06	0.07	0.09
Drapery				
10 oz/yd^2	0.04	0.11	0.17	0.24
14 oz/yd^2	0.31	0.49	0.75	0.70
18 oz/yd^2	0.35	0.55	0.72	0.70
Flooring				
Concrete or terrazzo	0.01	0.015	0.02	0.02
Resilient tile on concrete	0.03	0.03	0.03	0.03
Wood	0.08	0.08	0.06	0.06
Glass				
Large, heavy plate	0.06	0.04	0.03	0.02
Ordinary window	0.25	0.18	0.12	0.07
Gypsum board	0.10	0.05	0.04	0.07
Marble or glazed tile	0.01	0.01	0.01	0.02
Plaster				
On lath	0.10	0.06	0.05	0.04
On tile or brick	0.015	0.02	0.03	0.04
Plywood paneling	0.22	0.17	0.09	0.10
Seating				
Upholstered				
Audience seated[a]	0.74	0.88	0.96	0.93
Unocc., cloth-covered[a]	0.66	0.80	0.88	0.82
Unocc., leather-covered[a]	0.54	0.60	0.62	0.58
Wooden pews, occupied[a]	0.61	0.75	0.86	0.91
Metal or wood chairs[b]	0.19	0.22	0.39	0.38

[a] Per square foot of floor area.
[b] Per seat.

Irregularly shaped rooms with nonuniformly distributed acoustical absorption require a more complex calculation such as the *image theory*, which is practical only with the use of a computer. In this theory, the contributions of energy from the many sound waves reflected from the room surfaces are added to the direct sound wave from a source to calculate the total sound pressure level at any point in the room.

Reverberation and Reverberation Time

The support of sound by successive reflections is called *reverberation*. The degree of reverberation affects the aesthetic characteristics of the sounds heard and the general acoustical "feeling" of a space. Reverberation and reverberation time are important characteristics of room acoustics.

The average distance between reflections is called the *mean free path* of the sound wave. It has been found experimentally that the mean free path is approximately equal to four times the volume divided by the surface area.

A large amount of reflected sound is needed to maintain sound levels toward the rear of auditoriums. If, however, the reflected path is longer than the direct path by 60 to 65 feet (20 m), the reflected image may be manifested as an echo. In such cases, the reflection causes a masking of the sounds rather than a beneficial reinforcement of the sound pressure level. Thus, the ratio of reflected to direct sound should increase toward the rear. If the ratio is too high, there will be a loss of sense of direction indicating the position of the source.

Despite the loss of some directionality, the scattering of sound energy in a room from numerous reflecting surfaces is desirable because it helps make audibility more equal in all parts of a room. The greater the ratio of reflected to direct sound, the more live the space seems. The liveness of small rooms is limited by their small volume. Large wall splays and off-level ceilings help randomize the sound and increase liveness.

It is desirable for the sound producing the reverberation to also be well diffused because it causes the liveness to be somewhat "blurred." An overlay of distinct reflected sounds creates more confusion over subsequent direct sounds. A long, reflective corridor is an example of a reverberant space with poor diffusion.

In a diffuse environment, sound decays at a uniform rate. At each successive reflection, a portion of the sound energy is absorbed. The sound pressure level gradually decreases until it becomes inaudible. The time it takes for the sound to become inaudible is the *reverberation time*. The reverberation time is a function of the volume of the space and the absorptivity of the surfaces, people, and furnishings in it.

The quantitative definition of reverberation time is the time required for a given sound to undergo a reduction in pressure of 1/1,000th of its initial value. This corresponds to a sound pressure reduction of 60 dB, which is enough to reduce most sounds heard in enclosed spaces to inaudibility.

Sometimes a single reverberation time does not apply to an entire space. When spaces are subdivided by balconies or when a space is composed of a number of acoustically coupled smaller spaces, the reverberation time may be different for each subspace.

Room Resonance

An enclosed space with reflective surfaces excites certain frequencies of sound vibration, which depend on the relation between the sound wavelength and the space dimensions. The room is said to *resonate* at those frequencies.

The resonance raises the volume level for those frequencies and multiples of those frequencies, enabling them to be heard more predominantly. Each room usually has a number of these multiples within the range of human hearing. When organ builders install their instruments, they usually find it necessary to adjust the loudness of certain of the larger pipes whose notes coincide with the resonant frequencies of the space.

The effect of room resonance is emphasized if all the room dimensions are the same (cube shaped) or if the dimensions are related by simple ratios (e.g., 1:2 or 1:3). Resonant peaks can be lowered and made wider by introducing large irregularities in room shape or in the distribution of sound-absorbing materials. The difficulties arising from this phenomenon can be avoided by carefully selecting the dimensional relationships, making opposing walls not parallel, and using materials that absorb sound at the critical frequencies.

Public Assembly Spaces

The satisfactory use of large spaces such as assembly halls, auditoriums, lecture halls, and conference and music rooms depends on proper acoustic conditions. In auditoriums, the necessary analysis is of considerable complexity, so expert consultants are normally employed to advise on the acoustical design. The subject is far too detailed to be covered in its entirety here, but some general principles are presented below and in Figures 14.6 through 14.9.

Aside from proper insulation, the two principal aspects that govern satisfactory hearing are (1) the planning and shaping of the room in order to provide adequate paths for sound to travel from speakers or performers to the audience and (2) the rate at which sound dies away (reverberation time).

Direct Sound Path

Ideally, all occupants of a space should be able to hear by means of a direct path of sound from the speaker or performers. In conference, council, or board rooms, where any occupant may be called upon to speak, this is achieved by laying out the seating so that all the occupants can see one another. A circular board table is better for this purpose than a long, narrow one.

The intelligibility of speech depends to a large extent on the recognition of consonants. Unfortunately, consonants are weak compared to vowels. It is therefore extremely important to maintain the sound energy in lecture halls through the use of reflecting surfaces. In rooms where there is a large

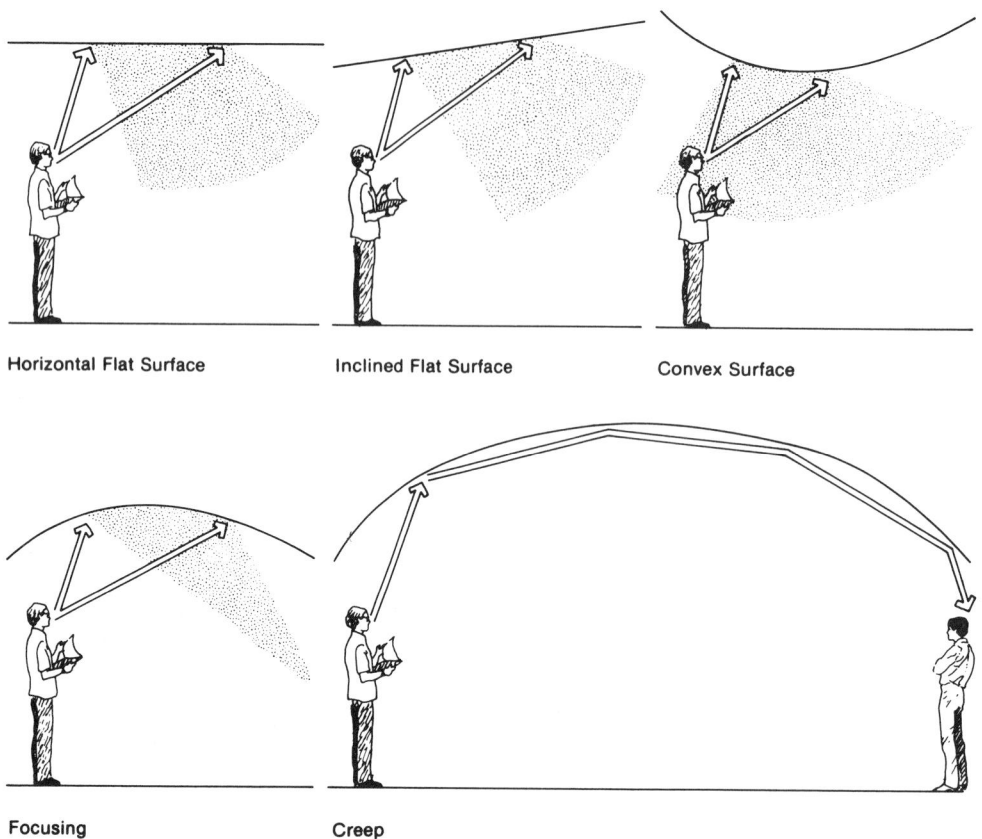

Horizontal Flat Surface   Inclined Flat Surface   Convex Surface

Focusing   Creep

FIGURE 14.6. Acoustical characteristics of ceilings. Sound-absorbing treatment may be used to eliminate sound focusing from a domed ceiling.

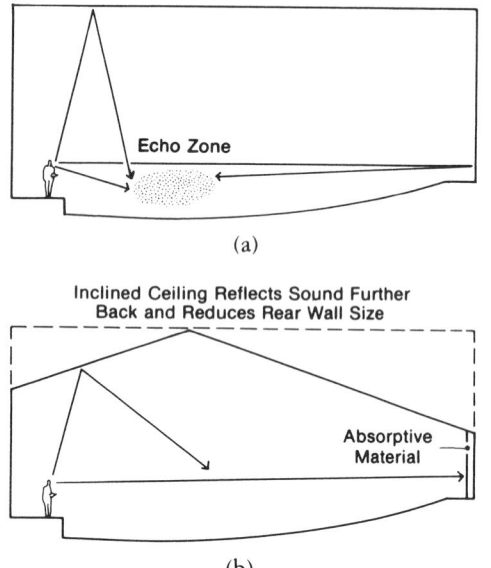

Echo Zone

(a)

Inclined Ceiling Reflects Sound Further
Back and Reduces Rear Wall Size

Absorptive
Material

(b)

FIGURE 14.7. (a) Echo zone and (b) uses of reflective and absorptive surfaces to avoid an echo zone.

passive audience, the direct path of sound is also improved by raising the speaker and by inclining the seating so that each member of the audience is less obstructed by those in front.

Reflected Sound

Reflected sound can, if properly directed, effectively augment the direct sound. Figure 14.10 shows a section through an auditorium where the stage or speaker's platform has been raised, the seating inclined, and portions of the ceiling arranged to reflect sound to specific areas of the audience. But to avoid echoes, the extra distance that first reflections travel farther than the direct path should not be over 60 feet (20 m).

Reflecting surfaces should generally be flat. Curved surfaces may sharply concentrate sounds in some places while leaving a lack of sound reinforcement elsewhere.

Just as properly arranged reflecting surfaces can assist in distributing sound, inappropriate surfaces can have an undesirable effect. For example, cross-beams projecting below the

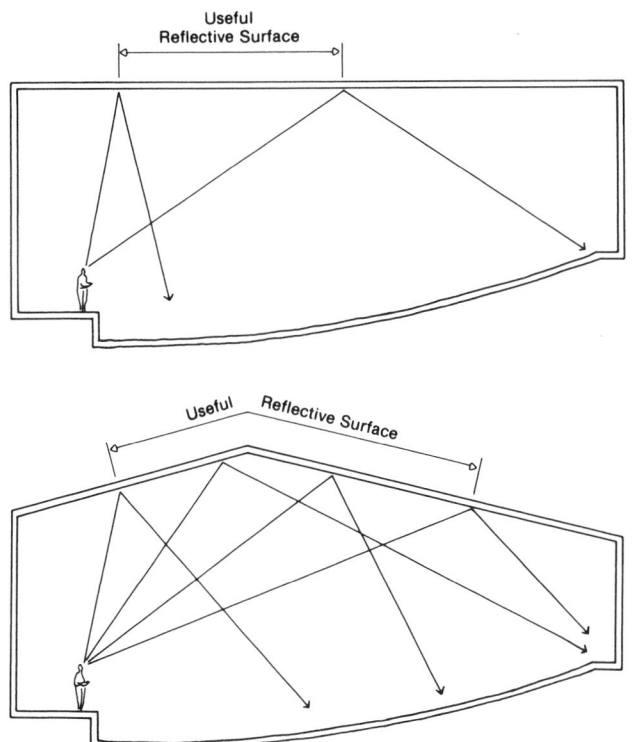

FIGURE 14.8. Usefulness of ceilings for acoustical reflection.

general ceiling level are very undesirable because they prevent a portion of the ceiling from reflecting sound to the audience.

Reflecting surfaces do not have to be part of the building structure. In very large halls, it is not unusual to suspend strategically placed decorative reflectors and sound traps from the ceiling.

Reverberation Time

Sound emitted from a source is normally reflected not once but several times within a space. Sounds that continue to be heard by reflection can persist long enough to mask subsequent sounds from the source.

Reverberation time can be altered by varying the absorption through the use of different surface finishes. A variety of absorbents is sometimes required to achieve a proper balance, since their performance varies greatly with frequency. Fiberboards and similar soft materials commonly used for absorbents are very efficient at high frequencies but not at low ones. A thin panel concealing a space lined with absorbent is effective for low frequencies.

SOUND ISOLATION

The acoustical concern for most spaces is to provide conditions that are favorable for promoting the audibility of sounds intended to be heard. In large spaces, this involves the proper use of reflective surfaces to sustain sound over long distances. If background noises are too high, the concern is the opposite: the suppression of noises that the occupants do not wish to hear. More specifically, the objective is to lower noise to a level that can be tolerated.

The importance of completely understanding how the building and its spaces will be used cannot be overemphasized. If noise levels are allowed to be too high, proper functioning within the space may be adversely affected, reducing the utility and value of the building. On the other hand, overly restricting noise levels can lead to needlessly large expenditures for acoustical designs and materials.

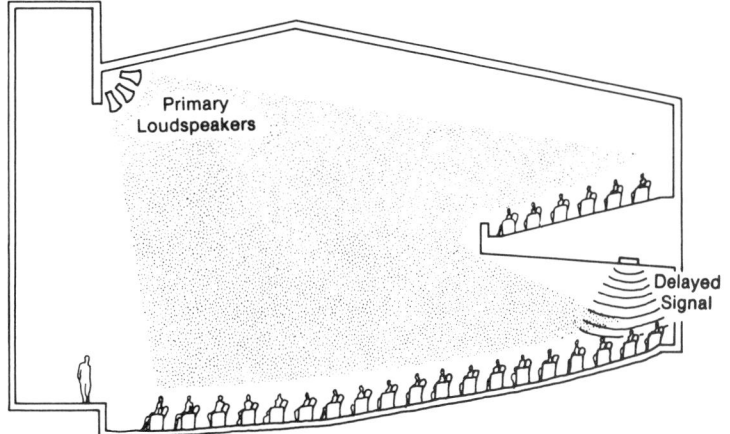

FIGURE 14.9. Spatial characteristics of electronic amplification.

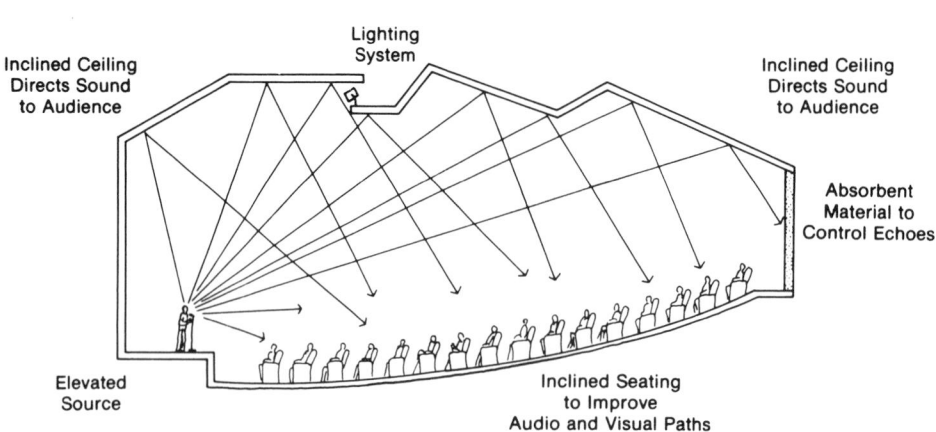

FIGURE 14.10. Auditorium shaping.

In an existing facility, measurements can be taken in order to define the character of a sound problem. Although it is more difficult, in a proposed design it is necessary to imagine what noise sources can be expected. All parts of the building as well as its surroundings must be considered as potential paths for sound travel, and a general acoustical consciousness should be fostered.

Noise Sources

Noise is generated at a source and travels through various paths to an auditor. Noise control requires an analysis of the noise generated with respect to the frequencies, loudness, directivity, and variations in sound pressure level with time. It is also important to know whether the sound is produced in the air (such as human voices or musical instruments), by impact (such as footfalls or other blows), by machinery vibration, or by flow of fluids (such as air in ducts or water in pipes).

The sound level in a space is not the sum of the actual sounds emitted. The combined effect of two sound sources, as discussed earlier, produces a maximum gain of 3 dB over that emitted by the louder source alone. And if the *difference* in sound level between two sources is greater than 10 dB, the addition of the lesser sound source does not add to the overall sound level in the space at all.

As an example, consider a room where the total measured sound level is 60 dB. A compressor operating within the room is contributing 53 dB to the overall sound level. If the compressor is turned off, the total sound level will be reduced only by 1 dB, to 59 dB. Therefore, noise problems cannot be solved by quieting only one piece of equipment unless it is the principal sound source.

Mechanical Equipment

All kinds of operating equipment can raise the noise level in spaces and can transmit impact sound over long distances through the structure. Motors, HVAC systems, and plumbing are commonplace examples in most buildings. Certain measures must be taken to ensure that their operation is not acoustically disturbing.

Operating machinery sets up vibrations that travel through the mounting or foundation to the building structure. The vibrating structure then delivers airborne noise to building occupants.

Traffic Noise

Vehicular traffic can generate a significant amount of intrusive noise in many locations. The sound level varies with traffic density, which in turn is a function of the use of the roadway. Arteries to urban centers have rush-hour peaks. Highways to recreation areas may have peak loads on weekends.

Roadside sound levels vary from about 54 dB for 1 vehicle per minute to 72 dB for 100 vehicles per minute. The sound level increases approximately logarithmically with the traffic density, so that a rate of only 10 vehicles per minute (1/10th of the difference between 1 and 100) produces 63 dB (halfway between 54 and 72 dB).

These sound levels are only approximate, since actual values are subject to fluctuation. Actual values also depend on the type of road surface and on the presence of trucks, stop-and-go movement, honking horns, or vehicles laboring up grades.

Around airports, the noise from air traffic can be a serious concern. A four-engine commercial jet at takeoff creates a sound pressure of 160 dB, well above the threshold of pain.

At a distance of 500 feet (150 m), the level is still about 110 dB. Buildings in the vicinity of airports must be especially well insulated against outside noise.

Community Noise

In addition to traffic noises, many other community sources of noise exist such as industrial plants, outdoor sports arenas, and playgrounds.

Two special characteristics of cities contribute to increased noise: hard surfaces reflect rather than absorb sound, and close parallel walls amplify noise between them rather than dissipating it. Although building surfaces are generally hard for durability and weatherability, softer and multiplaned materials (such as plants) would be more desirable from a public noise-reduction point of view. Hard paving should be kept to a minimum and grass and planting used as much as possible.

It may be desirable in some cases to interpose solid screens between noise sources and windows. For such screens to be effective, they must be close to either the source or to the receiver. In intermediate positions, sound may pass over them. While the upper floors of high buildings receive less sound from immediately adjacent ground-level areas, they receive more sound from distant sources which are blocked at lower heights.

It is important to recognize not only noise sources external to the site, but also sites and activities on the site that may generate an annoying noise level. These include children's play areas, refuse collection, and delivery or garage areas. Interior planning arrangements should attempt to group quiet rooms remote from external noise sources, and to cluster noisy and quiet rooms separately and in isolation from each other.

Degree of Noise Reduction

The desired degree of noise reduction is the difference between the sound level produced at the source and the level desired at the listeners' position. First, the level of tolerance of the listeners in any given circumstance must be established. From an economic point of view, it is usually not cost-effective to reduce the sound level any more than necessary. In fact, too much reduction can produce an undesirably low background noise level.

The determination of the optimum background noise level for any given occupancy may involve considerations of the damage to hearing organs from excessively high levels, the speech interference level, and the level that would cause annoyance. The first is usually associated with industrial problems.

Speech interference is influenced by the duration and distance of the communication. The briefer the conversation and the closer the talkers are to each other, the noisier the environment can be.

Annoyance depends on a number of factors. For critical listening tasks such as hearing music or theatrical performances, a high background noise level would be intolerable. Sudden noises in very quiet surroundings can be more disturbing than a somewhat higher continuous background noise level. Even very slight noises can disturb sleep, and even low-level conversations in a library can be intrusive. In these types of applications, a low limit on continuous background noise is just as desirable and important as a high limit.

Methods

The effective sound insulative quality of the various floor, wall, and ceiling constructions that enclose a particular space must be known in order to determine the degree of sound insulation provided. Once the expected noise level and desirable NC are established, the architect must decide on a procedure to lower predicted levels if they are higher than design values.

Noise reduction can be accomplished by (1) modifying the source or transmission paths, (2) modifying the relative positions of the source and listener, (3) equipping listeners with protective devices to wear, or (4) any combination of these techniques.

The modification of the source of machinery noise is generally the responsibility of the manufacturer. For example, developing a noiseless computer printer represents an improvement by the manufacturer.

The architect, through planning, can often control the source location. However, changing a plan arrangement in order to separate source and quiet areas may compromise other aspects of the design. The desirability of preserving a plan must be weighed against the increased cost of sound-insulating construction.

Buildings in which quiet environments are essential should be built at a distance from noise sources if possible. The additional cost of a quieter site may be less than the additional cost of more insulative construction at a noisier location.

Airborne and Structure-Borne Sound

The possible paths of transmission are through the air or through the solid parts of building construction and equipment. Noises originating within a room are generally *air-*

borne, while those intruding from elsewhere may be transmitted both through the air and through *solid structure.* All possible transmission paths must be considered, because when sound has a choice of paths to traverse, the greatest quantity flows through the path of least resistance.

For example, a partition between two offices may have adequate sound insulation, but for reasons of flexibility, it may terminate at a suspended ceiling of sound-absorbent panels. Not only can sound easily pass through gaps in the construction, it can also pass through the light, rigid ceiling, then over the partition and down through the ceiling of the adjacent office, as shown in Figure 14.11.

The sound-insulative function of the partition is negated unless the ceilings are also of adequate insulative construction. The ceiling can be made insulative using (1) a solid backing, (2) a resilient mounting, or (3) airtight construction with as much space as possible above and, in critical cases, additional absorbent. Otherwise, the partition may be extended up to the underside of the construction above in order to plug the sound leak.

Acoustically insulative construction reduces the amount of sound transmitted by a constant proportion, regardless of the magnitude of the noise level. Insulation values are expressed in decibels, and the reduction in noise level is found by subtracting the insulation value in decibels from the noise level on the other side of the construction.

Airborne Sound

If noise traveling through the air is to be significantly reduced in intensity before being heard at a given point,

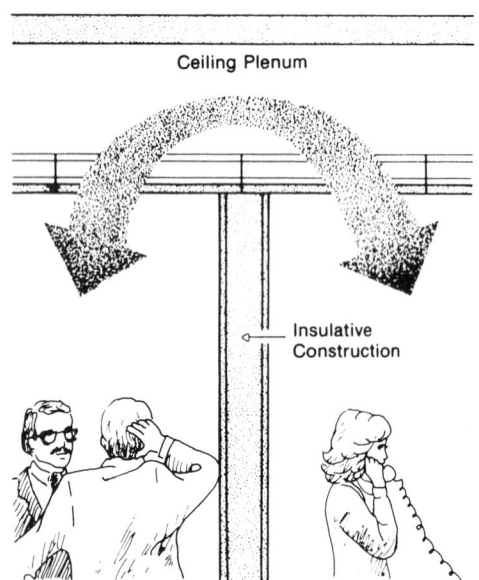

FIGURE 14.11. Sound transmission over partitions.

some barrier having the necessary sound insulation must be provided, or a sufficient distance must intervene between the noise source and the listener.

Site selection and development should locate the source of offending sounds as far as possible from listeners. According to the inverse square law, doubling the distance between source and listener causes a fourfold decrease in sound pressure level. Frequencies above 1,000 may be further attenuated by air absorption.

Outdoors, sound may be reflected by ground surfaces and other buildings as it travels. Reflected sound waves can intrude upon a building just as readily as direct waves.

An enclosed room has several kinds of construction surrounding it and acting as a barrier to outside noise entry. Identifying and then reinforcing the weakest links in the construction is of greatest importance.

When sound strikes a wall or barrier, it tries to make it vibrate. Wall vibration in turn sets up a sound wave on the other side of the wall. The heavier the wall, the greater its inertia and the less sound transmitted.

Porous materials absorb sound from a room and inhibit reflection of sound to farther parts of a room. But they also allow sound to pass through. Although some energy is absorbed during the passage, the sound level transmitted is approximately proportional to the inverse of the thickness. The thicker the material, the more the sound is attenuated.

Even though absorption has little direct effect on sound transmission through walls, floors, and ceilings, the intensity of sound within a space—whether it is transmitted to the space or generated from within—may still be reduced by using absorptive surfaces rather than reflective ones. This is because a reflective and reverberant acoustical environment enclosing the listener can maintain high sound levels, while an absorptive one does not. Even with absorptive treatment on the walls, however, the maximum possible reduction is limited to about 6 dB.

Absorptive treatment should not be used to lower the background noise level in spaces, such as auditoriums, that need to be live so that sounds generated within can be clearly heard. In those spaces, absorbents should be used only to control reverberation time and sound distribution. The surrounding construction must be able to exclude enough unwanted background noise so that the sounds that are intended to be heard are not masked.

In spaces such as restaurant dining rooms, theater lobbies, and clerical offices, objectionable noise is generated within the space. Since reflective surfaces sustain the noise, the only way to provide a quieter environment is by substituting absorptive surfaces for reflective ones. Since this treatment does not affect the direct component of the sound, in extreme cases, individual sources must be shielded locally.

Sound Leaks

Ordinary windows, doors, and other openings generally have a high transmittance rate, forming an acoustical bridge. They provide easy passage for sound through otherwise well-insulated walls and limit the level of sound insulation attainable. Any air passages, such as keyholes, cracks around doors and windows, or gaps between walls and floor constitute *sound leaks* which can nullify all other efforts to provide adequate sound-insulative construction.

Doors. Typical folding and sliding doors provide visual separation but very little acoustical privacy. Hollow-core doors do not provide much insulation either. A hollow hardboard door with normal crack widths gives about 15 dB reduction. This may not be critical if doors open to corridors rather than directly into other rooms.

If better insulation is required, heavier doors must close tightly against rubber gaskets on the stops, and be sealed at the threshold. In practice, TLs claimed by the manufacturer must be examined in terms of the edge seals that can be maintained in use. Generally, the TL of the material of which the door is made and that of the installed door are not the same.

Double doors (one behind the other) may be used for additional insulation if necessary. A short sound-absorptive vestibule between them can form a sound trap.

Windows. Whenever windows in external walls must be open, it is not possible to have much protection from outside airborne noise (Figure 14.12). An open window can provide only a 5- to 10-dB reduction, depending on whether it is fully or only slightly open, compared to the 50 dB or more reduction by a typical cavity wall. A closed single-glazed window with ⅛-inch (3-mm) glass provides only about a 20-dB reduction.

Better insulation is gained by a tight fit of sash and by heavier glass in double or triple panes. Sealing improves the effect up to about 25 dB, and the use of heavy glass can improve it slightly more. To gain any significant further increase, it is necessary to use double or triple glazing.

The sound insulation of glass is a function of its thickness and area. The highest degree of sound insulation is obtained by multiple sheets of heavy plate of different thickness mounted in resilient gaskets and not parallel to each other.

Specially constructed laminated glazing is also available for sound insulation through windows. A plastic interlayer dampens sound vibrations between glass faces on either side.

Other Openings. Inconspicuous holes in solid construction that otherwise provides high sound insulation must not be overlooked. In building construction, many perforations through the envelope—including pipe sleeves, electri-

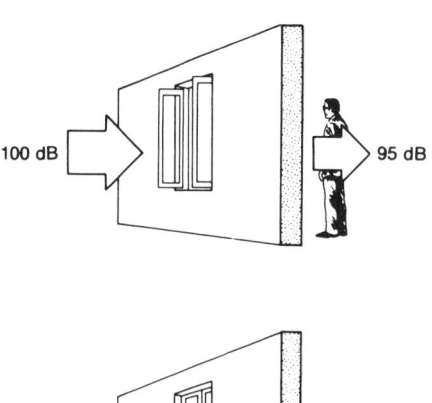
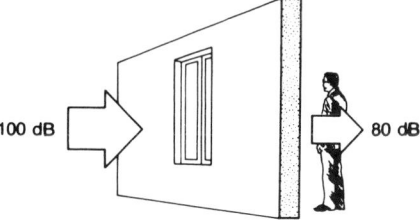
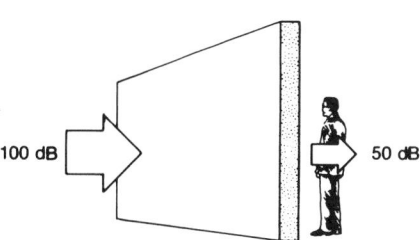

FIGURE 14.12. Sound transmission through a barrier.

cal raceways, back-to-back electrical outlet boxes in walls, recessed panelboards, and duct openings—are necessary. These must be tightly sealed against sound leaks.

Solid-Borne and Impact Noise

Vibrations and *impacts* can travel long distances through solid material, and it is possible for sound to be transmitted along structural members that link but do not divide rooms, as shown in Figure 14.13. The continuity of building structures, which is essential to their stability, makes controlling impact noise rather difficult.

Airborne sound may become solid-borne if it sets construction in motion. The construction in turn radiates sound energy as airborne sound to an auditor. Even when the structure is set in motion by direct contact with vibration or by impact from people, machinery, doors, and so on, the solid path usually terminates by radiating airborne sound to the auditor.

Discontinuity in construction, as shown in Figure 14.14, is an effective means of improving sound insulation against both airborne noise and solid-borne vibration. To achieve

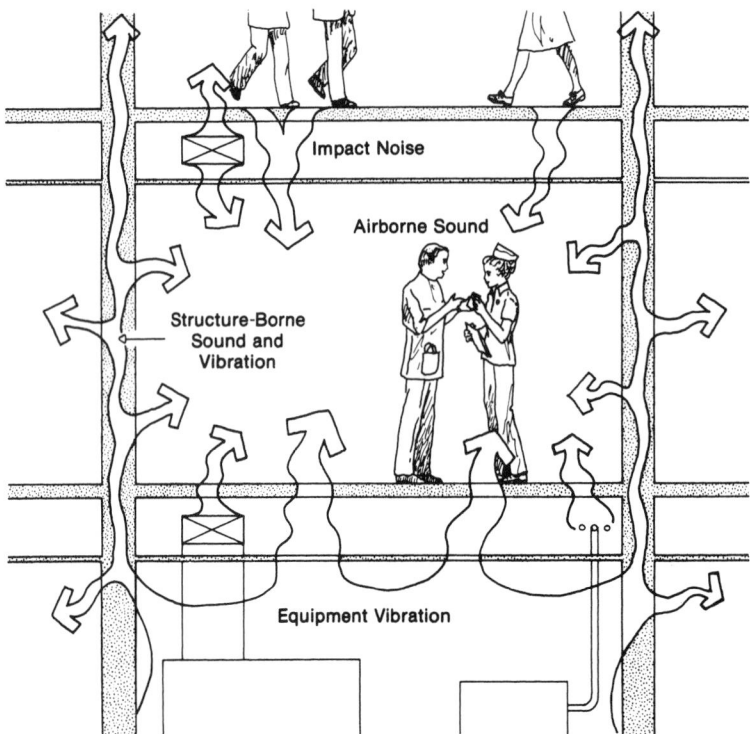

FIGURE 14.13. Solid-borne sound.

this, a wall is divided into two separate skins, and consequently vibrations are not easily transmitted from one to the other. If a wall consists of two walls with an air space between them, both walls must be set in motion in order to transmit sound across the construction. In general, two walls will decrease sound transmission more than a single wall equal to their combined thickness. It is less easy to make floors discontinuous.

It is very difficult to achieve complete discontinuity: ties across masonry cavity walls, top or bottom plates on staggered stud walls, or other structural continuity can transmit sound between the surfaces on either side of an air space. But the more discontinuity the better.

Resilient springs and cushioned mountings are sometimes used to increase the sound insulation of construction. These can be quite effective.

Resilient floor finishes provide varying degrees of impact insulation. In comparison to sound transmission by a bare concrete floor, thin composition tiles on concrete floors reduce transmitted sound 2 to 5 dB, cork tile 5/16 inch (0.8 cm) thick about 10 dB, and a thick carpet more than 20 dB.

The impact-sound insulation of wood floor finishes increases as their direct contact with the concrete base slabs

is lessened. A floor on battens will have an impact-sound insulation of about 6 or 7 dB. Resilient strips of 1-inch mineral wool or fiberglass under the battens approximately doubles the insulation value.

Floating floor construction is used for high-impact insulation. Such construction involves the use of a resilient pad or resilient chairs supporting screeds that break the solid contact between the floor finish and the structural support below. Glass or mineral wool blankets, cork, or fiberboards are placed over structural concrete slabs. A concrete slab is then placed over the blanket to support the floor finish. The insulation attainable ranges from 10 dB for ½-inch (1.27-cm) fiberboard to 24 dB for a 1-inch (2.54-cm) blanket. Resilient chairs and screeds are proprietary products for which insulation values should be available from the manufacturer.

It is of the utmost importance that there be no solid contact between the floating finish and any other part of the surrounding structure. Edges of blankets should turn up at walls. Baseboards on walls should not bridge the gap between floor slab and finish. Lack of attention to these details can result in an acoustical short-circuit which nullifies the value of the resilient construction.

It is also possible to provide separate structural systems

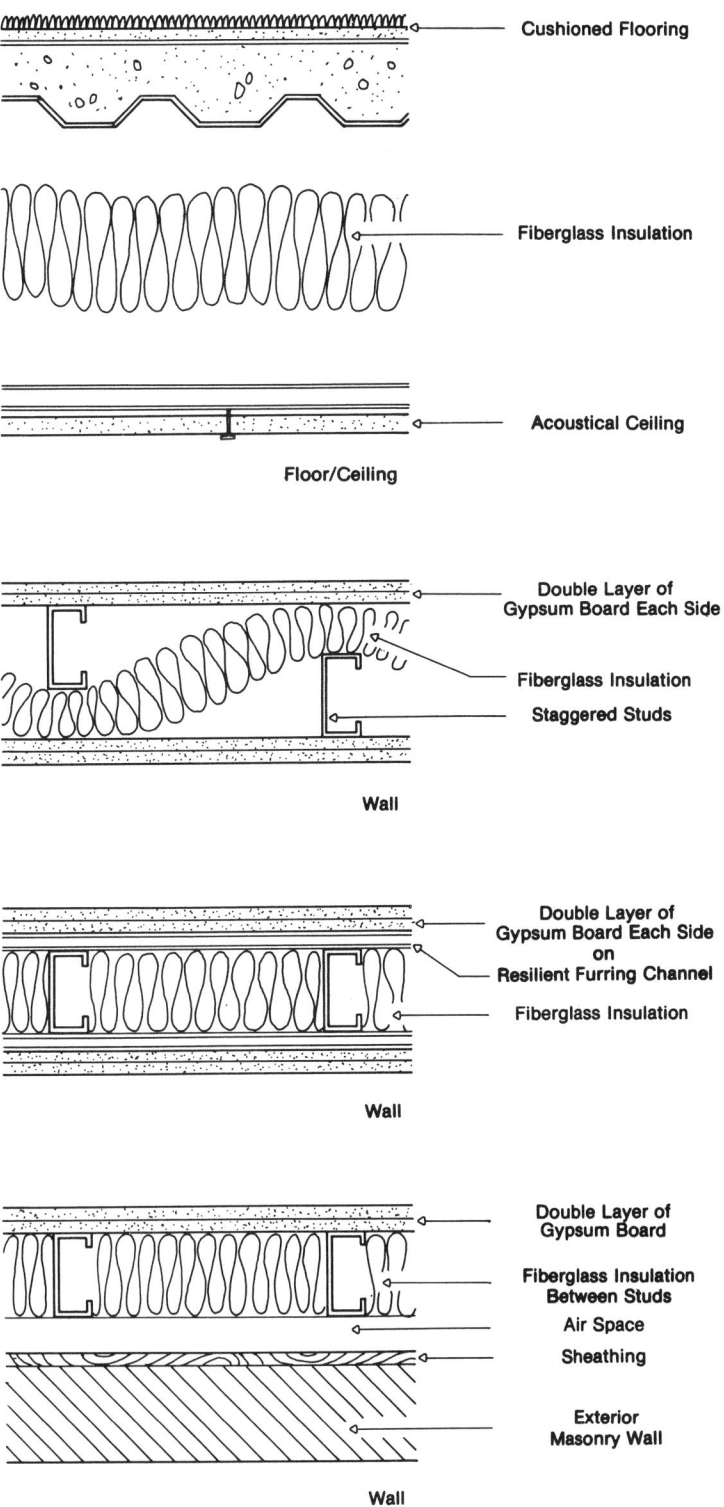

FIGURE 14.14. Insulative constructions.

for the support of floors and ceilings below. These eliminate the structural connections that readily transmit sound. Again, edge conditions must be considered carefully in order to avoid unintentional transmission. Resilient hangers are available for hung-ceiling construction.

Open Office Plans

The objective in the design of open office plans, such as that shown in Figure 14.15, is to keep sound from passing from one work station to another. Special acoustical partitions are used both to block the sound from continuing in its original direction and to absorb the sound in order to prevent it from ricocheting off the partitions and annoying other workers.

In order to be effective, acoustical partitions need a solid core, or *septum,* which acts as a barrier between sound sources and listeners. They also need a fibrous surface to admit sound and trap it. The best sound-absorbing material for covering a panel is an open-weave fabric.

A solid-core septum can provide a panel with an STC of at least 25, depending on the composition of the core. The barriers selected should have an NRC in excess of 0.7 on both sides.

Another critical element in an open office sound-control design is the ceiling. Ceilings, too, must be designed so as to prevent sound from reflecting back down into nearby work stations and other workers in the office. In arrangements that include traditional closed private offices, ceilings must also prevent sound from being carried through the plenum into or

FIGURE 14.15. Acoustical ceilings, carpeting, and acoustically controlled work stations such as this reduce sound problems in open office plans. (Photo courtesy of Steelcase Inc.)

out of closed offices. Since most sounds hit the ceiling at a 45° to 50° angle, ceiling materials must be able to capture noise from almost any angle.

Reflected sound may also be reduced with absorptive wall coverings. Standard paint and wallpaper are of negligible sound-absorptive value. There are, however, sprays and other finishes that are fairly effective.

Most acoustics experts believe that carpeted floors are relatively ineffective in absorbing unwanted conversation noise. Nevertheless, carpeting is a popular floor covering for a variety of other reasons, and it does provide for a quieter work environment simply by cushioning footsteps and the noises of sliding chairs.

Mechanical Equipment Noise Control

In addition to external noises and internal noises due to the occupants, considerable noise can be generated by the services in buildings. Fans, boilers, chillers, pumps, plumbing, and other equipment may produce noise.

Noise and vibration are generated by the compressor, and the great quantity of air that must be rapidly circulated through outdoor condensers can also be quite noisy.

In places of assembly, air-conditioning systems must not be heard. Concert halls and opera houses have the most critical noise requirements, followed by theaters for drama and dance, auditoriums, houses of worship, and sports arenas, in that order. Although these are the most critical, equipment noises must be controlled to some extent in all installations.

HVAC noise control is accomplished by breaking the path of travel of the vibration. The mechanical equipment should be isolated from the structure of the building so that vibrations are not transmitted (see Figure 14.16). An excellent solution for especially noisy apparatus such as refrigeration compressors is the use of a separate equipment foundation that rests on the ground and is independent of the building structure.

Often, equipment must be located on upper floors or there are other reasons why mounting it on the main structure is unavoidable. In those cases, the equipment may be supported on *vibration isolators* between the equipment and foundation. Isolator elements, which reduce the vibration transmitted from the machine to an acceptably low level, may be springs, rubber, and other resilient materials.

Besides being isolated from the building structure, machinery must not be in direct contact with other equipment or building elements. Electrical, piping, ductwork, and other service connections must be *flexible.*

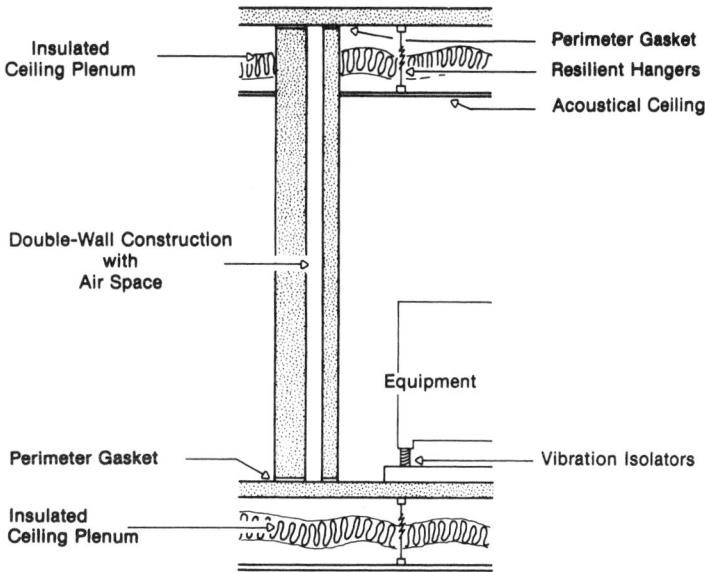

FIGURE 14.16. Isolation of sound and vibration from mechanical and electrical equipment.

Plumbing and heating piping can be a source of noise from flowing water or steam. Noise radiating from large piping can be reduced by wrapping pipes with a fibrous insulation and covering them with an impervious jacket.

In addition to isolation for control of solid-borne noise, heavy construction may be necessary for enclosing machinery rooms in order to insulate against the usually high airborne noise levels. Heavy slabs and masonry walls that may not be justified structurally can be essential to reduce sound transmission.

The mounting of machinery is normally the mechanical consultant's responsibility, but the enclosure around mechanical equipment may be the joint responsibility of the architect and engineer.

Duct Noise

Ducts are avenues through which sound can travel readily from central equipment to rooms and between rooms. Noise emanates from ducted air systems in two main ways.

First, vibration from the plant can be transmitted along the ductwork. This is avoided by the use of *flexible* canvas links joining the fan to all main distribution ducts.

Second, noise is generated in the ductwork or outlets by the passage of air. This noise can either emerge from the outlets or penetrate the wall of the duct along its length. To keep sound levels within acceptable limits, air velocity in ducts is limited. Air noise that might be transmitted through the duct system is reduced with sound-absorbent inner linings, plenums lined with absorbent material, or premanufactured sound traps designed for insertion in the ducts.

The fast-moving air causes high-velocity air distribution systems to be noisy. Abrupt changes in the size, shape, or direction of a duct exacerbate the problem and are to be avoided. In high-velocity systems, the ducts are enclosed behind sound-insulating construction and terminate in absorbent-lined, sound-attenuating boxes that also reduce the velocity at which the air is admitted into the room.

Control of air noise from central fans is accomplished by housing air handlers within rooms having adequate sound-insulating construction and by mounting them on vibration isolators. Large fans operating at slow speeds are quieter than small, high-speed fans.

Both ducts and pipes should be located far from quiet rooms in order to avoid the possibility that noise from them will cause annoyance. Firmly attaching such items to heavy walls minimizes transmission of noise to rooms.

The air discharge velocity through registers and diffusers determines the airflow noise level. Air outlets can be selected with any of a variety of discharge noise levels, and selection is based on maximum and minimum (for sound masking) desired noise levels, cost, necessary airflow, and space considerations.

Carefully designed sound-insulative construction between rooms may be rendered ineffective by direct air paths through air inlets or outlets connected to common ductwork.

Such acoustical links can easily transmit intelligible conversations between spaces. This phenomenon, known as *crosstalk,* can be avoided by proper duct layout and by sound-absorbent duct linings or other sound attenuators in the ductwork.

Electrical Equipment Noise

Some electrical equipment produces a hum that can be annoying under some circumstances. Large transformers can usually be located far away from quiet areas or enclosed in sound-insulative construction. In the latter case, special provisions for adequate ventilation must be arranged.

Fluorescent lamp ballasts can be objectionably noisy in quiet areas. In such spaces, ballasts with low noise levels should be selected, and they can even be mounted in a remote location if necessary.

SUMMARY

The subject of architectural acoustics actually encompasses two types of sound control: (1) the acoustical environment within a room and (2) isolation of unwanted sounds. The former has to do primarily with positioning sound sources with respect to the listeners and arranging appropriate absorptivity or reflectivity levels for all interior surfaces. The latter involves insulating building occupants from intrusive noise. In some spaces, a continuous low background noise level is desirable for masking distracting sounds.

Sound waves are air pressure pulses traveling outward from a source. When pulses at the appropriate frequency arrive at a human ear, they set up vibrations in the ear's receiving mechanism, which are then perceived as a sound.

Sounds are measured in the logarithmic decibel scale. Since some frequencies are more noticeable than others at the same energy level, other scales, such as NC, SIL, and dBA, have been developed to quantify noise levels for specific purposes.

When a wave front hits a barrier, it tries to make the barrier vibrate. Portions of the sound energy may be absorbed, transmitted, or reflected. The proportion of each depends on the material properties. Hard surfaces mostly reflect sound, while soft porous surfaces mostly absorb it. The absorption coefficient, α, represents the relative absorptivity of a material at a given sound frequency. The NRC is an average of α values over a range of frequencies.

The amount of sound transmitted is a function of the weight and discontinuity of the construction. The TL depends on the specific construction and represents the decibel reduction from one side of an element to the other.

The sound pressure level at any point in a space is a function of the absorptivity of its surfaces, which determines the reverberant quality of the space, and the effective TL, or insulative quality, of the enclosing construction, which impedes the flow of sound into the space.

Reverberation is the prolongation of sound energy by successive reflection. A certain amount of reverberation is necessary to enhance the sound in some spaces, but excessive reverberation is damaging to clarity. If the interreflections from a number of reflection paths continue for too long, they produce an echo effect, causing intelligibility to suffer.

Noise and vibration from outside traffic, other community sources, mechanical equipment and services, and louder functions within a building must often be isolated from spaces. This can be accomplished by separating the source and receiver and by employing insulative construction.

Heavy walls decrease airborne transmission. Absorptive surface treatments can produce limited sound reduction in the source room as well as reduction of sound reinforcement in the receiving rooms. In equipment rooms and other unconditioned spaces, absorptive materials must resist high humidities and be able to be cleaned easily. Vibrating equipment is commonly mounted on resilient vibration isolators or on concrete slabs that are discontinuous with other floors in order to reduce the transfer of solid-borne sound and vibration. Links to sound-generating mechanical equipment are made with flexible connections, such as lengths of flexible pipe or canvas duct connectors. Duct linings are used to limit air noise transmission.

KEY TERMS

Acoustics	Transmission	Background
Noise	loss (TL)	noise level
Sound	Decibel (dB)	Sound
Frequency of	Noise criteria	audibility
vibration	(NC)	Sound masking
Pitch	Speech	Speech
Tone	interference	intelligibility
Wavelength	level (SIL)	Geometric
Absorption	A-weighted	acoustics
coefficient	sound level	Physical
Noise reduction	(dBA)	acoustics
coefficient	Phon	Dead space
(NRC)	Sone	Live space

Absorptive materials	Sound isolation	Structural discontinuity
Acoustical materials	Airborne sound	Floating floor
Image theory	Structure-borne or solid-borne sound	Septum
Reverberation		Vibration isolation mountings
Mean free path	Acoustically insulative construction	
Reverberation time		Flexible connectors
Room resonance	Sound leak	Cross-talk
	Impact noise	

STUDY QUESTIONS

1. A member of an audience is 25 feet (8 m) from a speaker on stage. The sound pressure level at that point due to the direct sound path is 70 dB.
   a. What is the sound pressure level due to the direct sound path 50 feet (15 m) from the speaker?
   b. If the background noise level from conversation and the rustling of papers is 60 dB, would it interfere with the intelligibility of the speaker, considering the direct sound path only?
   c. What means could be used to reinforce the sound level so that the speaker would be more easily heard?
2. Indicate which of the following spaces should have a predominantly dead quality and which a live quality: auditorium, church, classroom, hospital patient room, hospital operating room, lecture hall, library, movie theater, office, restaurant, and theater lobby.

BIBLIOGRAPHY

American Society of Heating, Refrigerating, and Air-Conditioning Engineers, Inc. *Practical Guide to Noise and Vibration Control for HVAC Systems.*

Archibald, Claudia J. *Noise Control Directory.* Fairmont Press, Inc., 1979.

Beranek, Leo L. (editor). *Noise Reduction.* McGraw-Hill Book Company, 1960.

Beranek, Leo L. *Music, Acoustics and Architecture.* Krieger Publishing Company, 1979.

Beranek, Leo L., and István L. Vér (editors). *Noise and Vibration Control Engineering: Principles and Applications.* Wiley-Interscience, 1992.

Doelle, Leslie L. *Environmental Acoustics.* McGraw-Hill Book Company, 1972.

Egan, M. David. *Concepts in Architectural Acoustics.* McGraw-Hill Book Company, 1972.

Harris, C.M. *Handbook of Noise Control,* 2nd ed. McGraw-Hill Book Company, 1979.

Miller, Richard K., and Wayne V. Monotone. *Handbook of Acoustical Enclosures and Barriers.* Fairmont Press, Inc., 1978.

Peterson, A.P.G., and E.E. Gross, Jr. *Handbook of Noise Measurement.* GenRad, Inc., 1974.

Stein, Benjamin, and John S. Reynolds. *Mechanical and Electrical Equipment for Buildings,* 9th ed. John Wiley & Sons, Inc., 1999.

Yerges, Lyle F. *Sound, Noise, and Vibration Control,* 2nd ed. Krieger Publishing Company, 1978.

Chapter 15

Building Commissioning

THE COMMISSIONING PLAN

APPLICABLE SYSTEMS AND EQUIPMENT

Commissioning is a process for verifying and documenting that the performance of a building and its various systems meets the design intent and the owner's operational needs. Commissioning tests the operation of the building's mechanical, electrical, and controls systems and equipment to ensure that they operate as designed and at optimum performance, and can satisfactorily meet the needs of the whole building as intended. The systems must function interactively throughout the entire range of operating conditions to achieve their intended purpose.

Normally, a *commissioning agent* independent of the owner, design consultants, and contractors is employed to avoid conflicts of interest. The commissioning agent works on behalf of the owner, is an advocate for the owner's interests, and makes recommendations to the owner about the performance of the commissioned systems. The commissioning agent should also provide constructive input for the resolution of system deficiencies.

The commissioning agent coordinates the various testing, documentation, and other tasks. Some of these tasks may actually be performed by the building operations staff, a construction manager, the architect, various members of the engineering team, the general contractor, subcontractors, or equipment manufacturers' representatives. The overall process is a team effort led by the independent commissioning agent.

The commissioning process is intended to facilitate the orderly and efficient transfer of the systems to the owner. It is an inspection function that is not merely redundant to the contractor's obligation for startup, calibration, tuning, system adjustment, testing and balancing, proof of performance, or contract closeout. It is an enhancement to the traditional project delivery process.

To avoid wasting time, the contractor should be required to start up, test, adjust, and certify that all systems and equipment are fully operational before the commissioning process begins. It is also critical that operator familiarization and training, which are typically required by construction contracts, be undertaken prior to commissioning.

Another purpose of the building commissioning process is to verify that operation and maintenance manuals, as-built (record) documents, spare parts listings, and special tools listings are not only furnished by the contractor but are prepared in such a way as to be useful.

The costs of not commissioning are:

- Higher energy use
- Occupant discomfort and complaints until construction problems are resolved
- The high cost of investigation, diagnosis, and correction of problems after the building is occupied and under operation
- A lower market value for a poorly performing building

The scope of commissioning must be clearly defined in the commissioning contract. The roles and scope of all building team members in the commissioning process should be clearly defined in the design and engineering consultants' contracts; in the construction contract; in the General Conditions of the Specifications; in the divisions of the specifications covering work to be commissioned; and in the specifications for each system or component for which a supplier's support is required.

The scope of commissioning may include

- Ensuring appropriate product selection during design
- Reviewing specifications to ensure that product specifications are clear
- Overseeing the review of product submissions to ensure selections are acceptable
- Ensuring products are properly installed
- Reviewing operations and maintenance (O&M) manuals to ensure that adequate documentation is provided to enable facility staff to properly maintain systems and equipment
- Verifying that facility staff receives adequate training to operate and maintain systems and equipment
- Reviewing the O&M plan for completeness
- Ensuring that the building's indoor air quality meets the design objectives

Commissioning of existing systems may require the development of new functional criteria in order to address the owner's current systems performance requirements. For new construction, the commissioning agent should review systems installations for commissioning issues throughout construction.

THE COMMISSIONING PLAN

A *commissioning plan* must be produced to describe how the commissioning process will be carried out. It should identify the systems to be commissioned and the scope of the commissioning process. It is desirable to develop the commissioning plan during the design process, especially when very complex systems are involved. Integrating it throughout the design results in a higher-quality product.

A commissioning plan includes the following:

- A clear definition of the project goals
- A statement of the commissioning goals
- A summary description of the systems designed to meet the project goals
- A description of the roles, responsibilities and lines of communication of all parties involved in the commissioning process
- A list of all systems and equipment to be commissioned
- The scope of operational modes to be tested
- A schedule for the commissioning tasks
- A detailed description of each functional test to be performed and what verification techniques are to be used

It is important to demonstrate that all systems and installed equipment are fully operational before the commissioning tests begin. During the functional performance tests of the commissioning process, the systems and equipment are verified to perform as intended in all modes of operation. Tests should also include any manual override controls. Once the tests are successfully completed, all system setpoints that were changed for testing purposes need to be returned to their proper values. The results of all tests need to be completely documented.

The commissioning plan should include a test of battery backup systems, where applicable, and recovery of all systems from an interruption of power. Upon restoration of power, all systems and components should automatically restart without any loss of programming, setpoints, schedules, or other required functions.

A building commissioning plan must be tailored to each specific project. Since commissioning is a dynamic process, the plan may evolve as the project proceeds. Changes to the prepared commissioning procedures may be necessary due to unexpected field conditions. Any such changes should be approved by the party responsible for the particular system design.

The acceptance criteria or test results that are required before the mode of performance or inspection item in question will be considered acceptable must be specified. If an item fails its test, the problem must be corrected and a retest conducted. In the event of a second failure, the responsible party may be assessed a charge to cover each person's actual expenses for materials, equipment rental, and travel, as well as normal hourly billing rates for preparation, testing, and travel time.

At the conclusion of the process, the commissioning agent prepares and submits a project *commissioning report.* It includes summaries of the overall process, all deficiencies found, deficiency corrections, unresolved deficiencies, approved equipment and systems, and any implemented changes to the design intent. The report also includes the commissioning checklists, test documentation, and other commissioning documentation.

APPLICABLE SYSTEMS AND EQUIPMENT

The systems and equipment that may be included in a commissioning plan are:

- Air-handling and conditioning systems
- Heating-water systems
- Chilled-water systems
- HVAC control systems
- Smoke control systems
- Electrical systems
- Daylighting controls
- Occupancy sensor or other automatic lighting controls
- Telecommunications systems
- Security systems
- Backup power systems
- Plumbing fixtures
- Rainwater harvesting systems
- Graywater systems
- Electronic water controls
- Door hardware
- Other automated systems

SUMMARY

Building commissioning is the process of ensuring that building systems are designed, installed, functionally tested, and capable of being operated and maintained according to the owner's operational needs. Ideally, the commissioning agent should become involved in the conceptual stage of the project in order to provide input during the design phase.

Prior to a completed building's delivery to its owner, the commissioning agent performs a thorough review of all systems, including but not limited to HVAC, plumbing, fire protection, power, lighting, telecommunications, and elevators. The functional testing program is composed of written, repeatable test procedures.

The commissioning process concludes with the preparation of a report indicating both expected and actual results. The commissioning report should include an evaluation of the operating condition of each system, deficiencies that were discovered and measures that were taken to correct them, uncorrected operational deficiencies accepted by the owner, the test procedures and results, documentation of all commissioning activities, and a description and estimated schedule for any deferred testing.

Building commissioning is an important part of the project delivery process and reaps enormous benefits in the form of reduced operations and maintenance costs. It has the potential for generating tremendous savings for the building owner. Building commissioning can be even more effective if it reoccurs periodically throughout a building's life cycle, because complex systems tend to drift out of specification and even fail.

KEY TERMS

Commissioning
Commissioning agent
Commissioning plan
Commissioning report

STUDY QUESTIONS

1. What is the purpose of the commissioning process?
2. Can the same commissioning plan be used for two different projects?
3. Which should occur first, commissioning or operator training?

BIBLIOGRAPHY

American Society of Heating, Refrigerating, and Air-Conditioning Engineers, Inc. *Guideline 0—The Commissioning Process.*

American Society of Heating, Refrigerating, and Air-Conditioning Engineers, Inc. *Guideline 4—Preparation of Operating and Maintenance Documentation for Building Systems.*

American Society of Heating, Refrigerating, and Air-Conditioning Engineers, Inc. *Guideline 5—Commissioning Smoke Management Systems.*

Odom, J. David, and George Dubose (editors). *Commissioning Buildings in Hot Humid Climates: Design and Construction Guidelines.* Fairmont Press, 2004.

Chapter 16

Design Economics

PAYBACK PERIOD
SIMPLE PAYBACK PROCEDURE
LIMITATIONS
ADVANTAGES

LIFE-CYCLE COST
TIME VALUE OF MONEY
PRESENT WORTH METHOD
ANNUAL CASH-FLOW METHOD

RATE OF RETURN

COMPARATIVE VALUE ANALYSIS

OTHER CONSIDERATIONS
INFLATION
FUEL COST ESCALATION
POINT OF DIMINISHING RETURNS
CREATIVE FINANCING

It is important to understand how economics affects building decisions. For every design, alternative solutions are possible; it is the responsibility of the designer to select the one that is most economical as well as consistent with the degree of safety and aesthetic value desired. Each design must conform to the capital available to the client. If the project is one that is to yield monetary returns, such as rental property or industrial facilities, the effect on the anticipated volume of these returns must also be considered.

In many cases, the most appropriate decision is clear-cut, but uncertainty can arise if there are conflicts between the various criteria. One design may have aesthetic advantages, another may be more economical to build, and yet another may be more economical to operate. Figure 16.1 shows how initial and energy costs can differ between two alternative design options. An incorrect interpretation of the economic value of one design relative to another can result in expensive mistakes.

The final choice among the competing values requires an energy-conscious evaluation of the important factors. The first cost is often considered the dominant factor. But rising salaries and energy costs make operating and maintenance increasingly significant. All the facts must be laid out on the table.

A building is ultimately for the benefit of the owner. It is up to the designer to make design decisions that are in the client's best interests. Since the client is the one who will test the designs with large amounts of money, he or she should be in a position to make the decisions that have a major impact on cost. The designer must therefore be able to present the benefits of a design in such a way that the client can clearly see them.

It is up to the client to establish priorities and guidelines for trade-offs. In some cases, the client may want to reduce energy operating costs, while in others, only first costs may be important. In any case, the architect must be sensitive to

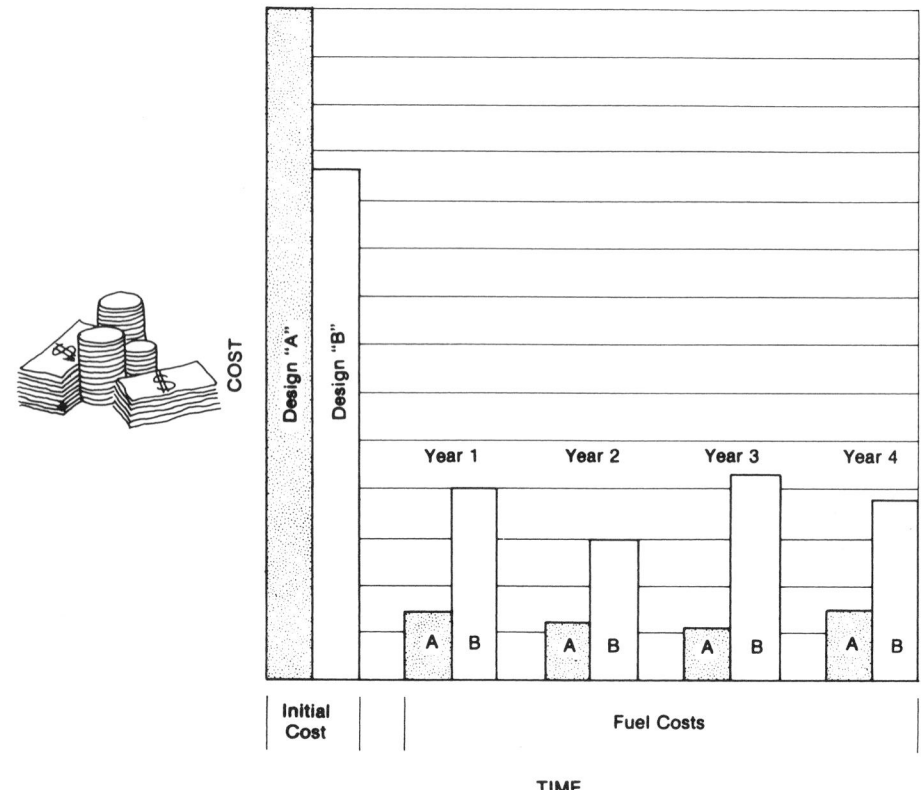

FIGURE 16.1. Initial cost versus energy costs for alternative design options.

the investment needs of the client. This involves understanding the energy trade-offs of alternative strategies and being able to evaluate their costs and benefits to determine the best value for the client. The architect or a consultant must be able to explain the advantages—tangible and otherwise—and the costs for those advantages in both capital and operating expenses.

Energy analyses are necessary when the situation arises in which one design is less expensive to install but another is more energy-efficient because it has more insulation, uses more efficient equipment, or utilizes solar energy. Examples of decisions that may need an energy evaluation are optimizing the quantity of insulation, double versus triple glazing, type of lighting, weighing aesthetic considerations against economic cost to the client, or whether or not to include solar energy utilization.

While there are many advantages in saving energy for its own sake, most clients are under the constraint of limited available financial resources, so dollar savings is their primary goal. Achieving energy savings is simply considered a way of attaining that goal. Usually, an energy-saving invest-

ment that cannot pay for itself with dollar savings will not be made.

The four alternative methods of design value analysis presented here are (1) payback period, (2) life-cycle cost, (3) return on investment, and (4) comparative value analysis. The first three are strictly economic, while the fourth can account for aesthetic preferences as well.

PAYBACK PERIOD

The simplest method of value analysis is payback. The *payback period* is defined as the length of time required for an investment to pay for itself, or, in other words, how long it takes to recoup the initial investment. When comparing non-revenue-producing projects, the payback period is the time required to regain from savings the additional cost of a retrofit or the difference in cost between competing projects.

Simple Payback Procedure

When the calculation makes no allowance for the time value of money, it is called *simple payback* or *payback ratio*. It is the simple ratio of initial cost to annual savings, as follows:

$$\text{Simple payback} = \frac{\text{capital cost}}{\text{annual savings}} \quad (16.1)$$

The reciprocal of the simple payback is called the *capital recovery factor* (CRF):

$$\text{CRF} = \frac{\text{annual savings}}{\text{capital cost}} \quad (16.2)$$

When expressed as a percentage, the CRF is sometimes used as an initial rate of "return on investment. This should not be confused with ordinary interest rates.

As an example, consider building alterations for solar energy or energy conservation. If the initial outlay is $10,000 and each year the net fuel cost savings averages $2,500, the payback period is $10,000/2,500 = 4 years. The costs and benefits on an annual basis are shown in Table 16.1.

Obviously, an acceptable payback period must be within the projected useful life of an improvement. The savings can include other owning and operating costs as well as energy costs. Alternatives must be converted to the same time basis before they can be compared: for example, if the replacement period of components for system A is 2 years, and that of system B is 3 years, total costs and savings should be compared on a 6-year cycle.

Limitations

The shortcomings of simple payback are that it does not take into account the time value of money or escalating energy costs. There is no provision for interest on the capital that is tied up while the payback occurs. Nor are credits considered for savings that occur after payback and before the end of the system lifetime. Tax deductions for interest payments are also a form of annual return. And there is no comparison of the cash flow prior to payback. These factors may be negligible when interest rates and fuel rates are low, but that is not always the case.

Tables 16.2, 16.3, and 16.4 demonstrate some of the drawbacks to the payback period method. Table 16.2 compares the annual savings of two alternative improvements, both of which require an initial expense of $10,000.

According to the simple payback method, investment A is preferable because it has a faster payback. However, as Table 16.3 shows, benefits beyond the payback period can sometimes have a greater economic impact.

TABLE 16.1 SIMPLE PAYBACK

Year	Initial Cost	Savings
1	$10,000	$2,500
2		2,500
3		2,500
4		2,500
	$10,000	$10,000

TABLE 16.2 PAYBACK PERIOD COMPARISON

Year	Annual Savings Investment A	Investment B
1	$ 2,500	$ 2,000
2	2,500	2,000
3	2,500	2,000
4	2,500	2,000
5	—	2,000
Payback	4 years	5 years

TABLE 16.3 COMPARISON BEYOND PAYBACK

Year	Annual Savings Investment A	Investment B
1	$ 2,500	$ 2,000
2	2,500	2,000
3	2,500	2,000
4	2,500[a]	2,000
5	2,500	2,000[a]
6	2,500	6,000
Payback	4 years	5 years

[a] Payback year.

TABLE 16.4 COMPARISON OF TIMING OF BENEFITS

Year	Annual Savings Investment A	Investment B
1	$ 1,000	$ 8,000
2	1,000	1,000
3	1,000	500
4	7,000	250
5	—	250
Payback	4 years	5 years

Although investment A has the faster payback, investment B provides greater financial benefit over 6 years. The timing of the financial benefits can also be more important than payback period, as indicated in Table 16.4.

The payback periods and the *average* savings in these three tables are the same—$2,500 for investment A and $2,000 for investment B—but the distribution of savings within the payback period is uneven. In this case, the larger initial savings for investment B may be more desirable.

Advantages

Despite these drawbacks, payback has some distinct advantages: (1) it is easy to understand, calculate, and communicate to others, (2) it can be used to screen out proposals, and (3) it can be used to assess the liquidity of a measure.

The first advantage is self-evident. The second involves rejecting a project if its payback period is longer than a specified time period. Even though there is no absolute basis for the acceptance or rejection of a given payback time, a screening policy can be established based on prior experience with similar circumstances. If a project passes this preliminary screening, additional criteria may be imposed before a decision is reached.

Liquidity refers to how readily invested funds are available for reuse. Some companies consider a short-payback policy the most flexible position during uncertain economic times or when their capital is limited.

LIFE-CYCLE COST

Life-cycle cost is the total cost of owning, operating, and maintaining a planned project over its useful life. That is, it includes not only the acquisition and energy costs, but also the repair/maintenance cost and the salvage value of an improvement. The life-cycle cost analysis method can be used to determine whether proceeding with an energy-conservation project is cost-effective, or to compare the economic consequences of alternative design solutions.

Life-cycle costs generally consist of the following major categories. These expenses and financial benefits should be carefully itemized for each alternative:

- *Acquisition cost* is the one-time purchasing cost or first cost and includes taxes, fees, delivery, installation, and initial start-up.
- *Annual owning and operating costs* include those for maintenance, energy, taxes, insurance, interest on borrowed money, and any other recurring costs over the useful life of the project. Some of these costs may vary with age and inflation.
- *Major repairs and component replacements* are intermittent costs.
- *Terminal value* is a one-time disposal cost, or deduction from costs if a salvage value, computed at the end of the useful life.

With sufficient research, each of the above costs can be estimated. As with anything else, experience brings more accurate judgment and may lead to shortcuts in computations. The reader is encouraged to wade through the laborious process thoroughly the first few times in order to learn it fully.

Projected future costs obviously depend on the assumptions that are made. The useful life of the project and the salvage value at the end of the useful life must be based on past averages.

Besides deciding which cost items should be included in a life-cycle cost analysis, the big question is what length of time to work with. The further into the future a building's life is projected, the less certain all costs become. For this reason, a client may request a life-cycle analysis to be calculated over an arbitrary period of time, such as the client's standard investment period. Usually, however, the anticipated useful life of the project is used.

The expected *useful life,* or life expectancy, is arbitrary for such things as major machinery components, piping systems, and building structures. The factors that determine useful life are the economics of maintenance and obsolescence.

A properly maintained building, like a well-maintained piece of machinery, will last indefinitely if a reliable maintenance program is employed to ensure the replacement of worn components prior to their total failure. If a nonmaintainable or nonserviceable compressor wears out, for example, the compressor can be replaced as a modular component of a larger subsystem, such as a chiller. Thus, if an adequate budget is allocated for maintenance and service, the machine will never wear out. Unlike animal and plant life, machinery and buildings are not organic; thus, the owner has total control over their life expectancy.

The only factor that truly limits the life of the building and the building systems is obsolescence. An item becomes obsolete when a more desirable item is available to perform the same function.

Many federal and state agencies require a life-cycle cost analysis for public projects and procurement. Designers must understand and be able to perform this type of energy/economic analysis in order to participate in such programs.

Time Value of Money

An amount of $100 in cash today is worth more than $100 next year. If in hand today, it can be deposited in a bank at 5 percent simple interest, and at the end of 12 months it will have grown to a value of $105. The process of growth from $100 today to $105 next year is called *compounding.*

The reverse process of finding out how much money must be deposited in the bank today at 5 percent in order to have $105 by the end of the year is called *discounting.* The expected future return is discounted, or reduced in value, to account for its lesser worth.

This concept of the *time value of money* is illustrated in Figure 16.2. An initial cost of $10,000 for some energy-conservation measure may save an estimated $2,500 per year in energy costs. If the owner could invest $2,500 today at 10 percent interest, it would be worth $2,750 in 1 year, $3,025 in 2 years, $3,327.50 in 3 years, and so on. The value of the money would grow at a rate of 10 percent per year. Thus, an economic benefit of $2,750 received

next year would have the same value as $2,500 in hand today.

Furthermore, suppose that the cost of energy is growing at 10 percent per year. Each year's energy savings, if discounted back to the present time, would be worth $2,500, even though it would be worth more when the benefit is received in the future. (Note that if the energy costs didn't increase over time, the savings would be worth only $2,500 each year when accrued, and the value today would be less than $2,500; the further in the future, the less the savings would be worth at the present time.)

Over 10 years, therefore, the total present-time value of the savings would be 10 × $2,500 per year = $25,000. If $5,000 were required in the fifth year for a partial replacement, its cost in present-time money (discounted back 5 years according to Equation 16.3 below) would be $3,105. If the salvage value of the material is $2,500 at the end of the 10-year period, that would be worth $964. In this way, all costs and savings can be placed in a comparable frame of reference, and the total value can be compared with the total cost.

When uneven savings occur, as in Table 16.4, the value of each investment cannot be directly compared. However,

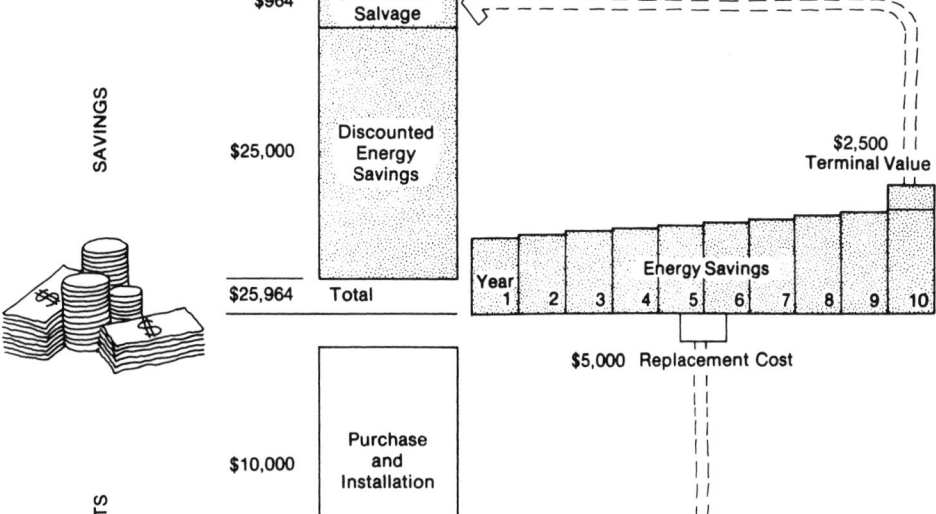

FIGURE 16.2. Time value of money.

using the discounting procedure, the value of the future savings over the entire life of each investment can be adjusted to a common base year, usually the present. The resulting amount is called the *present value.*

Net Present Value

The formula for present value (PV) is as follows:

$$PV = \frac{FV}{(1 + i)^n} \qquad (16.3)$$

where

FV = future value amount
i = rate of interest involved
n = number of years of discounting

Thus, if $1,000 is due 10 years from now, and the interest rate that the money could have been invested at is 15 percent, the future value is discounted to its present value as follows:

$$PV = \frac{\$1,000}{(1 + 0.15)^{10}} = \$614 \qquad (16.4)$$

This means that if $614 were invested in the bank now at 15 percent interest, it would have the same economic worth as $1,000 appearing in 10 years. The two amounts are interchangeable.

The *net present value* of a contemplated investment is the total present value minus the initial capital invested. As an example, consider a design option that requires an incremental increase in initial cost of $10,000 together with recurring incremental costs, incremental savings, and net cash flows as shown in Table 16.5.

For the sake of simplicity, let us say that cash flow occurs at the end of each year. Thus, the first year's $3,000 cash flow occurs 12 months from now. In order to determine how much it is worth today, the interest rate must be known. In this case, suppose it is 10 percent. The net cash gain of $3,000 at the end of the year would then be worth approximately $2,727 at the beginning of the year. In other words, if $2,727 were deposited in a bank at 10 percent interest, payable annually, it would grow to $3,000 by the end of the year.

The second year's $3,300 must be discounted back 2 years, resulting in a present value of $2,727. The discounted values for all 4 years is presented in Table 16.6.

The net present value would be $10,966 − $10,000 = $966. As long as the total present value is greater than the initial investment, the return is greater than the interest that was lost. In this example, the investment of $10,000 returned $10,966 in 4 years discounted at 10 percent. Therefore, since $10,966 is greater than $10,000, this investment would be viable if 10 percent is a satisfactory minimum rate of return. By incorporating this time value of money, the designer can assure his client that this investment wouldn't earn more in a bank.

Discounted Payback Period

Returning to the concept of payback, the $10,000 investment would have a simple payback period of 3 years based on actual cash flow. However, the payback period computed by using the discounted cash flow column is about $3\frac{2}{3}$ years. When the payback period is based on discounted cash flow figures, it is referred to as the *discounted payback period.*

Discount Rate

It is very important to choose the right *discount rate* for a project if the objective is cost minimization rather than profit maximization. The higher the *interest rate,* the less viable an energy-conservation measure is. Varying the interest rate a few percentage points can turn a potentially profitable energy-conservation measure into an unprofitable one.

Many developers use their own *cost of capital* as a starting point in deciding which interest rate to use as a basis for considering alternative investments. That is, they start with an interest rate that reflects their cost of acquiring financing.

TABLE 16.5   PRESENT VALUE EXAMPLE

Year	(1) Incremental Costs	(2) Incremental Savings	(3) = (2) − (1) Net Positive Cash Flow
1	$ 0	$ 3,000	$ 3,000
2	200	3,500	3,300
3	300	4,000	3,700
4	500	4,500	4,000
	$1000	$15,000	$14,000

TABLE 16.6   DISCOUNTED CASH FLOWS

(3) Net Positive Cash Flow	(4) Discounting Factor	(5) = (3)/(4) Discounted Cash Flow at 10%
3,000	$(1 + 0.1)^1 = 1.1000$	$ 2,727
3,300	$(1 + 0.1)^2 = 1.2100$	2,727
3,700	$(1 + 0.1)^3 = 1.3310$	2,780
4,000	$(1 + 0.1)^4 = 1.4641$	2,732
$ 14,000		$10,966

Construction Financing. Every construction project is financed, whether with cash or borrowed money. Most large construction projects are financed with long-term debt, such as borrowing money at 10 percent for 20 years. One type of long-term financing is a mortgage loan obtained from a lending institution. Large corporate building projects are often financed through the sale of corporate stocks and bonds.

Small energy-conservation retrofit projects are normally financed out of operating funds requiring a 1- or 2-year simple payback. However, a number of retrofit projects can be assembled together in a single long-term financing package.

When considering a component of a new building, such as extra insulation or an additional layer of glazing, that component should be analyzed as if it were financed in the same manner as the entire building. Setting an arbitrary simple payback limit on the extra insulation could result in substantially less insulation than an accurate financial analysis would justify.

When there are no taxes involved, the discount rate is the same as the interest rate on the construction financing. Thus, the interest rate for discounting is sometimes taken as the rate of return that the developer could receive on alternative investments. However, interest on investments is taxable, unlike fuel savings, and the after-tax return from an ordinary investment may be less than the posted rate.

Tax Effects. The appropriate discount rate for a taxpaying owner is the after-tax cost of raising the additional funds needed for the project. In addition to the tax bracket of the client, other tax considerations are interest deductions, depreciation, and investment tax credits. These have the benefit of reducing taxes in order to offset some of the increased cost of a project.

Interest payments on borrowed money are *tax deductible,* which effectively decreases the net interest rate. *Depreciation* is a method of spreading out the tax deduction on the expense of any capital equipment, building, or subsystem over its useful life. There are a number of different forms of depreciation, and the length and type of depreciation can greatly effect the profitability of an investment. This information should therefore be supplied by the client.

As an example, consider a $10,000 investment by a company with a taxable income of $100,000. If the investment is depreciated straight-line for 10 years, the depreciation for each year is $1,000. Supposing that the company's tax rate is 50 percent, the taxes paid with and without the investment would be as shown in Table 16.7.

The use of an accelerated depreciation method advances the benefits to the earlier years of the asset's life, which reduces the payback time.

TABLE 16.7 TAX EXAMPLE

	Without Investment	With Investment
Taxable income	$100,000	$100,000
Depreciation	—	1,000
Net taxable income	$100,000	$ 99,000
Taxes (50%)	50,000	49,500
Reduction in taxes	—	500

If it satisfies other criteria, an investment may also qualify for an investment *tax credit,* which is different from deductions such as depreciation. Referring to the example in Table 16.7, the $10,000 investment was entitled to an annual $1,000 tax deduction for 10 years due to depreciation. The $1,000 deduction would reduce taxes by $500.

That same investment might also qualify for a one-time investment tax credit of 10 percent. That would mean a $1,000 credit for a $10,000 investment. The $1,000 credit would be subtracted directly from the tax liability rather than from the taxable income. The tax liability would thus be reduced from $49,500 to $48,500. This represents a double benefit, although tax credits can be taken only once, while depreciation is taken repeatedly over the life of the improvement up to its original cost.

The types of property that qualify for an investment tax credit are tangible property such as machinery, office furniture, and computers. Buildings and their structural components, such as overhead lighting fixtures and central heating or cooling systems (including related motors, compressors, pipes, ducts, and plumbing) are not eligible.

Ordinarily, computers qualify for an investment tax credit unless they are part of an energy-management, security, or fire alarm system for the building (these systems are interpreted as structural components). The complexity of the investment tax credit is further demonstrated by the following situation: While furniture, such as a desk, qualifies for the credit, equipment to provide overhead lighting is considered a structural component, so its cost does not. If, however, a portion of the lighting in an area is provided by task lighting as part of the desk, the entire desk with its lighting component would be eligible for an investment tax credit.

Depreciation values and an item's suitability for a tax credit can be confirmed by a qualified tax accountant in critical cases. Other tax benefits may also be available, such as an additional tax credit on investments in energy-conserving and alternative (other than oil or natural gas) energy-utilization equipment.

Once the net tax rate (TR) is established, the discount rate, i_d, is found by the following formula:

$$i_d = i(1 - TR) \qquad (16.5)$$

For example the tax bracket in the example above is 50 percent. However, due to the benefits of depreciation and an investment credit, the actual tax liability is $48,500 on a $100,000 taxable income. Thus, TR = 48.5 percent. If the cost of money before taxes is 10 percent, $i_d = 0.10(1 - 0.485) = 0.05$.

This demonstrates how sensitive the discount rate is to taxes. In this particular example, depreciation and investment credit effects were negligible, but if the investment were a larger proportion of the taxable income, they could be significant.

Sensitivity Analysis.   When there is uncertainty about the appropriate discount rate, more than one may be calculated and the results compared in order to see how much effect different rates have and at what rate the project is no longer viable. Called a *sensitivity analysis,* this can involve modifying a number of variables such as interest rate, method of depreciation, period of financing, tax credits, and future escalation of energy prices so as to compare the overall cost effectiveness under these different conditions. The use of a computer is particularly helpful in performing these multiple calculations.

Present Worth Method

There are two basic methods of life-cycle cost analysis: (1) present worth and (2) annual cash flow. The *present-worth method* consists of determining the total of all initial costs and the present value of all subsequent costs.

The *annual cash-flow method* accounts for all cash flows at the time they occur. Both methods result in the same ranking order for alternative projects, and in most cases either will suffice for decision making.

There is a benefit, however, to the annual cash-flow method in cases where cost differentials between projects are relatively small and noneconomic factors are a consideration. For example, project A may have a present-worth cost $5,000 greater than project B, which amounts to an average annual cost of $100 greater.

Project A, on the other hand, has nonquantifiable benefits that are very desirable. Project A may be hard to justify economically because of the $5,000 additional cost, but the realization that the difference amounts to only $100 a year may make option A more acceptable considering the nontangible benefits.

The example in Table 16.8 illustrates the use of the net present-worth approach in choosing between two alternative projects. The tabulated results show that one project has a lower initial cost, while the other has a lower life-cycle cost.

Thus, while B would be more desirable based on first cost alone, A has a lower life-cycle cost on a present-value basis. Project A shows a lower life-cycle cost by $3,500,000 − $3,100,000 = $400,000.

Annual Cash-Flow Method

Lending institutions often study the annual cash flow of a project to ascertain the safety in making a loan. The annual cash-flow method compares the annual costs to the annual profits over the life of an investment. As long as the annual income or net savings exceeds the annual costs, enough money should be available to meet the annual bank payments. The prospective owners of buildings intended for generating income, such as rental property, also often prefer to see the cost in terms of annual cash flow.

The advantage over the present-worth method is that the annual cash-flow method considers when the profits are received relative to when the expenses occur. The present-worth method may show that an investment would make a large profit, but it doesn't show whether this occurs at the beginning, as annual payments, or as a lump sum at the end. If the profit came at the end, the owner might be "cash poor" and not have enough money to meet the annual payments.

The first step in a cash-flow analysis is to determine how the building is financed. This converts a single initial cost into a series of annual cash payments. These payments can then be compared with the annual operating savings generated by that investment.

Consideration must also be given to the time value of the cash flow. For example, a $100,000 annual cash flow next year is worth more than a $100,000 annual cash flow 20 years in the future.

Cash-flow analysis is basically a schedule of annual profit and loss. The calculations for each year are not difficult, but since they must be repeated for every year of the life of the system, a computer is useful.

TABLE 16.8   LIFE-CYCLE COST EXAMPLE

Alternative	Acquisition Cost	Present Value of Savings	Net Present Value
A	$600,000	$2,500,000	$3,100,000
B	500,000	3,000,000	3,500,000
Difference	$100,000	$ 500,000	$ 400,000

Consider the example of a $10,000 investment with a resultant $1,000 annual fuel saving. The simple-payback period would be 10 years, which would not be immediately acceptable to most investors. However, an annual cash-flow life-cycle cost analysis provides different results as follows.

The $10,000 would probably either be borrowed or come from savings; either way, there would be an interest rate involved. Suppose the $10,000 represents an incremental increase in the cost of a construction project that is financed at 15 percent over 20 years. Considering fuel savings in excess of the loan payment to be "profit," the incremental cost should make a profit in order to be a viable investment.

The annual payment, A, on a loan principal, P, to be paid off in N years at an interest rate of i is as follows:

$$A = \frac{P \times i}{1 - (1 + i)^{-N}}$$

$$= \frac{\$10,000 \times 0.15}{1 - (1 + 0.15)^{-20}} = \$1,598 \qquad (16.6)$$

The annual payment, at first impression, seems higher than the $1,000 annual energy savings. However, the interest on the loan for the first year is $10,000 × 0.15 = $1,500. If the client is in a 50 percent tax bracket, the tax savings will be $750 on this $1,500 deduction. Adding that to the $1,000 energy saving results in a saving (expense reduction) of $1,750 during the first year. That is a net cash gain the very first year. Tables 16.9 and 16.10 present the cash flow story for the full 20-year period.

Notice that there is an actual profit (positive cash flow) during each of the first 11 years. The cumulative profit reaches over $1,000. After that, the tax deduction benefit diminishes below the break-even point, and the annual loan payment exceeds the energy and tax savings. By the end of the 20 years, there is a cumulative net loss of nearly $1,000 before the loan is completely paid off.

However, the significant profits during the first few years can be invested in something else before being paid back out as later losses. This is reflected in the present-value columns. When the annual profits are converted into their present values at a 10 percent discount rate, the early gains are worth more than the later losses, and the cumulative present-value profits remain positive over the full 20 years. Thus, the investment not only pays for itself over the 20-year loan period, it also generates a net profit. After the 20-year period, the $1,000 per year energy savings are a net gain.

TABLE 16.9   CASH-FLOW EXAMPLE LOAN PAYMENTS AND TAX SAVINGS

(1) Year	(2) Initial Loan Principal	(3) Payment Due	(4) = i × (2) Interest Portion	(5) = TR × (4) Tax Saved by Interest Deduction	(6) = (3) − (5) Net Payment After Credit for Interest Deduction	(7) Residual Loan Principal
1	10,000	1,598	1,500	750	848	9,902
2	9,902	1,598	1,485	742	856	9,789
3	9,789	1,598	1,468	734	864	9,659
4	9,659	1,598	1,449	724	874	9,510
5	9,510	1,598	1,426	713	885	9,338
6	9,338	1,598	1,401	700	898	9,141
7	9,141	1,598	1,371	686	912	8,914
8	8,914	1,598	1,337	668	930	8,653
9	8,653	1,598	1,298	649	949	8,353
10	8,353	1,598	1,253	626	972	8,008
11	8,008	1,598	1,201	600	998	7,611
12	7,611	1,598	1,142	571	1,027	7,155
13	7,155	1,598	1,073	536	1,062	6,630
14	6,630	1,598	994	497	1,101	6,026
15	6,026	1,598	904	452	1,146	5,332
16	5,332	1,598	800	400	1,198	4,534
17	4,534	1,598	680	340	1,258	3,616
18	3,616	1,598	542	271	1,327	2,560
19	2,560	1,598	384	192	1,406	1,346
20	1,346	1,548	202	101	1,447	0

TABLE 16.10 CASH-FLOW EXAMPLE ANNUAL CASH FLOW

Year	End-of-Year Annual Fuel Savings	Net Annual Loan Payment	Annual Profit	Cumulative Profit	Present Value (@10%)	
					Annual Profit	Cumulative Profit
1	1,000	848	152	152	138	138
2	1,000	856	144	296	119	257
3	1,000	864	136	432	102	359
4	1,000	874	126	558	86	445
5	1,000	885	115	673	71	516
6	1,000	898	102	775	58	574
7	1,000	912	88	863	45	619
8	1,000	930	70	933	33	652
9	1,000	949	51	984	22	674
10	1,000	972	28	1,012	11	685
11	1,000	998	2	1,014	1	686
12	1,000	1,027	⟨27⟩	987	⟨9⟩	677
13	1,000	1,062	⟨62⟩	925	⟨18⟩	659
14	1,000	1,101	⟨101⟩	824	⟨27⟩	632
15	1,000	1,146	⟨146⟩	678	⟨35⟩	597
16	1,000	1,198	⟨198⟩	480	⟨43⟩	554
17	1,000	1,258	⟨258⟩	222	⟨51⟩	503
18	1,000	1,327	⟨327⟩	⟨105⟩	⟨59⟩	444
19	1,000	1,406	⟨406⟩	⟨511⟩	⟨66⟩	378
20	1,000	1,447	⟨447⟩	⟨958⟩	⟨66⟩	312

A number of assumptions were made in this example, and it is always wise to recognize assumptions:

1. Energy rates resulting in the cost savings were considered constant. In reality, energy costs have historically risen and are likely to continue to do so. This would result in a larger amount of savings than that shown.
2. The cost of money (construction financing) was set at 15 percent over a 20-year period. If the interest rate were lower or the period longer, the annual loan payments would be lower. If the interest rate were higher or the period shorter, the annual loan payments would be higher.
3. The time value of the client's money was set at 10 percent. This is the rate at which the client could presumably reinvest profits and the rate at which lost profits are magnified. A higher discount rate would tend to accentuate the earlier profits compared to the later losses, resulting in an even more profitable cumulative total. A lower discount rate would have the opposite effect.
4. The client's tax rate was figured as 50 percent. A lower rate would mean proportionately lower interest deductions and lower profits.

5. The project involved an incremental increase in the total project cost, and no down payment was attributed to the project. If the project were to have an initial cost associated with it in addition to annual payments, there would more likely be some negative cash flows in the first few years.

Ideally, all these factors should be given consideration for a given proposed project. Whenever there is uncertainty about one or more of these assumptions, they should be tested with a sensitivity analysis. In order to justify a project on a life-cycle basis, an architect must be armed with calculations showing what the economic consequences would be if such variables were altered.

RATE OF RETURN

An energy-conservation project is desirable as long as the net present worth of its cash flow is greater than zero. However, when there are alternative projects, the project with the highest net present worth should be chosen.

Another method of ranking alternative projects is to determine the rate of return on the investment. This is referred to either as the *rate-of-return* (ROR) or the *return-on-investment* (ROI) method.

Implicit in any investment is some element of risk. In theory, a profit from the investment is intended to be a compensation for that risk. Some investors are accustomed to selecting investments on the basis of the ROR versus the risk involved. They prefer to compare alternatives in terms of the rate of return they would receive.

Consider an investment of $100,000 that nets a uniform return of $15,000/year. The annual ROR on the investment would thus be

$$ROR = \frac{\$15,000}{\$100,000} = 15\% \qquad (16.7)$$

If the return is not a uniform amount each year, an average ROR may be calculated over a specified period of time, usually the loan period or the useful life of the project. Thus, a $100,000 investment that returns $1,000 the first year and an additional $1,000 each year thereafter over a period of 10 years would yield a total return of $55,000. This represents $5,500/year, and the average annual rate of return would be

$$ROR = \frac{\$5,500}{\$100,000} = 5.5\% \qquad (16.8)$$

Actually, when the ROI is extended that far into the future, it is appropriate to discount it to its present value. This can be demonstrated using the example in Table 16.6. An initial investment of $10,000 discounted at 10 percent resulted in a present worth of $10,966. If the discount rate were 14 percent instead, the total discounted cash flow over the 4-year period would approximately equal the initial investment of $10,000 ($10,036 to be exact). Therefore, it can be said that this investment approximates a 14 percent rate of return, or has an equivalent annual return of 14 percent.

The important question is, could the same money be invested anywhere else (in the building construction project or elsewhere) in order to earn more interest? The ROR can be computed for each alternative investment opportunity, and then all options can be ranked according to their rates of return.

COMPARATIVE VALUE ANALYSIS

The previous decision-making methods dealt only with the economic aspects of a project. But certain design features—such as convenience, appearance, added comfort, improved

worker efficiency, better acoustic control, reduced air or water pollution, conservation of nonrenewable resources, and a more pleasing environment—are difficult to quantify. Even though these considerations are not easily quantifiable, at times they must somehow be incorporated into the decision-making process.

In such cases, these important factors may be evaluated by the *comparative value analysis* method. First, the various key factors involved are listed. Each one is then weighted according to its relative importance. Then all of the competing designs are compared and the listed characteristics of each design are assigned a numerical value in comparison with the same characteristic of the other designs. An evaluation matrix is thus formed.

The weighting factor and the comparative evaluation factor are then multiplied together for each characteristic, and then the scores for all characteristics are added together for a total score on each design. If the weighting is done carefully and the comparative evaluations are properly assigned, the design with the highest score is the preferred alternative. The example in Table 16.11 illustrates this procedure.

Although difficult, it is extremely important to select the proper numerical values. It is helpful to keep them on a scale from 1 to 10.

TABLE 16.11 COMPARATIVE VALUE ANALYSIS EXAMPLE

Identification of Competing Factors	Design 1	Design 2
Initial cost	$ 100,000	$ 150,000
Annual heating/cooling cost	75,000	50,000
Life-cycle cost	20,000,000	10,000,000
Interior environment	more pleasant	less pleasant

Numerical Evaluation	Weighting Factor	Comparative Value	Total Score
Design 1			
Initial cost	5	3	15
Annual heating/cooling cost	8	2	16
Life-cycle cost	10	1	10
Interior environment	8	2	16
			57
Design 2			
Initial cost	5	2	10
Annual heating/cooling cost	8	3	24
Life-cycle cost	10	2	20
Interior environment	8	1	8
			62

The design features that are commonly considered when deciding whether or not to expend funds on building construction are:

- First cost
- Expected reliability (low maintenance cost and downtime)
- Relative energy efficiency
- Tax advantages
- Net after-tax cash flow
- Maintenance cost
- Payback period
- Life-cycle cost (or overall ROI)
- ROR on investment
- Length of construction period
- Usable space available
- Design life
- Other personal or aesthetic factors

The last item is sometimes the deciding (overriding) factor, but it is usually in the client's best interest to consider the entire economic picture because of the large costs involved. For a simple analysis, the single factor that best represents the economics is the life-cycle cost, since it can take all the others into account.

OTHER CONSIDERATIONS

Inflation

The above discussions have ignored the factor of *inflation*, which refers to a continual increase in the prices of goods and services. As prices rise, a fixed amount of money has less buying power and is said to have less value. This decreasing value of money over time in an inflationary economy offsets the increase in the potential time value of money due to interest on investments. This has the effect of discounting the value of future money at whatever the inflation rate is.

It also causes a double-discounting effect on present-value calculations. Thus, the present value of future money is discounted once to account for the time value of money and once again for the effect of inflation.

This is illustrated in Figure 16.3. The line defined by points A and B represents the rising time value of money at the applicable interest rate. Moving to the right on the line shows the effect of compounding. Moving back to the left then depicts the process of discounting.

The line connecting points A and D represents the decreasing value of money over time due to inflation. The combined

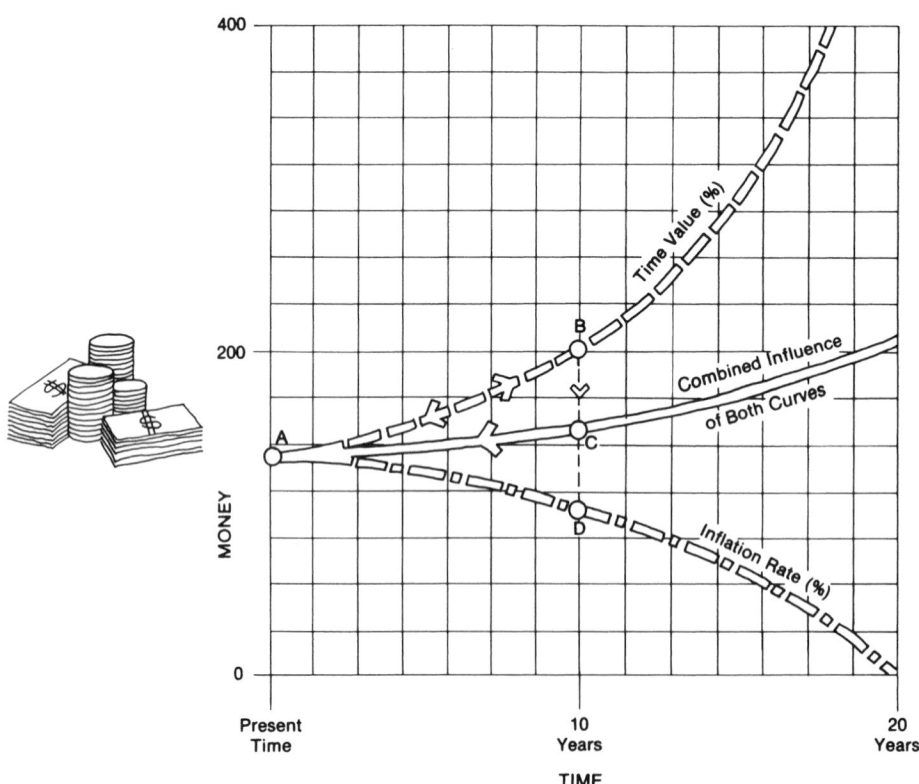

FIGURE 16.3. Effect of inflation on the time value of money.

effect of both influences is shown by the net buying-power line, AC.

At a given point in time, B, the influence of the inflation line is to decrease the value of cash flow to point C. Mathematically, this is expressed in the following formula, where i refers to the time value interest rate and inf represents the applicable rate of inflation:

$$FV = PV\{(1 + i)^n - (1 + inf)^n\} \qquad (16.9)$$

Likewise, to compare investment alternatives under the influence of inflation, the present value of a future cash flow can be calculated by the following formula:

$$PV = \frac{FV}{(1 + i)^n - (1 + inf)^n} \qquad (16.10)$$

Fuel Cost Escalation

At the same time that the prices of goods and services in general become inflated, energy costs also increase, and often at a different rate from the general inflation rate. The future value (FV) of energy savings is thus determined by four factors: (1) calculated annual cost savings at today's rates (PV), (2) the fuel escalation (FE) rate, (3) the time value of money (i), and (4) the inflation (inf) rate. Mathematically, FV is found as follows:

$$FV = PV \{(1 + FE)^n + (1 + i)^n - (1 + inf)^n\} \quad (16.11)$$

Typically, the cost of fossil fuel–based energy supplies increases at a rate faster than the general rate of inflation due to its limited availability as a resource and its increasing scarcity. The federal government usually projects future fuel escalation rates for the United States, and such estimates can be used in comparative evaluations.

Once the future value of the energy savings is found using Equation 16.11, it can be added to any other gains or savings, and any losses can be subtracted for a net cash flow at any given time in the future. That cash flow may then be treated as any other cash flow and analyzed as is, or discounted according to Equation 16.10 in order to obtain the present value for comparison with other alternative investments.

Point of Diminishing Returns

If an energy-conserving measure, such as added insulation, is determined to be a profitable investment, why not insulate the roof even more to obtain even greater savings? The answer lies in the procedure for calculating energy savings discussed in Chapter 4.

Suppose that the roof of a building is initially designed for an R-value of 10. The energy use is a function of the U-value of $\frac{1}{10}$ (0.1). Then suppose that doubling the R-value is considered. Twice the R-value is 20, with a corresponding U-value of $\frac{1}{20}$ (0.05). Having found that an R-20 is cost-effective, let us now consider an R-40. The U-value drops to $\frac{1}{40}$ (.025).

In going from R-10 to R-20, twice as much insulation material is used, which would double the cost. The energy savings is proportional to $0.1 - 0.05 = 0.05$. In going from R-20 to R-40, however, four times the amount of initial insulation is used, with a consequent quadrupling of the cost, but the savings is proportional only to $0.1 - 0.025 = 0.075$. The trend is shown in Table 16.12.

The cost of insulation is directly proportional to the physical amount used. Each additional increment adds as much to the cost as the previous increment. The energy savings, though, become smaller and smaller as more insulation is added. A point is eventually reached at which the cost for any additional R-value would save very little additional energy. At that point, referred to as the *point of diminishing returns,* it would no longer be economically justifiable to add more insulation. This is illustrated in Figure 16.4.

The top curve represents the savings as more and more insulation is added. It gradually tapers off to show that there is very little additional savings with excessive amounts of insulation. The straight line represents the cost of the insulation as a direct function of the quantity. The third line is a composite of the other two and represents the net savings.

At the point where this third line levels off, the net savings are at a maximum. Insulation levels should achieve this point as a minimum. Some investors see this as the optimum insulation level, but notice that a net savings occurs all the way up to the R-maximum point. If the dollar scale is expressed in present value of the net cash-flow savings, all points above the zero line would yield savings above the required investment interest rate and would therefore be a profitable investment.

From the point of view of energy usage, a given R-value of insulation takes a fixed amount of energy to manufacture.

TABLE 16.12 POINT-OF-DIMINISHING-RETURNS EXAMPLE

Insulation R-Value	U-Value	Energy Use	Energy Savings	Insulation Cost
R-10	0.10	100 units		10 units
R-20	0.05	50 units	50 units	20 units
R-40	0.025	25 units	25 units	40 units
R-80	0.0125	12.5 units	12.5 units	80 units

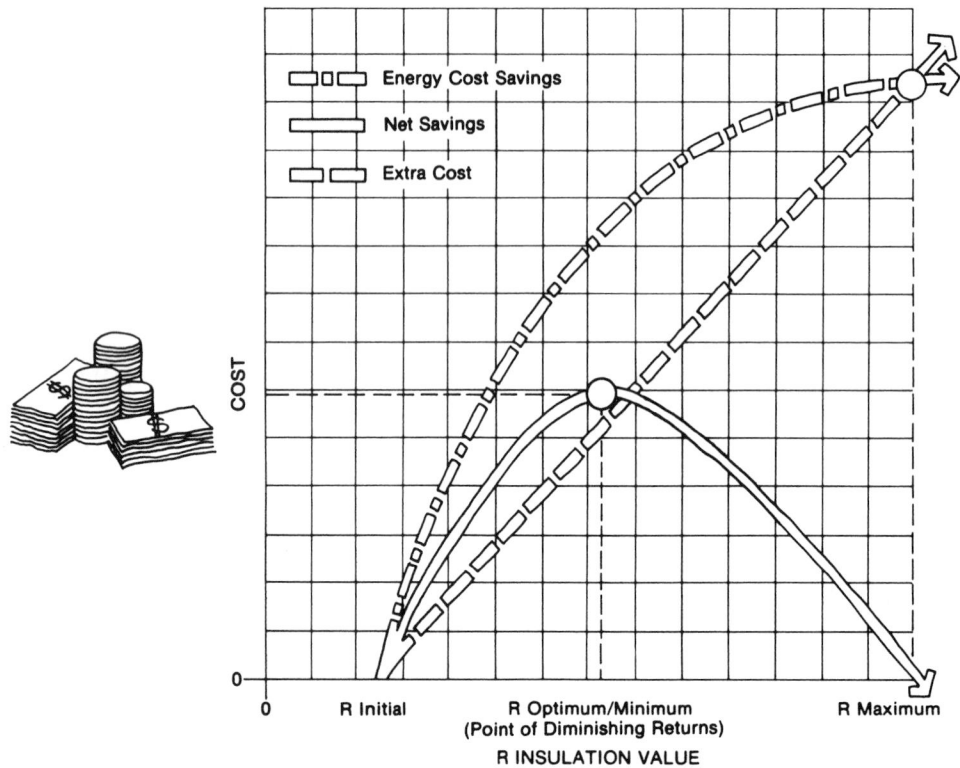

FIGURE 16.4. Point of diminishing returns.

If too much insulation is added, the amount of energy saved over the life of the building is less than the amount of energy it took to produce the insulation in the first place. In that event, rather than saving energy, more energy has been used up.

While the above example refers to insulation, the same concept applies to other energy-saving measures, such as additional layers of window glazing and solar energy utilization.

Creative Financing

In recent years, financing arrangements have become available that provide energy-conserving and alternative-energy equipment at no initial cost to the user. Under these arrangements, the venders or third-party investors finance the equipment and the installation at the client's site. The client repays the investors or venders in one of three ways: (1) by sharing the savings with the investor, (2) by making lease payments for a given period, or (3) by buying the energy produced by an alternative system. In all these arrangements, the user makes no capital investment and has the option of purchas-

ing the equipment at the end of the lease or contract at a predetermined price. Furthermore, the equipment is typically maintained by the vender or investor.

Under the third plan, the client contracts to purchase the energy, usually at prices guaranteed to be 10 to 20 percent lower than the current local utility cost. The price of the energy thus varies with the costs of oil, natural gas, or electricity in the area. The period of time for the agreement is usually about 10 years, with an option to renew the contract or purchase the equipment at the end of the period.

If a third-party investment company owns the equipment, it receives the 10 percent investment credit and can depreciate the equipment in 5 years. If a vender simply provides financing, the client retains all or a portion of the tax benefits.

SUMMARY

Economic evaluation methods are useful for comparing designs with differing installation costs and differing total costs over the life of the project. They can be used to make decisions if the choice is strictly economic, or to determine the cost of an aesthetic consideration or special need in terms of lifetime cost. They are applicable for any component that isn't required or a bare necessity. They are also useful for deciding between energy source options.

While construction decisions are often made on the basis of the lowest initial cost, a lower life-cycle cost is actually in the client's best economic interest. In order to accomplish a life-cycle cost analysis, all of the costs of each design alternative over the anticipated lifetime of the building or component must be estimated. Then the total cost of the design alternatives can be compared.

Life-cycle costs include:

1. Initial costs (design, materials, and labor/installation)
2. Owning and operating costs (maintenance and repairs, energy, taxes, interest on borrowed money, and insurance)
3. Salvage value or disposal cost at the end of the project's useful life

Life-cycle cost evaluations usually incorporate the concept of the time value of money, which refers to its capacity to increase in value over time due to interest on investments. In the present-worth method of computing life-cycle cost, sums of money in the future are discounted to the present time in order to compensate for the increase in value over time. These sums represent the incremental costs and savings over the useful life of the project.

The other method of life-cycle cost analysis is the annual cash-flow method. If all factors are considered, including taxes and interest rates, a good investment over the life cycle will yield a net positive cash-flow advantage soon after the investment. The cash flow is of paramount importance to some clients.

Other clients prefer to know the ROR on the investment. If the ROR isn't fast enough, some clients won't invest in a project, even though its eventual benefits are great. Regardless of whether the results are presented as ROI or as net annual cash flow, the life expectancy of the project must be established.

If there is doubt about the appropriate interest rate to use for discounting, more than one may be used for calculations and the results compared to see whether the effect is significant. The type and period of financing, tax benefits, energy-cost escalation rate, or any other factor that may be uncertain can also be varied to determine the sensitivity of the estimates and under what conditions the project is viable.

The payback method is an alternative that is much simpler than the others, but it also has limitations. Its major flaws are that it fails to consider either the timing of cash flows or the economic life of an investment, which are directly related to the worthiness of an expenditure.

The comparative evaluation method involves a decision-making matrix to compare alternative designs. This is the only method that can take noneconomic factors into account for evaluating any type of trade-off. The accuracy with which each parameter is determined and assigned a value affects the reliability of the decision and subsequent satisfaction with the installation.

Many decisions made in planning and design have consequences in terms of the energy use and economics of the building, affecting the owners and users over the life of the facility. The techniques discussed in this chapter can provide the designer with the tools needed to evaluate the consequences of their decisions.

KEY TERMS

Incremental costs	Time value of money	Annual cash-flow method
Incremental savings	Present value	Rate of return (ROR) or
Payback period	Net present value	Return on investment (ROI)
Simple payback (payback ratio)	Discounted payback period	
Capital recovery factor (CRF)	Discount rate	Comparative value analysis
	Interest rate	
	Cost of capital	Inflation
Life-cycle cost	Tax deduction	Fuel cost escalation
Useful life	Tax credit	
Compounding	Sensitivity analysis	Point of diminishing returns
Discounting	Present-worth method	

STUDY QUESTIONS

1. If fuel oil costs $3.00 per gallon (per liter) today and escalates in cost at the rate of 10 percent each year, what will it cost in 10 years? What will it cost in 10 years if the escalation rate is 15 percent?
2. If the discount rate is 15 percent, what is the present value of a $200 salvage value 10 years from now?

3. Two design options are being considered. Alternative A would require a probable replacement at $500 in 5 years. Alternative B would require replacement at $600 in 6 years. Using the present-worth method of evaluation, which option would be less expensive if the discount rate is 15 percent? Which would have a lower present value if the discount rate is 20 percent? If it is 25 percent?

4. Given a building with an uninsulated roof (R-3) and the following conditions:

   Area = 20,000 ft^2 (1,860 m^2)

   Insulation cost = $.10/ft^2/R ($6.11/m^2/R) installed

   Heating energy cost = $5/million Btu ($4.74/GJ)

   Cooling energy cost = $.06/kWh

   Heating climate = 5,000 degree days (2,800 °C-days) (M factor = 0.75)

   Cooling climate = 1,500 equivalent full-load hours

   Heating system efficiency = 60 percent (seasonal average)

   Cooling power input = 1.4 kW/ton (0.40 kW/kW)

   a. What would the total initial cost be to install insulation to a level of R-23 (R-4.05)?

   b. What would the annual energy cost savings be with this much insulation?

   c. What would be the simple payback?

   d. If the improvement is fully financed with a 12 percent loan for 10 years, what would the monthly payment be?

   e. Tabulate the annual net cash flow over the 10-year loan period.

   f. Tabulate the annual net cash flow over the 10-year loan period including the effect of a tax deduction based on a 20 percent tax rate and straight-line depreciation.

   g. What is the present worth life-cycle savings based on the results of (f)?

   h. What is the effective rate of return over the 10-year period?

5. A third glazing for a 4 × 5-ft (20-ft^2, 1.9-m^2) window costs $100. If the energy cost savings amounts to $15 per year, what would the simple payback be? What would the discounted payback be at a 12 percent discount rate?

BIBLIOGRAPHY

American Society of Heating, Refrigerating, and Air-Conditioning Engineers, Inc. *Guideline 8—Energy Cost Allocation for Multiple-Occupancy Residential Buildings.*

Dietrich, Norman L. (editor). *Kerr's Cost Data for Landscape Construction 1994: Unit Prices for Site Development* (*Landscape Architecture*), 14th ed. Van Nostrand Reinhold Company, 1994.

Khashab, A.E. *Heating, Ventilating, and Air-Conditioning Systems Estimating Manual,* 2nd ed. McGraw-Hill Book Company, 1984. (Charts, tables, and detailed discussions of equipment, materials, and estimating methods.)

Kolstad, C. Kenneth, and Gerald V. Kohnert. *Rapid Electrical Estimating and Pricing,* 5th ed. McGraw-Hill Book Company, 1993. (Charts and tables with adjustment factors for increasing prices.)

R.S. Means Company, Inc. *Building Construction Cost Data; MEANS Man-Hour Standards; Mechanical & Electrical Cost Data;* and *Repair & Remodeling Cost Data.* (Published each year.)

Taylor, George A. *Managerial and Engineering Economy: Economic Decision-Making.* Van Nostrand Reinhold Company, 1975.

Thumann, Albert, *Plant Engineers and Managers Guide to Energy Conservation,* 8th ed. Marcel Dekker, 2002.

Turner, Wayne C. (editor). *Energy Management Handbook,* 4th ed. Marcel Dekker, 2002.

U.S. Department of Commerce. *Energy Conservation in Buildings: An Economic Guidebook for Investment Decisions.* United States Government Printing Office, 1980.

West, Ronald E., and Frank Kreith (editors). *Economic Analysis of Solar Thermal Energy Systems (Solar Heat Technologies).* MIT Press, 1988.

Zimmerman, Larry, and Glen Hart. *Value Engineering: A Practical Approach for Owners, Designers, and Contractors.* Van Nostrand Reinhold Company, 1997. (Value engineering methods and applications to help reduce costs and increase value.)

Appendix

TABLE A1   UNIT CONVERSIONS

Multiply	By	To Obtain	Multiply	By	To Obtain
Length			barrels petroleum (bbl)	5.6146	cubic feet
centimeters	0.394	inches	barrels	42	gallons
feet	30.5	centimeters	barrels	159	liters
feet	0.305	meters	cubic centimeters	0.061	cubic inches
inches	2.54	centimeters	cubic centimeters	1	mililiters
inches	25.4	millimeters	cubic feet	1728	cubic inches
kilometers	0.6214	miles	cubic feet	0.0283	cubic meters
meters	3.28	feet	cubic feet	7.481	gallons
meters	39.37	inches	cubic meters	35.31	cubic feet
meters	1.0936	yards	cubic meters	1.308	cubic yards
miles	5,280	feet	cubic meters	264.17	gallons
miles	1.609	kilometers	cubic meters	1057	quarts
miles	1,609	meters	cubic yards	0.7646	cubic meters
miles	1,760	yards	gallons	0.1337	cubic feet
yards	0.9144	meters	gallons	231	cubic inches
Area			gallons	0.003785	cubic meters
acres	43560	square feet	gallons	3.785	liters
acres	4047	square meters	liters	1000	cubic centimeters
acres	4840	square yards	liters	0.0353	cubic feet
square centimeters	0.155	square inches	liters	0.2642	gallons
square feet	0.0929	square meters	liters	1.0567	quarts
square inches	6.45	square centimeters	quart	0.946	liter
square kilometers	247.1	acres	**Mass**		
square kilometers	0.3861	square miles	kilograms	2.205	pounds
square meters	10.76	square feet	ounce	28.35	grams
square meters	1.196	square yards	pounds	0.4536	kilograms
square miles	640	acres	tons	907.2	kilograms
square miles	2.59	square kilometers	tons	2,000	pounds
square yards	9	square feet	tonnes (metric tons)	1000	kilograms
square yards	0.836	square meters	tonnes	2,205	pounds
Volume			tonnes	1.1	tons
acre-feet	43,560	cubic feet			
acre-feet	1,234	cubic meters			
acre-feet	1,613	cubic yards			

TABLE A1   UNIT CONVERSIONS *(Continued)*

Multiply	By	To Obtain	Multiply	By	To Obtain
Speed			**Pressure**		
feet/minute	0.508	centimeters/second	atmospheres	33.9	feet of water
feet/minute	0.01829	kilometers/hour	atmospheres	760	millimeters of Hg
feet/minute	0.00508	meters/second	atmospheres	1033	centimeters of water
feet/minute	0.01136	miles/hour	atmospheres	29.92	inches of Hg
feet/second	0.3048	meters/second	atmospheres	407	inches of water
kilometers/hour	54.68	feet/minute	atmospheres	14.696	psi (pounds/sq inch)
meters/second	200	feet/minute	feet of water	2.989	kilopascals
meters/second	3.28	feet/second	feet of water	0.4335	psi
meters/second	2.24	miles/hour	inches of Hg	0.03342	atmospheres
miles/hour	44.704	centimeters/second	inches of Hg	1.133	feet of water
miles/hour	88	feet/minute	inches of Hg	3,377	pascals
miles/hour	1.609	kilometers/hour	inches of Hg	0.491	psi
miles/hour	0.447	meters/second	inches of water	249.1	pascals
Flow Rate			inches of water	0.03614	psi
cfm (cubic feet/ minute)	1.7	cubic meters/hour	kilogram/sq centimeter	14.7	psi
cubic feet/second	448.8	gallons/minute	millimeters of Hg	0.01934	psi
gpm (gallons/minute)	0.0631	liters/second	psf (pounds/sq ft)	47.88	pascals
			psi	0.06805	atmospheres
liters/minute	0.0353	cfm	psi	5.1715	centimeters of Hg
liters/minute	0.2642	gpm	psi	2.307	feet of water
meter³/second	15,873	gpm	psi	2.036	inches of Hg
meter³/hour	0.59	cfm	psi	27.68	inches of water
Energy			psi	0.068	kilograms/square centimeter
Btu	1,055	joules	psi	6.895	kilopascals
Btu	1.055	kilojoules	**Temperature**		
calories	0.003968	Btu	$°F = (1.8 \times °C) + 32$		
calories	0.00116	watt-hours			
foot-pounds	0.001286	Btu	$°C = \dfrac{°F - 32}{1.8}$		
foot-pounds	1.365	joule			
horsepower-hours	2546.5	Btu	**Mass/Unit Area**		
horsepower-hours	2.685	megajoules	pounds/square foot	4.882	kilograms/square meter
kilowatt-hours	3,413	Btu	**Density**		
kilowatt-hours	3.6	megajoules	pounds/cubic foot	16.02	kilograms/cubic meter
watt-hours	3.413	Btu	**Thermal Conductivity**		
Power			Btu-in./hr-sq ft-°F	0.1442	W/m-°K
boiler horsepower (BHP)	33,479	Btuh	**Thermal Conductance** (U-Value)		
Btuh	4.2	calories/minute	Btu/hr-sq ft-°F	5.678	W/sq m-°K
Btuh	777.65	foot-pounds/hour	**Thermal Resistance**		
Btuh	0.2931	watt	hr-sq ft-°F/Btu	0.1761	sq m-°K/W
horsepower	2,546	Btuh	**Thermal Resistivity**		
horsepower	550	foot-pounds/second	hr-sq ft-°F/Btu-in.	6.933	m-°K/W
horsepower	745.7	watts	**Heat Flow**		
ton refrigeration	12,000	Btuh	Btuh/sq ft	0.00451	cal/sq cm-min
ton refrigeration	3.516	kilowatt	Btuh/sq ft	3.1546	W/sq m
watts	3.413	Btuh			
watts	1	joules/second			

TABLE A1   UNIT CONVERSIONS *(Continued)*

Multiply	By	To Obtain
langleys/min	1	cal/sq cm-min
watts/sq cm	3172	Btuh/sq ft
Heat Capacity		
Btu/°F	1.899	kilojoule/°C
Specific Enthalpy		
Btu/pound	2.326	kilojoule/kilogram
Lighting		
candela/sq inch	452	footlambert
candlepower	1	candela
candlepower	12.566	lumens
footcandles	1	lumens/sq ft
footcandles	10.764	lux
footlambert	0.00221	candela/sq in.
footlambert	1.076	millilambert
lumens	0.079577	candlepower
lux	0.0929	footcandle
millilambert	0.929	footlambert

TABLE A2   STANDARD SI UNITS

Measure	Unit	Symbol
Length	meter	m
Area	square meter	m^2
Volume	cubic meter	m^3
Mass	kilogram	kg
Time	second	s
Frequency	hertz	Hz
Speed	meter per second	m/s
Energy, work, heat	joule	J
Power	watt	W
Force	newton	N
Pressure	pascal	Pa[a]
Temperature	degrees Celsius	°C
Density	kilogram per cubic meter	kg/m^3
Voltage	volt	V
Electrical current	ampere	A
Electric resistance	ohm	Ω

[a] 1 pascal (Pa) = 1 newton per square meter (N/m^2).

TABLE B1   COOLING LOAD CHECK FIGURES

Building Type	Btuh/sq ft			Sq ft/ton		
	Low	Avg.	High	Low	Avg.	High
Apartments						
Individual unit		26			450	
Corridors		22			550	
Overall, high-rise	27	30	34	450	400	350
Auditoriums and theaters[a]	30	48	133	400	250	90
Banks		50			240	
Bars and taverns		133			90	
Bowling alleys		68			175	
Churches[b]	30	48	133	400	250	90
Cocktail lounges		68			175	
Computer rooms		141			85	
Dental offices		52			230	
Dormitories						
Rooms		40			300	
Corridors		30			400	
Overall	34	40	55	350	300	220
Factories						
Assembly areas	50	80	133	240	150	90
Light manufacturing	60	80	120	200	150	100
Heavy manufacturing[c]	120	150	200	100	80	60
Partial cooling		40			300	
Hospitals						
Patient rooms	44	55	73	275	220	165
Public areas	69	86	109	175	140	110
Hotels						
Guest rooms		44			275	
Public spaces		55			220	
Corridors		30			400	
Overall	34	40	55	350	300	220
Libraries and museums	35	43	60	340	280	200
Medical centers		28			425	
Motels		28			425	
Office buildings						
High-rise, exterior rooms		46			263	
High-rise, interior rooms		37			325	
Low-rise, exterior rooms		38			320	
Low-rise, interior rooms		33			360	
Overall	33	43	63	360	280	190
Post offices						
Individual office		42			285	
Central area		46			260	
Residences						
Large	20	24	32	600	500	380
Medium	17	22	30	700	550	400
Cluster central plant	19	24	32	625	500	375
Restaurants						
Large	89	120	150	135	100	80
Medium	80	100	120	150	120	100

TABLE B1   COOLING LOAD CHECK FIGURES *(Continued)*

Building Type	Btuh/sq ft			Sq ft/ton		
	Low	Avg.	High	Low	Avg.	High
Retail shops						
Barber shops		48		250	250	240
Beauty parlors		66		240	180	105
Dept. stores, basement	34	42	53	350	285	225
Main floor	34	49	80	350	245	150
Upper floors	30	35	43	400	340	280
Dress shops	35	43	65	345	280	185
Drugstores	66	89	109	180	135	110
Five-and-ten-cent stores	35	55	100	345	220	120
Hat shops	38	44	65	315	270	185
Shoe stores	40	55	80	300	220	150
Specialty shops		60			200	
Supermarkets[d]		34			350	
Malls	33	52	75	365	230	160
Shopping centers (central plant)	36	45	60	330	265	200
Schools, colleges, and universities						
Individual buildings	46	65	80	260	185	150
Central plant	30	38	50	400	320	240
Urban district central plant	25	32	42	475	380	285

[a] 666 Btuh/person; 18 persons/ton.

[b] 600 Btuh/person; 20 persons/ton.

[c] Based on exhaust fans or other supplementary means to remove excessive heat.

[d] Does not include food refrigeration.

TABLE B2 CLIMATIC CONDITIONS FOR THE UNITED STATES

Col. 1	Col. 2		Col. 3		Col. 4	Col. 5 Winter,[b] °F		Col. 6 Summer,[c] °F			Col. 7	Col. 8			Col. 9 Prevailing Wind			Col. 10 Temp. °F	
	Lat.		Long.		Elev.	Design Dry-Bulb		Design Dry-Bulb and Coincident Wet-Bulb			Mean Daily	Design Wet-Bulb			Winter		Summer	Median of Annual Extr.	
State and Station[a]	°	'	°	'	Feet	99%	97.5%	1%	2.5%	5%	Range	1%	2.5%	5%	Knots[d]			Max.	Min.
ALABAMA																			
Alexander City	32	57	85	57	660	18	22	96/77	93/76	91/76	21	79	78	78					
Anniston AP	33	35	85	51	599	18	22	97/77	94/76	92/76	21	79	78	78	SW	5	SW	98.4	12.4
Auburn	32	36	85	30	652	18	22	96/77	93/76	91/76	21	79	78	78				99.8	14.6
Birmingham AP	33	34	86	45	620	17	21	96/74	94/75	92/74	21	78	77	76	NNW	8	WNW	98.5	12.9
Decatur	34	37	86	59	580	11	16	95/75	93/74	91/74	22	78	77	76					
Dothan AP	31	19	85	27	374	23	27	94/76	92/76	91/76	20	80	79	78					
Florence AP	34	48	87	40	581	17	21	97/74	94/74	92/74	22	78	77	76	NW	7	NW		
Gadsden	34	01	86	00	554	16	20	96/75	94/75	92/74	22	78	77	76	NNW	8	WNW		
Huntsville AP	34	42	86	35	606	11	16	95/75	93/74	91/74	23	78	77	76	N	9	SW		
Mobile AP	30	41	88	15	211	25	29	95/77	93/77	91/76	18	80	79	78	N	10	N		
Mobile Co	30	40	88	15	211	25	29	95/77	93/77	91/76	16	80	79	78				97.9	22.3
Montgomery AP	32	23	86	22	169	22	25	96/76	95/76	93/76	21	79	79	78	NW	7	W	98.9	18.2
Selma-Craig AFB	32	20	87	59	166	22	26	97/78	95/77	93/77	2ᒪ	81	80	79	N	9	SW	100.1	17.6
Talladega	33	27	86	06	565	18	22	97/77	94/76	92/76	21	79	78	78				99.6	11.2
Tuscaloosa AP	33	13	87	37	169	20	23	98/75	96/76	94/76	22	79	78	77	N	5	WNW		
ALASKA																			
Anchorage AP	61	10	150	01	114	−23	−18	71/59	68/58	66/56	15	60	59	57	SE	3	WNW		
Barrow (S)	71	18	156	47	31	−45	−41	57/53	53/50	49/47	12	54	50	47	SW	8	SE		
Fairbanks AP (S)	64	49	147	52	436	−51	−47	82/62	78/60	75/59	24	64	62	60	N	5	S		
Juneau AP	58	22	134	35	12	−4	1	74/60	70/58	67/57	15	61	59	58	N	7	W		
Kodiak	57	45	152	29	73	10	13	69/58	65/56	62/55	10	60	58	56	WNW	14	NW		
Nome AP	64	30	165	26	13	−31	−27	66/57	62/55	59/54	10	58	56	55	N	4	W		
ARIZONA																			
Douglas AP	31	27	109	36	4098	27	31	98/63	95/63	93/63	31	70	69	68				104.4	14.0
Flagstaff AP	35	08	111	40	7006	−2	4	84/55	82/55	80/54	31	61	60	59	NE	5	SW	90.0	−11.6
Fort Huachuca AP (S)	31	35	110	20	4664	24	28	95/62	92/62	90/62	27	69	68	67	SW	5	W		
Kingman AP	35	12	114	01	3539	18	25	103/65	100/64	97/64	30	70	69	69					
Nogales	31	21	110	55	3800	28	32	99/64	96/64	94/64	31	71	70	69	SW	5	W		
Phoenix AP (S)	33	26	112	01	1112	31	34	109/71	107/71	105/71	27	76	75	75	E	4	W	112.8	26.7
Prescott AP	34	39	112	26	5010	4	9	96/61	94/60	92/60	30	66	65	64					
Tucson AP (S)	32	07	110	56	2558	28	32	104/66	102/66	100/66	26	72	71	71	SE	6	WNW	108.9	.3
Winslow AP	35	01	110	44	4895	5	10	97/61	95/60	93/60	32	66	65	64	SW	6	WSW	102.7	−.4
Yuma AP	32	39	114	37	213	36	39	111/72	109/72	107/71	27	79	78	77	NNE	6	WSW	114.8	30.8
ARKANSAS																			
Blytheville AFB	35	57	89	57	264	10	15	96/78	94/77	91/76	21	81	80	78	N	8	SSW		
Camden	33	36	92	49	116	18	23	98/76	96/76	94/76	21	80	79	78					
El Dorado AP	33	13	92	49	277	18	23	98/76	96/76	94/76	21	80	79	78	S	6	SE	101.0	13.9
Fayetteville AP	36	00	94	10	1251	7	12	97/72	94/73	92/73	23	77	76	75	NE	9	SSW	99.4	−.3
Fort Smith AP	35	20	94	22	463	12	17	101/75	98/76	95/76	24	80	79	78	NW	8	SW	101.9	7.0
Hot Springs	34	29	93	06	535	17	23	101/77	97/77	94/77	22	80	79	78	N	8	SW	103.0	10.6
Jonesboro	35	50	90	42	345	10	15	96/78	94/77	91/76	21	81	80	78				101.7	7.3
Little Rock AP (S)	34	44	92	14	257	15	20	99/76	96/77	94/77	22	80	79	78	N	9	SSW	99.0	11.2
Pine Bluff AP	34	18	92	05	241	16	22	100/78	97/77	95/78	22	81	80	80	N	7	SW	102.2	13.1
Texarkana AP	33	27	93	59	389	18	23	98/76	96/77	93/76	21	80	79	78	WNW	9	SSW	104.8	14.0
CALIFORNIA																			
Bakersfield AP	35	25	119	03	475	30	32	104/70	101/69	98/68	32	73	71	70	ENE	5	WNW	109.8	25.3
Barstow AP	34	51	116	47	1927	26	29	106/68	104/68	102/67	37	73	71	70	WNW	7	W	110.4	17.4
Blythe AP	33	37	114	43	395	30	33	112/71	110/71	108/70	28	75	75	74				116.8	24.1
Burbank AP	34	12	118	21	775	37	39	95/68	91/68	88/67	25	71	70	69	NW	3	S		
Chico	39	48	121	51	238	28	30	103/69	100/68	98/67	36	71	70	68	NW	5	SSE	109.0	22.6
Concord	37	58	121	59	200	24	27	100/69	97/68	94/67	32	71	70	68	WNW	5	NW		
Covina	34	05	117	52	575	32	35	98/69	95/68	92/67	31	73	71	70					
Crescent City AP	41	46	124	12	40	31	33	68/60	65/59	63/58	18	62	60	59					

[a] AP or AFB following the station name designates airport or Air Force base temperature observations. Co designates office locations within an urban area that are affected by the surrounding area. Undesignated stations are semirural and may be compared to airport data.

[b] Winter design data are based on the 3-month period December through February.
[c] Summer design data are based on the 4-month period June through September.
[d] Mean wind speeds occurring coincidentally with the 99.5% dry-bulb winter design temperature.

TABLE B2 CLIMATIC CONDITIONS FOR THE UNITED STATES (*Continued*)

Col. 1	Col. 2		Col. 3		Col. 4	Col. 5 Winter,[b] °F		Col. 6 Summer,[c] °F			Col. 7	Col. 8			Col. 9 Prevailing Wind				Col. 10 Temp. °F	
	Lat.		Long.		Elev.	Design Dry-Bulb		Design Dry-Bulb and Coincident Wet-Bulb			Mean Daily	Design Wet-Bulb			Winter		Summer		Median of Annual Extr.	
State and Station[a]	°	′	°	′	Feet	99%	97.5%	1% Mean	2.5%	5%	Range	1%	2.5%	5%	Knots[d]				Max.	Min.
Downey	33	56	118	08	116	37	40	93/70	89/70	86/69	22	72	71	70						
El Cajon	32	49	116	58	367	42	44	83/69	80/69	78/68	30	71	70	68						
El Centro AP (S)	32	49	115	40	−43	35	38	112/74	110/74	108/74	34	81	80	78	W	6	SE			
Escondido	33	07	117	05	660	39	41	89/68	85/68	82/68	30	71	70	69						
Eureka/Arcata AP	40	59	124	06	218	31	33	68/60	65/59	63/58	11	62	60	59	E	5	NW		75.8	29.7
Fairfield-Travis AFB	38	16	121	56	62	29	32	99/68	95/67	91/66	34	70	68	67	N	5	WSW			
Fresno AP (S)	36	46	119	43	328	28	30	102/70	100/69	97/68	34	72	71	70	E	4	WNW		108.7	25.8
Hamilton AFB	38	04	122	30	3	30	32	89/68	84/66	80/65	28	72	69	67	N	4	SE			
Laguna Beach	33	33	117	47	35	41	43	83/68	80/68	77/67	18	70	69	68						
Livermore	37	42	121	57	545	24	27	100/69	97/68	93/67	24	71	70	68	WNW	4	NW			
Lompoc, Vandenberg AFB	34	43	120	34	368	35	38	75/61	70/61	67/60	20	63	61	60	ESE	5	NW			
Long Beach AP	33	49	118	09	30	41	43	83/68	80/68	77/67	22	70	69	68	NW	4	WNW			
Los Angeles AP (S)	33	56	118	24	97	41	43	83/68	80/68	77/67	15	70	69	68	E	4	WSW			
Los Angeles Co (S)	34	03	118	14	270	37	40	93/70	89/70	86/69	20	72	71	70	NW	4	NW		98.1	35.9
Merced-castle AFB	37	23	120	34	188	29	31	102/70	99/69	96/68	36	72	71	70	ESE	4	NW			
Modesto	37	39	121	00	91	28	30	101/69	98/68	95/67	36	71	70	69					105.8	26.2
Monterey	36	36	121	54	39	35	38	75/63	71/61	68/61	20	64	62	61	SE	4	NW			
Napa	38	13	122	17	56	30	32	100/69	96/68	92/67	30	71	69	68					103.1	25.8
Needles AP	34	46	114	37	913	30	33	112/71	110/71	108/70	27	75	75	74					116.4	26.7
Oakland AP	37	49	122	19	5	34	36	85/64	80/63	75/62	19	66	64	63	E	5	WNW		93.0	31.8
Oceanside	33	14	117	25	26	41	43	83/68	80/68	77/67	13	70	69	68						
Ontario	34	03	117	36	952	31	33	102/70	99/69	96/68	36	74	72	71	E	4	WSW			
Oxnard	34	12	119	11	49	34	36	83/66	80/64	77/63	19	70	68	67						
Palmdale AP	34	38	118	06	2542	18	22	103/65	101/65	98/64	35	69	67	66	SW	5	WSW			
Palm Springs	33	49	116	32	411	33	35	112/71	110/70	108/70	35	76	74	73						
Pasadena	34	09	118	09	864	32	35	98/69	95/68	92/67	29	73	71	70					102.8	30.4
Petaluma	38	14	122	38	16	26	29	94/68	90/66	87/65	31	72	70	68					102.0	24.2
Pomona Co	34	03	117	45	934	28	30	102/70	99/69	95/68	36	74	72	71	E	4	W		105.7	26.2
Redding AP	40	31	122	18	495	29	31	105/68	102/67	100/66	32	71	69	68					109.2	26.0
Redlands	34	03	117	11	1318	31	33	102/70	99/69	96/68	33	74	72	71					106.7	27.1
Richmond	37	56	122	21	55	34	36	85/64	80/63	75/62	17	66	64	63						
Riverside-March AFB (S)	33	54	117	15	1532	29	32	100/68	98/68	95/67	37	72	71	70	N	4	NW		107.6	26.6
Sacramento AP	38	31	121	30	17	30	32	101/70	98/70	94/69	36	72	71	70	NNW	6	SW		105.1	27.6
Salinas AP	36	40	121	36	75	30	32	74/61	70/60	67/59	24	62	61	59						
San Bernardino, Norton AFB	34	08	117	16	1125	31	33	102/70	99/69	96/68	38	74	72	71	E	3	W		109.3	25.3
San Diego AP	32	44	117	10	13	42	44	83/69	80/69	78/68	12	71	70	68	NE	3	WNW		91.2	37.4
San Fernando	34	17	118	28	965	37	39	95/68	91/68	88/67	38	71	70	69						
San Francisco AP	37	37	122	23	8	35	38	82/64	77/63	73/62	20	65	64	62	S	5	NW			
San Francisco Co	37	46	122	26	72	38	40	74/63	71/62	69/61	14	64	62	61	W	5	W		91.3	35.9
San Jose AP	37	22	121	56	56	34	36	85/66	81/65	77/64	26	68	67	65	SE	4	NNW		98.6	28.2
San Luis Obispo	35	20	120	43	250	33	35	92/69	88/70	84/69	26	73	71	70	E	4	W		99.8	29.3
Santa Ana AP	33	45	117	52	115	37	39	89/69	85/68	82/68	28	71	70	69	E	3	SW		101.0	29.9
Santa Barbara MAP	34	26	119	50	10	34	36	81/67	77/66	75/65	24	68	67	66	NE	3	SW		97.1	31.7
Santa Cruz	36	59	122	01	125	35	38	75/63	71/61	68/61	28	64	62	61					97.5	26.8
Santa Maria AP (S)	34	54	120	27	236	31	33	81/64	76/63	73/62	23	65	64	63	E	4	WNW			
Santa Monica Co	34	01	118	29	64	41	43	83/68	80/68	77/67	16	70	69	68						
Santa Paula	34	21	119	05	263	33	35	90/68	86/67	84/66	36	71	69	68						
Santa Rosa	38	31	122	49	125	27	29	99/68	95/67	91/66	34	70	68	67	N	5	SE		102.5	23.4
Stockton AP	37	54	121	15	22	28	30	100/69	97/68	94/67	37	71	70	68	WNW	4	NW		104.1	24.5
Ukiah	39	09	123	12	623	27	29	99/69	95/68	91/67	40	70	68	67					108.1	21.6
Visalia	36	20	119	18	325	28	30	102/70	100/69	97/68	38	72	71	70					108.4	25.1
Yreka	41	43	122	38	2625	13	17	95/65	92/64	89/63	38	67	65	64					102.8	7.1
Yuba City	39	08	121	36	80	29	31	104/68	101/67	99/66	36	71	69	68						
COLORADO																				
Alamosa AP	37	27	105	52	7537	−21	−16	84/57	82/57	80/57	35	62	61	60						
Boulder	40	00	105	16	5445	2	8	93/59	91/59	89/59	27	64	63	62					96.0	−8.4
Colorado Springs AP	38	49	104	43	6145	−3	2	91/58	88/57	86/57	30	63	62	61	N	9	S		92.3	−12.1
Denver AP	39	45	104	52	5283	−5	1	93/59	91/59	89/59	28	64	63	62	S	8	SE		96.8	−10.4
Durango	37	17	107	53	6550	−1	4	89/59	87/59	85/59	30	64	63	62					92.4	−11.2
Fort Collins	40	35	105	05	4999	−10	−4	93/59	91/59	89/59	28	64	63	62					95.2	−18.1
Grand Junction AP (S)	39	07	108	32	4843	2	7	96/59	94/59	92/59	29	64	63	62	ESE	5	WNW		99.9	−3.4
Greeley	40	26	104	38	4648	−11	−5	96/60	94/60	92/60	29	65	64	63						
Lajunta AP	38	03	103	30	4160	−3	3	100/68	98/68	95/67	31	72	70	69	W	8	S			
Leadville	39	15	106	18	10155	−8	−4	84/52	81/51	78/50	30	56	55	54					79.7	−17.8
Pueblo AP	38	18	104	29	4641	−7	0	97/61	95/61	92/61	31	67	66	65	W	5	SE		100.5	−12.2
Sterling	40	37	103	12	3939	−7	−2	95/62	93/62	90/62	30	67	66	65					100.3	−15.4
Trinidad AP	37	15	104	20	5740	−2	3	93/61	91/61	89/61	32	66	65	64	W	7	WSW		96.8	−10.5

TABLE B2   CLIMATIC CONDITIONS FOR THE UNITED STATES (*Continued*)

Col. 1	Col. 2 Lat.		Col. 3 Long.		Col. 4 Elev.	Col. 5 Winter,[b] °F Design Dry-Bulb		Col. 6 Summer,[c] °F Design Dry-Bulb and Mean Coincident Wet-Bulb			Col. 7 Mean Daily Range	Col. 8 Design Wet-Bulb			Col. 9 Prevailing Wind		Col. 10 Temp. °F Median of Annual Extr.	
State and Station[a]	°	′	°	′	Feet	99%	97.5%	1%	2.5%	5%	Range	1%	2.5%	5%	Winter (Knots[d])	Summer	Max.	Min.
CONNECTICUT																		
Bridgeport AP	41	11	73	11	25	6	9	86/73	84/71	81/70	18	75	74	73	NNW 13	WSW		
Hartford, Brainard Field	41	44	72	39	19	3	7	91/74	88/73	85/72	22	77	75	74	N 5	SSW	95.7	−4.4
New Haven AP	41	19	73	55	6	3	7	88/75	84/73	82/72	17	76	75	74	NNE 7	SW	93.0	−.2
New London	41	21	72	06	59	5	9	88/73	85/72	83/71	16	76	75	74				
Norwalk	41	07	73	25	37	6	9	86/73	84/71	81/70	19	75	74	73				
Norwich	41	32	72	04	20	3	7	89/75	86/73	83/72	18	76	75	74				
Waterbury	41	35	73	04	843	−4	2	88/83	85/71	82/70	21	75	74	72	N 8	SW		
Windsor Locks, Bradley Fld	41	56	72	41	169	0	4	91/74	88/72	85/71	22	76	75	73	N 8	SW		
DELAWARE																		
Dover AFB	39	08	75	28	28	11	15	92/75	90/75	87/74	18	79	77	76	W 9	SW	97.0	7.0
Wilmington AP	39	40	75	36	74	10	14	92/74	89/74	87/73	20	77	76	75	WNW 9	WSW	95.4	4.9
DISTRICT OF COLUMBIA																		
Andrews AFB	38	5	76	5	279	10	14	92/75	90/74	87/73	18	78	76	75				
Washington, National AP	38	51	77	02	14	14	17	93/75	91/74	89/74	18	78	77	76	WNW 11	S	97.6	7.4
FLORIDA																		
Belle Glade	26	39	80	39	16	41	44	92/76	91/76	89/76	16	79	78	78			94.7	30.9
Cape Kennedy AP	28	29	80	34	16	35	38	90/78	88/78	87/78	15	80	79	79				
Daytona Beach AP	29	11	81	03	31	32	35	92/78	90/77	88/77	15	80	79	78	NW 8			
E Fort Lauderdale	26	04	80	09	10	42	46	92/78	91/78	90/78	15	80	79	79	NW 9	ESE		
Fort Myers AP	26	35	81	52	15	41	44	93/78	92/78	91/77	18	80	79	79	NNE 7	W	94.9	34.9
Fort Pierce	27	28	80	21	25	38	42	91/78	90/78	89/78	15	80	79	79			96.1	34.0
Gainesville AP (S)	29	41	82	16	152	28	31	95/77	93/77	92/77	18	80	79	78	W 6	W	97.8	23.3
Jacksonville AP	30	30	81	42	26	29	32	96/77	94/77	92/76	19	79	79	78	NW 7	SW	97.5	25.4
Key West AP	24	33	81	45	4	55	57	90/78	90/78	89/78	09	80	79	79	NNE 12	SE	92.0	51.5
Lakeland Co (S)	28	02	81	57	214	39	41	93/76	91/76	89/76	17	79	78	78	NNW 9	SSW		
Miami AP (S)	25	48	80	16	7	44	47	91/77	90/77	89/77	15	79	79	78	NNW 8	SE	92.5	39.0
Miami Beach Co	25	47	80	17	10	45	48	90/77	89/77	88/77	10	79	79	78				
Ocala	29	11	82	08	89	31	34	95/77	93/77	92/76	18	80	79	78			98.6	24.8
Orlando AP	28	33	81	23	100	35	38	94/76	93/76	91/76	17	79	78	78	NNW 9	SSW		
Panama City, Tyndall AFB	30	04	85	35	18	29	33	92/78	90/77	89/77	14	81	80	79	N 8	WSW		
Pensacola Co	30	25	87	13	56	25	29	94/77	93/77	91/77	14	80	79	79	NNE 7	SW	96.3	23.3
St. Augustine	29	58	81	20	10	31	35	92/78	89/78	87/78	16	80	79	79	NW 7	W	97.6	25.8
St. Petersburg	27	46	82	80	35	36	40	92/77	91/77	90/76	16	79	79	78	N 8	W	94.8	35.6
Sanford	28	46	81	17	89	35	38	94/76	93/76	91/76	17	79	78	78				
Sarasota	27	23	82	33	26	39	42	93/77	92/77	90/76	17	79	79	78				
Tallahassee AP (S)	30	23	84	22	55	27	30	94/77	92/76	90/76	19	79	78	78	NW 6	NW	97.6	20.9
Tampa AP (S)	27	58	82	32	19	36	40	92/77	91/77	90/76	17	79	79	78	N 8	W	95.0	31.5
West Palm Beach AP	26	41	80	06	15	41	45	92/78	91/78	90/78	16	80	79	79	NW 9	ESE		
GEORGIA																		
Albany, Turner AFB	31	36	84	05	223	25	29	97/77	95/76	93/76	20	80	79	78	N	W	100.6	19.9
Americus	32	03	84	14	456	21	25	97/77	94/76	92/75	20	79	78	77			100.4	16.5
Athens	33	57	83	19	802	18	22	94/74	92/74	90/74	21	78	77	76	NW 9	WNW	98.7	13.5
Atlanta AP (S)	33	39	84	26	1010	17	22	94/74	92/74	90/73	19	77	76	75	NW 11	NW	95.7	11.9
Augusta AP	33	22	81	58	145	20	23	97/77	95/76	93/76	19	80	79	78	W 4	WSW	99.0	17.5
Brunswick	31	15	81	29	25	29	32	92/78	89/78	87/78	18	80	79	79			99.3	24.7
Columbus, Lawson AFB	32	31	84	56	242	21	24	95/76	93/76	91/75	21	79	78	77	NW 8	W		
Dalton	34	34	84	57	720	17	22	94/76	93/76	91/76	22	79	78	77				
Dublin	32	20	82	54	215	21	25	96/77	93/76	91/75	20	79	78	77			101.0	16.7
Gainesville	34	11	83	41	50	24	27	96/77	93/77	91/77	20	80	79	78	WNW 7	SW	98.7	21.9
Griffin	33	13	84	16	981	18	22	93/76	90/75	88/74	21	78	77	76				
LaGrange	33	01	85	04	709	19	23	94/76	91/75	89/74	21	78	77	76			12	98
Macon AP	32	42	83	39	354	21	25	96/77	93/76	91/75	22	79	78	77	NW 8	WNW	17	100
Marietta, Dobbins AFB	33	55	84	31	1068	17	21	94/74	92/74	90/74	21	78	77	76	NNW 12	NW		
Savannah	32	08	81	12	50	24	27	96/77	93/77	91/77	20	80	79	78	WNW 7	SW	22	99
Valdosta-Moody AFB	30	58	83	12	233	28	31	96/77	94/77	92/76	20	80	79	78	WNW 6	W		
Waycross	31	15	82	24	148	26	29	96/77	94/77	91/76	20	80	79	78			100.0	19.5
HAWAII																		
Hilo AP (S)	19	43	155	05	36	61	62	84/73	83/72	82/72	15	75	74	74	SW 6	NE		
Honolulu AP	21	20	157	55	13	62	63	87/73	86/73	85/72	12	76	75	74	ENE 12	ENE		
Kaneohe Bay MCAS	21	27	157	46	18	65	66	85/75	84/74	83/74	12	76	76	75	NNE 9	NE		
Wahiawa	21	03	158	02	900	58	59	86/73	85/72	84/72	14	75	74	73	WNW 5	E		

TABLE B2 CLIMATIC CONDITIONS FOR THE UNITED STATES *(Continued)*

State and Station[a]	Lat. ° '	Long. ° '	Elev. Feet	Winter,[b] °F Design Dry-Bulb 99%	97.5%	Summer,[c] °F Mean 1%	Design Dry-Bulb and Coincident Wet-Bulb 2.5%	5%	Mean Daily Range	Design Wet-Bulb 1%	2.5%	5%	Prevailing Wind Winter	Summer	Temp. °F Median of Annual Extr. Max.	Min.
IDAHO																
Boise AP (S)	43 34	116 13	2838	3	10	96/65	94/64	91/64	31	68	66	65	SE 6	NW	103.2	.6
Burley	42 32	113 46	4156	− 3	2	99/62	95/61	92/66	35	64	63	61			98.6	− 8.3
Coeur D'Alene AP	47 46	116 49	2972	− 8	− 1	89/62	86/61	83/60	31	64	63	61			99.9	− 4.5
Idaho Falls AP	43 31	112 04	4741	− 11	− 6	89/61	87/61	84/59	38	65	63	61	N 9	S	96.2	− 16.0
Lewiston AP	46 23	117 01	1413	− 1	6	96/65	93/64	90/63	32	67	66	64	W 3	WNW	105.9	2.7
Moscow	46 44	116 58	2660	− 7	0	90/63	87/62	84/61	32	65	64	62			98.0	− 5.9
Mountain Home AFB	43 02	115 54	2996	6	12	99/64	97/63	94/62	36	66	65	63	ESE 7	NW	103.2	− 6.5
Pocatello AP	42 55	112 36	4454	− 8	− 1	94/61	91/60	89/59	35	64	63	61	NE 5	W	97.9	− 11.4
Twin Falls AP (S)	42 29	114 29	4150	− 3	2	99/62	95/61	92/60	34	64	63	61	SE 6	NW	100.9	− 5.1
ILLINOIS																
Aurora	41 45	88 20	744	− 6	− 1	93/76	91/76	88/75	20	79	78	76			96.7	− 13.0
Belleville, Scott AFB	38 33	89 51	453	1	6	94/76	92/76	89/75	21	79	78	76	WNW 8	S		
Bloomington	40 29	88 57	876	− 6	− 2	92/75	90/74	88/73	21	78	76	75			98.4	− 9.6
Carbondale	37 47	89 15	417	2	7	95/77	93/77	90/76	21	80	79	77			100.9	− .8
Champaign/Urbana	40 02	88 17	777	− 3	2	95/75	92/74	90/73	21	78	77	75				
Chicago, Midway AP	41 47	87 45	607	− 5	0	94/74	91/73	88/72	20	77	75	74	NW 11	SW		
Chicago, O'Hare AP	41 59	87 54	658	− 8	− 4	91/74	89/74	86/72	20	77	76	74	WNW 9	SW		
Chicago Co	41 53	87 38	590	− 3	2	94/75	91/74	88/73	15	79	77	75			96.1	− 8.3
Danville	40 12	87 36	695	− 4	1	93/75	90/74	88/73	21	78	77	75	W 10	SSW	98.2	− 8.4
Decatur	39 50	88 52	679	− 3	2	94/75	91/74	88/73	21	78	77	75	NW 10	SW	99.0	− 8.1
Dixon	41 50	89 29	696	− 7	− 2	93/75	90/74	88/73	23	78	77	75			97.5	− 13.5
Elgin	42 02	88 16	758	− 7	− 2	91/75	88/74	86/73	21	78	77	75				
Freeport	42 18	89 37	780	− 9	− 4	91/74	89/73	87/72	24	77	76	74				
Galesburg	40 56	90 26	764	− 7	− 2	93/75	91/75	88/74	22	78	77	75	WNW 8	SW		
Greenville	38 53	89 24	563	− 1	4	94/76	92/75	89/74	21	79	78	76				
Joliet	41 31	88 10	582	− 5	0	93/75	90/74	88/73	20	78	77	75	NW 11	SW		
Kankakee	41 05	87 55	625	− 4	1	93/75	90/74	88/73	21	78	77	75				
La Salle/Peru	41 19	89 06	520	− 7	− 2	93/75	91/75	88/74	22	78	77	75				
Macomb	40 28	90 40	702	− 5	0	95/76	92/76	89/75	22	79	78	76				
Moline AP	41 27	90 31	582	− 9	− 4	93/75	91/75	88/74	23	78	77	75	WNW 8	SW	96.8	− 12.7
Mt Vernon	38 19	88 52	479	0	5	95/76	92/75	89/74	21	79	78	76			100.5	− 2.9
Peoria AP	40 40	89 41	652	− 8	− 4	91/75	89/74	87/73	22	78	76	75	WNW 8	SW	98.0	− 10.9
Quincy AP	39 57	91 12	769	− 2	3	96/76	93/76	90/76	22	80	78	77	NW 11	SSW	101.1	− 6.7
Rantoul, Chanute AFB	40 18	88 08	753	− 4	1	94/75	91/74	89/73	21	78	77	75	W 10	SSW		
Rockford	42 21	89 03	741	− 9	− 4	91/74	89/73	87/72	24	77	76	74			97.4	− 13.8
Springfield AP	39 50	89 40	588	− 3	2	94/75	92/74	89/74	21	79	77	76	NW 10	SW	98.1	− 7.2
Waukegan	42 21	87 53	700	− 6	− 3	92/76	89/74	87/73	21	78	76	75			96.5	− 10.6
INDIANA																
Anderson	40 06	85 37	919	0	6	95/76	92/75	89/74	22	79	78	76	W 9	SW	95.1	− 6.0
Bedford	38 51	86 30	670	0	5	95/76	92/75	89/74	22	79	78	76			97.5	− 4.4
Bloomington	39 08	86 37	847	0	5	95/76	92/75	89/74	22	79	78	76	W 9	SW	97.8	− 4.6
Columbus, Bakalar AFB	39 16	85 54	651	3	7	95/76	92/75	90/74	22	79	78	76	W 9	SW	98.3	− 6.4
Crawfordsville	40 03	86 54	679	− 2	3	94/75	91/74	88/73	22	79	77	76			98.4	− 7.6
Evansville AP	38 03	87 32	381	4	9	95/76	93/75	91/75	22	79	78	77	NW 9	SW	98.2	.2
Fort Wayne AP	41 00	85 12	791	− 4	1	92/73	89/72	87/72	24	77	75	74	WSW 10	SW		
Goshen AP	41 32	85 48	827	− 3	1	91/73	89/73	86/72	23	77	75	74			96.8	− 10.5
Hobart	41 32	87 15	600	− 4	2	91/73	88/73	85/72	21	77	75	74			98.5	− 8.5
Huntington	40 53	85 30	802	− 4	1	92/73	89/72	87/72	23	77	75	74			96.9	− 8.1
Indianapolis AP	39 44	86 17	792	− 2	2	92/74	90/74	87/73	22	78	76	75	WNW 10	SW	96	− 7
Jeffersonville	38 17	85 45	455	5	10	95/74	93/74	90/74	23	79	77	76			98	2
Kokomo	40 25	86 03	855	− 4	0	91/74	90/73	88/73	22	77	75	74			98.2	− 7.5
Lafayette	40 2	86 5	600	− 3	3	94/74	91/74	88/73	22	78	76	75				
La Porte	41 36	86 43	810	− 3	3	93/74	90/74	87/73	22	78	76	75			98.1	− 10.5
Marion	40 29	85 41	859	− 4	0	91/74	90/73	88/73	23	77	75	74			97.0	− 8.6
Muncie	40 11	85 21	957	− 3	2	92/74	90/73	87/73	22	76	75	74				
Peru, Grissom AFB	40 39	86 09	813	− 6	− 1	90/74	88/73	86/73	22	77	75	74	W 10	SW		
Richmond AP	39 46	84 50	1141	− 2	2	92/74	90/74	87/73	22	78	76	75			94.8	− 8.5
Shelbyville	39 31	85 47	750	− 1	3	93/74	91/74	88/73	22	78	76	75			97.7	− 6.0
South Bend AP	41 42	86 19	773	− 3	1	91/73	89/73	86/72	22	77	75	74	SW 11	SSW	96.2	− 9.2
Terre Haute AP	39 27	87 18	585	− 2	4	95/74	92/74	89/73	22	79	77	76	NNW 7	SSW	98.3	− 4.9
Valparaiso	41 31	87 02	801	− 3	3	93/74	90/74	87/73	22	78	76	75			95.5	− 11.0
Vincennes	38 41	87 32	420	1	6	95/75	92/74	90/73	22	79	77	76			100.3	− 2.8

TABLE B2   CLIMATIC CONDITIONS FOR THE UNITED STATES *(Continued)*

State and Station[a]	Lat. °	Lat. '	Long. °	Long. '	Elev. Feet	Winter 99%	Winter 97.5%	Summer Mean 1%	Summer 2.5%	Summer 5%	Mean Daily Range	Wet-Bulb 1%	Wet-Bulb 2.5%	Wet-Bulb 5%	Wind Winter	Knots	Wind Summer	Max.	Min.
IOWA																			
Ames (S)	42	02	93	48	1099	−11	− 6	93/75	90/74	87/73	23	78	76	75				97.4	−17.8
Burlington AP	40	47	91	07	692	− 7	− 3	94/74	91/75	88/73	22	78	77	75	NW	9	SSW	98.6	−11.0
Cedar Rapids AP	41	53	91	42	863	−10	− 5	91/76	88/75	86/74	23	78	77	75	NW	9	S	97.7	−15.6
Clinton	41	50	90	13	595	− 8	− 3	92/75	90/75	87/74	23	78	77	75				97.5	−13.8
Council Bluffs	41	20	95	49	1210	− 8	− 3	94/76	91/75	88/74	22	78	77	75					
Des Moines AP	41	32	93	39	938	−10	− 5	94/75	91/74	88/73	23	78	77	75	NW	11	S	98.2	−14.2
Dubuque	42	24	90	42	1056	−12	− 7	90/74	88/73	86/72	22	77	75	74	N	10	SSW	95.2	−15.0
Fort Dodge	42	33	94	11	1162	−12	− 7	91/74	88/74	86/72	23	77	75	74	NW	11	S	98.5	−19.1
Iowa City	41	38	91	33	661	−11	− 6	92/76	89/76	87/74	22	80	78	76	NW	9	SSW	97.4	−15.2
Keokuk	40	24	91	24	574	− 5	0	95/75	92/75	89/74	22	79	77	76				98.4	− 8.8
Marshalltown	42	04	92	56	898	−12	− 7	92/76	90/75	88/74	23	78	77	75				98.5	−13.4
Mason City AP	43	09	93	20	1213	−15	−11	90/74	88/74	85/72	24	77	75	74	NW	11	S	96.5	−21.7
Newton	41	41	93	02	936	−10	− 5	94/75	91/74	88/73	23	78	77	75				98.2	−14.7
Ottumwa AP	41	06	92	27	840	− 8	− 4	94/75	91/74	88/73	22	78	77	75				99.1	−12.0
Sioux City AP	42	24	96	23	1095	−11	− 7	95/74	92/74	89/73	24	78	77	75	NNW	9	S	99.9	−17.7
Waterloo	42	33	92	24	868	−15	−10	91/76	89/75	86/74	23	78	77	75	NW	9	S	97.7	−19.8
KANSAS																			
Atchison	39	34	95	07	945	− 2	2	96/77	93/76	91/76	23	81	79	77				100.5	− 8.8
Chanute AP	37	40	95	29	981	3	7	100/74	97/74	94/74	23	78	77	76	NNW	11	SSW	102.8	− 2.8
Dodge City AP (S)	37	46	99	58	2582	0	5	100/69	97/69	95/69	25	74	73	71	N	12	SSW	102.9	− 7.0
El Dorado	37	49	96	50	1282	3	7	101/72	98/73	96/73	24	77	76	75				103.5	− 5.0
Emporia	38	20	96	12	1210	1	5	100/74	97/74	94/73	25	78	77	76				102.4	− 6.4
Garden City AP	37	56	100	44	2880	− 1	4	99/69	96/69	94/69	28	74	73	71					
Goodland AP	39	22	101	42	3654	− 5	0	99/66	96/65	93/66	31	71	70	68	WSW	10	S	103.2	−10.4
Great Bend	38	21	98	52	1889	0	4	101/73	98/73	95/73	28	78	76	75					
Hutchinson AP	38	04	97	52	1542	4	8	102/72	99/72	97/72	28	77	75	74	N	14	S	105.3	− 6.1
Liberal	37	03	100	58	2870	2	7	99/68	96/68	94/68	28	73	72	71				105.8	− 3.8
Manhattan, Ft Riley (S)	39	03	96	46	1065	− 1	3	99/75	95/75	92/74	24	78	77	76	NNE	8	S	104.5	− 8.6
Parsons	37	20	95	31	899	5	9	100/74	97/74	94/74	23	79	77	76	NNW	11	SSW		
Russell AP	38	52	98	49	1866	0	4	101/73	98/73	95/73	29	78	76	75					
Salina	38	48	97	39	1272	0	5	103/74	100/74	97/73	26	78	77	75	N	8	SSW		
Topeka AP	39	04	95	38	877	0	4	99/75	96/75	93/74	24	79	78	76	NNW	10	S	101.8	− 6.4
Wichita AP	37	39	97	25	1321	3	7	101/72	98/73	96/73	23	77	76	75	NNW	12	SSW	102.5	− 2.8
KENTUCKY																			
Ashland	38	33	82	44	546	5	10	94/76	91/74	89/73	22	78	77	75	W	6	SW	97.4	.8
Bowling Green AP	35	58	86	28	535	4	10	94/77	92/75	89/74	21	79	77	76				99.9	1.2
Corbin AP	36	57	84	06	1174	4	9	94/73	92/73	89/72	23	77	76	75					
Covington AP	39	03	84	40	869	1	6	92/73	90/72	88/72	22	77	75	74	W	9	SW		
Hopkinsville, Ft Campbell	36	40	87	29	571	4	10	94/77	92/75	89/74	21	79	77	76	N	6	W	100.1	− .4
Lexington AP (S)	38	02	84	36	966	3	8	93/73	91/73	88/72	22	77	76	75	WNW	9	SW	95.3	− .5
Louisville AP	38	11	85	44	477	5	10	95/74	93/74	90/74	23	79	77	76	NW	8	SW	97.4	1.2
Madisonville	37	19	87	29	439	5	10	96/76	93/75	90/75	22	79	78	77					
Owensboro	37	45	87	10	407	5	10	97/76	94/75	91/75	23	79	78	77	NW	9	SW	98.0	− .2
Paducah AP	37	04	88	46	413	7	12	98/76	95/75	92/75	20	79	78	77					
LOUISIANA																			
Alexandria AP	31	24	92	18	92	23	27	95/77	94/77	92/77	20	80	79	78	N	7	S	100.1	− 5.7
Baton Rouge AP	30	32	91	09	64	25	29	95/77	93/77	92/77	19	80	80	79	ENE	8	W	98.0	21.4
Bogalusa	30	47	89	52	103	24	28	95/77	93/77	92/77	19	80	80	79				99.3	20.2
Houma	29	31	90	40	13	31	35	95/78	93/78	92/77	15	81	80	79				97.2	22.5
Lafayette AP	30	12	92	00	42	26	30	95/78	94/78	92/78	18	81	80	79	N	8	SW	98.2	22.6
Lake Charles AP (S)	30	07	93	13	9	27	31	95/77	93/77	92/77	17	80	79	79	N	9	SSW	99.2	20.5
Minden	32	36	93	18	250	20	25	99/77	96/76	94/76	20	79	79	78				101.7	− 4.9
Monroe AP	32	31	92	02	79	20	25	99/77	96/76	94/76	20	79	79	78	N	9	S	101.1	− 5.9
Natchitoches	31	46	93	05	130	22	26	97/77	95/77	93/77	20	80	79	78					
New Orleans AP	29	59	90	15	4	29	33	93/78	92/78	90/77	16	81	80	79	NNE	9	SSW	96.3	27.7
Shreveport AP (S)	32	28	93	49	254	20	25	99/77	96/76	94/76	20	79	79	78	N	9	S		
MAINE																			
Augusta AP	44	19	69	48	353	− 7	− 3	88/73	85/70	82/68	22	74	72	70	NNE	10	WNW		
Bangor, Dow AFB	44	48	68	50	192	−11	− 6	86/70	83/68	80/67	22	73	71	69	WNW	7	S		
Caribou AP (S)	46	52	68	01	624	−18	−13	84/69	81/67	78/66	21	71	69	67	WSW	10	SW		
Lewiston	44	02	70	15	200	− 7	− 2	88/73	85/70	82/68	22	74	72	70				94.0	−13.7

TABLE B2 CLIMATIC CONDITIONS FOR THE UNITED STATES *(Continued)*

	Col. 2		Col. 3		Col. 4	Col. 5 Winter,[b] °F		Col. 6 Summer,[c] °F			Col. 7	Col. 8			Col. 9 Prevailing Wind			Col. 10 Temp. °F	
Col. 1	Lat.		Long.		Elev.	Design Dry-Bulb		Design Dry-Bulb and Mean Coincident Wet-Bulb			Mean Daily	Design Wet-Bulb			Winter	Summer		Median of Annual Extr.	
State and Station[a]	°	′	°	′	Feet	99%	97.5%	1%	2.5%	5%	Range	1%	2.5%	5%	Knots[d]			Max.	Min.
Millinocket AP	45	39	68	42	413	−13	−9	87/69	83/68	80/66	22	72	70	68	WNW 11	WNW		92.4	−23.0
Portland (S)	43	39	70	19	43	−6	−1	87/72	84/71	81/69	22	74	72	70	W 7	S		93.5	−9.9
Waterville	44	32	69	40	302	−8	−4	87/72	84/69	81/68	22	74	72	70					
MARYLAND																			
Baltimore AP	39	11	76	40	148	10	13	94/75	91/75	89/74	21	78	77	76	W 9	WSW			
Baltimore Co	39	20	76	25	20	14	17	92/77	89/76	87/75	17	80	78	76	WNW 9	S		97.9	7.2
Cumberland	39	37	78	46	790	6	10	92/75	89/74	87/74	22	77	76	75	WNW 10	W			
Frederick AP	39	27	77	25	313	8	12	94/76	91/75	88/74	22	78	77	76	N 9	WNW			
Hagerstown	39	42	77	44	704	8	12	94/75	91/74	89/74	22	77	76	75	WNW 10	W			
Salisbury (S)	38	20	75	30	59	12	16	93/75	91/75	88/74	18	79	77	76				96.8	7.4
MASSACHUSETTS																			
Boston AP (S)	42	22	71	02	15	6	9	91/73	88/71	85/70	16	75	74	72	WNW 16	SW		95.7	−1.2
Clinton	42	24	71	41	398	−2	2	90/72	87/71	84/69	17	75	73	72				91.7	−8.5
Fall River	41	43	71	08	190	5	9	87/72	84/71	81/69	18	74	73	72	NW 10	SW		92.1	−1.0
Framingham	42	17	71	25	170	3	6	89/72	86/71	83/69	17	74	73	71				96.0	−7.7
Gloucester	42	35	70	41	10	2	5	89/73	86/71	83/70	15	75	74	72					
Greenfield	42	3	72	4	205	−7	−2	88/72	85/71	82/69	23	74	73	71					
Lawrence	42	42	71	10	57	−6	0	90/73	87/72	84/70	22	76	74	73	NW 8	WSW		95.2	−9.0
Lowell	42	39	71	19	90	−4	1	91/73	88/72	85/70	21	76	74	73				95.1	−8.5
New Bedford	41	41	70	58	79	5	9	85/72	82/71	80/69	19	74	73	72	NW 10	SW		91.4	2.2
Pittsfield AP	42	26	73	18	1194	−8	−3	87/71	84/70	81/68	23	73	72	70	NW 12	SW			
Springfield, Westover AFB	42	12	72	32	245	−5	0	90/72	87/71	84/69	19	75	73	72	N 8	SSW		95.7	−4.7
Taunton	41	54	71	04	20	5	9	89/73	86/72	83/70	18	75	74	73				92.9	−9.8
Worcester AP	42	16	71	52	986	0	4	87/71	84/70	81/68	18	73	72	70	W 14	W			
MICHIGAN																			
Adrian	41	55	84	01	754	−1	3	91/73	88/72	85/71	23	76	75	73				97.2	−7.0
Alpena AP	45	04	83	26	610	−11	−6	89/70	85/70	83/69	27	73	72	70	W 5	SW		93.9	−14.8
Battle Creek AP	42	19	85	15	941	1	5	92/74	88/72	85/70	23	76	74	73	SW 8	SW			
Benton Harbor AP	42	08	86	26	643	1	5	91/72	88/72	85/70	20	75	74	72	SSW 8	WSW			
Detroit	42	25	83	01	619	3	6	91/73	88/72	86/71	20	76	74	73	W 11	SW		95.1	−2.6
Escanaba	45	44	87	05	607	−11	−7	87/70	83/69	80/68	17	73	71	69				88.8	−16.1
Flint AP	42	58	83	44	771	−4	1	90/73	87/72	85/71	25	76	74	72	SW 8	SW		95.3	−9.9
Grand Rapids AP	42	53	85	31	784	1	5	91/72	88/72	85/70	24	75	74	72	WNW 8	WSW		95.4	−5.6
Holland	42	42	86	06	678	2		88/72	86/71	83/70	22	75	73	72				94.1	−6.8
Jackson AP	42	16	84	28	1020	1	5	92/74	88/72	85/70	23	76	74	73				96.5	−7.8
Kalamazoo	42	17	85	36	955	1	5	92/74	88/72	85/70	23	76	74	73				95.9	−6.7
Lansing AP	42	47	84	36	873	−3	1	90/73	87/72	84/70	24	75	74	72	SW 12	W		94.6	−11.0
Marquette Co	46	34	87	24	735	−12	−8	84/70	81/69	77/66	18	72	70	68				94.5	−11.8
Mt Pleasant	43	35	84	46	796	0	4	91/73	87/72	84/71	24	76	74	72				95.4	−11.1
Muskegon AP	43	10	86	14	625	2	6	86/72	84/70	82/70	21	75	73	72	E 8	SW			
Pontiac	42	40	83	25	981	0	4	90/73	87/72	85/71	21	76	74	73				95.0	−6.8
Port Huron	42	59	82	25	586	0	4	90/73	87/72	83/71	21	76	74	73	W 8	S			
Saginaw AP	43	32	84	05	667	0	4	91/73	87/72	84/71	23	76	74	72	WSW 7	SW		96.1	−7.6
Sault Ste. Marie AP (S)	46	28	84	22	721	−12	−8	84/70	81/69	77/66	23	72	70	68	E 7	SW		89.8	−21.0
Traverse City AP	44	45	85	35	624	−3	1	89/72	86/71	83/69	22	75	73	71	SSW 9	SW		95.4	−10.7
Ypsilanti	42	14	83	32	716	1	5	92/72	89/71	86/70	22	75	74	72	SW 10	SW			
MINNESOTA																			
Albert Lea	43	39	93	21	1220	−17	−12	90/74	87/72	84/71	24	77	75	73					
Alexandria AP	45	52	95	23	1430	−22	−16	91/72	88/72	85/70	24	76	74	72				95.1	−28.0
Bemidji AP	47	31	94	56	1389	−31	−26	88/69	85/69	81/67	24	73	71	69	N 8	S		94.5	−36.9
Brainerd	46	24	94	08	1227	−20	−16	90/73	87/71	84/70	24	75	73	71					
Duluth AP	46	50	92	11	1428	−21	−16	85/70	82/68	79/66	22	72	70	68	WNW 12	WSW		90.9	−27.4
Fairbault	44	18	93	16	940	17	−12	91/74	88/72	85/71	24	77	75	73				95.8	−24.3
Fergus Falls	46	16	96	04	1210	−21	−17	91/72	88/72	85/70	24	76	74	72				96.9	−27.8
International Falls AP	48	34	93	23	1179	−29	−25	85/68	83/68	80/66	26	71	70	68	N 9	S		93.4	−36.5
Mankato	44	09	93	59	1004	−17	−12	91/72	88/72	85/70	24	77	75	73					
Minneapolis/St. Paul AP	44	53	93	13	834	−16	−12	92/75	89/73	86/71	22	77	75	73	NW 8	S		96.5	−22.0
Rochester AP	43	55	92	30	1297	−17	−12	90/74	87/72	84/71	24	77	75	73	NW 9	SSW			
St. Cloud AP (S)	45	35	94	11	1043	−15	−11	91/74	88/72	85/70	24	76	74	72					
Virginia	47	30	92	33	1435	−25	−21	85/69	83/68	80/66	23	71	70	68				92.6	−33.0
Willmar	45	07	95	05	1128	−15	−11	91/74	88/72	85/71	24	76	74	72				96.8	−24.3
Winona	44	03	91	38	652	−14	−10	91/75	88/73	85/72	24	77	75	74					

TABLE B2   CLIMATIC CONDITIONS FOR THE UNITED STATES *(Continued)*

Col. 1	Col. 2		Col. 3		Col. 4	Col. 5 Design Dry-Bulb		Col. 6 Design Dry-Bulb and Coincident Wet-Bulb			Col. 7 Mean Daily	Col. 8 Design Wet-Bulb			Col. 9 Prevailing Wind		Col. 10 Median of Annual Extr.	
	Lat.		Long.		Elev.	Winter,[b] °F		Summer,[c] °F			Range				Winter	Summer	Temp. °F	
State and Station[a]	° ′		° ′		Feet	99%	97.5%	1%	2.5%	5%		1%	2.5%	5%	Knots[d]		Max.	Min.
MISSISSIPPI																		
Biloxi, Keesler AFB	30	25	88	55	26	28	31	94/79	92/79	90/78	16	82	81	80	N 8	S	23	98
Clarksdale	34	12	90	34	178	14	19	96/77	94/77	92/76	21	80	79	78			100.9	13.2
Columbus AFB	33	39	88	27	219	15	20	95/77	93/77	91/76	22	80	79	78	N 7	W	101.6	12.7
Greenville AFB	33	29	90	59	138	15	20	95/77	93/77	91/76	21	80	79	78			99.5	14.9
Greenwood	33	30	90	05	148	15	20	95/77	93/77	91/76	21	80	79	78			100.6	15.3
Hattiesburg	31	16	89	15	148	24	27	96/78	94/77	92/77	21	81	80	79			99.9	18.2
Jackson AP	32	19	90	05	310	21	25	97/76	95/76	93/76	21	79	78	78	NNW 6	NW	99.8	16.0
Laurel	31	40	89	10	236	24	27	96/78	94/77	92/77	21	81	80	79			99.7	17.8
Mccomb AP	31	15	90	28	469	21	26	96/77	94/76	92/76	18	80	79	78				
Meridian AP	32	20	88	45	290	19	23	97/77	95/76	93/76	22	80	79	78	N 6	WSW	98.3	15.7
Natchez	31	33	91	23	195	23	27	96/78	94/78	92/77	21	81	80	79			98.4	18.4
Tupelo	34	16	88	46	361	14	19	96/77	94/77	92/76	22	80	79	78			100.7	11.8
Vicksburg Co	32	24	90	47	262	22	26	97/78	95/78	93/77	21	81	80	79			96.9	18.0
MISSOURI																		
Cape Girardeau	37	14	89	35	351	8	13	98/76	95/75	92/75	21	79	78	77				
Columbia AP (S)	38	58	92	22	778	− 1	4	97/74	94/74	91/73	22	78	77	76	WNW 9	WSW	99.5	− 6.2
Farmington AP	37	46	90	24	928	3	8	96/76	93/75	90/74	22	78	77	75			99.9	− 2.1
Hannibal	39	42	91	21	489	− 2	3	96/76	93/76	90/76	22	80	78	77	NNW 11	SSW	98.4	− 7.6
Jefferson City	38	34	92	11	640	2	7	98/75	95/74	92/74	23	78	77	76			101.2	− 6.1
Joplin AP	37	09	94	30	980	6	10	100/73	97/73	94/73	24	78	77	76	NNW 12	SSW		
Kansas City AP	39	07	94	35	791	2	6	99/75	96/74	93/74	20	78	77	76	NW 9	S	100.2	− 4.3
Kirksville AP	40	06	92	33	964	− 5	0	96/74	93/74	90/73	24	78	77	76			98.3	− 10.8
Mexico	39	11	91	54	775	− 1	4	97/74	94/74	91/73	22	78	77	76			101.2	− 8.0
Moberly	39	24	92	26	850	− 2	3	97/74	94/74	91/73	23	78	77	76				
Poplar Bluff	36	46	90	25	380	11	16	98/78	95/76	92/76	22	81	79	78				
Rolla	37	59	91	43	1204	3	9	94/77	91/75	89/74	22	78	77	76			99.4	− 3.1
St. Joseph AP	39	46	94	55	825	− 3	2	96/77	93/76	91/76	23	81	79	77	NNW 9	S	100.6	− 8.0
St. Louis AP	38	45	90	23	535	2	6	97/75	94/75	91/74	21	78	77	76	NW 9	WSW		
St. Louis Co	38	39	90	38	462	3	8	98/75	94/75	91/74	18	78	77	76	NW 6	S	99.1	− 2.7
Sikeston	36	53	89	36	325	9	15	98/77	95/76	92/75	21	80	78	77			100.0	− 5.1
Sedalia, Whiteman AFB	38	43	93	33	869	− 1	4	95/76	92/76	90/75	22	79	78	76	NNW 7	SSW		
Sikeston	36	53	89	36	325	9	15	98/77	95/76	92/75	21	80	78	77				
Springfield AP	37	14	93	23	1268	3	9	96/73	93/74	91/74	23	78	77	75	NNW 10	S	97.2	− 2.4
MONTANA																		
Billings AP	45	48	108	32	3567	− 15	− 10	94/64	91/64	88/63	31	67	66	64	NE 9	SW	100.5	− 19.1
Bozeman	45	47	111	09	4448	− 20	− 14	90/61	87/60	84/59	32	63	62	60			92.2	− 23.2
Butte AP	45	57	112	30	5553	− 24	− 17	86/58	83/56	80/56	35	60	58	57	S 5	NW	91.8	− 26.3
Cut Bank AP	48	37	112	22	3838	− 25	− 20	88/61	85/61	82/60	35	64	62	61			94.7	− 30.9
Glasgow AP (S)	48	25	106	32	2760	− 22	− 18	92/64	89/63	85/62	29	68	66	64	E 8	S		
Glendive	47	08	104	48	2476	− 18	− 13	95/66	92/64	89/62	29	69	67	65			103.3	− 29.8
Great Falls AP (S)	47	29	111	22	3662	− 21	− 15	91/60	88/60	85/59	28	64	62	60	SW 7	WSW	98.0	− 25.1
Havre	48	34	109	40	2492	− 18	− 11	94/65	90/64	87/63	33	68	66	65			99.7	− 31.3
Helena AP	46	36	112	00	3828	− 21	− 16	90/60	88/60	85/59	32	64	62	61	N 12	WNW	95.6	− 23.7
Kalispell AP	48	18	114	16	2974	− 14	− 7	91/62	87/61	84/60	34	65	63	62			94.4	− 16.8
Lewiston AP	47	04	109	27	4122	− 22	− 16	90/62	87/61	83/60	30	65	63	62	NW 9	NW	96.2	− 27.7
Livingstown AP	45	42	110	26	4618	− 20	− 14	90/61	87/60	84/59	32	63	62	60			97.2	− 21.2
Miles City AP	46	26	105	52	2634	− 20	− 15	98/66	95/66	92/65	30	70	68	67	NW 7	SE	103.6	− 27.7
Missoula AP	46	55	114	05	3190	− 13	− 6	92/62	88/61	85/60	36	65	63	62	ESE 7	NW	98.6	− 13.9
NEBRASKA																		
Beatrice	40	16	96	45	1235	− 5	− 2	99/75	95/74	92/74	24	78	77	76			103.1	− 11.3
Chadron AP	42	50	103	05	3313	− 8	− 3	97/66	94/65	91/65	30	71	69	68				
Columbus	41	28	97	20	1450	− 6	− 2	98/74	95/73	92/73	25	77	76	75				
Fremont	41	26	96	29	1200	− 6	− 2	98/75	95/74	92/74	22	78	77	76				
Grand Island AP	40	59	98	19	1860	− 8	− 3	97/72	94/71	91/71	28	75	74	73	NNW 10	S	103.3	− 14.2
Hastings	40	36	98	26	1954	− 7	− 3	97/72	94/71	91/71	27	75	74	73	NNW 10	S	103.5	− 10.7
Kearney	40	44	99	01	2132	− 9	− 4	96/71	93/70	90/70	28	74	73	72			102.9	− 13.7
Lincoln Co (S)	40	51	96	45	1180	− 5	− 2	99/75	95/74	92/74	24	78	77	76	N 8	S	102.0	− 12.4
McCook	40	12	100	38	2768	− 6	− 2	98/69	95/69	91/69	28	74	72	71				
Norfolk	41	59	97	26	1551	− 8	− 4	97/74	93/74	90/73	30	78	77	75			102.0	− 20.0
North Platte AP (S)	41	08	100	41	2779	− 8	− 4	97/69	94/69	90/69	28	74	72	71	NW 9	SSE	100.8	− 15.8
Omaha AP	41	18	95	54	977	− 8	− 3	94/76	91/75	88/74	22	78	77	75	NW 8	S	100.2	− 13.2
Scottsbluff AP	41	52	103	36	3958	− 8	− 3	95/65	92/65	90/64	31	70	68	67	NW 9	SE	101.6	− 18.9
Sidney AP	41	13	103	06	4399	− 8	− 3	95/65	92/65	90/64	31	70	68	67				

TABLE B2 CLIMATIC CONDITIONS FOR THE UNITED STATES (*Continued*)

Col. 1	Col. 2 Lat.		Col. 3 Long.		Col. 4 Elev.	Col. 5 Winter,[b] °F Design Dry-Bulb		Col. 6 Summer,[c] °F Design Dry-Bulb and Mean Coincident Wet-Bulb			Col. 7 Mean Daily Range	Col. 8 Design Wet-Bulb			Col. 9 Prevailing Wind			Col. 10 Temp. °F Median of Annual Extr.	
State and Station[a]	°	′	°	′	Feet	99%	97.5%	1%	2.5%	5%	Range	1%	2.5%	5%	Winter		Summer	Max.	Min.
NEVADA																			
Carson City	39	10	119	46	4675	4	9	94/60	91/59	89/58	42	63	61	60	SSW 3		WNW	99.2	− 5.0
Elko AP	40	50	115	47	5050	− 8	− 2	94/59	92/59	90/58	42	63	62	60	E 4		SW		
Ely AP (S)	39	17	114	51	6253	− 10	− 4	89/57	87/56	85/55	39	60	59	58	S 9		SSW		
Las Vegas AP (S)	36	05	115	10	2178	25	28	108/66	106/65	104/65	30	71	70	69	ENE 7		SW		
Lovelock AP	40	04	118	33	3903	8	12	98/63	96/63	93/62	42	66	65	64				103.0	− 1.0
Reno AP (S)	39	30	119	47	4404	5	10	95/61	92/60	90/59	45	64	62	61	SSW 3		WNW		
Reno Co	39	30	119	47	4408	6	11	96/61	93/60	91/59	45	64	62	61				98.9	.2
Tonopah AP	38	04	117	05	5426	5	10	94/60	92/59	90/58	40	64	62	61	N 8		S		
Winnemucca AP	40	54	117	48	4301	− 1	3	96/60	94/60	92/60	42	64	62	61	SE 10		W	100.1	− 8.1
NEW HAMPSHIRE																			
Berlin	44	03	71	01	1110	− 14	− 9	87/71	84/69	81/68	22	73	71	70				93.2	− 24.7
Claremont	43	02	72	02	420	− 9	− 4	89/72	86/70	83/69	24	74	73	71					
Concord AP	43	12	71	30	342	− 8	− 3	90/72	87/70	84/69	26	74	73	71	NW 7		SW	94.8	− 16.0
Keene	42	55	72	17	490	− 12	− 7	90/72	87/70	83/69	24	74	73	71				94.6	− 18.9
Laconia	43	03	71	03	505	− 10	− 5	89/72	86/70	83/69	25	74	73	71					
Manchester, Grenier AFB	42	56	71	26	233	− 8	− 3	91/72	88/71	85/70	24	75	74	72	N 11		SW	93.7	− 12.6
Portsmouth, Pease AFB	43	04	70	49	101	− 2	2	89/73	85/71	83/70	22	75	74	72	W 8		W		
NEW JERSEY																			
Atlantic City Co	39	23	74	26	11	10	13	92/74	89/74	86/72	18	78	77	75	NW 11		WSW	93.0	7.5
Long Branch	40	19	74	01	15	10	13	93/74	90/73	87/72	18	78	77	75				95.9	4.3
Newark AP	40	42	74	10	7	10	14	94/74	91/73	88/72	20	77	76	75	WNW 11		WSW		
New Brunswick	40	29	74	26	125	6	10	92/74	89/73	86/72	19	77	76	75					
Paterson	40	54	74	09	100	6	10	94/74	91/73	88/72	21	77	76	75					
Phillipsburg	40	41	75	11	180	1	6	92/73	89/72	86/71	21	76	75	74				97.4	− .7
Trenton Co	40	13	74	46	56	11	14	91/75	88/74	85/73	19	78	76	75	W 9		SW	96.2	4.2
Vineland	39	29	75	00	112	8	11	91/75	89/74	86/73	19	78	76	75					
NEW MEXICO																			
Alamagordo, Holloman AFB	32	51	106	06	4093	14	19	98/64	96/64	94/64	30	69	68	67					
Albuquerque AP (S)	35	03	106	37	5311	12	16	96/61	94/61	92/61	27	66	65	64	N 7		W	98.1	5.1
Artesia	32	46	104	23	3320	13	19	103/67	100/67	97/67	30	72	71	70				105.5	3.7
Carlsbad AP	32	20	104	16	3293	13	19	103/67	100/67	97/67	28	72	71	70	N 6		SSE		
Clovis AP	34	23	103	19	4294	8	13	95/65	93/65	91/65	28	69	68	67				102.0	2.5
Farmington AP	36	44	108	14	5503	1	6	95/63	93/62	91/61	30	67	65	64	ENE 5		SW		
Gallup	35	31	108	47	6465	0	5	90/59	89/58	86/58	32	64	62	61					
Grants	35	10	107	54	6524	− 1	4	89/59	88/58	85/57	32	64	62	61					
Hobbs AP	32	45	103	13	3690	13	18	101/66	99/66	97/66	29	71	70	69					
Las Cruces	32	18	106	55	4544	15	20	99/64	96/64	94/64	30	69	68	67	SE 5		SE		
Los Alamos	35	52	106	19	7410	5	9	89/60	87/60	85/60	32	62	61	60				89.8	− 2.3
Raton AP	36	45	104	30	6373	− 4	1	91/60	89/60	87/60	34	65	64	63					
Roswell, Walker AFB	33	18	104	32	3676	13	18	100/66	98/66	96/66	33	71	70	69	N 6		SSE	103.0	2.7
Santa Fe Co	35	37	106	05	6307	6	10	90/61	88/61	86/61	28	63	62	61				90.1	− 1.2
Silver City AP	32	38	108	10	5442	5	10	95/61	94/60	91/60	30	66	64	63					
Socorro AP	34	03	106	53	4624	13	17	97/62	95/62	93/62	30	67	66	65					
Tucumcari AP	35	11	103	36	4039	8	13	99/66	97/66	95/65	28	70	69	68	NE 8		SW	102.7	1.1
NEW YORK																			
Albany AP (S)	42	45	73	48	275	− 6	− 1	91/73	88/72	85/70	23	75	74	72	WNW 8		S		
Albany Co	42	39	73	45	19	− 4	1	91/73	88/72	85/70	20	75	74	72				95.2	− 11.4
Auburn	42	54	76	32	715	− 3	2	90/73	87/71	84/70	22	75	73	72				92.4	− 9.5
Batavia	43	00	78	11	922	1	5	90/72	87/71	84/70	22	75	73	72				92.2	− 7.5
Binghamton AP	42	13	75	59	1590	− 2	1	86/71	83/69	81/68	20	73	72	70	WSW 10		WSW	92.9	− 9.3
Buffalo AP	42	56	78	44	705	2	6	88/71	85/70	83/69	21	74	73	72	W 10		SW	90.0	− 3.2
Cortland	42	36	76	11	1129	− 5	0	88/71	85/71	82/70	23	74	73	71				93.8	− 11.2
Dunkirk	42	29	79	16	692	4	9	88/73	85/72	83/71	18	75	74	72	SSW 10		WSW		
Elmira AP	42	10	76	54	955	− 4	1	89/71	86/71	83/70	24	74	73	71				96.2	− 6.7
Geneva (S)	42	45	76	54	613	− 3	2	90/73	87/71	84/70	22	75	73	72				96.1	− 6.5
Glens Falls	43	20	73	37	328	− 11	− 5	88/72	85/71	82/69	23	74	73	71	NNW 6		S		
Gloversville	43	02	74	21	760	− 8	− 2	89/72	86/71	83/69	23	75	74	72				93.2	− 14.6
Hornell	42	21	77	42	1325	− 4	0	88/71	85/70	82/69	24	74	73	72					
Ithaca (S)	42	27	76	29	928	− 5	0	88/71	85/71	82/70	24	74	73	71	W 6		SW		
Jamestown	42	07	79	14	1390	− 1	3	88/70	86/70	83/69	20	74	72	71	WSW 9		WSW		
Kingston	41	56	74	00	279	− 3	2	91/73	88/72	85/70	22	76	74	73					
Lockport	43	09	79	15	638	4	7	89/74	86/72	84/71	21	76	74	73	N 9		SW	92.2	− 4.8

TABLE B2 CLIMATIC CONDITIONS FOR THE UNITED STATES (*Continued*)

Col. 1	Lat.		Long.		Elev.	Winter,[b] °F Col. 5 Design Dry-Bulb		Summer,[c] °F Col. 6 Design Dry-Bulb and Coincident Wet-Bulb			Col. 7 Mean Daily Range	Col. 8 Design Wet-Bulb			Prevailing Wind Col. 9				Temp. °F Col. 10 Median of Annual Extr.	
State and Station[a]	°	′	°	′	Feet	99%	97.5%	1%	2.5%	5%	Range	1%	2.5%	5%	Winter		Summer		Max.	Min.
Massena AP	44	56	74	51	207	−13	−8	86/70	83/69	80/68	20	73	72	70						
Newburgh, Stewart AFB	41	30	74	06	471	−1	4	90/73	88/72	85/70	21	76	74	73	W	10	W			
NYC-Central Park (S)	40	47	73	58	157	11	15	92/74	89/73	87/72	17	76	75	74					94.9	3.8
NYC-Kennedy AP	40	39	3	47	13	12	15	90/73	87/72	84/71	16	76	75	74	WNW	4	SSW			
NYC-La Guardia AP	40	46	73	54	11	11	15	92/74	89/73	87/72	16	76	75	74	WNW	15	SW			
Niagara Falls AP	43	06	79	57	590	4	7	89/74	86/72	84/71	20	76	74	73	W	9	SW			
Olean	42	14	78	22	2119	−2	2	87/71	84/71	81/70	23	74	73	71						
Oneonta	42	31	75	04	1775	−7	−4	86/71	83/69	80/68	24	73	72	70						
Oswego Co	43	28	76	33	300	1	7	86/73	83/71	80/70	20	75	73	72	E	7	WSW		91.3	−7.4
Plattsburg AFB	44	39	73	28	235	−13	−8	86/70	83/69	80/68	22	73	72	70	NW	6	SE			
Poughkeepsie	41	38	73	55	165	0	6	92/74	89/74	86/72	21	77	75	74	NNE	6	SSW		98.1	−5.6
Rochester AP	43	07	77	40	547	1	5	91/73	88/71	85/70	22	75	73	72	WSW	11	WSW			
Rome, Griffiss AFB	43	14	75	25	514	−11	−5	88/71	85/70	83/69	22	75	73	71	NW	5	W			
Schenectady (S)	42	51	73	57	377	−4	1	90/73	87/72	84/70	22	75	74	72	WNW	8	S			
Suffolk County AFB	40	51	72	38	67	7	10	86/72	83/71	80/70	16	76	74	73	NW	9	SW			
Syracuse AP	43	07	76	07	410	−3	2	90/73	87/71	84/70	20	75	73	72	N	7	WNW		93.	−10.0
Utica	43	09	75	23	714	−12	−6	88/73	85/71	82/70	22	75	73	71	NW	12	W			
Watertown	43	59	76	01	325	−11	−6	86/73	83/71	81/70	20	75	73	72	E	7	WSW		91.7	−19.6
NORTH CAROLINA																				
Asheville AP	35	26	82	32	2140	10	14	89/73	87/72	85/71	21	75	74	72	NNW	12	NNW		91.9	5.8
Charlotte AP	35	13	80	56	736	18	22	95/74	93/74	91/74	20	77	76	76	NNW	6	SW		97.8	12.6
Durham	35	52	78	47	434	16	20	94/75	92/75	90/75	20	78	77	76					98.9	9.6
Elizabeth City AP	36	16	76	11	12	12	19	93/78	91/77	89/76	18	80	78	78	NW	8	SW			
Fayetteville, Pope AFB	35	10	79	01	218	17	20	95/76	92/76	90/75	20	79	78	77	N	6	SSW		99.1	13.1
Goldsboro, Seymour-Johnson	35	20	77	58	109	18	21	94/77	91/76	89/75	18	79	78	77	N	8	SW		99.8	13.0
Greensboro AP (S)	36	05	79	57	897	14	18	93/74	91/73	89/73	21	77	76	75	NE	7	SW		97.7	9.7
Greenville	35	37	77	25	75	18	21	93/77	91/76	89/75	19	79	78	77						
Henderson	36	22	78	25	480	12	15	95/77	92/76	90/76	20	79	78	77						
Hickory	35	45	81	23	1187	14	18	92/73	90/72	88/72	21	75	74	73					96.5	9.6
Jacksonville	34	50	77	37	95	20	24	92/78	90/78	88/77	18	80	79	78						
Lumberton	34	37	79	04	129	18	21	95/76	92/76	90/75	20	79	78	77						
New Bern AP	35	05	77	03	20	20	24	92/78	90/78	88/77	18	80	79	78					98.2	15.1
Raleigh/Durham AP (S)	35	52	78	47	434	16	20	94/75	92/75	90/75	20	78	77	76	N	7	SW		97.7	12.2
Rocky Mount	35	58	77	48	121	18	21	94/77	91/76	89/75	19	79	78	77						
Wilmington AP	34	16	77	55	28	23	26	93/79	91/78	89/77	18	81	80	79	N	8	SW		96.9	18.2
Winston-Salem AP	36	08	80	13	969	16	20	94/74	91/73	89/73	20	76	75	74	NW	8	WSW			
NORTH DAKOTA																				
Bismarck AP (S)	46	46	100	45	1647	−23	−19	95/68	91/68	88/67	27	73	71	70	WNW	7	S		100.3	−31.5
Devils Lake	48	07	98	54	1450	−25	−21	91/69	88/68	85/66	25	73	71	69					97.5	−30.4
Dickinson AP	46	48	102	48	2585	−21	−17	94/68	90/66	87/65	25	71	69	68	WNW	12	SSE		101.0	−31.3
Fargo AP	46	54	96	48	896	−22	−18	92/73	89/71	85/69	25	76	74	72	SSE	11	S		97.3	−29.7
Grand Forks AP	47	57	97	24	911	−26	−22	91/70	87/70	84/68	25	74	72	70	N	8	S		97.6	−29.0
Jamestown AP	46	55	98	41	1492	−22	−18	94/70	90/69	87/68	26	74	74	71					101.3	−27.9
Minot AP	48	25	101	21	1668	−24	−20	92/68	89/67	86/65	25	72	70	68	WSW	10	S			
Williston	48	09	103	35	1876	−25	−21	91/68	88/67	85/65	25	72	70	68					99.7	−32.9
OHIO																				
Akron-Canton AP	40	55	81	26	1208	1	6	89/72	86/71	84/70	21	75	73	72	SW	9	SW		94.4	−4.6
Ashtabula	41	51	80	48	690	4	9	88/73	85/72	83/71	18	75	74	72						
Athens	39	20	82	06	700	0	6	95/75	92/74	90/73	22	78	76	74						
Bowling Green	41	23	83	38	675	−2	2	92/73	89/73	86/71	23	76	75	73					96.7	−7.3
Cambridge	40	04	81	35	807	1	7	93/75	90/74	87/73	23	78	76	75						
Chillicothe	39	21	83	00	640	0	6	95/75	92/74	90/73	22	78	76	74	W	8	WSW		98.2	−2.1
Cincinnati Co	39	09	84	31	758	1	6	92/73	90/72	88/72	21	77	75	74	W	9	SW		97.2	−.2
Cleveland AP (S)	41	24	81	51	777	1	5	91/73	88/72	86/71	22	76	74	73	SW	12	N		94.7	−3.1
Columbus AP (S)	40	00	82	53	812	0	5	92/73	90/73	87/72	24	77	75	74	W	8	SSW		96.0	−3.4
Dayton AP	39	54	84	13	1002	−1	4	91/73	89/72	86/71	20	76	75	73	WNW	11	SW		96.6	−4.5
Defiance	41	17	84	23	700	−1	4	94/74	91/73	88/72	24	77	76	74						
Findlay AP	41	01	83	40	804	2	3	92/74	90/73	87/72	24	77	76	74					97.4	−7.4
Fremont	41	20	83	07	600	−3	1	90/73	88/73	85/71	24	76	75	73						
Hamilton	39	24	84	35	650	0	5	92/73	90/72	87/71	22	76	75	73					98.2	−2.8

TABLE B2 CLIMATIC CONDITIONS FOR THE UNITED STATES (*Continued*)

	Lat.		Long.		Elev.	Winter,[b] °F Design Dry-Bulb		Summer,[c] °F Design Dry-Bulb and Coincident Wet-Bulb			Mean Daily Range	Design Wet-Bulb			Prevailing Wind		Temp. °F Median of Annual Extr.	
								Mean							Winter	Summer		
State and Station[a]	°	′	°	′	Feet	99%	97.5%	1%	2.5%	5%	Range	1%	2.5%	5%	Knots[d]		Max.	Min.
Lancaster	39	44	82	38	860	0	5	93/74	91/73	88/72	23	77	75	74				
Lima	40	42	84	02	975	− 1	4	94/74	91/73	88/72	24	77	76	74	WNW 11	SW	96.0	− 6.5
Mansfield AP	40	49	82	31	1295	0	5	90/73	87/72	85/72	22	76	74	73	W 8	SW	93.8	− 10.7
Marion	40	36	83	10	920	0	5	93/74	91/73	88/72	23	77	76	74				
Middletown	39	31	84	25	635	0	5	92/73	90/72	87/71	22	76	75	73				
Newark	40	01	82	28	880	− 1	5	94/73	92/73	89/72	23	77	75	74	W 8	SSW	95.8	− 6.8
Norwalk	41	16	82	37	670	− 3	1	90/73	88/73	85/71	22	76	75	73			97.3	− 8.3
Portsmouth	38	45	82	55	540	5	10	95/76	92/74	89/73	22	78	77	75	W 8	SW	97.9	1.0
Sandusky Co	41	27	82	43	606	1	6	93/73	91/72	88/71	21	76	74	73			96.7	− 1.9
Springfield	39	50	83	50	1052	− 1	3	91/74	89/73	87/72	21	77	76	74	W 7	W		
Steubenville	40	23	80	38	992	1	5	89/72	86/71	84/70	22	74	73	72				
Toledo AP	41	36	83	48	669	− 3	1	90/73	88/73	85/71	25	76	75	73	WSW 8	SW	95.4	− 5.2
Warren	41	20	80	51	928	0	5	89/71	87/71	85/70	23	74	73	71				
Wooster	40	47	81	55	1020	1	6	89/72	86/71	84/70	22	75	73	72			94.0	− 7.7
Youngstown AP	41	16	80	40	1178	− 1	4	88/71	86/71	84/70	23	74	73	71	SW 10	SW		
Zanesville AP	39	57	81	54	900	1	7	93/75	90/74	87/73	23	78	76	75	W 6	WSW		
OKLAHOMA																		
Ada	34	47	96	41	1015	10	14	100/74	97/74	95/74	23	77	76	75				
Altus AFB	34	39	99	16	1378	11	16	102/73	100/73	98/73	25	77	76	75	N 10	S		
Ardmore	34	18	97	01	771	13	17	100/74	98/74	95/74	23	77	77	76				
Bartlesville	36	45	96	00	715	6	10	101/73	98/74	95/74	23	77	77	76				
Chickasha	35	03	97	55	1085	10	14	101/74	98/74	95/74	24	78	77	76				
Enid, Vance AFB	36	21	97	55	1307	9	13	103/74	100/74	97/74	24	79	77	76				
Lawton AP	34	34	98	25	1096	12	16	101/74	99/74	96/74	24	78	77	76				
McAlester	34	50	95	55	776	14	19	99/74	96/74	93/74	23	77	76	75	N 10	S		
Muskogee AP	35	40	95	22	610	10	15	101/74	98/75	95/75	23	79	78	77				
Norman	35	15	97	29	1181	9	13	99/74	96/74	94/74	24	77	76	75	N 10	S		
Oklahoma City AP (S)	35	24	97	36	1285	9	13	100/74	97/74	95/73	23	78	77	76	N 14	SSW		
Ponca City	36	44	97	06	997	5	9	100/74	97/74	94/74	24	77	76	76				
Seminole	35	14	96	40	865	11	15	99/74	96/74	94/73	23	77	76	75				
Stillwater (S)	36	10	97	05	984	8	13	100/74	96/74	93/74	24	77	76	75	N 12	SSW	103.7	1.6
Tulsa AP	36	12	95	54	650	8	13	101/74	98/75	95/75	22	79	78	77	N 11	SSW		
Woodward	36	36	99	31	2165	6	10	100/73	97/73	94/73	26	78	76	75			107.1	− 1.3
OREGON																		
Albany	44	38	123	07	230	18	22	92/67	89/66	86/65	31	69	67	66			97.5	16.6
Astoria AP (S)	46	09	123	53	8	25	29	75/65	71/62	68/61	16	65	63	62	ESE 7	NNW		
Baker AP	44	50	117	49	3372	− 1	6	92/63	89/61	86/60	30	65	63	61			97.5	− 6.8
Bend	44	04	121	19	3595	− 3	4	90/62	87/60	84/59	33	64	62	60			96.4	− 5.8
Corvallis (S)	44	30	123	17	246	18	22	92/67	89/66	86/65	31	69	67	66	N 6	N	98.5	17.1
Eugene AP	44	07	123	13	359	17	22	92/67	89/66	86/65	31	69	67	66	N 7	N		
Grants Pass	42	26	123	19	925	20	24	99/69	96/68	93/67	33	71	69	68	N 5	N	103.6	16.4
Klamath Falls AP	42	09	121	44	4092	4	9	90/61	87/60	84/59	36	63	61	60	N 4	W	96.3	.9
Medford AP (S)	42	22	122	52	1298	19	23	98/68	94/67	91/66	35	70	68	67	S 4	WMW	103.8	15.0
Pendleton AP	45	41	118	51	1482	− 2	5	97/65	93/64	90/62	29	66	65	63	NNW 6	WNW		
Portland AP	45	36	122	36	21	17	23	89/68	85/67	81/65	23	69	67	66	ESE 12	NW	96.6	18.3
Portland Co	45	32	122	40	75	18	24	90/68	86/67	82/65	21	69	67	66			97.6	20.5
Roseburg AP	43	14	123	22	525	18	23	93/67	90/66	87/65	30	69	67	66			99.6	19.5
Salem AP	44	55	123	01	196	18	23	92/68	88/66	84/65	31	69	68	66	N 6	N	98.9	15.9
The Dalles	45	36	121	12	100	13	19	93/69	89/68	85/66	28	70	68	67			105.1	7.9
PENNSYLVANIA																		
Allentown AP	40	39	75	26	387	4	9	92/73	88/72	86/72	22	76	75	73	W 11	SW		
Altoona Co	40	18	78	19	1504	0	5	90/72	87/71	84/70	23	74	73	72	WNW 11	WSW	93.7	− 5.2
Butler	40	52	79	54	1100	1	6	90/73	87/72	85/71	22	75	74	73				
Chambersburg	39	56	77	38	640	4	8	93/75	90/74	87/73	23	77	76	75			97.4	− .3
Erie AP	42	05	80	11	731	4	9	88/73	85/72	83/71	18	75	74	72	SSW 0	WSW	91.3	− 2.2
Harrisburg AP	40	12	76	46	308	7	11	94/75	91/74	88/73	21	77	76	75	NW 11	WSW	96.5	3.7
Johnstown	40	19	78	50	2284	− 3	2	86/70	83/70	80/68	23	72	71	70	WNW 8	WSW	96.4	− 1.8
Lancaster	40	07	76	18	403	4	8	93/75	90/74	87/73	22	77	76	75	NW 11	WSW		

TABLE B2   CLIMATIC CONDITIONS FOR THE UNITED STATES (*Continued*)

Col. 1	Col. 2		Col. 3		Col. 4	Col. 5 Design Dry-Bulb		Col. 6 Design Dry-Bulb and Coincident Wet-Bulb			Col. 7 Mean Daily	Col. 8 Design Wet-Bulb			Col. 9 Prevailing Wind			Col. 10 Median of Annual Extr.	
	Lat.		Long.		Elev.	Winter,°F		Summer,°F							Winter		Summer	Temp. °F	
State and Station	°	′	°	′	Feet	99%	97.5%	1%	2.5%	5%	Range	1%	2.5%	5%	Knots			Max.	Min.
Meadville	41	38	80	10	1065	0	4	88/71	85/70	83/69	21	73	72	71				93.2	−8.5
New Castle	41	01	80	22	825	2	7	91/73	88/72	86/71	23	75	74	73	WSW 10		WSW	94.7	−6.4
Philadelphia AP	39	53	75	15	5	10	14	93/75	90/74	87/72	21	77	76	75	WNW 10		WSW	96.4	5.9
Pittsburgh AP	40	30	80	13	1137	1	5	89/72	86/71	84/70	22	74	73	72	WSW 10		WSW		
Pittsburgh Co	40	27	80	00	1017	3	7	91/72	88/71	86/70	19	74	73	72				94.6	−1.1
Reading Co	40	20	75	38	266	9	13	92/73	89/72	86/72	19	76	75	73	W 11		SW	97.0	3.6
Scranton/Wilkes-Barre	41	20	75	44	930	1	5	90/72	87/71	84/70	19	74	73	72	SW 8		WSW	94.8	−2.2
State College (S)	40	48	77	52	1175	3	7	90/72	87/71	84/70	23	74	73	72	NNW 8		WSW	93.2	−3.6
Sunbury	40	53	76	46	446	2	7	92/73	89/72	86/70	22	75	74	73					
Uniontown	39	55	79	43	956	5	9	91/74	88/73	85/72	22	76	75	74				93.9	−2.5
Warren	41	51	79	08	1280	−2	4	89/71	86/71	83/70	24	74	73	72				93.3	−10.7
West Chester	39	58	75	38	450	9	13	92/75	89/74	86/72	20	77	76	75					
Williamsport AP	41	15	76	55	524	2	7	92/73	89/72	86/70	23	75	74	73	W 9		WSW	95.5	−4.6
York	39	55	76	45	390	8	12	94/75	91/74	88/73	22	77	76	75				97.0	−2.4
RHODE ISLAND																			
Newport (S)	41	30	71	20	10	5	9	88/73	85/72	82/70	16	76	75	73	WNW 10		SW		
Providence AP	41	44	71	26	51	5	9	89/73	86/72	83/70	19	75	74	73	WNW 11		SW	94.6	−.5
SOUTH CAROLINA																			
Anderson	34	30	82	43	774	19	23	94/74	92/74	90/74	21	77	76	75				99.5	13.3
Charleston AFB (S)	32	54	80	02	45	24	27	93/78	91/78	89/77	18	81	80	79	NNE 8		SW		
Charleston Co	32	54	79	58	3	25	28	94/78	92/78	90/77	13	81	80	79				97.8	21.4
Columbia AP	33	57	81	07	213	20	24	97/76	95/75	93/75	22	79	78	77	W 6		SW	100.6	16.2
Florence AP	34	11	79	43	147	22	25	94/77	92/77	90/76	21	80	79	78	N 7		SW	99.5	16.5
Georgetown	33	23	79	17	14	23	26	92/79	90/78	88/77	18	81	80	79	N 7		SSW	98.2	19.1
Greenville AP	34	54	82	13	957	18	22	93/74	91/74	89/74	21	77	76	75	NW 8		SW	97.3	12.6
Greenwood	34	10	82	07	620	18	22	95/75	93/74	91/74	21	78	77	76				99.5	14.1
Orangeburg	33	30	80	52	260	20	24	97/76	95/75	93/75	20	79	78	77				101.2	18.0
Rock Hill	34	59	80	58	470	19	23	96/75	94/74	92/74	20	78	77	76					
Spartanburg AP	34	58	82	00	823	18	22	93/74	91/74	89/74	20	77	76	75				99.5	13.9
Sumter, Shaw AFB	33	54	80	22	169	22	25	95/77	92/76	90/75	21	79	78	77	NNE 6		W	100.0	15.4
SOUTH DAKOTA																			
Aberdeen AP	45	27	98	26	1296	−19	−15	94/73	91/72	88/70	27	77	75	73	NNW 8		S	102.3	−28.1
Brookings	44	18	96	48	1637	−17	−13	95/73	92/72	89/71	25	77	75	73				97.8	−26.5
Huron AP	44	23	98	13	1281	−18	−14	96/73	93/72	90/71	28	77	75	73	NNW 8		S	101.5	−25.8
Mitchell	43	41	98	01	1346	−15	−10	96/72	93/71	90/70	28	76	75	73				103.0	−22.7
Pierre AP	44	23	100	17	1742	−15	−10	99/71	95/71	92/69	29	75	74	72	NW 11		SSE	105.7	−20.6
Rapid City AP (S)	44	03	103	04	3162	−11	−7	95/66	92/65	89/65	28	71	69	67	NNW 10		SSE	100.9	−19.0
Sioux Falls AP	43	34	96	44	1418	−15	−11	94/73	91/72	88/71	24	76	75	73	NW 8		S		
Watertown AP	44	55	97	09	1738	−19	−15	94/73	91/72	88/71	26	76	75	73				97.8	−26.5
Yankton	42	55	97	23	1302	−13	−7	94/73	91/72	88/71	25	77	76	74				100.8	−19.1
TENNESSEE																			
Athens	35	26	84	35	940	13	18	95/74	92/73	90/73	22	77	76	75					
Bristol-Tri City AP	36	29	82	24	1507	9	14	91/72	89/72	87/71	22	75	75	73	WNW 6		SW		
Chattanooga AP	35	02	85	12	665	13	18	96/75	93/74	91/74	22	78	77	76	NNW 8		WSW	97.2	9.8
Clarksville	36	33	87	22	382	6	12	95/76	93/74	90/74	21	78	77	76				99.8	3.7
Columbia	35	38	87	02	690	10	15	97/75	94/74	91/74	21	78	77	76					
Dyersburg	36	01	89	24	344	10	15	96/78	94/77	91/76	21	81	80	78					
Greenville	36	04	82	50	1319	11	16	92/73	90/72	88/72	22	76	75	74				95.6	.8
Jackson AP	35	36	88	55	423	11	16	98/76	95/75	92/75	21	79	78	77				99.2	6.6
Knoxville AP	35	49	83	59	980	13	19	94/74	92/73	90/73	21	77	76	75	NE 8		W	96.0	7.0
Memphis AP	35	03	90	00	258	13	18	98/77	95/76	93/76	21	80	79	78	N 10		SW	97.9	10.4
Murfreesboro	34	55	86	28	600	9	14	97/75	94/74	91/74	22	78	77	76				97.7	4.5
Nashville AP (S)	36	07	86	41	590	9	14	97/75	94/74	91/74	21	78	77	76	NW 8		WSW		
Tullahoma	35	23	86	05	1067	8	13	96/74	93/73	91/73	22	77	76	75	NW 9		WSW	96.7	3.7

TABLE B2 CLIMATIC CONDITIONS FOR THE UNITED STATES (*Continued*)

Col. 1	Col. 2		Col. 3		Col. 4	Col. 5 Winter,[b] °F		Col. 6 Summer,[c] °F			Col. 7	Col. 8			Col. 9		Col. 10	
	Lat.		Long.		Elev.	Design Dry-Bulb		Design Dry-Bulb and Mean Coincident Wet-Bulb			Mean Daily	Design Wet-Bulb			Prevailing Wind		Temp. °F Median of Annual Extr.	
State and Station[a]	°	′	°	′	Feet	99%	97.5%	1%	2.5%	5%	Range	1%	2.5%	5%	Winter	Summer	Max.	Min.
															Knots[d]			
TEXAS																		
Abilene AP	32	25	99	41	1784	15	20	101/71	99/71	97/71	22	75	74	74	N 12	SSE	103.6	10.4
Alice AP	27	44	98	02	180	31	34	100/78	98/77	95/77	20	82	81	79			104.9	24.8
Amarillo AP	35	14	101	42	3604	6	11	98/67	95/67	93/67	26	71	70	70	N 11	S	100.8	.9
Austin AP	30	18	97	42	597	24	28	100/74	98/74	97/74	22	78	77	77	N 11	S	101.6	19.7
Bay City	29	00	95	58	50	29	33	96/77	94/77	92/77	16	80	79	79				
Beaumont	29	57	94	01	16	27	31	95/79	93/78	91/78	19	81	80	80			99.7	23.5
Beeville	28	22	97	40	190	30	33	99/78	97/77	95/77	18	82	81	79	N 9	SSE	103.1	22.5
Big Spring AP (S)	32	18	101	27	2598	16	20	100/69	97/69	95/69	26	74	73	72			105.3	10.7
Brownsville AP (S)	25	54	97	26	19	35	39	94/77	93/77	92/77	18	80	79	79	NNW 13	SE	98.1	30.1
Brownwood	31	48	98	57	1386	18	22	101/73	99/73	96/73	22	77	76	75	N 9	S	105.3	13.0
Bryan AP	30	40	96	33	276	24	29	98/76	96/76	94/76	20	79	78	78				
Corpus Christi AP	27	46	97	30	41	31	35	95/78	94/78	92/78	19	80	80	79	N 12	SSE	97.0	27.2
Corsicana	32	05	96	28	425	20	25	100/75	98/75	96/75	21	79	78	77			104.2	15.2
Dallas AP	32	51	96	51	481	18	22	102/75	100/75	97/75	20	78	78	77	N 11	S		
Del Rio, Laughlin AFB	29	22	100	47	1081	26	31	100/73	98/73	97/73	24	79	77	76			103.8	23.0
Denton	33	12	97	06	630	17	22	101/74	99/74	97/74	22	78	77	76			104.5	11.8
Eagle Pass	28	52	100	32	884	27	32	101/73	99/73	98/73	24	78	78	77	NNW 9	ESE	107.7	22.1
El Paso AP (S)	31	48	106	24	3918	20	24	100/64	98/64	96/64	27	69	68	68	N 7	S	103.0	15.7
Fort Worth AP (S)	32	50	97	03	537	17	22	101/74	99/74	97/74	22	78	77	76	NW 11	S	103.2	13.5
Galveston AP	29	18	94	48	7	31	36	90/79	89/79	88/78	10	81	80	80	N 15	S	93.9	27.5
Greenville	33	04	96	03	535	17	22	101/74	99/74	97/74	21	78	77	76			103.6	11.7
Harlingen	26	14	97	39	35	35	39	96/77	94/77	93/77	19	80	79	79	NNW 10	SSE	102.3	29.3
Houston AP	29	58	95	21	96	27	32	96/77	94/77	92/77	18	80	79	79	NNW 11	S		
Houston Co	29	59	95	22	108	28	33	97/77	95/77	93/77	18	80	79	79			99.0	23.5
Huntsville	30	43	95	33	494	22	27	100/75	98/75	96/75	20	78	78	77			100.8	18.7
Killeen, Robert Gray AAF	31	05	97	41	850	20	25	99/73	97/73	95/73	22	77	76	75				
Lamesa	32	42	101	56	2965	13	17	99/69	96/69	94/69	26	73	72	71			105.5	8.9
Laredo AFB	27	32	99	27	512	32	36	102/73	101/73	99/73	23	78	77	77	N 8	SE		
Longview	32	28	94	44	330	19	24	99/76	97/76	95/76	20	80	79	78				
Lubbock AP	33	39	101	49	3254	10	15	98/69	96/69	94/69	26	73	72	71	NNE 10	SSE		
Lufkin AP	31	25	94	48	277	25	29	99/76	97/76	94/76	20	80	79	78	NNW 12	S		
Mcallen	26	12	98	13	122	35	39	97/77	95/77	94/77	21	80	79	79				
Midland AP (S)	31	57	102	11	2851	16	21	100/69	98/69	96/69	26	73	72	71	NE 9	SSE	103.6	10.8
Mineral Wells AP	32	47	98	04	930	17	22	101/74	99/74	97/74	22	78	77	76				
Palestine Co	31	47	95	38	600	23	27	100/76	98/76	96/76	20	79	79	78			101.2	16.3
Pampa	35	32	100	59	3250	7	12	99/67	96/67	94/67	26	71	70	70				
Pecos	31	25	103	30	2610	16	21	100/69	98/69	96/69	27	73	72	71				
Plainview	34	11	101	42	3370	8	13	98/68	96/68	94/68	26	72	71	70			102.7	3.1
Port Arthur AP	29	57	94	01	16	27	31	95/79	93/78	91/78	19	81	80	80	N 9	S	97.7	24.0
San Angelo, Goodfellow AFB	31	26	100	24	1877	18	22	101/71	99/71	97/70	24	75	74	73	NNE 10	SSE		
San Antonio AP (S)	29	32	98	28	788	25	30	99/72	97/73	96/73	19	77	76	76	N 8	SSE	101.3	21.1
Sherman, Perrin AFB	33	43	96	40	763	15	20	100/75	98/75	95/74	22	78	77	76	N 10	S	103.0	11.9
Snyder	32	43	100	55	2325	13	18	100/70	98/70	96/70	26	74	73	72				
Temple	31	06	97	21	700	22	27	100/74	99/74	97/74	22	78	77	77				
Tyler AP	32	21	95	16	530	19	24	99/76	97/76	95/76	21	80	79	78	NNE 23	S		
Vernon	34	10	99	18	1212	13	17	102/73	100/73	97/73	24	77	76	75				
Victoria AP	28	51	96	55	104	29	32	98/78	96/77	94/77	18	82	81	79			101.4	23.4
Waco AP	31	37	97	13	500	21	26	101/75	99/75	97/75	22	78′	78	77				
Wichita Falls AP	33	58	98	29	994	14	18	103/73	101/73	98/73	24	77	76	75	NNW 12	S		
UTAH																		
Cedar City AP	37	42	113	06	5617	− 2	5	93/60	91/60	89/59	32	65	63	62	SE 5	SW		
Logan	41	45	111	49	4785	− 3	2	93/62	91/61	88/60	33	65	64	63			95.5	− 7.8
Moab	38	36	109	36	3965	6	11	100/60	98/60	96/60	30	65	64	63				
Ogden AP	41	12	112	01	4455	1	5	93/63	91/61	88/61	33	66	65	64	S 6	SW	99.5	− 3.9
Price	39	37	110	50	5580	− 2	5	93/60	91/60	89/59	33	65	63	62				
Provo	40	13	111	43	4448	1	6	98/62	96/62	94/61	32	66	65	64	SE 5	SW		

TABLE B2 CLIMATIC CONDITIONS FOR THE UNITED STATES (*Continued*)

Col. 1	Col. 2 Lat.		Col. 3 Long.		Col. 4 Elev.	Col. 5 Winter,[b] °F Design Dry-Bulb		Col. 6 Summer,[c] °F Design Dry-Bulb and Coincident Wet-Bulb			Col. 7 Mean Daily	Col. 8 Design Wet-Bulb			Col. 9 Prevailing Wind			Col. 10 Temp. °F Median of Annual Extr.	
State and Station[a]	°	′	°	′	Feet	99%	97.5%	Mean 1%	2.5%	5%	Range	1%	2.5%	5%	Winter		Summer	Max.	Min.
Richfield	38	46	112	05	5270	− 2	5	93/60	91/60	89/59	34	65	63	62				98.1	− 10.5
St George Co	37	02	113	31	2900	14	21	103/65	101/65	99/64	33	70	68	67				109.3	11.1
Salt Lake City AP (S)	40	46	111	58	4220	3	8	97/62	95/62	92/61	32	66	65	64	SSE	6	N	99.4	− .1
Vernal AP	40	27	109	31	5280	− 5	0	91/61	89/60	86/59	32	64	63	62					
VERMONT																			
Barre	44	12	72	31	600	− 16	− 11	84/71	81/69	78/68	23	73	71	70					
Burlington AP (S)	44	28	73	09	332	− 12	− 7	88/72	85/70	82/69	23	74	72	71	E	7	SSW	92.4	− 16.9
Rutland	43	36	72	58	620	− 13	− 8	87/72	84/70	81/69	23	74	72	71				92.5	− 17.5
VIRGINIA																			
Charlottesville	38	02	78	31	870	14	18	94/74	91/74	88/73	23	77	76	75	NE	7	SW	97.4	8.0
Danville AP	36	34	79	20	590	14	16	94/74	92/73	90/73	21	77	76	75				100.1	9.2
Fredericksburg	38	18	77	28	100	10	14	96/76	93/75	90/74	21	78	77	76					
Harrisonburg	38	27	78	54	1370	12	16	93/72	91/72	88/71	23	75	74	73					
Lynchburg AP	37	20	79	12	916	12	16	93/74	90/74	88/73	21	77	76	75	NE	7	SW	97.2	7.6
Norfolk AP	36	54	76	12	22	20	22	93/77	91/76	89/76	18	79	78	77	NW	10	SW	97.2	15.3
Petersburg	37	11	77	31	194	14	17	95/76	92/76	90/75	20	79	78	77					
Richmond AP	37	30	77	20	164	14	17	95/76	92/76	90/75	21	79	78	77	N	6	SW	97.9	9.6
Roanoke AP	37	19	79	58	1193	12	16	93/72	91/72	88/71	23	75	74	73	NW	9	SW		
Staunton	38	16	78	54	1201	12	16	93/72	91/72	88/71	23	75	74	73	NW	9	SW	95.9	2.5
Winchester	39	12	78	10	760	6	10	93/75	90/74	88/74	21	77	76	75				97.3	3.7
WASHINGTON																			
Aberdeen	46	59	123	49	12	25	28	80/65	77/62	73/61	16	65	63	62	ESE	6	NNW	91.9	19.3
Bellingham AP	48	48	122	32	158	10	15	81/67	77/65	74/63	19	68	65	63	NNE	15	WSW	87.4	10.3
Bremerton	47	34	122	40	162	21	25	82/65	78/64	75/62	20	66	64	63	E	8	N		
Ellensburg AP	47	02	120	31	1735	2	6	94/65	91/64	87/62	34	66	65	63					
Everett, Paine AFB	47	55	122	17	596	21	25	80/65	76/64	73/62	20	67	64	63	ESE	6	NNW	84.9	15.2
Kennewick	46	13	119	08	392	5	11	99/68	96/67	92/66	30	70	68	67				103.4	2.0
Longview	46	10	122	56	12	19	24	88/68	85/67	81/65	30	69	67	66	ESE	9	NW	96.0	14.8
Moses Lake, Larson AFB	47	12	119	19	1185	1	7	97/66	94/65	90/63	32	67	66	64	N	8	SSW		
Olympia AP	46	58	122	54	215	16	22	87/66	83/65	79/64	32	67	66	64	NE	4	NE		
Port Angeles	48	07	123	26	99	24	27	72/62	69/61	67/60	18	64	62	61				83.5	19.4
Seattle-Boeing Field	47	32	122	18	23	21	26	84/68	81/66	77/65	24	69	67	65					
Seattle Co (S)	47	39	122	18	20	22	27	85/68	82/66	78/65	19	69	67	65	N	7	N	90.2	22.0
Seattle-Tacoma AP (S)	47	27	122	18	400	21	26	84/65	80/64	76/62	22	66	64	63	E	9	N	90.1	19.9
Spokane AP (S)	47	38	117	31	2357	− 6	2	93/64	90/63	87/62	28	65	64	62	NE	6	SW	98.8	− 4.9
Tacoma, McChord AFB	47	15	122	30	100	19	24	86/66	82/65	79/63	22	68	66	64	S	5	NNE	89.4	18.8
Walla Walla AP	46	06	118	17	1206	0	7	97/67	94/66	90/65	27	69	67	66	W	5	W	103.0	3.8
Wenatchee	47	25	120	19	632	7	11	99/67	96/66	92/64	32	68	67	65				101.1	1.0
Yakima AP	46	34	120	32	1052	− 2	5	96/65	93/65	89/63	36	68	66	65	W	5	NW		
WEST VIRGINIA																			
Beckley	37	47	81	07	2504	− 2	4	83/71	81/69	79/69	22	73	71	70	WNW	9	WNW		
Bluefield AP	37	18	81	13	2867	− 2	4	83/71	81/69	79/69	22	73	71	70					
Charleston AP	38	22	81	36	939	7	11	92/74	90/73	87/72	20	76	75	74	SW	8	SW	97.2	2.9
Clarksburg	39	16	80	21	977	6	10	92/74	90/73	87/72	21	76	75	74					
Elkins AP	38	53	79	51	1948	1	6	86/72	84/70	82/70	22	74	72	71	WNW	9	WNW	90.6	− 7.3
Huntington Co	38	25	82	30	565	5	10	94/76	91/74	89/73	22	78	77	75	W	6	SW	97.1	2.1
Martinsburg AP	39	24	77	59	556	6	10	93/75	90/74	88/74	21	77	76	75	WNW	10	W	99.0	1.1
Morgantown AP	39	39	79	55	1240	4	8	90/74	87/73	85/73	22	76	75	74					
Parkersburg Co	39	16	81	34	615	7	11	93/75	90/74	88/73	21	77	76	75	WSW	7	WSW	95.9	.7
Wheeling	40	07	80	42	665	1	5	89/72	86/71	84/70	21	74	73	72	WSW	10	WSW	97.5	− .6
WISCONSIN																			
Appleton	44	15	88	23	730	− 14	− 9	89/74	86/72	83/71	23	76	74	72				94.6	− 16.2
Ashland	46	34	90	58	650	− 21	− 16	85/70	82/68	79/66	23	72	70	68				94.1	− 26.8
Beloit	42	30	89	02	780	− 7	− 3	92/75	90/75	88/74	24	78	77	75					
Eau Claire AP	44	52	91	29	888	− 15	− 11	92/75	89/73	86/71	23	77	75	73					
Fond Du Lac	43	48	88	27	760	− 12	− 8	89/74	86/72	84/71	23	76	74	72				96.0	− 17.7
Green Bay AP	44	29	88	08	682	− 13	− 9	88/74	85/72	83/71	23	76	74	72	W	8	SW	94.3	− 17.9
La Crosse AP	43	52	91	15	651	− 13	− 9	91/75	88/73	85/72	22	77	75	74	NW	10	S	95.7	− 21.3
Madison AP (S)	43	08	89	20	858	− 11	− 7	91/74	88/73	85/71	22	77	75	73	NW	8	SW	93.6	− 16.8
Manitowoc	44	06	87	41	660	− 11	− 7	89/74	86/72	83/71	21	76	74	72				94.1	− 13.7

TABLE B2 CLIMATIC CONDITIONS FOR THE UNITED STATES *(Continued)*

Col. 1	Col. 2		Col. 3		Col. 4	Col. 5 Winter,[b] °F		Col. 6 Summer,[c] °F			Col. 7	Col. 8			Col. 9 Prevailing Wind		Col. 10 Temp. °F	
	Lat.		Long.		Elev.	Design Dry-Bulb		Design Dry-Bulb and Mean Coincident Wet-Bulb			Mean Daily	Design Wet-Bulb			Winter	Summer	Median of Annual Extr.	
State and Station[a]	°	′	°	′	Feet	99%	97.5%	1% Mean	2.5%	5%	Range	1%	2.5%	5%	Knots[d]		Max.	Min.
Marinette	45	06	87	38	605	−15	−11	87/73	84/71	82/70	20	75	73	71			95.9	−15.8
Milwaukee AP	42	57	87	54	672	−8	−4	90/74	87/73	84/71	21	76	74	73	WNW 10	SSW		
Racine	42	43	87	51	730	−6	−2	91/75	88/73	85/72	21	77	75	74				
Sheboygan	43	45	87	43	648	−10	−6	89/75	86/73	83/72	20	77	75	74			97.0	−12.4
Stevens Point	44	30	89	34	1079	−15	−11	92/75	89/73	86/71	23	77	75	73			95.3	−24.1
Waukesha	43	01	88	14	860	−9	−5	90/74	87/73	84/71	22	76	74	73			95.7	−14.3
Wausau AP	44	55	89	37	1196	−16	−12	91/74	88/72	85/70	23	76	74	72				
WYOMING																		
Casper AP	42	55	106	28	5338	−11	−5	92/58	90/57	87/57	31	63	61	60	NE 10	SW	97.3	−20.9
Cheyenne	41	09	104	49	6126	−9	−1	89/58	86/58	84/57	30	63	62	60	N 11	WNW	92.5	−15.9
Cody AP	44	33	109	04	4990	−19	−13	89/60	86/60	83/59	32	64	63	61			97.4	−21.9
Evanston	41	16	110	57	6780	−9	−3	86/55	84/55	82/54	32	59	58	57			89.2	−21.2
Lander AP (S)	42	49	108	44	5563	−16	−11	91/61	88/61	85/60	32	64	63	61	E 5	NW	94.9	−22.6
Laramie AP (S)	41	19	105	41	7266	−14	−6	84/56	81/56	79/55	28	61	60	59				
Newcastle	43	51	104	13	4265	−17	−12	91/64	87/63	84/63	30	69	68	66			99.4	−19.0
Rawlins	41	48	107	12	6740	−12	−4	86/57	83/57	81/56	40	62	61	60				
Rock Springs AP	41	36	109	04	6745	−9	−3	86/55	84/55	82/54	32	59	58	57	WSW 10	W		
Sheridan AP	44	46	106	58	3964	−14	−8	94/62	91/62	88/61	32	66	65	63	NW 7	N	99.8	−23.6
Torrington	42	05	104	13	4098	−14	−8	94/62	91/62	88/61	30	66	65	63			101.1	−20.7

TABLE B2 CLIMATIC CONDITIONS FOR CANADA

	Col. 2		Col. 3		Col. 4	Winter,[b] °F Col. 5		Summer,[c] °F Col. 6			Col. 7	Col. 8			Prevailing Wind Col. 9		
Col. 1	Lat.		Long.		Elev.	Design Dry-Bulb		Design Dry-Bulb and Mean Coincident Wet-Bulb			Mean Daily	Design Wet-Bulb			Winter	Summer	
State and Station[a]	°	′	°	′	Feet	99%	97.5%	1%	2.5%	5%	Range	1%	2.5%	5%	Knots[d]		
ALBERTA																	
Calgary AP	51	06	114	01	3540	−27	−23	84/63	81/61	79/60	25	65	63	62	NNW	8	SE
Edmonton AP	53	34	113	31	2219	−29	−25	85/66	82/65	79/63	23	68	66	65	E	9	SE
Grande Prairie AP	55	11	118	53	2190	−39	−33	83/64	80/63	78/61	23	66	64	62			
Jasper	52	53	118	04	3480	−31	−26	83/64	80/62	77/61	28	66	64	63			
Lethbridge AP (S)	49	38	112	48	3018	−27	−22	90/65	87/64	84/63	28	68	66	65			
McMurray AP	56	39	111	13	1216	−41	−38	86/67	82/65	79/64	26	69	67	65			
Medicine Hat AP	50	01	110	43	2365	−29	−24	93/66	90/65	87/64	28	70	68	66			
Red Deer AP	52	11	113	54	2965	−31	−26	84/65	81/64	78/62	25	67	66	64			
BRITISH COLUMBIA																	
Dawson Creek	55	44	120	11	2164	−37	−33	82/64	79/63	76/61	26	66	64	62			
Fort Nelson AP (S)	58	50	122	35	1230	−43	−40	84/64	81/63	78/62	23	67	65	64			
Kamloops Co	50	43	120	25	1133	−21	−15	94/66	91/65	88/64	29	68	66	65			
Nanaimo (S)	49	11	123	58	230	16	20	83/67	80/65	77/64	21	68	66	65			
New Westminster	49	13	122	54	50	14	18	84/68	81/67	78/66	19	69	68	66			
Penticton AP	49	28	119	36	1121	0	4	92/68	89/67	87/66	31	70	68	67			
Prince George AP (S)	53	53	122	41	2218	−33	−28	84/64	80/62	77/61	26	66	64	62	N	11	N
Prince Rupert Co	54	17	130	23	170	−2	2	64/59	63/57	61/56	12	60	58	57			
Trail	49	08	117	44	1400	−5	0	92/66	89/65	86/64	33	68	67	65			
Vancouver AP (S)	49	11	123	10	16	15	19	79/67	77/66	74/65	17	68	67	66	E	6	WNW
Victoria Co	48	25	123	19	228	20	23	77/64	73/62	70/60	16	64	62	60			
MANITOBA																	
Brandon	49	52	99	59	1200	−30	−27	89/72	86/70	83/68	25	74	72	70			
Churchill AP (S)	58	45	94	04	155	−41	−39	81/66	77/64	74/62	18	67	65	63	SE	11	S
Dauphin AP	51	06	100	03	999	−31	−28	87/71	84/70	81/68	23	74	72	70			
Flin Flon	54	46	101	51	1098	−41	−37	84/68	81/66	79/65	19	70	68	67			
Portage La Prairie AP	49	54	98	16	867	−28	−24	88/73	86/72	83/70	22	76	74	71			
The Pas AP (S)	53	58	101	06	894	−37	−33	85/68	82/67	79/66	20	71	69	68	W	8	W
Winnipeg AP (S)	49	54	97	14	786	−30	−27	89/73	86/71	84/70	22	75	73	71	W	8	S
NEW BRUNSWICK																	
Campbellton Co	48	00	66	40	25	−18	−14	85/68	82/67	79/66	21	72	70	68			
Chatham AP	47	01	65	27	112	15	−10	89/69	85/68	82/67	22	72	71	69			
Edmundston Co	47	22	68	20	500	−21	−16	87/70	83/68	80/67	21	73	71	69			
Fredericton AP (S)	45	52	66	32	74	−16	−11	89/71	85/69	82/68	23	73	71	70			
Moncton AP (S)	46	07	64	41	248	−12	−8	85/70	82/69	79/67	23	72	71	69			
Saint John AP	45	19	65	53	352	−12	−8	80/67	77/65	75/64	19	70	68	66			
NEWFOUNDLAND																	
Corner Brook	48	58	57	57	15	−5	0	76/64	73/63	71/62	17	67	66	65			
Gander AP	48	57	54	34	482	−5	−1	82/66	79/65	77/64	19	69	67	66	WNW	11	SW
Goose Bay AP (S)	53	19	60	25	144	−27	−24	85/66	81/64	77/63	19	68	66	64	N	9	SW
St John's AP (S)	47	37	52	45	463	3	7	77/66	75/65	73/64	18	69	67	66	N	20	WSW
Stephenville AP	48	32	58	33	44	−3	4	76/65	74/64	71/63	14	67	66	65	WNW	10	S
NORTHWEST TERRITORIES																	
Fort Smith AP(S)	60	01	111	58	665	−49	−45	85/66	81/64	78/63	24	68	66	65	NW	4	S
Frobisher AP (S)	63	45	68	33	68	−43	−41	66/53	63/51	59/50	14	54	52	51	NNW	9	NW
Inuvik (S)	68	18	133	29	200	−56	−53	79/62	77/60	75/59	21	64	62	61			
Resolute AP (S)	74	43	94	59	209	−50	−47	57/48	54/46	51/45	10	50	48	46			
Yellowknife AP	62	28	114	27	682	−49	−46	79/62	77/61	74/60	16	64	63	62	SSE	7	S

TABLE B2 CLIMATIC CONDITIONS FOR CANADA (*Continued*)

State and Station[a]	Col. 2 Lat. ° ′	Col. 3 Long. ° ′	Col. 4 Elev. Feet	Col. 5 Winter,[b] °F Design Dry-Bulb 99%	97.5%	Col. 6 Summer,[c] °F Design Dry-Bulb and Mean Coincident Wet-Bulb 1%	2.5%	5%	Col. 7 Mean Daily Range	Col. 8 Design Wet-Bulb 1% 2.5% 5%	Col. 9 Prevailing Wind Winter Summer Knots[d]
NOVA SCOTIA											
Amherst	45 49	64 13	65	− 11	− 6	84/69	81/68	79/67	21	72 70 68	
Halifax AP (S)	44 39	63 34	83	1	5	79/66	76/65	74/64	16	69 67 66	
Kentville (S)	45 03	64 36	40	− 3	1	85/69	83/68	80/67	22	72 71 69	
New Glasgow	45 37	62 37	317	− 9	− 5	81/69	79/68	77/67	20	72 70 69	
Sydney AP	46 10	60 03	197	− 1	3	82/69	80/68	77/66	19	71 70 68	
Truro Co	45 22	63 16	131	− 8	− 5	82/70	80/69	78/68	22	73 71 70	
Yarmouth AP	43 50	66 05	136	5	9	74/65	72/64	70/63	15	68 66 65	NW 11 S
ONTARIO											
Belleville	44 09	77 24	250	− 11	− 7	86/73	84/72	82/71	20	75 74 73	
Chatham	42 24	82 12	600	0	3	89/74	87/73	85/72	19	76 75 74	
Cornwall	45 01	74 45	210	− 13	− 9	89/73	87/72	84/71	21	75 74 72	
Hamilton	43 16	79 54	303	− 3	1	88/73	86/72	83/71	21	76 74 73	
Kapuskasing AP (S)	49 25	82 28	752	− 31	− 28	86/70	83/69	80/67	23	72 70 69	
Kenora AP	49 48	94 22	1345	− 32	− 28	84/70	82/69 ·	80/68	19	73 71 70	
Kingston	44 16	76 30	300	− 11	− 7	87/73	84/72	82/71	20	75 74 73	
Kitchener	43 26	80 30	1125	− 6	− 2	88/73	85/72	83/71	23	75 74 72	
London AP	43 02	81 09	912	− 4	0	87/74	85/73	83/72	21	76 74 73	
North Bay AP	46 22	79 25	1210	− 22	− 18	84/68	81/67	79/66	20	71 70 68	
Oshawa	43 54	78 52	370	− 6	− 3	88/73	86/72	84/71	20	75 74 73	
Ottawa AP (S)	45 19	75 40	413	− 17	− 13	90/73	87/71	84/70	21	75 73 72	
Owen Sound	44 34	80 55	597	− 6	− 2	84/71	82/70	80/69	21	73 72 70	
Peterborough	44 17	78 19	635	− 13	− 9	87/72	85/71	83/70	21	75 73 72	
St Catharines	43 11	79 14	325	− 1	3	87/73	85/72	83/71	20	76 74 73	
Sarnia	42 58	82 22	625	0	3	88/73	86/72	84/71	19	76 74 73	
Sault Ste Marie AP	46 32	84 30	675	− 7	− 13	85/71	82/69	79/68	22	73 71 70	
Sudbury AP	46 37	80 48	1121	− 22	− 19	86/69	83/67	81/66	22	72 70 68	
Thunder Bay AP	48 22	89 19	644	− 27	− 24	85/70	83/68	80/67	24	72 70 68	W 8 W
Timmins AP	48 34	81 22	965	− 33	− 29	87/69	84/68	81/66	25	72 70 68	
Toronto AP (S)	43 41	79 38	578	− 5	− 1	90/73	87/72	85/71	20	75 74 73	N 10 SW
Windsor AP	42 16	82 58	637	0	4	90/74	88/73	86/72	20	77 75 74	
PRINCE EDWARD ISLAND											
Charlottetown AP (S)	46 17	63 08	186	− 7	− 4	80/69	78/68	76/67	16	71 70 68	
Summerside AP	46 26	63 50	78	− 8	− 4	81/69	79/68	77/67	16	72 70 68	
QUEBEC											
Bagotville AP	48 20	71 00	536	− 28	− 23	87/70	83/68	80/67	21	72 70 68	
Chicoutimi	48 25	71 05	150	− 26	− 22	86/70	83/68	80/67	20	72 70 68	
Drummondville	45 53	72 29	270	− 18	− 14	88/72	85/71	82/69	21	75 73 71	
Granby	45 23	72 42	550	− 19	− 14	88/72	85/71	83/70	21	75 73 72	
Hull	45 26	75 44	200	− 18	− 14	90/72	87/71	84/70	21	75 73 72	
Megantic AP	45 35	70 52	1362	− 20	− 16	86/71	83/70	81/69	20	74 72 71	
Montreal AP (S)	45 28	73 45	98	− 16	− 10	88/73	85/72	83/71	17	75 74 72	
Quebec AP	46 48	71 23	245	− 19	− 14	87/72	84/70	81/68	20	74 72 70	
Rimouski	48 27	68 32	117	− 16	− 12	83/68	79/66	76/65	18	71 69 67	
St Jean	45 18	73 16	129	− 15	− 11	88/73	86/72	84/71	20	75 74 72	
St Jerome	45 48	74 01	556	− 17	− 13	88/72	86/71	83/70	23	75 73 72	
Sept. Iles AP (S)	50 13	66 16	190	− 26	− 21	76/63	73/61	70/60	17	67 65 63	
Shawinigan	46 34	72 43	306	− 18	− 14	86/72	84/70	82/69	21	74 72 71	
Sherbrooke Co	45 24	71 54	595	− 25	− 21	86/72	84/71	81/69	20	74 73 71	
Thetford Mines	46 04	71 19	1020	− 19	− 14	87/71	84/70	81/69	21	74 72 71	
Trois Rivieres	46 21	72 35	50	− 17	− 13	88/72	85/70	82/69	23	74 72 71	
Val D'or AP	48 03	77 47	1108	− 32	− 27	85/70	83/68	80/67	22	72 70 68	
Valleyfield	45 16	74 06	150	− 14	− 10	89/73	86/72	84/71	20	75 74 72	

TABLE B2   CLIMATIC CONDITIONS FOR CANADA _(Continued)_

Col. 1	Col. 2		Col. 3		Col. 4	Winter,[b] °F Col. 5 Design Dry-Bulb		Summer,[c] °F Col. 6 Design Dry-Bulb and			Col. 7 Mean Daily	Col. 8 Design Wet-Bulb			Prevailing Wind Col. 9		
	Lat.		Long.		Elev.			Mean Coincident		Wet-Bulb	Range				Winter		Summer
State and Station[a]	°	′	°	′	Feet	99%	97.5%	1%	2.5%	5%		1%	2.5%	5%	Knots[d]		
SASKATCHEWAN																	
Estevan AP	49	04	103	00	1884	−30	−25	92/70	89/68	86/67	26	72	70	69			
Moose Jaw AP	50	20	105	33	1857	−29	−25	93/69	89/67	86/66	27	71	69	68			
North Battleford AP	52	46	108	15	1796	−33	−30	88/67	85/66	82/65	23	69	68	66			
Prince Albert AP	53	13	105	41	1414	−42	−35	87/67	84/66	81/65	25	70	68	67			
Regina AP	50	26	104	40	1884	−33	−29	91/69	88/68	84/67	26	72	70	68			
Saskatoon AP (S)	52	10	106	41	1645	−35	−31	89/68	86/66	83/65	26	70	68	67			
Swift Current AP (S)	50	17	107	41	2677	−28	−25	93/68	90/66	87/65	25	70	69	67			
Yorkton AP	51	16	102	28	1653	−35	−30	87/69	84/68	80/66	23	72	70	68			
YUKON TERRITORY																	
Whitehorse AP (S)	60	43	135	04	2289	−46	−43	80/59	77/58	74/56	22	61	59	58	NW	5	SE

[a] AP following the station name designates airport temperature observations. Co designates office locations within an urban area that are affected by the surrounding area. Undesignated stations are semirural and may be compared to airport data.

[b] Winter design data are based on the month of January only.

[c] Summer design data are based on the month of July only. See also Boughner, C. C. Canadian Meteorological Memoirs No. 5. Meteorological Branch, Department of Transport, Toronto. (Now Environment Canada, Ontario), 1960.

[d] Mean wind speeds occurring coincidentally with the 99.5% dry-bulb winter design temperature.

TABLE B2   CLIMATIC CONDITIONS FOR OTHER COUNTRIES

Col. 1 Country and Station	Col. 2 Lat. °	′	Long. °	′	Col. 3 Elevation, ft	Winter, °F Mean of Annual Extremes	Col. 4 99%	97.5%	Summer, °F Col. 5 Design Dry-Bulb 1%	2.5%	5%	Col. 6 Mean Daily Range, °F	Col. 7 Design Wet-Bulb 1%	2.5%	5%	Prevailing Winds Winter		Summer
AFGHANISTAN																		
Kabul	34	35N	69	12E	5955	2	6	9	98	96	93	32	66	65	64	N	4	N
ALGERIA																		
Algiers	36	46N	3	03E	194	38	43	45	95	92	89	14	77	76	75			
ARGENTINA																		
Buenos Aires	34	35S	58	29W	89	27	32	34	91	89	86	22	77	76	75	SW	9	NNE
Cordoba	31	22S	64	15W	1388	21	28	32	100	96	93	27	76	75	74			
Tucuman	26	50S	65	10W	1401	24	32	36	102	99	96	23	76	75	74			
AUSTRALIA																		
Adelaide	34	56S	138	35E	140	36	38	40	98	94	91	25	72	70	68	NE	5	NW
Alice Springs	23	48S	133	53E	1795	28	34	37	104	102	100	27	75	74	72	N	6	SE
Brisbane	27	28S	153	02E	137	39	44	47	91	88	86	18	77	76	75	N	7	NNE
Darwin	12	28S	130	51E	88	60	64	66	94	93	91	16	82	81	81	E	10	WNW
Melbourne	37	49S	144	58E	114	31	35	38	95	91	86	21	71	69	68			
Perth	31	57S	115	51E	210	38	40	42	100	96	93	22	76	74	73	N	6	E
Sydney	33	52S	151	12E	138	38	40	42	89	84	80	13	74	73	72	N	8	NE
AUSTRIA																		
Vienna	48	15N	16	22E	644	−2	6	11	88	86	83	16	71	69	67	W	13	SSE
AZORES																		
Lajes (Terceira)	38	45N	27	05W	170	42	46	49	80	78	77	11	73	72	71	W	9	NW
BAHAMAS																		
Nassau	25	05N	77	21W	11	55	61	63	90	89	88	13	80	80	79			
BANGLADESH																		
Chittagong	22	21N	91	50E	87	48	52	54	93	91	89	20	82	81	81			
BELGIUM																		
Brussels	50	48N	4	21E	328	13	15	19	83	79	77	19	70	68	67	NE	8	ENE
BERMUDA																		
Kindley AFB	33	22N	64	41W	129	47	53	55	87	86	85	12	79	78	78	NW	16	S
BOLIVIA																		
La Paz	16	30S	68	09W	12001	28	31	33	71	69	68	24	58	57	56			
BRAZIL																		
Belem	1	27S	48	29W	42	67	70	71	90	89	87	19	80	79	78	SE	5	E
Belo Horizonte	19	56S	43	57W	3002	42	47	50	86	84	83	18	76	75	75			
Brasilia	15	52S	47	55W	3442	46	49	51	89	88	86	17	76	75	75	N	6	E
Curitiba	25	25S	49	17W	3114	28	34	37	86	84	82	21	75	74	74			
Fortaleza	3	46S	38	33W	89	66	69	70	91	90	89	17	79	78	78			
Porto Alegre	30	02S	51	13W	33	32	37	40	95	92	89	20	76	76	75			
Recife	8	04S	34	53W	97	67	69	70	88	87	86	10	78	77	77	S	7	ESE
Rio De Janeiro	22	55S	43	12W	201	56	58	60	94	92	90	11	80	79	78	N	5	S
Salvador	13	00S	38	30W	154	65	67	68	88	87	86	12	79	79	78			
Sao Paulo	23	33S	46	38W	2608	36	42	46	86	84	82	18	75	74	74	N	6	W
BELIZE																		
Belize	17	31N	88	11W	17	55	60	62	90	90	89	13	82	82	81			
BULGARIA																		
Sofia	42	42N	23	20E	1805	−2	3	8	89	86	84	26	71	70	69			
BURMA (Myanmar)																		
Mandalay	21	59N	96	06E	252	50	54	56	104	102	101	30	81	80	80			
Rangoon	16	47N	96	09E	18	59	62	63	100	98	95	25	83	82	82	W	6	W
CAMBODIA																		
Phnom Penh	11	33N	104	51E	36	62	66	68	98	96	94	19	83	82	82	N	4	W
CHILE																		
Punta Arenas	53	10S	70	54W	26	22	25	27	68	66	64	14	56	55	54			
Santiago	33	27S	70	42W	1706	27	32	35	90	89	88	32	71	70	69	N	4	SW
Valparaiso	33	01S	71	38W	135	39	43	46	81	79	77	16	67	66	65			
CHINA																		
Chongquing	29	33N	106	33E	755	34	37	39	99	97	95	18	81	80	79			
Shanghai	31	12N	121	26E	23	16	23	26	94	92	90	16	81	81	80	WNW	6	S
COLOMBIA																		
Baranquilla	10	59N	74	48W	44	66	70	72	95	94	93	17	83	82	82			
Bogota	4	36N	74	05W	8406	42	45	46	72	70	69	19	60	59	58	E	8	E
Cali	3	25N	76	30W	3189	53	57	58	84	82	79	15	70	69	68			
Medellin	6	13N	75	36W	4650	48	53	55	87	85	84	25	73	72	72			
Republic of the Congo																		
Brazzaville	4	15S	15	15E	1043	54	60	62	93	92	91	21	81	81	80			

TABLE B2　CLIMATIC CONDITIONS FOR OTHER COUNTRIES *(Continued)*

Col. 1 Country and Station	Lat. °	Lat. ′	Long. °	Long. ′	Col. 3 Elevation, ft	Winter, °F Mean of Annual Extremes	Col. 4 99%	Col. 4 97.5%	Col. 5 Design Dry-Bulb 1%	Col. 5 Design Dry-Bulb 2.5%	Col. 5 Design Dry-Bulb 5%	Col. 6 Mean Daily Range, °F	Col. 7 Design Wet-Bulb 1%	Col. 7 Design Wet-Bulb 2.5%	Col. 7 Design Wet-Bulb 5%	Prevailing Winds Winter (Knots)	Prevailing Winds Winter (Knots)	Prevailing Winds Summer (Knots)
CUBA																		
Guantanamo Bay	19	54N	75	09W	21	60	64	66	94	93	92	16	82	81	80	N	6	ESE
Havana	23	08N	82	21W	80	54	59	62	92	91	89	14	81	81	80	N	11	E
CZECH REPUBLIC																		
Prague	50	05N	14	25E	662	3	4	9	88	85	83	16	66	65	64			
DENMARK																		
Copenhagen	55	41N	12	33E	43	11	16	19	79	76	74	17	68	66	64	NE	11	N
DOMINICAN REPUBLIC																		
Santo Domingo	18	29N	69	54W	57	61	63	65	92	90	88	16	81	80	80	NNE	6	SE
EQUADOR																		
Guayaquil	2	10S	79	53W	20	61	64	65	92	91	89	20	80	80	79			
Quito	0	13S	78	32W	9446	30	36	39	73	72	71	32	63	62	62	N	3	N
EGYPT																		
Cairo	29	52N	31	20E	381	39	45	46	102	100	98	26	76	75	74	N	9	NNW
EL SALVADOR																		
San Salvador	13	42N	89	13W	2238	51	54	56	98	96	95	32	77	76	75	N	7	S
ETHIOPIA																		
Addis Ababa	9	02N	38	45E	7753	35	39	41	84	82	81	28	66	65	64	E	10	S
Asmara	15	17N	38	55E	7628	36	40	42	83	81	80	27	65	64	63	E	9	WNW
FINLAND																		
Helsinki	60	10N	24	57E	30	−11	−7	−1	77	74	72	14	66	65	63	E	4	S
FRANCE																		
Lyon	45	42N	4	47E	938	−1	10	14	91	89	86	23	71	70	69	N	7	W
Marseilles	43	18N	5	23E	246	23	25	28	90	87	84	22	72	71	69	SE	14	W
Nantes	47	15N	1	34W	121	17	22	26	86	83	80	21	70	69	67	NNE	6	E
Nice	43	42N	7	16E	39	31	34	37	87	85	83	15	73	72	72			
Paris	48	49N	2	29E	164	16	22	25	89	86	83	21	70	68	67	NE	7	E
Strasbourg	48	35N	7	46E	465	9	11	16	86	83	80	20	70	69	67			
FRENCH GUIANA																		
Cayenne	4	56N	52	27W	20	69	71	72	92	91	90	17	83	83	82	ENE	5	E
GERMANY																		
Berlin	52	27N	13	18E	187	6	7	12	84	81	78	19	68	67	66	E	6	E
Hamburg	53	33N	9	58E	66	10	12	16	80	76	73	13	68	66	65			
Hannover	52	24N	9	40E	561	7	16	20	82	78	75	17	68	67	65	E	8	E
Mannheim	49	34N	8	28E	359	2	8	11	87	85	82	18	71	69	68	N	5	S
Munich	48	09N	11	34E	1729	−1	5	9	86	83	80	18	68	66	64	S	4	N
GHANA																		
Accra	5	33N	0	12W	88	65	68	69	91	90	89	13	80	79	79	WSW	5	SW
GIBRALTAR																		
Gibraltar	36	09N	5	22W	11	38	42	45	92	89	86	14	76	75	74			
GREECE																		
Athens	37	58N	23	43E	351	29	33	36	96	93	91	18	72	71	71	N	9	NNE
Thessaloniki	40	37N	22	57E	78	23	28	32	95	93	91	20	77	76	75			
GREENLAND																		
Narsarssuaq	61	11N	45	25W	85	−23	−12	−8	66	63	61	20	56	54	52			
GUATEMALA																		
Guatemala City	14	37N	90	31W	4855	45	48	51	83	82	81	24	69	68	67	N	9	S
GUYANA																		
Georgetown	6	50N	58	12W	6	70	72	73	89	88	87	11	80	79	79			
HAITI																		
Port Au Prince	18	33N	72	20W	121	63	65	67	97	95	93	20	82	81	80	N	6	ESE
HONDURAS																		
Tegucigalpa	14	06N	87	13W	3094	44	47	50	89	87	85	28	73	72	71	N	8	E
HONG KONG																		
Hong Kong	22	18N	114	10E	109	43	48	50	92	91	90	10	81	80	80	N	9	W
HUNGARY																		
Budapest	47	31N	19	02E	394	8	10	14	90	86	84	21	72	71	70	N	5	S
ICELAND																		
Reykjavik	64	08N	21	56E	59	8	14	17	59	58	56	16	54	53	53	E	12	E
INDIA																		
Ahmenabad	23	02N	72	35E	163	49	53	56	109	107	105	28	80	79	78			
Bangalore	12	57N	77	37E	3021	53	56	58	96	94	93	26	75	74	74			
Bombay	18	54N	72	49E	37	62	65	67	96	94	92	13	82	81	81	NW	6	NW
Calcutta	22	32N	88	20E	21	49	52	54	98	97	96	22	83	82	82	N	4	S
Madras	13	04N	80	15E	51	61	64	66	104	102	101	19	84	83	83	W	3	W
Nagpur	21	09N	79	07E	1017	45	51	54	110	108	107	30	79	79	78			
New Delhi	28	35N	77	12E	703	35	39	41	110	107	105	26	83	82	82	N	6	NW

TABLE B2 CLIMATIC CONDITIONS FOR OTHER COUNTRIES (*Continued*)

Col. 1 Country and Station	Col. 2 Lat. °	′	Long. °	′	Col. 3 Elevation, ft	Winter, °F Mean of Annual Extremes	Col. 4 99%	97.5%	Summer, °F Col. 5 Design Dry-Bulb 1%	2.5%	5%	Col. 6 Mean Daily Range, °F	Col. 7 Design Wet-Bulb 1%	2.5%	5%	Prevailing Winds Winter		Summer
INDONESIA																		
Djakarta	6	11S	106	50E	26	69	71	72	90	89	88	14	80	79	78	N	11	N
Kupang	10	10S	123	34E	148	63	66	68	94	93	92	20	81	80	80			
Makassar	5	08S	119	28E	61	64	66	68	90	89	88	17	80	80	79			
Medan	3	35N	98	41E	77	66	69	71	92	91	90	17	81	80	79			
Palembang	3	00S	104	46E	20	67	70	71	92	91	90	17	80	79	79			
Surabaya	7	13S	112	43E	10	64	66	68	91	90	89	18	80	79	79			
IRAN																		
Abadan	30	21N	48	16E	7	32	39	41	116	113	110	32	82	81	81	W	6	WNW
Meshed	36	17N	59	36E	3104	3	10	14	99	96	93	29	68	67	66			
Tehran	35	41N	51	25E	4002	15	20	24	102	100	98	27	75	74	73	W	5	SE
IRAQ																		
Baghdad	33	20N	44	24E	111	27	32	35	113	111	108	34	73	72	72	WNW	5	WNW
Mosul	36	19N	43	09E	730	23	29	32	114	112	110	40	73	72	72			
IRELAND																		
Dublin	53	22N	6	21W	155	19	24	27	74	72	70	16	65	64	62	W	9	SW
Shannon	52	41N	8	55W	8	19	25	28	76	73	71	14	65	64	63	SE	4	W
IRIAN BARAT																		
Manokwari	0	52S	134	05E	62	70	71	72	89	88	87	12	82	81	81			
ISRAEL																		
Jerusalem	31	47N	35	13E	2485	31	36	38	95	94	92	24	70	69	69	W	12	NW
Tel Aviv	32	06N	34	47E	36	33	39	41	96	93	91	16	74	73	72	N	8	W
ITALY																		
Milan	45	27N	9	17E	341	12	18	22	89	87	84	20	76	75	74	W	4	SW
Naples	40	53N	14	18E	220	28	34	36	91	88	86	19	74	73	72	N	6	SSW
Rome	41	48N	12	36E	377	25	30	33	94	92	89	24	74	73	72	E	6	WSW
IVORY COAST																		
Abidjan	5	19N	4	01W	65	64	67	69	91	90	88	15	83	82	81	WSW	5	SW
JAPAN																		
Fukuoka	33	35N	130	27E	22	26	29	31	92	90	89	20	82	80	79			
Sapporo	43	04N	141	21E	56	−7	1	5	86	83	80	20	76	74	72	SE	3	SE
Tokyo	35	41N	139	46E	19	21	26	28	91	89	87	14	81	80	79	SW	10	S
JORDAN																		
Amman	31	57N	35	57E	2548	29	33	36	97	94	92	25	70	69	68	N	6	NNW
KENYA																		
Nairobi	1	16S	36	48E	5971	45	48	50	81	80	78	24	66	65	65	E	13	ENE
KOREA																		
Pyongyang	39	02N	125	41E	186	−10	−2	3	89	87	85	21	77	76	76			
Seoul	37	34N	126	58E	285	−1	7	9	91	89	87	16	81	79	78	NW	7	W
LEBANON																		
Beirut	33	54N	35	28E	111	40	42	45	93	91	90	15	78	77	76	N	7	SW
LIBERIA																		
Monrovia	6	18N	10	48W	75	64	68	69	90	89	88	19	82	82	81	E	3	WSW
LIBYA																		
Benghazi	32	06N	20	04E	82	41	46	48	97	94	91	13	77	76	75	SSE	8	S
MADAGASCAR																		
Tananarive	18	55S	47	33E	4531	39	43	46	86	84	83	23	73	72	71			
MALAYSIA																		
Kuala Lumpur	3	07N	101	42E	127	67	70	71	94	93	92	20	82	82	81	N	4	W
Penang	5	25N	100	19E	17	69	72	73	93	93	92	18	82	81	80			
MARTINIQUE																		
Fort De France	14	37N	61	05W	13	62	64	66	90	89	88	14	81	81	80			
MEXICO																		
Guadalajara	20	41N	103	20W	5105	35	39	42	93	91	89	29	68	67	66	N	7	W
Merida	20	58N	89	38W	72	56	59	61	97	95	94	21	80	79	77	E	11	E
Mexico City	19	24N	99	12W	7575	33	37	39	83	81	79	25	61	60	59	N	8	N
Monterrey	25	40N	100	18W	1732	31	38	41	98	95	93	20	79	78	77			
Vera Cruz	19	12N	96	08W	184	55	60	62	91	89	88	12	83	83	82			
MOROCCO																		
Casablanca	33	35N	7	39W	164	36	40	42	94	90	86	50	73	72	70			
NEPAL																		
Katmandu	27	42N	85	12E	4388	30	33	35	89	87	86	25	78	77	76	W	4	NW
NETHERLANDS																		
Amsterdam	52	23N	4	55E	5	17	20	23	79	76	73	10	65	64	63	S	8	E
NEW ZEALAND																		
Auckland	36	51S	174	46E	140	37	40	42	78	77	76	14	67	66	65			
Christchurch	43	32S	172	37E	32	25	28	31	82	79	76	17	68	67	66	W	4	NNW

TABLE B2 CLIMATIC CONDITIONS FOR OTHER COUNTRIES (*Continued*)

Col. 1	Lat.		Long.		Col. 3 Elevation, ft	Winter, °F Mean of Annual Extremes	Col. 4 99%	97.5%	Col. 5 Design Dry-Bulb 1%	2.5%	5%	Col. 6 Mean Daily Range, °F	Col. 7 Design Wet-Bulb 1%	2.5%	5%	Prevailing Winds Winter		Summer
Country and Station	°	′	°	′												Knots		
Wellington	41	17S	174	46E	394	32	35	37	76	74	72	14	66	65	64	NE	6	NNE
NICARAGUA																		
Managua	12	10N	86	15W	135	62	65	67	94	93	92	21	81	80	79	E	9	E
NIGERIA																		
Lagos	6	27N	3	24E	10	67	70	71	92	91	90	12	82	82	81	WSW	8	S
NORWAY																		
Bergen	60	24N	5	19E	141	14	17	20	75	74	73	21	67	66	65			
Oslo	59	56N	10	44E	308	−2	0	4	79	77	74	17	67	66	64	N	10	S
PAKISTAN																		
Karachi	24	48N	66	59E	13	45	49	51	100	98	95	14	82	82	81	N	4	SSW
Lahore	31	35N	74	20E	702	32	35	37	109	107	105	27	83	82	81	NW	3	SE
Peshwar	34	01N	71	35E	1164	31	35	37	109	106	103	29	81	80	79	W	5	
NE PANAMA and CANAL ZONE																		
Panama City	8	58N	79	33W	21	69	72	73	93	92	91	18	81	81	80			
PAPUA NEW GUINEA																		
Port Moresby	9	29S	147	09E	126	62	67	69	92	91	90	14	80	80	79			
PARAGUAY																		
Asuncion	25	17S	57	30W	456	35	43	46	100	98	96	24	81	81	80	NE	7	NE
PERU																		
Lima	12	05S	77	03W	394	51	53	55	86	85	84	17	76	75	74	N	10	S
PHILIPPINES																		
Manila	14	35N	120	59E	47	69	73	74	94	92	91	20	82	81	81	N	3	ESE
POLAND																		
Krakow	50	04N	19	57E	723	−2	2	6	84	81	78	19	68	67	66			
Warsaw	52	13N	21	02E	394	−3	3	8	84	81	78	19	71	70	68	E	7	SE
PORTUGAL																		
Lisbon	38	43N	9	08W	313	32	37	39	89	86	83	16	69	68	67	ENE	5	N
PUERTO RICO																		
San Juan	18	29N	66	07W	82	65	67	68	89	88	87	11	81	80	79	ENE	10	E
RUMANIA																		
Bucharest	44	25N	26	06E	269	−2	3	8	93	91	89	26	72	71	70			
SAUDI ARABIA																		
Dhahran	26	17N	50	09E	80	39	45	48	111	110	108	32	86	85	84	N	8	N
Jedda	21	28N	39	10E	20	52	57	60	106	103	100	22	85	84	83			
Riyadh	24	39N	46	42E	1938	29	37	40	110	108	106	32	78	77	76	N	8	N
SENEGAL																		
Dakar	14	42N	17	29W	131	58	61	62	95	93	91	13	81	80	80	N	8	NW
SINGAPORE																		
Singapore	1	18N	103	50E	33	69	71	72	92	91	90	14	82	81	80	N	4	SE
SOMALIA																		
Mogadiscio	2	02N	49	19E	39	67	69	70	91	90	89	12	82	82	81	SSW	16	E
SOUTH AFRICA																		
Cape Town	33	56S	18	29E	55	36	40	42	93	90	86	20	72	71	70			
Johannesburg	26	11S	78	03E	5463	26	31	34	85	83	81	24	70	69	69			
Pretoria	25	45S	28	14E	4491	27	32	35	90	87	85	23	70	69	68	N	4	W
SOUTH YEMEN																		
Aden	12	50N	45	02E	10	63	68	70	102	100	98	11	83	82	82			
FORMER SOVIET UNION																		
Alma Ata	43	14N	76	53E	2543	−18	−10	−6	88	86	83	21	69	68	67			
Archangel	64	33N	40	32E	22	−29	−23	−18	75	71	68	13	60	58	57			
Kaliningrad	54	43N	20	30E	23	−3	1	6	83	80	77	17	67	66	65			
Krasnoyarsk	56	01N	92	57E	498	−41	−23	−27	84	80	76	12	64	62	60			
Kiev	50	27N	30	30E	600	−12	−5	1	87	84	81	22	69	68	67			
Kharkov	50	00N	36	14E	472	−19	−10	−3	87	84	82	23	69	68	67			
Kuibyshev	53	11N	50	06E	190	−23	−19	−13	89	85	81	20	69	67	66			
Leningrad	59	56N	30	16E	16	−14	−9	−5	78	75	72	15	65	64	63			
Minsk	53	54N	27	33E	738	−19	−11	−4	80	77	74	16	67	66	65			
Moscow	55	46N	37	40E	505	−19	−11	−6	84	81	78	21	69	67	65	SW	11	S
Odessa	46	29N	30	44E	214	−1	4	8	87	84	82	14	70	69	68			
Petropavlovsk	52	53N	158	42E	286	−9	−3	0	70	68	65	13	58	57	56			
Rostov on Don	47	13N	39	43E	159	−9	−2	4	90	87	84	20	70	69	68			
Sverdlovsk	56	49N	60	38E	894	−34	−25	−20	80	76	72	16	63	62	60			
Tashkent	41	20N	69	18E	1569	−4	3	8	95	93	90	29	71	70	69			
Tbilisi	41	43N	44	48E	1325	12	18	22	87	85	83	18	68	67	66			
Vladivostok	43	07N	131	55E	94	−15	−10	−7	80	77	74	11	70	69	68			
Volgograd	48	42N	44	31E	136	−21	−13	−7	93	89	86	19	71	70	69			

TABLE B2   CLIMATIC CONDITIONS FOR OTHER COUNTRIES *(Continued)*

Col. 1 Country and Station	Lat. °	Lat. ′	Long. °	Long. ′	Col. 3 Elevation, ft	Mean of Annual Extremes	Col. 4 99%	Col. 4 97.5%	Col. 5 Design Dry-Bulb 1%	2.5%	5%	Col. 6 Mean Daily Range, °F	Col. 7 Design Wet-Bulb 1%	2.5%	5%	Prevailing Winds Winter Knots		Summer Knots
SPAIN																		
Barcelona	41	24N	2	09E	312	31	33	36	88	86	84	13	75	74	73	N	10	S
Madrid	40	25N	3	41W	2188	22	25	28	93	91	89	25	71	69	67	NNE	5	W
Valencia	39	28N	0	23W	79	31	33	37	92	90	88	14	75	74	73	W	7	ESE
SRI LANKA																		
Colombo	6	54N	79	52E	24	65	69	70	90	89	88	15	81	80	80	W	6	W
SUDAN																		
Khartoum	15	37N	32	33E	1279	47	53	56	109	107	104	30	77	76	75	N	6	NW
SURINAM																		
Paramaribo	5	49N	55	09W	12	66	68	70	93	92	90	18	82	82	81	NE	9	E
SWEDEN																		
Stockholm	59	21N	18	04E	146	3	5	8	78	74	72	15	64	62	60	W	4	S
SWITZERLAND																		
Zurich	47	23N	8	33E	1617	4	9	14	84	81	78	21	68	67	66			
SYRIA																		
Damascus	33	30N	36	20E	2362	25	29	32	102	100	98	35	72	71	70			
TAIWAN																		
Tainan	22	57N	120	12E	70	40	46	49	92	91	90	14	84	83	82	N	10	W
Taipei	25	02N	121	31E	30	41	44	47	94	92	90	16	83	82	81	E	7	E
TANZANIA																		
Dar es Salaam	6	50S	39	18E	47	62	64	65	90	89	88	13	82	81	81			
THAILAND																		
Bangkok	13	44N	100	30E	39	57	61	63	97	95	93	18	82	82	81	N	4	S
TRINIDAD																		
Port of Spain	10	40N	61	31W	67	61	64	66	91	90	89	16	80	80	79			
TUNISIA																		
Tunis	36	47N	10	12E	217	35	39	41	102	99	96	22	77	76	74	W	10	E
TURKEY																		
Adana	36	59N	35	18E	82	25	33	35	100	97	95	22	79	78	77			
Ankara	39	57N	32	53E	2825	2	9	12	94	92	89	28	68	67	66	N	8	W
Istanbul	40	58N	28	50E	59	23	28	30	91	88	86	16	75	74	73	N	10	NE
Izmir	38	26N	27	10E	16	24	27	29	98	96	94	23	75	74	73	NNE	8	N
UNITED KINGDOM																		
Belfast	54	36N	5	55W	24	19	23	26	74	72	69	16	65	64	62			
Birmingham	52	29N	1	56W	535	21	24	27	79	76	73	15	66	64	63			
Cardiff	51	28N	3	10W	203	21	24	27	79	76	73	14	64	63	62			
Edinburgh	55	55N	3	11W	441	22	25	28	73	70	68	13	64	62	61	WSW	6	WSW
Glasgow	55	52N	4	17W	85	17	21	24	74	71	68	13	64	63	61			
London	51	29N	0	00	149	20	24	26	82	79	76	16	68	66	65	W	7	E
URUGUAY																		
Montevideo	34	51S	56	13W	72	34	37	39	90	88	85	21	73	72	71	N	11	NNE
VENEZUELA																		
Caracas	10	30N	66	56W	3418	49	52	54	84	83	81	21	70	69	69	E	8	ENE
Maracaibo	10	39N	71	36W	20	69	72	73	97	96	95	17	84	83	83			
VIETNAM																		
Da Nang	16	04N	108	13E	23	56	60	62	97	95	93	14	86	86	85	NW	5	N
Hanoi	21	02N	105	52E	53	46	50	53	99	97	95	16	85	85	84			
Ho Chi Minh City (Saigon)	10	47N	106	42E	30	62	65	67	93	91	89	16	85	84	83			
YUGOSLAVIA																		
Belgrade	44	48N	20	28E	453	4	9	13	92	89	86	23	74	73	72	ESE	9	SE
ZAIRE																		
Kinshasa (Leopoldville)	4	20S	15	18E	1066	54	60	62	92	91	90	19	81	80	80	NNW	7	W
Kisangani (Stanleyville)	0	26S	15	14E	1370	65	67	68	92	91	90	19	81	80	80			

Source: © 1989. Reprinted by permission from ASHRAE *Handbook of Fundamentals 1989.*

TABLE B3.1 SOLAR INTENSITY AND SOLAR HEAT GAIN FACTORS FOR 0° NORTH LATITUDE, CONVENTIONAL UNITS

Date	Solar Time (am)	Direct Normal (Btuh/ft²)	N	NNE	NE	ENE	E	ESE	SE	SSE	S	SSW	SW	WSW	W	WNW	NW	NNW	HOR	Solar Time (pm)
Jan 21	7	218	10	12	78	149	193	205	188	139	65	11	10	10	10	10	10	10	40	5
	8	288	19	20	88	177	234	254	235	179	91	20	19	19	19	19	19	19	121	4
	9	315	25	26	67	153	212	238	227	182	104	30	25	25	25	25	25	25	194	3
	10	328	30	31	40	99	158	192	195	168	112	48	31	30	30	30	30	30	250	2
	11	334	33	33	34	43	82	122	143	142	117	76	40	34	33	33	33	33	284	1
	12	335	34	34	34	35	36	52	83	109	118	109	83	52	36	35	34	34	296	12
	HALF DAY TOTALS		135	140	329	647	909	1053	1045	875	551	237	162	139	135	135	135	135	1037	
Feb 21	7	219	11	32	112	172	203	203	171	112	31	11	10	10	10	10	10	10	43	5
	8	287	20	39	132	205	245	247	210	141	47	20	19	19	19	19	19	19	128	4
	9	313	27	35	110	182	222	228	199	138	56	27	26	26	26	26	26	26	203	3
	10	325	32	33	70	127	166	178	162	120	62	33	31	31	31	31	31	31	260	2
	11	331	35	36	38	59	87	105	107	92	66	42	36	35	34	34	34	34	295	1
	12	333	36	36	36	37	37	42	53	63	67	63	53	42	37	37	36	36	306	12
	HALF DAY TOTALS		142	194	488	773	954	997	891	643	297	161	146	141	140	140	140	139	1081	
Mar 21	7	206	12	69	138	182	197	182	138	69	12	11	11	11	11	11	11	11	43	5
	8	275	22	87	170	223	242	223	170	87	22	20	20	20	20	20	20	20	128	4
	9	302	29	80	156	204	221	204	156	80	29	27	27	27	27	27	27	27	202	3
	10	314	34	64	116	154	167	154	116	64	34	33	32	32	32	32	32	33	259	2
	11	320	37	46	65	82	86	82	65	46	37	36	36	35	35	35	36	36	293	1
	12	322	38	38	38	38	38	38	38	38	38	38	38	38	38	38	38	38	303	12
	HALF DAY TOTALS		153	369	672	874	944	874	672	369	153	146	145	145	144	145	145	146	1075	
Apr 21	7	177	31	98	147	172	171	144	94	28	11	11	11	11	11	11	11	12	40	5
	8	249	49	132	193	224	221	184	118	36	21	21	21	21	21	21	21	22	118	4
	9	278	59	134	188	213	206	168	102	35	28	28	28	28	28	28	28	29	188	3
	10	291	66	119	156	169	157	119	67	35	33	33	33	33	33	33	33	35	241	2
	11	298	70	94	106	103	85	59	40	37	36	36	36	36	36	37	38	45	274	1
	12	300	71	67	57	45	39	38	38	38	37	38	38	38	39	45	56	67	284	12
	HALF DAY TOTALS		311	615	826	911	864	697	442	191	147	147	147	148	148	150	155	174	1002	
May 21	7	156	53	109	145	159	149	116	63	14	12	12	12	12	12	12	12	13	37	5
	8	231	83	156	202	218	201	154	80	24	21	21	21	21	21	21	21	24	109	4
	9	261	97	164	203	212	190	139	65	30	28	28	28	28	28	28	28	34	174	3
	10	275	106	154	178	175	146	94	43	34	33	33	33	33	33	33	35	51	224	2
	11	282	111	133	134	115	81	47	38	36	36	36	36	36	36	38	43	75	255	1
	12	284	113	104	82	54	40	38	37	37	37	37	37	38	40	54	82	104	265	12
	HALF DAY TOTALS		507	771	909	912	789	571	310	157	149	149	149	149	150	155	176	246	931	
Jun 21	7	146	60	109	141	150	137	103	51	13	12	12	12	12	12	12	12	14	35	5
	8	221	95	162	202	212	191	140	66	23	21	21	21	21	21	21	21	27	104	4
	9	252	112	173	206	209	182	126	53	30	28	28	28	28	28	28	29	42	167	3
	10	267	122	166	184	175	140	84	38	34	33	33	33	33	33	33	36	63	215	2
	11	274	127	146	143	119	79	43	38	36	36	36	36	36	36	38	50	90	245	1
	12	276	129	119	93	59	40	38	37	37	37	37	37	38	40	58	92	119	255	12
	HALF DAY TOTALS		580	818	927	900	750	517	265	154	149	149	149	149	151	158	189	293	892	
Jul 21	7	146	53	105	139	151	141	109	59	14	12	12	12	12	12	12	12	13	36	5
	8	221	84	154	198	213	195	149	77	24	22	22	22	22	22	22	22	25	107	4
	9	252	99	164	201	209	186	135	63	31	29	29	29	29	29	29	29	36	171	3
	10	266	108	155	177	173	144	92	43	35	34	34	34	34	34	34	36	53	219	2
	11	274	113	134	135	116	81	47	39	37	37	37	37	37	37	39	45	78	250	1
	12	276	115	107	84	56	41	39	38	38	38	38	38	39	41	56	84	107	260	12
	HALF DAY TOTALS		516	767	896	894	769	553	300	159	152	152	152	152	153	159	182	256	912	
Aug 21	7	159	32	93	136	159	157	132	85	26	12	12	12	12	12	12	12	12	39	5
	8	233	52	131	187	216	212	175	112	35	22	22	22	22	22	22	22	23	115	4
	9	262	63	134	185	208	200	162	98	36	29	29	29	29	29	29	29	31	182	3
	10	277	70	121	156	167	154	116	65	37	34	34	34	34	34	34	35	37	234	2
	11	284	74	97	109	104	85	59	41	39	37	37	37	37	37	38	40	49	266	1
	12	286	75	71	60	47	41	40	39	39	38	39	39	40	41	47	60	71	276	12
	HALF DAY TOTALS		329	615	809	882	830	666	422	192	154	154	154	154	155	157	163	186	973	
Sep 21	7	184	13	65	127	167	181	167	127	65	13	11	11	11	11	11	11	11	42	5
	8	256	23	84	163	213	231	213	163	84	23	21	21	21	21	21	21	21	123	4
	9	284	30	80	152	198	214	198	152	80	30	28	28	28	28	28	28	28	195	3
	10	298	35	64	114	150	163	150	114	64	35	34	33	33	33	33	33	34	250	2
	11	304	39	48	66	82	88	82	66	48	39	38	37	36	36	36	37	38	283	1
	12	306	40	40	40	40	40	40	40	40	40	40	40	40	40	40	40	40	293	12
	HALF DAY TOTALS		160	362	646	836	902	836	646	362	160	152	150	150	150	150	150	152	1039	
Oct 21	7	202	11	32	106	161	190	190	160	104	30	11	11	11	11	11	11	11	42	5
	8	273	20	40	129	199	236	238	202	135	45	20	20	20	20	20	20	20	125	4
	9	300	27	37	110	179	217	222	194	134	55	28	27	27	27	27	27	27	198	3
	10	313	33	34	71	126	164	174	158	117	61	34	32	32	32	32	32	32	254	2
	11	319	36	36	40	60	87	103	105	90	65	42	37	36	35	35	35	35	288	1
	12	321	37	37	37	38	38	43	53	63	66	62	53	43	38	38	37	37	299	12
	HALF DAY TOTALS		146	199	481	751	921	960	856	618	290	164	149	145	144	143	143	143	1056	
Nov 21	7	210	10	12	77	145	187	199	182	135	62	11	10	10	10	10	10	10	39	5
	8	282	19	20	88	175	230	250	230	176	89	20	19	19	19	19	19	19	120	4
	9	309	25	27	67	152	210	235	224	179	102	30	25	25	25	25	25	25	192	3
	10	322	30	31	40	99	157	190	192	165	111	48	31	30	30	30	30	30	247	2
	11	328	34	34	35	44	82	121	142	140	115	75	40	35	34	34	34	34	282	1
	12	330	35	35	35	35	36	52	82	107	117	107	82	52	36	35	35	35	293	12
	HALF DAY TOTALS		136	141	329	640	895	1035	1025	857	540	235	162	141	137	136	136	136	1027	
Dec 21	7	215	10	11	64	138	186	203	190	147	77	11	10	10	10	10	10	10	38	5
	8	287	18	19	71	164	226	253	240	191	108	23	18	18	18	18	18	18	117	4
	9	314	25	25	52	140	206	239	235	196	123	39	25	25	25	25	25	25	189	3
	10	327	29	30	33	87	153	194	204	183	132	62	31	29	29	29	29	29	243	2
	11	333	33	33	33	39	79	127	155	159	137	93	46	34	33	33	33	33	277	1
	12	335	34	34	34	34	35	56	96	127	138	127	95	56	35	34	34	34	288	12
	HALF DAY TOTALS		131	134	274	592	879	1059	1089	949	648	288	172	138	132	131	131	131	1008	
			N	NNW	NW	WNW	W	WSW	SW	SSW	S	SSE	SE	ESE	E	ENE	NE	NNE	HOR	PM

TABLE B3.1 SOLAR INTENSITY AND SOLAR HEAT GAIN FACTORS FOR 8° NORTH LATITUDE, CONVENTIONAL UNITS

Date	Solar Time (am)	Direct Normal (Btuh/ft²)	N	NNE	NE	ENE	E	ESE	SE	SSE	S	SSW	SW	WSW	W	WNW	NW	NNW	HOR	Solar Time (pm)
Jan 21	7	141	5	6	44	92	124	134	126	96	49	6	5	5	5	5	5	5	14	5
	8	262	14	15	55	147	210	240	233	189	114	25	14	14	14	14	14	14	79	4
	9	300	21	21	32	122	200	244	251	219	152	58	22	21	21	21	21	21	150	3
	10	317	26	26	27	66	150	209	233	223	178	102	31	26	26	26	26	26	203	2
	11	325	29	29	29	31	77	148	195	210	194	146	75	31	29	29	29	29	236	1
	12	327	30	30	30	30	32	73	139	184	199	184	138	72	32	30	30	30	248	12
	HALF DAY TOTALS		110	112	196	461	760	1000	1096	1020	781	426	211	127	111	110	110	110	805	
Feb 21	7	182	8	17	84	138	169	172	150	103	36	8	8	8	8	8	8	8	25	5
	8	273	17	19	96	180	231	247	224	166	77	18	17	17	17	17	17	17	101	4
	9	305	23	24	64	153	214	242	233	188	110	30	23	23	23	23	23	23	174	3
	10	319	28	29	33	92	161	202	211	188	134	61	30	28	28	28	28	28	229	2
	11	326	32	32	32	37	83	136	167	172	149	102	49	33	32	32	32	32	263	1
	12	328	33	33	33	33	34	60	107	142	154	142	106	60	34	33	33	33	275	12
	HALF DAY TOTALS		124	137	321	609	865	1023	1034	885	582	287	174	132	124	124	124	124	930	
Mar 21	7	201	11	53	124	172	192	183	145	82	15	10	10	10	10	10	10	10	40	5
	8	272	20	50	140	205	239	235	195	123	35	19	19	19	19	19	19	19	120	4
	9	299	26	35	109	179	218	225	197	138	57	27	26	26	26	26	26	26	192	3
	10	312	31	33	61	120	165	182	172	134	76	34	32	31	31	31	31	31	247	2
	11	318	34	35	36	53	87	114	125	116	89	55	36	35	34	34	34	34	280	1
	12	320	35	35	36	36	37	47	69	87	93	86	68	47	37	36	36	35	291	12
	HALF DAY TOTALS		141	226	494	755	928	975	879	643	319	187	153	142	139	139	139	139	1025	
Apr 21	6	14	2	8	12	14	14	12	8	2	1	1	1	1	1	1	1	1	1	6
	7	197	24	94	153	187	191	167	117	45	14	13	13	13	13	13	13	13	53	5
	8	256	27	99	172	216	227	204	150	69	24	22	22	22	22	22	22	22	131	4
	9	280	31	79	149	193	208	193	147	77	31	29	29	29	29	29	29	29	197	3
	10	293	35	54	102	141	158	151	120	73	37	34	33	33	33	33	33	33	249	2
	11	299	38	40	54	72	86	88	78	60	43	38	38	36	36	36	36	37	279	1
	12	301	39	39	39	40	40	41	43	45	45	45	43	41	40	39	39	39	289	12
	HALF DAY TOTALS		179	403	674	859	922	851	653	352	174	159	157	156	155	155	155	156	1057	
May 21	6	44	14	30	41	45	43	34	19	4	3	3	3	3	3	3	3	3	5	6
	7	193	50	120	168	191	185	150	92	24	16	16	16	16	16	16	16	17	62	5
	8	244	52	132	189	218	215	179	115	38	25	24	24	24	24	24	24	25	135	4
	9	268	49	116	171	198	197	167	109	45	32	30	30	30	30	30	30	32	197	3
	10	280	47	89	130	151	150	126	84	44	37	35	35	35	35	35	35	37	245	2
	11	286	47	63	79	87	83	70	52	41	40	39	38	38	38	39	39	41	273	1
	12	288	46	46	44	43	42	41	41	41	41	41	41	41	42	43	44	46	282	12
	HALF DAY TOTALS		283	575	804	916	897	748	493	217	172	167	167	167	167	168	169	176	1058	
Jun 21	6	53	20	39	52	55	51	39	20	4	4	4	4	4	4	4	4	4	7	6
	7	188	62	128	172	190	179	141	80	20	16	16	16	16	16	16	16	18	64	5
	8	238	66	142	194	217	207	167	99	31	25	25	25	25	25	25	25	27	135	4
	9	261	63	130	178	198	190	154	93	37	31	31	31	31	31	31	31	33	194	3
	10	273	59	104	140	154	145	115	70	39	37	36	36	36	36	36	36	38	241	2
	11	279	57	76	90	92	82	63	46	41	40	39	39	39	39	40	41	43	268	1
	12	281	57	55	50	45	43	42	41	41	41	41	41	41	42	42	45	50	55	277	12
	HALF DAY TOTALS		356	648	850	929	876	700	430	194	174	171	171	171	172	173	176	190	1049	
Jul 21	6	41	14	29	39	42	40	31	18	4	3	3	3	3	3	3	3	3	6	6
	7	184	51	118	164	185	179	145	88	23	16	16	16	16	16	16	16	17	62	5
	8	236	55	132	187	214	210	174	111	37	25	25	25	25	25	25	25	26	133	4
	9	259	52	117	170	196	193	163	106	44	32	31	31	31	31	31	31	33	194	3
	10	272	50	92	131	151	148	123	81	44	38	36	36	36	36	36	36	38	241	2
	11	278	49	66	81	88	83	69	52	42	41	40	39	39	39	40	40	42	269	1
	12	279	49	48	46	44	43	42	42	42	42	42	42	42	43	44	46	48	277	12
	HALF DAY TOTALS		296	580	799	903	878	729	478	215	176	172	171	171	171	172	173	182	1043	
Aug 21	6	11	2	7	10	12	12	10	6	2	1	1	1	1	1	1	1	1	1	6
	7	180	26	92	145	176	180	156	109	42	15	14	14	14	14	14	14	14	53	5
	8	240	30	100	168	209	219	196	143	65	25	23	23	23	23	23	23	23	128	4
	9	266	33	82	148	190	203	187	142	74	33	30	30	30	30	30	30	30	193	3
	10	279	37	58	104	140	155	147	117	71	39	36	35	35	35	35	35	35	243	2
	11	285	40	43	57	75	86	87	76	59	44	40	39	38	38	38	38	39	273	1
	12	287	41	41	41	42	42	43	44	45	46	45	44	43	42	41	41	41	282	12
	HALF DAY TOTALS		191	410	666	837	891	817	624	339	180	167	165	164	163	163	163	164	1033	
Sep 21	7	179	12	50	114	158	176	168	133	76	15	11	11	11	11	11	11	11	39	5
	8	253	21	49	134	196	227	224	186	119	36	20	20	20	20	20	20	20	116	4
	9	281	28	36	106	173	211	217	191	134	57	28	27	27	27	27	27	27	185	3
	10	295	32	34	61	118	161	178	168	132	76	35	33	32	32	32	32	32	238	2
	11	302	35	36	37	54	86	113	123	114	88	56	38	36	35	35	35	35	271	1
	12	304	36	36	37	38	39	49	69	86	93	86	69	48	39	38	37	36	282	12
	HALF DAY TOTALS		146	226	475	722	885	931	842	622	319	192	159	148	145	144	144	144	991	
Oct 21	7	166	8	18	79	128	156	159	139	95	33	9	8	8	8	8	8	8	25	5
	8	259	17	20	95	174	223	237	215	159	74	19	17	17	17	17	17	17	99	4
	9	292	24	25	65	150	209	235	225	182	106	31	24	24	24	24	24	24	170	3
	10	307	29	30	34	92	158	197	205	183	130	60	31	29	29	29	29	29	224	2
	11	314	32	32	33	39	83	133	163	167	145	100	49	34	32	32	32	32	258	1
	12	316	33	33	33	34	35	60	105	139	150	138	104	60	35	34	33	33	270	12
	HALF DAY TOTALS		127	141	318	592	836	986	996	852	563	283	175	136	128	127	127	127	911	
Nov 21	7	134	5	6	43	89	119	129	120	92	47	6	5	5	5	5	5	5	14	5
	8	255	14	15	55	145	206	235	228	185	111	25	15	15	15	15	15	15	78	4
	9	295	21	21	33	121	197	241	247	215	150	57	22	21	21	21	21	21	149	3
	10	312	26	26	28	67	147	206	230	220	176	100	31	26	26	26	26	26	201	2
	11	320	29	29	29	31	77	146	192	207	191	144	74	31	29	29	29	29	234	1
	12	322	30	30	30	30	32	72	137	181	196	181	137	72	32	30	30	30	246	12
	HALF DAY TOTALS		112	113	197	456	749	983	1077	1001	767	420	210	128	112	112	112	112	799	
Dec 21	7	118	4	5	30	72	101	112	107	85	48	7	4	4	4	4	4	4	10	5
	8	255	13	14	41	132	198	233	231	193	124	33	13	13	13	13	13	13	69	4
	9	297	20	20	25	108	191	241	254	227	165	72	21	20	20	20	20	20	138	3
	10	315	25	25	26	56	144	208	239	233	192	117	35	25	25	25	25	25	191	2
	11	323	28	28	28	29	73	150	202	221	207	161	86	30	28	28	28	28	223	1
	12	325	29	29	29	29	30	77	149	197	212	196	149	76	30	29	29	29	234	12
	HALF DAY TOTALS		104	105	159	402	710	975	1099	1050	836	484	228	125	105	104	104	104	748	
			N	NNW	NW	WNW	W	WSW	SW	SSW	S	SSE	SE	ESE	E	ENE	NE	NNE	HOR	PM

481

TABLE B3.1   SOLAR INTENSITY AND SOLAR HEAT GAIN FACTORS FOR 16° NORTH LATITUDE, CONVENTIONAL UNITS

Solar Heat Gain Factors (Btuh/ft²)

Date	Solar Time (am)	Direct Normal (Btuh/ft²)	N	NNE	NE	ENE	E	ESE	SE	SSE	S	SSW	SW	WSW	W	WNW	NW	NNW	HOR	Solar Time (pm)
Jan 21	7	141	5	6	44	92	124	134	126	96	49	6	5	5	5	5	5	5	14	5
	8	262	14	15	55	147	210	240	233	189	114	25	14	14	14	14	14	14	79	4
	9	300	21	21	32	122	200	244	251	219	152	58	22	21	21	21	21	21	150	3
	10	317	26	26	27	66	150	209	233	223	178	102	31	26	26	26	26	26	203	2
	11	325	29	29	29	31	77	148	195	210	194	146	75	31	29	29	29	29	236	1
	12	327	30	30	30	30	30	73	139	184	199	184	138	72	32	30	30	30	248	12
	HALF DAY TOTALS		110	112	196	461	760	1000	1096	1020	781	426	211	127	111	110	110	110	805	
Feb 21	7	182	8	17	84	138	169	172	150	103	36	8	8	8	8	8	8	8	25	5
	8	273	17	19	96	180	231	247	224	166	77	18	17	17	17	17	17	17	101	4
	9	305	23	24	64	153	214	242	233	188	110	30	23	23	23	23	23	23	174	3
	10	319	28	29	33	92	161	202	211	188	134	61	30	28	28	28	28	28	229	2
	11	326	32	32	32	37	83	136	167	172	149	102	49	33	32	32	32	32	263	1
	12	328	33	33	33	33	34	60	107	142	154	142	106	60	34	33	33	33	275	12
	HALF DAY TOTALS		124	137	321	609	865	1023	1034	885	582	287	174	132	124	124	124	124	930	
Mar 21	7	201	11	53	124	172	192	183	145	82	15	10	10	10	10	10	10	10	40	5
	8	272	20	50	140	205	239	235	195	123	35	19	19	19	19	19	19	19	120	4
	9	299	26	35	109	179	218	225	197	138	57	27	26	26	26	26	26	26	192	3
	10	312	31	33	61	120	165	182	172	134	76	34	32	31	31	31	31	31	247	2
	11	318	34	35	36	53	87	114	125	116	89	55	36	35	34	34	34	34	280	1
	12	320	35	35	36	36	37	47	69	87	93	86	68	47	37	36	36	35	291	12
	HALF DAY TOTALS		141	226	494	755	928	975	879	643	319	187	153	142	139	139	139	139	1025	
Apr 21	6	14	2	8	12	14	14	12	8	2	1	1	1	1	1	1	1	1	1	6
	7	197	24	94	153	187	191	167	117	45	14	13	13	13	13	13	13	13	53	5
	8	256	27	99	172	216	227	204	150	69	24	22	22	22	22	22	22	22	131	4
	9	280	31	79	149	193	208	193	147	77	31	29	29	29	29	29	29	29	197	3
	10	293	35	54	102	141	158	151	120	73	37	34	33	33	33	33	33	33	249	2
	11	299	38	40	54	72	86	88	78	60	43	38	38	36	36	36	36	37	279	1
	12	301	39	39	39	40	40	41	43	45	45	45	43	41	40	39	39	39	289	12
	HALF DAY TOTALS		179	403	674	859	922	851	653	352	174	159	157	156	155	155	155	156	1057	
May 21	6	44	14	30	41	45	43	34	19	4	3	3	3	3	3	3	3	3	5	6
	7	193	50	120	168	191	185	150	92	24	16	16	16	16	16	16	16	17	62	5
	8	244	52	132	189	218	215	179	115	38	25	24	24	24	24	24	24	25	135	4
	9	268	49	116	171	198	197	167	109	45	32	30	30	30	30	30	30	32	197	3
	10	280	47	89	130	151	150	126	84	44	37	35	35	35	35	35	35	37	245	2
	11	286	47	63	79	87	83	70	52	41	40	39	38	38	38	39	39	41	273	1
	12	288	46	46	44	43	42	41	41	41	41	41	41	41	42	43	44	46	282	12
	HALF DAY TOTALS		283	575	804	916	897	748	493	217	172	167	167	167	167	168	169	176	1058	
Jun 21	6	53	20	39	52	55	51	39	20	4	4	4	4	4	4	4	4	4	7	6
	7	188	62	128	172	190	179	141	80	20	16	16	16	16	16	16	16	18	64	5
	8	238	66	142	194	217	207	167	99	31	25	25	25	25	25	25	25	27	135	4
	9	261	63	130	178	198	190	154	93	37	31	31	31	31	31	31	31	33	194	3
	10	273	59	104	140	154	145	115	70	39	37	36	36	36	36	36	36	38	241	2
	11	279	57	76	90	92	82	63	46	41	40	39	39	39	39	39	41	43	268	1
	12	281	57	55	50	45	43	42	41	41	41	41	41	41	42	42	45	50	277	12
	HALF DAY TOTALS		356	648	850	929	876	700	430	194	174	171	171	171	171	172	173	176	190	1049	
Jul 21	6	41	14	29	39	42	40	31	18	4	3	3	3	3	3	3	3	3	6	6
	7	184	51	118	164	185	179	145	88	23	16	16	16	16	16	16	16	17	62	5
	8	236	55	132	187	214	210	174	111	37	25	25	25	25	25	25	25	26	133	4
	9	259	52	117	170	196	193	163	106	44	32	31	31	31	31	31	31	33	194	3
	10	272	50	92	131	151	148	123	81	44	38	36	36	36	36	36	36	38	241	2
	11	278	49	66	81	88	83	69	52	42	41	40	39	39	39	40	40	42	269	1
	12	279	49	48	46	44	43	42	42	42	42	42	42	42	43	44	46	48	277	12
	HALF DAY TOTALS		296	580	799	903	878	729	478	215	176	172	171	171	171	172	173	182	1043	
Aug 21	6	11	2	7	10	12	12	10	6	2	1	1	1	1	1	1	1	1	1	6
	7	180	26	92	145	176	180	156	109	42	15	14	14	14	14	14	14	14	53	5
	8	240	30	100	168	209	219	196	143	65	25	23	23	23	23	23	23	23	128	4
	9	266	33	82	148	190	203	187	142	74	33	30	30	30	30	30	30	30	193	3
	10	279	37	58	104	140	155	147	117	71	39	36	35	35	35	35	35	35	243	2
	11	285	40	43	57	75	86	87	76	59	44	40	39	38	38	38	38	39	273	1
	12	287	41	41	42	42	42	43	44	45	46	45	44	43	42	41	41	41	282	12
	HALF DAY TOTALS		191	410	666	837	891	817	624	339	180	167	165	164	163	163	163	164	1033	
Sep 21	7	179	12	50	114	158	176	168	133	76	15	11	11	11	11	11	11	11	39	5
	8	253	21	49	134	196	227	224	186	119	36	20	20	20	20	20	20	20	116	4
	9	281	28	36	106	173	211	217	191	134	57	28	27	27	27	27	27	27	185	3
	10	295	32	34	61	118	161	178	168	132	76	35	33	32	32	32	32	32	238	2
	11	302	35	35	37	54	86	113	123	114	88	56	38	36	35	35	35	35	271	1
	12	304	36	36	37	38	39	49	69	86	93	86	69	48	39	38	37	36	282	12
	HALF DAY TOTALS		146	226	475	722	885	931	842	622	319	192	159	148	145	144	144	144	991	
Oct 21	7	166	8	18	79	128	156	159	139	95	33	9	8	8	8	8	8	8	25	5
	8	259	17	20	95	174	223	237	215	159	74	19	17	17	17	17	17	17	99	4
	9	292	24	25	65	150	209	235	225	182	106	31	24	24	24	24	24	24	170	3
	10	307	29	30	34	92	158	197	205	183	130	60	31	29	29	29	29	29	224	2
	11	314	32	32	33	39	83	133	163	167	145	100	49	34	32	32	32	32	258	1
	12	316	33	33	33	34	35	60	105	139	150	138	104	60	35	34	33	33	270	12
	HALF DAY TOTALS		127	141	318	592	836	986	996	852	563	283	175	136	128	127	127	127	911	
Nov 21	7	134	5	6	43	89	119	129	120	92	47	6	5	5	5	5	5	5	14	5
	8	255	15	15	55	145	206	235	228	185	111	25	15	15	15	15	15	15	78	4
	9	295	21	21	33	121	197	241	247	215	150	57	22	21	21	21	21	21	149	3
	10	312	26	26	28	67	147	206	230	220	176	100	31	26	26	26	26	26	201	2
	11	320	29	29	29	31	77	146	192	207	191	144	74	31	29	29	29	29	234	1
	12	322	30	30	30	30	32	72	137	181	196	181	137	72	32	30	30	30	246	12
	HALF DAY TOTALS		112	113	197	456	749	983	1077	1001	767	420	210	128	112	112	112	112	799	
Dec 21	7	118	4	5	30	72	101	112	107	85	48	7	4	4	4	4	4	4	10	5
	8	255	13	14	41	132	198	233	231	193	124	33	13	13	13	13	13	13	69	4
	9	297	20	20	25	108	191	241	254	227	165	72	21	20	20	20	20	20	138	3
	10	315	25	25	26	56	144	208	239	233	192	117	35	25	25	25	25	25	191	2
	11	323	28	28	28	29	73	150	202	221	207	161	86	30	28	28	28	28	223	1
	12	325	29	29	29	29	30	77	149	197	212	196	149	76	30	29	29	29	234	12
	HALF DAY TOTALS		104	105	159	402	710	975	1099	1050	836	484	228	125	105	104	104	104	748	
			N	NNW	NW	WNW	W	WSW	SW	SSW	S	SSE	SE	ESE	E	ENE	NE	NNE	HOR	PM

TABLE B3.1 SOLAR INTENSITY AND SOLAR HEAT GAIN FACTORS FOR 24° NORTH LATITUDE, CONVENTIONAL UNITS

Date	Solar Time (am)	Direct Normal (Btuh/ft²)	N	NNE	NE	ENE	E	ESE	SE	SSE	S	SSW	SW	WSW	W	WNW	NW	NNW	HOR	Solar Time (pm)
Jan 21	7	71	2	3	21	45	62	67	63	49	25	3	2	2	2	2	2	2	5	5
	8	239	12	12	41	128	190	221	218	181	114	28	12	12	12	12	12	12	55	4
	9	288	18	18	23	106	190	240	253	227	166	73	19	18	18	18	18	18	121	3
	10	308	23	23	24	53	144	211	245	241	200	125	38	24	23	23	23	23	172	2
	11	317	26	26	26	27	73	156	211	234	220	173	95	29	26	26	26	26	204	1
	12	320	27	27	27	27	29	82	160	210	227	210	160	81	29	27	27	27	214	12
	HALF DAY TOTALS		95	96	148	372	671	942	1076	1039	840	505	241	120	96	95	95	95	664	
Feb 21	7	133	6	12	67	114	141	145	128	90	33	6	6	6	6	6	6	6	17	5
	8	262	15	16	80	165	220	240	224	172	89	17	15	15	15	15	15	15	83	4
	9	297	21	22	46	138	208	244	243	205	133	42	22	21	21	21	21	21	153	3
	10	314	26	26	28	76	157	209	228	213	165	87	28	26	26	26	26	26	205	2
	11	321	29	29	29	31	80	148	191	203	185	137	68	31	29	29	29	29	238	1
	12	323	30	30	30	30	32	70	134	177	192	177	133	70	32	30	30	30	249	12
	HALF DAY TOTALS		113	119	257	527	806	1011	1072	965	699	374	200	127	113	113	113	113	820	
Mar 21	7	194	11	45	115	164	186	180	145	86	17	10	10	10	10	10	10	10	36	5
	8	267	18	35	124	195	234	237	204	138	48	19	18	18	18	18	18	18	112	4
	9	295	25	27	85	165	215	232	214	163	82	27	25	25	25	25	25	25	180	3
	10	309	30	31	41	103	162	194	195	168	112	47	31	30	30	30	30	30	232	2
	11	315	33	33	34	42	85	129	154	155	139	86	43	34	33	33	33	33	264	1
	12	317	34	34	34	34	35	56	96	126	137	126	95	56	35	34	34	34	275	12
	HALF DAY TOTALS		133	189	422	693	906	1011	970	778	458	249	169	139	133	133	133	133	962	
Apr 21	6	40	6	21	33	39	39	33	22	7	2	2	2	2	2	2	2	2	4	6
	7	203	20	88	151	189	197	176	127	55	15	14	14	14	14	14	14	14	58	5
	8	256	24	80	159	209	228	212	164	88	24	22	22	22	22	22	22	22	132	4
	9	280	30	54	126	181	208	203	169	105	39	29	28	28	28	28	28	28	195	3
	10	292	34	37	75	125	157	165	148	107	56	35	33	33	33	33	33	33	244	2
	11	298	36	37	40	59	85	103	106	94	70	45	38	37	36	36	36	36	274	1
	12	299	37	37	38	38	39	46	59	70	75	70	58	45	39	38	38	37	283	12
	HALF DAY TOTALS		168	339	607	826	940	924	773	494	244	180	163	157	155	155	154	154	1048	
May 21	6	86	25	57	79	87	84	66	38	8	6	6	6	6	6	6	6	6	13	6
	7	203	43	117	171	199	196	163	105	32	17	17	17	17	17	17	17	18	73	5
	8	248	38	114	178	214	218	190	132	54	26	25	25	25	25	25	25	26	142	4
	9	269	35	88	150	188	198	179	132	66	33	31	31	31	31	31	31	31	201	3
	10	280	38	59	103	137	150	141	111	67	39	36	35	35	35	35	35	36	247	2
	11	286	40	43	55	72	83	84	75	58	44	40	39	38	38	38	38	39	274	1
	12	288	41	41	41	41	42	43	44	46	46	46	44	43	42	41	41	41	282	12
	HALF DAY TOTALS		238	492	749	909	943	840	614	308	187	176	174	173	172	172	172	175	1089	
Jun 21	6	97	36	70	93	101	94	73	39	8	7	7	7	7	7	7	7	8	17	6
	7	201	55	127	177	199	192	155	94	26	18	18	18	18	18	18	18	20	77	5
	8	242	50	126	184	214	212	179	117	43	27	26	26	26	26	26	26	27	145	4
	9	263	43	102	158	189	192	168	116	53	34	32	32	32	32	32	33	33	201	3
	10	274	41	72	113	140	146	131	96	55	39	36	36	36	36	36	36	38	245	2
	11	279	42	50	65	77	82	77	64	49	42	41	40	39	39	39	40	41	271	1
	12	281	43	43	43	43	43	43	43	43	43	43	43	43	43	43	43	43	279	12
	HALF DAY TOTALS		284	562	802	933	932	797	544	255	187	181	180	179	179	179	180	187	1096	
Jul 21	6	81	26	56	76	84	80	63	36	8	6	6	6	6	6	6	6	7	13	6
	7	195	45	116	168	194	190	158	101	31	18	18	18	18	18	18	18	19	73	5
	8	239	41	115	176	210	213	185	128	52	27	26	26	26	26	26	26	26	141	4
	9	261	37	90	150	186	195	175	129	64	34	32	32	32	32	32	32	32	198	3
	10	272	39	62	104	137	149	139	108	65	39	37	36	36	36	36	36	37	243	2
	11	278	41	44	58	73	83	83	73	57	44	41	40	39	39	39	39	40	270	1
	12	280	42	42	42	43	43	44	45	46	46	46	45	43	43	42	42	42	278	12
	HALF DAY TOTALS		247	498	746	897	925	820	595	300	191	181	178	177	177	177	177	181	1076	
Aug 21	6	35	6	20	30	35	35	30	19	6	2	2	2	2	2	2	2	2	4	6
	7	186	22	87	144	179	186	165	119	51	16	15	15	15	15	15	15	15	58	5
	8	241	26	82	156	203	220	204	157	84	26	24	24	24	24	24	24	24	130	4
	9	265	32	57	126	178	202	197	162	101	39	31	30	30	30	30	30	30	191	3
	10	278	36	40	78	125	155	161	143	103	55	37	35	35	35	35	35	35	239	2
	11	284	38	39	42	61	85	101	104	91	68	46	40	38	37	37	37	39	268	1
	12	286	38	39	40	41	47	47	58	69	72	68	58	47	41	40	40	39	277	12
	HALF DAY TOTALS		179	347	601	806	910	889	740	473	243	186	171	165	164	163	163	162	1028	
Sep 21	8	248	19	36	119	185	222	225	194	132	48	20	19	19	19	19	19	19	108	4
	9	278	26	28	84	160	207	223	206	158	81	28	26	26	26	26	26	26	174	3
	10	292	31	32	42	101	158	188	190	163	110	48	32	31	31	31	31	31	224	2
	11	299	34	34	35	43	84	127	151	151	128	86	44	35	34	34	34	34	256	1
	12	301	35	35	35	36	37	57	95	124	134	124	94	57	37	36	35	35	266	12
	HALF DAY TOTALS		139	190	406	661	863	964	927	749	451	251	174	145	139	138	138	138	930	
Oct 21	7	138	6	12	62	104	129	133	117	82	31	7	6	6	6	6	6	6	17	5
	8	247	16	17	79	159	211	230	214	164	85	17	16	16	16	16	16	16	82	4
	9	284	22	23	47	135	202	237	235	198	128	41	23	22	22	22	22	22	150	3
	10	301	27	27	29	77	154	204	222	207	160	85	29	27	27	27	27	27	201	2
	11	309	30	30	30	33	80	145	186	198	180	133	67	32	30	30	30	30	233	1
	12	311	31	31	31	31	33	70	131	173	187	172	130	69	33	31	31	31	244	12
	HALF DAY TOTALS		116	123	255	512	778	974	1032	929	675	367	200	131	117	116	116	116	804	
Nov 21	7	67	2	3	20	43	59	64	60	46	24	3	2	2	2	2	2	2	5	5
	8	232	12	13	42	126	186	216	213	177	111	28	12	12	12	12	12	12	55	4
	9	282	19	19	23	106	187	236	249	223	163	71	20	19	19	19	19	19	120	3
	10	303	23	23	24	53	143	209	241	237	197	123	37	24	23	23	23	23	171	2
	11	312	26	26	26	28	73	154	209	230	217	171	93	29	26	26	26	26	202	1
	12	315	27	27	27	27	29	81	158	207	224	207	158	80	29	27	27	27	213	12
	HALF DAY TOTALS		97	97	149	368	661	926	1056	1020	825	497	239	121	98	97	97	97	659	
Dec 21	7	30	1	1	7	18	25	28	27	21	12	2	1	1	1	1	1	1	2	5
	8	225	10	10	29	112	174	208	209	178	118	35	11	10	10	10	10	10	44	4
	9	281	17	17	19	93	180	234	252	231	174	84	18	17	17	17	17	17	107	3
	10	304	22	22	22	44	137	209	247	247	209	137	44	22	22	22	22	22	157	2
	11	314	25	25	25	26	69	156	216	241	230	183	104	29	25	25	25	25	188	1
	12	317	26	26	26	26	27	85	167	219	237	219	167	84	27	26	26	26	199	12
	HALF DAY TOTALS		88	88	118	313	611	899	1054	1042	868	550	257	117	89	88	88	88	598	
			N	NNW	NW	WNW	W	WSW	SW	SSW	S	SSE	SE	ESE	E	ENE	NE	NNE	HOR	PM

483

TABLE B3.1 SOLAR INTENSITY AND SOLAR HEAT GAIN FACTORS FOR 32° NORTH LATITUDE, CONVENTIONAL UNITS

Solar Heat Gain Factors (Btuh/ft²)

Date	Solar Time (am)	Direct Normal (Btuh/ft²)	N	NNE	NE	ENE	E	ESE	SE	SSE	S	SSW	SW	WSW	W	WNW	NW	NNW	HOR	Solar Time (pm)
Jan 21	7	1	0	0	0	1	1	1	1	1	1	0	0	0	0	0	0	0	0	5
	8	203	9	9	29	105	160	189	189	159	103	28	9	9	9	9	9	9	32	4
	9	269	15	15	17	91	175	229	246	225	169	82	17	15	15	15	15	15	88	3
	10	295	20	20	20	41	135	209	249	250	212	141	46	20	20	20	20	20	136	2
	11	306	23	23	23	24	68	159	221	249	238	191	110	29	23	23	23	23	166	1
	12	310	24	24	24	24	25	88	174	228	246	228	174	88	25	24	24	24	176	12
	HALF DAY TOTALS		79	79	107	284	570	856	1015	1014	853	553	264	112	80	79	79	79	512	
Feb 21	7	112	4	7	47	82	102	106	95	67	26	4	4	4	4	4	4	4	9	5
	8	245	13	14	65	149	205	228	216	170	95	17	13	13	13	13	13	13	64	4
	9	287	19	19	32	122	199	242	248	216	149	55	20	19	19	19	19	19	127	3
	10	305	24	24	25	62	151	213	241	232	189	112	31	24	24	24	24	24	176	2
	11	314	26	26	26	28	76	156	208	227	212	165	87	28	26	26	26	26	207	1
	12	316	27	27	27	27	29	79	155	204	221	204	155	79	29	27	27	27	217	12
	HALF DAY TOTALS		100	103	201	445	735	978	1080	1010	780	452	228	122	100	100	100	100	691	
Mar 21	7	185	10	37	105	153	176	173	142	88	20	9	9	9	9	9	9	9	32	5
	8	260	17	25	107	183	227	237	209	150	62	18	17	17	17	17	17	17	100	4
	9	290	23	25	64	151	210	237	227	183	107	30	23	23	23	23	23	23	164	3
	10	304	28	28	30	87	158	202	215	195	144	70	29	28	28	28	28	28	211	2
	11	311	31	31	31	34	82	142	179	188	168	120	59	32	31	31	31	31	242	1
	12	313	32	32	32	32	33	66	122	162	176	162	122	66	33	32	32	32	252	12
	HALF DAY TOTALS		124	162	359	629	875	1033	1041	888	589	326	193	136	125	124	124	124	874	
Apr 21	6	66	9	35	54	65	66	56	38	12	4	3	3	3	3	3	3	3	7	6
	7	206	17	80	146	188	200	182	136	65	16	14	14	14	14	14	14	14	61	5
	8	255	23	61	144	200	227	219	177	107	30	22	22	22	22	22	22	22	129	4
	9	278	28	36	103	168	206	212	187	133	58	29	28	28	28	28	28	28	188	3
	10	290	32	34	52	108	155	177	172	141	87	39	33	32	32	32	32	32	233	2
	11	295	35	35	36	47	83	118	135	132	108	70	40	36	35	35	35	35	262	1
	12	297	36	36	36	37	38	53	82	106	115	106	82	53	38	37	36	36	271	12
	HALF DAY TOTALS		161	296	550	792	952	992	889	645	360	228	177	157	153	152	152	152	1015	
May 21	6	119	33	77	108	121	116	94	56	13	8	8	8	8	8	8	8	9	21	6
	7	211	36	111	170	202	204	174	118	42	19	18	18	18	18	18	18	19	81	5
	8	250	29	94	165	208	220	199	149	73	27	25	25	25	25	25	25	25	146	4
	9	269	33	61	128	177	198	190	155	93	37	32	31	31	31	31	31	31	201	3
	10	280	36	40	76	121	150	156	138	99	54	37	35	35	35	35	35	35	243	2
	11	285	38	39	42	59	83	99	102	90	68	47	40	39	37	37	37	37	269	1
	12	286	38	39	40	40	41	47	59	70	74	70	59	47	41	40	39	38	277	12
	HALF DAY TOTALS		222	438	702	900	985	933	747	447	250	199	183	177	175	174	174	175	1098	
Jun 21	6	131	44	92	123	135	127	99	55	12	10	10	10	10	10	10	10	11	28	6
	7	210	47	122	176	204	201	168	108	35	20	20	20	20	20	20	20	21	88	5
	8	245	36	106	171	208	214	189	135	60	28	27	27	27	27	27	27	27	151	4
	9	264	35	74	137	178	193	180	139	77	35	32	32	32	32	32	32	32	204	3
	10	274	38	47	86	125	146	145	123	83	45	38	36	36	36	36	36	36	244	2
	11	279	40	41	47	64	82	91	89	75	56	43	41	40	39	39	39	39	269	1
	12	280	41	41	41	42	42	46	52	58	60	58	52	46	42	42	41	41	276	12
	HALF DAY TOTALS		261	504	762	935	985	897	678	372	225	197	189	185	184	184	183	186	1122	
Jul 21	6	113	34	76	105	117	113	90	53	12	9	9	9	9	9	9	9	9	22	6
	7	203	38	111	167	198	198	169	114	41	20	19	19	19	19	19	19	19	81	5
	8	241	31	95	163	204	215	194	145	70	28	26	26	26	26	26	26	26	145	4
	9	261	34	64	129	175	195	186	150	90	37	32	32	32	32	32	32	32	198	3
	10	271	37	42	78	121	148	153	134	96	53	38	36	36	36	36	36	36	240	2
	11	277	39	40	43	60	83	98	99	88	66	47	41	40	38	38	38	38	265	1
	12	279	40	40	41	41	42	48	58	68	72	68	58	48	42	41	41	40	273	12
	HALF DAY TOTALS		231	444	701	890	967	912	726	433	248	202	187	182	180	179	179	180	1088	
Aug 21	6	59	10	33	50	60	60	51	34	11	4	4	4	4	4	4	4	4	8	6
	7	190	19	79	141	179	190	172	128	61	17	15	15	15	15	15	15	15	61	5
	8	240	25	63	141	195	219	210	170	102	31	23	23	23	23	23	23	23	128	4
	9	263	30	39	104	166	200	206	181	127	57	31	29	29	29	29	29	29	185	3
	10	276	34	36	55	109	153	173	167	136	84	40	35	34	34	34	34	34	229	2
	11	282	36	37	39	50	84	116	131	127	104	69	41	38	36	36	36	36	256	1
	12	284	37	37	37	39	40	54	81	103	111	103	81	54	40	39	37	37	265	12
	HALF DAY TOTALS		171	303	546	774	922	955	854	618	352	231	184	166	162	161	160	160	999	
Sep 21	7	163	10	35	96	139	159	156	128	80	20	10	10	10	10	10	10	10	31	5
	8	240	18	26	103	173	215	224	198	143	60	19	18	18	18	18	18	18	96	4
	9	272	24	26	64	146	202	227	218	177	105	31	24	24	24	24	24	24	158	3
	10	287	29	29	32	86	154	196	208	189	141	70	31	29	29	29	29	29	204	2
	11	294	32	32	32	36	81	139	174	182	163	118	59	34	32	32	32	32	234	1
	12	296	33	33	33	33	35	66	120	158	171	158	120	66	35	33	33	33	244	12
	HALF DAY TOTALS		130	164	345	598	831	982	993	852	574	325	197	142	130	129	129	129	845	
Oct 21	7	99	4	7	43	74	92	96	85	60	24	5	4	4	4	4	4	4	10	5
	8	229	13	15	63	143	195	217	206	162	90	17	13	13	13	13	13	13	63	4
	9	273	20	20	33	120	193	234	239	208	144	54	21	20	20	20	20	20	125	3
	10	293	24	24	26	62	147	207	234	225	183	109	32	24	24	24	24	24	173	2
	11	302	27	27	27	29	76	152	203	221	207	160	85	29	27	27	27	27	203	1
	12	304	28	28	28	28	30	78	151	199	215	199	151	78	30	28	28	28	213	12
	HALF DAY TOTALS		103	106	200	433	708	941	1038	972	753	441	226	125	104	103	103	103	679	
Nov 21	7	2	0	0	0	1	1	1	1	1	1	1	0	0	0	0	0	0	0	5
	8	196	9	9	29	103	156	184	184	155	100	27	9	9	9	9	9	9	32	4
	9	263	16	16	17	90	173	225	241	221	166	80	17	16	16	16	16	16	88	3
	10	289	20	20	21	41	134	206	245	246	209	138	45	21	20	20	20	20	136	2
	11	301	23	23	23	24	67	157	218	245	234	188	109	29	23	23	23	23	165	1
	12	304	24	24	24	24	25	87	171	224	243	224	171	87	25	24	24	24	175	12
	HALF DAY TOTALS		80	81	108	282	561	841	996	995	838	544	261	113	81	80	80	80	509	
Dec 21	8	176	7	7	19	84	135	163	166	143	97	31	7	7	7	7	7	7	22	4
	9	257	14	14	15	77	162	218	238	222	171	89	15	14	14	14	14	14	72	3
	10	288	18	18	18	34	127	204	246	251	216	148	52	19	18	18	18	18	119	2
	11	301	21	21	21	22	63	157	222	252	243	197	116	29	21	21	21	21	148	1
	12	304	22	22	22	23	23	89	177	232	252	232	177	89	23	22	22	22	158	12
	HALF DAY TOTALS		71	71	84	227	500	792	965	986	852	578	275	107	71	71	71	71	440	
			N	NNW	NW	WNW	W	WSW	SW	SSW	S	SSE	SE	ESE	E	ENE	NE	NNE	HOR	PM

TABLE B3.1   SOLAR INTENSITY AND SOLAR HEAT GAIN FACTORS FOR 40° NORTH LATITUDE, CONVENTIONAL UNITS

Date	Solar Time (am)	Direct Normal (Btuh/ft²)	N	NNE	NE	ENE	E	ESE	SE	SSE	S	SSW	SW	WSW	W	WNW	NW	NNW	HOR	Solar Time (pm)
Jan 21	8	142	5	5	17	71	111	132	133	114	75	22	6	5	5	5	5	5	14	4
	9	239	12	12	13	74	154	205	224	209	160	82	13	12	12	12	12	12	55	3
	10	274	16	16	16	31	124	199	241	246	213	146	51	17	16	16	16	16	96	2
	11	289	19	19	19	20	61	156	222	252	244	198	118	28	19	19	19	19	124	1
	12	294	20	20	20	20	21	90	179	234	254	234	179	90	21	20	20	20	133	12
	HALF DAY TOTALS		61	61	73	199	452	734	904	932	813	561	273	101	62	61	61	61	354	
Feb 21	7	55	2	3	23	40	51	53	47	34	14	2	2	2	2	2	2	2	4	5
	8	219	10	11	50	129	183	206	199	160	94	18	10	10	10	10	10	10	43	4
	9	271	16	16	22	107	186	234	245	218	157	66	17	16	16	16	16	16	98	3
	10	294	21	21	21	49	143	211	246	243	203	129	38	21	21	21	21	21	143	2
	11	304	23	23	23	24	71	160	219	244	231	184	103	27	23	23	23	23	171	1
	12	307	24	24	24	24	25	86	170	222	241	222	170	86	25	24	24	24	180	12
	HALF DAY TOTALS		84	86	152	361	648	916	1049	1015	821	508	250	114	85	84	84	84	548	
Mar 21	7	171	9	29	93	140	163	161	135	86	22	8	8	8	8	8	8	8	26	5
	8	250	16	18	91	169	218	232	211	157	74	17	16	16	16	16	16	16	85	4
	9	282	21	22	47	136	203	238	236	198	128	40	22	21	21	21	21	21	143	3
	10	297	25	25	27	72	153	207	229	216	171	195	29	25	25	25	25	25	186	2
	11	305	28	28	28	30	78	151	198	213	197	150	77	30	28	28	28	28	213	1
	12	307	29	29	29	29	31	75	145	191	206	191	145	75	31	29	29	29	223	12
	HALF DAY TOTALS		114	139	302	563	832	1035	1087	968	694	403	220	132	114	113	113	113	764	
Apr 21	6	89	11	46	72	87	88	76	52	18	5	5	5	5	5	5	5	5	11	6
	7	206	16	71	140	185	201	186	143	75	16	14	14	14	14	14	14	14	61	5
	8	252	22	44	128	190	224	223	188	124	41	22	21	21	21	21	21	21	123	4
	9	274	27	29	80	155	202	219	203	156	83	29	27	27	27	27	27	27	177	3
	10	286	31	31	37	92	152	187	193	170	121	56	32	31	31	31	31	41	217	2
	11	292	33	33	34	39	81	130	160	166	146	102	52	35	33	33	33	33	243	1
	12	293	34	34	34	34	36	62	108	142	154	142	108	62	36	34	34	34	252	12
	HALF DAY TOTALS		154	265	501	758	957	1051	994	782	488	296	199	157	148	147	147	147	957	
May 21	5	1	0	1	1	1	1	1	1	0	0	0	0	0	0	0	0	0	0	7
	6	144	36	90	128	145	141	115	71	18	10	10	10	10	10	10	10	11	31	6
	7	216	28	102	165	202	209	184	131	54	20	19	19	19	19	19	19	19	87	5
	8	250	27	73	149	199	220	208	164	93	29	25	25	25	25	25	25	25	146	4
	9	267	31	42	105	164	197	200	175	121	53	32	30	30	30	30	30	30	195	3
	10	277	34	36	54	105	148	168	163	133	83	40	35	34	34	34	34	34	234	2
	11	283	36	36	38	48	81	113	130	127	105	70	42	38	36	36	36	36	257	1
	12	284	37	37	37	38	40	54	82	104	113	104	82	54	40	38	37	37	265	12
	HALF DAY TOTALS		215	404	666	893	1024	1025	881	601	358	247	200	180	176	175	174	175	1083	
Jun 21	5	22	10	17	21	22	20	14	6	2	1	1	1	1	1	1	1	2	3	7
	6	155	48	104	143	159	151	121	70	17	13	13	13	13	13	13	13	14	40	6
	7	216	37	113	172	205	207	178	122	46	22	21	21	21	21	21	21	21	97	5
	8	246	30	85	156	201	216	199	152	80	29	27	27	27	27	27	27	27	153	4
	9	263	33	51	114	166	192	190	161	105	45	33	32	32	32	32	32	32	201	3
	10	272	35	38	63	109	145	158	148	116	69	39	36	35	35	35	35	35	238	2
	11	277	38	39	40	52	81	105	116	110	88	60	41	39	38	38	38	38	260	1
	12	279	38	38	38	40	41	52	72	89	95	89	72	52	41	40	38	38	267	12
	HALF DAY TOTALS		253	470	734	941	1038	999	818	523	315	236	204	191	188	187	186	188	1126	
Jul 21	5	2	1	2	2	2	2	1	1	0	0	0	0	0	0	0	0	0	0	7
	6	138	37	89	125	142	137	112	68	18	11	11	11	11	11	11	11	12	32	6
	7	208	30	102	163	198	204	179	127	53	20	20	20	20	20	20	20	20	88	5
	8	241	28	75	148	196	216	203	160	90	30	26	25	26	26	26	26	26	145	4
	9	259	32	44	106	163	193	196	170	118	52	33	31	31	31	31	31	31	194	3
	10	269	35	37	56	106	146	165	159	129	81	41	36	35	35	35	35	35	231	2
	11	275	37	38	40	50	81	111	127	123	102	69	43	39	37	37	37	37	254	1
	12	276	38	38	38	40	41	55	80	101	109	101	80	55	41	40	38	38	262	12
	HALF DAY TOTALS		223	411	666	885	1008	1003	858	584	352	248	204	186	181	180	180	181	1076	
Aug 21	6	81	12	44	68	81	82	71	48	17	6	5	5	5	5	5	5	5	12	6
	7	191	17	71	135	177	191	177	135	70	17	16	16	16	16	16	16	16	62	5
	8	237	24	47	126	185	216	214	180	118	41	23	23	23	23	23	23	23	122	4
	9	260	28	31	82	153	197	212	196	151	80	31	28	28	28	28	28	28	174	3
	10	272	32	33	40	93	150	182	187	165	116	56	34	32	32	32	32	32	214	2
	11	278	35	35	36	41	81	128	156	160	141	99	52	37	35	35	35	35	239	1
	12	280	35	35	35	36	38	63	106	138	149	138	106	63	38	36	35	35	247	12
	HALF DAY TOTALS		164	273	498	741	928	1013	956	751	474	296	205	166	157	156	156	156	946	
Sep 21	7	149	9	27	84	125	146	144	121	77	21	9	9	9	9	9	9	9	25	5
	8	230	17	19	87	160	205	218	199	148	71	18	17	17	17	17	17	17	82	4
	9	263	22	23	47	131	194	227	226	190	124	41	23	22	22	22	22	22	138	3
	10	280	27	27	28	71	148	200	221	209	165	93	30	27	27	27	27	27	180	2
	11	287	29	29	29	31	78	147	192	207	191	146	77	31	29	29	29	29	206	1
	12	290	30	30	30	30	32	75	142	185	200	185	142	75	32	30	30	30	215	12
	HALF DAY TOTALS		119	142	291	534	787	980	1033	925	672	396	222	137	119	118	118	118	738	
Oct 21	7	48	2	3	20	36	45	47	42	30	12	2	2	2	2	2	2	2	4	5
	8	204	11	12	49	123	173	195	188	151	89	18	11	11	11	11	11	11	43	4
	9	257	17	17	23	104	180	225	235	209	151	64	18	17	17	17	17	17	97	3
	10	280	21	21	22	50	139	205	238	235	196	125	38	22	21	21	21	21	140	2
	11	291	24	24	24	25	71	156	212	236	224	178	101	28	24	24	24	24	168	1
	12	294	25	25	25	25	27	85	165	216	234	216	165	85	27	25	25	25	177	12
	HALF DAY TOTALS		88	89	152	351	623	878	1006	974	791	493	247	117	89	88	88	88	540	
Nov 21	8	136	5	5	18	69	108	128	129	110	72	21	6	5	5	5	5	5	14	4
	9	232	12	12	13	73	151	201	219	204	156	80	13	12	12	12	12	12	55	3
	10	268	16	16	16	31	122	196	237	242	209	143	50	17	16	16	16	16	96	2
	11	283	19	19	19	20	61	154	218	248	240	194	116	28	19	19	19	19	123	1
	12	288	20	20	20	20	21	89	176	231	250	231	176	89	21	20	20	20	132	12
	HALF DAY TOTALS		63	63	75	198	445	721	887	914	798	551	269	101	63	63	63	63	354	
Dec 21	8	89	3	3	8	41	67	82	84	73	50	17	3	3	3	3	3	3	6	4
	9	217	10	10	11	60	135	185	205	194	151	83	13	10	10	10	10	10	39	3
	10	261	14	14	14	25	113	188	232	239	210	146	55	15	14	14	14	14	77	2
	11	280	17	17	17	17	56	151	217	249	242	198	120	28	17	17	17	17	104	1
	12	285	18	18	18	18	19	89	178	233	253	233	178	89	19	18	18	18	113	12
	HALF DAY TOTALS		52	52	56	146	374	649	822	867	775	557	276	94	53	52	52	52	282	
			N	NNW	NW	WNW	W	WSW	SW	SSW	S	SSE	SE	ESE	E	ENE	NE	NNE	HOR	PM

TABLE B3.1 SOLAR INTENSITY AND SOLAR HEAT GAIN FACTORS FOR 48° NORTH LATITUDE, CONVENTIONAL UNITS

Date	Solar Time (am)	Direct Normal (Btuh/ft²)	N	NNE	NE	ENE	E	ESE	SE	SSE	S	SSW	SW	WSW	W	WNW	NW	NNW	HOR	Solar Time (pm)
Jan 21	8	37	1	1	4	18	29	34	35	30	20	6	1	1	1	1	1	1	2	4
	9	185	8	8	8	53	118	160	176	166	129	69	10	8	8	8	8	8	25	3
	10	239	12	12	12	22	106	175	216	223	195	136	50	12	12	12	12	12	55	2
	11	261	14	14	14	15	53	144	208	239	233	190	116	26	14	14	14	14	77	1
	12	267	15	15	15	15	16	86	171	226	245	226	171	86	16	15	15	15	85	12
	HALF DAY TOTALS		43	43	46	117	316	567	729	776	701	512	259	85	43	43	43	43	203	
Feb 21	7	4	0	0	1	3	3	3	3	2	1	0	0	0	0	0	0	0	0	5
	8	180	8	8	36	103	149	170	166	136	82	17	8	8	8	8	8	8	25	4
	9	247	13	13	16	90	168	216	230	209	155	71	14	13	13	13	13	13	66	3
	10	275	17	17	17	38	131	203	242	244	207	138	44	18	17	17	17	17	105	2
	11	288	19	19	19	20	65	158	221	249	239	192	113	27	19	19	19	19	130	1
	12	292	20	20	20	20	22	89	176	231	250	231	176	89	22	20	20	20	138	12
	HALF DAY TOTALS		68	68	107	274	541	816	968	967	813	531	261	104	68	68	68	68	395	
Mar 21	7	153	7	22	80	123	145	145	123	80	23	7	7	7	7	7	7	7	20	5
	8	236	14	15	76	154	204	222	206	158	82	15	14	14	14	14	14	14	68	4
	9	270	19	19	3	121	193	234	239	207	142	52	20	19	19	19	19	19	118	3
	10	287	23	23	24	58	146	208	237	231	189	115	33	23	23	23	23	23	156	2
	11	295	25	25	25	26	74	156	210	232	218	172	94	28	25	25	25	25	180	1
	12	298	26	26	26	26	27	83	161	211	228	211	161	83	27	26	26	26	188	12
	HALF DAY TOTALS		100	118	250	494	775	1012	1100	1014	767	465	244	126	101	100	100	100	636	
Apr 21	6	108	12	53	86	105	107	93	64	23	6	6	6	6	6	6	6	6	15	6
	7	205	15	61	132	180	199	189	148	84	18	14	14	14	14	14	14	14	60	5
	8	247	20	32	111	179	219	225	196	138	55	21	20	20	20	20	20	20	114	4
	9	268	25	26	60	141	197	223	215	176	106	33	25	25	25	25	25	25	161	3
	10	280	28	28	31	77	148	193	209	194	150	80	31	28	28	28	28	28	196	2
	11	286	31	31	31	33	78	140	181	193	177	133	69	33	31	31	31	31	218	1
	12	288	31	31	31	31	34	71	131	172	186	172	131	71	34	31	31	31	226	12
	HALF DAY TOTALS		147	242	461	724	957	1098	1081	895	605	370	226	156	141	140	140	140	875	
May 21	5	41	17	31	40	42	39	29	14	3	3	3	3	3	3	3	3	3	5	7
	6	162	35	97	141	162	160	133	85	24	12	12	12	12	12	12	12	13	40	6
	7	219	23	90	158	200	212	191	142	68	21	19	19	19	19	19	19	19	91	5
	8	248	26	54	132	190	218	214	178	113	38	25	25	25	25	25	25	25	142	4
	9	264	29	32	82	151	194	208	192	147	77	32	29	29	29	29	29	29	185	3
	10	274	33	34	39	90	145	178	184	163	116	57	35	33	33	33	33	33	219	2
	11	279	35	35	36	40	79	126	155	160	142	101	54	37	35	35	35	35	240	1
	12	280	35	35	35	36	38	63	107	139	150	139	107	63	38	36	36	35	247	12
	HALF DAY TOTALS		215	388	645	893	1065	1114	1007	749	483	316	225	184	174	173	173	174	1045	
Jun 21	5	77	35	61	76	80	72	53	24	6	5	5	5	5	5	5	5	8	12	7
	6	172	46	110	155	175	169	138	84	22	14	14	14	14	14	14	14	16	51	6
	7	220	29	101	165	204	211	187	135	60	23	21	21	21	21	21	21	21	103	5
	8	246	29	64	139	191	215	206	168	101	34	27	27	27	27	27	27	27	152	4
	9	261	31	36	91	153	190	199	180	133	66	33	31	31	31	31	31	31	193	3
	10	269	34	36	45	94	143	169	171	148	101	50	36	34	34	34	34	34	225	2
	11	274	36	36	38	44	79	118	142	145	126	88	49	38	36	36	36	36	246	1
	12	275	37	37	37	38	40	60	96	124	134	124	96	60	40	38	37	37	252	12
	HALF DAY TOTALS		257	459	722	955	1095	1102	955	678	436	299	228	197	189	188	188	191	1108	
Jul 21	5	43	18	33	42	45	41	30	15	3	3	3	3	3	3	3	3	4	6	7
	6	156	37	96	138	159	156	129	82	24	13	13	13	13	13	13	13	14	41	6
	7	211	25	90	156	196	207	186	138	66	22	20	20	20	20	20	20	20	92	5
	8	240	27	56	132	187	214	209	174	110	38	26	26	26	26	26	26	26	142	4
	9	256	30	34	83	149	191	204	187	143	75	33	30	30	30	30	30	30	184	3
	10	266	34	35	41	90	143	174	180	158	113	56	36	34	34	34	34	34	217	2
	11	271	36	36	42	79	124	151	156	138	99	54	38	36	36	36	36	237	1	1
	12	272	36	36	36	37	39	63	104	136	146	136	104	63	39	37	36	36	244	12
	HALF DAY TOTALS		223	395	646	886	1050	1092	983	730	474	315	229	190	181	179	179	180	1042	
Aug 21	6	99	13	51	81	98	100	87	60	22	7	7	7	7	7	7	7	7	16	6
	7	190	17	61	128	172	190	179	141	79	19	15	15	15	15	15	15	15	61	5
	8	232	22	34	110	174	211	216	188	132	53	23	22	22	22	22	22	22	114	4
	9	154	27	28	63	139	192	216	108	169	102	34	27	27	27	27	27	27	159	3
	10	266	30	30	33	78	145	188	203	188	144	78	33	30	30	30	30	30	193	2
	11	272	32	32	32	36	78	137	175	187	171	129	68	35	32	32	32	32	215	1
	12	274	33	33	33	33	36	71	128	167	189	167	128	71	36	33	33	33	223	12
	HALF DAY TOTALS		157	251	459	709	929	1060	1040	862	587	366	231	165	151	149	149	149	869	
Sep 21	7	131	8	21	71	108	128	128	108	71	21	8	7	7	7	7	7	7	20	5
	8	215	15	16	72	144	191	207	193	148	77	16	15	15	15	15	15	15	65	4
	9	251	20	20	34	116	184	223	227	197	136	52	21	20	20	20	20	20	114	3
	10	269	24	24	25	58	141	200	228	221	182	112	34	24	24	24	24	24	151	2
	11	278	26	26	26	28	73	151	203	223	210	166	92	29	26	26	26	26	174	1
	12	280	27	27	27	27	29	82	156	204	220	204	156	82	29	27	27	27	182	12
	HALF DAY TOTALS		105	121	240	465	729	953	1040	963	737	453	243	131	106	105	105	105	614	
Oct 21	7	0	0	0	2	3	4	4	3	2	1	0	0	0	0	0	0	0	0	5
	8	165	8	9	35	96	139	159	155	126	77	16	8	8	8	8	8	8	25	4
	9	233	14	14	16	88	161	207	220	199	148	68	15	14	14	14	14	14	66	3
	10	262	18	18	18	39	128	196	233	234	199	133	43	18	18	18	18	18	104	2
	11	274	20	20	20	21	64	153	213	241	231	186	109	27	20	20	20	20	128	1
	12	278	21	21	21	21	23	87	171	223	242	223	171	87	23	21	21	21	136	12
	HALF DAY TOTALS		71	71	108	266	519	780	925	925	779	513	256	106	72	71	71	71	391	
Nov 21	8	36	1	1	4	18	29	34	35	30	20	6	1	1	1	1	1	1	2	4
	9	179	8	8	8	52	115	156	171	161	125	67	10	8	8	8	8	8	26	3
	10	233	12	12	12	22	104	172	212	218	191	133	49	13	12	12	12	12	55	2
	11	255	15	15	15	15	52	142	204	234	228	186	114	26	15	15	15	15	77	1
	12	261	15	15	15	15	17	85	168	222	240	222	168	85	17	15	15	15	85	12
	HALF DAY TOTALS		44	44	47	117	310	555	713	760	686	502	255	85	44	44	44	44	204	
Dec 21	9	140	5	5	6	36	86	120	133	127	100	56	8	5	5	5	5	5	13	3
	10	214	10	10	10	16	91	156	194	201	179	126	49	10	10	10	10	10	38	2
	11	242	12	12	12	13	46	134	195	225	220	180	111	25	12	12	12	12	57	1
	12	250	13	13	13	13	14	81	163	215	233	215	168	81	14	13	13	13	65	12
	HALF DAY TOTALS		33	33	34	73	233	458	610	665	616	468	247	76	34	33	33	33	141	
			N	NNW	NW	WNW	W	WSW	SW	SSW	S	SSE	SE	ESE	E	ENE	NE	NNE	HOR	PM

TABLE B3.1 SOLAR INTENSITY AND SOLAR HEAT GAIN FACTORS FOR 56° NORTH LATITUDE, CONVENTIONAL UNITS

Date	Solar Time (am)	Direct Normal (Btuh/ft²)	N	NNE	NE	ENE	E	ESE	SE	SSE	S	SSW	SW	WSW	W	WNW	NW	NNW	HOR	Solar Time (pm)
Jan 21	9	78	3	3	3	21	49	67	74	70	55	30	4	3	3	3	3	3	5	3
	10	170	7	7	7	13	74	126	156	162	143	100	38	7	7	7	7	7	21	2
	11	207	9	9	9	10	40	116	169	194	190	156	96	21	9	9	9	9	34	1
	12	217	10	10	10	10	11	71	144	190	205	190	144	71	11	10	10	10	40	12
	HALF DAY TOTALS		23	23	24	46	163	343	468	517	487	378	206	61	24	23	23	23	80	
Feb 21	8	115	4	4	21	64	95	109	107	88	55	12	4	4	4	4	4	4	10	4
	9	203	10	10	11	71	139	183	197	182	136	66	10	10	10	10	10	10	36	3
	10	246	13	13	13	28	115	184	223	227	196	133	45	14	13	13	13	13	65	2
	11	262	15	15	15	16	57	148	210	239	232	188	112	25	15	15	15	15	84	1
	12	267	16	16	16	16	17	86	171	225	244	225	171	86	17	16	16	16	91	12
	HALF DAY TOTALS		49	50	66	182	409	666	821	846	737	509	253	89	50	49	49	49	241	
Mar 21	7	128	6	16	65	101	121	122	105	70	21	6	6	6	6	6	6	6	14	5
	8	215	12	13	61	136	185	205	194	152	84	15	12	12	12	12	12	12	49	4
	9	253	16	16	23	105	179	224	207	148	61	17	16	16	16	16	16	89	3	
	10	272	19	19	20	46	136	203	238	236	198	128	39	20	19	19	19	19	122	2
	11	282	21	21	21	22	68	156	215	241	230	184	106	27	21	21	21	21	142	1
	12	284	22	22	22	22	24	86	170	222	241	222	170	86	24	22	22	22	149	12
	HALF DAY TOTALS		85	97	200	419	699	956	1071	1016	800	502	258	118	86	85	85	85	491	
Apr 21	6	122	13	58	95	118	121	107	75	29	7	7	7	7	7	7	7	7	18	6
	7	201	15	51	123	173	195	188	152	91	21	14	14	14	14	14	14	14	56	5
	8	239	19	23	95	167	211	223	201	148	68	20	19	19	19	19	19	19	101	4
	9	260	23	24	44	126	190	223	223	189	126	44	24	23	23	23	23	23	140	3
	10	272	26	26	27	63	142	196	220	212	171	102	33	26	26	26	26	26	170	2
	11	278	28	28	28	30	74	147	195	213	200	156	86	31	28	28	28	28	189	1
	12	280	28	28	28	28	31	79	149	194	210	194	149	79	31	28	28	28	195	12
	HALF DAY TOTALS		139	226	430	694	951	1132	1147	982	699	437	252	154	132	131	131	131	772	
May 21	5	93	36	68	89	95	88	66	33	7	6	6	6	6	6	6	6	7	14	7
	6	175	33	99	148	174	173	147	97	31	14	14	14	14	14	14	14	14	48	6
	7	219	21	77	149	195	212	197	152	81	22	19	19	19	19	19	19	19	92	5
	8	244	25	38	115	179	215	218	189	131	52	25	24	24	24	24	24	24	135	4
	9	259	28	30	62	136	189	213	206	168	102	36	28	28	28	28	28	28	171	3
	10	268	31	31	33	75	141	185	200	187	145	80	33	31	31	31	31	31	199	2
	11	273	32	32	32	35	76	135	174	187	172	131	71	35	32	32	32	32	216	1
	12	275	33	33	33	33	36	71	129	168	181	168	129	71	36	33	33	33	222	12
	HALF DAY TOTALS		222	391	644	906	1112	1202	1120	878	604	392	256	187	172	170	170	173	986	
Jun 21	4	21	13	19	22	21	18	11	3	1	1	1	1	1	1	1	2	5	3	8
	5	122	53	94	119	126	115	85	40	10	9	9	9	9	9	9	9	12	25	7
	6	185	42	111	160	185	182	152	97	30	16	16	16	16	16	16	16	17	62	6
	7	222	25	86	156	199	213	195	147	74	24	22	22	22	22	22	22	22	105	5
	8	243	27	46	122	181	213	213	181	122	46	27	26	26	26	26	26	26	146	4
	9	257	30	32	69	139	187	206	196	156	91	34	30	30	30	30	30	30	181	3
	10	265	33	33	36	79	139	178	190	174	132	71	35	33	33	33	33	33	208	2
	11	269	34	34	35	38	76	129	164	174	159	119	65	37	34	34	34	34	225	1
	12	271	35	35	35	35	38	68	119	155	168	155	119	68	38	35	35	35	231	12
	HALF DAY TOTALS		275	473	738	989	1162	1207	1082	822	562	376	160	103	190	189	189	196	1070	
Jul 21	5	91	37	69	89	95	88	66	33	8	7	7	7	7	7	7	7	8	16	7
	6	169	34	98	145	170	170	143	95	31	15	14	14	14	14	14	14	15	50	6
	7	212	23	77	147	192	208	193	148	79	23	20	20	20	20	20	20	20	93	5
	8	237	26	40	115	177	211	214	185	128	51	26	25	25	25	25	25	25	135	4
	9	252	29	31	63	135	186	209	201	164	99	36	29	29	29	29	29	29	171	3
	10	261	32	32	34	76	139	181	196	182	142	78	35	32	32	32	32	32	198	2
	11	265	33	33	33	37	76	133	171	183	168	128	70	36	33	33	33	33	215	1
	12	267	34	34	34	34	37	71	126	164	177	164	126	71	37	34	34	34	221	12
	HALF DAY TOTALS		231	398	646	901	1097	1180	1096	859	593	390	259	193	179	177	177	180	987	
Aug 21	5	1	0	1	1	1	1	1	0	0	0	0	0	0	0	0	0	0	0	7
	6	112	14	56	91	111	114	101	71	28	8	8	8	8	8	8	8	8	20	6
	7	187	16	51	119	165	186	179	144	86	22	15	15	15	15	15	15	15	58	5
	8	225	20	25	94	162	203	214	192	142	66	22	20	20	20	20	20	20	101	4
	9	246	25	26	46	124	184	216	215	182	121	44	26	25	25	25	25	25	140	3
	10	258	28	28	30	65	139	191	213	204	165	99	34	28	28	28	28	28	169	2
	11	264	30	30	30	32	74	143	189	206	193	152	84	33	30	30	30	30	187	1
	12	266	30	30	30	30	30	78	145	188	203	188	145	78	33	30	30	30	198	12
	HALF DAY TOTALS		149	235	429	680	923	1092	1104	946	678	431	256	163	142	140	140	141	771	
Sep 21	7	107	6	15	56	87	104	105	90	60	19	6	6	6	6	6	6	6	14	5
	8	194	12	14	58	126	171	189	179	140	78	16	12	12	12	12	12	12	48	4
	9	233	17	17	24	100	170	211	220	195	140	59	18	17	17	17	17	17	86	3
	10	253	20	20	21	46	131	194	227	225	189	123	39	21	20	20	20	20	118	2
	11	263	22	22	22	24	67	150	206	230	220	176	103	28	22	22	22	22	137	1
	12	266	23	23	23	23	25	85	163	213	231	213	163	85	25	23	23	23	144	12
	HALF DAY TOTALS		89	99	191	391	652	893	1004	958	761	484	255	121	90	89	89	89	474	
Oct 21	8	104	4	5	20	59	87	100	98	81	50	11	4	4	4	4	4	4	10	4
	9	193	10	10	11	68	132	173	186	171	129	63	11	10	10	10	10	10	37	3
	10	231	14	14	14	28	111	176	213	216	186	127	44	14	14	14	14	14	64	2
	11	248	16	16	16	17	56	142	202	229	222	180	108	25	16	16	16	16	84	1
	12	253	16	16	16	16	18	83	164	216	234	216	164	83	18	16	16	16	91	12
	HALF DAY TOTALS		52	52	68	177	390	633	779	804	702	487	246	90	53	52	52	52	240	
Nov 21	9	76	3	3	3	21	48	66	72	69	54	29	4	3	3	3	3	3	6	3
	10	165	7	7	7	13	72	122	152	157	139	98	37	7	7	7	7	7	21	2
	11	201	9	9	9	10	39	113	165	190	186	152	94	21	9	9	9	9	35	1
	12	211	10	10	10	10	11	70	140	186	200	186	140	70	11	10	10	10	40	12
	HALF DAY TOTALS		24	24	24	47	161	336	457	505	475	369	202	61	24	24	24	24	81	
Dec 21	9	5	0	0	0	1	3	4	5	5	4	2	0	0	0	0	0	0	0	3
	10	113	4	4	4	7	47	82	103	107	96	68	27	4	4	4	4	4	9	2
	11	166	6	6	6	7	30	92	135	156	154	127	78	17	6	6	6	6	19	1
	12	180	7	7	7	7	8	59	120	159	171	159	120	59	8	7	7	7	23	12
	HALF DAY TOTALS		14	14	14	20	88	217	311	354	343	277	163	47	15	14	14	14	40	
			N	NNW	NW	WNW	W	WSW	SW	SSW	S	SSE	SE	ESE	E	ENE	NE	NNE	HOR	PM

487

TABLE B3.1 SOLAR INTENSITY AND SOLAR HEAT GAIN FACTORS FOR 64° NORTH LATITUDE, CONVENTIONAL UNITS

Date	Solar Time (am)	Direct Normal (Btuh/ft²)	N	NNE	NE	ENE	E	ESE	SE	SSE	S	SSW	SW	WSW	W	WNW	NW	NNW	HOR	Solar Time (pm)
Jan 21	10	22	1	1	1	1	9	16	20	21	19	13	5	1	1	1	1	1	1	2
	11	81	3	3	3	3	15	45	67	77	75	62	38	8	3	3	3	3	6	1
	12	100	3	3	3	3	4	33	67	89	96	89	67	33	4	3	3	3	8	12
	HALF DAY TOTALS		5	5	5	6	25	79	121	142	141	119	75	23	5	5	5	5	11	
Feb 21	8	18	1	1	3	10	15	17	17	14	9	2	1	1	1	1	1	1	1	4
	9	134	5	5	6	43	89	118	128	119	90	45	6	5	5	5	5	5	13	3
	10	190	8	8	8	18	87	144	176	180	157	108	38	9	8	8	8	8	28	2
	11	215	10	10	10	11	44	122	177	202	197	160	97	20	10	10	10	10	41	1
	12	222	11	11	11	11	12	73	147	194	210	194	147	73	12	11	11	11	45	12
	HALF DAY TOTALS		29	30	33	89	244	446	578	617	560	411	212	66	30	29	29	29	106	
Mar 21	7	95	4	11	47	74	90	91	79	53	17	4	4	4	4	4	4	4	9	5
	8	185	9	10	46	113	158	177	170	135	78	14	9	9	9	9	9	9	32	4
	9	227	13	13	16	88	159	203	215	194	143	64	14	13	13	13	13	13	59	3
	10	249	16	16	16	35	122	190	226	228	194	130	42	16	16	16	16	16	84	2
	11	260	17	17	17	18	60	148	209	236	228	184	109	25	17	17	17	17	99	1
	12	263	18	18	18	18	19	85	168	221	239	221	168	85	19	18	18	18	105	12
	HALF DAY TOTALS		68	74	150	334	596	854	984	958	779	504	257	104	68	68	68	68	335	
Apr 21	5	27	8	18	24	27	26	20	12	2	1	1	1	1	1	1	1	1	2	7
	6	133	12	59	102	127	132	118	84	35	8	8	8	8	8	8	8	8	21	6
	7	194	14	41	113	163	189	185	153	96	25	13	13	13	13	13	13	13	51	5
	8	228	17	19	79	153	201	217	201	153	79	19	17	17	17	17	17	17	85	4
	9	248	21	21	32	111	180	219	225	197	138	55	22	21	21	21	21	21	116	3
	10	260	23	23	24	51	134	194	225	221	185	118	38	24	23	23	23	23	140	2
	11	266	24	24	24	26	68	148	202	225	214	171	99	29	24	24	24	24	155	1
	12	268	25	25	25	25	27	83	159	208	224	208	159	83	27	25	25	25	160	12
	HALF DAY TOTALS		131	218	410	671	943	1150	1186	1036	763	487	273	149	121	120	120	120	651	
May 21	4	51	30	44	51	51	43	28	8	3	3	3	3	3	3	3	3	10	6	8
	5	132	48	95	125	135	125	96	50	11	9	9	9	9	9	9	9	11	26	7
	6	185	28	97	150	181	183	158	109	40	15	15	15	15	15	15	15	15	55	6
	7	218	21	63	138	189	211	201	161	94	24	19	19	19	19	19	19	19	90	5
	8	239	23	28	97	167	209	220	198	146	68	25	23	23	23	23	23	23	124	4
	9	252	26	27	45	122	183	215	215	184	123	46	27	26	26	26	26	26	152	3
	10	261	28	28	30	61	135	188	212	205	167	102	36	28	28	28	28	28	174	2
	11	265	30	30	30	32	72	141	188	207	195	154	87	33	30	30	30	30	188	1
	12	267	30	30	30	30	33	78	146	189	204	189	146	78	33	30	30	30	192	12
	HALF DAY TOTALS		247	425	680	950	1177	1291	1218	985	708	465	288	191	169	168	168	176	911	
Jun 21	4	93	53	83	96	94	78	50	14	7	7	7	7	7	7	7	7	21	16	8
	5	154	62	114	148	158	145	110	55	14	12	12	12	12	12	12	12	14	39	7
	6	194	36	107	162	191	192	163	110	39	18	17	17	17	17	17	17	18	71	6
	7	221	24	71	145	193	213	200	158	89	25	22	22	22	22	22	22	22	105	5
	8	239	25	33	104	170	208	216	192	139	62	27	25	25	25	25	25	25	137	4
	9	251	28	29	51	124	181	210	208	175	115	43	29	28	28	28	28	28	165	3
	10	258	30	30	32	65	134	183	204	195	157	94	36	30	30	30	30	30	186	2
	11	262	32	32	32	34	72	137	180	184	144	82	35	32	32	32	32		199	1
	12	263	32	32	32	32	35	76	138	179	193	179	138	76	35	32	32	32	203	12
	HALF DAY TOTALS		322	533	801	1061	1253	1317	1195	946	679	455	296	211	192	190	191	213	1021	
Jul 21	4	53	32	47	55	54	46	29	9	4	4	4	4	4	4	4	4	11	8	8
	5	128	49	94	123	133	124	95	50	11	10	10	10	10	10	10	10	11	28	7
	6	179	30	96	148	177	180	155	106	39	16	15	15	15	15	15	15	15	57	6
	7	211	22	64	137	186	207	197	157	92	25	20	20	20	20	20	20	20	92	5
	8	231	24	30	97	165	205	215	193	142	67	26	24	24	24	24	24	24	124	4
	9	245	27	28	47	121	180	211	211	179	120	46	28	27	27	27	27	27	152	3
	10	253	29	29	31	62	134	185	208	200	164	100	37	29	29	29	29	29	174	2
	11	257	31	31	31	33	72	139	185	202	191	151	86	34	31	31	31	31	187	1
	12	259	31	31	31	31	34	78	143	185	200	185	143	78	34	31	31	31	192	12
	HALF DAY TOTALS		258	434	684	946	1163	1269	1193	965	697	462	292	198	177	175	175	185	918	
Aug 21	5	29	9	20	27	30	28	22	13	2	2	2	2	2	2	2	2	3	3	7
	6	123	13	58	97	121	125	111	80	34	9	9	9	9	9	9	9	9	23	6
	7	181	15	42	109	157	180	176	145	92	26	14	14	14	14	14	14	14	53	5
	8	214	19	21	78	148	193	208	192	147	76	21	19	19	19	19	19	19	87	4
	9	234	22	22	34	109	174	211	217	189	133	55	23	22	22	22	22	22	117	3
	10	246	25	25	26	52	131	188	217	214	178	114	39	25	25	25	25	25	140	2
	11	252	26	26	26	28	69	144	196	217	207	166	97	31	26	26	26	26	154	1
	12	254	27	27	27	27	29	82	155	201	217	201	155	82	29	27	27	27	159	12
	HALF DAY TOTALS		142	226	410	657	914	1109	1141	997	740	478	275	158	131	130	130	130	656	
Sep 21	7	77	4	10	39	62	74	75	65	44	15	4	4	4	4	4	4	4	8	5
	8	163	10	10	43	103	143	160	154	123	71	14	10	10	10	10	10	10	31	4
	9	206	14	14	17	83	148	189	200	181	133	61	15	14	14	14	14	14	57	3
	10	229	16	16	17	35	116	179	213	214	183	123	41	17	16	16	16	16	81	2
	11	240	18	18	18	19	59	141	198	224	216	174	104	26	18	18	18	18	96	1
	12	244	19	19	19	19	21	82	160	209	227	209	160	82	21	19	19	19	101	12
	HALF DAY TOTALS		71	77	142	307	547	787	910	891	731	480	249	106	72	71	71	71	324	
Oct 21	8	17	1	1	3	10	14	16	16	13	8	2	1	1	1	1	1	1	1	4
	9	122	5	5	6	40	82	109	118	110	83	42	6	5	5	5	5	5	13	3
	10	176	9	9	9	18	83	135	165	169	147	102	36	9	9	9	9	9	29	2
	11	201	11	11	11	11	43	116	167	191	186	152	92	20	11	11	11	11	41	1
	12	208	11	11	11	11	13	70	140	184	199	184	140	70	13	11	11	11	46	12
	HALF DAY TOTALS		31	31	34	86	231	420	542	580	527	388	202	66	32	31	31	31	108	
Nov 21	10	23	1	1	1	1	10	17	21	22	20	14	5	1	1	1	1	1	1	2
	11	79	3	3	3	3	15	44	65	75	74	61	37	8	3	3	3	3	6	1
	12	97	4	4	4	4	4	32	66	87	93	87	66	32	4	4	4	4	8	12
	HALF DAY TOTALS		5	5	5	6	26	79	120	141	140	117	74	23	6	5	5	5	11	
Dec 21	11	4	0	0	0	0	1	2	3	4	4	3	2	0	0	0	0	0	0	1
	12	16	0	0	0	0	1	5	11	14	15	14	11	5	1	0	0	0	0	12
	HALF DAY TOTALS		0	0	0	0	1	5	9	11	11	10	7	3	0	0	0	0		
			N	NNW	NW	WNW	W	WSW	SW	SSW	S	SSE	SE	ESE	E	ENE	NE	NNE	HOR	PM

TABLE B3.1   SI UNITS—SOLAR INTENSITY AND SOLAR HEAT GAIN FACTORS FOR 0° NORTH LATITUDE

Solar Heat Gain Factors (W/m²)

Date	Solar Time (am)	Direct Normal (W/m²)	N	NNE	NE	ENE	E	ESE	SE	SSE	S	SSW	SW	WSW	W	WNW	NW	NNW	HOR	Solar Time (pm)
Jan 21	7	686	31	37	247	469	608	648	592	439	204	34	31	31	31	31	31	31	126	5
	8	909	58	63	278	559	737	801	740	565	287	263	58	58	58	58	58	58	383	4
	9	994	79	83	211	483	670	751	718	575	329	96	79	79	79	79	79	79	613	3
	10	1033	95	97	125	311	500	605	614	529	354	153	98	95	95	95	95	95	787	2
	11	1052	105	105	109	137	259	385	452	447	369	239	125	108	105	105	105	105	897	1
	12	1058	108	108	108	111	163	262	344	374	374	344	262	163	113	111	108	108	933	12
	HALF DAY TOTALS		424	441	1037	2042	2867	3320	3296	2759	1739	748	510	440	427	425	424	424	3272	
Feb 21	7	689	34	100	354	541	640	640	540	352	99	34	33	33	33	33	33	33	135	5
	8	906	62	122	415	646	771	778	664	444	147	62	61	61	61	61	61	61	403	4
	9	987	84	112	348	575	700	719	629	436	177	85	82	82	82	82	82	82	639	3
	10	1025	100	104	222	402	524	562	512	378	196	103	98	98	98	98	98	98	820	2
	11	1044	110	112	121	188	274	331	337	291	208	132	113	110	109	109	109	109	910	1
	12	1049	113	114	114	116	117	133	168	200	212	200	168	133	117	116	114	114	964	12
	HALF DAY TOTALS		447	613	1541	2440	3008	3144	2810	2029	938	509	460	445	442	441	440	440	3409	
Mar 21	7	649	38	219	436	575	621	575	436	219	38	34	34	34	34	34	34	34	136	5
	8	868	68	273	538	704	763	704	538	273	68	63	63	63	63	63	63	63	402	4
	9	951	90	253	493	645	697	645	493	253	90	85	85	85	85	85	85	85	636	3
	10	991	107	201	365	485	526	485	365	201	107	103	101	101	101	101	101	103	816	2
	11	1009	117	146	206	258	278	258	206	146	117	115	113	111	111	111	113	115	924	1
	12	1015	121	121	121	121	121	121	121	121	121	121	121	121	121	121	121	121	956	12
	HALF DAY TOTALS		482	1164	2121	2758	2979	2758	2121	1164	482	460	457	456	456	456	457	460	3391	
Apr 21	7	558	99	309	463	543	540	454	296	87	36	36	36	36	36	36	36	37	126	5
	8	786	155	418	610	707	697	581	372	115	65	65	65	65	65	65	65	68	373	4
	9	876	187	422	593	670	648	530	320	112	87	87	87	27	87	87	87	92	592	3
	10	919	208	374	493	534	494	377	210	111	103	103	103	103	103	103	103	110	760	2
	11	939	220	296	336	325	267	185	125	118	113	113	113	113	113	116	119	143	863	1
	12	945	224	211	178	141	124	121	120	119	116	119	120	121	124	141	178	211	895	12
	HALF DAY TOTALS		982	1940	2606	2874	2725	2198	1394	603	463	464	464	465	467	472	490	550	3160	
May 21	7	493	167	343	458	500	470	366	200	43	36	36	36	36	36	36	36	41	116	5
	8	728	260	491	637	688	635	486	253	74	67	67	67	67	67	67	67	75	344	4
	9	822	307	518	641	670	600	439	207	95	89	89	89	89	89	89	89	108	549	3
	10	868	335	486	561	552	461	297	136	108	104	104	104	104	104	104	110	160	706	2
	11	890	350	418	423	364	255	147	120	114	114	114	114	114	114	119	137	237	804	1
	12	896	355	329	258	171	126	121	117	117	117	117	117	121	126	171	258	329	836	12
	HALF DAY TOTALS		1599	2431	2867	2877	2490	1801	976	494	469	469	469	470	474	488	556	776	2935	
Jun 21	7	459	189	345	444	473	433	325	161	41	36	36	36	36	36	36	36	43	110	5
	8	696	299	510	638	670	601	443	207	73	68	68	68	68	68	68	68	85	328	4
	9	793	353	546	651	661	574	398	167	94	89	89	89	89	89	89	92	133	526	3
	10	841	384	522	579	553	443	265	120	107	105	105	105	105	105	105	112	198	677	2
	11	864	401	461	452	375	248	137	119	114	114	114	114	114	114	121	158	283	772	1
	12	870	407	376	292	185	127	121	117	117	117	117	117	121	127	185	292	376	804	12
	HALF DAY TOTALS		1831	2581	2925	2838	2365	1631	835	487	471	471	471	472	476	497	597	923	2814	
Jul 21	7	462	167	331	438	476	446	345	187	43	37	37	37	37	37	37	37	42	114	5
	8	697	264	485	625	671	616	470	243	76	69	69	69	69	69	69	69	78	337	4
	9	794	313	516	634	659	587	427	200	97	91	91	91	91	91	91	91	114	538	3
	10	841	342	488	558	547	454	291	135	111	107	107	107	107	107	107	113	168	691	2
	11	863	358	423	426	364	254	148	123	116	116	116	116	116	116	122	143	246	788	1
	12	870	363	336	264	176	129	124	119	119	119	119	119	124	129	176	264	336	820	12
	HALF DAY TOTALS		1626	2421	2828	2819	2425	1746	947	503	479	479	479	480	484	501	573	807	2876	
Aug 21	7	500	102	292	430	501	496	415	269	80	38	38	38	38	38	38	38	39	122	5
	8	734	164	412	591	681	667	553	353	112	69	69	69	69	69	69	69	73	362	4
	9	828	198	424	585	655	630	512	308	113	92	92	92	92	92	92	92	98	575	3
	10	873	220	381	492	528	485	367	205	116	108	108	108	108	108	108	110	117	738	2
	11	895	233	307	343	328	268	186	129	123	118	118	118	118	118	121	125	154	839	1
	12	901	237	223	188	149	130	127	123	123	121	123	125	127	130	149	188	223	870	12
	HALF DAY TOTALS		1037	1939	2551	2783	2619	2101	1332	606	484	484	485	487	488	495	516	586	3070	
Sep 21	7	581	41	205	401	528	570	528	401	205	41	36	36	36	36	36	36	36	131	5
	8	808	72	266	515	672	728	672	515	266	72	66	66	66	66	66	66	66	388	4
	9	896	95	251	479	624	674	624	479	251	95	88	88	88	88	88	88	88	615	3
	10	939	112	202	360	474	513	474	360	202	112	107	105	105	105	105	105	107	789	2
	11	959	122	150	208	258	277	258	208	150	122	119	117	115	115	115	117	119	894	1
	12	965	125	125	125	125	125	125	125	125	125	125	125	125	125	125	125	125	925	12
	HALF DAY TOTALS		505	1143	2039	2637	2844	2637	2039	1143	505	479	475	474	473	474	475	479	3279	
Oct 21	7	636	35	100	336	509	600	599	505	329	93	35	34	34	34	34	34	34	131	5
	8	861	64	126	407	626	746	750	639	427	143	65	63	63	63	63	63	63	393	4
	9	947	86	117	347	564	684	700	611	423	173	88	85	85	85	85	85	85	624	3
	10	988	103	108	225	399	516	550	499	368	193	106	101	101	101	101	101	101	801	2
	11	1008	113	115	125	191	273	326	330	284	205	132	115	113	111	111	111	111	910	1
	12	1013	116	116	117	118	120	135	168	197	209	197	168	135	120	118	117	116	943	12
	HALF DAY TOTALS		439	629	1516	2370	2906	3028	2701	1950	914	518	471	457	453	452	451	451	3330	
Nov 21	7	622	32	38	243	457	591	629	574	424	196	35	32	32	32	32	32	32	124	5
	8	889	59	64	278	553	726	787	727	554	280	64	59	59	59	59	59	59	379	4
	9	976	80	84	213	480	664	742	708	566	323	96	80	80	80	80	80	80	607	3
	10	1017	96	99	127	311	497	598	607	521	349	151	99	96	96	96	96	96	780	2
	11	1036	106	106	110	139	259	381	447	441	363	236	125	109	106	106	106	106	889	1
	12	1042	109	109	109	112	114	163	259	339	368	339	259	163	114	112	109	109	924	12
	HALF DAY TOTALS		428	446	1037	2020	2824	3264	3235	2705	1704	742	512	444	431	429	428	428	3240	
Dec 21	7	678	30	33	203	435	586	641	601	462	242	35	30	30	30	30	30	30	120	5
	8	905	57	60	224	518	714	798	759	601	340	72	57	57	57	57	57	57	370	4
	9	991	78	80	163	441	651	754	742	618	387	122	79	78	78	78	78	78	595	3
	10	1032	93	94	104	274	484	613	644	578	415	195	97	93	93	93	93	93	766	2
	11	1051	103	103	105	122	251	400	490	501	431	292	146	106	103	103	103	103	874	1
	12	1057	106	106	106	108	112	177	302	400	436	400	302	177	112	108	106	106	910	12
	HALF DAY TOTALS		414	423	864	1869	2772	3342	3434	2993	2045	909	543	436	417	414	414	414	3179	

| | | | N | NNW | NW | WNW | W | WSW | SW | SSW | S | SSE | SE | ESE | E | ENE | NE | NNE | HOR | PM |

TABLE B3.1 SI UNITS—SOLAR INTENSITY AND SOLAR HEAT GAIN FACTORS FOR 8° NORTH LATITUDE

Solar Heat Gain Factors (W/m²)

Date	Solar Time (am)	Direct Normal (W/m²)	N	NNE	NE	ENE	E	ESE	SE	SSE	S	SSW	SW	WSW	W	WNW	NW	NNW	HOR	Solar Time (pm)
Jan 21	7	590	25	27	196	394	521	561	519	392	192	27	25	25	25	25	25	25	82	5
	8	876	52	56	223	513	706	788	748	591	332	67	52	52	52	52	52	52	320	4
	9	975	74	76	149	433	653	766	762	642	414	133	75	74	74	74	74	74	550	3
	10	1020	89	89	95	258	489	637	683	626	469	234	94	89	89	89	89	89	722	2
	11	1041	99	99	101	111	252	430	542	566	502	354	173	103	99	99	99	99	831	1
	12	1048	102	102	102	104	107	195	354	471	512	471	354	195	107	104	102	102	868	12
HALF DAY TOTALS			389	397	805	1744	2648	3267	3421	3041	2156	1040	577	420	392	389	389	389	2937	
Feb 21	7	642	29	76	312	494	595	602	517	347	110	30	29	29	29	29	29	29	108	5
	8	888	57	83	359	608	754	783	691	491	197	60	57	57	57	57	57	57	367	4
	9	977	79	84	271	529	690	745	687	522	262	83	79	79	79	79	79	79	602	3
	10	1018	95	98	149	346	518	603	595	493	309	125	97	95	95	95	95	95	780	2
	11	1037	105	106	109	148	269	382	437	422	338	212	117	108	105	105	105	105	892	1
	12	1043	108	108	108	111	113	158	246	320	348	320	246	158	113	111	108	108	928	12
HALF DAY TOTALS			419	501	1254	2180	2883	3207	3064	2443	1388	661	488	432	421	419	419	419	3210	
Mar 21	7	645	37	194	417	563	618	581	450	242	41	34	34	34	34	34	34	34	133	5
	8	866	66	212	491	678	761	726	579	334	80	62	62	62	62	62	62	62	397	4
	9	949	88	170	419	607	695	680	561	346	117	86	84	84	84	84	84	84	628	3
	10	989	103	125	273	433	524	533	458	309	147	104	100	100	100	100	100	100	806	2
	11	1008	113	116	139	209	277	310	297	241	167	119	114	112	110	110	110	110	914	1
	12	1013	116	117	117	118	120	129	148	167	174	167	148	129	118	118	117	117	947	12
HALF DAY TOTALS			466	886	1821	2575	2966	2929	2449	1570	639	484	461	454	452	451	451	451	3351	
Apr 21	6	2	0	1	2	2	2	2	1	0	0	0	0	0	0	0	0	0	0	6
	7	596	88	308	478	572	577	495	334	112	41	39	39	39	39	39	39	39	149	5
	8	800	114	369	580	698	710	614	424	161	71	68	68	68	68	68	68	68	398	4
	9	883	125	337	535	643	655	570	394	166	94	90	90	90	90	90	90	92	615	3
	10	923	132	265	409	492	498	428	291	147	110	105	105	105	105	105	105	109	781	2
	11	942	136	188	245	276	270	228	169	127	120	118	115	115	115	115	118	121	882	1
	12	948	138	136	132	128	126	125	124	124	124	124	124	125	126	128	132	136	912	12
HALF DAY TOTALS			672	1569	2364	2803	2829	2443	1703	784	501	484	482	482	482	483	486	498	3283	
May 21	6	18	6	12	17	18	17	14	8	1	1	1	1	1	1	1	1	2		6
	7	560	169	370	506	562	536	425	246	57	43	43	43	43	43	43	43	47	156	5
	8	753	215	460	624	694	660	527	307	89	72	72	72	72	72	72	72	78	390	4
	9	836	228	448	595	652	613	483	272	104	93	93	93	93	93	93	93	101	592	3
	10	878	232	387	490	518	469	347	190	116	108	108	108	108	108	111	120	748		2
	11	898	233	303	336	320	260	180	129	123	118	118	118	118	118	122	126	156	843	1
	12	904	233	220	186	148	130	127	125	123	121	123	125	127	130	148	186	220	873	12
HALF DAY TOTALS			1215	2133	2718	2899	2670	2077	1236	556	500	500	500	502	504	510	531	610	3171	
Jun 21	6	27	11	20	26	28	26	20	10	2	2	2	2	2	2	2	2	3		6
	7	537	201	387	507	549	510	391	207	51	45	45	45	45	45	45	45	51	157	5
	8	728	260	487	632	683	631	486	258	82	74	74	74	74	74	74	74	82	382	4
	9	811	278	485	611	648	590	443	226	102	95	95	95	95	95	95	95	107	577	3
	10	853	283	432	516	523	453	313	159	116	110	110	110	110	110	110	114	140	726	2
	11	874	284	352	371	334	254	164	127	122	119	119	119	119	119	124	130	192	819	1
	12	880	284	266	217	160	131	127	125	122	122	122	125	127	131	160	217	266	849	12
HALF DAY TOTALS			1475	2336	2826	2901	2573	1913	1065	539	507	507	507	509	512	522	561	702	3093	
Jul 21	6	16	5	11	15	16	16	12	7	1	1	1	1	1	1	1	1	2		6
	7	530	171	362	490	542	514	406	233	56	45	45	45	45	45	45	45	49	155	5
	8	725	222	457	614	679	643	511	295	89	74	74	74	74	74	74	74	81	383	4
	9	809	237	450	591	643	601	471	263	106	96	96	96	96	96	96	96	104	582	3
	10	851	241	393	491	315	463	340	185	119	111	111	111	111	111	111	114	125	734	2
	11	872	242	312	341	322	260	179	131	125	120	120	120	120	124	129	164	828		1
	12	878	243	229	193	153	133	129	127	123	123	123	127	129	133	153	193	229	857	12
HALF DAY TOTALS			1255	2139	2692	2849	2608	2018	1198	563	511	511	512	513	515	522	547	635	3116	
Aug 21	6	1	0	1	1	1	1	1	1	0	0	0	0	0	0	0	0	0		6
	7	540	93	296	451	535	537	458	308	104	44	42	42	42	42	42	42	43	147	5
	8	749	124	368	565	675	682	587	403	154	75	72	72	72	72	72	72	74	388	4
	9	836	136	343	530	630	637	552	378	162	99	94	94	94	94	94	94	97	599	3
	10	878	143	277	413	488	489	417	283	146	115	110	110	110	110	110	110	115	761	2
	11	899	147	201	255	282	271	227	169	131	126	123	120	120	120	120	124	128	859	1
	12	905	149	146	141	135	132	131	130	129	129	129	130	131	132	135	141	146	888	12
HALF DAY TOTALS			725	1586	2325	2722	2726	2341	1628	768	525	507	505	505	505	507	511	530	3200	
Sep 21	7	577	39	182	383	516	566	532	414	225	43	36	36	36	36	36	36	36	129	5
	8	805	70	209	471	647	725	692	554	323	84	66	66	66	66	66	66	66	383	4
	9	894	92	172	408	587	672	657	544	339	121	90	88	88	88	88	88	88	607	3
	10	937	108	130	271	424	512	520	448	306	150	109	104	104	104	104	104	104	780	2
	11	957	117	120	144	210	276	307	295	242	170	123	118	116	114	114	114	114	884	1
	12	963	120	121	122	123	124	134	153	170	177	170	153	134	124	123	122	121	916	12
HALF DAY TOTALS			487	879	1752	2461	2831	2799	2351	1529	657	504	480	472	469	468	468	468	3241	
Oct 21	7	589	30	76	295	462	555	560	480	322	103	31	30	30	30	30	30	30	106	5
	8	843	59	87	353	589	728	754	664	471	190	62	59	59	59	59	59	59	358	4
	9	957	81	87	272	520	674	725	666	505	254	86	81	81	81	81	81	81	588	3
	10	981	97	101	154	344	509	590	580	480	301	126	100	97	97	97	97	97	763	2
	11	1001	107	109	112	151	268	376	428	411	330	208	119	110	107	107	107	107	873	1
	12	1007	111	111	111	114	117	159	244	313	340	313	243	159	117	114	111	111	908	12
HALF DAY TOTALS			430	515	1237	2119	2786	3091	2949	2351	1345	661	497	443	432	430	430	430	3140	
Nov 21	7	566	25	28	192	382	504	542	501	377	185	28	25	25	25	25	25	25	81	5
	8	855	53	57	223	507	694	774	734	579	325	67	53	53	53	53	53	53	317	4
	9	957	75	77	151	431	646	756	751	632	407	132	76	75	75	75	75	75	545	3
	10	1003	90	90	96	258	486	630	674	618	462	230	95	90	90	90	90	90	716	2
	11	1025	100	100	102	113	252	426	535	558	494	349	171	104	100	100	100	100	824	1
	12	1032	103	103	103	105	109	195	349	465	505	465	349	195	109	105	103	103	861	12
HALF DAY TOTALS			393	402	806	1725	2609	3212	3359	2984	2115	1027	576	424	396	393	393	393	2911	
Dec 21	7	560	22	24	152	350	481	531	531	394	215	30	22	22	22	22	22	22	71	5
	8	864	50	52	172	470	677	777	757	618	377	88	50	50	50	50	50	50	298	4
	9	968	71	72	112	391	631	764	780	677	465	173	73	71	71	71	71	71	523	3
	10	1016	86	86	90	223	472	641	707	667	522	285	95	86	86	86	86	86	692	2
	11	1038	96	96	97	102	242	442	573	611	555	408	204	100	96	96	96	96	798	1
	12	1044	99	99	99	100	104	210	392	522	566	522	392	210	104	100	99	99	835	12
HALF DAY TOTALS			374	378	661	1562	2520	3240	3498	3208	2403	1234	624	415	376	374	374	374	2796	
			N	NNW	NW	WNW	W	WSW	SW	SSW	S	SSE	SE	ESE	E	ENE	NE	NNE	HOR	PM

TABLE B3.1 SI UNITS—SOLAR INTENSITY AND SOLAR HEAT GAIN FACTORS FOR 16° NORTH LATITUDE

Solar Heat Gain Factors (W/m²)

Date	Solar Time (am)	Direct Normal (W/m²)	N	NNE	NE	ENE	E	ESE	SE	SSE	S	SSW	SW	WSW	W	WNW	NW	NNW	HOR	Solar Time (pm)
Jan 21	7	445	17	19	138	291	390	424	397	303	155	19	17	17	17	17	17	17	43	5
	8	827	45	48	174	463	662	757	734	596	359	79	45	45	45	45	45	45	249	4
	9	948	67	67	102	384	630	770	791	690	481	183	69	67	67	67	67	67	472	3
	10	1001	82	82	86	209	474	658	737	704	563	321	97	82	82	82	82	82	640	2
	11	1025	92	92	92	96	242	467	614	663	612	462	236	96	92	92	92	92	745	1
	12	1032	95	95	95	95	100	228	438	580	628	580	438	228	100	95	95	95	782	12
HALF DAY TOTALS			348	352	618	1453	2398	3153	3458	3217	2465	1344	666	401	350	348	348	348	2539	
Feb 21	7	575	24	55	265	435	532	544	474	326	113	26	24	24	24	24	24	24	80	5
	8	862	53	60	304	567	729	778	706	525	244	56	53	53	53	53	53	53	319	4
	9	961	74	77	202	482	676	763	733	592	347	96	74	74	74	74	74	74	549	3
	10	1006	90	91	104	292	508	636	665	593	423	193	94	90	90	90	90	90	722	2
	11	1027	99	99	102	118	262	428	527	542	471	323	154	103	99	99	99	99	831	1
	12	1034	103	103	103	105	108	189	336	448	487	448	336	189	108	105	103	103	868	12
HALF DAY TOTALS			390	431	1013	1922	2730	3228	3263	2792	1836	906	547	417	393	390	390	390	2933	
Mar 21	7	634	36	167	393	544	606	578	458	260	47	33	33	33	33	33	33	33	126	5
	8	857	63	157	442	648	752	741	615	390	111	61	61	61	61	61	61	61	380	4
	9	943	84	110	343	565	689	709	622	435	180	86	82	82	82	82	82	82	606	3
	10	983	98	103	191	379	519	575	543	424	240	107	100	98	98	98	98	98	778	2
	11	1003	108	110	113	166	273	361	395	366	281	173	114	110	108	108	108	108	885	1
	12	1008	111	111	112	114	117	149	216	273	295	273	216	149	117	114	112	111	919	12
HALF DAY TOTALS			443	712	1558	2383	2926	3076	2773	2028	1006	588	483	448	440	483	437	437	3233	
Apr 21	6	44	2	24	37	43	43	37	24	7	2	2	2	2	2	2	12	2	4	6
	7	622	75	298	482	589	604	528	369	141	45	42	42	42	42	42	42	42	169	5
	8	807	85	312	543	682	718	644	473	217	74	70	70	70	70	70	70	70	413	4
	9	885	97	248	469	610	657	608	465	244	97	90	90	90	90	90	90	90	623	3
	10	924	112	171	321	444	499	476	380	231	118	109	166	106	106	106	106	106	784	2
	11	942	120	127	169	228	270	276	245	189	136	121	118	115	115	115	115	118	882	1
	12	948	123	123	124	125	126	129	135	141	143	141	135	129	126	125	124	123	911	12
HALF DAY TOTALS			565	1272	2127	2711	2909	2684	2059	1111	547	503	494	491	490	489	490	492	3333	
May 21	6	138	43	94	128	141	134	106	59	11	9	9	9	9	9	9	9	10	195	6
	7	608	157	378	531	603	583	474	290	76	49	49	49	49	49	49	49	55	17	5
	8	771	165	415	598	689	677	564	362	121	78	76	76	76	76	76	76	80	425	4
	9	845	156	366	539	626	622	526	344	141	100	96	96	96	96	96	96	100	621	3
	10	883	149	281	410	478	474	398	264	139	116	111	111	111	111	111	111	116	772	2
	11	902	147	198	248	273	262	220	166	130	126	123	120	120	120	122	124	128	862	1
	12	907	146	144	140	134	132	131	130	129	129	129	130	131	132	134	140	144	890	12
HALF DAY TOTALS			893	1813	2537	2891	2829	2360	1555	685	541	528	527	526	527	529	533	557	3338	
Jun 21	6	168	64	124	163	175	162	123	64	13	12	12	12	12	12	12	12	13	24	6
	7	593	195	404	543	598	565	445	252	63	52	52	52	52	52	52	52	57	203	5
	8	750	209	449	612	684	653	526	313	98	79	79	79	79	79	79	79	84	425	4
	9	823	199	409	560	626	601	487	294	118	98	98	98	98	98	98	98	105	613	3
	10	861	187	329	441	486	459	363	222	125	116	113	113	113	113	113	113	121	759	2
	11	879	181	241	283	290	258	200	146	130	126	122	122	122	122	125	128	136	847	1
	12	885	179	173	158	142	134	132	130	129	129	129	130	132	134	142	158	173	873	12
HALF DAY TOTALS			1122	2043	2683	2930	2763	2207	1357	612	548	540	540	540	542	545	555	599	3308	
Jul 21	6	128	43	90	122	134	127	99	55	11	9	9	9	9	9	9	9	10	17	6
	7	579	161	373	518	585	564	456	277	74	51	51	51	51	51	51	51	55	194	5
	8	743	173	415	590	676	661	549	349	110	78	78	78	78	78	78	78	83	419	4
	9	818	165	371	537	618	610	513	334	138	102	99	99	99	99	99	99	104	611	3
	10	857	157	289	413	475	467	389	257	138	118	113	113	113	113	113	113	120	759	2
	11	876	154	207	255	277	262	218	164	133	128	125	122	122	122	125	128	131	848	1
	12	882	154	151	145	138	135	134	133	132	132	132	133	134	135	138	145	151	875	12
HALF DAY TOTALS			933	1829	2521	2848	2770	2298	1507	680	554	541	539	539	540	542	547	575	3289	
Aug 21	6	136	7	21	31	37	36	31	20	6	2	2	2	2	2	2	2	2	4	6
	7	569	81	289	458	555	567	493	343	131	48	45	45	45	45	45	45	45	167	5
	8	757	94	315	531	660	691	617	451	206	79	74	74	74	74	74	74	74	404	4
	9	838	104	258	467	598	640	589	448	235	103	95	95	95	95	95	95	95	608	3
	10	879	118	183	328	443	490	464	368	224	122	114	110	110	110	110	110	110	766	2
	11	899	127	136	180	235	271	273	240	185	138	127	124	120	120	120	120	124	860	1
	12	905	129	130	130	131	132	134	139	143	145	143	139	134	132	131	130	130	889	12
HALF DAY TOTALS			603	1294	2099	2640	2810	2578	1969	1069	568	528	519	516	515	514	515	518	3258	
Sep 21	7	565	38	157	360	497	554	529	419	240	48	35	35	35	35	35	35	35	122	5
	8	797	67	156	424	618	716	705	587	374	113	64	64	64	64	64	64	64	367	4
	9	887	87	114	335	547	665	684	602	424	181	90	86	86	86	86	86	86	585	3
	10	931	101	107	193	372	507	560	529	415	240	111	103	101	101	101	101	101	752	2
	11	952	111	114	118	169	272	356	389	361	279	176	118	114	111	111	111	111	856	1
	12	958	114	114	116	118	121	153	217	272	293	272	217	153	121	118	116	114	889	12
HALF DAY TOTALS			461	712	1500	2276	2791	2937	2658	1963	1007	605	501	466	457	455	454	454	3126	
Oct 21	7	524	25	56	249	404	492	502	437	300	105	25	25	25	25	25	25	25	79	5
	8	816	55	64	299	548	702	747	677	502	234	59	55	55	55	55	55	55	313	4
	9	920	76	80	204	473	659	741	711	573	336	97	76	76	76	76	76	76	537	3
	10	968	92	94	108	291	499	621	647	577	412	189	97	92	92	92	92	92	707	2
	11	991	102	105	122	261	420	515	528	459	315	154	106	102	102	102	102	102	814	1
	12	997	105	105	105	107	111	189	330	437	474	437	330	189	111	107	105	105	850	12
HALF DAY TOTALS			402	444	1002	1869	2637	3110	3141	2689	1776	891	553	428	404	402	402	402	2872	
Nov 21	7	423	17	19	134	280	375	406	379	289	147	19	17	17	17	17	17	17	43	5
	8	806	46	49	174	456	651	742	719	583	350	78	46	46	46	46	46	46	247	4
	9	929	67	67	105	382	623	779	779	679	472	180	70	67	67	67	67	67	468	3
	10	984	83	83	87	210	470	651	727	694	554	316	98	83	83	83	83	83	635	2
	11	1009	92	92	92	98	242	462	607	654	603	455	234	98	92	92	92	92	740	1
	12	1016	96	96	96	96	101	227	432	572	618	572	432	227	101	96	96	96	775	12
HALF DAY TOTALS			352	357	620	1438	2363	3101	3396	3159	2421	1324	664	404	355	352	352	352	2520	
Dec 21	7	372	13	14	93	228	318	354	339	268	150	21	13	13	13	13	13	13	31	5
	8	803	42	43	129	416	625	734	728	609	390	105	42	42	42	42	42	42	219	4
	9	936	63	63	78	341	604	761	800	716	520	226	66	63	63	63	63	63	436	3
	10	993	78	78	81	178	445	657	753	735	605	369	112	79	78	78	78	78	602	2
	11	1019	87	87	87	92	231	474	638	698	654	506	270	94	87	87	87	87	704	1
	12	1026	91	91	91	91	95	241	469	620	670	620	469	241	95	91	91	91	739	12
HALF DAY TOTALS			328	330	501	1269	2241	3076	3468	3312	2638	1527	721	393	331	328	328	328	2361	
			N	NNW	NW	WNW	W	WSW	SW	SSW	S	SSE	SE	ESE	E	ENE	NE	NNE	HOR	PM

TABLE B3.1 SI UNITS—SOLAR INTENSITY AND SOLAR HEAT GAIN FACTORS FOR 24° NORTH LATITUDE

Solar Heat Gain Factors (W/m²)

Date	Solar Time (am)	Direct Normal (W/m²)	N	NNE	NE	ENE	E	ESE	SE	SSE	S	SSW	SW	WSW	W	WNW	NW	NNW	HOR	Solar Time (pm)
Jan 21	7	223	7	8	65	143	195	212	200	154	80	9	7	7	7	7	7	7	7	15	5
	8	754	37	39	131	404	600	698	689	571	360	89	37	37	37	37	37	37	37	174	4
	9	908	58	58	72	335	598	757	799	716	523	229	61	58	58	58	58	58	58	381	3
	10	973	973	73	75	166	454	667	772	759	631	395	119	74	73	73	73	73	73	543	2
	11	1001	82	82	82	86	229	492	667	737	695	546	299	91	82	82	82	82	82	643	1
	12	1010	85	85	85	85	90	257	504	663	717	663	504	257	90	85	85	85	85	677	12
	HALF DAY TOTALS		301	303	466	1172	2118	2971	3393	3276	2649	1592	759	379	303	301	301	301	301	2094	
Feb 21	7	483	19	38	212	359	445	459	405	283	105	20	19	19	19	19	19	19	53	5
	8	825	47	51	252	521	695	758	706	542	281	52	47	47	47	47	47	47	263	4
	9	938	68	70	144	434	695	770	767	646	419	132	69	68	68	68	68	68	482	3
	10	989	83	83	87	241	494	659	720	673	521	276	89	83	83	83	83	83	647	2
	11	1012	92	92	92	99	252	465	601	641	584	431	213	97	92	92	92	92	750	1
	12	1019	95	95	95	95	100	221	421	558	605	558	421	221	100	95	95	95	785	12
	HALF DAY TOTALS		355	375	809	1663	2541	3191	3383	3043	2204	1180	630	402	358	355	355	355	2586	
Mar 21	7	613	33	141	364	517	586	567	457	272	55	31	31	31	31	31	31	31	115	5
	8	843	58	111	390	614	738	748	643	437	152	60	58	58	58	58	58	58	352	4
	9	932	79	84	269	521	678	732	675	514	260	84	79	79	79	79	79	79	568	3
	10	974	94	96	130	325	511	610	616	528	352	148	97	94	94	94	94	94	731	2
	11	994	103	103	106	132	267	408	486	488	411	272	135	107	103	103	103	103	834	1
	12	1000	106	106	106	106	108	112	178	301	397	432	397	301	178	112	108	106	106	869	12
	HALF DAY TOTALS		421	597	1330	2187	2859	3190	3060	2453	1446	785	534	439	421	418	418	418	3034	
Apr 21	6	126	19	67	104	123	124	105	70	21	7	6	6	6	6	6	6	6	12	6
	7	639	62	278	476	595	622	554	400	172	47	44	44	44	44	44	44	44	183	5
	8	808	76	252	500	659	719	669	519	277	77	70	70	70	70	70	70	70	416	4
	9	883	94	169	397	572	655	641	532	333	122	92	90	90	90	90	90	90	615	3
	10	920	107	116	236	394	496	521	466	338	177	110	104	104	104	104	104	104	769	2
	11	939	113	118	125	185	267	325	336	296	220	143	119	116	113	113	113	113	863	1
	12	945	116	118	119	121	123	143	184	221	235	221	184	143	123	121	119	118	893	12
	HALF DAY TOTALS		531	1070	1915	2607	2965	2914	2438	1559	770	569	515	495	490	488	487	486	3307	
May 21	6	270	80	181	249	276	264	209	121	25	18	18	18	18	18	18	18	20	40	6
	7	642	136	370	540	627	618	515	332	101	54	54	54	54	54	54	54	56	229	5
	8	782	121	359	562	676	688	598	417	169	82	78	78	78	78	78	78	80	449	4
	9	849	112	277	475	594	625	565	418	209	104	98	98	98	98	98	98	98	634	3
	10	884	119	186	323	432	475	446	349	212	122	115	111	111	111	111	111	114	778	2
	11	902	127	134	175	227	263	266	235	184	139	127	124	120	120	120	120	124	864	1
	12	907	129	129	130	131	132	134	134	140	146	144	140	134	131	130	129	129	890	12
	HALF DAY TOTALS		750	1553	2361	2867	2973	2651	1936	973	591	556	548	544	543	542	543	552	3437	
Jun 21	6	307	112	222	294	318	298	229	122	26	22	22	22	22	22	22	22	26	54	6
	7	633	175	402	557	628	605	490	297	82	57	57	57	57	57	57	57	62	244	5
	8	765	157	397	580	674	668	563	370	135	85	82	82	82	82	82	82	86	456	4
	9	829	136	322	500	597	607	529	367	168	106	101	101	101	101	101	101	104	635	3
	10	864	129	227	357	442	462	412	301	172	122	114	114	114	114	114	114	118	774	2
	11	881	132	157	204	244	259	242	201	154	132	128	126	122	122	122	126	129	856	1
	12	887	135	135	135	135	135	135	135	135	136	135	135	135	135	135	135	135	880	12
	HALF DAY TOTALS		897	1772	2530	2942	2939	2513	1718	803	590	572	567	565	565	567	567	589	3456	
Jul 21	6	255	81	176	241	266	253	200	114	25	19	19	19	19	19	19	19	21	42	6
	7	615	142	367	529	611	600	498	319	98	56	56	56	56	56	56	56	59	229	5
	8	755	128	362	556	664	673	583	404	163	85	81	81	81	81	81	81	83	444	4
	9	823	117	284	474	588	615	553	406	202	107	100	100	100	100	100	100	100	626	3
	10	859	123	194	328	431	469	437	340	206	124	117	114	114	114	114	114	117	767	2
	11	876	130	140	182	232	263	262	230	180	139	130	127	122	122	122	122	127	851	1
	12	882	133	133	133	134	135	137	141	145	146	145	141	137	135	134	133	133	876	12
	HALF DAY TOTALS		779	1572	2352	2830	2917	2586	1878	947	602	570	562	559	557	557	558	570	3395	
Aug 21	6	109	19	62	93	110	110	93	62	19	7	7	7	7	7	7	7	7	13	6
	7	588	69	273	455	565	586	521	375	161	51	47	47	47	47	47	47	47	183	5
	8	759	82	257	491	640	693	642	496	264	82	74	74	74	74	74	74	74	409	4
	9	837	100	181	398	563	638	621	513	319	123	97	95	95	95	95	95	95	603	3
	10	876	113	125	245	394	488	507	451	325	174	116	109	109	109	109	109	109	753	2
	11	896	121	124	134	193	269	320	327	287	214	145	125	121	118	118	118	118	844	1
	12	902	121	124	125	127	130	148	184	216	229	216	184	148	130	127	125	124	872	12
	HALF DAY TOTALS		566	1094	1896	2544	2870	2804	2336	1493	767	588	539	521	516	514	513	512	3243	
Sep 21	7	545	35	133	333	471	533	516	417	250	55	33	33	33	33	33	33	33	111	5
	8	782	61	112	374	584	701	711	612	418	151	64	61	61	61	61	61	61	340	4
	9	875	82	89	265	504	654	705	651	498	256	88	82	82	82	82	82	82	549	3
	10	921	97	100	134	320	498	593	599	515	346	151	101	97	97	97	97	97	707	2
	11	942	106	106	111	136	266	400	476	477	404	271	139	111	106	106	106	106	807	1
	12	949	110	110	110	112	117	180	298	391	424	391	298	180	117	112	110	110	840	12
	HALF DAY TOTALS		438	601	1280	2085	2722	3041	2926	2364	1424	793	550	457	438	435	435	435	2932	
Oct 21	7	434	20	39	197	329	407	419	369	258	97	21	20	20	20	20	20	20	53	5
	8	778	49	54	248	503	666	726	674	517	268	54	49	49	49	49	49	49	258	4
	9	897	70	72	148	426	637	747	742	624	405	130	72	70	70	70	70	70	473	3
	10	950	85	85	91	242	484	643	701	654	506	269	92	85	85	85	85	85	634	2
	11	975	95	95	95	103	252	456	586	624	568	420	210	100	95	95	95	95	735	1
	12	982	98	98	98	98	104	220	412	544	589	544	412	220	104	98	98	98	770	12
	HALF DAY TOTALS		366	388	803	1617	2454	3073	3255	2931	2130	1156	631	412	369	366	366	366	2536	
Nov 21	7	210	7	8	63	137	185	202	190	146	76	9	7	7	7	7	7	7	15	5
	8	733	38	40	131	398	588	683	673	557	351	87	38	38	38	38	38	38	173	4
	9	889	59	59	74	333	590	746	785	704	513	225	62	59	59	59	59	59	378	3
	10	955	74	74	77	167	450	659	762	748	621	389	118	75	74	74	74	74	539	2
	11	984	83	83	83	88	229	486	659	727	685	538	295	92	83	83	83	83	638	1
	12	993	86	86	86	86	91	255	498	654	707	654	498	255	91	86	86	86	672	12
	HALF DAY TOTALS		305	307	469	1161	2086	2921	3332	3217	2601	1568	754	382	308	305	305	305	2080	
Dec 21	7	94	3	3	22	56	79	89	85	68	38	5	3	3	3	3	3	3	5	5
	8	710	33	33	92	352	548	655	659	561	372	111	34	33	33	33	33	33	139	4
	9	887	54	54	59	293	567	738	794	728	547	266	58	54	54	54	54	54	336	3
	10	959	68	68	70	139	432	660	779	780	660	432	139	70	68	68	68	68	496	2
	11	991	77	77	77	81	217	494	681	761	726	579	328	92	77	77	77	77	594	1
	12	1000	81	81	81	81	85	267	527	691	747	691	527	267	85	81	81	81	627	12
	HALF DAY TOTALS		277	278	371	988	1928	2835	3323	3288	2738	1734	810	369	280	277	277	277	1887	
			N	NNW	NW	WNW	W	WSW	SW	SSW	S	SSE	SE	ESE	E	ENE	NE	NNE	HOR	PM

TABLE B3.1   SI UNITS—SOLAR INTENSITY AND SOLAR HEAT GAIN FACTORS FOR 32° NORTH LATITUDE

Solar Heat Gain Factors (W/m²)

Date	Solar Time (am)	Direct Normal (W/m²)	N	NNE	NE	ENE	E	ESE	SE	SSE	S	SSW	SW	WSW	W	WNW	NW	NNW	HOR	Solar Time (pm)
Jan 21	7	4	0	0	1	3	4	4	4	3	2	0	0	0	0	0	0	0	0	5
	8	640	28	29	93	330	505	597	596	502	326	88	29	28	28	28	28	28	102	4
	9	849	48	48	53	286	553	721	775	711	534	258	52	48	48	48	48	48	278	3
	10	931	63	63	64	129	427	659	784	788	670	444	144	64	63	63	63	63	430	2
	11	967	71	71	71	75	213	502	698	784	750	602	347	90	71	71	71	71	524	1
	12	977	74	74	74	74	79	277	548	718	777	718	548	277	79	74	74	74	556	12
	HALF DAY TOTALS		249	250	338	897	1797	2700	3201	3198	2692	1746	831	353	252	249	249	249	1614	
Feb 21	7	352	13	23	148	258	323	336	299	212	83	14	13	13	13	13	13	13	30	5
	8	771	41	43	204	469	646	719	683	538	300	53	41	41	41	41	41	41	200	4
	9	905	60	60	101	386	627	764	783	680	471	174	63	60	60	60	60	60	402	3
	10	964	75	75	78	195	475	670	760	733	595	352	99	75	75	75	75	75	556	2
	11	990	83	83	83	88	240	491	656	717	670	519	275	89	83	83	83	83	652	1
	12	998	86	86	86	86	91	250	489	644	696	643	489	250	91	86	86	86	684	12
	HALF DAY TOTALS		314	324	635	1405	2318	3086	3408	3188	2461	1426	718	384	316	314	314	314	2181	
Mar 21	7	583	30	116	332	483	556	545	447	276	63	29	29	29	29	29	29	29	100	5
	8	821	54	78	339	576	717	746	661	473	195	57	54	54	54	54	54	54	315	4
	9	914	74	77	203	475	662	746	716	578	339	95	74	74	74	74	74	74	516	3
	10	959	88	88	96	273	499	637	677	615	456	221	93	88	88	88	88	88	667	2
	11	980	96	96	98	108	258	447	564	592	529	380	185	101	96	96	96	96	762	1
	12	987	99	99	99	99	105	209	386	511	554	511	386	209	105	99	99	99	795	12
	HALF DAY TOTALS		393	512	1131	1985	2761	3258	3284	2801	1858	1030	609	429	394	391	391	391	2757	
Apr 21	6	210	29	110	172	205	207	177	119	38	11	11	11	11	11	11	11	11	23	6
	7	649	53	253	462	593	631	575	428	204	49	45	45	45	45	45	45	45	191	5
	8	804	73	192	453	632	715	689	559	337	95	69	69	69	69	69	69	69	408	4
	9	876	89	113	324	532	649	669	591	418	183	92	87	87	87	87	87	87	593	3
	10	913	101	106	165	342	490	599	543	445	275	123	104	101	101	101	101	101	736	2
	11	932	109	109	115	149	263	371	426	415	340	222	126	113	109	109	109	109	825	1
	12	937	112	112	112	116	120	167	260	335	363	335	260	167	120	116	112	112	854	12
	HALF DAY TOTALS		508	932	1734	2500	3003	3129	2806	2033	1135	721	559	496	482	479	479	479	3203	
May 21	6	374	104	244	340	381	367	295	175	39	26	26	26	26	26	26	26	28	67	6
	7	666	112	350	535	638	643	550	374	134	60	57	57	57	57	57	57	59	256	5
	8	787	93	295	519	655	694	629	470	229	86	80	80	80	80	80	80	80	461	4
	9	849	104	194	404	558	625	601	488	294	117	100	97	97	97	97	97	97	633	3
	10	882	114	127	240	383	473	491	435	313	170	117	110	110	110	110	110	110	766	2
	11	898	121	124	132	186	260	312	322	285	215	147	126	122	118	118	118	118	847	1
	12	904	121	123	125	127	130	148	186	220	223	220	186	148	130	127	125	123	873	12
	HALF DAY TOTALS		700	1382	2214	2840	3106	2943	2356	1409	788	627	577	558	551	550	548	551	3464	
Jun 21	6	412	140	289	388	425	401	314	174	39	32	32	32	32	32	32	32	35	89	6
	7	662	148	384	556	644	635	529	342	109	62	62	62	62	62	62	62	65	279	5
	8	773	115	335	540	656	677	597	427	188	89	84	84	84	84	84	84	84	476	4
	9	831	110	234	431	563	609	567	440	244	111	101	101	101	101	101	101	101	642	3
	10	863	120	150	272	395	461	458	387	261	143	120	114	114	114	114	114	114	770	2
	11	880	126	130	148	202	258	288	280	237	177	135	128	125	122	122	122	122	847	1
	12	885	128	129	130	132	134	144	164	183	191	183	164	144	134	132	130	129	871	12
	HALF DAY TOTALS		824	1589	2403	2950	3108	2829	2137	1174	710	621	595	585	582	579	579	586	3538	
Jul 21	6	358	106	240	332	370	355	285	168	39	27	27	27	27	27	27	27	30	70	6
	7	640	118	349	526	623	626	534	361	129	62	59	59	59	59	59	59	61	257	5
	8	761	98	300	515	645	680	613	456	221	89	82	82	82	82	82	82	82	457	4
	9	823	107	202	405	553	615	588	475	284	117	100	100	100	100	100	100	100	626	3
	10	856	118	133	247	383	467	481	424	303	167	120	113	113	113	113	113	113	757	2
	11	873	124	128	137	191	261	308	314	277	210	147	129	125	121	121	121	121	836	1
	12	879	126	127	128	130	133	150	184	214	226	214	184	150	133	130	128	127	861	12
	HALF DAY TOTALS		728	1402	2210	2809	3052	2877	2290	1365	783	637	591	574	568	566	565	568	3431	
Aug 21	6	187	30	103	159	188	189	162	108	35	12	12	12	12	12	12	12	12	25	6
	7	599	59	250	444	565	598	542	402	192	53	49	49	49	49	49	49	49	192	5
	8	756	79	200	446	614	690	662	536	321	96	74	74	74	74	74	74	74	403	4
	9	830	95	124	328	523	632	648	570	402	179	98	93	93	93	93	93	93	583	3
	10	869	106	113	175	344	482	544	526	429	266	126	110	106	106	106	106	106	722	2
	11	889	115	117	122	157	264	365	414	401	328	217	131	119	115	115	115	115	808	1
	12	894	117	117	117	122	126	170	255	324	350	324	255	170	126	122	117	117	837	12
	HALF DAY TOTALS		539	957	1721	2443	2910	3014	2694	1950	1109	729	581	524	510	507	506	506	3152	
Sep 21	7	514	32	109	302	437	502	492	405	252	62	30	30	30	30	30	30	30	97	5
	8	758	57	81	324	546	679	706	626	450	190	61	57	57	57	57	57	57	304	4
	9	857	77	82	201	459	637	717	688	557	330	99	77	77	77	77	77	77	498	3
	10	905	91	91	101	270	485	617	655	596	444	220	97	91	91	91	91	91	645	2
	11	928	100	100	102	113	257	437	549	576	516	373	187	106	100	100	100	100	737	1
	12	935	103	103	103	103	110	210	379	499	540	499	379	210	110	103	103	103	769	12
	HALF DAY TOTALS		409	518	1089	1888	2621	3097	3132	2689	1812	1025	620	446	411	408	408	408	2664	
Oct 21	7	312	13	24	136	233	291	301	268	190	75	14	13	13	13	13	13	13	30	5
	8	724	42	46	200	450	616	684	649	511	285	54	42	42	42	42	42	42	198	4
	9	862	63	63	104	378	608	738	755	655	454	170	65	63	63	63	63	63	395	3
	10	923	77	77	81	197	465	652	737	711	577	342	101	77	77	77	77	77	546	2
	11	952	86	86	86	91	239	481	639	697	651	505	269	93	86	86	86	86	639	1
	12	960	89	89	89	89	95	247	478	626	677	626	478	247	95	89	89	89	671	12
	HALF DAY TOTALS		325	335	632	1365	2234	2968	3275	3067	2376	1390	714	394	328	325	325	325	2143	
Nov 21	7	5	0	0	1	3	4	5	4	3	2	0	0	0	0	0	0	0	0	5
	8	619	29	29	93	323	493	581	579	488	316	86	29	29	29	29	29	29	102	4
	9	829	49	49	55	284	545	709	761	697	523	253	49	49	49	49	49	49	277	3
	10	912	64	64	65	131	423	650	772	775	659	437	142	65	64	64	64	64	428	2
	11	949	72	72	72	76	213	496	688	773	739	593	342	91	72	72	72	72	521	1
	12	959	75	75	75	75	80	274	541	708	766	708	541	274	80	75	75	75	553	12
	HALF DAY TOTALS		253	254	342	890	1769	2653	3143	3138	2643	1717	824	355	256	253	253	253	1607	
Dec 21	8	556	22	22	59	265	426	515	523	452	305	98	23	22	22	22	22	22	69	4
	9	812	43	43	46	244	512	686	751	701	539	282	48	43	43	43	43	43	228	3
	10	908	57	57	57	106	402	642	777	792	683	466	163	59	57	57	57	57	375	2
	11	949	66	66	66	69	200	497	700	795	766	620	367	92	66	66	66	66	468	1
	12	960	69	69	69	69	73	281	559	733	794	733	559	281	73	69	69	69	500	12
	HALF DAY TOTALS		223	223	265	717	1577	2499	3043	3111	2687	1823	868	339	225	223	223	223	1389	
			N	NNW	NW	WNW	W	WSW	SW	SSW	S	SSE	SE	ESE	E	ENE	NE	NNE	HOR	PM

TABLE B3.1 SI UNITS—SOLAR INTENSITY AND SOLAR HEAT GAIN FACTORS FOR 40° NORTH LATITUDE

Solar Heat Gain Factors (W/m²)

Date	Solar Time (am)	Direct Normal (W/m²)	N	NNE	NE	ENE	E	ESE	SE	SSE	S	SSW	SW	WSW	W	WNW	NW	NNW	HOR	Solar Time (pm)
Jan 21	8	446	17	17	55	223	350	417	420	358	236	60	17	17	17	17	17	17	44	4
	9	753	37	37	41	233	485	648	706	658	504	260	42	37	37	37	37	37	173	3
	10	865	51	51	51	97	390	627	761	776	671	460	161	53	51	51	51	51	303	2
	11	913	59	59	59	62	193	493	699	796	769	623	372	89	59	59	59	59	390	1
	12	926	62	62	62	66	293	563	740	802	740	563	283	62	62	62	62	62	419	12
	HALF DAY TOTALS		194	194	231	628	1426	2316	2852	2941	2566	1770	860	318	196	194	194	194	1117	
Feb 21	7	175	6	10	71	127	160	167	150	107	43	6	6	6	6	6	6	6	11	5
	8	692	33	35	158	407	576	651	628	505	296	56	33	33	33	33	33	33	136	4
	9	854	52	52	70	337	587	738	773	689	496	209	55	52	52	52	52	52	309	3
	10	926	65	65	67	155	450	666	777	768	641	408	120	66	65	65	65	65	451	2
	11	958	73	73	73	77	224	504	690	769	730	579	325	86	73	73	73	73	538	1
	12	967	76	76	76	76	80	271	536	702	760	702	536	271	80	76	76	76	568	12
	HALF DAY TOTALS		267	271	478	1140	2043	2888	3308	3202	2591	1602	790	361	269	267	267	267	1730	
Mar 21	5	540	27	92	295	441	514	509	425	271	69	26	26	26	26	26	26	26	83	5
	8	789	50	57	288	534	686	732	665	494	232	54	50	50	50	50	50	50	268	4
	9	889	67	70	147	429	640	749	744	625	404	127	69	67	67	67	67	67	450	3
	10	938	80	80	85	226	482	653	722	682	539	299	91	80	80	80	80	80	587	2
	11	961	88	88	88	94	247	476	623	673	622	473	244	94	88	88	88	88	673	1
	12	968	91	91	91	91	97	238	458	602	650	602	458	238	97	91	91	91	702	12
	HALF DAY TOTALS		358	440	954	1777	2626	3265	3429	3055	2191	1270	694	417	360	337	337	337	2411	
Apr 21	6	282	36	144	228	275	279	241	164	56	16	15	15	15	15	15	15	15	34	6
	7	651	50	223	442	584	633	588	451	255	51	45	45	45	45	45	45	45	193	5
	8	795	69	140	402	601	706	703	594	391	130	69	67	67	67	67	67	67	389	4
	9	865	84	91	254	488	638	691	640	494	260	91	84	84	84	84	84	84	557	3
	10	901	96	99	117	291	480	589	608	538	381	177	101	96	96	96	96	96	685	2
	11	920	104	107	107	122	255	411	506	522	459	323	164	109	104	104	104	104	766	1
	12	926	106	106	106	109	114	196	341	448	486	448	341	196	114	109	106	106	794	12
	HALF DAY TOTALS		487	835	1580	2390	3020	3314	3135	2466	1539	935	628	495	467	464	464	464	3020	
May 21	5	3	1	2	3	3	3	2	1	0	0	0	0	0	0	0	0	0	0	7
	6	453	113	284	403	458	446	364	223	56	33	33	33	33	33	33	33	35	96	6
	7	681	90	320	520	638	659	580	412	172	63	59	59	59	59	59	59	59	276	5
	8	787	86	230	471	629	694	655	519	295	92	80	80	80	80	80	80	80	461	4
	9	844	99	131	330	518	620	632	551	382	168	101	96	96	96	96	96	96	616	3
	10	875	107	114	171	332	467	529	513	419	262	127	111	107	107	107	107	107	737	2
	11	891	115	115	121	152	256	357	409	400	331	222	133	120	115	115	115	115	812	1
	12	896	117	117	117	121	126	171	258	329	355	329	258	171	126	121	117	117	836	12
	HALF DAY TOTALS		679	1275	2102	2818	3231	3232	2778	1897	1129	780	630	568	554	550	550	552	3418	
Jun 21	5	68	32	54	68	70	63	46	20	5	4	4	4	4	4	4	4	7	8	7
	6	488	150	329	450	501	478	380	222	53	39	39	39	39	39	39	39	43	126	6
	7	681	118	355	543	648	654	562	385	145	68	65	65	65	65	65	65	67	306	5
	8	776	94	268	492	633	680	626	480	252	93	85	85	85	85	85	85	85	484	4
	9	829	105	160	359	524	607	601	507	332	142	104	100	100	100	100	100	100	633	3
	10	859	112	121	197	345	457	500	468	366	218	122	115	112	112	112	112	112	750	2
	11	874	119	123	128	166	254	332	367	347	279	188	130	124	119	119	119	119	821	1
	12	879	121	121	121	127	131	163	227	281	301	281	227	163	131	127	121	121	844	12
	HALF DAY TOTALS		799	1483	2314	2968	3275	3151	2580	1649	995	743	642	602	592	588	587	594	3551	
Jul 21	5	7	3	5	7	7	6	5	2	0	0	0	0	0	0	0	0	1	1	7
	6	435	116	281	395	447	433	352	216	55	34	34	34	34	34	34	34	37	100	6
	7	656	95	321	513	625	643	564	400	166	66	62	62	62	62	62	62	62	278	5
	8	762	90	236	468	620	680	639	505	285	94	83	83	83	83	83	83	83	459	4
	9	818	102	138	333	513	610	618	537	371	165	104	99	99	99	99	99	99	611	3
	10	850	110	118	177	333	462	519	501	407	255	129	114	110	110	110	110	110	729	2
	11	866	117	120	125	157	256	352	400	389	321	217	135	123	117	117	117	117	802	1
	12	871	120	120	120	125	130	172	253	320	345	320	253	172	130	125	120	120	826	12
	HALF DAY TOTALS		705	1296	2102	2792	3180	3164	2707	1842	1110	783	643	586	572	568	567	570	3395	
Aug 21	6	255	38	137	214	256	259	223	151	52	18	17	17	17	17	17	17	17	38	6
	7	603	55	223	426	557	602	557	426	222	55	49	49	49	49	49	49	49	196	5
	8	747	75	149	397	584	681	676	569	374	128	74	72	72	72	72	72	72	386	4
	9	819	89	97	259	481	621	669	618	475	251	97	89	89	89	89	89	89	549	3
	10	857	102	105	126	294	472	574	590	519	567	173	107	102	102	102	102	102	674	2
	11	876	109	109	113	130	257	403	492	505	443	313	165	116	109	109	109	109	753	1
	12	882	112	112	112	115	120	197	333	434	470	434	333	197	120	115	112	112	780	12
	HALF DAY TOTALS		518	861	1571	2338	2929	3196	3015	2370	1496	932	647	524	496	492	492	492	2983	
Sep 21	7	472	28	87	265	395	460	456	381	244	66	27	27	27	27	27	27	27	80	5
	8	725	52	61	275	504	646	689	626	467	224	57	52	52	52	52	52	52	258	4
	9	830	71	73	148	413	613	717	712	599	391	129	73	71	71	71	71	71	434	3
	10	882	84	84	89	224	468	631	697	659	522	293	96	84	84	84	84	84	567	2
	11	906	92	92	92	99	245	463	652	652	603	461	242	99	92	92	92	92	651	1
	12	914	95	95	95	95	101	237	446	584	631	584	446	237	101	95	95	95	679	12
	HALF DAY TOTALS		374	447	917	1683	2484	3092	3257	2918	2119	1250	699	432	376	373	373	373	2329	
Oct 21	7	153	6	10	64	113	142	148	132	94	38	7	6	6	6	6	6	38	12	5
	8	644	35	37	155	387	545	615	592	476	280	56	35	35	35	35	35	35	136	4
	9	811	54	54	73	329	567	710	743	661	476	202	57	54	54	54	54	54	305	3
	10	884	67	67	70	157	439	646	752	742	619	395	120	69	67	67	67	67	443	2
	11	917	76	76	76	80	223	491	670	745	707	562	317	89	76	76	76	76	529	1
	12	927	78	78	78	78	84	267	521	681	737	681	521	267	84	78	78	78	558	12
	HALF DAY TOTALS		277	282	479	1106	1965	2771	3173	3074	2494	1555	780	369	280	277	277	277	1704	
Nov 21	8	430	17	17	55	217	339	404	406	346	228	66	18	17	17	17	17	17	44	4
	9	733	38	38	42	231	476	634	691	643	492	254	43	38	38	38	38	38	173	3
	10	846	52	52	52	99	385	617	748	763	659	454	159	54	52	52	52	52	303	2
	11	894	60	60	60	63	192	486	688	783	757	613	367	89	60	60	60	60	388	1
	12	908	63	63	63	63	67	280	555	728	789	728	555	281	67	63	63	63	418	12
	HALF DAY TOTALS		197	198	235	623	1403	2273	2797	2884	2516	1738	850	319	199	197	197	197	1115	
Dec 21	8	279	9	9	25	129	212	259	264	230	157	52	10	9	9	9	9	9	20	4
	9	685	31	31	33	188	427	584	646	611	477	260	40	31	31	31	31	31	124	3
	10	825	45	45	45	78	358	594	732	755	661	462	173	47	45	45	45	45	244	2
	11	882	53	53	53	55	177	477	685	786	765	624	379	90	53	53	53	53	327	1
	12	898	56	56	56	56	60	281	560	736	798	736	560	281	60	56	56	56	356	12
	HALF DAY TOTALS		165	165	178	461	1180	2046	2594	2736	2446	1757	870	298	167	165	165	165	891	
			N	NNW	NW	WNW	W	WSW	SW	SSW	S	SSE	SE	ESE	E	ENE	NE	NNE	HOR	PM

TABLE B3.1 SI UNITS—SOLAR INTENSITY AND SOLAR HEAT GAIN FACTORS FOR 48° NORTH LATITUDE

Solar Heat Gain Factors (W/m²)

Date	Solar Time (am)	Direct Normal (W/m²)	N	NNE	NE	ENE	E	ESE	SE	SSE	S	SSW	SW	WSW	W	WNW	NW	NNW	HOR	Solar Time (pm)
Jan 21	8	116	4	4	13	57	90	109	110	94	63	18	4	4	4	4	4	4	7	4
	9	584	24	24	26	168	371	505	555	523	406	217	30	24	24	24	24	24	79	3
	10	754	37	37	37	69	333	554	682	702	615	429	159	39	37	37	37	37	174	2
	11	823	45	45	45	47	166	455	656	753	734	598	365	83	45	45	45	45	244	1
	12	842	48	48	48	48	51	271	541	713	772	713	541	271	51	48	48	48	269	12
	HALF DAY TOTALS		134	134	144	368	996	1788	2298	2448	2212	1616	817	267	136	134	134	134	639	
Feb 21	7	11	0	1	5	8	10	11	10	7	0	0	0	0	0	0	0	0	1	5
	8	568	24	25	114	324	470	537	524	428	259	52	24	24	24	24	24	24	78	4
	9	780	42	42	49	284	530	683	727	660	488	225	45	42	42	42	42	42	210	3
	10	869	54	54	55	120	415	641	764	768	653	434	137	55	54	54	54	54	331	2
	11	908	61	61	61	64	205	498	696	787	755	607	355	84	61	61	61	61	409	1
	12	920	63	63	63	63	68	280	555	728	790	728	555	280	68	63	63	63	435	12
	HALF DAY TOTALS		214	216	337	864	1708	2575	3054	3051	2565	1677	825	328	216	214	214	214	1246	
Mar 21	7	482	23	71	253	387	458	458	388	253	71	23	22	22	22	22	22	22	64	5
	8	744	44	48	239	486	644	701	651	498	257	49	44	44	44	44	44	44	214	4
	9	853	60	60	104	381	609	738	753	652	449	164	62	60	60	60	60	60	371	3
	10	906	72	72	75	183	460	656	749	728	596	363	104	72	72	72	72	72	493	2
	11	932	79	79	79	83	232	492	663	731	689	541	297	88	79	79	79	79	568	1
	12	939	81	81	81	81	86	261	500	666	720	666	509	261	86	81	81	81	594	12
	HALF DAY TOTALS		317	372	790	1558	2446	3193	3471	3199	2420	1466	769	399	319	316	316	316	2006	
Apr 21	6	340	39	167	271	330	337	294	203	74	20	19	19	19	19	19	19	19	46	6
	7	646	49	191	417	567	628	595	468	264	56	45	45	45	45	45	45	45	189	5
	8	779	64	99	350	566	690	709	619	435	173	67	64	64	64	64	64	64	359	4
	9	847	79	83	191	444	621	703	679	554	335	105	79	79	79	79	79	79	507	3
	10	884	90	90	97	244	466	610	660	613	472	251	97	90	90	90	90	90	618	2
	11	902	97	97	97	105	245	443	570	609	558	419	217	103	97	97	97	97	689	1
	12	908	99	99	99	99	106	224	414	543	587	543	414	224	106	99	99	99	713	12
	HALF DAY TOTALS		463	765	1454	2285	3019	3465	3411	2825	1908	1169	713	492	445	442	442	442	2761	
May 21	5	129	52	97	125	133	122	91	44	9	8	8	8	8	8	8	8	10	16	7
	6	511	112	305	443	512	504	418	267	75	38	38	38	38	38	38	38	40	125	6
	7	690	73	283	498	631	668	604	448	214	66	61	61	61	61	61	61	61	287	5
	8	782	83	170	418	599	689	675	562	358	120	79	79	79	79	79	79	79	449	4
	9	834	93	101	259	475	611	656	605	463	243	100	93	93	93	93	93	93	585	3
	10	864	103	106	124	283	458	561	579	513	366	179	109	103	103	103	103	103	690	2
	11	879	109	109	113	127	249	396	488	505	447	320	170	116	109	109	109	109	756	1
	12	884	111	111	111	114	120	198	336	439	474	439	336	198	120	114	111	111	778	12
	HALF DAY TOTALS		679	1224	2035	2817	3360	3515	3176	2363	1525	997	709	579	550	546	546	549	3297	
Jun 21	5	243	111	192	241	252	228	166	75	19	17	17	17	17	17	17	17	24	38	7
	6	544	146	348	488	552	534	434	266	70	46	46	46	46	46	46	46	49	162	6
	7	693	93	317	521	642	467	591	427	188	72	67	67	67	67	67	67	67	324	5
	8	775	90	203	440	604	678	651	529	320	107	84	84	84	84	84	84	84	479	4
	9	822	98	114	286	482	600	629	567	418	208	105	98	98	98	98	98	98	610	3
	10	849	108	113	142	296	450	534	540	465	319	157	114	108	108	108	108	108	711	2
	11	864	114	114	120	138	248	373	449	456	396	279	156	121	114	114	114	114	775	1
	12	869	116	116	116	120	126	189	303	391	423	391	303	189	126	120	116	116	796	12
	HALF DAY TOTALS		811	1446	2279	3013	3454	3477	3013	2140	1376	942	718	620	597	593	592	602	3495	
Jul 21	5	135	57	104	133	141	129	96	46	11	9	9	9	9	9	9	9	12	18	7
	6	492	116	302	436	501	492	407	259	75	40	40	40	40	40	40	40	43	130	6
	7	666	78	285	492	619	653	588	436	207	69	63	63	63	63	63	63	63	290	5
	8	757	86	176	416	590	675	660	547	348	119	81	81	81	81	81	81	81	448	4
	9	809	96	106	263	471	601	643	591	450	237	104	96	96	96	96	96	96	582	3
	10	839	106	109	130	285	453	550	566	500	356	177	112	106	106	106	106	106	684	2
	11	855	112	112	117	132	249	390	477	492	435	312	169	119	112	112	112	112	749	1
	12	859	114	114	114	117	124	198	329	428	462	428	329	198	124	117	114	114	771	12
	HALF DAY TOTALS		705	1247	2039	2795	3311	3446	3101	2303	1496	993	721	598	570	565	565	569	3287	
Aug 21	6	311	42	161	256	310	316	273	190	70	22	21	21	21	21	21	21	21	51	6
	7	599	53	193	403	543	598	565	444	250	59	49	49	49	49	49	49	49	193	5
	8	732	69	108	347	549	665	681	593	417	168	73	69	69	69	69	69	69	358	4
	9	801	84	90	198	437	605	681	655	534	323	108	84	84	84	84	84	84	502	3
	10	839	95	95	104	247	453	593	639	592	456	245	104	95	95	95	95	95	610	2
	11	858	102	102	102	112	247	433	553	589	539	406	215	110	102	102	102	102	679	1
	12	864	104	104	104	104	113	224	404	526	568	526	404	224	113	104	104	104	702	12
	HALF DAY TOTALS		496	790	1449	2237	2929	3343	3282	2720	1852	1156	728	521	475	471	471	471	2741	
Sep 21	7	414	24	66	224	342	403	403	342	224	66	24	23	23	23	23	23	23	62	5
	8	678	46	51	228	455	602	654	608	467	244	52	46	46	46	46	46	46	206	4
	9	792	63	63	107	366	581	702	716	621	430	163	66	63	63	63	63	63	358	3
	10	848	75	75	79	182	444	630	719	699	573	353	107	75	75	75	75	75	476	2
	11	876	82	82	82	88	230	476	639	704	664	524	292	93	82	82	82	82	549	1
	12	884	84	84	84	84	91	257	494	643	695	643	494	257	91	84	84	84	574	12
	HALF DAY TOTALS		332	381	758	1467	2300	3007	3280	3039	2324	1429	766	412	334	331	331	331	1937	
Oct 21	7	12	0	1	5	9	11	12	11	8	3	0	0	0	0	0	0	0	1	5
	8	522	25	27	111	304	439	501	489	398	242	51	25	25	25	25	25	25	79	4
	9	734	44	44	52	276	508	652	693	629	466	216	47	44	44	44	44	44	208	3
	10	825	56	56	58	122	403	618	736	739	628	418	136	58	56	56	56	56	327	2
	11	866	64	64	64	67	203	483	673	760	729	587	345	86	64	64	64	64	403	1
	12	878	66	66	66	66	71	274	538	704	763	704	538	274	71	66	66	66	429	12
	HALF DAY TOTALS		223	225	340	838	1637	2461	2918	2918	2459	1619	808	334	226	223	223	223	1233	
Nov 21	8	115	4	4	13	57	90	108	109	93	62	18	4	4	4	4	4	4	7	4
	9	565	25	25	27	165	363	492	540	509	394	211	31	25	25	25	25	25	81	3
	10	735	38	38	38	70	328	543	668	688	602	420	156	40	38	38	38	38	175	2
	11	804	46	46	46	48	165	448	645	739	720	587	358	83	46	46	46	46	244	1
	12	823	48	48	48	48	53	267	531	700	758	700	531	267	53	48	48	48	269	12
	HALF DAY TOTALS		138	138	147	368	979	1752	2250	2396	2165	1583	804	268	140	138	138	138	642	
Dec 21	9	442	16	16	18	113	272	378	420	401	316	176	25	16	16	16	16	16	42	3
	10	676	30	30	30	51	286	491	613	636	563	398	156	32	30	30	30	30	119	2
	11	765	38	38	38	40	146	421	615	708	694	569	351	80	38	38	38	38	181	1
	12	789	41	41	41	41	44	256	514	679	734	679	514	256	44	41	41	41	204	12
	HALF DAY TOTALS		105	105	107	231	735	1444	1924	2095	1944	1477	778	239	107	105	105	105	444	
			N	NNW	NW	WNW	W	WSW	SW	SSW	S	SSE	SE	ESE	E	ENE	NE	NNE	HOR	PM

TABLE B3.1   SI UNITS—SOLAR INTENSITY AND SOLAR HEAT GAIN FACTORS FOR 56° NORTH LATITUDE

Solar Heat Gain Factors (W/m²)

Date	Solar Time (am)	Direct Normal (W/m²)	N	NNE	NE	ENE	E	ESE	SE	SSE	S	SSW	SW	WSW	W	WNW	NW	NNW	HOR	Solar Time (pm)
Jan 21	9	245	8	8	9	66	54	212	234	222	173	94	12	8	8	8	8	8	17	3
	10	538	21	21	21	40	232	396	493	509	451	317	121	23	21	21	21	21	65	2
	11	653	29	29	29	30	125	365	533	613	600	492	303	65	29	29	29	29	108	1
	12	684	31	31	31	31	34	225	225	453	647	599	453	225	34	31	31	31	125	12
	HALF DAY TOTALS		73	73	74	145	514	1082	1477	1632	1536	1192	651	193	75	73	73	73	251	
Feb 21	8	363	13	14	65	202	298	344	338	279	172	36	13	13	13	13	13	13	31	4
	9	655	30	30	33	223	440	578	621	573	431	210	33	30	30	30	30	30	115	3
	10	775	41	41	41	88	363	580	703	715	617	420	142	43	41	41	41	41	204	2
	11	827	47	47	47	50	178	465	662	755	731	592	355	78	47	47	47	47	266	1
	12	843	50	50	50	50	54	270	539	710	769	710	539	270	54	50	50	50	288	12
	HALF DAY TOTALS		156	156	209	575	1290	2102	2589	2669	2326	1605	798	281	158	156	156	156	759	
Mar 21	7	404	18	51	204	320	383	386	330	220	67	18	18	18	18	18	18	18	45	5
	8	679	37	40	192	429	585	647	612	479	264	47	37	37	37	37	37	37	156	4
	9	799	51	51	73	331	566	706	735	652	467	193	54	51	51	51	51	51	282	3
	10	859	61	61	63	145	429	640	750	745	624	403	122	63	61	61	61	61	384	2
	11	889	67	67	67	71	214	491	677	759	724	579	334	85	67	67	67	67	448	1
	12	897	69	69	69	69	75	273	536	701	759	701	536	273	75	69	69	69	470	12
	HALF DAY TOTALS		269	305	632	1321	2206	3017	3379	3207	2524	1584	815	372	271	268	268	268	1549	
Apr 21	5	1	0	1	1	1	1	1	1	0	0	0	0	0	0	0	0	0	0	7
	6	386	39	181	301	371	382	337	237	92	-24	22	22	22	22	22	22	22	57	6
	7	634	46	160	388	545	616	594	479	287	67	43	43	43	43	43	43	43	178	5
	8	755	60	73	299	527	666	704	634	467	216	64	60	60	60	60	60	60	318	4
	9	821	73	75	139	398	598	705	704	598	396	138	75	73	73	73	73	73	443	3
	10	858	82	82	87	200	447	619	695	668	540	321	103	82	82	82	82	82	537	2
	11	877	88	88	88	94	232	462	615	672	631	493	271	97	88	88	88	88	595	1
	12	883	90	90	90	90	97	248	470	613	662	613	470	248	97	90	90	90	615	12
	HALF DAY TOTALS		437	714	1356	2189	3001	3571	3617	3097	2205	1380	796	485	416	413	413	413	2437	
May 21	5	292	114	215	280	300	277	209	104	22	19	19	19	19	19	19	19	23	45	7
	6	553	103	311	466	548	547	463	306	99	43	43	43	43	43	43	43	44	152	6
	7	672	68	242	469	616	670	622	480	257	68	61	61	61	61	61	61	61	290	5
	8	771	78	121	363	565	678	689	597	414	165	80	76	76	76	76	76	76	425	4
	9	819	88	93	195	430	597	672	648	529	322	112	88	88	88	88	88	88	539	3
	10	847	97	97	105	236	445	583	632	589	458	252	106	97	97	97	97	97	627	2
	11	862	102	102	102	111	239	427	550	590	543	413	223	110	102	102	102	102	682	1
	12	867	104	104	104	104	112	225	407	530	572	530	407	225	112	104	104	104	701	12
	HALF DAY TOTALS		701	1232	2032	2859	3507	3791	3535	2771	1905	1238	807	590	542	538	538	544	3110	
Jun 21	4	67	42	60	69	67	55	35	9	4	4	4	4	4	4	4	5	16	8	8
	5	385	167	296	377	398	362	268	127	33	29	29	29	29	29	29	29	37	78	7
	6	585	133	350	506	585	576	479	307	93	51	51	51	51	51	51	51	53	195	6
	7	699	78	272	492	629	673	614	465	235	75	68	68	68	68	68	68	68	332	5
	8	767	86	146	384	571	670	670	571	383	146	86	83	83	83	83	83	83	462	4
	9	810	94	101	218	438	588	650	617	493	287	109	94	94	94	94	94	94	572	3
	10	835	103	103	113	249	439	561	598	550	417	224	111	103	103	103	103	103	656	2
	11	849	108	108	110	119	239	407	517	549	500	375	204	116	108	108	108	108	709	1
	12	853	110	110	110	110	119	216	376	490	529	490	376	216	119	110	110	110	727	12
	HALF DAY TOTALS		867	1494	2328	3119	3666	3809	3415	2592	1774	1185	819	640	600	595	595	617	3376	
Jul 21	4	1	1	1	1	1	1	1	0	0	0	0	0	0	0	0	0	0	0	8
	5	288	117	217	281	300	277	208	104	24	21	21	21	21	21	21	21	25	50	7
	6	534	107	309	458	537	535	452	299	97	47	45	45	45	45	45	45	47	158	6
	7	668	71	244	464	605	656	607	468	250	71	64	64	64	64	64	64	64	294	5
	8	747	81	127	362	557	665	674	583	404	162	83	79	79	79	79	79	79	426	4
	9	794	91	97	200	427	587	659	634	516	314	114	91	91	91	91	91	91	538	3
	10	822	100	100	109	238	440	572	618	576	446	247	109	100	100	100	100	100	623	2
	11	837	105	105	105	115	239	420	539	576	530	403	220	114	105	105	105	105	677	1
	12	842	107	107	107	107	116	225	399	518	559	518	399	225	116	107	107	107	696	12
	HALF DAY TOTALS		730	1257	2039	2841	3461	3722	3458	2709	1871	1230	817	610	564	559	559	567	3114	
Aug 21	5	4	1	2	3	4	4	3	2	0	0	0	0	0	0	0	0	0	0	7
	6	355	43	176	287	351	361	318	223	88	26	24	24	24	24	24	24	24	63	6
	7	589	51	162	375	521	587	565	455	273	69	47	47	47	47	47	47	47	183	5
	8	708	65	80	296	511	642	676	607	448	209	70	65	65	65	65	65	65	320	4
	9	775	78	81	146	392	581	681	679	575	382	138	81	78	78	78	78	78	441	3
	10	813	87	87	93	204	438	601	673	645	522	312	108	87	87	87	87	87	532	2
	11	832	93	93	93	101	233	452	596	651	610	478	266	103	93	93	93	93	589	1
	12	839	95	95	95	95	103	246	457	594	640	594	457	246	103	95	95	95	608	12
	HALF DAY TOTALS		470	740	1353	2144	2912	3445	3482	2984	2139	1359	807	514	447	443	443	443	2433	
Sep 21	7	339	18	47	177	276	329	332	284	190	60	19	18	18	18	18	18	18	44	5
	8	611	39	43	182	397	540	596	565	443	247	49	39	39	39	39	39	39	151	4
	9	736	54	54	76	316	535	666	693	615	442	188	58	54	54	54	54	54	272	3
	10	799	64	64	67	145	412	611	715	709	595	387	123	66	64	64	64	64	371	2
	11	830	71	71	71	75	211	473	649	727	693	556	324	89	71	71	71	71	433	1
	12	839	73	73	73	73	79	267	515	672	727	672	515	267	79	73	73	73	454	12
	HALF DAY TOTALS		282	313	604	1233	2056	2818	3168	3021	2401	1528	803	382	285	282	282	282	1496	
Oct 21	8	328	14	15	63	186	274	315	310	255	158	35	14	14	14	14	14	14	33	4
	9	608	32	32	36	214	416	545	585	540	406	199	35	32	32	32	32	32	116	3
	10	729	43	43	43	90	349	555	671	682	588	401	138	45	43	43	43	43	203	2
	11	783	50	50	50	52	176	448	636	724	701	568	342	80	50	50	50	50	264	1
	12	798	52	52	52	52	57	263	519	682	738	682	519	263	52	52	52	52	286	12
	HALF DAY TOTALS		164	164	215	557	1230	1997	2458	2536	2214	1536	775	284	166	164	164	164	757	
Nov 21	9	238	8	8	9	66	151	207	228	217	169	92	12	8	8	8	8	8	18	3
	10	521	22	22	22	41	227	386	480	496	439	309	118	24	22	22	22	22	66	2
	11	634	30	30	30	31	123	357	521	599	586	481	296	65	30	30	30	30	109	1
	12	665	32	32	32	32	35	221	443	585	632	585	443	221	35	32	32	32	126	12
	HALF DAY TOTALS		76	76	77	147	507	1059	1443	1594	1499	1164	637	192	77	76	76	76	255	
Dec 21	9	16	0	0	1	4	10	14	15	15	12	7	1	0	0	0	0	0	1	3
	10	356	12	12	12	22	148	259	325	339	302	215	85	13	12	12	12	12	29	2
	11	523	20	20	20	21	94	289	427	493	485	400	247	53	20	20	20	20	60	1
	12	568	23	23	23	23	26	187	378	501	540	501	378	187	26	23	23	23	73	12
	HALF DAY TOTALS		45	45	45	64	276	683	982	1118	1082	874	514	149	47	45	45	45	127	

N	NNW	NW	WNW	W	WSW	SW	SSW	S	SSE	SE	ESE	E	ENE	NE	NNE	HOR	PM

TABLE B3.1 SI UNITS—SOLAR INTENSITY AND SOLAR HEAT GAIN FACTORS FOR 64° NORTH LATITUDE

Solar Heat Gain Factors (W/m²)

Date	Solar Time (am)	Direct Normal (W/m²)	N	NNE	NE	ENE	E	ESE	SE	SSE	S	SSW	SW	WSW	W	WNW	NW	NNW	HOR	Solar Time (pm)
Jan 21	10	60	2	2	2	4	29	51	63	66	58	41	16	2	2	2	2	2	4	2
	11	256	9	9	9	9	46	143	210	242	238	196	120	24	9	9	9	9	18	1
	12	316	11	11	11	11	12	104	211	280	302	280	211	104	12	11	11	11	24	12
	HALF DAY TOTALS		16	16	16	18	80	250	382	449	446	374	237	72	17	16	16	16	33	
Feb 21	8	56	2	2	9	31	46	53	53	44	27	6	2	2	2	2	2	2	3	4
	9	422	16	16	18	136	280	373	403	375	285	143	18	16	16	16	16	16	41	3
	10	601	26	26	26	57	276	453	554	567	495	341	119	28	26	26	26	26	90	2
	11	679	32	32	32	34	139	386	558	638	622	506	307	64	32	32	32	32	128	1
	12	701	34	34	34	34	37	231	464	613	662	613	464	231	37	34	34	34	143	12
	HALF DAY TOTALS		93	93	103	279	770	1408	1822	1947	1767	1298	668	210	95	93	93	93	334	
Mar 21	7	300	12	34	147	235	284	288	249	168	54	13	12	12	12	12	12	12	27	5
	8	582	29	31	147	357	499	559	536	427	246	44	29	29	29	29	29	29	100	4
	9	717	41	41	51	277	502	642	679	613	450	203	44	41	41	41	41	41	187	3
	10	786	49	49	50	111	386	598	714	719	613	410	132	51	49	49	49	49	264	2
	11	821	54	54	54	57	190	468	658	746	718	579	344	80	54	54	54	54	314	1
	12	831	56	56	56	56	61	269	530	696	754	696	530	269	61	56	56	56	331	12
	HALF DAY TOTALS		213	234	473	1052	1879	2695	3103	3021	2457	1591	811	330	215	213	213	213	1057	
Apr 21	5	85	24	56	77	85	81	64	36	6	4	4	4	4	4	4	4	5	8	7
	6	419	38	187	320	400	416	371	266	111	26	24	24	24	24	24	24	24	67	6
	7	613	43	129	355	516	595	584	482	303	80	41	41	41	41	41	41	41	162	5
	8	720	54	60	249	483	633	685	634	484	250	60	54	54	54	54	54	54	269	4
	9	783	65	65	100	351	567	691	710	620	436	174	68	65	65	65	65	65	367	3
	10	820	72	72	76	161	421	613	709	698	582	372	119	74	72	72	72	72	441	2
	11	839	77	77	77	82	215	468	637	710	676	540	312	93	77	77	77	77	448	1
	12	846	79	79	79	79	85	262	503	655	708	655	503	262	85	79	79	79	504	12
	HALF DAY TOTALS		413	687	1294	2116	2974	3628	3741	3268	2408	1537	861	471	381	378	378	378	2053	
May 21	4	160	94	139	162	161	135	87	25	10	10	10	10	10	10	10	11	31	20	8
	5	416	152	298	393	425	395	302	158	34	29	29	29	29	29	29	29	33	81	7
	6	584	90	304	474	570	579	499	343	126	49	46	46	46	46	46	46	46	175	6
	7	688	65	199	436	596	665	634	507	297	78	60	60	60	60	60	60	60	284	5
	8	754	72	89	307	527	660	694	623	459	215	78	72	72	72	72	72	72	390	4
	9	796	82	85	143	384	577	678	679	579	389	147	85	82	82	82	82	82	481	3
	10	822	89	89	94	193	427	593	668	645	528	322	114	89	89	89	89	89	549	2
	11	836	93	93	93	100	226	446	594	652	614	485	275	105	93	93	93	93	592	1
	12	841	95	95	95	95	103	248	460	597	644	597	460	248	103	95	95	95	606	12
	HALF DAY TOTALS		780	1341	2145	2998	3712	4073	3841	3108	2234	1467	909	603	533	529	530	554	2875	
Jun 21	4	294	181	263	302	297	247	156	43	21	21	21	21	21	21	21	23	67	50	8
	5	485	195	360	466	498	458	346	175	44	39	39	39	39	39	39	39	45	124	7
	6	614	113	338	510	603	605	516	347	122	57	55	55	55	55	55	55	57	223	6
	7	698	74	225	457	610	672	632	498	282	80	68	68	68	68	68	68	68	331	5
	8	754	79	105	327	535	657	682	605	438	197	85	79	79	79	79	79	79	433	4
	9	791	89	93	161	393	572	662	655	552	362	137	92	89	89	89	89	89	520	3
	10	814	96	96	102	205	423	577	642	615	497	297	114	96	96	96	96	96	585	2
	11	827	100	100	100	108	228	431	568	619	581	456	258	110	100	100	100	100	627	1
	12	831	101	101	101	101	110	240	436	566	610	566	436	240	110	101	101	101	641	12
	HALF DAY TOTALS		1016	1680	2526	3347	3953	4154	3769	2985	2142	1436	935	667	605	601	603	673	3219	
Jul 21	4	168	101	149	173	172	144	93	27	12	12	12	12	12	12	12	12	35	24	8
	5	405	154	297	389	420	390	298	156	36	32	32	32	32	32	32	32	36	88	7
	6	564	94	303	467	560	567	488	335	124	51	49	49	49	49	49	49	49	181	6
	7	665	69	201	431	585	652	620	495	290	80	63	63	63	63	63	63	63	289	5
	8	730	75	94	307	520	648	680	609	449	211	81	75	75	75	75	75	75	393	4
	9	771	85	88	148	382	567	665	664	566	380	146	88	85	85	85	85	85	481	3
	10	797	92	92	98	196	422	583	655	631	516	315	116	92	92	92	92	92	548	2
	11	812	96	96	96	104	226	439	582	638	601	475	271	108	96	96	96	96	590	1
	12	816	98	98	98	98	107	246	452	585	630	585	452	246	107	98	98	98	604	12
	HALF DAY TOTALS		814	1370	2157	2985	3669	4004	3763	3046	2198	1457	920	626	557	553	554	582	2985	
Aug 21	5	92	28	62	85	94	89	71	40	8	6	6	6	6	6	6	6	6	11	7
	6	388	42	182	306	380	395	352	252	107	30	27	27	27	27	27	27	27	73	6
	7	570	48	132	344	494	567	555	458	289	81	45	45	45	45	45	45	45	168	5
	8	675	59	66	247	468	609	657	607	464	241	66	59	59	59	59	59	59	273	4
	9	737	70	70	107	345	549	666	683	597	420	172	74	70	70	70	70	70	368	3
	10	775	78	78	82	165	412	594	685	674	562	361	122	80	78	78	78	78	440	2
	11	794	82	82	82	89	216	456	617	686	653	523	305	99	82	82	82	82	485	1
	12	801	84	84	84	84	92	260	489	635	684	633	489	260	92	84	84	84	501	12
	HALF DAY TOTALS		447	714	1293	2073	2884	3498	3600	3147	2333	1509	869	500	413	409	409	410	2069	
Sep 21	7	242	12	30	122	194	234	238	206	139	47	13	12	12	12	12	12	12	26	5
	8	513	30	33	136	324	451	505	484	387	225	45	30	30	30	30	30	30	97	4
	9	651	43	43	54	261	468	596	631	570	420	193	47	43	43	43	43	43	181	3
	10	722	52	52	53	111	366	563	672	677	577	388	131	54	52	52	52	52	255	2
	11	758	57	57	57	61	185	445	623	706	680	549	329	83	57	57	57	57	303	1
	12	769	59	59	59	59	65	260	504	660	715	660	504	260	65	59	59	59	320	12
	HALF DAY TOTALS		224	241	448	968	1727	2484	2871	2810	2305	1513	787	336	227	224	224	224	1021	
Oct 21	8	54	2	2	10	30	45	52	51	42	26	6	2	2	2	2	2	2	4	4
	9	383	17	17	19	127	259	345	372	346	263	133	19	17	17	17	17	17	42	3
	10	556	28	28	28	58	261	426	520	532	465	321	115	29	28	28	28	28	91	2
	11	633	34	34	34	36	135	367	528	603	588	479	292	64	34	34	34	34	130	1
	12	655	36	36	36	36	40	222	441	581	628	581	441	222	40	36	36	36	144	12
	HALF DAY TOTALS		98	98	109	273	728	1324	1711	1829	1661	1225	638	209	100	98	98	98	339	
Nov 21	10	72	2	2	2	5	31	53	67	69	62	43	17	3	2	2	2	2	4	2
	11	250	9	9	9	10	46	140	207	238	234	192	118	24	9	9	9	9	19	1
	12	307	11	11	11	11	13	102	207	274	295	274	207	102	13	11	11	11	25	12
	HALF DAY TOTALS		17	17	17	19	81	248	378	444	440	369	234	71	18	17	17	17	35	
Dec 21	11	12	0	0	0	0	2	7	10	12	11	9	6	1	0	0	0	0	1	1
	12	51	1	1	1	1	2	16	34	45	48	45	34	16	2	1	1	1	3	12
	HALF DAY TOTALS		1	1	1	1	3	16	28	35	36	32	22	8	1	1	1	1	2	
			N	NNW	NW	WNW	W	WSW	SW	SSW	S	SSE	SE	ESE	E	ENE	NE	NNE	HOR	PM

[1] Based on a ground reflectance of 0.20.

[2] Underlined values = monthly maximums.

[3] Boxed values = yearly maximums.

[4] For steel sash or no sash, multiply by 1.17.

[5] Altitude correction = +0.7% per 1000 ft. elevation.

[6] Dew point correction:

+7% per 10°F below 67°F D.P.

−7% per 10°F above 67°F D.P.

[7] In southern latitudes in December or January, add 7%.

Source: Reprinted from ASHRAE *Handbook of Fundamentals 1981* by permission of the American Society of Heating, Refrigerating and Air-Conditioning Engineers, Inc.

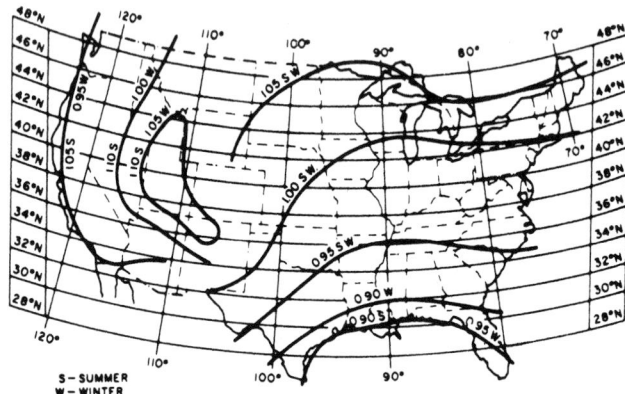

FIGURE B3.1. Estimated atmospheric clearness numbers in the United States for nonindustrial localities.

TABLE B3.2 SOLAR REFLECTANCES OF VARIOUS FOREGROUND SURFACES

Foreground Surface	Incident Angle (deg.)					
	20	30	40	50	60	70
Asphalt parking lot	0.09	0.09	0.10	0.10	0.11	0.12
Concrete, new	0.31	0.31	0.32	0.32	0.33	0.34
Concrete, old	0.22	0.22	0.22	0.23	0.23	0.25
Gravel and bitumen roof	0.14	0.14	0.14	0.14	0.14	0.14
Green grass or other vegetation	0.21	0.22	0.23	0.25	0.28	0.31
Red brick	0.45	0.45	0.45	0.45	0.45	0.45
Sand, dry	0.18	0.18	0.18	0.18	0.18	0.18
Snow	0.75	0.75	0.75	0.75	0.75	0.75
Tar paper (black)	0.07	0.07	0.08	0.08	0.08	0.09
Water	0.06	0.24	0.41	0.59	0.76	0.94

TABLE B3.3 SHADING COEFFICIENTS (SC) FOR GLASS

| Glazing Material | | Nominal Pane Thickness | | | Shading Coefficient | | | | | | | |
| | | | | | Indoor Shading | | | | | Shading Between Glazing | | |
Outer	Inner	in.	mm	No shading	Medium Venetian Blinds	Light Venetian Blinds	Opaque Roller Shades (Dark)	Opaque Roller Shades (White)	Translucent Roller Shade	Medium Venetian Blinds	Light Venetian Blinds	Louvered Sun Screen
Single-pane glass												
		1/8	3	1.0								
Clear		1/4	6	.94	.64	.55	.59	.25	.39			
glass		3/8	10	.91								
		1/2	12	.88								
Heat-absorbing glass[a]		1/8	3	.84	.57	.53	.45	.30	.36			
		1/4	6	.71								
		3/8	10	.62	.54	.52	.40	.28	.32			
		1/2	12	.56	.42	.40	.36	.28	.31			
Reflective coated glass				.30	.25	.23						
				.40	.33	.29						
				.50	.42	.38						
				.60	.50	.44						
Painted glass												
Light color				.28								
Medium color				.39								
Dark color				.50								
Stained glass												
Amber				.70								
Dark red				.56								
Dark blue				.60								
Dark green				.32								
Greyed green				.46								
Light opalescent				.43								
Dark opalescent				.37								
Double-pane glass												
Clear	Clear	1/8	3	.88	.57	.51	.60	.25	.37	.36	.33	.43
Glass	Glass	1/4	6	.82								.49
Heat-absorbing glass[a]	Clear Glass	1/4	6	.56	.39	.36	.40	.22	.30	.30	.28	.39
Reflective coated glass	Clear Glass			.20	.19	.18						
				.30	.27	.26						
				.40	.34	.33						
Triple-pane glass												
Clear	Clear	1/8	3	.80								
Clear	Clear	1/4	6	.71								

Shading devices fully drawn except roller shades. For fully drawn roller shades, multiply light colors by .73, medium colors by .95, and dark colors by 1.08.

[a] Refers to gray-, bronze-, and green-tinted heat-absorbing float glass.

TABLE B3.4   SHADING COEFFICIENTS (SC) FOR PLASTIC (SINGLE PANE)

	Nominal Glazing Thickness		
Glazing Material	in.	mm	No Shading
Clear acrylic			.98
Gray tint acrylic			.52
			.63
			.74
			.80
			.89
Bronze tint acrylic			.46
			.58
			.75
			.80
			.90
Reflective acrylic[a]			.21
Clear polycarbonate	⅛	3	.98
Gray polycarbonate	⅛	3	.74
Bronze polycarbonate	⅛	3	.74

[a] Aluminum-metallized polyester film on plastic.

TABLE B3.5   SHADING COEFFICIENTS (SC) FOR GLASS WITH EXTERIOR SHADING

	Group 1 Solar Altitude				Group 2 Solar Altitude				Group 3 Solar Altitude				Group 4 Solar Altitude				
Glazing Material	10	20	30	40	10	20	30	40	10	20	30	40	10	20	30	40	Outside Awning[a]
Single-Pane																	
Clear glass	.35	.17	.15	.15	.33	.23	.21	.20	.51	.42	.31	.18	.59	.50	.38	.30	.25
Heat-absorbing glass																	.18
Double-Pane																	
Clear glass	.27	.11	.10	.10	.45	.35	.26	.13									.22
Heat-absorbing glass																	.13
Triple-Pane																	
Clear glass																	.18

Group 1. Black, 23 louvers per inch, width over spacing ratio 1.15.
Group 2. Light color, high reflectance, otherwise same as Group 1.
Group 3. Black or dark color, 17 louvers per inch, w/s ratio 0.85.
Group 4. Light color or unpainted aluminum, high reflectance; otherwise, the same as Group 3.

U-value = 0.85 Btu/(hr · ft² · F) for Groups 1 through 4, when used with single glazing.
[a] With vented sides and top. When awning is tight against building on sides and top, multiply SC by 1.4.

TABLE B3.6 SHADING COEFFICIENTS FOR SINGLE AND INSULATING GLASS WITH DRAPERIES

Glazing	Glass Trans.	Glass SC[b]	SC for Index Letters in Chart Below									
			A	B	C	D	E	F	G	H	I	J
Single Glass												
1/4 in. Clear	0.80	0.95	0.80	0.75	0.70	0.65	0.60	0.55	0.50	0.45	0.40	0.35
1/2 in. Clear	0.71	0.88	0.74	0.70	0.66	0.61	0.56	0.52	0.48	0.43	0.39	0.35
1/4 in. Heat Abs.	0.46	0.67	0.57	0.54	0.52	0.49	0.46	0.44	0.41	0.38	0.36	0.33
1/2 in. Heat Abs.	0.24	0.50	0.43	0.42	0.40	0.39	0.38	0.36	0.34	0.33	0.32	0.30
Reflective Coated	—	0.60	0.57	0.54	0.51	0.49	0.46	0.43	0.41	0.38	0.36	0.33
(see manufacturers' literature	—	0.50	0.46	0.44	0.42	0.41	0.39	0.38	0.36	0.34	0.33	0.31
for exact values)	—	0.40	0.36	0.35	0.34	0.33	0.32	0.30	0.29	0.28	0.27	0.26
	—	0.30	0.25	0.24	0.24	0.23	0.23	0.23	0.22	0.21	0.21	0.20
Insulating Glass 1/2-in. Air												
Space Clear Out and Clear In	0.64	0.83	0.66	0.62	0.58	0.56	0.52	0.48	0.45	0.42	0.37	0.35
Heat Abs. Out and Clear In	0.37	0.55	0.49	0.47	0.45	0.43	0.41	0.39	0.37	0.35	0.33	0.32
Reflective Coated	—	0.40	0.38	0.37	0.37	0.36	0.34	0.32	0.31	0.29	0.28	0.28
(see manufacturers' literature	—	0.30	0.29	0.28	0.27	0.27	0.26	0.26	0.25	0.25	0.24	0.24
for exact values)	—	0.20	0.19	0.19	0.18	0.18	0.17	0.17	0.16	0.16	0.15	0.15

[b] For glass alone, with no drapery.

SHADING COEFFICIENT INDEX LETTER ►
GLAZING INDICATED IN TABLE ABOVE
DRAPERIES ARE 100% FULLNESS
(Fabric width two times draped width)

Notes:
1. Shading Coefficients are for draped fabrics.
2. Other properties are for fabrics in flat orientation.
3. Use Fabric Reflectance and Transmittance to obtain accurate Shading Coefficients.
4. Use Openness and Yarn Reflectance or Openness and Fabric Reflectance to obtain the Various Environmental Characteristics, or to obtain Approximate Shading Coefficients.

CLASSIFICATION OF FABRICS
I = Open Weave
II = Semi-open Weave
III = Closed Weave

D = Dark "Color"
M = Medium "Color"
L = Light "Color"

Source: Reprinted from ASHRAE *Handbook of Fundamentals 1981* by permission of the American Society of Heating, Refrigerating and Air-Conditioning Engineers, Inc.

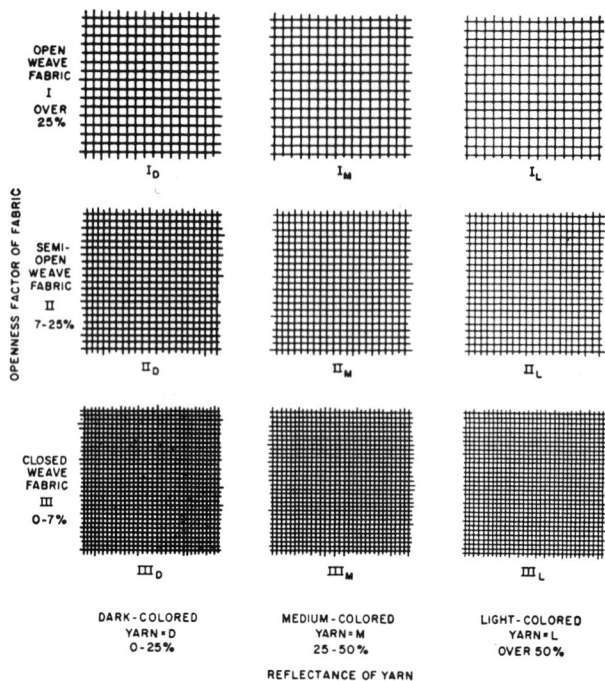

OPENNESS FACTOR OF FABRIC

OPEN WEAVE FABRIC I OVER 25%

SEMI-OPEN WEAVE FABRIC II 7-25%

CLOSED WEAVE FABRIC III 0-7%

I_D I_M I_L

II_D II_M II_L

III_D III_M III_L

DARK-COLORED YARN = D 0-25%

MEDIUM-COLORED YARN = M 25-50%

LIGHT-COLORED YARN = L OVER 50%

REFLECTANCE OF YARN

Note 1. Designators I_M, III_L indicate *open weave, medium colored yarn,* and *closed weave light colored yarn, and so forth.*

Note 2. Classes may be approximated by eye. With *closed* fabrics, no objects are visible through the material, and large light or dark areas may show. *Semi-open* fabrics do not permit details to be seen, and large objects are clearly defined. *Open fabrics* allow details to be seen, and the general view is relatively clear with no confusion of vision. Light, medium, and dark fabrics may be identified by eye; keep in mind it is the yarn *color* or shade of light or dark which is being observed.

FIGURE B3.2. Classification of drapery fabrics. Reprinted from ASHRAE *Handbook of Fundamentals 1981* by permission of American Society of Heating, Refrigerating and Air-Conditioning Engineer, Inc.

TABLE B3.7   SOLAR HEAT GAIN FACTORS FOR GLASS BLOCK WITH AND WITHOUT SHADING DEVICES

	Multiplying Factors for Glass Block			
Exposure in North Latitudes	Instantaneous Transmission Factor (B_i)	Absorption Transmission		Exposure in South Latitudes
		Factor (B_a)	Time Lag (hours)	
North	.27	.24	3.0	South
Northeast	.27	.24	3.0	Southeast
East	.39	.21	3.0	East
Southeast	.35	.22	3.0	Northeast
South				North
Summer	.27	.24	3.0	Summer
Winter	.39	.22	3.0	Winter
Southwest	.35	.22	3.0	Northwest
West	.39	.21	3.0	West
Northwest	.27	.24	3.0	Southwest
Horizontal	.14	.13	3.0	Horizontal

Equations:

Solar load without shading devices

$$= (B_i \times SC_i) + (B_a \times SC_a)$$

Solar load with outdoor shading devices

$$= (B_i \times SC_i + B_a \times SC_a) \times .25$$

Solar load with inside shading devices

$$= (B_i \times SC_i + B_a \times SC_a) \times .90$$

where:

B_i = instantaneous transmission factor from Table B3.7.
B_a = absorption transmission factor from Table B3.7.
SC_i = solar heat gain value from Table B3.1 for the desired time and wall facing.
SC_a = solar heat gain value from Table B3.1 for 3 hours earlier than SC_i and same wall facing.

Source: Reprinted with permission from *Handbook of Air-Conditioning System Design* by the Carrier Air Conditioning Company, McGraw-Hill Book Company, 1965.

TABLE B3.8 MAXIMUM SOLAR HEAT GAIN FACTOR FOR EXTERNALLY SHADED GLASS, W/M² (BASED ON GROUND REFLECTANCE OF 0.2)

Use for latitudes 0–24 deg.
For latitudes greater than 24, use north orientation, Table B3.1.
For horizontal glass in shade, use the tabulated values for all latitudes.

	N	NNE/ NNW	NE/ NW	ENE/ WNW	E/ W	ESE/ WSW	SE/ SW	SSE/ SSW	S	(ALL LATIT.) HOR
Jan.	98	98	98	101	107	114	117	117	120	50
Feb.	107	107	107	110	114	117	120	120	123	50
Mar.	114	114	117	120	123	126	126	123	123	60
Apr.	126	126	130	133	133	133	129	126	126	76
May	137	139	142	145	142	136	129	126	126	88
June	[142]	[145]	[148]	[148]	145	139	129	126	126	[98]
July	142	142	145	148	[148]	[142]	133	129	129	98
Aug.	133	133	136	142	145	142	136	[133]	[133]	88
Sept.	117	117	120	126	129	133	[133]	129	129	73
Oct.	107	107	107	114	120	123	126	126	126	60
Nov.	101	101	101	101	107	114	120	120	123	54
Dec.	95	95	95	98	101	107	114	117	117	47

MAXIMUM SOLAR HEAT GAIN FACTOR FOR EXTERNALLY SHADED GLASS, BTU/HR-FT² (BASED ON GROUND REFLECTANCE OF 0.2)

Use for latitudes 0–24 deg.
For latitudes greater than 24, use north orientation, Table B3.1.
For horizontal glass in shade, use the tabulated values for all latitudes.

	N	NNE/ NNW	NE/ NW	ENE/ WNW	E/ W	ESE/ WSW	SE/ SW	SSE/ SSW	S	(ALL LATIT.) HOR
Jan.	31	31	31	32	34	36	37	37	38	16
Feb.	34	34	34	35	36	37	38	38	39	16
Mar.	6	36	37	38	39	40	40	39	39	19
Apr.	40	40	41	42	42	42	41	40	40	24
May	43	44	45	46	45	43	41	40	40	28
June	[45]	[46]	[47]	[47]	46	44	41	40	40	[31]
July	45	45	46	47	[47]	[45]	42	41	41	31
Aug.	42	42	43	45	46	45	[43]	[42]	[42]	28
Sept.	37	37	38	40	41	42	42	41	41	23
Oct.	34	34	34	36	38	39	40	40	40	19
Nov.	32	32	32	32	34	36	38	38	39	17
Dec.	30	30	30	31	32	34	36	37	37	15

Source: Reprinted from ASHRAE *Handbook of Fundamentals 1981* by permission of the American Society of Heating, Refrigerating and Air-Conditioning Engineers, Inc.

TABLE B3.9 SHADOW LENGTHS AND SHADOW WIDTHS FOR BUILDING EXTERIOR PROJECTIONS

Horizontal Projection — Shadow Length, foot per foot (meter per meter) Projection, 0.0° North Latitude

Date	Solar Time (am)	N	NNE	NE	ENE	E	ESE	SE	SSE	S	SSW	SW	WSW	ALT	AZ
DEC	7			0.7	0.4	0.3	0.2	0.3	0.3	0.6	8.1			14	66
	8			1.7	0.8	0.6	0.5	0.6	0.7	1.2	7.1			27	63
	9			3.7	1.4	1.0	0.9	0.9	1.0	1.6	5.7			40	58
	10			19.1	2.9	1.7	1.4	1.3	1.5	2.0	4.0	8.1		53	49
	11				14.3	3.7	2.4	2.0	2.0	2.2	3.3	8.1		62	31
	12					6.3	3.3	2.5	2.4		3.3	6.3		67	0
JAN/NOV	7		8.1	0.6	0.3	0.3	0.2	0.3	0.4	0.7				14	69
	8			1.4	0.8	0.6	0.5	0.6	0.8	1.4	57.3			28	67
	9			2.9	1.4	1.0	0.9	0.9	1.2	2.0	11.4			42	63
	10			9.5	2.7	1.7	1.4	1.4	1.7	2.4	6.3			54	54
	11				9.5	3.7	2.6	2.1	2.2	2.6	4.0	14.3		65	35
	12						7.1	4.0	2.9	2.7	2.9	4.0	7.1	70	0
FEB/OCT	7		1.3	0.5	0.3	0.2	0.3	0.3	0.5	1.4				15	79
	8		3.3	1.0	0.7	0.6	0.6	0.7	1.0	2.6				29	78
	9		7.1	2.0	1.2	1.0	1.0	1.1	1.6	3.7				44	75
	10		57.3	4.0	2.2	1.7	1.6	1.8	2.4	4.7				58	69
	11			19.1	5.7	3.7	3.1	3.1	3.5	5.1	11.4			72	54
	12					14.3	7.1	7.1	5.7	5.1	5.7	7.1	14.3	79	0
MAR/SEPT	7		0.7	0.4	0.3	0.3	0.3	0.4	0.7					15	90
	8		1.5	0.8	0.6	0.6	0.6	0.8	1.5					30	90
	9		2.6	1.4	1.1	1.0	1.1	1.4	2.6					45	90
	10		4.7	2.5	1.9	1.7	1.9	2.5	4.7					60	90
	11		9.5	5.1	4.0	3.7	4.0	5.1	9.5					75	90
	12					14.3	7.1					7.1	14.3	90	90
APR/AUG	7	1.3	1.5	0.5	0.3	0.3	0.3	0.5	1.4					15	102
	8	2.5	1.0	0.6	0.6	0.6	0.7	1.1	3.5					29	103
	9	3.5	1.5	1.1	1.0	1.0	1.2	2.0	8.1					44	106
	10	4.3	2.2	1.7	1.6	1.7	2.2	4.0						58	112
	11	4.7	3.3	2.9	3.1	3.7	6.3	28.6						71	128
	12	4.7	5.1	7.1	14.3									78	180
MAY/JUL	7	0.7	0.4	0.3	0.3	0.3	0.3	0.6	8.1					14	111
	8	1.4	0.8	0.6	0.6	0.6	0.8	1.4						28	113
	9	2.0	1.2	0.9	1.0	1.0	1.4	2.9						42	117
	10	2.4	1.7	1.4	1.4	1.7	2.7	9.5						54	126
	11	2.6	2.2	2.1	2.6	3.7	9.5							65	145
	12	2.7	2.9	4.0	7.1									70	180
JUN	7	0.6	0.3	0.3	0.3	0.3	0.4	0.7						14	114
	8	1.2	0.7	0.6	0.6	0.6	0.8	1.7						27	117
	9	1.6	1.0	0.9	1.0	1.0	1.4	3.7						40	122
	10	2.0	1.5	1.3	1.4	1.7	2.9	19.1						53	131
	11	2.2	2.0	2.0	2.4	3.7	14.3							62	149
	12	2.4	2.5	3.3	6.3									67	180

Side Projection — Shadow Width, foot per foot (meter per meter) Projection, 0.0° North Latitude

Date	Time (pm)	N	NNE	NE	ENE	E	ESE	SE	SSE	S	SSW	SW	WSW	ALT	AZ
DEC	5			2.6	1.1	0.4	0.0	0.4	0.9	2.2	34.0			14	66
	4			3.0	1.2	0.5	0.1	0.3	0.9	2.0	13.9			27	63
	3			4.2	1.4	0.6	0.2	0.2	0.7	1.6	6.3			40	58
	2			14.1	2.0	0.9	0.3	0.1	0.5	1.2	3.0			53	49
	1				6.8	1.7	0.7	0.3	0.1	0.6	1.3	4.0		62	31
	12						2.4	1.0	0.4	0.0	0.4	1.0	2.4	67	0
JAN/NOV	5	30.9		2.2	0.9	0.4	0.0	0.5	1.1	2.7				14	69
	4			2.4	1.0	0.4	0.0	0.4	1.0	2.4				28	67
	3			3.1	1.2	0.5	0.1	0.3	0.8	1.9	12.1			42	63
	2			6.4	1.6	0.7	0.2	0.2	0.6	1.4	4.1			54	54
	1				4.4	1.4	0.6	0.2	0.2	0.7	1.6	5.9		65	35
	12						2.4	1.0	0.4	0.0	0.4	1.0	2.4	70	0
FEB/OCT	5	5.0		1.5	0.7	0.2	0.2	0.7	1.5	5.1				15	79
	4	5.6		1.6	0.7	0.2	0.2	0.6	1.4	4.5				29	78
	3	7.7		1.7	0.8	0.3	0.1	0.6	1.3	3.7				44	75
	2	35.4		2.2	0.9	0.4	0.0	0.4	1.1	2.6				58	69
	1			6.6	1.7	0.7	0.2	0.2	0.6	1.4	4.0			72	54
	12						2.4	1.0	0.4	0.0	0.4	1.0	2.4	79	0
MAR/SEPT	5	2.4		1.0	0.4	0.0	0.4	1.0	2.4					15	90
	4	2.4		1.0	0.4	0.0	0.4	1.0	2.4					30	90
	3	2.4		1.0	0.4	0.0	0.4	1.0	2.4					45	90
	2	2.4		1.0	0.4	0.0	0.4	1.0	2.4					60	90
	1	2.4		1.0	0.4	0.0	0.4	1.0	2.4					75	90
	12	2.4		1.0	0.4	0.0	0.4	1.0	2.4					90	90
APR/AUG	5	4.7	1.5	0.6	0.2	0.2	0.7	1.5	5.4					15	102
	4	4.2	1.4	0.6	0.2	0.2	0.7	1.6	6.2					29	103
	3	3.4	1.2	0.6	0.1	0.3	0.8	1.8	9.0					44	106
	2	2.4	1.0	0.4	0.0	0.4	1.0	2.4						58	112
	1	1.3	0.6	0.1	0.31	0.8	1.8	8.7						71	128
	12	0.0	0.4	1.0	2.4									78	180
MAY/JUL	5	2.7	1.1	0.5	0.4	0.4	0.9	2.2	30.9					14	111
	4	2.4	1.0	0.4	0.01	0.4	1.0	2.4						28	113
	3	1.9	0.8	0.3	0.1	0.5	1.2	3.1						42	117
	2	1.4	0.6	0.2	0.2	0.7	1.6	6.4						54	126
	1	0.7	0.2	0.2	0.6	1.4	4.4							65	145
	12	0.0	0.4	1.0	2.4									70	180
JUN	5	2.2	0.9	0.4	0.0	0.4	1.1	2.6						14	114
	4	2.0	0.9	0.3	0.1	0.5	1.2	3.0						27	117
	3	1.6	0.7	0.2	0.2	0.6	1.4	4.2						40	122
	2	1.2	0.5	0.1	0.3	0.9	2.0	14.1						53	131
	1	0.6	0.1	0.31	0.7	1.7	6.8							62	149
	12	0.0	0.4	1.0	2.4									67	180

Source: Reprinted from ASHRAE *Cooling and Heating Load Calculation Manual* by permission of the American Society of Heating, Refrigerating and Air-Conditioning Engineers, Inc.

505

TABLE B3.9 SHADOW LENGTHS AND SHADOW WIDTHS FOR BUILDING EXTERIOR PROJECTIONS (Continued)

Horizontal Projection — 8.0° North Latitude
Shadow Length, foot per foot (meter per meter) Projection

Date	Solar Time (am)	N	NNE	NE	ENE	E	ESE	SE	SSE	S	SSW	SW	WSW
DEC	7		57.3	0.6	0.3	0.2	0.2	0.2	0.2	0.4	3.3		
	8			1.7	0.7	0.5	0.4	0.4	0.6	0.9	3.3		
	9			5.1	1.4	0.9	0.8	0.8	0.8	1.2	2.9		
	10			57.3	3.1	1.6	1.2	1.1	1.2	1.4	2.5	19.1	
	11				57.3	3.5	2.0	1.5	1.4	1.6	2.1	4.0	
	12						4.3	2.4	1.8	1.7	1.8	2.4	4.3
JAN/NOV	7		57.3	0.5	0.3	0.2	0.2	0.2	0.3	0.5			
	8			1.4	0.7	0.5	0.5	0.5	0.6	1.0	7.1		
	9			3.7	1.4	0.9	0.8	0.8	0.9	1.4	4.3		
	10			57.3	2.9	1.6	1.2	1.2	1.3	1.7	3.3	19.1	
	11				19.1	3.5	2.1	1.7	1.6	1.8	2.5	5.7	
	12						4.7	2.6	2.1	1.9	2.1	2.6	4.7
FEB/OCT	7		1.4	0.4	0.3	0.2	0.2	0.3	0.4	1.0			
	8		5.1	1.1	0.7	0.5	0.5	0.6	0.8	1.8			
	9			2.2	1.2	1.0	0.9	1.0	1.2	2.4			
	10			6.3	2.5	1.7	1.4	1.4	1.7	2.7	9.5	19.1	
	11				8.1	3.5	2.6	2.2	2.4	2.9	4.7	4.0	
	12						8.1	4.0	3.3	2.9	3.3	4.0	8.1
MAR/SEPT	7		0.8	0.4	0.3	0.3	0.3	0.4	0.6	7.1			
	8		1.9	0.9	0.6	0.6	0.6	0.8	1.2	7.1			
	9		4.0	1.6	1.2	1.0	1.0	1.4	2.0	7.1			
	10		11.4	3.3	2.1	1.7	1.7	2.0	2.9	7.1	28.6		
	11			11.4	5.1	3.7	3.3	3.5	4.3	7.1	8.1	9.5	
	12						19.1	9.5	8.1	7.1	8.1	9.5	19.1
APR/AUG	6	0.1	0.1	0.0	0.0	0.0	0.0	0.1	0.1	0.1			
	7	1.7	0.6	0.4	0.3	0.3	0.3	0.5	1.3	1.0			
	8	4.0	1.2	0.8	0.6	0.6	0.7	1.0	2.5	2.5			
	9	7.1	2.0	1.3	1.0	1.0	1.2	1.7	4.3	4.3			
	10	11.4	3.3	2.1	1.8	1.8	2.1	3.5	8.1	8.1			
	11	14.3	6.3	4.3	3.7	4.0	6.3	57.3	28.6	28.6			
	12	14.3	19.1	19.1	57.3								
MAY/JUL	6	0.1	0.1	0.1	0.1	0.1	0.1	0.1	1.0	1.0			
	7	0.9	0.5	0.3	0.3	0.3	0.4	0.7	4.3	4.3			
	8	1.9	0.9	0.7	0.6	0.6	0.8	1.3	8.1	8.1			
	9	2.9	1.5	1.1	1.0	1.1	1.4	2.4	28.6	28.6			
	10	3.7	2.2	1.7	1.7	1.8	2.5	5.1					
	11	4.3	3.3	2.9	3.1	4.0	6.3	57.3					
	12	4.7	5.1	6.3	11.4								
JUN	6	0.1	0.1	0.1	0.1	0.1	0.1	0.2	1.0	1.0		57.3	
	7	0.8	0.4	0.3	0.3	0.3	0.3	0.8	4.3	1.5			
	8	1.5	0.8	0.6	0.6	0.6	0.8	1.5	8.1	2.9			
	9	2.2	1.3	1.0	1.0	1.1	1.4	2.9	28.6	7.1			
	10	2.9	2.0	1.6	1.6	1.8	2.6	7.1					
	11	3.5	2.7	2.6	2.9	4.0	7.1	57.3					
	12	3.7	4.0	5.1	9.5	8.1							
		N	NNW	NW	WNW	W	WSW	SW	SSW	S	SSE	SE	ESE

Side Projection — 8.0° North Latitude
Shadow Width, foot per foot (meter per meter) Projection

Date	N	NNE	NE	ENE	E	ESE	SE	SSE	S	SSW	SW	WSW	Time (pm)	ALT	AZ
DEC			2.9	1.1	0.5	0.1	0.3	0.9	2.1	17.7			5	10	64
			3.7	1.3	0.6	0.1	0.3	0.8	1.7	7.6			4	24	60
			6.9	1.7	0.7	0.3	0.1	0.6	1.3	3.9			3	36	53
				2.8	1.1	0.5	0.0	0.4	0.9	2.1	21.0		2	47	42
					2.2	0.9	0.4	0.0	0.5	1.1	2.7		1	55	25
						2.4	1.0	0.4	0.0	0.4	1.0	2.4	12	59	0
JAN/NOV			2.4	1.0	0.4	0.0	0.4	1.0	2.4	14.6			5	11	68
			3.0	1.1	0.5	0.1	0.3	0.9	2.0	5.4			4	25	64
			4.7	1.5	0.6	0.2	0.2	0.7	1.5	2.6			3	38	57
			51.5	2.3	1.0	0.4	0.0	0.4	1.0	1.2			2	49	46
				11.2		0.8	0.3	0.1	0.5	0.4			1	58	28
						2.4	1.0	0.4	0.0	0.4			12	62	0
FEB/OCT		6.0	1.6	0.7	0.2	0.2	0.6	1.4	4.3				5	13	77
		9.7	1.9	0.8	0.3	0.1	0.5	1.2	3.4				4	27	73
			2.4	1.0	0.4	0.2	0.4	1.0	2.5				3	41	68
			4.2	1.4	0.6	0.2	0.2	0.7	1.6	6.1	9.0		2	55	58
				3.4	1.2	0.6	0.1	0.3	0.8	1.8			1	66	39
						2.4	1.0	0.4	0.0	0.4	1.0	2.4	12	71	0
MAR/SEPT		2.7	1.1	0.5	0.0	0.4	0.9	2.2	26.8				5	15	88
		3.1	1.2	0.5	0.1	0.3	0.8	2.0	12.4				4	30	85
		3.8	1.3	0.6	0.1	0.3	0.6	1.7	7.2				3	44	82
		6.4	1.6	0.7	0.2	0.2	0.6	1.4	4.1	11.6			2	59	76
			3.2	1.2	0.5	0.1	0.3	0.8	1.9	8.1			1	73	63
						2.4	1.0	0.4	0.0	0.4	1.0	2.4	12	82	0
APR/AUG	4.9	1.5	0.7	0.2	0.2	0.7	1.5	5.1	5.1				6	2	101
	5.8	1.6	0.7	0.2	0.2	0.6	1.4	4.4	1.3				5	16	100
	6.5	1.6	0.7	0.2	0.2	0.6	1.4	4.1	2.5				4	31	99
	6.7	1.7	0.7	0.3	0.1	0.6	1.3	4.0	4.3				3	46	98
	6.0	1.6	0.7	0.2	0.2	0.6	1.3	4.3	8.1				2	60	99
	3.8	1.3	0.6	0.1	0.3	0.8	1.7	7.5	28.6				1	75	105
	0.0	0.4	1.0	2.4									12	86	180
MAY/JUL	2.8	1.1	0.5	0.0	0.4	0.9	2.1	21.4	1.0				6	3	110
	3.0	1.1	0.5	0.1	0.3	0.9	2.0	14.5	4.3				5	17	109
	3.0	1.1	0.5	0.1	0.3	0.9	2.0	14.5	8.1				4	31	109
	2.7	1.1	0.5	0.0	0.4	0.9	2.2	26.4	28.6				3	45	110
	2.1	0.9	0.4	0.1	0.5	1.1	2.8						2	59	116
	1.1	0.5	0.1	0.3	0.9	2.0	14.8						1	71	131
	0.0	0.4	1.0	2.4									12	78	180
JUN	2.3	1.0	0.4	0.0	0.4	1.0	2.5			57.3			6	3	113
	2.5	1.0	0.4	0.0	0.4	1.0	2.4			1.5			5	17	112
	2.4	1.0	0.4	0.0	0.4	1.0	2.4			2.9			4	31	113
	2.1	0.9	0.4	0.4	0.5	1.1	2.8			7.1			3	44	115
	1.6	0.7	0.2	0.2	0.6	1.4	4.2						2	57	122
	0.9	0.3	0.1	0.5	1.1	2.9							1	69	139
	0.0	0.4	1.0	2.4									12	75	180
	N	NNW	NW	WNW	W	WSW	SW	SSW	S	SSE	SE	ESE	PM		

Horizontal Projection — 16.0° North Latitude
Shadow Length, foot per foot (meter per meter) Projection

Date	Solar Time (am)	N	NNE	NE	ENE	E	ESE	SE	SSE	S	SSW	SW	WSW
DEC	7			0.4	0.2	0.1	0.1	0.1	0.2	0.3	1.6		
	8			1.7	0.6	0.4	0.4	0.4	0.4	0.6	2.0		
	9			8.1	1.3	0.8	0.6	0.6	0.7	0.9	1.9		
	10				3.5	1.4	1.0	0.9	0.9	1.1	1.7	6.3	
	11					3.1	1.6	1.2	1.1	1.2	1.5	2.7	28.6
	12						3.3	1.7	1.3	1.2	1.3	1.7	3.3
JAN/NOV	7			0.4	0.2	0.2	0.1	0.2	0.2	0.3	8.1		
	8			1.4	0.6	0.4	0.4	0.4	0.5	0.8	3.3		
	9			5.1	1.3	0.8	0.7	0.6	0.8	1.1	2.5		
	10				3.1	1.5	1.1	1.0	1.0	1.2	2.1	11.4	
	11					3.3	1.7	1.3	1.2	1.3	1.7	3.3	
	12						3.5	2.0	1.5	1.4	1.5	2.0	3.5
FEB/OCT	7		1.5	0.4	0.2	0.2	0.2	0.2	0.3	0.8			
	8		11.4	1.1	0.6	0.5	0.5	0.5	0.7	1.3			
	9			2.7	1.2	0.9	0.8	0.8	1.0	1.7	8.1		
	10			14.3	2.6	1.5	1.2	1.2	1.3	1.9	4.0		
	11				14.3	3.5	1.7	1.7	1.7	2.0	2.7	6.3	
	12						5.1	2.7	2.1	2.0	2.1	2.7	5.1
MAR/SEPT	7		0.8	0.4	0.3	0.3	0.3	0.3	0.6	3.5			
	8		2.4	0.9	0.6	0.6	0.6	0.7	1.0	3.5			
	9		7.1	1.9	1.2	1.0	0.9	1.1	1.5	3.5			
	10			4.3	2.2	1.7	1.5	1.6	2.1	3.5	28.6		
	11				7.1	3.5	2.7	2.5	2.7	3.5	6.3	5.1	
	12						9.5	5.1	3.7	3.5	3.7	5.1	9.5
APR/AUG	6	0.3	0.1	0.1	0.1	0.1	0.1	0.1	0.3				
	7	2.5	0.6	0.4	0.3	0.3	0.4	0.5	1.2				
	8	9.5	1.4	0.8	0.6	0.6	0.7	1.2	2.0				
	9		2.7	1.5	1.1	1.0	1.1	1.5	2.7	19.1			
	10		5.7	2.7	1.9	1.8	1.9	2.4	4.0	14.3			28.6
	11		28.6	7.1	3.7	3.7	3.7	4.3	6.3	14.3	19.1	19.1	28.6
	12			19.1	28.6			19.1	14.3	14.3	14.3	19.1	28.6
MAY/JUL	6	0.3	0.1	0.1	0.1	0.1	0.1	0.2	1.7				
	7	1.2	0.6	0.4	0.3	0.4	0.4	0.7	3.1				
	8	2.7	1.1	0.8	0.6	0.7	0.8	1.2	4.3				
	9	5.1	1.9	1.3	1.1	1.1	1.3	2.0	6.3				
	10	8.1	3.3	2.1	1.9	1.9	2.2	3.5	11.4				
	11	11.4	5.7	4.3	3.7	4.0	5.1	11.4	57.3				
	12	14.3	14.3	19.1	28.6				57.3				
JUN	6	0.3	0.2	0.1	0.1	0.1	0.2	0.3	7.1				
	7	1.1	0.5	0.4	0.4	0.4	0.5	0.8	8.1				
	8	2.1	1.0	0.8	0.6	0.7	0.9	1.4	11.4				
	9	3.5	1.7	1.2	1.1	1.2	1.4	2.4	28.6				
	10	5.1	2.6	2.0	1.8	1.9	2.5	4.3					
	11	7.1	4.3	3.7	3.5	4.0	5.7	14.3					
	12	8.1	8.1	11.4	19.1								28.6
		N	NNW	NW	WNW	W	WSW	SW	SSW	S	SSE	SE	ESE

Side Projection — 16.0° North Latitude
Shadow Width, foot per foot (meter per meter) Projection

N	NNE	NE	ENE	E	ESE	SE	SSE	S	SSW	SW	WSW	Time (pm)	ALT	AZ
		3.0	1.2	0.5	0.1	0.3	0.9	2.0	13.2			5	7	63
		4.6	1.4	0.6	0.2	0.2	0.7	1.6	5.6			4	19	57
		13.9	2.0	0.9	0.3	0.1	0.5	1.2	3.0			3	31	49
			3.8	1.3	0.6	0.1	0.3	0.8	1.7	7.4		2	41	37
				2.6	1.1	0.5	0.0	0.4	0.9	2.2	32.7	1	48	21
					2.4	1.0	0.4	0.0	0.4	1.0	2.4	12	51	0
		2.5	1.0	0.4	0.0	0.4	1.0	2.3	54.0			5	8	66
		3.6	1.3	0.6	0.1	0.3	0.8	1.8	8.3			4	21	61
		7.7	1.7	0.8	0.3	0.1	0.6	1.3	3.7			3	33	52
			3.1	1.2	0.5	0.1	0.3	0.8	1.9	12.3		2	43	40
				2.4	1.0	0.4	0.0	0.4	1.0	2.4		1	51	23
					2.4	1.0	0.4	0.0	0.4	1.0	2.4	12	54	0
	7.4	1.7	0.8	0.3	0.1	0.6	1.3	3.8				5	11	75
	26.6	2.2	0.9	0.4	0.0	0.5	1.1	2.7				4	25	70
		3.3	1.2	0.5	0.1	0.3	0.8	1.9	10.1			3	38	62
		11.8	1.9	0.8	0.3	0.1	0.5	1.2	3.1			2	50	50
			7.7	1.7	0.8	0.3	0.1	0.6	1.3	3.7		1	59	30
					2.4	1.0	0.4	0.0	0.4	1.0	2.4	12	63	0
	3.0	1.2	0.5	0.1	0.3	0.9	2.0	13.5				5	14	86
	4.2	1.4	0.6	0.2	0.2	0.7	1.6	6.3				4	29	81
	8.0	1.8	0.8	0.3	0.1	0.6	1.3	3.6				3	43	75
		2.8	1.1	0.5	0.1	0.4	0.9	2.1	19.0			2	56	64
			2.5	1.0	0.4	0.0	0.4	1.0	2.3	70.7		1	68	44
					2.4	1.0	0.4	0.0	0.4	1.0	2.4	12	74	0
5.1	1.5	0.7	0.2	0.2	0.7	1.5	5.0					6	3	101
7.7	1.7	0.8	0.3	0.1	0.6	1.3	3.7					5	17	97
14.6	2.0	0.9	0.3	0.1	0.5	1.1	3.0					4	32	94
	2.4	1.0	0.4	0.0	0.4	1.0	2.4					3	46	90
	3.1	1.2	0.5	0.1	0.3	0.8	1.9	12.1				2	61	85
	7.5	1.7	0.8	0.3	0.1	0.6	1.3	3.8				1	75	75
0.0	0.4	1.0	2.4			1.0	0.4	0.0	0.4	1.0	2.4	12	86	0
2.9	1.1			0.3	0.9	2.1	17.8					6	5	109
3.5	1.3			0.3	0.8	1.8	8.9					5	19	106
4.1	1.4			0.2	0.7	1.6	6.5					4	33	104
4.6	1.4			0.2	0.7	1.6	5.6					3	47	102
4.5	1.4			0.2	0.7	1.6	5.7					2	61	103
3.1	1.2			0.3	0.9	2.0	12.4					1	75	108
0.0	0.4	1.0	2.4			1.0	0.4	0.0				12	86	180
2.4	1.0			0.4	1.0	2.4	20.4					6	6	113
2.8	1.1			0.4	0.9	2.1	12.3					5	20	110
3.1	1.2			0.3	0.9	2.0	11.3					4	33	108
3.2	1.2			0.3	0.8	1.9	11.3					3	47	107
2.8	1.1			0.4	0.9	2.1	19.9					2	61	110
1.7	0.8	0.3	0.6	1.3	3.8							1	74	120
0.0	0.4	1.0	2.4									12	83	180
N	NNW	NW	WNW	W	WSW	SW	SSW	S	SSE	SE	ESE	PM		

507

TABLE B3.9 SHADOW LENGTHS AND SHADOW WIDTHS FOR BUILDING EXTERIOR PROJECTIONS (*Continued*)

Horizontal Projection — 24.0° North Latitude
Shadow Length, foot per foot (meter per meter) Projection

Date	Solar Time (am)	N	NNE	NE	ENE	E	ESE	SE	SSE	S	SSW	SW	WSW
DEC	7		0.2		0.1	0.1	0.1	0.1	0.1	0.5	0.6		
	8			1.5	0.5	0.3	0.3	0.3	0.3	0.7	1.3		
	9			28.6	1.2	0.7	0.5	0.5	0.5	1.0	1.3		
	10				3.5	1.2	0.8	0.7	0.7	0.8	1.2	3.5	
	11					2.7	1.3	1.2	0.8	0.9	1.1	1.9	11.4
	12						2.4	1.3	1.0	0.9	1.0	1.3	2.4
JAN/NOV	7		0.2		0.1	0.1	0.1	0.1	0.1	0.2	2.6		
	8			1.3	0.5	0.4	0.3	0.3	0.4	0.6	1.9		
	9			8.1	1.2	0.7	0.5	0.5	0.6	0.8	1.7		
	10				3.3	1.3	0.9	0.8	0.8	0.9	1.5	5.1	
	11					2.9	1.4	1.1	1.0	1.0	1.3	2.2	19.1
	12						2.7	1.5	1.1	1.0	1.1	1.5	2.7
FEB/OCT	7		1.4		0.3	0.2	0.2	0.2	0.2	0.6			
	8			1.1	0.6	0.4	0.4	0.4	0.6	1.0	19.1		
	9			3.3	1.2	0.8	0.7	0.7	0.8	1.2	3.7		
	10				2.7	1.4	1.1	1.0	1.0	1.4	2.5	28.6	
	11					3.1	1.7	1.4	1.3	1.4	1.9	3.7	
	12						3.7	2.1	1.5	1.4	1.5	2.1	3.7
MAR/SEPT	7		0.9	0.4	0.3	0.2	0.2	0.3	0.5	2.2			
	8		3.3	1.0	0.6	0.5	0.5	0.6	0.9	2.2			
	9			2.1	1.2	0.9	0.8	0.9	1.2	2.2	5.7		
	10			7.1	2.5	1.6	1.3	1.3	1.5	2.2	3.3	9.5	
	11				9.5	3.5	2.2	1.9	1.9	2.2	2.5	3.7	5.7
	12						5.7	3.3	2.5	2.2	2.5	3.3	5.7
APR/AUG	6	0.4	0.2	0.1	0.1	0.1	0.1	0.1	0.4				
	7	4.0	0.7	0.4	0.3	0.3	0.4	0.5	1.1				
	8		1.7	0.9	0.7	0.6	0.7	0.9	1.5	28.6			
	9		4.0	1.7	1.2	1.0	1.1	1.3	2.1	7.1			
	10		19.1	3.7	2.1	1.7	1.7	1.9	2.6	5.1			
	11			28.6	6.3	3.7	3.1	2.9	3.3	4.7	6.3	19.1	
	12					28.6	11.4	5.7	5.1	4.7	5.1	5.7	11.4
MAY/JUL	6	0.4	0.2	0.2	0.1	0.1	0.2	0.3	2.0				
	7	1.7	0.7	0.4	0.3	0.4	0.5	0.7	2.4				
	8	4.7	1.3	0.9	0.6	0.7	0.8	1.2	2.9				
	9	19.1	2.6	1.5	1.1	1.2	1.2	1.7	3.5				
	10		5.7	2.7	1.9	1.9	2.1	2.6	4.3	57.3			
	11		28.6	7.1	4.0	4.0	4.3	4.7	6.3	19.1			
	12						28.6	19.1	14.3	14.3	14.3	19.1	28.6
JUN	6	0.4	0.2	0.2	0.1	0.2	0.2	0.4	11.4				
	7	1.4	0.6	0.5	0.3	0.4	0.5	0.9	4.0				
	8	3.3	1.2	0.8	0.6	0.7	0.9	1.3	4.0				
	9	7.1	2.2	1.4	1.2	1.2	1.3	2.0	4.7				
	10	19.1	4.3	2.5	2.1	2.0	2.1	3.1	6.3				
	11		9.5	5.7	4.3	4.0	4.3	5.7	11.4	57.3			
	12							57.3	57.3	57.3	57.3	57.3	
		N	NNW	NW	WNW	W	WSW	SW	SSW	S	SSE	SE	ESE

Side Projection — 24.0° North Latitude
Shadow Width, foot per foot (meter per meter) Projection

Date	N	NNE	NE	ENE	E	ESE	SE	SSE	S	SSW	SW	WSW	Time (pm)	ALT	AZ
DEC			3.2	1.2	0.5	0.1	0.3	0.8	1.9	11.6			5	3	63
			5.5	1.6	0.7	0.2	0.2	0.7	1.4	4.6			4	15	55
			59.9	2.3	1.0	0.4	0.0	0.4	1.0	2.5			3	26	46
				5.0	1.5	0.7	0.2	0.2	0.7	1.5	5.0		2	34	34
					3.0	1.2	0.5	0.1	0.3	0.9	2.0	13.2	1	40	18
						2.4	1.0	0.4	0.0	0.4	1.0	2.4	12	43	0
JAN/NOV			2.7	1.1	0.5	0.0	0.4	0.9	2.2	30.5			5	5	66
			4.2	1.4	0.6	0.2	0.2	0.7	1.6	6.1			4	17	58
			15.3	2.0	0.9	0.3	0.1	0.5	1.1	2.9			3	28	49
				4.1	1.4	0.6	0.2	0.2	0.7	1.6	6.4		2	37	36
					2.8	1.1	0.5	0.1	0.4	0.9	2.1	20.0	1	44	20
						2.4	1.0	0.4	0.0	0.4	1.0	2.4	12	46	0
FEB/OCT		9.0	1.8	0.8	0.3	0.1	0.6	1.3	3.5	54.0			5	9	74
			2.5	1.0	0.4	0.0	0.4	1.0	2.3	5.3			4	22	66
			4.8	1.5	0.7	0.2	0.2	0.7	1.5	2.2			3	34	57
				2.6	1.1	0.4	0.0	0.4	0.9	1.1	38.7		2	45	44
					2.2	0.9	0.4	0.0	0.5	1.1	2.7		1	52	25
						2.4	1.0	0.4	0.0	0.4	1.0	2.4	12	55	0
MAR/SEPT		3.4	1.2	0.5	0.1	0.3	0.8	1.8	9.2				5	14	84
		6.1	1.6	0.7	0.2	0.2	0.6	1.4	4.3				4	27	77
			2.4	1.0	0.4	0.0	0.4	1.0	2.5	4.5			3	40	68
			5.8	1.6	0.7	0.2	0.2	0.6	1.4		4.9		2	52	55
				5.2	1.5	0.7	0.2	0.7	0.7	1.5		2.4	1	62	33
						2.4	1.0	0.4	0.0	0.4	1.0	2.4	12	66	0
APR/AUG	5.3	1.5	0.7	0.2	0.2	0.2	1.5	4.8					6	5	101
	11.7	1.9	0.8	0.3	0.1	0.1	1.2	3.1					5	18	95
		2.5	1.0	0.4	0.0	0.0	1.0	2.3	54.7				4	32	89
		3.9	1.3	0.6	0.1	0.1	0.8	1.7	7.1				3	46	82
		13.4	2.0	0.9	0.3	0.3	1.1	1.2	3.0				2	59	72
			8.7	1.8	0.8		0.6		1.3	3.5			1	71	52
									0.0				12	78	0
MAY/JUL	3.0	1.2	0.5	0.3	0.3	0.9	2.0	13.9					6	8	108
	4.3	1.4	0.6	0.2	0.2	0.7	1.6	6.1					5	21	103
	6.7	1.7	0.7	0.1	0.1	0.6	1.4	4.0					4	35	98
	15.7	2.0	0.9	0.1	0.1	0.5	1.1	2.9					3	48	94
		2.7	1.1	0.5	0.0	0.4	0.9	2.2	25.3				2	62	88
		6.1	1.6	0.7	0.2	0.2	0.6	1.4	4.3				1	76	77
								0.4	0.0	0.4			12	86	0
JUN	2.5	1.0	0.4	0.4	0.4	1.0	2.3	64.9					6	9	112
	3.3	1.2	0.5	0.3	0.3	0.8	1.9	10.0					5	22	107
	4.5	1.4	0.6	0.2	0.2	0.7	1.6	5.7					4	36	103
	6.5	1.6	0.7	0.2	0.2	0.6	1.4	4.1					3	49	99
	11.4	1.9	0.8	0.1	0.1	0.5	1.2	3.2					2	63	95
	76.2	2.3	1.0	0.0	0.0	0.4	1.0	2.5					1	76	91
									0.0	0.4	1.0	2.4	12	89	0
	N	NNW	NW	WNW	W	WSW	SW	SSW	S	SSE	SE	ESE	PM		

Side Projection and Horizontal Projection — 32.0° North Latitude

Horizontal Projection, 32.0° North Latitude
Shadow Length, foot per foot (meter per meter) Projection

Date	Solar Time (am)	N	NNE	NE	ENE	E	ESE	SE	SSE	S	SSW	SW	WSW
DEC	8			1.2	0.3	0.2	0.2	0.2	0.2	0.3	0.8		
	9				1.0	0.5	0.4	0.4	0.4	0.5	0.9	14.3	
	10				3.5	1.0	0.6	0.5	0.5	0.6	0.9	2.1	
	11					2.2	1.0	0.7	0.6	0.7	0.8	1.3	6.3
	12						1.8	1.0	0.8	0.7	0.8	1.0	1.8
JAN/NOV	7			0.1	0.0	0.0	0.0	0.0	0.0	0.1	0.6		
	8			1.1	0.4	0.3	0.2	0.2	0.3	0.4	1.2		
	9			28.6	1.0	0.6	0.4	0.4	0.4	0.6	1.1	2.9	
	10				3.3	1.1	0.7	0.6	0.6	0.7	1.0	1.6	
	11					2.5	1.2	0.8	0.7	0.8	1.0	1.6	8.1
	12						2.1	1.1	0.8	0.8	0.8	1.1	2.1
FEB/OCT	7		1.3	0.2	0.2	0.1	0.1	0.2	0.2	0.4			
	8			1.0	0.5	0.4	0.3	0.4	0.4	0.8	5.1		
	9			4.0	1.1	0.7	0.6	0.6	0.6	0.9	2.2	7.1	
	10				2.7	1.3	0.9	0.8	0.8	1.0	1.7	2.5	
	11					2.7	1.4	1.1	1.0	1.1	1.4	1.5	57.3
	12						2.7	1.5	1.2	1.1	1.2		2.7
MAR/SEPT	7		0.9	0.4	0.3	0.2	0.2	0.3	0.4	1.6			
	8		4.7	1.0	0.6	0.5	0.5	0.5	0.7	1.6	8.1		
	9			2.6	1.2	0.8	0.8	0.8	1.0	1.6	3.1		
	10			28.6	2.6	1.5	1.2	1.1	1.2	1.6	2.1	4.7	
	11				19.1	3.1	1.9	1.5	1.4	1.6	2.1	2.2	7.1
	12						4.3	2.2	1.7	1.6	1.7		4.3
APR/AUG	6	0.6	0.2	0.1	0.1	0.1	0.1	0.2	0.4	0.5			
	7	9.5	0.8	0.5	0.4	0.3	0.4	0.5	0.7	1.0			
	8		2.1	1.0	0.7	0.6	0.6	0.8	1.1	1.3	5.7		
	9		8.1	2.0	1.2	1.0	1.0	1.1	1.5	1.9	3.5		
	10			5.7	2.4	1.7	1.5	1.4	1.9	2.9	4.3	19.1	
	11				6.3	3.5	2.5	2.1	2.2	2.7	4.3	3.7	7.1
	12						7.1	3.7	2.9	2.7	2.9		7.1
MAY/JUL	6	0.6	0.3	0.2	0.2	0.2	0.2	0.4	0.5	0.5			
	7	2.4	0.8	0.5	0.4	0.4	0.5	0.7	1.0	1.0			
	8	14.3	1.7	1.0	0.8	0.7	0.8	1.1	2.1	11.4			
	9		3.7	1.7	1.3	1.1	1.2	1.4	2.4	6.3			
	10		19.1	3.7	2.2	1.9	1.8	2.1	2.7	5.1	4.3		
	11			28.6	6.3	4.0	3.3	3.1	3.5	5.1	11.4	19.1	
	12						11.4	6.3	5.1	4.7	5.1	6.3	11.4
JUN	5	0.0	0.0	0.0	0.0	0.0	0.0	0.0	0.5	5.1			
	6	0.6	0.3	0.2	0.2	0.2	0.2	0.3	0.5	2.9			
	7	2.0	0.8	0.5	0.5	0.5	0.6	0.9	0.9	2.7			
	8	6.3	1.5	1.0	0.8	0.8	0.9	1.2	2.4	3.1			
	9		3.1	1.7	1.3	1.2	1.3	1.7	3.1	11.4			
	10		9.5	3.3	2.2	2.0	2.0	2.4	3.5	7.1	28.6		
	11			14.3	5.7	4.0	3.5	3.7	4.7	7.1			
	12						19.1	9.5	7.1	6.3	7.1	9.5	19.1
		N	NNW	NW	WNW	W	WSW	SW	SSW	S	SSE	SE	ESE

Side Projection, 32.0° North Latitude
Shadow Width, foot per foot (meter per meter) Projection

Date	N	NNE	NE	ENE	E	ESE	SE	SSE	S	SSW	SW	WSW	Time (pm)	ALT	AZ
DEC			6.4	1.6	0.7	0.2	0.0	0.6	1.4	4.1	40.9		4	10	54
				2.6	1.1	0.4	0.0	0.4	1.0	2.3	4.1		3	20	44
				6.6	1.7	0.7	0.2	0.2	0.6	1.4	1.8		2	28	31
					3.4	1.2	0.5	0.1	0.3	0.8	1.0	9.3	1	33	16
						2.4	1.0	0.4	0.0	0.4	1.0	2.4	12	35	0
JAN/NOV			2.7	1.1	0.5	0.0	0.4	0.9	2.2	25.2			5	1	65
			4.9	1.5	0.7	0.2	0.2	0.7	1.5	5.1			4	13	56
			58.6	2.3	1.0	0.4	0.0	0.4	1.3	2.5			3	22	46
				5.4	1.5	0.7	0.2	0.2	0.7	1.5	4.7		2	31	33
					3.2	1.2	0.5	0.1	0.3	0.8	1.9	11.4	1	36	18
						2.4	1.0	0.4	0.0	0.4	1.0	2.4	12	38	0
FEB/OCT		10.8	1.9	0.8	0.3	0.1	0.5	1.2	3.2	15.3			5	7	73
			2.9	1.1	0.5	0.1	0.3	0.9	2.0	3.8			4	18	64
			7.3	1.7	0.8	0.3	0.1	0.6	1.3	1.8			3	29	53
				3.4	1.2	0.5	0.1	0.3	0.8	1.8	9.3		2	38	39
					2.6	1.1	0.4	0.0	0.4	0.9	2.2	38.6	1	45	21
						2.4	1.0	0.4	0.0	0.4	1.0	2.4	12	47	0
MAR/SEPT		3.9	1.3	0.6	0.1	0.3	0.8	1.7	7.0				5	13	82
		10.4	1.9	0.8	0.3	0.1	0.5	1.2	3.3	10.5			4	25	73
			3.3	1.2	0.5	0.1	0.3	0.8	1.9	2.7			3	37	62
			23.3	2.1	0.9	0.4	0.0	0.5	1.1	1.2	3.0		2	47	47
				13.2	2.0	0.9	0.3	0.1	0.5	1.2			1	55	27
						2.4	1.0	0.4	0.0	0.4	1.0	2.4	12	58	0
APR/AUG	5.7		0.7	0.7	0.2	0.6	1.4	4.5					6	6	100
	26.2		1.2	0.4	0.0	0.5	1.1	2.7	5.7				5	19	92
			1.8	0.5	0.1	0.3	0.8	1.8	3.5	9.5			4	31	84
			3.6	0.8	0.3	0.1	0.6	1.3	2.9	3.5	7.6		3	44	74
				1.3	0.6	0.1	0.3	0.8	1.8	1.7	1.7	7.9	2	56	60
				3.7	1.3	0.6	0.1	0.4	0.8	0.8	1.0	1.7	1	65	37
						2.4	1.0	0.4	0.0	0.4	1.0	2.4	12	70	0
MAY/JUL	3.2		0.5	0.1	0.3	0.8	1.9	10.7					6	10	107
	5.6		0.7	0.0	0.2	0.6	1.4	4.5	10.7				5	23	100
	19.8		0.9	0.2	0.1	0.5	1.1	2.8	3.3	10.7			4	35	93
			1.2	0.5	0.3	0.3	0.8	1.9	1.3	3.3	3.6		3	48	85
			1.9	0.8	0.4	0.1	0.5	1.2	0.6	1.3			2	61	73
			8.3	1.8	0.8	0.3	0.1	0.6		0.6	1.0		1	72	52
						2.4	1.0	0.4	0.0	0.4	1.0	2.4	12	78	0
JUN	1.9		0.3	0.1	0.5	1.2	3.2	24.9					7	1	118
	2.7		0.6	0.0	0.4	0.9	2.2	6.3	24.9				6	12	110
	4.2		0.8	0.2	0.2	0.7	1.6	3.6	6.3				5	24	103
	8.4		0.8	0.3	0.1	0.6	1.3	2.3	3.6				4	37	97
			1.0	0.4	0.0	0.4	1.0	2.3	5.5				3	50	89
			1.4	0.6	0.2	0.2	0.7	1.6	4.7	8.7			2	62	80
			3.5	1.3	0.6	0.1	0.3	0.8	1.8	0.4	1.0		1	74	61
						2.4	1.0	0.4	0.0	0.4	1.0	2.4	12	81	0
	N	NNW	NW	WNW	W	WSW	SW	SSW	S	SSE	SE	ESE	PM		

TABLE B3.9   SHADOW LENGTHS AND SHADOW WIDTHS FOR BUILDING EXTERIOR PROJECTIONS (*Continued*)

40.0° North Latitude

Horizontal Projection — Shadow Length, foot per foot (meter per meter) Projection

Date	Time (am)	N	NNE	NE	ENE	E	ESE	SE	SSE	S	SSW	SW	WSW
DEC	8			0.7	0.2	0.1	0.1	0.1	0.1	0.2	0.4		
	9	0.8			0.8	0.4	0.3	0.2	0.3	0.3	0.6	4.7	
	10			3.1	3.1	0.8	0.5	0.4	0.4	0.4	0.6	1.4	
	11				1.8	1.8	0.8	0.5	0.5	0.5	0.6	0.9	3.7
	12						1.3	1.3	0.7	0.7	0.5	0.7	1.3
JAN/NOV	8	0.8		0.8	0.3	0.2	0.1	0.1	0.2	0.2	0.7		
	9				0.8	0.4	0.3	0.3	0.3	0.4	0.8	19.1	
	10			3.1	3.1	0.9	0.6	0.4	0.4	0.5	0.7	1.8	
	11					2.0	0.9	0.6	0.6	0.6	0.7	1.1	4.7
	12						1.5	0.8	0.8	0.6	0.6	0.8	1.5
FEB/OCT	7	0.9		0.2	0.1	0.1	0.1	0.1	0.1	0.2			
	8			0.9	0.4	0.3	0.3	0.3	0.3	0.6	2.6		
	9			5.7	1.0	0.6	0.5	0.4	0.5	0.7	1.5	3.7	
	10				2.4	1.1	0.7	0.6	0.6	0.8	1.2	1.7	11.4
	11				2.7	1.1	1.2	0.8	0.8	0.8	1.0	1.2	2.1
	12					2.4	2.1	1.2	0.9	0.8	0.9	1.2	
MAR/SEPT	7	0.9		0.3	0.2	0.2	0.2	0.2	0.2	0.4	1.2		
	8	11.4		1.0	0.6	0.4	0.4	0.4	0.5	0.6	1.2		
	9			3.1	1.1	0.8	0.6	0.6	0.6	0.8	1.2	3.7	
	10				2.6	1.3	1.0	0.9	0.9	1.2	2.1	19.1	
	11					1.6	1.5	1.2	1.1	1.2	1.5	2.9	
	12					2.9	3.1	1.7	1.3	1.2	1.3	1.7	3.1
APR/AUG	6	0.8		0.2	0.1	0.1	0.2	0.2	0.4	0.6			
	7			0.5	0.4	0.3	0.4	0.5	0.9	1.1	28.6		
	8			1.0	0.7	0.6	0.6	0.7	1.1	2.2	3.1		
	9	2.1		1.1	1.2	0.7	0.8	0.9	1.0	2.0	4.3		
	10	7.1		2.6	2.6	1.6	1.3	1.2	1.4	1.7	2.0	4.3	14.3
	11			9.5	3.3	1.8	2.1	1.7	1.7	1.9	2.6	4.3	2.6
	12			14.3		3.7	4.7	2.6	2.0	1.9	2.0	2.6	4.7
MAY/JUL	5	0.1	0.1	0.0	0.0	0.0	0.1	0.1	0.1	0.2			
	6	0.8	0.4	0.3	0.2	0.2	0.3	0.5	0.5		1.9		
	7	3.7	0.9	0.6	0.5	0.4	0.5	0.7	0.7		1.6		
	8		2.1	1.1	0.8	0.7	0.8	1.0	1.0		1.7	14.3	
	9		7.1	2.1	1.3	1.1	1.1	1.2	1.4		4.3	4.3	
	10			5.7	2.5	1.8	1.6	1.6	2.0	3.3	3.3	19.1	
	11				3.7	2.4	2.6	2.2	2.4	2.9	4.3	4.0	
	12					4.0	7.1	4.0	2.9	2.7			7.1
JUN	5	0.2	0.1	0.1	0.1	0.1	0.1	0.1	0.2				
	6	0.8	0.4	0.3	0.3	0.3	0.3	0.6	0.6		3.7		
	7	2.9	0.9	0.6	0.5	0.5	0.6	0.8	0.8		2.2		
	8	57.3	2.0	1.1	0.8	0.8	0.8	1.1	1.1		2.1	7.1	
	9		5.1	2.0	1.4	1.2	1.2	1.4	1.8		2.4	57.3	
	10			4.7	2.5	1.9	1.7	1.6	2.2	2.6	3.5	6.3	57.3
	11				8.1	4.0	2.9	2.9	3.3	2.7	3.3	3.7	4.7
	12						9.5	9.5	4.7	3.3			9.5
		N	NNW	NW	WNW	W	WSW	SW	SSW	S	SSE	SE	ESE

Side Projection — Shadow Width, foot per foot (meter per meter) Projection

Date	N	NNE	NE	ENE	E	ESE	SE	SSE	S	SSW	SW	WSW	Time (pm)	ALT	AZ
DEC			7.2	1.7	0.8	0.3	0.1	0.6	1.3	3.9	18.7		4	5	53
				2.8	1.1	0.5	0.1	0.4	0.9	2.1	3.6		3	14	42
				8.3	1.8	0.8	0.3	0.1	0.6	1.3	1.7	7.8	2	21	29
						1.3	0.6	0.1	0.3	0.8	1.0	2.4	1	25	15
					3.7	2.4	1.0	0.4	0.0	0.4			12	27	0
JAN/NOV			5.5	1.6	0.7	0.2	0.2	0.6	1.4	4.6	54.9		4	8	55
				2.5	1.0	0.4	0.0	0.4	1.0	2.3	4.0		3	17	44
				6.8	1.7	0.7	0.3	0.1	0.6	1.3	1.8	8.8	2	24	31
						1.3	0.6	0.1	0.3	0.8	1.0	2.4	1	28	16
					3.5	2.4	1.0	0.4	0.0	0.4			12	30	0
FEB/OCT	12.5		2.0	0.9	0.3	0.1	0.5	1.2	3.1	9.7			5	4	72
			3.3	1.2	0.5	0.3	0.3	0.8	1.9	3.1			4	15	62
			12.3	1.9	0.8	0.3	0.1	0.5	1.2	1.6	5.9		3	24	50
				4.4	1.4	0.6	0.2	0.2	0.7	0.9	2.0	14.8	2	32	35
					3.0	1.1	0.5	0.1	0.3	0.4	1.0	2.4	1	37	19
						2.4	1.0	0.4	0.0	0.4	1.0		12	39	0
MAR/SEPT	4.4		1.4	0.6	0.2	0.2	0.7	1.6	5.8				5	11	80
	26.8		2.2	0.9	0.4	0.0	0.5	1.1	2.7				4	23	70
			4.6	1.4	0.6	0.2	0.2	0.7	1.6	5.5	18.6		3	33	57
				2.8	1.1	0.5	0.1	0.4	0.9	2.1	2.4		2	42	42
						1.0	0.4	0.0	0.4	1.0	1.0	2.4	1	48	23
					2.4	2.4	1.0	0.4	0.0	0.4	1.0		12	50	0
APR/AUG	6.4	1.6	0.7	0.2	0.2	0.6	1.4	4.1					6	7	99
		2.5	1.0	0.4	0.0	0.4	1.0	2.4					5	19	89
		4.8	1.5	0.7	0.2	0.2	0.7	1.5	5.3				4	30	79
			2.5	1.0	0.4	0.0	0.4	1.0	2.4				3	41	67
			8.9	1.8	0.8	0.3	0.1	0.6	1.3	3.5	3.5		2	51	51
				8.5	1.8	0.8	0.3	0.1	0.6	1.3	1.0		1	59	29
						2.4	1.0	0.4	0.0	0.4	1.0	2.4	12	62	0
MAY/JUL	2.2	0.9	0.4	0.0	0.5	1.1	2.7						7	2	115
	3.6	1.3	0.6	0.1	0.3	0.8	1.8	8.2					6	13	106
	8.6	1.8	0.8	0.3	0.1	0.6	1.3	3.5					5	24	97
		2.8	1.1	0.5	0.0	0.4	0.9	2.1	20.3				4	35	87
		6.7	1.7	0.7	0.2	0.2	0.6	1.4	4.0	8.7			3	47	76
			3.5	1.3	0.6	0.0	0.3	0.8	1.8	1.8	7.2		2	57	61
				3.8	1.3	0.6	0.1	0.3	0.8	1.7	1.0		1	66	37
						2.4	1.0	0.4	0.0	0.4			12	70	0
JUN	1.9	0.8	0.3	0.1	0.5	1.2	3.1						7	4	117
	3.0	1.2	0.5	0.1	0.3	0.9	2.0	13.9					6	15	108
	5.8	1.6	0.7	0.2	0.2	0.6	1.4	4.4					5	26	100
	79.5	2.3	1.0	0.4	0.0	0.4	1.0	2.5					4	37	91
		4.4	1.4	0.6	0.2	0.2	0.7	1.6	5.8				3	49	80
			2.6	1.1	0.4	0.0	0.4	0.9	2.2	34.3			2	60	66
				2.8	1.1	0.5	0.1	0.3	0.9	2.1	18.4		1	69	42
						2.4	1.0	0.4	0.0	0.4	1.0	2.4	12	73	0
	N	NNW	NW	WNW	W	WSW	SW	SSW	S	SSE	SE	ESE	PM		

Tables for 48.0° North Latitude — Horizontal Projection (left) and Side Projection with Solar Position (right). Shadow length / width expressed in foot per foot (meter per meter).

Horizontal Projection — 48.0° North Latitude
Shadow Length, foot per foot (meter per meter) Projection

Date	Solar Time (am)	N	NNE	NE	ENE	E	ESE	SE	SSE	S	SSW	SW	WSW
DEC	8		0.1			0.0	0.0	0.0	0.0	0.1	0.0	0.0	0.0
	9		0.4			0.2	0.3	0.2	0.1	0.2	0.3	2.0	
	10				2.5	0.5	0.5	0.4	0.3	0.2	0.3	0.8	
	11					1.2	1.4	0.5	0.5	0.4	0.4	0.6	2.2
	12						2.0	0.5	0.4	0.3	0.4	0.5	0.9
JAN/NOV	8		0.1			0.1	0.1	0.1	0.1	0.1	0.3		
	9		0.6			0.3	0.3	0.3	0.2	0.3	0.5	4.7	
	10				2.5	0.6	0.5	0.4	0.3	0.3	0.5	1.1	
	11					1.4	0.4	0.4	0.4	0.4	0.5	0.8	2.9
	12						2.0	0.6	0.4	0.4	0.5	0.6	1.1
FEB/OCT	7		0.4	0.1	0.0	0.0	0.0	0.0	0.1	0.1	0.1		
	8			0.8	0.3	0.2	0.2	0.2	0.2	0.4	1.5		
	9			8.1	0.8	0.5	0.4	0.3	0.4	0.5	1.0	2.2	
	10				2.6	0.9	0.6	0.5	0.5	0.6	0.8	1.2	
	11					2.0	0.9	0.6	0.6	0.6	0.7	0.9	6.3
	12						1.6	1.1	0.9	0.6	0.6	0.9	1.6
MAR/SEPT	7		0.9	0.3	0.2	0.2	0.2	0.2	0.2	0.3	0.9		
	8			1.0	0.5	0.4	0.4	0.4	0.4	0.5	0.9	28.6	
	9			3.7	1.0	0.7	0.6	0.5	0.6	0.6	0.9	2.2	
	10				2.7	1.2	0.8	0.7	0.7	0.9	1.4	2.0	5.7
	11					2.7	1.3	0.9	1.1	0.9	1.2	1.3	19.1
	12						2.4	1.3	1.5	0.9	1.0		2.4
APR/AUG	6	1.1	0.3	0.2	0.2	0.2	0.2	0.2	0.6				
	7		1.0	0.5	0.4	0.3	0.4	0.4	0.8	5.7			
	8		4.3	1.1	0.7	0.6	0.6	0.7	0.9	2.1			
	9			2.7	1.2	0.9	0.8	0.9	1.0	1.6	7.1		
	10				2.7	1.5	1.1	1.1	1.1	1.4	2.6	3.5	
	11				14.3	2.7	1.7	1.4	1.3	1.4	1.8	1.9	
	12		57.3			3.1	3.5	1.9	1.5	1.4	1.5	1.5	3.5
MAY/JUL	5	0.2	0.1	0.1	0.1	0.1	0.1	0.1	0.3				
	6	1.1	0.4	0.3	0.3	0.3	0.3	0.3	0.5	1.7			
	7	8.1	1.1	0.6	0.5	0.5	0.5	0.7	1.4	4.7			
	8		2.9	1.2	0.8	0.7	0.7	0.9	1.3	2.6			
	9		57.3	2.5	1.4	1.0	1.0	1.1	1.4	2.1	4.7		
	10			11.4	2.7	1.7	1.4	1.5	1.3	2.0	2.7	6.3	
	11				14.3	3.5	2.1	1.8	1.8	1.9	3.3	2.6	
	12						5.7	2.6	2.1		2.4		4.7
JUN	5	0.3	0.2	0.1	0.1	0.2	0.2	0.2	0.4				
	6	1.1	0.5	0.3	0.3	0.3	0.4	0.6	0.8	2.7			
	7	5.1	1.1	0.6	0.5	0.5	0.6	0.8	1.6	1.8			
	8		2.6	1.2	0.9	0.8	0.8	1.0	1.6	8.1			
	9		14.3	2.4	1.4	1.1	1.1	1.2	1.6	3.5	6.3		
	10			8.1	2.7	1.8	1.5	1.5	1.7	2.6	3.3	8.1	
	11				14.3	3.7	2.4	2.0	2.0	2.2	2.7	3.1	
	12						5.7	3.1	2.4	2.1	2.4		5.7
		N	NNW	NW	WNW	W	WSW	SW	SSW	S	SSE	SE	ESE

Side Projection — 48.0° North Latitude
Shadow Width, foot per foot (meter per meter) Projection — with Solar Position (ALT, AZ)

N	NNE	NE	ENE	E	ESE	SE	SSE	S	SSW	SW	WSW	Time (pm)	ALT	AZ
		7.5	1.7	0.8	0.3	0.1	0.6	1.3	3.8	14.0		4	1	53
			3.0	1.2	0.5	0.1	0.3	0.9	2.0	3.3		3	8	41
			10.1	1.9	0.8	0.3	0.1	0.5	1.2	1.7	7.0	2	14	28
				3.9	1.3	0.6	0.1	0.3	0.8	1.7		1	17	14
					2.4	1.0	0.4	0.0	0.4	1.0	2.4	12	19	0
		5.9	1.6	0.7	0.2	0.2	0.6	1.4	4.4	23.8		4	3	55
			2.7	1.1	0.5	0.0	0.4	0.9	2.2	3.6		3	11	43
			8.3	1.8	0.8	0.3	0.1	0.6	1.3	1.7	7.7	2	17	29
				3.7	1.3	0.6	0.1	0.3	0.8	1.7		1	21	15
					2.4	1.0	0.4	0.0	0.4	1.0	2.4	12	22	0
	13.7	2.0	0.9	0.3	0.1	0.5	1.2	3.0	7.6			5	2	72
		3.7	1.3	0.6	0.1	0.3	0.8	1.7	2.7	4.7		4	11	60
		25.2	2.2	0.9	0.4	0.0	0.5	1.1	1.5			3	19	47
			5.4	1.5	0.7	0.2	0.2	0.6	0.8	1.9	10.4	2	25	33
				3.3	1.2	0.5	0.1	0.3	0.8	1.9		1	30	17
					2.4	1.0	0.4	0.0	0.4	1.0	2.4	12	31	0
	5.0	1.5	0.7	0.2	0.2	0.7	1.5	5.0	79.4			5	10	79
		2.5	1.0	0.4	0.0	0.4	1.0	2.3	4.0	8.0		4	20	67
		6.8	1.7	0.7	0.3	0.1	0.6	1.3	1.8	2.1	21.4	3	28	53
			3.6	1.3	0.6	0.1	0.3	0.8	0.9	2.1		2	35	38
				2.8	1.1	0.5	0.0	0.4	1.2	2.1		1	40	20
					2.4	1.0	0.4	0.0	0.4	1.0	2.4	12	42	0
7.3	1.7	0.8	0.3	0.1	0.6	1.3	3.8	17.6				6	9	98
	2.9	1.1	0.5	0.1	0.3	0.9	2.1	3.7	9.1			5	19	87
		1.7	0.8	0.3	0.1	0.6	1.3	1.8	2.4			4	29	75
18.9		3.4	1.2	0.5	0.1	0.3	0.8	1.0	1.1	2.6		3	38	61
	75.2		2.5	1.0	0.4	0.0	0.4	0.4				2	46	45
			37.4	2.2	0.9	0.4	0.0	0.4				1	51	24
					2.4	1.0	0.4	0.0	0.4	1.0	2.4	12	54	0
2.2	1.7	0.4	0.0	0.5	1.1	2.6	6.5					7	5	114
4.1	1.4	0.6	0.2	0.3	0.7	1.6	2.8	6.8				6	15	104
18.9	2.1	0.9	0.4	0.1	0.5	1.1	1.7	2.5				5	25	93
	4.0	1.3	0.6	0.1	0.3	0.7	1.0	1.2	3.5			4	35	82
	9.0	2.3	1.0	0.4	0.0	0.4	0.6	1.2	3.4			3	44	68
			1.8	0.8	0.3	0.1	0.5	2.0				2	53	51
	75.2	9.3		1.8	0.8	0.3	0.1	0.0			2.4	1	59	29
					2.4	1.0	0.4	0.0	0.4	1.0	2.4	12	62	0
2.0	0.9	0.3	0.1	0.5	1.2	3.0	9.0	10.6				7	8	117
3.4	1.2	0.6	0.1	0.3	0.8	1.8	3.3	3.0				6	17	106
9.9	1.9	0.8	0.3	0.1	0.5	1.2	1.9	1.2				5	27	96
	3.2	1.2	0.5	0.1	0.3	0.8	1.9	10.6				4	37	85
	14.0	2.0	0.9	0.3	0.1	0.5	1.2	3.0	4.4			3	47	72
		5.8	1.6	0.7	0.2	0.2	0.6	1.4	1.4	4.1		2	56	55
			6.5	1.7	0.7	0.2	0.2	0.6	0.4	4.1		1	63	31
					2.4	1.0	0.4	0.0	0.4	1.0	2.4	12	65	0
N	NNW	NW	WNW	W	WSW	SW	SSW	S	SSE	SE	ESE	PM		

TABLE B3.9 SHADOW LENGTHS AND SHADOW WIDTHS FOR BUILDING EXTERIOR PROJECTIONS *(Continued)*

Horizontal Projection — Shadow Length, foot per foot (meter per meter) Projection — 56.0° North Latitude

Date	Solar Time (am)	N	NNE	NE	ENE	E	ESE	SE	SSE	S	SSW	SW	WSW
DEC	9				0.1	0.1	0.0	0.0	0.0	0.0	0.1	0.4	
	10				1.3	0.2	0.2	0.1	0.1	0.1	0.2	0.4	
	11					0.7	0.3	0.2	0.2	0.2	0.2	0.3	1.1
	12						0.5	0.3	0.2	0.2	0.2	0.3	0.5
JAN/NOV	9				0.3	0.1	0.1	0.1	0.1	0.1	0.2	1.6	
	10				1.7	0.4	0.2	0.2	0.2	0.2	0.3	0.6	
	11					0.9	0.4	0.3	0.2	0.2	0.3	0.4	1.7
	12						0.6	0.3	0.3	0.2	0.3	0.3	0.6
FEB/OCT	8			0.5	0.2	0.1	0.1	0.1	0.1	0.2	0.8		
	9			28.6	0.6	0.3	0.2	0.2	0.3	0.3	0.6		
	10				2.2	0.6	0.4	0.3	0.3	0.4	0.6	1.4	
	11					1.5	0.6	0.5	0.4	0.4	0.5	0.8	3.5
	12						1.1	0.6	0.5	0.4	0.5	0.6	1.1
MAR/SEPT	7		0.8	0.3	0.2	0.2	0.1	0.2	0.2	0.7			
	8			0.9	0.4	0.3	0.3	0.3	0.4	0.7	5.7		
	9			4.7	0.9	0.6	0.4	0.4	0.5	0.7	1.5		
	10				2.6	1.0	0.6	0.6	0.6	0.7	1.0	3.1	
	11					2.1	1.0	0.7	0.6	0.7	0.8	1.4	8.1
	12						1.7	1.0	0.7	0.7	0.7	1.0	1.7
APR/AUG	5	0.1	0.0	0.0	0.0	0.0	0.0	0.1	0.4				
	6	1.5	0.3	0.2	0.2	0.2	0.2	0.3	0.6				
	7		1.2	0.5	0.4	0.3	0.3	0.4	0.7	3.1			
	8		8.1	1.1	0.6	0.5	0.5	0.6	0.7	1.5			
	9			3.5	1.2	0.8	0.7	0.7	0.8	1.2	3.5		
	10				2.9	1.3	0.9	0.8	0.9	1.1	1.8	9.5	
	11					2.7	1.4	1.1	1.0	1.0	1.3	2.4	28.6
	12						2.6	1.4	1.1	1.0	1.1	1.4	2.6
MAY/JUL	4	0.0	0.0	0.0	0.0	0.0	0.0	0.1					
	5	0.4	0.2	0.2	0.1	0.2	0.2	0.4					
	6	1.5	0.5	0.4	0.3	0.3	0.4	0.5	1.5				
	7		1.2	0.6	0.5	0.5	0.5	0.6	1.2	57.3			
	8		4.3	1.2	0.8	0.7	0.6	0.8	1.1	2.7			
	9			3.1	1.4	1.0	0.9	0.9	1.1	1.8	8.1		
	10				2.9	1.6	1.2	1.1	1.2	1.5	2.7	57.3	
	11				57.3	3.3	1.8	1.4	1.3	1.4	1.9	3.5	
	12						3.5	2.0	1.5	1.4	1.5	2.0	3.5
JUN	4	0.1	0.1	0.1	0.1	0.1	0.1	0.5					
	5	0.5	0.3	0.2	0.2	0.2	0.3	0.6					
	6	1.5	0.6	0.4	0.3	0.4	0.4	0.7	2.2				
	7	19.1	1.3	0.7	0.6	0.5	0.6	0.8	1.5				
	8		3.7	1.3	0.9	0.7	0.7	0.9	1.3	3.7			
	9			2.9	1.4	1.1	1.0	1.0	1.3	2.2	14.3		
	10			57.3	3.1	1.7	1.3	1.2	1.3	1.8	3.5		
	11				28.6	3.5	2.0	1.5	1.5	1.6	2.1	4.3	
	12						4.0	2.2	1.7	1.5	1.7	2.2	4.0
		N	NNW	NW	WNW	W	WSW	SW	SSW	S	SSE	SE	ESE

Side Projection — Shadow Width, foot per foot (meter per meter) Projection — 56.0° North Latitude

Date	N	NNE	NE	ENE	E	ESE	SE	SSE	S	SSW	SW	WSW	Time (pm)	Solar Position ALT	Solar Position AZ
DEC				3.1	1.2	0.5	0.1	0.3	0.9	2.0	12.6		3	2	40
				11.4	1.9	0.8	0.3	0.1	0.5	1.2	3.2		2	7	27
					4.0	1.4	0.6	0.2	0.2	0.7	1.7	6.6	1	10	14
						2.4	1.0	0.4	0.0	0.4	1.0	2.4	12	11	0
JAN/NOV				2.8	1.1	0.5	0.1	0.4	0.9	2.1	18.1		3	5	42
				9.5	1.8	0.8	0.3	0.1	0.5	1.2	3.4		2	10	28
					3.9	1.3	0.6	0.1	0.3	0.8	1.7	7.1	1	13	14
						2.4	1.0	0.4	0.0	0.4	1.0	2.4	12	14	0
FEB/OCT			4.0	1.4	0.6	0.2	0.2	0.7	1.7	6.7			4	7	59
			98.5	2.3	1.0	0.4	0.0	0.4	1.0	2.5			3	13	46
				6.5	1.6	0.7	0.2	0.2	0.6	1.4	4.1		2	19	31
					3.5	1.3	0.6	0.1	0.3	0.8	1.8	8.7	1	22	16
						2.4	1.0	0.4	0.0	0.4	1.0	2.4	12	23	0
MAR/SEPT		5.7	1.6	0.7	0.2	0.2	0.6	1.4	4.5				5	8	77
			2.8	1.1	0.5	0.1	0.4	0.9	2.1	18.6			4	16	64
			10.7	1.9	0.8	0.3	0.1	0.5	1.2	3.2			3	23	50
				4.6	1.4	0.6	0.2	0.2	0.7	1.6	5.6		2	29	35
					3.1	1.2	0.5	0.1	0.3	0.9	2.0	12.5	1	33	18
						2.4	1.0	0.4	0.0	0.4	1.0	2.4	12	34	0
APR/AUG	2.9	1.1	0.5	0.1	0.3	0.9	2.0	15.6					7	1	109
	8.7	1.8	0.8	0.3	0.1	0.6	1.3	3.5					6	10	97
		3.4	1.2	0.5	0.1	0.3	0.8	1.8	9.7				5	18	84
		16.8	2.1	0.9	0.3	0.1	0.5	1.1	2.9				4	26	71
			5.0	1.5	0.7	0.2	0.2	0.7	1.5	5.1			3	34	56
				3.2	1.2	0.5	0.1	0.3	0.8	1.9	10.7		2	40	40
					2.7	1.1	0.5	0.0	0.4	0.9	2.2	31.3	1	44	21
						2.4	1.0	0.4	0.0	0.4	1.0	2.4	12	46	0
MAY/JUL	1.4	0.6	0.2	0.2	0.7	1.6	6.0						8	1	126
	2.3	1.0	0.4	0.0	0.4	1.0	2.5						7	8	113
	4.9	1.5	0.7	0.2	0.2	0.7	1.5	5.1					6	16	102
		2.5	1.0	0.4	0.0	0.4	1.0	2.3	87.5				5	25	89
		6.5	1.6	0.7	0.2	0.2	0.6	1.4	4.1				4	33	76
			3.4	1.2	0.5	0.1	0.3	0.8	1.8	9.6			3	41	62
				2.5	1.0	0.4	0.0	0.4	1.0	2.3	70.3		2	48	44
				61.7	2.3	1.0	0.4	0.0	0.4	1.0	2.5		1	52	23
						2.4	1.0	0.4	0.0	0.4	1.0	2.4	12	54	0
JUN	1.3	0.6	0.1	0.3	0.8	1.7	7.3						8	4	127
	2.1	0.9	0.4	0.0	0.5	1.1	2.8						7	11	115
	4.1	1.4	0.6	0.2	0.2	0.7	1.6	6.4					6	28	92
	34.5	2.2	0.9	0.4	0.0	0.4	1.1	2.6					5	28	92
		5.0	1.5	0.7	0.2	0.2	0.7	1.5	5.0				4	36	79
			2.9	1.1	0.5	0.1	0.3	0.9	2.1	16.7			3	44	64
			39.7	2.3	1.0	0.4	0.0	0.4	1.1	2.6			2	51	46
				24.1	2.2	0.9	0.4	0.0	0.5	1.1	2.7		1	56	25
						2.4	1.0	0.4	0.0	0.4	1.0	2.4	12	57	0
	N	NNW	NW	WNW	W	WSW	SW	SSW	S	SSE	SE	ESE	PM		

64.0° North Latitude — Sun Shadow Tables

Solar Position

Date	Solar Time (am)		Time (pm)	ALT	AZ
DEC	11		1	2	14
	12		12	3	0
JAN/NOV	10		2	3	28
	11		1	5	14
	12		12	6	0
FEB/OCT	8		4	3	58
	9		3	8	45
	10		2	12	30
	11		1	14	15
	12		12	15	0
MAR/SEPT	7		5	7	76
	8		4	13	63
	9		3	18	48
	10		2	22	33
	11		1	25	17
	12		12	26	0
APR/AUG	5		7	4	108
	6		6	10	95
	7		5	17	82
	8		4	23	67
	9		3	29	52
	10		2	34	36
	11		1	37	18
	12		12	38	0
MAY/JUL	4		8	6	125
	5		7	12	112
	6		6	18	99
	7		5	24	86
	8		4	31	71
	9		3	37	56
	10		2	42	39
	11		1	45	20
	12		12	46	0
JUN	4		8	9	126
	5		7	15	114
	6		6	21	101
	7		5	28	87
	8		4	34	73
	9		3	40	58
	10		2	45	40
	11		1	48	21
	12		12	49	0

Horizontal Projection — Shadow Length, foot per foot (meter per meter) Projection

Orientation columns: N | NNE | NE | ENE | E | ESE | SE | SSE | S | SSW | SW | WSW

Date	am	N	NNE	NE	ENE	E	ESE	SE	SSE	S	SSW	SW	WSW
DEC	11					0.1	0.1	0.0	0.0	0.0	0.0	0.1	0.2
	12						0.1	0.1	0.1	0.1	0.1	0.1	0.1
JAN/NOV	10				0.5	0.1	0.1	0.1	0.1	0.1	0.1	0.2	
	11					0.4	0.2	0.1	0.1	0.1	0.1	0.2	0.6
	12						0.3	0.1	0.1	0.1	0.1	0.1	0.3
FEB/OCT	8			0.2		0.1	0.1	0.1	0.1	0.1	0.3	14.3	
	9	0.2	0.7		0.1	0.2	0.1	0.1	0.1	0.2	0.3	0.8	2.1
	10				0.4	0.4	0.3	0.2	0.2	0.2	0.3	0.5	0.7
	11				1.6	1.0	0.4	0.3	0.2	0.3	0.3	0.4	
	12						0.7	0.4	0.3	0.3	0.3		
MAR/SEPT	7		0.7	0.2	0.1	0.1	0.1	0.1	0.2	0.5		2.0	4.7
	8			0.8	0.3	0.2	0.2	0.2	0.3	0.5	2.6	1.0	1.3
	9			6.3	0.8	0.4	0.3	0.3	0.4	0.5	1.0	0.7	
	10				2.4	0.8	0.5	0.4	0.4	0.5	0.7		
	11					1.7	0.8	0.5	0.5	0.5	0.6		
	12						1.3	0.7	0.7	0.5	0.5		
APR/AUG	5	0.2	0.1	0.1	0.1	0.1	0.1	0.2	1.0			4.3	9.5
	6	2.1	0.4	0.2	0.2	0.2	0.2	0.3	0.6	2.1	2.1	1.7	2.1
	7		1.2	0.5	0.3	0.3	0.3	0.4	0.6	1.1	1.3	1.1	
	8			1.1	0.5	0.5	0.4	0.5	0.6	0.9	1.0		
	9			4.3	1.1	0.7	0.6	0.7	0.6	0.8	0.8		
	10				2.9	1.1	0.8	0.8	0.7	0.8			
	11					2.4	1.2	0.8	0.8	0.8			
	12						2.1	1.1	0.8	0.8			
MAY/JUL	4	0.2	0.1	0.1	0.1	0.1	0.2	0.6	1.0				28.6
	5	0.6	0.3	0.2	0.2	0.2	0.3	0.5	28.6				2.7
	6	2.1	0.6	0.4	0.3	0.3	0.4	0.6	1.4	6.3		3.7	
	7		1.5	0.7	0.5	0.5	0.5	0.6	1.0	1.9		1.9	
	8		8.1	1.3	0.8	0.7	0.7	0.7	0.9	1.3	3.7		
	9			4.0	1.4	0.9	0.8	0.8	0.9	1.2	1.9		
	10				3.1	1.4	1.0	0.9	0.9	1.1	1.3		
	11					2.9	1.5	1.1	1.0	1.1	1.1		
	12						2.7	1.5	1.1	1.0			
JUN	4	0.3	0.2	0.1	0.1	0.2	0.3	1.1					57.3
	5	0.6	0.4	0.3	0.2	0.3	0.4	0.7	1.9				3.1
	6	2.1	0.7	0.5	0.4	0.4	0.5	0.7	1.2	1.9			
	7		1.5	0.8	0.6	0.5	0.6	0.7	1.1	2.4	11.4		
	8		6.3	1.4	0.9	0.7	0.7	0.8	1.0	1.6	1.6	11.4	
	9			3.7	1.4	1.0	0.8	0.9	1.0	1.3	2.1	2.7	
	10				3.3	1.5	1.1	1.0	1.0	1.2	1.5	1.7	
	11					3.1	1.6	1.2	1.1	1.2	1.3		
	12						3.1	1.7	1.3	1.2			

Bottom orientation labels (read for morning surfaces): N · NNW · NW · WNW · W · WSW · SW · SSW · S · SSE · SE · ESE

Side Projection — Shadow Width, foot per foot (meter per meter) Projection

Orientation columns: N | NNE | NE | ENE | E | ESE | SE | SSE | S | SSW | SW | WSW

Date	N	NNE	NE	ENE	E	ESE	SE	SSE	S	SSW	SW	WSW	pm
DEC					4.1	1.4	0.6	0.2	0.2	0.7	1.6	6.5	1
						2.4	1.0	0.4	0.0	0.4	1.0	2.4	12
JAN/NOV				10.3	1.9	0.8	0.3	0.1	0.5	1.2	3.3	6.8	2
					4.0	1.3	0.6	0.1	0.3	0.7	1.7	2.4	1
						2.4	1.0	0.4	0.0	0.4	1.0	2.4	12
FEB/OCT			4.2	1.4	0.6	0.2	0.2	0.7	1.6	6.2			4
				2.5	1.0	0.4	0.0	0.4	1.0	2.4	3.8	7.8	3
			7.5	1.7	0.8	0.8	0.3	0.1	0.6	1.3	1.7		2
					3.7	1.3	0.6	0.1	0.3	0.8	1.7		1
						2.4	1.0	0.4	0.0	0.4	1.0	2.4	12
MAR/SEPT		6.3	1.6	0.7	0.2	0.2	0.6	1.4	4.2	11.6		9.7	5
			3.2	1.2	0.5	0.1	0.3	0.8	1.9	2.8	4.6	2.4	4
			18.8	2.1	0.9	0.4	0.2	0.5	1.1	1.4	1.8		3
				5.5	1.6	0.7	0.2	0.6	0.6	0.8			2
					3.4	1.2	0.5	0.1	0.3				1
						2.4	1.0	0.4	0.0	0.4	1.0	2.4	12
APR/AUG	3.0	1.2	0.5	0.1	0.3	0.9	2.0	14.2		11.6			7
	11.1	1.9	0.8	0.3	0.1	0.5	1.2	3.2	6.8	2.8	6.3	13.9	6
		4.0	1.3	0.6	0.3	0.3	0.7	1.7	2.4	1.4	2.0	2.4	5
			2.4	1.0	0.4	0.0	0.4	1.0	1.3	0.8			4
				1.7	0.8	0.3	0.1	0.3	0.7		3.7		3
				4.2	1.4	0.6	0.2	0.6	1.6	0.9			2
					3.0	1.2	0.5	0.1	0.7	0.4			1
						2.4	1.0	0.4	0.0	0.4	1.0	2.4	12
MAY/JUL	1.4	0.6	0.2	0.2	0.7	1.6	5.7			4.9	9.4	23.5	8
	2.5	1.0	0.4	0.0	0.4	1.0	2.4	4.2		1.8	2.2	2.4	7
	6.3	1.6	0.7	0.2	0.2	0.6	1.4	2.0	13.2	0.9	1.0		6
	3.0	1.2	0.5	0.1	0.3	0.9	1.2	3.0		0.4			5
	14.3	1.5	0.9	0.3	0.2	0.5	1.5			0.0			4
		3.4	1.5	0.7	0.2	0.5	0.8	0.7	0.8				3
			2.7	1.2	0.5	0.1	0.3	0.4					2
				2.4	1.1	0.5	0.0	0.3					1
						2.4	1.0	0.4	0.0	0.4	1.0	2.4	12
JUN	1.4	0.6	0.2	0.2	0.7	1.7	6.6			5.8	12.3	35.7	8
	2.3	1.0	0.4	0.0	0.4	1.0	2.6	4.8		2.0	2.2	2.4	7
	5.3	1.5	0.7	0.2	0.2	0.7	1.5	2.1	22.7	0.9	1.0		6
		2.8	1.1	0.5	0.0	0.4	0.9	3.3	3.3	0.4			5
		9.8	1.9	0.8	0.3	0.1	0.5	1.2	1.6	0.0			4
			4.4	1.4	0.6	0.2	0.2	0.7	0.9				3
				3.1	1.2	0.5	0.1	0.3	0.4				2
					2.6	1.1	0.4	0.0	0.0				1
						2.4	1.0	0.4	0.0	0.4	1.0	2.4	12

Bottom orientation labels: N · NNW · NW · WNW · W · WSW · SW · SSW · S · SSE · SE · ESE · PM
Second bottom row: N · NE · E · SE · S · SW · W · NW · SW · SSE · SE · ESE

Surface Orientations and Azimuths, Measured from the South

Orientation	N	NNE	NE	E	SE	S	SW	W	NW
Surface azimuth (deg)	180	135	90	45	0	45	90	135	

513

TABLE B3.10 THERMAL LAG FACTORS FOR GLASS SOLAR LOAD WITHOUT INTERIOR SHADING

North Latitude Exposure	Room Construction	Solar Time (hours)																								South Latitude Exposure
		1	2	3	4	5	6	7	8	9	10	11	12	13	14	15	16	17	18	19	20	21	22	23	24	
N or Shaded	L	0.17	0.14	0.11	0.09	0.08	0.33	0.42	0.48	0.56	0.63	0.71	0.76	0.80	0.82	0.82	0.79	0.79	0.84	0.61	0.48	0.38	0.31	0.25	0.20	S or Shaded
	M	0.23	0.20	0.18	0.16	0.14	0.34	0.41	0.46	0.53	0.59	0.65	0.70	0.74	0.75	0.76	0.74	0.75	0.79	0.61	0.50	0.42	0.36	0.31	0.27	
	H	0.25	0.23	0.21	0.20	0.19	0.38	0.45	0.49	0.55	0.60	0.65	0.69	0.72	0.72	0.72	0.70	0.70	0.75	0.57	0.46	0.39	0.34	0.31	0.28	
NNE	L	0.06	0.05	0.04	0.03	0.03	0.26	0.43	0.47	0.44	0.41	0.40	0.39	0.39	0.38	0.36	0.33	0.30	0.26	0.20	0.16	0.13	0.10	0.08	0.07	SSE
	M	0.09	0.08	0.07	0.06	0.06	0.24	0.38	0.42	0.39	0.37	0.37	0.36	0.36	0.36	0.34	0.33	0.30	0.27	0.22	0.18	0.16	0.14	0.12	0.10	
	H	0.11	0.10	0.09	0.09	0.08	0.26	0.39	0.42	0.39	0.36	0.35	0.34	0.34	0.33	0.32	0.31	0.28	0.25	0.21	0.18	0.16	0.14	0.13	0.12	
NE	L	0.04	0.04	0.03	0.02	0.02	0.23	0.41	0.51	0.51	0.45	0.39	0.36	0.33	0.31	0.28	0.26	0.23	0.19	0.15	0.12	0.10	0.08	0.06	0.05	SE
	M	0.07	0.06	0.06	0.05	0.04	0.21	0.36	0.44	0.45	0.40	0.36	0.33	0.31	0.30	0.28	0.26	0.24	0.21	0.17	0.15	0.13	0.11	0.09	0.08	
	H	0.09	0.08	0.08	0.07	0.07	0.23	0.37	0.44	0.44	0.39	0.34	0.31	0.29	0.27	0.26	0.24	0.22	0.20	0.17	0.14	0.13	0.12	0.11	0.10	
ENE	L	0.04	0.03	0.03	0.02	0.02	0.21	0.40	0.52	0.57	0.53	0.45	0.39	0.34	0.31	0.28	0.25	0.22	0.18	0.14	0.12	0.09	0.08	0.06	0.05	ESE
	M	0.07	0.06	0.05	0.05	0.04	0.20	0.35	0.45	0.49	0.47	0.41	0.36	0.33	0.30	0.28	0.26	0.23	0.20	0.17	0.14	0.12	0.11	0.09	0.08	
	H	0.09	0.09	0.08	0.07	0.07	0.22	0.36	0.46	0.49	0.45	0.38	0.33	0.30	0.27	0.25	0.23	0.21	0.19	0.16	0.14	0.13	0.12	0.11	0.10	
E	L	0.04	0.03	0.03	0.02	0.02	0.19	0.37	0.51	0.57	0.57	0.50	0.42	0.37	0.32	0.29	0.25	0.22	0.19	0.15	0.12	0.10	0.08	0.06	0.05	E
	M	0.07	0.06	0.06	0.05	0.05	0.18	0.33	0.44	0.50	0.51	0.46	0.39	0.35	0.31	0.29	0.26	0.23	0.21	0.17	0.15	0.13	0.11	0.10	0.08	
	H	0.09	0.09	0.08	0.08	0.07	0.20	0.34	0.45	0.49	0.49	0.43	0.36	0.32	0.29	0.26	0.24	0.22	0.19	0.17	0.15	0.13	0.12	0.11	0.10	
ESE	L	0.05	0.04	0.03	0.03	0.02	0.17	0.34	0.49	0.58	0.61	0.57	0.48	0.41	0.36	0.32	0.28	0.24	0.20	0.16	0.13	0.10	0.09	0.07	0.06	ENE
	M	0.08	0.07	0.06	0.05	0.05	0.16	0.31	0.43	0.51	0.54	0.51	0.44	0.39	0.35	0.32	0.29	0.26	0.22	0.19	0.16	0.14	0.12	0.11	0.09	
	H	0.10	0.09	0.09	0.08	0.08	0.19	0.32	0.43	0.50	0.52	0.49	0.41	0.36	0.32	0.29	0.26	0.24	0.21	0.18	0.16	0.14	0.13	0.12	0.11	
SE	L	0.05	0.04	0.04	0.03	0.03	0.13	0.28	0.43	0.55	0.62	0.63	0.57	0.48	0.42	0.37	0.33	0.28	0.24	0.19	0.15	0.12	0.10	0.08	0.07	NE
	M	0.09	0.08	0.07	0.06	0.05	0.14	0.26	0.38	0.48	0.54	0.56	0.51	0.45	0.40	0.36	0.33	0.29	0.25	0.21	0.18	0.16	0.14	0.12	0.10	
	H	0.11	0.10	0.10	0.09	0.08	0.17	0.28	0.40	0.49	0.53	0.53	0.48	0.41	0.36	0.33	0.30	0.27	0.24	0.20	0.18	0.16	0.14	0.13	0.12	
SSE	L	0.07	0.05	0.04	0.04	0.03	0.06	0.15	0.29	0.43	0.55	0.63	0.64	0.60	0.52	0.45	0.40	0.35	0.29	0.23	0.18	0.15	0.12	0.10	0.08	NNE
	M	0.11	0.09	0.08	0.07	0.06	0.08	0.16	0.26	0.38	0.48	0.55	0.57	0.54	0.48	0.43	0.39	0.35	0.30	0.25	0.21	0.18	0.16	0.14	0.12	
	H	0.12	0.11	0.11	0.10	0.09	0.12	0.19	0.29	0.40	0.49	0.54	0.55	0.51	0.44	0.39	0.35	0.31	0.27	0.23	0.20	0.18	0.16	0.15	0.13	
S	L	0.08	0.07	0.05	0.04	0.04	0.06	0.09	0.14	0.22	0.34	0.48	0.59	0.65	0.65	0.59	0.50	0.43	0.36	0.28	0.22	0.18	0.15	0.12	0.10	N
	M	0.12	0.11	0.09	0.08	0.07	0.08	0.11	0.14	0.21	0.31	0.42	0.52	0.57	0.58	0.53	0.47	0.41	0.35	0.29	0.25	0.21	0.18	0.16	0.14	
	H	0.13	0.12	0.12	0.11	0.10	0.11	0.14	0.17	0.24	0.33	0.43	0.51	0.56	0.55	0.50	0.43	0.37	0.32	0.26	0.22	0.20	0.18	0.16	0.15	

Solar cooling load factors — room constructions (to be used with Table B3.1)

Direction		Hourly values																								Direction
SSW	L	0.10	0.08	0.07	0.06	0.05	0.06	0.09	0.11	0.15	0.19	0.27	0.39	0.52	0.62	0.67	0.65	0.58	0.46	0.36	0.28	0.23	0.19	0.15	0.12	NNW
	M	0.14	0.12	0.11	0.09	0.08	0.09	0.11	0.13	0.15	0.18	0.25	0.35	0.46	0.55	0.59	0.59	0.53	0.44	0.35	0.30	0.25	0.22	0.19	0.16	
	H	0.15	0.14	0.13	0.12	0.11	0.12	0.14	0.16	0.18	0.21	0.27	0.37	0.46	0.53	0.57	0.55	0.49	0.40	0.32	0.26	0.23	0.20	0.18	0.16	
SW	L	0.12	0.10	0.08	0.06	0.05	0.06	0.08	0.10	0.12	0.14	0.16	0.24	0.36	0.49	0.60	0.66	0.66	0.58	0.43	0.33	0.27	0.22	0.18	0.14	NW
	M	0.15	0.14	0.12	0.10	0.09	0.09	0.10	0.12	0.13	0.15	0.17	0.23	0.33	0.44	0.53	0.59	0.58	0.53	0.41	0.33	0.28	0.24	0.21	0.18	
	H	0.15	0.14	0.13	0.12	0.11	0.12	0.13	0.14	0.16	0.17	0.19	0.25	0.34	0.44	0.52	0.56	0.56	0.49	0.37	0.30	0.25	0.21	0.19	0.17	
WSW	L	0.12	0.10	0.08	0.07	0.05	0.06	0.07	0.09	0.10	0.12	0.13	0.17	0.26	0.40	0.52	0.62	0.66	0.61	0.44	0.34	0.27	0.22	0.18	0.15	WNW
	M	0.15	0.13	0.12	0.10	0.09	0.09	0.10	0.11	0.12	0.13	0.14	0.17	0.24	0.35	0.46	0.54	0.58	0.55	0.42	0.34	0.28	0.24	0.21	0.18	
	H	0.15	0.14	0.13	0.12	0.11	0.11	0.12	0.13	0.14	0.15	0.16	0.19	0.26	0.36	0.46	0.53	0.56	0.51	0.38	0.30	0.25	0.21	0.19	0.17	
W	L	0.12	0.10	0.08	0.06	0.05	0.06	0.07	0.08	0.10	0.11	0.12	0.14	0.20	0.32	0.45	0.57	0.64	0.61	0.44	0.34	0.27	0.22	0.18	0.14	W
	M	0.15	0.13	0.11	0.10	0.09	0.09	0.09	0.10	0.11	0.12	0.13	0.14	0.19	0.29	0.40	0.50	0.56	0.55	0.41	0.33	0.27	0.23	0.20	0.17	
	H	0.14	0.13	0.12	0.11	0.10	0.12	0.12	0.13	0.14	0.14	0.15	0.16	0.21	0.30	0.40	0.49	0.54	0.52	0.38	0.30	0.24	0.21	0.18	0.16	
WNW	L	0.12	0.10	0.08	0.06	0.05	0.06	0.07	0.09	0.10	0.12	0.13	0.15	0.17	0.26	0.40	0.53	0.63	0.62	0.44	0.34	0.27	0.22	0.18	0.14	WSW
	M	0.15	0.13	0.11	0.10	0.09	0.09	0.10	0.11	0.12	0.13	0.14	0.15	0.17	0.24	0.35	0.47	0.55	0.55	0.41	0.33	0.27	0.23	0.20	0.17	
	H	0.14	0.13	0.12	0.11	0.10	0.11	0.12	0.13	0.14	0.15	0.16	0.17	0.18	0.25	0.36	0.46	0.53	0.52	0.38	0.30	0.24	0.20	0.18	0.16	
NW	L	0.11	0.09	0.08	0.06	0.05	0.06	0.08	0.10	0.12	0.14	0.16	0.17	0.19	0.23	0.33	0.47	0.59	0.60	0.42	0.33	0.26	0.21	0.17	0.14	SW
	M	0.14	0.12	0.11	0.09	0.08	0.09	0.10	0.11	0.13	0.15	0.16	0.17	0.18	0.21	0.30	0.42	0.51	0.54	0.39	0.32	0.26	0.22	0.19	0.16	
	H	0.14	0.12	0.11	0.10	0.10	0.10	0.12	0.13	0.15	0.16	0.18	0.18	0.19	0.22	0.30	0.41	0.50	0.51	0.36	0.29	0.23	0.20	0.17	0.15	
NNW	L	0.12	0.09	0.08	0.06	0.05	0.07	0.11	0.14	0.18	0.22	0.25	0.27	0.29	0.30	0.33	0.44	0.57	0.62	0.44	0.33	0.26	0.21	0.17	0.14	SSW
	M	0.15	0.13	0.11	0.10	0.09	0.10	0.12	0.15	0.18	0.21	0.23	0.26	0.28	0.31	0.39	0.51	0.56	0.51	0.41	0.33	0.27	0.23	0.20	0.17	
	H	0.14	0.13	0.12	0.11	0.10	0.12	0.15	0.17	0.20	0.23	0.25	0.26	0.28	0.31	0.38	0.49	0.53	0.49	0.38	0.30	0.25	0.21	0.18	0.16	
HOR.	L	0.11	0.09	0.07	0.06	0.05	0.07	0.14	0.24	0.36	0.48	0.58	0.66	0.72	0.74	0.73	0.67	0.59	0.47	0.37	0.29	0.24	0.19	0.16	0.13	HOR.
	M	0.16	0.14	0.12	0.11	0.09	0.11	0.16	0.24	0.33	0.43	0.52	0.59	0.64	0.67	0.66	0.62	0.56	0.47	0.38	0.32	0.28	0.24	0.21	0.18	
	H	0.17	0.16	0.15	0.14	0.13	0.15	0.20	0.28	0.36	0.45	0.52	0.59	0.62	0.64	0.62	0.58	0.51	0.42	0.35	0.29	0.26	0.23	0.21	0.19	

1. To be used only with the maximum (underscored) values in Table B3.1.
2. Room Constructions:
L—Light construction: frame exterior wall, 2-in. (50.8-mm) concrete floor slab, approximately 30 lb (146 kg) of material/ft² (m²) of floor area.
M—Medium construction: 4-in. (101.6-mm) concrete exterior wall, 4-in. (101.6-mm) concrete floor slab, approximately 70 lb (341 kg) of building material/ft² (m²) of floor area.
H—Heavy construction: 6-in. (152.4-mm) concrete exterior wall, 6-in. (152.4-mm) concrete floor slab, approximately 130 lb (635 kg) of building materials/ft² (m²) of floor area.

TABLE B3.11 THERMAL LAG FACTORS FOR GLASS SOLAR LOAD WITH INTERIOR SHADING (ALL ROOM CONSTRUCTIONS)

North Latitude Exposure	1	2	3	4	5	6	7	8	9	10	11	12	13	14	15	16	17	18	19	20	21	22	23	24	South Latitude Exposure
													Solar Time (hours)												
N	0.08	0.07	0.06	0.06	0.07	0.73	0.66	0.65	0.73	0.80	0.86	0.89	0.89	0.86	0.82	0.75	0.78	0.91	0.24	0.18	0.15	0.13	0.11	0.10	S
NNE	0.03	0.03	0.02	0.02	0.03	0.64	0.77	0.62	0.42	0.37	0.37	0.37	0.36	0.35	0.32	0.28	0.23	0.17	0.08	0.07	0.06	0.05	0.04	0.04	SSE
NE	0.03	0.02	0.02	0.02	0.02	0.56	0.76	0.74	0.58	0.37	0.29	0.27	0.26	0.24	0.22	0.20	0.16	0.12	0.06	0.05	0.04	0.04	0.03	0.03	SE
ENE	0.03	0.02	0.02	0.02	0.02	0.52	0.76	0.80	0.71	0.52	0.31	0.26	0.24	0.22	0.20	0.18	0.15	0.11	0.06	0.05	0.04	0.04	0.03	0.03	ESE
E	0.03	0.02	0.02	0.02	0.02	0.47	0.72	0.80	0.76	0.62	0.41	0.27	0.24	0.22	0.20	0.17	0.14	0.11	0.06	0.05	0.04	0.04	0.03	0.03	E
ESE	0.03	0.03	0.02	0.02	0.02	0.41	0.67	0.79	0.80	0.72	0.54	0.34	0.27	0.24	0.21	0.19	0.15	0.12	0.07	0.06	0.05	0.04	0.04	0.03	ENE
SE	0.03	0.03	0.02	0.02	0.02	0.30	0.57	0.74	0.81	0.79	0.68	0.49	0.33	0.28	0.25	0.22	0.18	0.13	0.08	0.07	0.06	0.05	0.04	0.04	NE
SSE	0.04	0.03	0.03	0.03	0.02	0.12	0.31	0.54	0.72	0.81	0.81	0.71	0.54	0.38	0.32	0.27	0.22	0.16	0.09	0.08	0.07	0.06	0.05	0.04	NNE
S	0.04	0.03	0.03	0.03	0.03	0.09	0.16	0.23	0.38	0.58	0.75	0.83	0.80	0.68	0.50	0.35	0.27	0.19	0.11	0.09	0.08	0.07	0.06	0.05	N
SSW	0.05	0.04	0.03	0.03	0.03	0.09	0.14	0.18	0.22	0.27	0.43	0.63	0.78	0.84	0.80	0.66	0.46	0.25	0.13	0.11	0.09	0.08	0.07	0.06	NNW
SW	0.05	0.05	0.04	0.04	0.03	0.07	0.11	0.14	0.16	0.19	0.22	0.38	0.59	0.75	0.83	0.81	0.69	0.45	0.16	0.12	0.10	0.09	0.07	0.06	NW
WSW	0.05	0.05	0.04	0.04	0.03	0.07	0.10	0.12	0.14	0.16	0.17	0.23	0.44	0.64	0.78	0.84	0.78	0.55	0.16	0.12	0.10	0.09	0.07	0.06	WNW
W	0.05	0.05	0.04	0.04	0.03	0.06	0.09	0.11	0.13	0.15	0.16	0.17	0.31	0.53	0.72	0.82	0.81	0.61	0.16	0.12	0.10	0.08	0.07	0.06	W
WNW	0.05	0.05	0.04	0.03	0.03	0.07	0.10	0.12	0.14	0.16	0.17	0.18	0.22	0.43	0.65	0.80	0.84	0.66	0.16	0.12	0.10	0.08	0.07	0.06	WSW
NW	0.05	0.04	0.04	0.03	0.03	0.07	0.11	0.14	0.17	0.19	0.20	0.21	0.22	0.30	0.52	0.73	0.82	0.69	0.16	0.12	0.10	0.08	0.07	0.06	SW
NNW	0.05	0.05	0.03	0.03	0.03	0.11	0.17	0.22	0.26	0.30	0.32	0.33	0.34	0.34	0.39	0.61	0.82	0.76	0.17	0.12	0.10	0.08	0.07	0.06	SSW
HOR.	0.06	0.05	0.04	0.04	0.03	0.12	0.27	0.44	0.59	0.72	0.81	0.85	0.85	0.81	0.71	0.58	0.42	0.25	0.14	0.12	0.10	0.08	0.07	0.06	HOR.

Source: Table B3.10 and Table B3.11 reprinted from ASHRAE *Handbook of Fundamentals 1981* by permission of the American Society of Heating, Refrigerating and Air-Conditioning Engineers, Inc.

TABLE B4.1   R-VALUES OF AIR

Surface Air Films

Still Air	Surface	Direction of Heat Flow	Surface Reflective Quality	
			Nonreflective	Bright Aluminum Foil
	Horizontal	Up	0.61	1.32
	Vertical	Horizontal	0.68	1.70
	Horizontal	Down	0.92	4.55

Moving Air (any position)		7½ mph	15 mph	
	Smooth glass	0.29	0.20	
	Smooth plaster, wood	0.29	0.18	
	Concrete, brick, rough plaster	0.22	0.14	
	Stucco	0.18	0.11	
	General purpose	0.25	0.17	

Air Spaces

Surface	Direction of Heat Flow	Surface Reflective Quality	
		Nonreflective	Bright Aluminum Foil
Horizontal	Up		
	½″ air space winter	0.84	2.05
	summer	0.74	1.80
	¾″ air space winter	0.87	2.21
	summer	0.76	1.90
	1½″ air space winter	0.89	2.40
	summer	0.78	2.10
	3½″ air space winter	0.93	2.66
	summer	0.82	2.30
Vertical	Horizontal		
	½″ air space winter	0.91	2.54
	summer	0.77	2.34
	¾″ air space winter	1.01	3.46
	summer	0.84	3.24
	1½″ air space winter	1.02	3.55
	summer	0.87	3.66
	3½″ air space winter	1.01	3.40
	summer	0.85	3.40
Horizontal	Down		
	½″ air space winter	0.92	2.55
	summer	0.77	2.34
	¾″ air space winter	1.02	3.59
	summer	0.85	3.29
	1½″ air space winter	1.15	5.90
	summer	0.94	5.35
	3½″ air space winter	1.24	9.27
	summer	1.00	8.19

TABLE B4.1   R-VALUES OF AIR *(Continued)*

Attic Spaces					Ventilation Rate (cfm/ft²)				
(Summer condition) Ventilation	(No Ventilation) 0		(Natural Ventilation) 0.1		0.5		(Power Ventilation) 1.0		1.5	
				Ceiling R-Value					
Air Temp. (°F)	**10**	**20**	**10**	**20**	**10**	**20**	**10**	**20**	**10**	**20**
Nonreflective Surfaces										
80	1.9	1.9	2.8	3.5	6.5	10	9.8	17	12	21
90	1.9	1.9	2.6	3.1	5.2	7.9	7.6	12	8.6	15
100	1.9	1.9	2.4	2.7	4.2	6.1	5.8	8.7	6.5	10
Reflective surfaces										
80	6.5	6.5	8.2	9.0	14	18	18	26	20	31
90	6.5	6.5	7.7	8.3	12	15	14	10	16	22
100	6.5	6.5	7.3	7.8	10	12	11	15	12	16

1. A surface cannot take credit for both an air space resistance value and a surface resistance value. Take no credit for air spaces of less than 0.5 inch.

2. For horizontal surfaces serving as ceilings, the direction of heat flow is normally *up* in winter and *down* in summer. For floors, the direction of heat flow is *down* in winter and *up* in summer. Heat flow through vertical walls is in a horizontal direction.

3. Values for attic spaces with a reflective surface are for bright reflective aluminum foil at the ceiling level facing the attic space.

Source: National Bureau of Standards Housing Research Paper (HRP) No. 32, Housing and Home Finance Agency, 1954, U.S. Government Printing Office, Washington, D.C. 20402

TABLE B4.2   R-VALUES OF TYPICAL BUILDING MATERIALS

Material	Thickness in.	Thickness cm	Conventional Per inch Thickness (1/k)	Conventional For Thickness Listed (1/c)	SI $\left(\dfrac{m \cdot °C}{W}\right)$	SI $\left(\dfrac{m^2 \cdot °C}{W}\right)$	Density (lb/ft³)	Specific heat (Btu/lb-°F)
				Masonry				
Brick (common)	4	10	0.20	0.80	1.39	0.14	120	0.19
Brick (face)	4	10	0.11	0.44	0.76	0.08	130	0.19
Brick (fireclay)			0.14		0.97		112	0.20
Hollow clay tile								
1 cell deep	3	8		0.80		0.14	60	0.21
1 cell deep	4	10		1.11		0.20	48	0.21
2 cells deep	6	15		1.52		0.27	50	0.21
2 cells deep	8	20		1.85		0.33	45	0.21
2 cells deep	10	25		2.22		0.39	42	0.21
3 cells deep	12	30		2.50		0.44	40	0.21
Concrete blocks (3-oval core)								
Sand and	4	10		0.71		0.13	69	0.22
gravel	8	20		1.11		0.20	64	0.22
aggregate	12	30		1.28		0.23	63	0.22
Cinder	3	8		0.86		0.15	68	0.21
aggregate	4	10		1.11		0.20	60	0.21
	8	20		1.72		0.30	56	0.21
	12	30		1.89		0.33	53	0.21
Light-weight	3	8		1.27		0.22	60	0.21
aggregate[1]	4	10		1.50		0.26	51	0.21
	8	20		2.00		0.35	48	0.21
	12	30		2.27		0.40	43	0.21
Concrete blocks (rectangular core)								
Sand and gravel aggregate								
2 Core	8	20		1.04		0.18	65	0.22
2 Core filled[2]	8	20		1.93		0.34	65	0.22
Lightweight aggregate								
3 Core	6	15		1.65		0.29	46	0.21
3 Core filled[2]	6	15		2.99		0.53	46	0.21
2 Core	8	20		2.18		0.38	43	0.21
2 Core filled[2]	8	20		5.03		0.89	43	0.21
3 Core	12	30		2.48		0.44	46	0.21
3 Core filled[2]	12	30		5.82		1.02	46	0.21
Stone (lime or sand)			0.08		0.55		150	0.19
Gypsum partition tile (12″ × 30″)								
Solid	3	8		1.26		0.22	45	0.19
4-cell	3	8		1.35		0.24	35	0.19
3-cell	4	10		1.67		0.29	38	0.19

TABLE B4.2   R-VALUES OF TYPICAL BUILDING MATERIALS *(Continued)*

Material	Thickness in.	Thickness cm	R-Value Conventional Per inch Thickness (1/k)	R-Value Conventional For Thickness Listed (1/c)	R-Value SI $\left(\dfrac{m \cdot {}^\circ C}{W}\right)$	R-Value SI $\left(\dfrac{m^2 \cdot {}^\circ C}{W}\right)$	Density (lb/ft³)	Specific Heat (Btu/lb-°F)
Concrete								
Sand and gravel								
aggregate			0.08		0.55		140	0.22
Lightweight								
aggregates³			0.19		1.32		120	0.21
			0.28		1.94		100	0.21
			0.40		2.77		80	0.21
			0.59		4.09		60	0.21
			0.86		5.96		40	0.21
			1.11		7.69		30	0.21
			1.43		9.91		20	0.21
Perlite (expanded)			1.08		7.48		40	0.32
			1.41		9.77		30	0.32
			2.00		13.86		20	0.32
Stucco			0.20		1.39			
			Woods					
Plywood (Douglas fir)			1.25		8.66		34	0.29
Hardboard								
Med. density			1.37		9.49		50	0.31
High density			1.00		6.93		63	0.32
Particleboard			1.31		9.08		40	0.29
Wood siding			0.67		4.65		40	0.28
Drop (8″)	1	2.5		0.79		0.14		
Lapped (8″)	1/2	1.27		0.81		0.14		
Lapped (10″)	3/4	1.9		1.05		0.18		
Hardwoods⁴			0.91		6.31		45	0.30
Softwoods⁵			1.25		8.66		32	0.33
			Insulation					
Blanket								
Fiberglass			3.45		23.92			
Mineral wool			3.25		22.53			
Cotton fiber			3.85		26.69			
Loose Fill								
Fiberglass			3.00		20.80			
Mineral wool			3.75		26.00			
Cellulosic fiber⁶			3.50		24.27			0.33
Vermiculite								
(exfoliated)			2.13		14.76		7.0	3.20
Perlite (expanded)			2.70		18.71			0.26
Sawdust			2.22		15.39			0.33

Material	Thickness		R-Value				Density (lb/ft³)	Specific Heat (Btu/lb-°F)
			Conventional		SI			
	in.	cm	Per inch Thickness (1/k)	For Thickness Listed (1/c)	$\left(\dfrac{m \cdot °C}{W}\right)$	$\left(\dfrac{m^2 \cdot °C}{W}\right)$		
Rigid Board								
Polyurethane			6.5		45.07			
Polystyrene								
Styrofoam			4.5		31.20			
Bead board			3.57		24.75			
Cellular glass			2.50		17.33		8.5	0.24
Glass fiber			4.00		27.72		9.5	0.23
Cork board			3.57		24.75			
Fiberboard			3.56		24.68			
Foamed in-place								
Polyurethane			6.75		46.80			
Urea formaldehyde			4.80		33.28			
Roof insulation[7]								
Fesco board			2.78		19.28			
Fesco foam	1–1/2	3.8		6.67		1.17		
	1–5/8	4.13		7.69		1.35		
	2	5		10.00		1.76		
	2–1/4	5.7		12.50		2.20		
	2–1/2	6.4		14.29		2.52		
	2–3/4	7.0		16.67		2.94		
	3–1/4	8.3		20.00		3.52		
Polyisocyanurate			7.2		49.92			
Aged			6.0		41.60			
Phenolic foam board								
Closed cell			8.33		57.76			
Open cell			5.0		34.67			
			Roofing					
Built-up roofing	3/8	0.95	0.88	0.33	6.10	0.06	70	0.35
Vapor seal[8]				0.12		0.02		
Asphalt sheathing	1/2	1.27		1.46		0.26		
Asphalt roll roofing				0.15		0.03	70	0.36
Asphalt shingles				0.44		0.08	70	0.30
Asbestos-cement								
shingles				0.21		0.04	120	0.24
Slate	1/2	1.27		0.05		0.01	201	0.30
Wood shingles				0.94		0.17	40	0.31
Sheet metal				Negl.		Negl.		

TABLE B4.2   R-VALUES OF TYPICAL BUILDING MATERIALS *(Continued)*

Material	Thickness		R-Value				Density (lb/ft³)	Specific Heat (Btu/lb-°F)
			Conventional		SI			
	in.	cm	Per inch Thickness (1/k)	For Thickness Listed (1/c)	$\left(\dfrac{m \cdot {}^\circ C}{W}\right)$	$\left(\dfrac{m^2 \cdot {}^\circ C}{W}\right)$		
					Sidings			
Sheathing (Vegetable Fiber Board)								
Regular density	1/2	1.27		1.32		0.23	18	0.31
	25/32	1.98		2.06		0.36	18	0.31
Intermediate density	1/2	1.27		1.22		0.21	22	0.31
Nail-base	1/2	1.27		1.14		0.20	25	0.31
Shingle backer	3/8	0.95		0.94		0.17	18	0.31
	5/16	0.79		0.78		0.14	18	0.31
Sound deadening board	1/2	1.27		1.35		0.24	15	0.30
Wood Shingles[9]								
16″ (7½″ exposure)				0.87		0.15		0.31
16″ Double (12″ exposure)				1.19		0.21		0.28
Plus insul. backer board	5/16	0.79		1.40		0.25		0.31
Panels								
Asphalt roll siding				0.15		0.03		0.35
Asphalt insulating siding	1/2	1.27		1.46		0.26		0.35
Aluminum or steel over sheathing								
Hollow-backed				0.61		0.11		0.29
Insulating-board backed	3/8	0.95		1.82		0.32		0.32
Insulating-board backed (foil backed)	3/8	0.95		2.96		0.52		0.32
Architectural glass				0.10		0.02		0.20
Asbestos-cement board	1/8	0.32	0.25	0.03	1.73	0.005	120	0.24
	1/4	0.64	0.25	0.06	1.73	0.01	120	0.24
					PLASTER			
Gypsum or plaster board	3/8	0.95		0.32		0.06	50	0.26
	1/2	1.27		0.45		0.08	50	0.26
	5/8	1.59		0.56		0.10	50	0.26
Cement plaster, sand aggregate	3/8	0.95	0.20	0.08	1.39	0.01	116	0.20
	3/4	1.9	0.20	0.15	1.39	0.03	116	0.20

| Material | Thickness | | R-Value | | | | Density (lb/ft³) | Specific Heat (Btu/lb-°F) |
| | in. | cm | Conventional | | SI | | | |
			Per inch Thickness (1/k)	For Thickness Listed (1/c)	$\left(\dfrac{m \cdot {}^\circ C}{W}\right)$	$\left(\dfrac{m^2 \cdot {}^\circ C}{W}\right)$		
Gypsum plaster, lightweight aggregate	1/2	1.27	0.64	0.32	4.44	0.06	45	
	5/8	1.59	0.64	0.39	4.44	0.07	45	
(On metal lath)	3/4	1.9	0.63	0.47	4.44	0.08	45	
Perlite aggregate			0.67		4.64		45	0.32
Sand aggregate	1/2	1.27	0.18	0.09	1.25	0.02	105	0.20
	5/8	1.59	0.18	0.11	1.25	0.02	105	0.20
(On metal lath)	3/4	1.9	0.17	0.13	1.18	0.02	105	0.20
	3/4	1.9		0.40		0.07	105	0.20
Vermiculite aggregate			0.59		4.09		45	0.20

				Finish Flooring and Ceilings				
Hardwood	3/4	1.9		0.68		0.12	45	
Tile								
Asphalt	1/8	0.32		0.04		0.01	120	0.30
Linoleum	1/8	0.32		0.08		0.01	80	0.30
Rubber or vinyl	1/8	0.32		0.02		0.004	110	0.30
Cork	1/8	0.32	2.22	0.28	15.39	0.05	25	0.48
Ceramic	1	2.5	0.08	0.08	0.55	0.01		0.19
Terrazzo	1	2.5	0.08	0.08	0.55	0.01	140	0.19
Carpet								
On fibrous pad				2.08		0.37		0.34
On foam rubber pad				1.23		0.22		0.33
Lay-in ceiling tile[10]	1/2	1.27	2.50	1.25	17.33	0.22	18	0.14
	3/4	1.9	2.50	1.89	17.33	0.33	18	0.14

				Miscellaneous				
Asphalt			0.19		1.35		132	0.22
Cotton (fiber)			3.47		23.81		95	0.319
Earth								
Dry, loose			0.33		2.29			
Average			0.14		0.97			
Damp, packed			0.05		0.35			
Felt			2.78		20.00		21	
Leather (sole)			0.91		6.25		62	
Linen			1.67		11.11			
Metals			Negl.		Negl.			
Paints			0.56		3.85		63	
Paper			1.11		7.69		58	0.32
Corrugated cardboard			2.13		14.77		3.7	

TABLE B4.2 R-VALUES OF TYPICAL BUILDING MATERIALS *(Continued)*

| Material | Thickness | | R-Value | | | | Density (lb/ft³) | Specific Heat (Btu/lb-°F) |
| | in. | cm | Conventional | | SI | | | |
			Per inch Thickness (1/k)	For Thickness Listed (1/c)	$\left(\dfrac{m \cdot °C}{W}\right)$	$\left(\dfrac{m^2 \cdot °C}{W}\right)$		
Laminated paperboard			2.00		13.86		30	0.33
Homogeneous board from repulped paper			2.00		13.86		30	0.28
Porcelain			0.06		0.45		162	0.18
Rubber (vulcanized)								
Soft			1.04		10.00		69	0.48
Hard			0.91		6.25		74	0.48
Sand			0.44		3.03		95	0.191
Snow (freshly fallen)			0.24		1.67		7	
(at 32°F)			0.06		0.45		31	
Water			0.18		1.25			
Wool								
Fiber							82	0.32
Fabric			2.25-3.97		15.87-27.78		6.9-20.6	0.32

[1]Lightweight aggregate includes expanded shale, clay, slate or slag, pumice.

[2]Cores filled with vermiculite, perlite, or mineral wool insulation.

[3]Lightweight aggregates include expanded shale, clay or slate; expanded slag; cinders; pumice; vermiculite; cellular concretes.

[4]Maple, oak, and similar hardwoods.

[5]Fir, pine, and similar softwoods.

[6]Recycled newspaper.

[7]Rigid board insulations are also used for roofing applications.

[8]Two layers of mopped 15-lb felt.

[9]Some roofing materials are also used for siding.

[10]Plain or acoustic. Insulating value of acoustical tile varies, depending on density of the board and on the type, size, and depth of perforations.

[11]Face and common brick do not always have the densities listed. When the density is different from that listed, the insulation value will be different.

[12]Since metal itself has practically no insulation value, the R-value for metal siding varies widely, depending on the amount of infiltration into the air space behind the siding; whether the air space is reflective or nonreflective; and on the thickness, type, and application of insulating backing used. Departures of ± 50% or more from the values listed may occur.

[13]Heat capacity (Btu/ft³ - °F) = specific heat × density.

[14]SI unit conversions:

Density lb/ft³ × 16.018 = kg/m³

Specific heat Btu/lb - °F × 4,186.8 = J/kg - °C

TABLE B4.3A OVERALL COEFFICIENTS[a] OF HEAT TRANSMISSION (U-FACTOR) OF WINDOWS AND SKYLIGHTS, BTU/(HR-FT²-F)

Description	Exterior Vertical Panels				Exterior Horizontal Panels (Skylights)	
	Summer**		Winter*		Summer[j]	Winter[i]
	No Indoor Shade	Indoor Shade***	No Indoor Shade	Indoor Shade***		
Flat glass[b]						
Single glass	1.04	0.81	1.10	0.83	0.83	1.23
Insulating glass, double[c]						
³⁄₁₆ in. air space[d]	0.65	0.58	0.62	0.52	0.57	0.70
¼ in. air space[d]	0.61	0.55	0.58	0.48	0.54	0.65
½ in. air space[e]	0.56	0.52	0.49	0.42	0.49	0.59
½ in. air space, low emittance coating[f]						
$e = 0.20$	0.38	0.37	0.32	0.30	0.36	0.48
$e = 0.40$	0.45	0.44	0.38	0.35	0.42	0.52
$e = 0.60$	0.51	0.48	0.43	0.38	0.46	0.56
Insulating glass, triple[c]						
¼ in. air space[d]	0.44	0.40	0.39	0.31		
½ in. air space[g]	0.39	0.36	0.31	0.26		
Storm windows						
1 in. to 4 in. air spaces[d]	0.50	0.48	0.50	0.42		
Plastic bubbles[k]						
Single walled					0.80	1.15
Double walled					0.46	0.70

TABLE B4.3B ADJUSTMENT FACTORS FOR VARIOUS WINDOW AND SLIDING PATIO DOOR TYPES (MULTIPLY U-VALUES IN PART A BY THESE FACTORS)

Description	Single Glass	Double or Triple Glass	Storm Windows
Windows			
All glass[h]	1.00	1.00	1.00
Wood sash; 80% glass	0.90	0.95	0.90
Wood sash; 60% glass	0.80	0.85	0.80
Metal sash; 80% glass	1.00	1.20[m]	1.20[m]
Sliding patio doors			
Wood frame	0.95	1.00	—
Metal frame	1.00	1.10[m]	—

[a] See Table B4.3B for adjustments for various windows and sliding patio doors.

[b] Emittance of uncoated glass surface = 0.84.

[c] Double and triple refer to number of lights of glass.

[d] 0.125-in. glass.

[e] 0.25-in. glass.

[f] Coating on either glass surface facing air space; all other glass surfaces uncoated.

[g] Window design: 0.25-in. glass, 0.125-in. glass, 0.25-in. glass.

[h] Refers to windows with negligible opaque areas.

[i] For heat flow up.

[j] For heat flow down.

[k] Based on area of opening, not total surface area.

[m] Values will be less than these when metal sash and frame incorporate thermal breaks. In some thermal break designs U-values will be equal to or less than those for the glass. Window manufacturers should be consulted for specific data.

* 15 mph outdoor air velocity; 0°F outdoor air; 70°F inside air temp natural convection.

** 7.5 mph outdoor air velocity; 89°F outdoor air; 75°F inside air natural convection; solar radiation 248.3 Btu/(hr-ft²)

*** Values apply to tightly closed venetian and vertical blinds, draperies, and roller shades.

The reciprocal of the above U-factors is the thermal resistance, R, for each type of glazing. If tightly drawn drapes (heavy close weave), closed Venetian blinds, or closely fitted roller shades are used internally, the additional R is approximately 0.29 (hr·ft²·F)/Btu. If miniature louvered solar screens are used in close proximity to the outer fenestration surface, the additional R is approximately 0.24 (hr·ft²·F)/Btu.

Source: Reprinted from ASHRAE *Cooling and Heating Load Calculation Manual* by permission of the American Society of Heating, Refrigerating and Air-Conditioning Engineers, Inc

TABLE B4.4   COEFFICIENTS OF TRANSMISSION (U) FOR SLAB DOORS (BTU PER (HR-FT^2-F))

Thickness[a]	Winter			Summer
	Solid Wood, No Storm Door	Storm Door[b]		No Storm Door
		Wood	Metal	
1 in.	0.64	0.30	0.39	0.61
1.25 in.	0.55	0.28	0.34	0.53
1.5 in.	0.49	0.27	0.33	0.47
2 in.	0.43	0.24	0.29	0.42
	Steel Door			
1.75 in.				
A[c]	0.59	—	—	0.58
B[d]	0.19	—	—	0.18
C[e]	0.47	—	—	0.46

[a] Nominal thickness.

[b] Values for wood storm doors are for approximately 50% glass; for metal storm door values apply for any percent of glass.

[c] A = Mineral fiber core (2 lb/ft^3).

[d] B = Solid urethane foam core with thermal break.

[e] C = Solid polystyrene core with thermal break.

Source: Reprinted from ASHRAE *Cooling and Heating Load Calculation Manual* by permission of the American Society of Heating, Refrigerating and Air-Conditioning Engineers, Inc.

TABLE B4.5   TABLE OF RECIPROCALS ($U = 1/R$)

When a R- or U-value falls between the numbers in the table, use the closest number to find the corresponding U- or R-value.

R-Value (or U)	U (or R-Value)	R-Value (or U)	U (or R-Value)
1.00	1.00	7.00	0.14
1.05	0.95	8.00	0.13
1.10	0.91	9.00	0.11
1.20	0.83	10.00	0.10
1.40	0.71	11.00	0.09
1.50	0.66	12.00	0.08
2.00	0.50	15.00	0.06
2.50	0.40	20.00	0.05
3.00	0.33	25.00	0.04
4.00	0.25	33.00	0.03
5.00	0.20	50.00	0.02
6.00	0.16	100.00	0.01

TABLE B4.6   CONVERSION TABLE FOR WALL COEFFICIENT U FOR VARIOUS WIND VELOCITIES

U for 15 mph[a]	U for 0 to 30 mph Wind Velocities						U for 15 mph[a]	U for 0 to 30 mph Wind Velocities					
	0	5	10	20	25	30		0	5	10	20	25	30
0.050	0.049	0.050	0.050	0.050	0.050	0.050	0.290	0.257	0.278	0.286	0.293	0.295	0.296
0.060	0.059	0.059	0.060	0.060	0.060	0.060	0.310	0.273	0.296	0.305	0.313	0.315	0.317
0.070	0.068	0.069	0.070	0.070	0.070	0.070	0.330	0.288	0.314	0.324	0.333	0.336	0.338
0.080	0.078	0.079	0.080	0.080	0.080	0.080	0.350	0.303	0.332	0.344	0.354	0.357	0.359
0.090	0.087	0.089	0.090	0.090	0.091	0.091	0.370	0.318	0.350	0.363	0.375	0.378	0.380
0.100	0.096	0.099	0.100	0.100	0.101	0.101	0.390	0.333	0.368	0.382	0.395	0.399	0.401
0.110	0.105	0.108	0.109	0.110	0.111	0.111	0.410	0.347	0.385	0.402	0.416	0.420	0.422
0.130	0.123	0.127	0.129	0.131	0.131	0.131	0.430	0.362	0.403	0.421	0.436	0.441	0.444
0.150	0.141	0.147	0.149	0.151	0.151	0.152	0.450	0.376	0.420	0.439	0.457	0.462	0.465
0.170	0.158	0.166	0.169	0.171	0.172	0.172	0.500	0.410	0.464	0.487	0.509	0.514	0.518
0.190	0.175	0.184	0.188	0.191	0.192	0.193	0.600	0.474	0.548	0.581	0.612	0.620	0.626
0.210	0.192	0.203	0.208	0.212	0.213	0.213	0.700	0.535	0.631	0.675	0.716	0.728	0.736
0.230	0.209	0.222	0.227	0.232	0.233	0.234	0.800	0.592	0.711	0.766	0.821	0.836	0.847
0.250	0.226	0.241	0.247	0.252	0.253	0.254	0.900	0.645	0.789	0.858	0.927	0.946	0.960
0.270	0.241	0.259	0.266	0.273	0.274	0.275	1.000	0.695	0.865	0.949	1.034	1.058	1.075

[a] U in first column is from previous tables or as calculated for 15 mph wind velocity.

Source: Reprinted from ASHRAE *Handbook of Fundamentals 1981* by permission of the American Society of Heating, Refrigerating and Air-Conditioning Engineers, Inc.

TABLE B4.7A SOL-AIR TEMPERATURES FOR JULY 21, 40°N LATITUDE (°F)

Time	Air Temperature, (°F)	Light Colored								
		N	NE	E	SE	S	SW	W	NW	HOR
1	76	76	76	76	76	76	76	76	76	69
2	76	76	76	76	76	76	76	76	76	69
3	75	75	75	75	75	75	75	75	75	68
4	74	74	74	74	74	74	74	74	74	67
5	74	74	74	74	74	74	74	74	74	67
6	74	82	95	97	86	75	75	75	75	74
7	75	82	103	109	97	78	78	78	78	85
8	77	82	103	114	105	83	81	81	81	96
9	80	85	101	114	110	92	85	85	85	106
10	83	89	96	110	112	100	89	89	89	115
11	87	93	94	104	111	108	96	93	93	123
12	90	96	96	97	107	112	107	97	96	127
13	93	99	99	99	102	114	117	110	100	129
14	94	100	100	100	100	111	123	121	107	126
15	95	100	100	100	100	107	125	129	116	121
16	94	99	98	98	98	100	122	131	120	113
17	93	100	96	96	96	96	115	127	121	103
18	91	99	92	92	92	92	103	114	112	91
19	87	87	87	87	87	87	87	87	87	80
20	85	85	85	85	85	85	85	85	85	78
21	83	83	83	83	83	83	83	83	83	76
22	81	81	81	81	81	81	81	81	81	74
23	79	79	79	79	79	79	79	79	79	72
24	77	77	77	77	77	77	77	77	77	70
Avg.	**83**	**86**	**89**	**91**	**90**	**89**	**90**	**91**	**89**	**91**
		Dark Colored								
1	76	76	76	76	76	76	76	76	76	69
2	76	76	76	76	76	76	76	76	76	69
3	75	75	75	75	75	75	75	75	75	68
4	74	74	74	74	74	74	74	74	74	67
5	74	74	74	74	74	74	74	74	74	67
6	74	90	117	121	99	77	77	77	77	81
7	75	90	131	144	120	82	82	82	82	102
8	77	87	130	151	134	89	86	86	86	122
9	80	91	122	148	141	105	91	91	91	140
10	83	95	109	137	141	118	96	95	95	155
11	87	100	101	122	136	129	105	100	100	166
12	90	103	103	104	125	134	125	104	103	172
13	93	106	106	106	111	135	142	128	107	172
14	94	106	106	106	107	129	152	148	120	166
15	95	106	106	106	106	120	156	163	137	155
16	94	104	103	103	103	106	151	168	147	139
17	93	108	100	100	100	100	138	162	149	120
18	91	107	94	94	94	94	116	138	134	98
19	87	87	87	87	87	87	87	87	87	80
20	85	85	85	85	85	85	85	85	85	78
21	83	83	83	83	83	83	83	83	83	76
22	81	81	81	81	81	81	81	81	81	74
23	79	79	79	79	79	79	79	79	79	72
24	77	77	77	77	77	77	77	77	77	70
Avg.	**83**	**89**	**95**	**100**	**99**	**95**	**99**	**100**	**95**	**107**

Source: Reprinted from ASHRAE *Handbook of Fundamentals 1981* by permission of the American Society of Heating, Refrigerating and Air-Conditioning Engineers, Inc.

TABLE B4.7B SOL-AIR TEMPERATURES FOR JULY 21, 40°N LATITUDE (°C)

Time	Air Temperature, (°C)	N	NE	E	SE	S	SW	W	NW	HOR
					Light Colored					
1	24.4	24.4	24.4	24.4	24.4	24.4	24.4	24.4	24.4	20.5
2	24.4	24.4	24.4	24.4	24.4	24.4	24.4	24.4	24.4	20.5
3	23.8	23.8	23.8	23.8	23.8	23.8	23.8	23.8	23.8	20.0
4	23.3	23.3	23.3	23.3	23.3	23.3	23.3	23.3	23.3	17.4
5	23.3	23.3	23.3	23.3	23.3	23.3	23.3	23.3	23.3	19.4
6	23.3	27.7	35.0	36.1	30.0	23.8	23.8	23.8	23.8	23.3
7	23.8	27.7	39.4	42.7	36.1	25.5	25.5	25.5	25.5	29.4
8	25.0	27.7	39.4	45.5	40.5	28.3	27.2	27.2	27.2	35.5
9	26.6	29.4	38.3	45.5	43.3	33.3	29.4	29.4	29.4	41.1
10	28.3	31.6	35.5	43.3	44.4	37.7	31.6	31.6	31.6	46.1
11	30.5	33.8	34.4	40.0	43.8	42.2	35.5	33.8	33.8	50.5
12	32.2	35.5	35.5	36.1	41.6	44.4	41.6	36.1	35.5	52.7
13	33.8	37.2	37.2	37.2	38.8	45.5	47.2	43.3	37.7	53.8
14	34.4	37.7	37.7	37.7	37.7	43.8	50.5	49.4	41.6	52.2
15	35.0	37.7	37.7	37.7	37.7	41.6	51.6	53.8	46.6	49.4
16	34.4	37.2	36.6	36.6	36.6	37.7	50.0	55.0	48.8	45.0
17	33.8	37.7	35.5	35.5	35.5	35.5	46.1	52.7	49.4	39.4
18	32.7	37.2	33.3	33.3	33.3	33.3	39.4	45.5	44.4	32.7
19	30.5	30.5	30.5	30.5	30.5	30.5	30.5	30.5	30.5	26.6
20	29.4	29.4	29.4	29.4	29.4	29.4	29.4	29.4	29.4	25.5
21	38.3	28.3	28.3	28.3	28.3	28.3	28.3	28.3	28.3	24.4
22	27.2	27.2	27.2	27.2	27.2	27.2	27.2	27.2	27.2	23.3
23	26.1	26.1	26.1	26.1	26.1	26.1	26.1	26.1	26.1	22.2
24	25.0	25.0	25.0	25.0	25.0	25.0	25.0	25.0	25.0	21.1
Avg.	28.3	30.0	31.6	32.7	32.2	31.6	32.2	32.7	31.6	32.7
					Dark Colored					
1	24.4	24.4	24.4	24.4	24.4	24.4	24.4	24.4	24.4	20.5
2	24.4	24.4	24.4	24.4	24.4	24.4	24.4	24.4	24.4	20.5
3	23.8	23.8	23.8	23.8	23.8	23.8	23.8	23.8	23.8	20.0
4	23.3	23.3	23.3	23.3	23.3	23.3	23.3	23.3	23.3	19.4
5	23.3	23.3	23.3	23.3	23.3	23.3	23.3	23.3	23.3	19.4
6	23.3	32.2	47.2	49.4	37.2	25.0	25.0	25.0	25.0	27.2
7	23.8	32.2	55.0	62.2	48.8	27.7	27.7	27.7	27.7	38.8
8	25.0	30.5	54.4	66.1	56.6	31.6	30.0	30.0	30.0	50.0
9	26.6	32.7	50.0	64.4	60.5	40.5	32.7	32.7	32.7	60.0
10	28.3	35.0	42.7	58.3	60.5	47.7	35.5	35.0	35.0	68.3
11	30.5	37.7	38.3	50.0	57.7	53.8	40.5	37.7	37.7	74.4
12	32.2	39.4	39.4	40.0	51.6	56.6	51.6	40.0	39.4	77.7
13	33.8	41.1	41.1	41.1	43.8	57.2	61.1	53.3	41.6	77.7
14	34.4	41.1	41.1	41.1	41.6	53.8	66.6	64.4	48.8	74.4
15	35.0	41.1	41.1	41.1	41.1	48.8	68.8	72.7	58.3	68.3
16	34.4	40.0	39.4	39.4	39.4	41.1	66.1	75.5	63.8	59.4
17	33.8	42.2	37.7	37.7	37.7	37.7	58.8	72.2	65.0	48.8
18	32.7	41.6	34.4	34.4	34.4	34.4	46.6	58.8	56.6	36.6
19	30.5	30.5	30.5	30.5	30.5	30.5	30.5	30.5	30.5	26.6
20	29.4	29.4	29.4	29.4	29.4	29.4	29.4	29.4	29.4	25.5
21	28.3	28.3	28.3	28.3	28.3	28.3	28.3	28.3	28.3	24.4
22	27.2	27.2	27.2	27.2	27.2	27.2	27.2	27.2	27.2	23.3
23	26.1	26.1	26.1	26.1	26.1	26.1	26.1	26.1	26.1	22.2
24	25.0	25.0	25.0	25.0	25.0	25.0	25.0	25.0	25.0	21.1
Avg.	28.3	31.6	35.0	37.7	37.2	35.0	37.2	37.7	35.0	41.6

TABLE B4.8   PERCENTAGE OF THE DAILY RANGE

Time (hr)	%	Time (hr)	%	Time (hr)	%	Time (hr)	%
1	87	7	93	13	11	19	34
2	92	8	84	14	3	20	47
3	96	9	71	15	0	21	58
4	99	10	56	16	3	22	68
5	100	11	39	17	10	23	76
6	98	12	23	18	21	24	82

Source: Reprinted from ASHRAE *Handbook of Fundamentals 1981* by permission of the American Society of Heating, Refrigerating and Air-Conditioning Engineers, Inc.

TABLE B4.9   HEAT LOSS THROUGH BELOW GRADE WALLS

Heat Loss (W/m² · C)				Depth		Path Length through Soil		Heat Loss (Btu/hr-ft²-F)			
Uninsulated	R = 0.73	R = 1.47	R = 2.20	(m)	(ft)	(m)	(ft)	Uninsulated	R = 4.17	R = 8.34	R = 12.51
2.33	0.86	0.53	0.38	0–0.30	(0–1)	0.20	(0.68)	0.410	0.152	0.093	0.067
1.26	0.66	0.45	0.36	0.30–0.61	(1–2)	0.69	(2.27)	0.222	0.116	0.079	0.059
0.88	0.53	0.38	0.30	0.61–0.91	(2–3)	1.18	(3.88)	0.155	0.094	0.068	0.053
0.67	0.45	0.34	0.27	0.91–1.22	(3–4)	1.68	(5.52)	0.119	0.079	0.060	0.048
0.54	0.39	0.30	0.25	1.22–1.52	(4–5)	2.15	(7.05)	0.096	0.069	0.053	0.044
0.45	0.34	0.27	0.23	1.52–1.83	(5–6)	2.64	(8.65)	0.079	0.060	0.048	0.040
0.39	0.30	0.25	0.21	1.83–2.13	(6–7)	3.13	(10.28)	0.069	0.054	0.044	0.037

Source: Reprinted from ASHRAE *Handbook of Fundamentals 1981* by permission of the American Society of Heating, Refrigerating and Air-Conditioning Engineers, Inc.

TABLE B4.10A   HEAT LOSS THROUGH BASEMENT FLOORS (FOR FLOORS MORE THAN 3 FT (0.91 M) BELOW GRADE)[a]

Depth of Floor Below Grade		Narrowest Width of House							
		20 ft (6.1 m)		24 ft (7.3 m)		28 ft (8.5 m)		32 ft (9.8 m)	
(ft)	(m)	Btu/(hr-ft²-F)	W/(m²·C)	Btu/(hr-ft²-F)	W/(m²·C)	Btu/(hr-ft²-F)	W/(m²·C)	Btu/(hr-ft²-F)	W/(m²·C)
4	1.22	0.035	0.198	0.032	0.182	0.027	0.153	0.024	0.136
5	1.52	0.032	0.182	0.029	0.165	0.026	0.148	0.023	0.131
6	1.83	0.030	0.170	0.027	0.153	0.025	0.142	0.022	0.125
7	2.13	0.029	0.165	0.026	0.148	0.023	0.131	0.021	0.119

[a] For a depth below grade of 3 ft or less, treat as a slab on grade.
Source: Reprinted from ASHRAE *Cooling and Heating Load Calculation Manual* by permission of the American Society of Heating, Refrigerating and Air-Conditioning Engineers, Inc.

TABLE B4.10B   HEAT LOSS THROUGH CONCRETE FLOORS LESS THAN 3 FT (0.91 M) BELOW GRADE

Heat Loss per Unit Length of Exposed Edge

Outdoor Design		$R = 5$		$R = 2.5$		None		(hr-ft^2-°F)/Btu
Temperature			$R = 0.88$		$R = 0.44$			(m^2·C)/W
(°F)	(°C)	Btu/(hr-ft)	W/m	Btu/(hr-ft)	W/m	Btu/(hr-ft)	W/m	
−20 to −30	−29 to −34	50	48	60	58	75	72	
−10 to −20	−23 to −29	45	43	55	53	65	62	
0 to −10	−18 to −23	40	38	50	48	60	58	
+10 to 0	−12 to −18	35	34	45	43	55	53	
+20 to +10	−7 to −12	30	29	40	38	50	48	

[a] Insulation is assumed to extend 2 ft (0.61 m) either horizontally under slab or vertically along foundation wall.

Source: Reprinted from ASHRAE *Cooling and Heating Load Calculation Manual* by permission of the American Society of Heating, Refrigerating and Air-Conditioning Engineers, Inc

TABLE B4.11   THERMAL LAG FACTORS, WALLS AND ROOF

Latitude	Month	N	NE NW	E W	SE SW	S	HOR
0	Dec	.40	.28	.32	.45	.55	.33
	Jan/Nov	.40	.30	.33	.43	.51	.33
	Feb/Oct	.40	.35	.33	.40	.36	.35
	Mar/Sept	.40	.42	.33	.34	.19	.35
	Apr/Aug	.80	.47	.32	.28	.19	.32
	May/July	1.05	.51	.30	.25	.19	.30
	June	1.15	.51	.30	.23	.19	.29
8	Dec	.35	.26	.30	.47	.62	.29
	Jan/Nov	.40	.26	.32	.45	.57	.30
	Feb/Oct	.40	.33	.33	.42	.45	.33
	Mar/Sept	.40	.37	.33	.36	.28	.35
	Apr/Aug	.65	.44	.33	.30	.21	.33
	May/July	.90	.49	.32	.26	.21	.32
	June	1.00	.49	.32	.25	.21	.32
16	Dec	.35	.21	.29	.47	.64	.24
	Jan/Nov	.35	.23	.29	.47	.62	.26
	Feb/Oct	.40	.28	.32	.43	.55	.30
	Mar/Sept	.40	.35	.33	.40	.36	.33
	Apr/Aug	.50	.37	.33	.34	.23	.35
	May/July	.75	.47	.33	.30	.21	.35
	June	.85	.49	.33	.28	.21	.35
24	Dec	.30	.19	.24	.45	.64	.19
	Jan/Nov	.35	.21	.25	.45	.64	.21
	Feb/Oct	.35	.26	.30	.45	.57	.26
	Mar/Sept	.40	.33	.33	.42	.45	.31
	Apr/Aug	.45	.40	.33	.38	.30	.35
	May/July	.60	.44	.35	.34	.23	.36
	June	.70	.47	.35	.32	.23	.36
32	Dec	.30	.16	.22	.43	.62	.14
	Jan/Nov	.30	.19	.22	.43	.62	.17
	Feb/Oct	.35	.23	.29	.47	.60	.23
	Mar/Sept	.40	.30	.32	.45	.51	.29
	Apr/Aug	.45	.37	.35	.40	.38	.33
	May/July	.60	.42	.35	.38	.30	.36
	June	.60	.44	.35	.36	.28	.37
40	Dec	.25	.16	.19	.40	.57	.10
	Jan/Nov	.30	.16	.21	.42	.60	.12
	Feb/Oct	.30	.21	.25	.45	.62	.18
	Mar/Sept	.35	.28	.30	.47	.57	.25
	Apr/Aug	.45	.35	.35	.43	.45	.31
	May/July	.55	.40	.35	.40	.38	.36
	June	.60	.42	.37	.40	.34	.37

Latitude	Month	N	NE NW	E W	SE SW	S	HOR
48	Dec	.25	.14	.14	.34	.49	.05
	Jan/Nov	.25	.14	.17	.38	.53	.06
	Feb/Oct	.30	.16	.22	.42	.60	.13
	Mar/Sept	.35	.26	.29	.47	.60	.21
	Apr/Aug	.40	.33	.33	.47	.51	.29
	May/July	.55	.40	.37	.45	.45	.35
	June	.60	.44	.38	.43	.43	.37
56	Dec	.20	.12	.10	.23	.30	.01
	Jan/Nov	.25	.14	.13	.28	.40	.02
	Feb/Oct	.25	.16	.19	.40	.55	.08
	Mar/Sept	.30	.23	.27	.47	.62	.17
	Apr/Aug	.40	.30	.33	.49	.55	.25
	May/July	.55	.40	.38	.49	.51	.32
	June	.65	.44	.40	.47	.49	.36
64	Dec	.20	.12	.08	.09	.11	.01
	Jan/Nov	.20	.12	.10	.15	.19	.01
	Feb/Oct	.25	.14	.14	.32	.45	.04
	Mar/Sept	.30	.19	.24	.43	.60	.11
	Apr/Aug	.40	.30	.33	.49	.60	.21
	May/July	.60	.42	.40	.51	.57	.31
	June	.65	.44	.41	.51	.55	.35

For South latitudes, replace January through December by July through June.

TABLE B5.1 RATES OF HEAT GAIN FROM OCCUPANTS OF CONDITIONED SPACES[a,b,c]

Degree of Activity	Typical Application	Total Heat Adults (Watts)	(Male, Btu/hr)	Total Heat Adjusted[d] (Watts)	(Btu/hr)	Sensible Heat (Watts)	(Btu/hr)	Latent Heat (Watts)	(Btu/hr)
Seated at theater	Theater—Matinee	115	390	95	330	65	225	30	105
Seated at theater	Theater—Evening	115	390	105	350	70	245	30	105
Seated, very light work	Offices, hotels, apartments	130	450	115	400	70	245	45	155
Moderately active office work	Offices, hotels, apartments	140	475	130	450	75	250	60	200
Standing, light work; walking	Department store, retail store	160	550	130	450	75	250	60	200
Walking; standing	Drug store, bank	160	550	145	500	75	250	75	250
Seated, eating	Restaurant[e]	145	490	160	550	80	275	80	275
Light bench work	Factory	235	800	220	750	80	275	140	475
Moderate dancing	Dance hall	265	900	250	850	90	305	160	545
Walking 3 mph; light machine work	Factory	295	1000	295	1000	110	375	185	625
Bowling[f]	Bowling alley	440	1500	425	1450	170	580	255	870
Heavy work	Factory	440	1500	425	1450	170	580	255	870
Heavy machine work; lifting	Factory	470	1600	470	1600	185	635	285	965
Athletics	Gymnasium	585	2000	525	1800	210	710	320	1090

[a]Tabulated values are based on 75°F room dry-bulb temperature. For 80°F room dry-bulb, the total heat remains the same, but the sensible heat values should be decreased by approximately 20%, and the latent heat values increased accordingly.

[b]Also refer to Table 1.2, Chapter 1.

[c]All values are rounded to nearest 5 Btu/hr.

[d]Adjusted heat gain is based on normal percentage of men, women, and children for the application listed, with the postulate that the gain from an adult female is 85% of that for an adult male, and the gain from a child is 75% of that for an adult male.

[e]Adjusted total heat gain for Seated, eating, Restaurant, includes 60 Btu/hr for food per individual (30 Btu/hr sensible and 30 Btu/hr latent).

[f]For Bowling, figure one person per alley actually bowling, and all others as sitting (400 Btu/hr) or standing and walking slowly (550 Btu/hr).

Source: © 1989. Reprinted by permission from ASHRAE Handbook of Fundamentals 1989.

TABLE B5.2 BUILDING STORAGE FACTORS FOR PEOPLE

Total Hours in Space	Hours after Each Entry Into Space 1	2	3	4	5	6	7	8	9	10	11	12	13	14	15	16	17	18	19	20	21	22	23	24
2	0.49	0.58	0.17	0.13	0.10	0.08	0.07	0.06	0.05	0.04	0.04	0.03	0.03	0.02	0.02	0.02	0.02	0.01	0.01	0.01	0.01	0.01	0.01	0.01
4	0.49	0.59	0.66	0.71	0.27	0.21	0.16	0.14	0.11	0.10	0.08	0.07	0.06	0.06	0.05	0.04	0.04	0.03	0.03	0.03	0.02	0.02	0.02	0.01
6	0.50	0.60	0.67	0.72	0.76	0.79	0.34	0.26	0.21	0.18	0.15	0.13	0.11	0.10	0.08	0.07	0.06	0.06	0.05	0.04	0.04	0.03	0.03	0.03
8	0.51	0.61	0.67	0.72	0.76	0.80	0.82	0.84	0.38	0.30	0.25	0.21	0.18	0.15	0.13	0.12	0.10	0.09	0.08	0.07	0.06	0.05	0.05	0.04
10	0.53	0.62	0.69	0.74	0.77	0.80	0.83	0.85	0.87	0.89	0.42	0.34	0.28	0.23	0.20	0.17	0.15	0.13	0.11	0.10	0.09	0.08	0.07	0.06
12	0.55	0.64	0.70	0.75	0.79	0.81	0.84	0.86	0.88	0.89	0.91	0.92	0.45	0.36	0.30	0.25	0.21	0.19	0.16	0.14	0.12	0.11	0.09	0.08
14	0.58	0.66	0.72	0.77	0.80	0.83	0.85	0.87	0.89	0.90	0.91	0.92	0.93	0.94	0.47	0.38	0.31	0.26	0.23	0.20	0.17	0.15	0.13	0.11
16	0.62	0.70	0.75	0.79	0.82	0.85	0.87	0.88	0.90	0.91	0.92	0.93	0.94	0.95	0.95	0.96	0.49	0.39	0.33	0.28	0.24	0.20	0.18	0.16
18	0.66	0.74	0.79	0.82	0.85	0.87	0.89	0.90	0.92	0.93	0.94	0.94	0.95	J.96	0.96	0.97	0.97	0.97	0.50	0.40	0.33	0.28	0.24	0.21

Source: Sensible Heat Cooling Load Factors for People. Reprinted from ASHRAE Handbook of Fundamentals 1981 by permission of the American Society of Heating, Refrigerating and Air-Conditioning Engineers, Inc.

TABLE B5.3 AVERAGE VALUES OF BALLAST FACTOR, FOR FLUORESCENT LIGHTS

Specialty equipment, energy savings ballasts, and other lighting system conditions can produce significantly different results from these values.

Lamp Wattage	No. of Lamps per Fixture	Ballast Factor
35 / 40	1	1.30
35 / 40	2	1.20
60 / 75	1	1.30
60 / 75	2	1.20
110	1	1.25
110	2	1.07
160	1	1.15
160	2	1.08
185 / 215	1	1.08
185 / 215	2	1.06
32	1 (high-output fixture on 227 V rapid-start)	2.19

Source: Reprinted from ASHRAE *Cooling and Heating Load Calculation Manual* by permission of the American Society of Heating, Refrigerating and Air-Conditioning Engineers, Inc.

TABLE B5.4　LIGHTING WATTS/SQ FT

Building Type	Watts/sq ft			Watts/sq m		
	Low	Avg.	High	Low	Avg.	High
Auditoriums, theaters, town halls	1.0	2.0	3.0	10.8	21.5	32.3
Banks	2.5	3.0	5.0	26.9	32.3	53.8
Barber shops and beauty parlors[a]	3.0	5.0	9.0	32.3	53.8	96.9
Churches	1.0	1.8	3.0	10.8	19.4	32.3
Clubs and lodge rooms	1.5	2.0		16.1	21.5	
Court rooms		2.0			21.5	
Department stores						
Basement	2.0	3.0	5.0	21.5	32.3	53.8
Main floors	3.5	6.0[a]	9.0[a]	37.6	64.6	96.9
Upper floors	2.0	2.5	3.5[a]	21.5	26.9	37.6
Women's wear	3.0		5.0	32.3		53.8
House furnishings	2.0		3.0	21.5		32.3
Jewelry displays			8.0			86.1
Ancillary spaces[b]		2.0			21.5	
Factories[c]						
Assembly areas[a]	3.0	4.5	6.0	32.3	48.4	64.6
Light manufacturing[a]	9.0	10.0	12.0	96.9	107.6	129.2
Heavy manufacturing[a]	15.0	45.0	60.0	161.5	484.4	645.8
Garages (commercial)		0.5			5.4	
Hospitals[d]	1.0	1.5	2.0	10.8	16.1	21.5
Hotels and motels	1.0	2.0	3.0	10.8	21.5	32.3
Libraries and museums	1.0	1.5	3.0	10.8	16.1	32.3
Medical clinic		2.5			26.9	
Office buildings						
Private offices	2.0	5.8	8.0	21.5	62.4	86.1
Stenographic dept.[a]	5.0	7.5	10.0	53.8	80.7	107.6
Post office		3.0			32.3	
Radio/TV studio		3.8			40.9	
Residential	1.0	2.0	4.0	10.8	21.5	43.1
Restaurants	1.5	2.0	2.5	16.1	21.5	26.9
Retail stores						
Dress shops	1.0	2.0	4.0	10.8	21.5	43.1
Drive-in donut shops		3.0			32.3	
Drugstores	1.0	2.0	3.0	10.8	21.5	32.3
5 and 10 cent stores	1.5	3.0	5.0	16.1	32.3	53.8
Hat shops	1.0	2.0	3.0	10.8	21.5	32.3
Shoe stores	1.0	2.0	3.0	10.8	21.5	32.3
Supermarkets		3.0			32.3	
Malls	1.0	1.5	2.0	10.8	16.1	21.5

Building Type	Watts/sq ft			Watts/sq m		
	Low	Avg.	High	Low	Avg.	High
Schools, colleges, universities	1.5		6.0	16.1		64.6
Elementary		3.0			32.3	
Junior high		3.0			32.3	
Senior high		2.3			24.8	
College dormitory		1.5			16.1	
College library		2.5			26.9	
College physical education center		2.0			21.5	
Warehouses	¼	1.0		2.7	10.8	

[a] Includes other loads.

[b] Stock rooms, receiving, marking, toilet, and restroom areas.

[c] Food processing 3 W/ft^2 (32.3 W/m^2)
Apparel Manufacturing 2 W/ft^2 (21.5 W/m^2)
Tools Manufacturing 4 W/ft^2 (43.1 W/m^2)

[d] Pediatric Hospitals 3 W/ft^2 (32.3 W/m^2)

Note: In any of the above building types—except single-family dwellings and individual residential units in multifamily dwellings, the following spaces have average lighting loads as indicated:

Assembly halls and auditoriums	1 W/ft^2	(10.8 W/m^2)
Halls, corridors, closets	½ W/ft^2	(5.4 W/m^2)
Storage spaces	¼ W/ft^2	(2.7 W/m^2)

Note: Lighting limits for energy conservation:

Type of Use	Max. Watts/Sq Ft
Interior	
Category A: Classrooms, office areas, automotive mechanical areas, museums, conference rooms, drafting rooms, clerical areas, laboratories, merchandising areas, kitchens, examining rooms, book stacks, athletic facilities	3.00
Category B: Auditoriums, waiting areas, spectator areas, restrooms, dining areas, transportation terminals, working corridors in prisons and hospitals, book storage areas, active inventory storage, hospital bedrooms, hotel and motel bedrooms, enclosed shopping mall concourse areas, stairways	1.00
Category C: Corridors, lobbies, elevators, inactive storage areas	0.50
Category D: Indoor parking	0.25
Exterior	
Category E: Building perimeter: wallwash, facade, canopy	5.00 (per linear foot)
Category F: Outdoor parking	0.10

TABLE B5.5 BUILDING STORAGE FACTORS

When Lights Are on for 8 Hours

"a" Classification	"b" Classification	Number of Hours After Lights are Turned on																							
		0	1	2	3	4	5	6	7	8	9	10	11	12	13	14	15	16	17	18	19	20	21	22	23
0.45	A	0.02	0.46	0.57	0.65	0.72	0.77	0.82	0.85	0.88	0.46	0.37	0.30	0.24	0.19	0.15	0.12	0.10	0.08	0.06	0.05	0.04	0.03	0.03	0.02
	B	0.07	0.51	0.56	0.61	0.65	0.68	0.71	0.74	0.77	0.34	0.31	0.28	0.25	0.22	0.20	0.18	0.16	0.15	0.13	0.12	0.11	0.10	0.09	0.08
	C	0.11	0.55	0.58	0.60	0.63	0.65	0.67	0.69	0.71	0.28	0.26	0.25	0.23	0.22	0.20	0.19	0.18	0.17	0.16	0.15	0.14	0.13	0.12	0.12
	D	0.14	0.58	0.60	0.61	0.62	0.63	0.64	0.65	0.66	0.22	0.22	0.21	0.20	0.20	0.19	0.19	0.18	0.18	0.17	0.16	0.16	0.16	0.15	0.15
0.55	A	0.01	0.56	0.65	0.72	0.77	0.82	0.85	0.88	0.90	0.37	0.30	0.24	0.19	0.16	0.13	0.10	0.08	0.07	0.05	0.04	0.03	0.03	0.02	0.02
	B	0.06	0.60	0.64	0.68	0.71	0.74	0.76	0.79	0.81	0.28	0.25	0.23	0.20	0.18	0.16	0.15	0.13	0.12	0.11	0.10	0.09	0.08	0.07	0.06
	C	0.09	0.63	0.66	0.68	0.70	0.71	0.73	0.75	0.76	0.23	0.21	0.20	0.19	0.18	0.17	0.16	0.15	0.14	0.13	0.12	0.11	0.11	0.10	0.10
	D	0.11	0.66	0.67	0.68	0.69	0.70	0.71	0.72	0.72	0.18	0.18	0.17	0.17	0.16	0.16	0.15	0.15	0.14	0.14	0.13	0.13	0.13	0.12	0.12
0.65	A	0.01	0.66	0.73	0.78	0.82	0.86	0.88	0.91	0.93	0.29	0.23	0.19	0.15	0.12	0.10	0.08	0.06	0.05	0.04	0.03	0.03	0.02	0.02	0.01
	B	0.04	0.69	0.72	0.75	0.77	0.80	0.82	0.84	0.85	0.22	0.19	0.18	0.16	0.14	0.13	0.12	0.10	0.09	0.08	0.08	0.07	0.06	0.06	0.05
	C	0.07	0.72	0.73	0.75	0.76	0.78	0.79	0.80	0.82	0.18	0.17	0.16	0.15	0.14	0.13	0.12	0.11	0.11	0.10	0.10	0.09	0.08	0.08	0.07
	D	0.09	0.73	0.74	0.75	0.76	0.77	0.77	0.78	0.79	0.14	0.14	0.13	0.13	0.13	0.12	0.12	0.11	0.11	0.11	0.10	0.10	0.10	0.10	0.09
0.75	A	0.01	0.76	0.80	0.84	0.87	0.90	0.92	0.93	0.95	0.21	0.17	0.13	0.11	0.09	0.07	0.06	0.05	0.04	0.03	0.02	0.02	0.02	0.01	0.01
	B	0.03	0.78	0.80	0.82	0.84	0.85	0.87	0.88	0.89	0.15	0.14	0.13	0.11	0.10	0.09	0.08	0.07	0.07	0.06	0.05	0.05	0.04	0.04	0.04
	C	0.05	0.80	0.81	0.82	0.83	0.84	0.85	0.86	0.87	0.13	0.12	0.11	0.10	0.10	0.09	0.09	0.08	0.08	0.07	0.07	0.06	0.06	0.06	0.05
	D	0.06	0.81	0.82	0.82	0.83	0.83	0.84	0.84	0.85	0.10	0.10	0.10	0.09	0.09	0.09	0.08	0.08	0.08	0.08	0.07	0.07	0.07	0.07	0.07

When Lights Are on for 10 Hours

"a" Classification	"b" Classification	Number of Hours After Lights are Turned on																							
		0	1	2	3	4	5	6	7	8	9	10	11	12	13	14	15	16	17	18	19	20	21	22	23
0.45	A	0.03	0.47	0.58	0.66	0.73	0.78	0.82	0.86	0.88	0.91	0.93	0.49	0.39	0.32	0.26	0.21	0.17	0.13	0.11	0.09	0.07	0.06	0.05	0.04
	B	0.10	0.54	0.59	0.63	0.66	0.70	0.73	0.76	0.78	0.80	0.82	0.39	0.35	0.32	0.28	0.26	0.23	0.21	0.19	0.17	0.15	0.14	0.12	0.11
	C	0.15	0.59	0.61	0.64	0.66	0.68	0.70	0.72	0.73	0.75	0.76	0.33	0.31	0.29	0.27	0.26	0.24	0.23	0.21	0.20	0.19	0.18	0.17	0.16
	D	0.18	0.62	0.63	0.64	0.66	0.67	0.68	0.69	0.69	0.70	0.71	0.27	0.26	0.26	0.25	0.24	0.23	0.23	0.22	0.21	0.21	0.20	0.19	0.19
0.55	A	0.02	0.57	0.65	0.72	0.78	0.82	0.85	0.88	0.91	0.92	0.94	0.40	0.32	0.26	0.21	0.17	0.14	0.11	0.09	0.07	0.06	0.05	0.04	0.03
	B	0.08	0.62	0.66	0.69	0.73	0.75	0.78	0.80	0.82	0.84	0.85	0.32	0.29	0.26	0.23	0.21	0.19	0.17	0.15	0.14	0.12	0.11	0.10	0.09
	C	0.12	0.66	0.68	0.70	0.72	0.74	0.75	0.77	0.78	0.79	0.81	0.27	0.25	0.24	0.22	0.21	0.20	0.19	0.17	0.16	0.15	0.14	0.14	0.13
	D	0.15	0.69	0.70	0.71	0.72	0.73	0.73	0.74	0.75	0.76	0.76	0.22	0.22	0.21	0.20	0.20	0.19	0.18	0.18	0.17	0.17	0.16	0.16	0.15
0.65	A	0.02	0.66	0.73	0.78	0.83	0.86	0.89	0.91	0.93	0.94	0.95	0.31	0.25	0.20	0.16	0.13	0.11	0.08	0.07	0.05	0.04	0.04	0.03	0.02
	B	0.06	0.71	0.74	0.76	0.79	0.81	0.83	0.84	0.86	0.87	0.89	0.25	0.22	0.20	0.18	0.16	0.15	0.13	0.12	0.11	0.10	0.09	0.08	0.07
	C	0.09	0.74	0.75	0.77	0.78	0.80	0.81	0.82	0.83	0.84	0.85	0.21	0.20	0.18	0.17	0.16	0.15	0.14	0.14	0.13	0.12	0.11	0.11	0.10
	D	0.11	0.76	0.77	0.77	0.78	0.79	0.79	0.80	0.81	0.81	0.82	0.17	0.17	0.16	0.16	0.15	0.15	0.14	0.14	0.14	0.13	0.13	0.12	0.12
0.75	A	0.01	0.76	0.81	0.84	0.88	0.90	0.92	0.93	0.95	0.96	0.97	0.22	0.18	0.14	0.12	0.09	0.08	0.06	0.05	0.04	0.03	0.03	0.02	0.02
	B	0.04	0.79	0.81	0.83	0.85	0.86	0.88	0.89	0.90	0.91	0.92	0.18	0.16	0.14	0.13	0.12	0.10	0.09	0.08	0.08	0.07	0.06	0.06	0.05
	C	0.07	0.81	0.82	0.83	0.84	0.85	0.86	0.87	0.88	0.89	0.89	0.15	0.14	0.13	0.12	0.12	0.11	0.10	0.10	0.09	0.09	0.08	0.08	0.07
	D	0.08	0.83	0.83	0.84	0.84	0.85	0.85	0.86	0.86	0.87	0.87	0.12	0.12	0.12	0.11	0.11	0.11	0.10	0.10	0.10	0.09	0.09	0.09	0.09

When Lights Are on for 12 Hours

"a" Classification	"b" Classification	Number of Hours After Lights are Turned on																							
		0	1	2	3	4	5	6	7	8	9	10	11	12	13	14	15	16	17	18	19	20	21	22	23
0.45	A	0.05	0.49	0.59	0.67	0.73	0.78	0.83	0.86	0.89	0.91	0.93	0.94	0.95	0.51	0.41	0.33	0.27	0.22	0.17	0.14	0.11	0.09	0.07	0.06
	B	0.13	0.57	0.61	0.65	0.69	0.72	0.75	0.77	0.79	0.82	0.83	0.85	0.87	0.43	0.39	0.35	0.31	0.28	0.25	0.23	0.21	0.18	0.17	0.15
	C	0.19	0.63	0.65	0.67	0.69	0.71	0.73	0.74	0.76	0.77	0.79	0.80	0.81	0.37	0.35	0.33	0.31	0.29	0.27	0.26	0.24	0.23	0.21	0.20
	D	0.22	0.66	0.67	0.68	0.69	0.70	0.71	0.72	0.73	0.74	0.74	0.75	0.76	0.32	0.31	0.30	0.29	0.28	0.27	0.26	0.26	0.25	0.24	0.23
0.55	A	0.04	0.58	0.66	0.73	0.78	0.82	0.86	0.89	0.91	0.93	0.94	0.95	0.96	0.42	0.34	0.27	0.22	0.18	0.14	0.11	0.09	0.07	0.06	0.05
	B	0.11	0.65	0.68	0.72	0.74	0.77	0.79	0.81	0.83	0.85	0.86	0.88	0.89	0.35	0.32	0.28	0.26	0.23	0.21	0.19	0.17	0.15	0.14	0.12
	C	0.15	0.69	0.71	0.73	0.75	0.76	0.78	0.79	0.80	0.81	0.83	0.84	0.85	0.30	0.29	0.27	0.25	0.24	0.22	0.21	0.20	0.19	0.17	0.16
	D	0.18	0.72	0.73	0.74	0.75	0.76	0.76	0.77	0.78	0.78	0.79	0.80	0.80	0.26	0.25	0.24	0.24	0.23	0.22	0.22	0.21	0.20	0.20	0.19
0.65	A	0.03	0.67	0.74	0.79	0.83	0.86	0.89	0.91	0.93	0.94	0.95	0.96	0.97	0.33	0.26	0.21	0.17	0.14	0.11	0.09	0.07	0.06	0.05	0.04
	B	0.09	0.73	0.75	0.78	0.80	0.82	0.84	0.85	0.87	0.88	0.89	0.90	0.91	0.27	0.25	0.22	0.20	0.18	0.16	0.15	0.13	0.12	0.11	0.10
	C	0.12	0.76	0.78	0.79	0.80	0.81	0.83	0.84	0.85	0.86	0.86	0.87	0.88	0.24	0.22	0.21	0.20	0.19	0.17	0.16	0.15	0.14	0.14	0.13
	D	0.14	0.79	0.79	0.80	0.80	0.81	0.82	0.82	0.83	0.83	0.84	0.84	0.85	0.20	0.20	0.19	0.18	0.18	0.17	0.17	0.16	0.16	0.15	0.15
0.75	A	0.02	0.77	0.81	0.85	0.88	0.90	0.92	0.94	0.95	0.96	0.97	0.97	0.98	0.23	0.19	0.15	0.12	0.10	0.08	0.06	0.05	0.04	0.03	0.03
	B	0.06	0.81	0.82	0.84	0.86	0.87	0.88	0.90	0.91	0.92	0.92	0.93	0.94	0.19	0.18	0.16	0.14	0.13	0.12	0.10	0.09	0.08	0.08	0.07
	C	0.09	0.83	0.84	0.85	0.86	0.87	0.88	0.88	0.89	0.90	0.90	0.91	0.91	0.17	0.16	0.15	0.14	0.13	0.12	0.12	0.11	0.10	0.10	0.09
	D	0.10	0.85	0.85	0.86	0.86	0.86	0.87	0.87	0.88	0.88	0.88	0.89	0.89	0.14	0.14	0.14	0.13	0.13	0.12	0.12	0.12	0.11	0.11	0.11

When Lights Are on for 14 Hours

"a" Classification	"b" Classification	Number of Hours After Lights are Turned on																							
		0	1	2	3	4	5	6	7	8	9	10	11	12	13	14	15	16	17	18	19	20	21	22	23
0.45	A	0.07	0.51	0.61	0.68	0.74	0.79	0.83	0.87	0.89	0.91	0.93	0.94	0.95	0.96	0.97	0.53	0.42	0.34	0.27	0.22	0.18	0.14	0.12	0.09
	B	0.18	0.61	0.65	0.68	0.72	0.74	0.77	0.79	0.81	0.83	0.85	0.86	0.88	0.89	0.90	0.46	0.41	0.37	0.34	0.30	0.27	0.24	0.22	0.20
	C	0.24	0.67	0.69	0.71	0.73	0.74	0.76	0.77	0.79	0.80	0.81	0.82	0.83	0.84	0.85	0.41	0.39	0.36	0.34	0.32	0.30	0.28	0.27	0.25
	D	0.26	0.71	0.72	0.72	0.73	0.74	0.75	0.76	0.77	0.78	0.78	0.79	0.80	0.80	0.80	0.36	0.35	0.34	0.33	0.32	0.31	0.30	0.29	0.28
0.55	A	0.06	0.69	0.68	0.74	0.79	0.83	0.86	0.89	0.91	0.93	0.94	0.95	0.96	0.97	0.98	0.43	0.35	0.28	0.22	0.18	0.15	0.12	0.09	0.08
	B	0.15	0.68	0.71	0.74	0.77	0.79	0.81	0.83	0.85	0.86	0.88	0.89	0.90	0.91	0.92	0.38	0.34	0.31	0.27	0.25	0.22	0.20	0.18	0.16
	C	0.19	0.73	0.75	0.76	0.78	0.79	0.80	0.81	0.83	0.84	0.85	0.86	0.86	0.87	0.88	0.34	0.32	0.30	0.28	0.26	0.25	0.23	0.22	0.21
	D	0.22	0.76	0.77	0.77	0.78	0.79	0.79	0.80	0.81	0.81	0.82	0.82	0.83	0.83	0.84	0.29	0.28	0.28	0.27	0.26	0.25	0.24	0.24	0.23
0.65	A	0.05	0.69	0.75	0.80	0.84	0.87	0.89	0.92	0.93	0.95	0.96	0.96	0.97	0.98	0.98	0.34	0.27	0.22	0.17	0.14	0.11	0.09	0.07	0.06
	B	0.11	0.75	0.78	0.80	0.82	0.64	0.85	0.87	0.88	0.89	0.90	0.91	0.92	0.93	0.94	0.29	0.26	0.24	0.21	0.19	0.17	0.16	0.14	0.13
	C	0.15	0.79	0.80	0.82	0.83	0.84	0.85	0.86	0.86	0.87	0.88	0.89	0.89	0.90	0.91	0.26	0.25	0.23	0.22	0.20	0.19	0.18	0.17	0.16
	D	0.17	0.81	0.82	0.82	0.83	0.83	0.84	0.84	0.85	0.85	0.86	0.86	0.87	0.87	0.87	0.23	0.22	0.21	0.21	0.20	0.20	0.19	0.18	0.18
0.75	A	0.03	0.78	0.82	0.86	0.88	0.91	0.92	0.94	0.95	0.96	0.97	0.97	0.98	0.98	0.99	0.24	0.19	0.16	0.12	0.10	0.08	0.07	0.05	0.04
	B	0.08	0.82	0.84	0.86	0.87	0.88	0.90	0.91	0.92	0.92	0.93	0.94	0.94	0.95	0.96	0.21	0.19	0.17	0.15	0.14	0.12	0.11	0.10	0.09
	C	0.11	0.85	0.86	0.87	0.88	0.88	0.89	0.90	0.90	0.91	0.91	0.92	0.92	0.93	0.93	0.19	0.18	0.17	0.16	0.15	0.14	0.13	0.12	0.11
	D	0.12	0.87	0.87	0.87	0.88	0.88	0.89	0.89	0.89	0.90	0.90	0.90	0.90	0.91	0.91	0.16	0.16	0.15	0.15	0.14	0.14	0.14	0.13	0.13

When Lights Are on for 16 Hours

"a" Classification	"b" Classification	Number of Hours After Lights are Turned on																						
		0	1	2	3	4	5	6	7	8	9	10	11	12	13	14	15	16	17	18	19	20	21	22	23
0.45	A	0.12	0.54	0.63	0.70	0.76	0.81	0.85	0.88	0.90	0.92	0.94	0.95	0.96	0.97	0.97	0.98	0.98	0.54	0.43	0.35	0.28	0.23	0.18	0.15
	B	0.23	0.66	0.69	0.72	0.75	0.78	0.80	0.82	0.84	0.85	0.87	0.88	0.89	0.90	0.91	0.92	0.93	0.49	0.44	0.39	0.35	0.32	0.29	0.26
	C	0.29	0.72	0.74	0.75	0.77	0.78	0.80	0.81	0.82	0.83	0.84	0.85	0.86	0.87	0.88	0.88	0.89	0.45	0.42	0.39	0.37	0.35	0.33	0.31
	D	0.31	0.75	0.76	0.77	0.77	0.78	0.79	0.79	0.80	0.81	0.81	0.82	0.82	0.83	0.83	0.84	0.84	0.40	0.39	0.37	0.36	0.35	0.34	0.33
0.55	A	0.10	0.63	0.70	0.76	0.81	0.84	0.87	0.90	0.92	0.93	0.95	0.96	0.97	0.97	0.98	0.98	0.99	0.44	0.35	0.28	0.23	0.18	0.15	0.12
	B	0.19	0.72	0.75	0.77	0.80	0.82	0.84	0.85	0.87	0.88	0.89	0.90	0.91	0.92	0.93	0.94	0.94	0.40	0.36	0.32	0.29	0.26	0.24	0.21
	C	0.24	0.77	0.79	0.80	0.81	0.82	0.83	0.84	0.85	0.86	0.87	0.88	0.88	0.89	0.90	0.90	0.91	0.37	0.34	0.32	0.30	0.29	0.27	0.25
	D	0.26	0.80	0.80	0.81	0.82	0.82	0.83	0.83	0.84	0.84	0.85	0.85	0.86	0.86	0.86	0.87	0.87	0.33	0.32	0.31	0.30	0.29	0.28	0.27
0.65	A	0.07	0.71	0.77	0.81	0.85	0.88	0.90	0.92	0.94	0.95	0.96	0.97	0.97	0.98	0.98	0.99	0.99	0.34	0.27	0.22	0.18	0.14	0.12	0.09
	B	0.15	0.78	0.81	0.82	0.84	0.86	0.87	0.88	0.90	0.91	0.92	0.92	0.93	0.94	0.94	0.95	0.96	0.31	0.28	0.25	0.23	0.20	0.18	0.16
	C	0.18	0.82	0.83	0.84	0.85	0.86	0.87	0.88	0.89	0.89	0.90	0.90	0.91	0.92	0.92	0.93	0.93	0.28	0.27	0.25	0.24	0.22	0.21	0.20
	D	0.20	0.84	0.85	0.85	0.86	0.86	0.87	0.87	0.87	0.88	0.88	0.88	0.89	0.89	0.89	0.90	0.90	0.25	0.25	0.24	0.23	0.22	0.22	0.21
0.75	A	0.05	0.79	0.83	0.87	0.89	0.91	0.93	0.94	0.95	0.96	0.97	0.98	0.98	0.98	0.99	0.99	0.99	0.24	0.20	0.16	0.13	0.10	0.08	0.07
	B	0.11	0.85	0.86	0.87	0.89	0.90	0.91	0.92	0.93	0.93	0.94	0.95	0.95	0.96	0.96	0.96	0.97	0.22	0.20	0.18	0.16	0.15	0.13	0.12
	C	0.13	0.87	0.88	0.89	0.89	0.90	0.91	0.91	0.92	0.92	0.93	0.93	0.94	0.94	0.94	0.95	0.95	0.20	0.19	0.18	0.17	0.16	0.15	0.14
	D	0.14	0.89	0.89	0.89	0.90	0.90	0.90	0.91	0.91	0.91	0.91	0.92	0.92	0.92	0.92	0.93	0.93	0.18	0.18	0.17	0.17	0.16	0.16	0.15

"a" CLASSIFICATION FOR LIGHTS

This table is based on rooms having an average amount of furnishings.

"a"	Light Fixture and Ventilation Arrangements
0.45	Recessed lights which are not vented Low air supply rate—less than 0.5 cfm/ft^2 of floor area Supply and return diffusers below ceiling
0.55	Recessed lights which are not vented Medium to high air supply rate—more than 0.5 cfm/ft^2 of floor area Supply and return diffusers below ceiling or through ceiling space and grill
0.65	Vented light fixtures Medium to high air supply rate—more than 0.5 cfm/ft^2 of floor area Supply air through ceiling or wall but return air flows around light fixtures and through ceiling space
0.75	Vented or free hanging lights Supply air through ceiling or wall but return air flows around light fixtures and through a ducted return

TABLE B5.5 BUILDING STORAGE FACTORS *(Continued)*

"b" CLASSIFICATION FOR LIGHTS

This table is based on floor covered with carpet and rubber pad. For floor covered with floor tile use letter designation in next row down with the same floor weight.

Room Air Circulation and Type of Supply and Return	Floor Construction and Floor Weight in Pounds Per Square Foot of Floor Area				
	2 in. Wooden Floor 10 lb/ft²	3 in. Concrete Floor 40 lb/ft²	6 in. Concrete Floor 75 lb/ft²	8 in. Concrete Floor 120 lb/ft²	12 in. Concrete Floor 160 lb/ft²
Low ventilation rate—minimum required to handle cooling load. Supply through floor, wall or ceiling diffuser. Ceiling space not vented.	B	B	C	D	D
Medium ventilation rate. Supply through floor, wall or ceiling diffuser. Ceiling space not vented.	A	B	C	D	D
High room air circulation induced by primary air of induction unit or by fan coil unit. Return through ceiling space.	A	B	C	C	D
Very high room air circulation used to minimize room temperature gradients. Return through ceiling space.	A	A	B	C	D

Source: Data reprinted from ASHRAE *Handbook of Fundamentals 1981* by permission of the American Society of Heating, Refrigerating and Air-Conditioning Engineers, Inc.

TABLE B5.6A RECOMMENDED RATE OF HEAT GAIN FROM SELECTED RESTAURANT EQUIPMENT[a]

Appliance	Size	Input Rating (Btu/hr) Max.	Input Rating (Btu/hr) Standby[b]	Heat Gain (Btu/hr) Without Hood Sens.	Without Hood Latent	Without Hood Total	With Hood Sensible	Input Rating (Watts) Max.	Input Rating (Watts) Standby[b]	Heat Gain (Watts) Without Hood Sens.	Without Hood Latent	Without Hood Total	With Hood Sensible
Electric, No Hood Required													
Blender, per quart of capacity	1 to 4 qt	1550		1000	520	1520	480	450		290	150	440	140
Cabinet (large hot holding)	16.2 to 17.3 ft3	7100		610	340	960	290	2080		180	100	280	85
Cabinet (small hot holding)	3.2 to 6.4 ft3	37		270	140	410	130	900		80	40	120	40
Coffee brewer	12 cups/2 brnrs	5660		3750	1910	5660	1810	1660		1100	560	1660	530
Coffee brewing urn (large), per quart of capacity	23 to 40 qt	2130		1420	710	2130	680	620		415	210	625	200
Coffee heater, per warming burner	1 to 2 brnrs	340		230	110	340	110	100		70	30	100	30
Dishwasher (hood type chemical sanitizing), per 100 dishes/hr	950 to 2000 dishes/hr	1300		170	370	540	170	380		50	110	160	50
Dishwasher (conveyor type water sanitizing), per 100 disher/hr	5000 to 9000 dishes/hr	1160		150	370	520	170	340		40	110	150	50
Display case (refrigerated), per ft3 of interior	6 to 67 ft3	154		62	0	62	0	45		20	0	20	0
Food warmer (infrared bulb), per lamp	1 to 6 bulbs	850		850	0	850	850	250		250	0	250	250
Food warmer (well type), per ft3 of well	0.7 to 2.5 ft3	3620		1200	610	1810	580	1060		350	180	530	170
Freezer (large)	73 ft3	4570		1840	0	1840	0	1340		540	0	540	0
Griddle/grill (large), per ft2 of cooking surface	4.6 to 11.8 ft2	9200		620	340	960	340	2700		180	100	280	100
Hot plate (high speed double burner)		16720		7810	5430	13240	6240	4900		2290	1590	3880	1830
Ice maker (large)	220 lb/day	3720		9320	0	9320	0	1090		2730	0	2730	0
Microwave oven (heavy duty commercial)	0.7 ft3	8970		8970	0	8970	0	2630		2630	0	2630	0
Mixer (large), per quart of capacity	80 qt	94		94	0	94	0	28		28	0	28	0
Refrigerator (large), per 100 ft3 of space	25 to 74 ft3	750		300	0	300	0	220		90	0	90	0
Rotisserie	300	10920		7200	3720	10920	3480	3200		2110	1090	3200	1020
Serving cart (hot), per ft3 of well	1.8 to 3.2 ft3	2050		680	340	1020	330	600		200	100	300	95
Steam kettle (large), per quart of capacity	80 to 320 qt	300		23	16	40	13	90		7	5	12	4
Toaster (large pop-up)	10 slice	18080		9590	8500	18080	5800	5300		2810	2490	5300	1700
Electric, Exhaust Hood Required													
Charbroiler, per ft2 of cooking surface	1.5 to 4.6 ft2	7320					3310	2145					970
Fryer (deep fat), per lb of fat capacity	15 to 70 lb	1270					14	370					4
Fryer (pressurized), per lb of fat capacity	233	1570					59	460					17
Oven (large convection), per ft3 of oven space	7 to 19 ft3	4450					180	1305					55
Oven (small convection), per ft3 of oven space	1.4 to 5.3 ft3	10340					150	3030					45
Range (burners), per 2 burner section	2 to 10 burners	7170					2660	2100					780
Gas, No Hood Required													
Broiler, per ft2 of broiling area	2.7 ft2	14770	61	5310	2860	8170	1220	1330	18	1555	840	2395	355
Dishwasher (hood type chemical sanitizing), per 100 dishes/hr	950 to 2000 dishes/hr	1740	660b	510	200	710	230	510	195	150	60	210	65
Dishwasher (conveyor type water sanitizing), per 100 disher/hr	5000 to 9000 dishes/hr	1370	660b	370	80	450	140	400	195	110	25	130	40
Griddle/grill (large), per ft2 of cooking surface	4.6 to 11.8 ft2	17000	330	1140	610	1750	460	4980	95	335	180	515	135
Oven (pizza), per ft2 of hearth	6.4 to 12.9 ft2	4740	61b	620	220	840	84	1390	18	180	65	245	25
Gas, Exhaust Hood Required													
Braising pan, per quart of capacity	105 to 140 qt	9840	620				2430	2885	180				710
Charbroiler (large), per ft2 of cooking area	4.6 to 11.8 ft2	16440	510				790	4815	150				230
Fryer (deep fat), per lb of fat capacity	11 to 70 lb	2270	300b				160	665	90				45
Oven (convection), per ft3 of oven space	7.4 to 19.4 ft3	8670	19b				250	2540	6				75
Oven (pizza), per ft2 of oven hearth	9.3 to 25.8 ft2	7240	61b				130	2120	18				40
Range (burners), per 2 burner section	2 to 10 burners	33600	1325				6590	9845	390				1930
Range (hot top/fry top), per ft2 of cooking surface		11800	330				3390	3455	95				995
Steam													
Compartment steamer, per lb of food/hr	46 to 450 lb	280		22	14	36	11	80		6	4	10	3
Dishwasher (hood type chemical sanitizing), per 100 dishes/hr	950 to 2000 dishes/hr	3150		880	380	1260	410	925		260	110	370	120
Dishwasher (conveyor water sanitizing), per 100 dishes/hr	5000 to 9000 dishes/hr	1180		150	370	520	170	345		45	110	150	50
Steam kettle, per quart of capacity	13 to 32 qt	500		39	25	64	19	145		11	7	18	6

[a] In cases where heat gain is given per unit of capacity the heat gain is calculated by multiplying the capacity by the recommended heat gain per unit of capacity.

[b] Standby input rating is for the entire appliance regardless of size.

Source: © 1989. Reprinted by permission from ASHRAE Handbook of Fundamentals 1989.

TABLE B5.6B   RECOMMENDED RATE OF HEAT GAIN FROM SELECTED OFFICE EQUIPMENT

Appliance	Size	Maximum Input		Standby Input		Recommended Rate of Heat Gain	
		Watts	Btu/hr	Watts	Btu/hr	Watts	Btu/hr
Computer Devices							
Communication/ transmission		1800-4600	6140-15700	1640-2810	5600-9600	1640-2810	5600-9600
Disk drives/mass storage		1000-10000	3400-34100	1000-6600	3400-22400	1000-6600	3400-22400
Microcomputer/ wordprocessor	16-640 kbytes[a]	100-600	340-2050	90-530	300-1800	90-530	300-1800
Minicomputer		2200-6600	7500-15000	2200-6600	7500-15000	2200-6600	7500-15000
Printer (laser)	8 pages/min	870	3000	180	600	300	1000
Printer (Line, high speed	5000-more pages/min	1000-5300	3400-18000	500-2550	2160-9040	730-3800	2500-13000
Tape drives		1200-6500	4100-22200	1000-4700	3500-15000	1000-4700	3500-15000
Terminal		90-200	300-700	80-180	270-600	80-180	270-600
Copiers/Typesetters							
Blue print		1150-12500	3900-42700	500-5000	1700-17000	1150-12500	3900-42700
Copiers (large)	30-67[a] copies/min.	5800-22500	1700-6600	5800-22500	900	3100	1700-6600
Copiers	6-30[a] copies/min.	1570-5800	460-1700	1570-5800	300-900	1000-3100	460-1700
Phototypesetter		1725	5900			1520	5200
Mailprocessing							
Inserting machine 3600–6800 pieces/hr		600-3300	2000-11300			390-2150	1300-7300
Labeling machine 1500–30,000 pieces/hr		600-6600	2000-22500			390-4300	1300-14700
Miscellaneous							
Cash register		60	200			48	160
Cold food/beverage		1150-1920	3900-6600			575-960	1960-3280
Coffee maker	10 cup	1500	5120		sensible latent	1050 450	3580 1540
Microwave oven	1 ft^3	600	2050			400	1360
Paper shredder		250-3000	850-10200			200-2420	680-8250
Water cooler	8 gal/h	700	2400			1750	6000

[a] Input is not proportional to capacity.

Source: © 1989. Reprinted by permission from ASHRAE *Handbook of Fundamentals 1989.*

TABLE B5.6C RATE OF HEAT GAIN FROM MISCELLANEOUS APPLIANCES

Miscellaneous Data[a]	Manufacturer's Rating Watts	Sensible	Recommended Rate of Heat Gain, Watts Latent	Total	Appliance	Miscellaneous Data[a]	Manufacturer's Rating Btu/hr	Sensible	Recommended Rate of Heat Gain, Btu/hr Latent	Total
					Electrical Appliances					
Blower type	1580	675	120	785	Hair dryer	Blower type	5400	2300	400	2700
Helmet type	700	550	100	650	Hair dryer	Helmet type	2400	1870	330	2200
60 heaters @ 25 W					Permanent	60 heaters @ 25 W				
91.44-cm normal use	1500	250	50	300	wave machine	36-in. normal use	5000	850	150	1000
1.27-cm diameter		28		28	Neon sign, per	0.5-in., diameter		30		30
0.95-cm diameter		56		56	linear meter of tube	0.375-in. diameter		60		60
					Sterilizer,					
	1100	190	350	540	instrument		3750	650	1200	1850
					Magnetic cord					
					typewriter		690	350	0	350
Running	1760	1760	0	1760	Small copier	Running	6000	6000	0	6000
Standby	880	880	0	880		Standby	3000	3000	0	3000
Running	3515	3515	0	3515	Large copier	Running	12 000	12 000	0	12 000
Standby	1760	1760	0	1760		Standby	6000	6000	0	6000
					Gas-Burning Appliances					
					Lab burners					
1.1-cm barrel	880	495	125	620	Bunsen	0.4375-in barrel	3000	1680	430	2100
3.8-cm wide	1465	820	205	1025	Fishtail	1.5-in. wide	5000	2800	700	3500
2.54-cm diameter	1760	985	245	1230	Meeker	1-in. diameter	6000	3360	840	4200
Mantle type	585	530	60	590	Gas light, per	Mantle type	2000	1800	200	2000
					burner					
Continuous flame	730	265	30	295	Cigar lighter	Continuous flame	2500	900	100	1000

[a]English (Btu/hr) values are per linear foot of tube.

Source: Reprinted from ASHRAE *Handbook of Fundamentals 1981* by permission of the American Society of Heating, Refrigerating and Air-Conditioning Engineers, Inc.

TABLE B5.7A BUILDING STORAGE FACTORS FOR HOODED APPLIANCES

Total Operation (hours)	Hours After Appliances Are on 1	2	3	4	5	6	7	8	9	10	11	12	13	14	15	16	17	18	19	20	21	22	23	24
2	0.27	0.40	0.25	0.18	0.14	0.11	0.09	0.08	0.07	0.06	0.05	0.04	0.04	0.03	0.03	0.03	0.02	0.02	0.02	0.02	0.01	0.01	0.01	0.01
4	0.28	0.41	0.51	0.59	0.39	0.30	0.24	0.19	0.16	0.14	0.12	0.10	0.09	0.08	0.07	0.06	0.05	0.05	0.04	0.04	0.03	0.03	0.02	0.02
6	0.29	0.42	0.52	0.59	0.65	0.70	0.48	0.37	0.30	0.25	0.21	0.18	0.16	0.14	0.12	0.11	0.09	0.08	0.07	0.06	0.05	0.05	0.04	0.04
8	0.31	0.44	0.54	0.61	0.66	0.71	0.75	0.78	0.55	0.43	0.35	0.30	0.25	0.22	0.19	0.16	0.14	0.13	0.11	0.10	0.08	0.07	0.06	0.06
10	0.33	0.46	0.55	0.62	0.68	0.72	0.76	0.79	0.81	0.84	0.60	0.48	0.39	0.33	0.28	0.24	0.21	0.18	0.16	0.14	0.12	0.11	0.09	0.08
12	0.36	0.49	0.58	0.64	0.69	0.74	0.77	0.80	0.82	0.85	0.87	0.88	0.64	0.51	0.42	0.36	0.31	0.26	0.23	0.20	0.18	0.15	0.13	0.12
14	0.40	0.52	0.61	0.67	0.72	0.76	0.79	0.82	0.84	0.86	0.88	0.89	0.91	0.92	0.67	0.54	0.45	0.38	0.32	0.28	0.24	0.21	0.19	0.16
16	0.45	0.57	0.65	0.70	0.75	0.78	0.81	0.84	0.86	0.87	0.89	0.90	0.92	0.93	0.94	0.94	0.69	0.56	0.46	0.39	0.34	0.29	0.25	0.22
18	0.52	0.63	0.70	0.75	0.79	0.82	0.84	0.86	0.88	0.89	0.91	0.92	0.93	0.94	0.95	0.95	0.96	0.96	0.71	0.58	0.48	0.41	0.35	0.30

TABLE B5.7B BUILDING STORAGE FACTORS FOR UNHOODED APPLIANCES AND EQUIPMENT

Total Operation (hours)	Hours After Appliances Are on 1	2	3	4	5	6	7	8	9	10	11	12	13	14	15	16	17	18	19	20	21	22	23	24
2	0.56	0.64	0.15	0.11	0.08	0.07	0.06	0.05	0.04	0.04	0.03	0.03	0.02	0.02	0.02	0.02	0.01	0.01	0.01	0.01	0.01	0.01	0.01	0.01
4	0.57	0.65	0.71	0.75	0.23	0.18	0.14	0.12	0.10	0.08	0.07	0.06	0.05	0.05	0.04	0.04	0.03	0.03	0.02	0.02	0.02	0.02	0.01	0.01
6	0.57	0.65	0.71	0.76	0.79	0.82	0.29	0.22	0.18	0.15	0.13	0.11	0.10	0.08	0.07	0.06	0.06	0.05	0.04	0.04	0.03	0.03	0.03	0.02
8	0.58	0.66	0.72	0.76	0.80	0.82	0.85	0.87	0.33	0.26	0.21	0.18	0.15	0.13	0.11	0.10	0.09	0.08	0.07	0.06	0.05	0.04	0.04	0.03
10	0.60	0.68	0.73	0.77	0.81	0.83	0.85	0.87	0.89	0.90	0.36	0.29	0.24	0.20	0.17	0.15	0.13	0.11	0.10	0.08	0.07	0.07	0.06	0.05
12	0.62	0.69	0.75	0.79	0.82	0.84	0.86	0.88	0.89	0.91	0.92	0.93	0.38	0.31	0.25	0.21	0.18	0.16	0.14	0.12	0.11	0.09	0.08	0.07
14	0.64	0.71	0.76	0.80	0.83	0.85	0.87	0.89	0.90	0.92	0.93	0.93	0.94	0.95	0.40	0.32	0.27	0.23	0.19	0.17	0.15	0.13	0.11	0.10
16	0.67	0.74	0.79	0.82	0.85	0.87	0.89	0.90	0.91	0.92	0.93	0.94	0.95	0.96	0.96	0.97	0.42	0.34	0.28	0.24	0.20	0.18	0.15	0.13
18	0.71	0.78	0.82	0.85	0.87	0.89	0.90	0.92	0.93	0.94	0.94	0.95	0.96	0.96	0.97	0.97	0.97	0.98	0.43	0.35	0.29	0.24	0.21	0.18

Source: Sensible Heat Cooling Load Factors. Reprinted from ASHRAE *Handbook of Fundamentals 1981* by permission of the American Society of Heating, Refrigerating and Air-Conditioning Engineers, Inc.

TABLE B5.8 TYPICAL DIVERSITY FACTORS FOR LARGE BUILDINGS

Type of Application	Diversity Factor	
	People	Lights
Office	0.75 to 0.90	0.70 to 0.85
Apartment, Hotel	0.40 to 0.60	0.30 to 0.50
Department Store	0.80 to 0.90	0.90 to 1.0
Industrial*	0.85 to 0.95	0.80 to 0.90

Source: Reprinted with permission from *Handbook of Air-Conditioning System Design* by the Carrier Air Conditioning Company, McGraw-Hill Book Company, 1965.

TABLE B6.1 VOLUME OF AIR INFILTRATION AT A WIND SPEED OF 15 MPH (6.72 M/S)

	cfm per ft of Crack	cfm per sq ft Opening	L/s per m of Crack	L/s per m^2 Opening
Wood frame window				
Nonweatherstripped	0.42		0.65	
Weatherstripped	0.25		0.38	
Metal frame double-hung window				
Nonweatherstripped	0.83		1.28	
Weatherstripped	0.36		0.55	
Metal framed window or glazed door (nonweatherstripped)				
Industrial, pivoted or awning	2.36		3.63	
Residential, casement or hinged	0.35		0.54	
Sliding glass door		0.40		2.03
Hinged door, wood or metal				
Nonweatherstripped	1.50		2.31	
Weatherstripped	0.36		0.55	

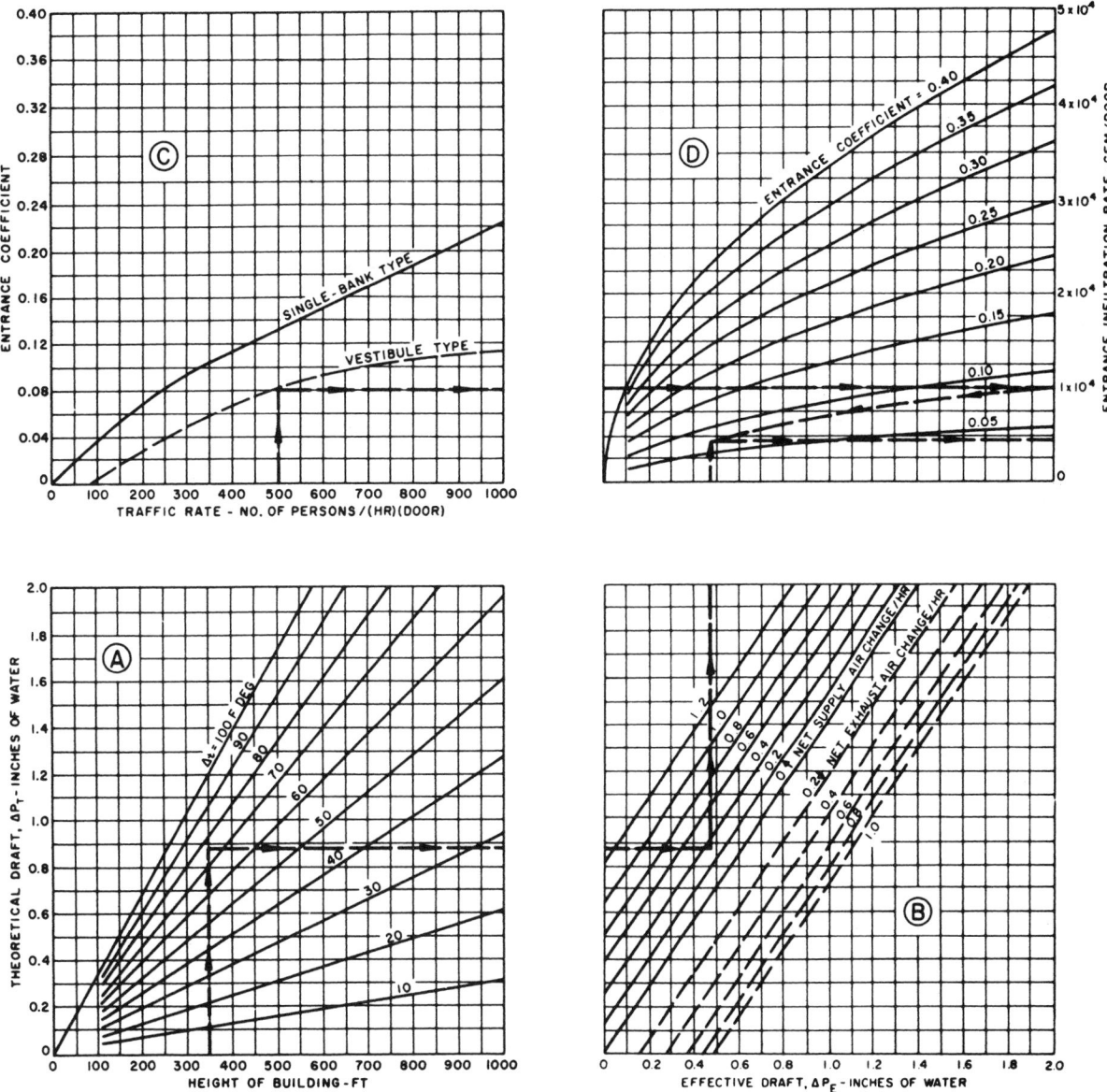

FIGURE B6.1. Design chart for evaluating air leakage rate through swinging door entrances under winter heating conditions. Reprinted from ASHRAE *Handbook of Fundamentals 1981* by permission of the American Society of Heating, Refrigerating & Air-Conditioning Engineers, Inc.

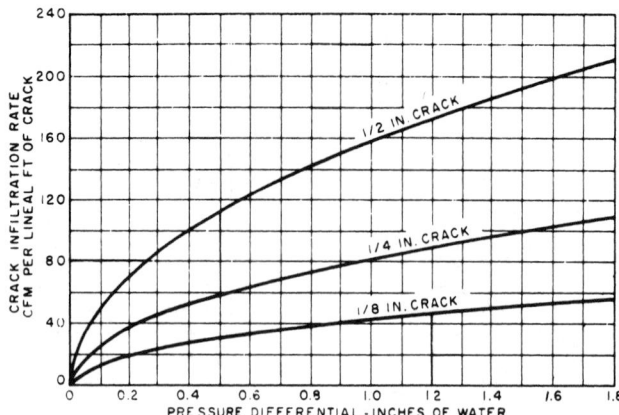

FIGURE B6.2.   Leakage rate through swinging door cracks. Reprinted from ASHRAE *Handbook of Fundamentals 1981* by permission of the American Society of Heating, Refrigerating & Air-Conditioning Engineers, Inc.

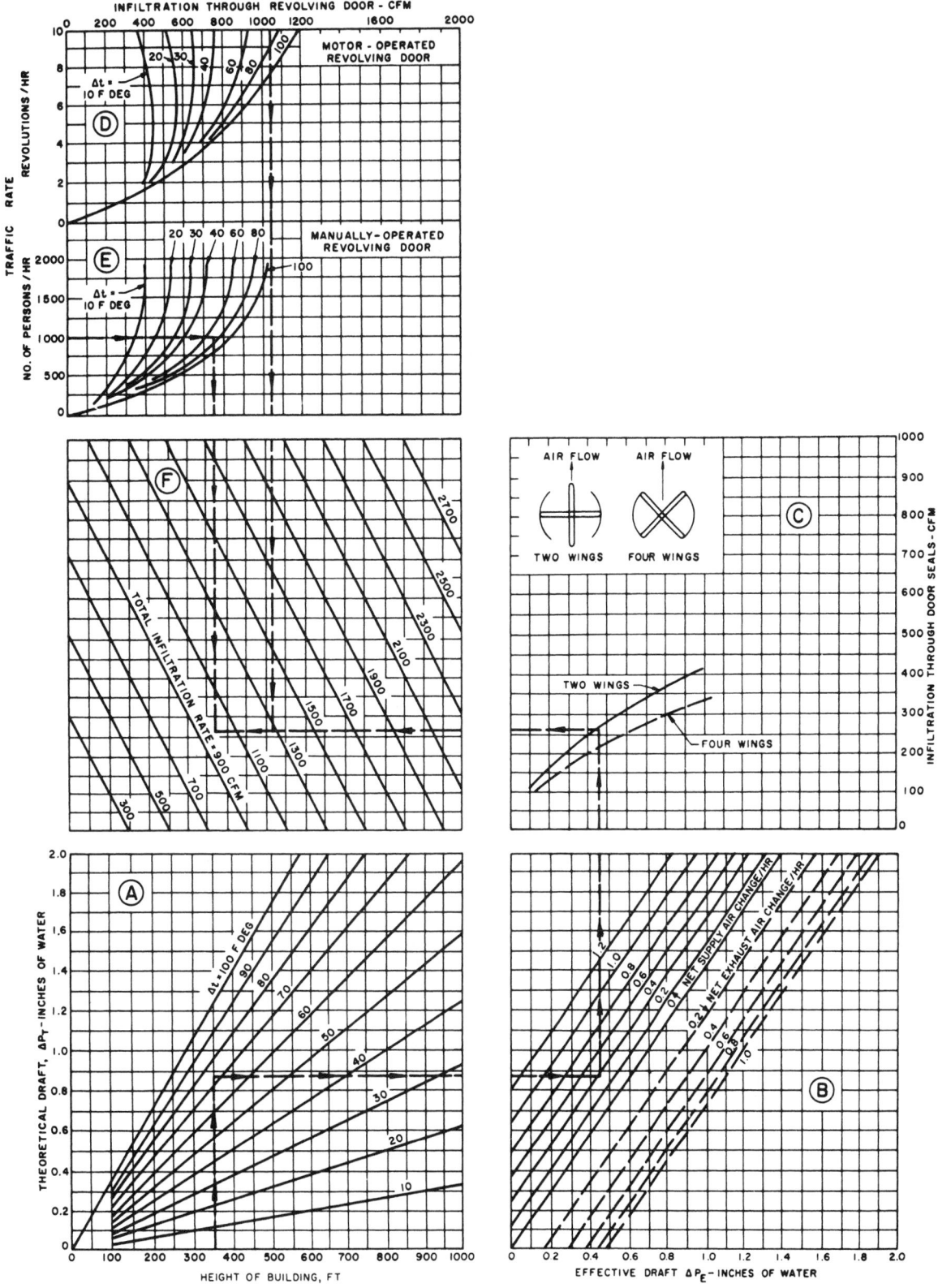

FIGURE B6.3. Design chart for evaluating air leakage rate through revolving doors. Reprinted from ASHRAE *Handbook of Fundamentals 1981* by permission of the American Society of Heating, Refrigerating & Air-Conditioning Engineers, Inc.

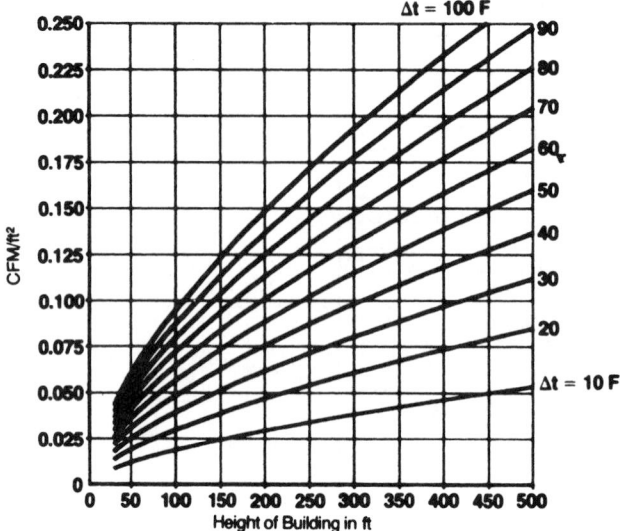

FIGURE B6.4. Infiltration through curtain wall for entire building due to stack effect and zero pressurization.

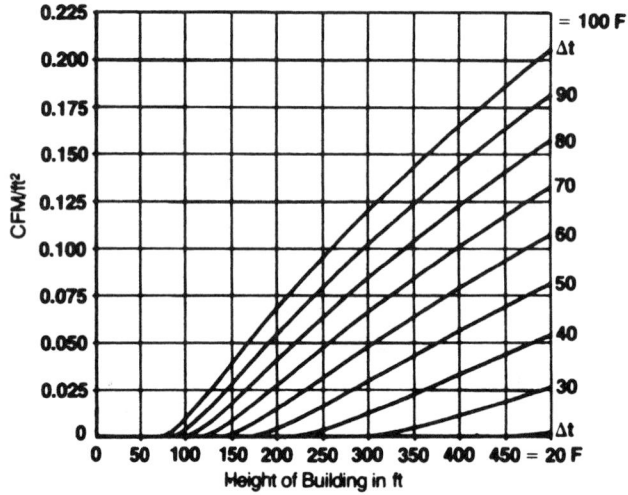

FIGURE B6.5. Infiltration through curtain wall for entire building due to stack effect, and a pressurization of 0.10 in. of water. Reprinted from ASHRAE *Cooling & Heating Load Calculation Manual* by permission of the American Society of Heating, Refrigerating & Air-Conditioning Engineers, Inc.

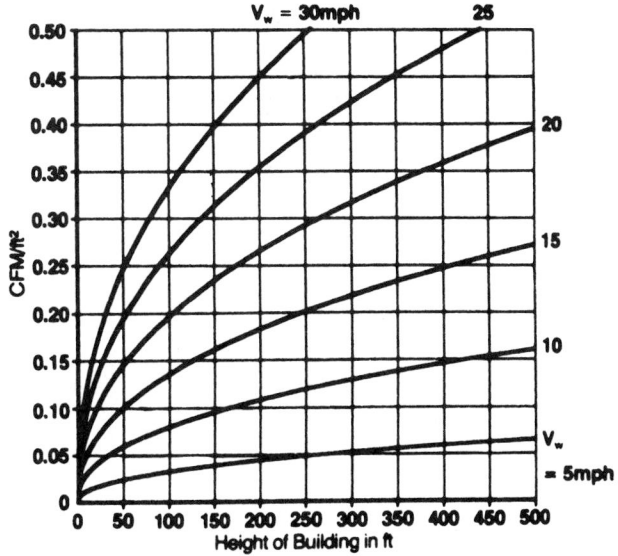

FIGURE B6.6. Infiltration through curtain wall for entire building due to wind and zero pressurization. Reprinted from ASHRAE *Cooling & Heating Load Calculation Manual* by permission of the American Society of Heating, Refrigerating & Air-Conditioning Engineers, Inc.

TABLE B6.2 AIR CHANGES/HOUR INFILTRATION

Building or Room Type	No. of Walls[a] with Windows or Exterior Doors	Winter (15 mph Winds)			Summer (7.5 mph Winds)		
		Loose	Medium	Tight	Loose	Medium	Tight
A	None	0.8	0.5	0.3	0.4	0.3	0.3
	1	1.3	1.0	0.7	0.6	0.5	0.4
	2	2.2	1.5	1.0	1.0	0.8	0.6
	3–4	3.0	2.0	1.3	1.2	1.0	0.8
B	Any	3.0	2.0	1.3	1.2	1.0	0.8
C	Any	3.0	2.0	1.3	1.2	1.0	0.8
D	Any	1.5	1.0	0.7	0.6	0.5	0.4
E	Any	3.0	1.5	1.0	1.0	0.8	0.6

A = Multistory offices, apartments, hotels, etc.

B = Entrance halls or vestibules

C = Industrial buildings

D = Houses (except vestibules)

E = Public or institutional buildings

Tight = New buildings where there is close supervision of workmanship and special precautions are taken to prevent infiltration; tight windows and doors.

Medium = Conventional construction procedures; average fitting windows and doors.

Loose = Buildings constructed with poor workmanship or older buildings where joints have separated; loose-fitting windows and doors.

[a]Number of walls in individual rooms with windows or exterior doors. Does not refer to number of walls of overall building with windows or doors.

TABLE B6.3 OUTDOOR AIR REQUIREMENTS FOR VENTILATION[a] COMMERCIAL FACILITIES (OFFICES, STORES, SHOPS, HOTELS, SPORTS FACILITIES)

Application	Estimated Maximum[b] Occupancy (P/1,000 ft² or 100 m²)	Outdoor Air Requirements				Comments
		cfm/ person	L/s· person	cfm/ft²	L/s·m²	
Dry Cleaners, Laundries						Dry-cleaning processes may require more air.
Commercial laundry	10	25	13			
Commercial dry cleaner	30	30	15			
Storage, pick up	30	35	18			
Coin-operated laundries	20	15	8			
Coin-operated dry cleaner	20	15	8			
Food and Beverage Service						
Dining rooms	70	20	10			
Cafeteria, fast food	100	20	10			
Bars, cocktail lounges	100	30	15			Supplementary smoke-removal equipment may be required.
Kitchens (cooking)	20	15	8			Makeup air for hood exhaust may require more ventilating air. The sum of the outdoor air and transfer air of acceptable quality from adjacent spaces shall be sufficient to provide an exhaust rate of not less than 1.5 cfm/ft² (7.5 L/s·m²).
Garages, Repair, Service Stations						
Enclosed parking garage				1.50	7.5	Distribution among people must consider worker location and concentration of running engines; stands where engines are run must incorporate systems for positive engine exhaust withdrawal. Contaminant sensors may be used to control ventilation.
Auto repair rooms				1.50	7.5	
Hotels, Motels, Resorts, Dormitories				cfm/room	L/s·room	Independent of room size.
Bedrooms				30	15	
Living rooms				30	15	
Baths				35	18	Installed capacity for intermittent use.
Lobbies	30	15	8			
Conference rooms	50	20	10			
Assembly rooms	120	15	8			
Dormitory sleeping areas	20	15	8			See also food and beverage services, merchandising, barber and beauty shops, garages.
Gambling casinos	120	30	15			Supplementary smoke-removal equipment may be required.
Offices						
Office space	7	20	10			Some office equipment may require local exhaust.
Reception areas	60	15	8			
Telecommunication centers and data entry areas	60	20	10			
Conference rooms	50	20	10			Supplementary smoke-removal equipment may be required.
Public Spaces				cfm/ft²	L/s·m²	
Corridors and utilities				0.05	0.25	
Public restrooms, cfm/wc or urinal		50	25			Mechanical exhaust with no recirculation is recommended.
Locker and dressing rooms				0.5	2.5	
Smoking lounge	70	60	30			Normally supplied by transfer air, local mechanical exhaust, with no recirculation recommended.
Elevators				1.00	5.0	Normally supplied by transfer air.

[a] Table B6.3 prescribes supply rates of acceptable outdoor air required for acceptable indoor air quality. These values have been chosen to control carbon dioxide and other contaminants with an adequate margin of safety and to account for health variations among people, varied activity levels, and a moderate amount of smoking.

[b] Net occupiable space.

TABLE B6.3 OUTDOOR AIR REQUIREMENTS FOR VENTILATION[a]
COMMERCIAL FACILITIES (OFFICES, STORES, SHOPS, HOTELS, SPORTS FACILITIES) *(Continued)*

Application	Estimated Maximum[b] Occupancy (P/1,000 ft² or 100 m²)	Outdoor Air Requirements				Comments
		cfm/ person	L/s· person	cfm/ft²	L/s·m²	
Retail Stores, Sales Floors, and Showroom Floors						
Basement and street	30			0.30	1.50	
Upper floors	20			0.20	1.00	
Storage rooms	15			0.15	0.75	
Dressing rooms				0.20	1.00	
Malls and arcades	20			0.20	1.00	
Shipping and receiving	10			0.15	0.75	
Warehouses	5			0.05	0.25	
Smoking lounge	70	60	30			Normally supplied by transfer air, local mechanical exhaust; exhaust with no recirculation recommended.
Specialty Shops						
Barber	25	15	8			
Beauty	25	25	13			
Reducing salons	20	15	8			
Florists	8	15	8			Ventilation to optimize plant growth may dictate requirements.
Clothiers, furniture				0.30	1.50	
Hardware, drugs, fabric	8	15	8			
Supermarkets	8	15	8			
Pet shops				1.00	5.00	
Sports and Amusement						
Spectator areas	150	15	8			When internal combustion engines are operated for maintenance of playing surfaces, increased ventilation rates may be required.
Game rooms	70	25	13			
Ice arenas (playing areas)				0.50	2.50	
Swimming pools (pool and deck area)				0.50	2.50	Higher values may be required for humidity control.
Playing floors (gymnasium)	30	20	10			
Ballrooms and discos	100	25	13			
Bowling alleys (seating areas)	70	25	13			
Theaters						Special ventilation will be needed to eliminate special stage effects (e.g., dry ice vapors, mists.)
Ticket booths	60	20	10			
Lobbies	150	20	10			
Auditorium	150	15	8			
Stages, studios	70	15	8			
Transportation						Ventilation within vehicles may require special considerations.
Waiting rooms	100	15	8			
Platforms	100	15	8			
Vehicles	150	15	8			
Workrooms						
Meat processing	10	15	8			Spaces maintained at low temperatures (−10°F to +50°F, or −23°C to +10°C) are not covered by these requirements unless the occupancy is continuous. Ventilation from adjoining spaces is permissible. When the occupancy is intermittent, infiltration will normally exceed the ventilation requirement.
Photo studios	10	15	8			
Darkrooms	10			0.50	2.50	
Pharmacy	20	15	8			
Bank vaults	5	15	8			

TABLE B6.3   OUTDOOR AIR REQUIREMENTS FOR VENTILATION[a]
COMMERCIAL FACILITIES (OFFICES, STORES, SHOPS, HOTELS, SPORTS FACILITIES) *(Continued)*

Application	Estimated Maximum[b] Occupancy ($P/1{,}000$ ft² or 100 m²)	Outdoor Air Requirements				Comments
		cfm/ person	L/s· person	cfm/ft²	L/s·m²	
Duplicating, printing				0.50	2.50	Installed equipment must incorporate positive exhaust and control (as required) of undesirable contaminants (toxic or otherwise).
		INSTITUTIONAL FACILITIES				
Education						
Classroom	50	15	8			
Laboratories	30	20	10			Special contaminant control
Training shop	30	20	10			systems may be required for
Music rooms	50	15	8			processes or functions including
Libraries	20	15	8			laboratory animal occupancy.
Locker rooms				0.50	2.50	
Corridors				0.10	0.50	
Auditoriums	150	15	8			
Smoking lounges	70	60	30			Normally supplied by transfer air. Local mechanical exhaust with no recirculation recommended.
Hospitals, Nursing and Convalescent Homes						
Patient rooms	10	25	13			Special requirements or codes and
Medical procedure	20	15	8			pressure relationships may deter-
Operating rooms	20	30	15			mine minimum ventilation rates
Recovery and ICU	20	15	8			and filter efficiency. Procedures generating contaminants may require higher rates.
Autopsy rooms				0.50	2.50	Air shall not be recirculated into other spaces.
Physical Therapy	20	15	8			
Correctional Facilities						
Cells	20	20	10			
Dining halls	100	15	8			
Guard stations	40	15	8			

[a] Table 2 prescribes supply rates of acceptable outdoor air required for acceptable indoor air quality. These values have been chosen to control carbon dioxide and other contaminants with an adequate margin of safety and to account for health variations among people, varied activity levels, and a moderate amount of smoking.

[b] Net occupiable space.

Source: © 1989. Reprinted by permission from ASHRAE Standard 62-1989 *Ventilation for Acceptable Indoor Air Quality.*

TABLE B6.4 OUTDOOR REQUIREMENTS FOR VENTILATION OF RESIDENTIAL FACILITIES[a] (PRIVATE DWELLINGS, SINGLE, MULTIPLE)

Applications	Outdoor Requirements	Comments
Living areas	0.35 air changes per hour but not less than 15 cfm (7.5 L/s) per person	For calculating the air changes per hour, the volume of the living spaces shall include all areas within the conditioned space. The ventilation is normally satisfied by infiltration and natural ventilation. Dwellings with tight enclosures may require supplemental ventilation supply for fuel-burning appliances, including fireplaces and mechanically exhausted appliances. Occupant loading shall be based on the number of bedrooms as follows: first bedroom, two persons; each additional bedroom, one person. Where higher occupant loadings are known, they shall be used.
Kitchens[b]	100 cfm (50 L/s) intermittent or 25 cfm (12 L/s) continuous or openable windows	Installed mechanical exhaust capacity.[c] Climatic conditions may affect choice of the ventilation system.
Baths, Toilets[b]	50 cfm (25 L/s) intermittent or 20 cfm (10 L/s) continuous or openable windows	Installed mechanical exhaust capacity.[c]
Garages: Separate for each dwelling unit	100 cfm (50 L/s) per car	Normally satisfied by infiltration or natural ventilation.
Common for several units	1.5 cfm/ft² (7.5 L/s·m²)	See "Enclosed parking garages."

[a] In using this table, the outdoor air is assumed to be acceptable.

[b] Climatic conditions may affect choice of ventilation option chosen.

[c] The air exhausted from kitchens, bath, and toilet rooms may utilize air supplied through adjacent living areas to compensate for the air exhausted. The air supplied shall be of sufficient quantities to meet the requirements of this table.

Source: © 1989. Reprinted by permission from ASHRAE Standard 62-1989 *Ventilation for Acceptable Indoor Air Quality.*

TABLE B7.1 AVERAGE MONTHLY AND YEARLY DEGREE DAYS FOR CITIES IN THE UNITED STATES AND CANADA[a,b,c] (BASE 65°F)

State	Station	Avg. Winter Temp[d]	July	Aug.	Sept.	Oct.	Nov.	Dec.	Jan.	Feb.	Mar.	Apr.	May	June	Yearly Total
Ala.	Birmingham .... A	54.2	0	0	6	93	363	555	592	462	363	108	9	0	2551
	Huntsville........................... A	51.3	0	0	12	127	426	663	694	557	434	138	19	0	3070
	Mobile.................................. A	59.9	0	0	0	22	213	357	415	300	211	42	0	0	1560
	Montgomery........................ A	55.4	0	0	0	68	330	527	543	417	316	90	0	0	2291
Alaska	Anchorage .... A	23.0	245	291	516	930	1284	1572	1631	1316	1293	879	592	315	10864
	Fairbanks .... A	6.7	171	332	642	1203	1833	2254	2359	1901	1739	1068	555	222	14279
	Juneau.................................. A	32.1	301	338	483	725	921	1135	1237	1070	1073	810	601	381	9075
	Nome .... A	13.1	481	496	693	1094	1455	1820	1879	1666	1770	1314	930	573	14171
Ariz.	Flagstaff .... A	35.6	46	68	201	558	867	1073	1169	991	911	651	437	180	7152
	Phoenix .... A	58.5	0	0	0	22	234	415	474	328	217	75	0	0	1765
	Tucson .... A	58.1	0	0	0	25	231	406	471	344	242	75	6	0	1800
	Winslow .... A	43.0	0	0	6	245	711	1008	1054	770	601	291	96	0	4782
	Yuma .... A	64.2	0	0	0	0	108	264	307	190	90	15	0	0	974
Ark.	Fort Smith .... A	50.3	0	0	12	127	450	704	781	596	456	144	22	0	3292
	Little Rock .... A	50.5	0	0	9	127	465	716	756	577	434	126	9	0	3219
	Texarkana .... A	54.2	0	0	0	78	345	561	626	468	350	105	0	0	2533
Calif.	Bakersfield .... A	55.4	0	0	0	37	282	502	546	364	267	105	19	0	2122
	Bishop .... A	46.0	0	0	48	260	576	797	874	680	555	306	143	36	4275
	Blue Canyon........................ A	42.2	28	37	108	347	594	781	896	795	806	597	412	195	5596
	Burbank .... A	58.6	0	0	6	43	177	301	366	277	239	138	81	18	1646
	Eureka C	49.9	270	257	258	329	414	499	546	470	505	438	372	285	4643
	Fresno.................................. A	53.3	0	0	0	84	354	577	605	426	335	162	62	6	2611
	Long Beach .... A	57.8	0	0	9	47	171	316	397	311	264	171	93	24	1803
	Los Angeles .... A	57.4	28	28	42	78	180	291	372	302	288	219	158	81	2061
	Los Angeles C	60.3	0	0	6	31	132	229	310	230	202	123	68	18	1349
	Mt. Shasta C	41.2	25	34	123	406	696	902	983	784	738	525	347	159	5722
	Oakland .... A	53.5	53	50	45	127	309	481	527	400	353	255	180	90	2870
	Red Bluff .... A	53.8	0	0	0	53	318	555	605	428	341	168	47	0	2515
	Sacramento .... A	53.9	0	0	0	56	321	546	583	414	332	178	72	0	2502
	Sacramento.......................... C	54.4	0	0	0	62	312	533	561	392	310	173	76	0	2419
	Sandberg C	46.8	0	0	30	202	480	691	778	661	620	426	264	57	4209
	San Diego............................. A	59.5	9	0	21	43	135	236	298	235	214	135	90	42	1458
	San Francisco....................... A	53.4	81	78	60	143	306	462	508	395	363	279	214	126	3015
	San Francisco....................... C	55.1	192	174	102	118	231	388	443	336	319	279	239	180	3001
	Santa Maria .... A	54.3	99	93	96	146	270	391	459	370	363	282	233	165	2967
Colo.	Alamosa .... A	29.7	65	99	279	639	1065	1420	1476	1162	1020	696	440	168	8529
	Colorado Springs .... A	37.3	9	25	132	456	825	1032	1128	938	893	582	319	84	6423
	Denver .... A	37.6	6	9	117	428	819	1035	1132	938	887	558	288	66	6283
	Denver C	40.8	0	0	90	366	714	905	1004	851	800	492	254	48	5524
	Grand Junction .... A	39.3	0	0	30	313	786	1113	1209	907	729	387	146	21	5641
	Pueblo.................................. A	40.4	0	0	54	326	750	986	1085	871	772	429	174	15	5462
Conn.	Bridgeport .... A	39.9	0	0	66	307	615	986	1079	966	853	510	208	27	5617
	Hartford .... A	37.3	0	12	117	394	714	1101	1190	1042	908	519	205	33	6235
	New Haven .... A	39.0	0	12	87	347	648	1011	1097	991	871	543	245	45	5897
Del.	Wilmington........................... A	42.5	0	0	51	270	588	927	980	874	735	387	112	6	4930
D.C.	Washington .... A	45.7	0	0	33	217	519	834	871	762	626	288	74	0	4224
Fla.	Apalachicola......................... C	61.2	0	0	0	16	153	319	347	260	180	33	0	0	1308
	Daytona Beach...................... A	64.5	0	0	0	0	75	211	248	190	140	15	0	0	879
	Fort Myers .... A	68.6	0	0	0	0	24	109	146	101	62	0	0	0	442
	Jacksonville .... A	61.9	0	0	0	12	144	310	332	246	174	21	0	0	1239
	Key West .... A	73.1	0	0	0	0	0	28	40	31	9	0	0	0	108
	Lakeland............................... C	66.7	0	0	0	0	57	164	195	146	99	0	0	0	661
	Miami................................... A	71.1	0	0	0	0	0	65	74	56	19	0	0	0	214

[a]Data for United States cities from a publication of the United States Weather Bureau, *Monthly Normals of Temperature, Precipitation and Heating Degree Days,* 1962, are for the period 1931 to 1960 inclusive. These data also include information from the 1963 revisions to this publication, where available.
[b]Data for airport stations, A, and city stations, C, are both given where available.
[c]Data for Canadian cities were computed by the Climatology Division, Department of Transport from normal monthly mean temperatures, and the monthly values of heating degree days data were obtained using the National Research Council computer and a method devised by H. C. S. Thom of the United States Weather Bureau. The heating degree days are based on the period from 1931 to 1960.
[d]For period October to April, inclusive.

State	Station		Avg. Winter Temp[d]	July	Aug.	Sept.	Oct.	Nov.	Dec.	Jan.	Feb.	Mar.	Apr.	May	June	Yearly Total
Fla. (Cont'd)	Miami Beach	C	72.5	0	0	0	0	0	40	56	36	9	0	0	0	141
	Orlando	A	65.7	0	0	0	0	72	198	220	165	105	6	0	0	766
	Pensacola	A	60.4	0	0	0	19	195	353	400	277	183	36	0	0	1463
	Tallahassee	A	60.1	0	0	0	28	198	360	375	286	202	36	0	0	1485
	Tampa	A	66.4	0	0	0	0	0	60	171	202	148	102	0	0	683
	West Palm Beach	A	68.4	0	0	0	0	6	65	87	64	31	0	0	0	253
Ga.	Athens	A	51.8	0	0	12	115	405	632	642	529	431	141	22	0	2929
	Atlanta	A	51.7	0	0	18	124	417	648	636	518	428	147	25	0	2961
	Augusta	A	54.5	0	0	0	78	333	552	549	445	350	90	0	0	2397
	Columbus	A	54.8	0	0	0	87	333	543	552	434	338	96	0	0	2383
	Macon	A	56.2	0	0	0	71	297	502	505	403	295	63	0	0	2136
	Rome	A	49.9	0	0	24	161	474	701	710	577	468	177	34	0	3326
	Savannah	A	57.8	0	0	0	47	246	437	437	353	254	45	0	0	1819
	Thomasville	C	60.0	0	0	0	25	198	366	394	305	208	33	0	0	1529
Hawaii	Lihue	A	72.7	0	0	0	0	0	0	0	0	0	0	0	0	0
	Honolulu	A	74.2	0	0	0	0	0	0	0	0	0	0	0	0	0
	Hilo	A	71.9	0	0	0	0	0	0	0	0	0	0	0	0	0
Idaho	Boise	A	39.7	0	0	132	415	792	1017	1113	854	722	438	245	81	5809
	Lewiston	A	41.0	0	0	123	403	756	933	1063	815	694	426	239	90	5542
	Pocatello	A	34.8	0	0	172	493	900	1166	1324	1058	905	555	319	141	7033
Ill.	Cairo	C	47.9	0	0	36	164	513	791	856	680	539	195	47	0	3821
	Chicago (O'Hare)	A	35.8	0	12	117	381	807	1166	1265	1086	939	534	260	72	6639
	Chicago (Midway)	A	37.5	0	0	81	326	753	1113	1209	1044	890	480	211	48	6155
	Chicago	C	38.9	0	0	66	279	705	1051	1150	1000	868	489	226	48	5882
	Moline	A	36.4	0	9	99	335	774	1181	1314	1100	918	450	189	39	6408
	Peoria	A	38.1	0	6	87	326	759	1113	1218	1025	849	426	183	33	6025
	Rockford	A	34.8	6	9	114	400	837	1221	1333	1137	961	516	236	60	6830
	Springfield	A	40.6	0	0	72	291	696	1023	1135	935	769	354	136	18	5429
Ind.	Evansville	A	45.0	0	0	66	220	606	896	955	767	620	237	68	0	4435
	Fort Wayne	A	37.3	0	9	105	378	783	1135	1178	1028	890	471	189	39	6205
	Indianapolis	A	39.6	0	0	90	316	723	1051	1113	949	809	432	177	39	5699
	South Bend	A	36.6	0	6	111	372	777	1125	1221	1070	933	525	239	60	6439
Iowa	Burlington	A	37.6	0	0	93	322	768	1135	1259	1042	859	426	177	33	6114
	Des Moines	A	35.5	0	6	96	363	828	1225	1370	1137	915	438	180	30	6588
	Dubuque	A	32.7	12	31	156	450	906	1287	1420	1204	1026	546	260	78	7376
	Sioux City	A	34.0	0	9	108	369	867	1240	1435	1198	989	483	214	39	6951
	Waterloo	A	32.6	12	19	138	428	909	1296	1460	1221	1023	531	229	54	7320
Kans.	Concordia	A	40.4	0	0	57	276	705	1023	1163	935	781	372	149	18	5479
	Dodge City	A	42.5	0	0	33	251	666	939	1051	840	719	354	124	9	4986
	Goodland	A	37.8	0	6	81	381	810	1073	1166	955	884	507	236	42	6141
	Topeka	A	41.7	0	0	57	270	672	980	1122	893	722	330	124	12	5182
	Wichita	A	44.2	0	0	33	229	618	905	1023	804	645	270	87	6	4620
Ky.	Covington	A	41.4	0	0	75	291	669	983	1035	893	756	390	149	24	5265
	Lexington	A	43.8	0	0	54	239	609	902	946	818	685	325	105	0	4683
	Louisville	A	44.0	0	0	54	248	609	890	930	818	682	315	105	9	4660
La.	Alexandria	A	57.5	0	0	0	56	273	431	471	361	260	69	0	0	1921
	Baton Rouge	A	59.8	0	0	0	31	216	369	409	294	208	33	0	0	1560
	Lake Charles	A	60.5	0	0	0	19	210	341	381	274	195	39	0	0	1459
	New Orleans	A	61.0	0	0	0	19	192	322	363	258	192	39	0	0	1385
	New Orleans	C	61.8	0	0	0	12	165	291	344	241	177	24	0	0	1254
	Shreveport	A	56.2	0	0	0	47	297	477	552	426	304	81	0	0	2184
Me.	Caribou	A	24.4	78	115	336	682	1044	1535	1690	1470	1308	858	468	183	9767
	Portland	A	33.0	12	53	195	508	807	1215	1339	1182	1042	675	372	111	7511
Md.	Baltimore	A	43.7	0	0	48	264	585	905	936	820	679	327	90	0	4654
	Baltimore	C	46.2	0	0	27	189	486	806	859	762	629	288	65	0	4111
	Frederick	A	42.0	0	0	66	307	624	955	995	876	741	384	127	12	5087
Mass.	Boston	A	40.0	0	9	60	316	603	983	1088	972	846	513	208	36	5634
	Nantucket	A	40.2	12	22	93	332	573	896	992	941	896	621	384	129	5891
	Pittsfield	A	32.6	25	59	219	524	831	1231	1339	1196	1063	660	326	105	7578
	Worcester	A	34.7	6	34	147	450	774	1172	1271	1123	998	612	304	78	6969

TABLE B7.1 AVERAGE MONTHLY AND YEARLY DEGREE DAYS FOR CITIES IN THE UNITED STATES AND CANADA[a,b,c] **(BASE 65°F)** (Continued)

State	Station	Avg. Winter Temp[d]	July	Aug.	Sept.	Oct.	Nov.	Dec.	Jan.	Feb.	Mar.	Apr.	May	June	Yearly Total
Mich.	Alpena .... A	29.7	68	105	273	580	912	1268	1404	1299	1218	777	446	156	8506
	Detroit (City) .... A	37.2	0	0	87	360	738	1088	1181	1058	936	522	220	42	6232
	Detroit (Wayne) .... A	37.1	0	0	96	353	738	1088	1194	1061	933	534	239	57	6293
	Detroit (Willow Run) .... A	37.2	0	0	90	357	750	1104	1190	1053	921	519	229	45	6258
	Escanaba C	29.6	59	87	243	539	924	1293	1445	1296	1203	777	456	159	8481
	Flint .... A	33.1	16	40	159	465	843	1212	1330	1198	1066	639	319	90	7377
	Grand Rapids..................... A	34.9	9	28	135	434	804	1147	1259	1134	1011	579	279	75	6894
	Lansing .... A	34.8	6	22	138	431	813	1163	1262	1142	1011	579	273	69	6909
	Marquette C	30.2	59	81	240	527	936	1268	1411	1268	1187	771	468	177	8393
	Muskegon .... A	36.0	12	28	120	400	762	1088	1209	1100	995	594	310	78	6696
	Sault Ste. Marie .... A	27.7	96	105	279	580	951	1367	1525	1380	1277	810	477	201	9048
Minn.	Duluth .... A	23.4	71	109	330	632	1131	1581	1745	1518	1355	840	490	198	10000
	Minneapolis .... A	28.3	22	31	189	505	1014	1454	1631	1380	1166	621	288	81	8382
	Rochester .... A	28.8	25	34	186	474	1005	1438	1593	1366	1150	630	301	93	8295
Miss.	Jackson .... A	55.7	0	0	0	65	315	502	546	414	310	87	0	0	2239
	Meridian .... A	55.4	0	0	0	81	339	518	543	417	310	81	0	0	2289
	Vicksburg C	56.9	0	0	0	53	279	462	512	384	282	69	0	0	2041
Mo.	Columbia .... A	42.3	0	0	54	251	651	967	1076	874	716	324	121	12	5046
	Kansas City .... A	43.9	0	0	39	220	612	905	1032	818	682	294	109	0	4711
	St. Joseph .... A	40.3	0	6	60	285	708	1039	1172	949	769	348	133	15	5484
	St. Louis .... A	43.1	0	0	60	251	627	936	1026	848	704	312	121	15	4900
	St. Louis C	44.8	0	0	36	202	576	884	977	801	651	270	87	0	4484
	Springfield......................... A	44.5	0	0	45	223	600	877	973	781	660	291	105	6	4900
Mont.	Billings .... A	34.5	6	15	186	487	897	1135	1296	1100	970	570	285	102	7049
	Glasgow .... A	26.4	31	47	270	608	1104	1466	1711	1439	1187	648	335	150	8996
	Great Falls .... A	32.8	28	53	258	543	921	1169	1349	1154	1063	642	384	186	7750
	Havre .... A	28.1	28	53	306	595	1065	1367	1584	1364	1181	657	338	162	8700
	Havre C	29.8	19	37	252	539	1014	1321	1528	1305	1116	612	304	135	8182
	Helena............................... A	31.1	31	59	294	601	1002	1265	1438	1170	1042	651	381	195	8129
	Kalispell .... A	31.4	50	99	321	654	1020	1240	1401	1134	1029	639	397	207	8191
	Miles City .... A	31.2	6	6	174	502	972	1296	1504	1252	1057	579	276	99	7723
	Missoula .... A	31.5	34	74	303	651	1035	1287	1420	1120	970	621	391	219	8125
Neb.	Grand Island .... A	36.0	0	6	108	381	834	1172	1314	1089	908	462	211	45	6530
	Lincoln C	38.8	0	6	75	301	726	1066	1237	1016	834	402	171	30	5864
	Norfolk .... A	34.0	9	0	111	397	873	1234	1414	1179	983	498	233	48	6979
	North Platte .... A	35.5	0	6	123	440	885	1166	1271	1039	930	519	248	57	6684
	Omaha .... A	35.6	0	12	105	357	828	1175	1355	1126	939	465	208	42	6612
	Scottsbluff .... A	35.9	0	0	138	459	876	1128	1231	1008	921	552	285	75	6673
	Valentine........................... A	32.6	9	12	165	493	942	1237	1395	1176	1045	579	288	84	7425
Nev.	Elko .... A	34.0	9	34	225	561	924	1197	1314	1036	911	621	409	192	7433
	Ely .... A	33.1	28	43	234	592	939	1184	1308	1075	977	672	456	225	7733
	Las Vegas.......................... A	53.5	0	0	0	78	387	617	688	487	335	111	6	0	2709
	Reno .... A	39.3	43	87	204	490	801	1026	1073	823	729	510	357	189	6332
	Winnemucca .... A	36.7	0	34	210	536	876	1091	1172	916	837	573	363	153	6761
N.H.	Concord .... A	33.0	6	50	177	505	822	1240	1358	1184	1032	636	298	75	7383
	Mt. Washington Obsv............	15.2	493	536	720	1057	1341	1742	1820	1663	1652	1260	930	603	13817
N.J.	Atlantic City .... A	43.2	0	0	39	251	549	880	936	848	741	420	133	15	4812
	Newark .... A	42.8	0	0	30	248	573	921	983	876	729	381	118	0	4589
	Trenton.............................. C	42.4	0	0	57	264	576	924	989	885	753	399	121	12	4980
N.M.	Albuquerque....................... A	45.0	0	0	12	229	642	868	930	703	595	288	81	0	4348
	Clayton .... A	42.0	0	6	66	310	699	899	986	812	747	429	183	21	5158
	Raton .... A	38.1	9	28	126	431	825	1048	1116	904	834	543	301	63	6228
	Roswell .... A	47.5	0	0	18	202	573	806	840	641	481	201	31	0	3793
	Silver City .... A	48.0	0	0	6	183	525	729	791	605	518	261	87	0	3705
N.Y.	Albany .... A	34.6	0	19	138	440	777	1194	1311	1156	992	564	239	45	6875
	Albany C	37.2	0	9	102	375	699	1104	1218	1072	908	498	186	30	6201
	Binghamton .... A	33.9	22	65	201	471	810	1184	1277	1154	1045	645	313	99	7286
	Binghamton C	36.6	0	28	141	406	732	1107	1190	1081	949	543	229	45	6451
	Buffalo .... A	34.5	19	37	141	440	777	1156	1256	1145	1039	645	329	78	7062
	New York (Cent. Park)....... C	42.8	0	0	30	233	540	902	986	885	760	408	118	9	4871
	New York (La Guardia) ...... A	43.1	0	0	27	223	528	887	973	879	750	414	124	6	4811

State	Station	Avg. Winter Temp[d]	July	Aug.	Sept.	Oct.	Nov.	Dec.	Jan.	Feb.	Mar.	Apr.	May	June	Yearly Total
	New York (Kennedy) .... A	41.4	0	0	36	248	564	933	1029	935	815	480	167	12	5219
	Rochester .... A	35.4	9	31	126	415	747	1125	1234	1123	1014	597	279	48	6748
	Schenectady C	35.4	0	22	123	422	756	1159	1283	1131	970	543	211	30	6650
	Syracuse .... A	35.2	6	28	132	415	744	1153	1271	1140	1004	570	248	45	6756
N.C.	Asheville................................ C	46.7	0	0	48	245	555	775	784	683	592	273	87	0	4042
	Cape Hatteras	53.3	0	0	0	78	273	521	580	518	440	177	25	0	2612
	Charlotte............................. A	50.4	0	0	6	124	438	691	691	582	481	156	22	0	3191
	Greensboro.......................... A	47.5	0	0	33	192	513	778	784	672	552	234	47	0	3805
	Raleigh .... A	49.4	0	0	21	164	450	716	725	616	487	180	34	0	3393
	Wilmington.......................... A	54.6	0	0	0	74	291	521	546	462	357	96	0	0	2347
	Winston-Salem .... A	48.4	0	0	21	171	483	747	753	652	524	207	37	0	3595
N.D.	Bismarck............................. A	26.6	34	28	222	577	1083	1463	1708	1442	1203	645	329	117	8851
	Devils Lake C	22.4	40	53	273	642	1191	1634	1872	1579	1345	753	381	138	9901
	Fargo .... A	24.8	28	37	219	574	1107	1569	1789	1520	1262	690	332	99	9226
	Williston .... A	25.2	31	43	261	601	1122	1513	1758	1473	1262	681	357	141	9243
Ohio	Akron-Canton .... A	38.1	0	9	96	381	726	1070	1138	1016	871	489	202	39	6037
	Cincinnati........................... C	45.1	0	0	39	208	558	862	915	790	642	294	96	6	4410
	Cleveland .... A	37.2	9	25	105	384	738	1088	1159	1047	918	552	260	66	6351
	Columbus .... A	39.7	0	6	84	347	714	1039	1088	949	809	426	171	27	5660
	Columbus C	41.5	0	0	57	285	651	977	1032	902	760	396	136	15	5211
	Dayton .... A	39.8	0	6	78	310	696	1045	1097	955	809	429	167	30	5622
	Mansfield .... A	36.9	9	22	114	397	768	1110	1169	1042	924	543	245	60	6403
	Sandusky C	39.1	0	6	66	313	684	1032	1107	991	868	495	198	36	5796
	Toledo .... A	36.4	0	16	117	406	792	1138	1200	1056	924	543	242	60	6494
	Youngstown .... A	36.8	6	19	120	412	771	1104	1169	1047	921	540	248	60	6417
Okla.	Oklahoma City..................... A	48.3	0	0	15	164	498	766	868	664	527	189	34	0	3725
	Tulsa .... A	47.7	0	0	18	158	522	787	893	683	539	213	47	0	3860
Ore.	Astoria .... A	45.6	146	130	210	375	561	679	753	622	636	480	363	231	5186
	Burns C	35.9	12	37	210	515	867	1113	1246	988	856	570	366	177	6957
	Eugene .... A	45.6	34	34	129	366	585	719	803	627	589	426	279	135	4726
	Meacham .... A	34.2	84	124	288	580	918	1091	1209	1005	983	726	527	339	7874
	Medford .... A	43.2	0	0	78	372	678	871	918	697	642	432	242	78	5008
	Pendleton .... A	42.6	0	0	111	350	711	884	1017	773	617	396	205	63	5127
	Portland .... A	45.6	25	28	114	335	597	735	825	644	586	396	245	105	4635
	Portland C	47.4	12	16	75	267	534	679	769	594	536	351	198	78	4109
	Roseburg............................. A	46.3	22	16	105	329	567	713	766	608	570	405	267	123	4491
	Salem .... A	45.4	37	31	111	338	594	729	822	647	611	417	273	144	4754
Pa.	Allentown............................ A	38.9	0	0	90	353	693	1045	1116	1002	849	471	167	24	5810
	Erie .... A	36.8	0	25	102	391	714	1063	1169	1081	973	585	288	60	6451
	Harrisburg .... A	41.2	0	0	63	298	648	992	1045	907	766	396	124	12	5251
	Philadelphia .... A	41.8	0	0	60	297	620	965	1016	889	747	392	118	40	5144
	Philadelphia C	44.5	0	0	30	205	513	856	924	823	691	351	93	0	4486
	Pittsburgh........................... A	38.4	0	9	105	375	726	1063	1119	1002	874	480	195	39	5987
	Pittsburgh........................... C	42.2	0	0	60	291	615	930	983	885	763	390	124	12	5053
	Reading............................... C	42.4	0	0	54	257	597	939	1001	885	735	372	105	0	4945
	Scranton .... A	37.2	0	19	132	434	762	1104	1156	1028	893	498	195	33	6254
	Williamsport .... A	38.5	0	9	111	375	717	1073	1122	1002	856	468	177	24	5934
R.I.	Block Island .... A	40.1	0	16	78	307	594	902	1020	955	877	612	344	99	5804
	Providence .... A	38.8	0	16	96	372	660	1023	1110	988	868	534	236	51	5954
S.C.	Charleston .... A	56.4	0	0	0	59	282	471	487	389	291	54	0	0	2033
	Charleston C	57.9	0	0	0	34	210	425	443	367	273	42	0	0	1794
	Columbia .... A	54.0	0	0	0	84	345	577	570	470	357	81	0	0	2484
	Florence .... A	54.5	0	0	0	78	315	552	552	459	347	84	0	0	2387
	Greenville-Spartenburg .... A	51.6	0	0	6	121	399	651	660	546	446	132	19	0	2980
S.D.	Huron .... A	28.8	9	12	165	508	1014	1432	1628	1355	1125	600	288	87	8223
	Rapid City .... A	33.4	22	12	165	481	897	1172	1333	1145	1051	615	326	126	7345
	Sioux Falls .... A	30.6	19	25	168	462	972	1361	1544	1285	1082	573	270	78	7839
Tenn.	Bristol................................. A	46.2	0	0	51	236	573	828	828	700	598	261	68	0	4143
	Chattanooga .... A	50.3	0	0	18	143	468	698	722	577	453	150	25	0	3254
	Knoxville .... A	49.2	0	0	30	171	489	725	732	613	493	198	43	0	3494
	Memphis .... A	50.5	0	0	18	130	447	698	729	585	456	147	22	0	3232

TABLE B7.1 AVERAGE MONTHLY AND YEARLY DEGREE DAYS FOR CITIES IN THE UNITED STATES AND CANADA[a,b,c] (BASE 65°F) *(Continued)*

State or Prov.	Station	Avg. Winter Temp[d]	July	Aug.	Sept.	Oct.	Nov.	Dec.	Jan.	Feb.	Mar.	Apr.	May	June	Yearly Total
	Memphis C	51.6	0	0	12	102	396	648	710	568	434	129	16	0	3015
	Nashville .... A	48.9	0	0	30	158	495	732	778	644	512	189	40	0	3578
	Oak Ridge C	47.7	0	0	39	192	531	772	778	669	552	228	56	0	3817
Tex.	Abilene .... A	53.9	0	0	0	99	366	586	642	470	347	114	0	0	2624
	Amarillo .... A	47.0	0	0	18	205	570	797	877	664	546	252	56	0	3985
	Austin .... A	59.1	0	0	0	31	225	388	468	325	223	51	0	0	1711
	Brownsville .... A	67.7	0	0	0	0	66	149	205	106	74	0	0	0	600
	Corpus Christi .... A	64.6	0	0	0	0	120	220	291	174	109	0	0	0	914
	Dallas .... A	55.3	0	0	0	62	321	524	601	440	319	90	6	0	2363
	El Paso .... A	52.9	0	0	0	84	414	648	685	445	319	105	0	0	2700
	Fort Worth .... A	55.1	0	0	0	65	324	536	614	448	319	99	0	0	2405
	Galveston .... A	62.2	0	0	0	6	147	276	360	263	189	33	0	0	1274
	Galveston C	62.0	0	0	0	0	138	270	350	258	189	30	0	0	1235
	Houston .... A	61.0	0	0	0	6	183	307	384	288	192	36	0	0	1396
	Houston C	62.0	0	0	0	0	165	288	363	258	174	30	0	0	1278
	Laredo .... A	66.0	0	0	0	0	105	217	267	134	74	0	0	0	797
	Lubbock .... A	48.8	0	0	18	174	513	744	800	613	484	201	31	0	3578
	Midland .... A	53.8	0	0	0	87	381	592	651	468	322	90	0	0	2591
	Port Arthur .... A	60.5	0	0	0	22	207	329	384	274	192	39	0	0	1447
	San Angelo .... A	56.0	0	0	0	68	318	536	567	412	288	66	0	0	2255
	San Antonio .... A	60.1	0	0	0	31	204	363	428	286	195	39	0	0	1546
	Victoria .... A	62.7	0	0	0	6	150	270	344	230	152	21	0	0	1173
	Waco .... A	57.2	0	0	0	43	270	456	536	389	270	66	0	0	2030
	Wichita Falls .... A	53.0	0	0	0	99	381	632	698	518	378	120	6	0	2832
Utah	Milford .... A	36.5	0	0	99	443	867	1141	1252	988	822	519	279	87	6497
	Salt Lake City .... A	38.4	0	0	81	419	849	1082	1172	910	763	459	233	84	6052
	Wendover .... A	39.1	0	0	48	372	822	1091	1178	902	729	408	177	51	5778
Vt.	Burlington .... A	29.4	28	65	207	539	891	1349	1513	1333	1187	714	353	90	8269
Va.	Cape Henry C	50.0	0	0	0	112	360	645	694	633	536	246	53	0	3279
	Lynchburg .... A	46.0	0	0	51	223	540	822	849	731	605	267	78	0	4166
	Norfolk .... A	49.2	0	0	0	136	408	698	738	655	533	216	37	0	3421
	Richmond .... A	47.3	0	0	36	214	495	784	815	703	546	219	53	0	3865
	Roanoke .... A	46.1	0	0	51	229	549	825	834	722	614	261	65	0	4150
Wash.	Olympia .... A	44.2	68	71	198	422	636	753	834	675	645	450	307	177	5236
	Seattle-Tacoma .... A	44.2	56	62	162	391	633	750	828	678	657	474	295	159	5145
	Seattle C	46.9	50	47	129	329	543	657	738	599	577	396	242	117	4424
	Spokane .... A	36.5	9	25	168	493	879	1082	1231	980	834	531	288	135	6655
	Walla Walla C	43.8	0	0	87	310	681	843	986	745	589	342	177	45	4805
	Yakima .... A	39.1	0	12	144	450	828	1039	1163	868	713	435	220	69	5941
W. Va.	Charleston .... A	44.8	0	0	63	254	591	865	880	770	648	300	96	9	4476
	Elkins .... A	40.1	9	25	135	400	729	992	1008	896	791	444	198	48	5675
	Huntington .... A	45.0	0	0	63	257	585	856	880	764	636	294	99	12	4446
	Parkersburg C	43.5	0	0	60	264	606	905	942	826	691	339	115	6	4754
Wisc.	Green Bay .... A	30.3	28	50	174	484	924	1333	1494	1313	1141	654	335	99	8029
	La Crosse .... A	31.5	12	19	153	437	924	1339	1504	1277	1070	540	245	69	7589
	Madison .... A	30.9	25	40	174	474	930	1330	1473	1274	1113	618	310	102	7863
	Milwaukee .... A	32.6	43	47	174	471	876	1252	1376	1193	1054	642	372	135	7635
Wyo.	Casper .... A	33.4	6	16	192	524	942	1169	1290	1084	1020	657	381	129	7410
	Cheyenne .... A	34.2	28	37	219	543	909	1085	1212	1042	1026	702	428	150	7381
	Lander .... A	31.4	6	19	204	555	1020	1299	1417	1145	1017	654	381	153	7870
	Sheridan .... A	32.5	25	31	219	539	948	1200	1355	1154	1051	642	366	150	7680
Alta.	Banff C	—	220	295	498	797	1185	1485	1624	1364	1237	855	589	402	10551
	Calgary .... A	—	109	186	402	719	1110	1389	1575	1379	1268	798	477	291	9703
	Edmonton .... A	—	74	180	411	738	1215	1603	1810	1520	1330	765	400	222	10268
	Lethbridge .... A	—	56	112	318	611	1011	1277	1497	1291	1159	696	403	213	8644
B.C.	Kamloops .... A	—	22	40	189	546	894	1138	1314	1057	818	462	217	102	6799
	Prince George* .... A	—	236	251	444	747	1110	1420	1612	1319	1122	747	468	279	9755
	Prince Rupert C	—	273	248	339	539	708	868	936	808	812	648	493	357	7029
	Vancouver* .... A	—	81	87	219	456	657	787	862	723	676	501	310	156	5515
	Victoria* .... A	—	136	140	225	462	663	775	840	718	691	504	341	204	5699
	Victoria C	—	172	184	243	426	607	723	805	668	660	487	354	250	5579

State or Prov.	Station		Avg. Winter Temp^d	July	Aug.	Sept.	Oct.	Nov.	Dec.	Jan.	Feb.	Mar.	Apr.	May	June	Yearly Total
Man.	Brandon*	A	—	47	90	357	747	1290	1792	2034	1737	1476	837	431	198	11036
	Churchill	A	—	360	375	681	1082	1620	2248	2558	2277	2130	1569	1153	.675	16728
	The Pas	C	—	59	127	429	831	1440	1981	2232	1853	1624	969	508	228	12281
	Winnipeg	A	—	38	71	322	683	1251	1757	2008	1719	1465	813	405	147	10679
N.B.	Fredericton*	A	—	78	68	234	592	915	1392	1541	1379	1172	753	406	141	8671
	Moncton	C	—	62	105	276	611	891	1342	1482	1336	1194	789	468	171	8727
	St. John	C	—	109	102	246	527	807	1194	1370	1229	1097	756	490	249	8219
Nfld.	Argentia	A	—	260	167	294	564	750	1001	1159	1085	1091	879	707	483	8440
	Corner Brook	C	—	102	133	324	642	873	1194	1358	1283	1212	885	639	333	8978
	Gander	A	—	121	152	330	670	909	1231	1231	1370	1266	1243	939	657	366	9254
	Goose*	A	—	130	205	444	843	1227	1745	1947	1689	1494	1074	741	348	11887
	St. John's*	A	—	186	180	342	651	831	1113	1262	1170	1187	927	710	432	8991
N.W.T.	Aklavik	C	—	273	459	807	1414	2064	2530	2632	2336	2282	1674	1063	483	18017
	Fort Norman	C	—	164	341	666	1234	1959	2474	2592	2209	2058	1386	732	294	16109
	Resolution Island	C	—	843	831	900	1113	1311	1724	2021	1850	1817	1488	1181	942	16021
N.S.	Halifax	C	—	58	51	180	457	710	1074	1213	1122	1030	742	487	237	7361
	Sydney	A	—	62	71	219	518	765	1113	1262	1206	1150	840	567	276	8049
	Yarmouth	A	—	102	115	225	471	696	1029	1156	1065	1004	726	493	258	7340
Ont.	Cochrane	C	—	96	180	405	760	1233	1776	1978	1701	1528	963	570	222	11412
	Fort William	A	—	90	133	366	694	1140	1597	1792	1557	1380	876	543	237	10405
	Kapuskasing	C	—	74	171	405	756	1245	1807	2037	1735	1562	978	580	222	11572
	Kitchener	C	—	16	59	177	505	855	1234	1342	1226	1101	663	322	66	7566
	London	A	—	12	43	159	477	837	1206	1305	1198	1066	648	332	66	7349
	North Bay	C	—	37	90	267	608	990	1507	1680	1463	1277	780	400	120	9219
	Ottawa	C	—	25	81	222	567	936	1469	1624	1441	1231	708	341	90	8735
	Toronto	C	—	7	18	151	439	760	1111	1233	1119	1013	616	298	62	6827
P.E.I.	Charlottetown	C	—	40	53	198	518	804	1215	1380	1274	1169	813	496	204	8164
	Summerside	C	—	47	84	216	546	840	1246	1438	1291	1206	841	518	216	8488
Que.	Arvida	C	—	102	136	327	682	1074	1659	1879	1619	1407	891	521	231	10528
	Montreal*	A	—	9	43	165	521	882	1392	1566	1381	1175	684	316	69	8203
	Montreal	C	—	16	28	165	496	864	1355	1510	1328	1138	657	288	54	7899
	Quebec*	A	—	56	84	273	636	996	1516	1665	1477	1296	819	428	126	9372
	Quebec	C	—	40	68	243	592	972	1473	1612	1418	1228	780	400	111	8937
Sasks	Prince Albert	A	—	81	136	414	797	1368	1872	2108	1763	1559	867	446	219	11630
	Regina	A	—	78	93	360	741	1284	1711	1965	1687	1473	804	409	201	10806
	Saskatoon	C	—	56	87	372	750	1302	1758	2006	1689	1463	798	403	186	10870
Y.T.	Dawson	C	—	164	326	645	1197	1875	2415	2561	2150	1838	1068	570	258	15067
	Mayo Landing	C	—	208	366	648	1135	1794	2325	2427	1992	1665	1020	580	294	14454

*The data for these normals were from the full ten-year period 1951–1960, adjusted to the standard normal period 1931–1960.

Source: Reprinted from ASHRAE *Handbook of Fundamentals 1981* by permission of the American Society of Heating, Refrigerating and Air-Conditioning Engineers, Inc.

TABLE B7.2 DEGREE DAY MODIFICATION FACTORS

Outdoor design temp. °F	−20	−10	0	+10	+20
Factor M	0.57	0.64	0.71	0.79	0.89

Source: Reprinted from ASHRAE *Handbook of Systems 1976* by permission of the American Society of Heating, Refrigerating and Air-Conditioning Engineers, Inc.

TABLE B7.3 PART LOAD CORRECTION FACTORS

Percent oversizing	0	20	40	60	80
Factor P	1.36	1.56	1.79	2.04	2.32

P = 1.0 for electric resistance heating.

Source: Reprinted from ASHRAE *Handbook of Systems 1976* by permission of the American Society of Heating, Refrigerating and Air-Conditioning Engineers, Inc.

TABLE B7.4 EFFICIENCIES AND FUEL HEATING VALUES

Fuel	Fuel Heating Value (V)	Full Load Efficiency
Electricity	3413 Btu/kwh	
	(3,600 J/kwh)	.95
Natural gas	1,030 Btu/ft^3	.65 (old)
	(38.38 MJ/m^3)	.85 (new)
No. 2 oil	138,690 Btu/gallon	.55 (old)
	(38.66 MJ/L)	.65 (new)
No. 6 oil	149,690 Btu/gallon	
	(41.72 MJ/L)	
Butane/propane	95,475 Btu/gallon	.65 (old)
	(26.61 MJ/L)	.85 (new)
Anthracite coal	30,090,000 Btu/ton	
	(35 MJ/kg)	
Bituminous coal	25,800,000 Btu/ton	
	(30 MJ/kg)	
Coke	24,080,000 Btu/ton	
	(28 MJ/kg)	
Coal (unspecified)	24,500,000 Btu/ton	
	(28.5 MJ/kg)	
Steam	1,390 Btu/lb	
	(3.23 MJ/kg)	0.95

TABLE 7.5   APPROXIMATE POWER INPUTS

System	Compressor kW/kW (kW/ton)	Auxiliaries kW/kW (kW/ton)
Window units	0.415 (1.46)	0.091 (0.32)
Through-wall units	0.466 (1.64)	0.085 (0.30)
Dwelling unit, central air-cooled	0.423 (1.49)	0.040 (0.14)
Central, group, or bldg. cooling plants		
(3 to 25 tons) air-cooled	0.341 (1.20)	0.057 (0.20)
(25 to 100 tons) air-cooled	0.330 (1.18)	0.060 (0.21)
(25 to 100 tons) water-cooled	0.267 (0.94)	0.048 (0.17)
(Over 100 tons) water-cooled	0.225 (0.79)	0.057 (0.20)

Source: Reprinted from ASHRAE *Handbook of Fundamentals 1981* by permission of the American Society of Heating, Refrigerating and Air-Conditioning Engineers, Inc.

TABLE B7.6   ESTIMATED EQUIVALENT RATED FULL LOAD HOURS OF OPERATION FOR PROPERLY SIZED EQUIPMENT DURING NORMAL COOLING SEASON

Albuquerque, NM	800–2200	Indianapolis, IN	600–1600
Atlantic City, NJ	500–800	Little Rock, AR	1400–2400
Birmingham, AL	1200–2200	Minneapolis, MN	400–800
Boston, MA	400–1200	New Orleans, LA	1400–2800
Burlington, VT	200–600	New York, NY	500–1000
Charlotte, NC	700–1100	Newark, NJ	400–900
Chicago, IL	500–1000	Oklahoma City, OK	1100–2000
Cleveland, OH	400–800	Pittsburgh, PA	900–1200
Cincinnati, OH	1000–1500	Rapid City, SD	800–1000
Columbia, SC	1200–1400	St. Joseph, MO	1000–1600
Corpus Christi, TX	2000–2500	St. Petersburg, FL	1500–2700
Dallas, TX	1200–1600	San Diego, CA	800–1700
Denver, CO	400–800	Savannah, GA	1200–1400
Des Moines, IA	600–1000	Seattle, WA	400–1200
Detroit, MI	700–1000	Syracuse, NY	200–1000
Duluth, MN	300–500	Trenton, NJ	800–1000
El Paso, TX	1000–1400	Tulsa, OK	1500–2200
Honolulu, HI	1500–3500	Washington, DC	700–1200

Source: Reprinted from ASHRAE *Handbook of Fundamentals 1981* by permission of the American Society of Heating, Refrigerating and Air-Conditioning Engineers, Inc.

TABLE B7.7 HOURLY WEATHER OCCURRENCES

Location	Outdoor Temperature (°F)

Location	72	67	62	57	52	47	42	37	32	27	22	17	12	7	2	-3	-8	-13	-18
Albany, NY	588	733	740	708	652	625	647	769	793	574	404	278	184	110	63	32	10	5	4
Albuquerque, NM	767	831	719	651	687	734	741	689	552	346	154	66	21	4	1	1			
Atlanta, GA	1185	926	823	784	735	676	598	468	271	112	44	19	8	2					
Bakersfield, CA	831	898	966	977	908	746	541	247	77	7									
Birmingham, AL	1138	908	805	742	668	614	528	433	292	143	69	17	6	3					
Bismark, ND	454	566	614	606	563	520	518	604	653	550	474	371	338	292	278	208	131	77	80
Boise, ID	492	575	643	702	786	798	878	829	522	307	148	53	26	14	6	2			
Boston, MA	676	819	804	781	766	757	828	848	674	429	256	151	74	35	4	9	1		
Buffalo, NY	646	772	760	700	666	624	647	756	849	602	426	267	170	815		24	2		
Burlington, VT	573	670	703	694	655	603	637	716	752	561	491	336	272	216	135	81	39	17	8
Casper, WY	423	532	592	642	606	670	782	831	806	683	495	325	200	116	73	45	30	15	5
Charleston, SC	1267	1090	889	787	651	576	434	321	192	79	27	5							
Charleston, WV	912	949	767	689	661	667	607	633	630	356	252	135	73	22	7	1			
Charlotte, NC	1115	908	839	752	730	684	634	515	360	166	64	23	5	2					
Chattanooga, TN	1021	895	775	722	713	679	642	553	414	228	113	45	4	4	2				
Chicago, IL	762	769	653	592	569	543	591	800	822	551	335	196	117	85	59	25	12	3	
Cincinnati, OH	879	843	726	639	611	599	627	698	711	460	249	131	68	44	18	8	2		
Cleveland, OH	763	831	723	641	638	607	620	754	806	578	355	201	111	47	22	11	2		
Columbus, OH	774	820	720	648	622	603	658	730	772	502	280	169	94	40	20	10	4	1	
Corpus Christi, TX	1175	1041	748	551	444	302	180	83	27	9	3								
Dallas, TX	831	795	693	656	629	576	504	371	231	91	34	17	4	1					
Denver, CO	549	684	783	731	678	704	692	717	721	553	359	216	119	78	36	22	6	1	1
Des Moines, IA	707	751	681	600	585	512	510	627	747	557	405	281	211	152	104	59	23	8	1
Detroit, MI	721	783	695	633	592	566	595	808	884	618	377	248	131	61	17	4	1		
El Paso, TX	933	839	749	760	687	611	494	369	233	104	34	10	2						
Ft. Wayne, IN	728	777	699	608	569	552	601	725	905	596	381	205	124	69	40	19	6	1	
Fresno, CA	709	803	921	1006	1036	952	673	426	168	34									
Grand Rapids, MI	634	739	712	647	571	565	554	742	938	690	469	293	172	78	31	10	1	1	
Great Falls, MT	407	520	636	754	822	830	832	813	698	533	355	218	167	136	118	101	68	51	62
Harrisburg, PA	807	824	737	692	635	659	722	888	749	427	222	125	52	18	4	1			
Hartford, CT	617	755	751	752	649	575	683	807	825	552	370	233	153	77	33	11	3	2	
Houston, TX	1172	980	772	681	570	452	291	141	64	18	4	2							
Indianapolis, IN	821	815	722	585	586	579	605	712	791	551	293	152	97	60	35	13	3	2	
Jackson, MS	1168	922	790	677	618	605	484	367	224	103	41	6	2	2	1				
Jacksonville, FL	1334	975	879	692	530	355	288	154	83	24	2								
Kansas City, MO	761	723	601	572	553	562	628	625	591	407	265	175	99	51	21	4			
Knoxville, TN	1056	889	746	675	672	689	648	590	456	217	101	41	21	7	2				
Las Vegas, NV	651	644	699	786	769	716	591	396	194	44	7	1							
Little Rock, AR	940	803	725	672	638	669	605	509	363	172	50	25	5	1					
Los Angeles, CA	881	1654	2193	1904	1054	428	107	10											

Source: Reprinted from ASHRAE *Handbook of Fundamentals 1981* by permission of the American Society of Heating, Refrigerating and Air-Conditioning Engineers, Inc.

Location	Outdoor Temperature (°F)																		
	72	67	62	57	52	47	42	37	32	27	221	7	12	7	2	-	-8	-13	-15
Louisville, KY	869	758	693	654	619	634	649	703	631	332	169	97	45	25	8	3	1		
Lubbock, TX	833	829	688	700	642	618	620	546	490	346	180	86	33	7	5	1			
Memphis, TN	977	798	715	690	618	633	614	532	374	196	74	25	10	4					
Miami, FL	1705	810	452	277	147	71	26	4											
Milwaukee, WI	597	753	749	634	585	591	611	774	913	659	421	285	176	116	83	47	18	4	3
Minneapolis, MN	621	690	695	602	588	482	500	560	632	6009	514	383	311	246	186	119	62	31	16
Mobile, AL	1411	1038	882	698	609	506	377	214	109	49	7	3							
Nashville, TN	933	838	738	697	637	619	627	565	463	263	132	67	28	9	3	1	1		
New Orleans, LA	1189	987	850	692	621	449	282	128	47	9	2								
New York, NY	926	877	754	745	722	796	838	858	603	330	188	2	26	10	1				
Oklahoma City, OK	881	769	717	643	645	611	641	570	468	287	173	77	36	12	3	1			
Omaha, NB	726	721	606	558	539	543	543	655	663	511	390	287	189	135	93	40	15	1	
Philadelphia, PA	863	809	735	710	663	701	758	818	654	335	189	100	32	9					
Phoenix, AZ	762	776	767	769	659	540	391	182	57	8									
Pittsburgh, PA	722	910	799	678	637	587	631	688	569	774	360	233	159	60	30	7	1		
Portland, ME	407	627	780	808	760	748	722	839	820	599	408	293	190	109	60	29	15	5	1
Portland, OR	373	581	1001	1316	1274	1271	1238	772	343	123	40	10	4	1					
Raleigh, NC	1087	937	848	762	707	672	638	527	410	236	103	38	11	1					
Reno, NV	418	477	572	690	845	909	890	829	733	530	387	277	101	37	15	4	1		
Richmond, VA	953	850	784	745	690	673	699	632	478	285	138	67	19	2	1				
Sacramento, CA	630	773	1071	1329	1298	1049	701	355	93	8									
Salt Lake City, UT	569	615	614	635	682	685	755	831	798	564	328	158	80	41	16	2			
San Antonio, TX	1086	943	789	669	569	445	387	190	94	31	11	4	1	1					
San Francisco, CA	285	665	1264	2341	2341	1153	449	99	10										
Seattle, WA	258	448	750	1272	1462	1445	1408	914	427	104	39	20	3						
Shreveport, LA	1063	886	772	679	619	609	516	361	200	72	23	6	2						
Sioux Falls, SD	566	684	669	605	522	498	501	625	712	585	520	448	293	208	152	102	59	43	18
St. Louis, MO	823	728	646	575	585	578	620	671	650	411	219	134	77	40	15	7	1		
Syracuse, NY	627	735	723	717	656	641	651	720	830	547	392	282	190	102	55	23	5	2	2
Tampa, FL	1387	1187	877	570	345	216	137	48	10	1									
Waco, TX	909	830	701	622	651	558	501	354	216	84	24	3	1						
Washington, DC	960	766	740	673	690	684	790	744	542	254	138	54	17	2					
Wichita, KS	758	709	641	603	589	592	611	584	607	426	273	161	85	45	14	3	1		

Index

Absolute humidity, 18
Absolute temperature, 41
Absolute zero, 41, 166
Absorptance (of light), 254
Absorption:
    chiller, 164–166, 168
    coefficient, 413, 417, 418, 430
Absorptivity, thermal, 5, 9
ACES, 202–203
Acoustics, 411
Acoustically absorptive materials, 417–418, 420, 421
Acoustically insulative construction, see Sound, insulation
Actuator, 204, 388
Airborne living organisms, 30, 31
Air change method, 104, 105–106
Air-conditioning system, 34, 35, 61–62, 140, 171–183
Air conditioning, process, 4
Air-cooled condenser, 164, 170, 168, 169, 170
Air-handling unit (AHU), 142, 143, 171, 172–180
Air infiltration:
    changes/hour (Appendix Table B6.2), 549
    rate:
        through curtain walls (Appendix Figures B6.4–B6.6),
            548–549
        through revolving doors (Appendix Figure B6.3), 547
        through swinging door cracks (Appendix Figure B6.2),
            546
        through swinging door entrances (Appendix Figure B6.1),
            545
        through windows and doors (Appendix Table B6.1),
            544
Air mass, 87–88, 153–154
Air sterilization, 31
Albedo, 61, 88
Alternating current (AC), 290, 293, 294
Ampacity, 302
Amperage, 289
Amperes, 289, 312

Angle of incidence:
    of light, 254–255
    of solar radiation, 46, 90
    of sound, 413, 416
Annual cash-flow method, 444–446
Annual Cycle Energy System (ACES), 202–203
Annunciation, 388
Antifreeze system, 402
Areaway drain, 375, 376
ASHRAE, 4
Aspect ratio, 182, 206, 220
Audio indicators, 335–336
Auditorium, acoustic design of, 416, 419–421, 422, 424
Automatic fire detection, 336–337, 385–387
Automatic fire-suppression system, 399–405
Automatic sprinkler system, 399–404
Average trip time, elevator, see Elevator, travel time
Avoided costs, 316

Backflow preventer, 365
Background mechanical noise, 141, 205–206, 422
Backwater valve, 373
Balance point temperature (BPT), 64, 107, 236–237
Ballast:
    factor, 102
    factors, for fluorescent lights (Appendix Table B5.3), 535
    lighting, 257, 265–266, 267
Bidet, 361,
Bin method, 108, 109
Blackfields, 125
Blackwater, 129
Blowdown, 149
Body heat balance, 7–8, 9, 12–13
Boiler, 146–150, 171
    firetube, 149
    portable, 149
    waste heat, see Waste heat boiler
    watertube, 149

Bottoming cycle, 320
Branch circuit, electrical, 307, 308
Branch circuit panelboard, *see* Subpanel
Branch drains (branch lines), 370
Brightness, 255, 258, 260
Brownfields, 125
Btu (British Thermal Unit), 6, 150
Btuh (Btu/hr), 6, 150
Buffer space, 224
Building:
    drain, 370
    population, 33, 101–102, 103, 106, 343
    sewer, 370
Building storage factors (BSF), 1101–103
    for appliances (Appendix Table B5.7), 543
    for lights (Appendix Table B5.5), 538–540
    for people (Appendix Table B5.2), 534
Buoyancy, 44, 61
Busbar, 303. *See also* Conductor, electrical
Bus duct, 303, 307
Busway, *see* Bus duct

Cable, 303. *See also* Conductor, electrical
Cablebus, 303, 307
Cable tray, 303, 304, 305
Cable-type elevator, *see* Elevator, electric-type
Canal, 330
Candela, 253, 254
Candlepower, 253, 254, 256
Capacitor, 312
Capacity:
    defined, 61
    heat pump, 166
    HVAC equipment, 61, 141–142, 146
    solar energy system, 158
    water heater storage tank, 357–358
Capital recovery factor (CRF), 439
Car capacity, elevator, 343, 346
Car loading, elevator, *see* Passenger load, elevator
Carbon dioxide fire-suppression system, automatic, 404
Carousel toilet, 369
Catch basin, 376
Ceiling diffuser, 185–187
Ceiling raceway, 303, 307
Cellular floor raceway, 305, 306
Celsius, *see* Temperature scales
Central systems, HVAC, 141, 143, 179–180
Cesspool, 355
CFC, 119–121
CFM, defined, 44
Check valve, 365
Chilled water system:
    for air conditioning, 62, 162, 163, 164–166
    drinking fountain, 359

Chiller, 62, 163, 164–166, 170
Circuit:
    breakers, 292, 297, 300
    electrical, 290
Cistern, 131–132, 377–378
Clean-out, 373
Clearness correction, 89
Clerestories, 65, 224, 279, 280
Climate, design impacts, 64–65, 71, 214–216, 218–219, 221. *See also* Microclimate
Climatic conditions:
    for Canada (Appendix Table B2), 472–474
    for other countries (Appendix Table B2), 475–479
    for the United States (Appendix Table B2), 458–471
Clivus Multrum, 369
Clo, 14–16, 17, 25
Closed circuit:
    electrical, 290
    television (CCTV), 335, 338–339
Closed switch, 299
Clustering of buildings, 221, 222
Coal, xiii–xiv, 120, 144, 147–148
Coefficient:
    of beam utilization (CBU), 257
    of transmission, *see* Luminous transmittance
    of utilization (CU), 256–257, 271, 273
Cogeneration, 319–323
Coil, heating or cooling, 52, 172, 173–174
Collector panel, *see* Solar collector
Color rendition, 257, 264, 265, 267
Combination system, HVAC, 171
Combined heat and power (CHP) system, *see* Cogeneration
Comfort:
    ASHRAE standard on, 25
    chart, 22–25
    conditioning, 4
    conditions affecting, 4–5, 14–21, 35
    defined, 4
    design considerations, 10, 25–27, 66, 85
    envelope, 22–24
    indices of, 21–22
    nonthermal aspects, 4, 5, 30
    physiological, 7, 12, 35
    and productivity, 4, 7
    radiant, 10, 11
    thermal, 4–5
Commissioning, 433–435
Comparative value analysis, 447–448
Compartmentation, 391–393, 397
Composting toilet, 369
Compounding, 441, 448
Compressor, 162, 163, 168, 369
Computer room unit, air-conditioning, 179, 180
Concentration ratio, 327

Condensate, 52–53, 146, 163
Condensate drain, 52–53, 146, 163, 194
Condensation, 12, 23, 50, 52, 55, 72–75, 76, 146, 162
  concealed, 55, 72, 74–75, 76
  visible, 23, 72, 73–74, 76
Condenser:
  air-conditioning system, 162, 163, 164–165, 166, 168–170
  electrical, *see* Capacitor
Condensing combustion process, 146
Condensing unit, 170, 180
Conductance, thermal, 42–43, 44
Conduction, thermal, 41–43
  defined, 5, 41–42, 55
  sensation, 5, 11–12
Conductivity:
  electrical, 290, 302
  thermal (k), 7, 42–43
Conductor:
  electrical, 18, 290, 292, 302–303, 312
  thermal, 11, 14
Conduit, 303, 304
Connected load, 292, 297
Contrast, 258. *See also* Glare
Convection, 43–44
  and comfort, 8, 9, 11–12
  defined, 5, 41, 43–44, 55
  impact of air motion on, 11, 20–21
Convector, 142, 188. *See also* Finned-tube radiation
Convenience outlet, *see* Receptacle
Conveyors, 349
Cooling, 6, 50–53, 55, 141–142, 161–171
Cooling degree day method, 108–109
Cooling load, 61–63
  calculation of, 80–106, 113
  example of calculation, 110, 112
Cooling load check figures (Appendix Table B1), 456–457
Cooling approximate power inputs (Appendix Table B7.5), 589
Cooling system equivalent full load hours of operation (Appendix Table B7.6), 561
Cooling tower, 164, 168–170
COP (coefficient of performance), 166
Cost of capital, 442, 444, 446
Counteraction, 33
Crack length method, 104–105
Cross-contamination, 365
Cross-talk, 429–430
Current, electrical, 289, 312

Dampers, 174, 176, 179, 182, 185
Daylighting, 152, 153, 224, 235, 276–285
dBA (A-weighted sound level) scale, 415
DDC (direct digital control) system, 204–205
Dead space, acoustically, 417
Decibel (dB), 414–415, 422

Deciduous vegetation, 216, 217
Degree day:
  modification factors (Appendix Table B7.2), 560
  part load correction factors (Appendix Table B7.3), 560
  values for the United States and Canada (Appendix Table B7.1), 554–559
Degree day method, 107–108, 110
Degree of saturation, 18
Deluge system, 403
Demand charge, 310
Demand control, 310–311
Demand leveling, 310–311
Depreciation factor, *see* Maintenance factor
Design conditions, 81, 83, 85–87
Dew-point temperature, 6, 18, 49, 73
Diffuse lighting, *see* Indirect lighting
Diffuser, air, 142, 185–187
Diffuse reflection:
  of light, 255
  of solar radiation, 88
Diffuse solar radiation, 88, 89, 153, 155
Diffuse transmission of light, 255, 256
Direct Current (DC), 290, 293, 294
Direct effect, 120–121
Direct expansion (DX), 162, 163, 174, 180, 195
Direct gain, passive solar heating, 238, 239, 280
Direct lighting, 268–270
Direct solar radiation, 88, 89, 155–156
Direct transmission of light, 255, 256
Direct water heater, 358
Discounted payback period, 442
Discounting, 441, 449
Discount rate, 442, 444
Disposal field, 374–375
District heating and cooling, 145, 150, 206–207
Diurnal (day-night) cycle, 69
Diversity, 101, 103
  factor, 103
Diversity factors, typical, for large buildings (Appendix Table B5.8), 544
Domestic hot water (DHW), 357
Double-bundle condenser, 200
Downfeed distribution, 357, 379
Downspout, 375, 376, 377
Draft, flue, 145
Drain field, 374, 375
Drift, temperature, 17
Drinking fountain, 359, 360, 366–368
Dry air, 49, 53
Dry-bulb (DB) temperature, 6, 16–18, 21–22, 49
Dry chemical fire-suppression system, 405
Dry pipe system, 402
Dry well, 375, 377, 378
Dual duct system, 175, 177, 178

Ducted system, 171
Dumbwaiters, 349
Dust, 30
Dust control, 31, 173
DX, *see* Direct expansion
Dynamic envelope, 65, 66, 214, 224, 225, 226, 247

Earth sheltering, 228, 234–235
Echoes, 420, 430
Economizer cycle:
    using ventilation air, 68, 175, 176, 179
    water-side, *see* Water-side economizer cycle
EER (energy efficiency ratio), 294–295
Efficacy of light sources, 256, 263, 264, 265, 266, 267
Efficiencies and fuel-heating values (Appendix Table B7.4), 560
Egress, means of, 285, 389–391
Electric:
    control system for HVAC, 203–205
    stairway, *see* Escalator
Electrical:
    closets, 307–308, 337–338
    current, *see* Current, electrical
    energy, 289, 290–291, 312
    panel, 300, 307–308
    power, 290–291, 310–311, 312
    resistance, *see* Resistance, electrical
    supervision, 388
    symbols, 312, 336
    system space requirements, 297–299, 301, 307–308, 337–338
Electrolysis, 316, 326
Electromagnetic spectrum, 45, 153
Elevator:
    average trip time, 344
    car capacity, 343, 346
    car loading, 343, 344
    electric-type, 340, 341, 342, 346–347, 350
    freight, 347
    hydraulic-type, 340, 341–342, 343, 347, 350
    recall (capture) system, 391
    shaft, *see* Hoistway
    travel time, 344
    zone, 345
Embodied energy, 123–124, 132, 136
EMCS (energy management control system), 311, 340
Emergency electrical systems, 308–310, 319, 388
Emergency lighting, 285
Emissivity, 5, 67
EMT, 303, 304. *See also* Conduit
Energy:
    in building materials, 220. *See also* Embodied energy
    -conscious design, xiii–xv, 1, 24–25, 39–40, 122–126, 127, 134–136, 224–225
    conversion efficiency, 13, 58. *See also* Efficiencies and fuel-heating values

for cooling in the U.S., 241
of electromagnetic waves, 45
for lighting, 257, 262, 263
resources, xiii–xv, 141–146, 151, 310
savings and clothing, 16
for ventilation, 33–34
Engine-generator set, 317. *See also* Internal combustion engine
Enthalpy (total heat), 5–6, 49, 52
Environmental amenity, 132
Equipment load, 81
Equivalent full-load hours, 107, 109, 110
Escalator, 347–348, 350
Estimated loads, 297–298
Evaporative condenser, 168, 169, 170
Evaporative cooler, 53, 170–171
Evaporative cooling:
    of body, 7–9, 12–13, 16, 18, 20–21, 23
    passive systems, 245
    principle of, 53, 55
    by water bodies, 59–61
Evapotranspiration, 61, 127, 218
Exfiltration, 69
Exhaust-air heat recovery, 34, 199–200
Expectancy, 27

Fahrenheit, *see* Temperature scales
Fan, 172–173
Fan coil unit (FCU), 142, 172, 191–194
Feeder, electrical, 300, 307–308
Fenestration, 65–66, 67, 229–234, 238, 242, 243–244
    for daylighting, 214, 224, 279, 280, 281
    heating and cooling loads due to, 63, 65–66, 87–95
Film coefficient, *see* Surface conductance
Film conductance, *see* Surface conductance
Finned-tube radiation, 172, 184, 188
Fire alarms, 336–337, 387–389
Fire barrier, 391
Fire compartmentation, 391–393, 397
Fire damper, 182, 398
Fire detectors, 335–337, 384–387, 398
Fire door, 392
Fire extinguisher, portable, 405
Fire-fighters' communication system, 339, 389
Fire hose:
    cabinet (F.H.C.), 405
    standpipe (S.P.), 405
Fire-rated access panel, 392
Fire-resistance rating, 391–392
Fire sprinkler:
    head, 399–401
    system, automatic, 399–404
Flame detector, 385, 387
Flexible connectors, 206, 428, 429, 430
Flexible cord, 309–310

Floating floor, 426, 428
Floor drain, 164, 365, 373
Flow rate, river, 330–331
Flues, 142, 145, 146, 149–150, 151
Fluorescent lamp, 102, 256, 263, 264, 265–266, 277
Flush valve, 360, 361
FM-200, 404
Fog, 31, 403
Footcandle, 253–254, 255, 258, 259
    meter, 254
Footing drain, 377, 378
Footlambert, 255–256
Forced convection, 20, 44
Foreground reflectance correction, 88, 89
Four-pipe system, 183, 192–193
Free convection, *see* Natural convection
Frequency:
    of electrical power, 290, 293
    of sound waves, 412
Fresh-air vent (fresh-air inlet), plumbing, 371
Fuel:
    alternatives, xiii–xiv, 142–148, 296
    cell, 316, 317, 322, 326
    cost escalation, 449
Fumes, 30
Furnace, 146, 147, 179, 181
Fuses, 292, 297, 300

Gas turbine, 317–318, 321
Geometric acoustics, 416–417, 418
GFCI (Ground fault circuit interrupter), 308
Glare, 256, 258, 260, 279–285
Glass blocks, 92–93
Global warming, 120–122
Globe thermometer temperature, 22
Grains of moisture, 49–50
Grayfields, 125
Graywater, 129
Green design, 122–136
Greenfields, 125
Greenhouse effect, 120, 216
Greenhouse, passive solar heating, 239, 240–241
Green roof (eco-roof), 127, 128
Grilles, 141, 185, 186, 195
Grounding, 303
Gutter, 375–376, 377

Handling capacity, elevator, 343
Head, 330–331
Heat:
    addition rate, 81
    balance, body, 7–13
    capacity, *see* Thermal capacity
    content, *see* Enthalpy

defined, 5
detector, 385, 386
exchanger, 150, 162, 168, 169, 173, 199–200
extraction rate, 81
gain:
    body, 7–12, 25, 28
    building, 46, 58–69, 80–81, 86
    latent, *see* Latent heat gain
    sensible, *see* Sensible heat gain
    solar, 87–95
loss:
    body, 7–13, 18–20, 23, 25–26, 28–30
    building, 46, 58–69, 80–81, 85
pipe, 200
pump, 108, 145, 162, 166–168, 180, 196–197
    closed-loop system, 167, 196, 200
    hydronic, 166–167, 196–197
    solar-assisted, 157
recovery incinerator, 151, 152, 201–202
specific, *see* Specific heat
stroke, 28
wheel, 199–200
Heat gain, rate of:
    from miscellaneous appliances (Appendix Table B5.6C), 543
    from occupants of conditioned spaces (Appendix Table B5.1), 534
    from office equipment (Appendix Table B5.6B), 542
    from restaurant equipment (Appendix Table B5.6A), 541
Heating load:
    calculation of, 81–83, 85–86, 106, 110–113
    categorized, 62–63
    defined, 62, 81
    example of calculation, 110–111
Heating water converter, 141, 142, 146, 150–151, 162
Heat island, 127
Heat loss:
    through basement floors (Appendix Table B4.10A), 558
    through below grade walls (Appendix Table B4.9), 558
    through concrete floors on grade (Appendix Table B4.10B), 559
Hertz, 290, 293, 322, 412
H.I.D. (high-intensity discharge lamp), 263, 264, 266–267, 277, 309
High-expansion foam fire-suppression system, 404
High-pressure sodium (HPS) lamp, 263, 264, 267
Hoistway, 340–341, 342, 346
Home run, 307
Horsepower (hp):
    boiler, 150
    motor, 293
Hose bib, 365
Hose standpipe, *see* Fire hose standpipe
Hot gas, 162
Hot refrigerant gas heat recovery, 200

Hourly computer simulation of energy use, 109–110
Hourly weather occurrences (bins) (Appendix Table B7.7), 562–563
Humidity ratio, 18, 49, 72–73
Humus type toilet, 369
HVAC (heating, ventilation, and air conditioning), 25
HVAC system, 25, 27, 139–209
   for smoke control, 395–398
   symbols and abbreviations, 144
   See also Air-conditioning system
Hydraulic-type elevator, *see* Elevator, hydraulic-type
Hydronic heating, 183
Hydropower (hydroelectric power generation), 328–332

IES (Illuminating Engineering Society of North America), 257
Illumination level, *see* Lighting, level
Image theory, 418
Immersed water heater, *see* Direct water heater
Impact noise, *see* Noise, impact
Incandescent lamp, 262–264, 277
Incineration:
   heat recovery, 151, 152
   toilet, 369
Incremental costs, 445, 451
Incremental savings, 451
Incremental unit (for heating and cooling), 195–197
Indicator devices, 204, 335, 336–337, 388–389
Indirect gain, passive solar heating, 238–241
Indirect lighting, 268–269
Indirect water heater, 358
Indoor air quality (IAQ) problem, 34, 133–134
Induction unit, 179, 195
Induction system, 175, 179, 195
Inductive loads, 293, 311
Infiltration, 44, 62–63, 68–69, 104–106
   at entrances, 104–105, 224
   through unit ventilators, 194
inflation, 448–449
Infrared heater, 189, 190
Initiating device, 336
In-line water heater, 357–359
Inrush current, 293, 308
Instantaneous water heater, *see* In-line water heater
Insulating glass, 66, 67
Insulation:
   of clothing, 14–16, 17
   and condensation, 73–75, 225–229
   of ductwork, 181
   of electrical conductors, 303
   and MRT, 10, 20
   of piping, 184, 206–207
   principle of, 44
   R-value of, 42–43, 46–47

   of slab-on-grade floors, 98–99
   window, *see* Operable thermal barrier
   and zoning, 172
Insulator, electrical, 290
Interceptor, 373, 379
Interest rate, 441, 442, 445
Internal combustion engine, 168, 317, 321
Internal heat gains, 63, 69, 83–84, 100–103, 235
Internal-load-dominated building:
   defined, 214, 220
   design for, 221, 222, 225, 228, 236–238, 241
Internal loads, 63, 100–103, 235
Interval, elevator, 344
Intrusion detector, 339
Ionization detector, 385, 386
Irrigation quality (IQ) water, 129
Isolated gain, passive solar heating, 241

Jack, 341, 342
Joule, 6

Kelvin, *see* Temperature scales,
Known loads, 297

Lambert, 256
Lamp efficacy, 256, 262–267
Lamps, 256, 260, 262–267
Latent heat:
   defined, 5, 49, 51
   and evaporative cooling, 6, 53
   loads, 8–9, 100–103
   loss from body, 12–13, 25–26
   on psychrometric chart, 51–52, 53
Latent heat of fusion, 5, 201
Latent heat gain, 62–63, 80–83
Latent heat of vaporization, 9, 170–171
Lavatory, 360, 362, 366–368
Leader, 376, 377
LEED, 123
Life-cycle cost, 129, 142, 158, 440–446, 451
   of lighting, 262, 263
Lighting:
   fixture, *see* Luminaire
   heat from:
      energy implications of, 262, 263, 276–277
      internal load due to, 63, 67–68, 102
      luminaire mounting concerns due to, 266
      and MRT, 10
      and thermal storage, 69
   heat recovery, 200
   level, 258, 259
   watts/square foot (Appendix Table B5.4), 536–537
Light-loss factor, *see* Maintenance factor
Light pollution, 126

Light sources, 256, 262–267, 276
Light trespass, 126
Liquid line, 162
Live space, acoustically, 417, 419
Load:
    electrical, 290, 291, 292–296, 312
    thermal, 61–64, 79–106
Load center substation, *see* Unit substation
Load factor, 310
Load shedding, 311
Lobby dispatch time, *see* Interval, elevator
Local water heater, *see* Point-of-use water heater
Locked rotor current, 293
Low-pressure sodium (LPS) lamp, 263, 264, 267
Lumen, 253, 254, 255–260
Luminaire, 256, 267–276
    effect on cooling load, 102
Luminaire efficiency, 256
Luminance, *see* Brightness
Luminance ratio, *see* Contrast
Luminous ceiling, 270, 272
Luminous flux, 253, 256
Luminous intensity, 253, 254, 255–256
Luminous transmittance, 255–256
Lux, 253–254

Main switch, *see* Service disconnect
Maintenance factor, 270
Make-up air unit, 179, 180, 181, 197
Mass, thermal storage, 6–72, 228
MCM wire gauge, 303
Mean coincident wet-bulb temperature, 86–87
Mean daily temperature range, 87
Mean free path, 418–419
Mean radiant temperature (MRT), 6, 9–10, 12, 18–20, 22
Mean time, 87
Mercury vapor (M-V) lamp, 263, 264, 266–267
Metabolism, 6, 8, 9, 12, 13–14
Metal-halide lamp, 263, 264, 267
Metering, electrical, 301–302
Met units, 13–14, 25
Microclimate, 58, 126, 214, 215, 216
Micro-hydro, 328, 330
Millilambert, 256
Mist, 30–31
MIUS (modular integrated utility system), 322–323
Moisture content, *see* Humidity ratio
Monitors, daylight, 279, 280, 281
Motion detector, 338
Motors, electric, 293–296, 308, 311
Moving ramp, 349, 350
Moving stairway, *see* Escalator
Moving walk (moving sidewalk), 349, 350
Multiple-step methods, 107, 109–110

Multi-zone system, 176–177, 180
Municipal (public) sewer, 355, 370, 374

National Electrical Code (NEC), 291
National Electrical Manufacturers' Association (NEMA), 293
National Fire Protection Association (NFPA), 291
Natural convection, 44, 241
Natural ventilation, 126, 224, 225, 242–245, 246
NC (noise criteria), 205, 415
NEII (National Elevator Industry, Inc.), 342
Negative pressure, 69, 104, 397. *See also* Pressurization
Net present value, 442, 444
New effective temperature (ET*), 22, 23
Night sky radiation, 73, 214–215, 241–242
Noise:
    background, 205–206, 266, 415–416, 421–423
    defined, 411
    impact, 422, 425–428
    mechanical equipment, 141, 205–206, 422
        control of, 206, 428–430
        layout considerations of, 141, 163, 206
        sources of, 163, 181, 422, 429–430
NPT (number of passengers per trip), elevator, *see* Passenger load, elevator
NRC (noise reduction coefficient), 413, 417, 430

Occupied zone, 17, 185, 186
Odor masking, 33
Odors, 21, 31, 32–33
Office acoustics, 416, 424, 428
Off-peak electricity, 202, 311, 357
Ohm's Law, 290
One-pipe system, 183
Opaque material, 45
Open circuit, electrical, 290
Openness factor, 91
Open switch, 299
Operable thermal barrier, 231
Operable shading, 231–234
Operative temperature, 15, 22, 24
Optimum operative temperature, 15, 16
Outside air loads, 81–83, 104–106, 197–199
Overcurrent protection, 300, 308
Overshot water wheel, *see* Water wheel
Ozone depletion, 119–120

Panelboard, *see* Electrical panel
Passenger load, elevator, 343
Passive cooling, 241–245
Passive solar heating, 63, 153, 222, 235–241
    orientation for, 221, 222
    summer shading, 223
Payback period, 438–440
Payback ratio, *see* Simple payback

Percentage humidity, 6, 18
Perforated ceiling panel, 185, 186
Perimeter heat, 68, 172, 183, 187
PERS (public emergency reporting system), 389
Phase-change materials, 156, 201
Phon, 415
Photoelectric detectors, see Smoke detectors
Photovoltaic solar cell, 152, 326–328
Physical acoustics, 416
Piped system, 171, 183–184
Pitch, 412
Plumbing:
    core, 372, 379
    symbols, 379
Plunger, elevator, see Jack
Pneumatic tubes, 349
Point of diminishing returns, 229, 449–450
Point-of-use water heater, 358
Positive pressure, 68–69, 106, 397–398. *See also* Pressurization
Power, relationship to energy, 290
Power factor, 310, 311–312
Power pole, 307
Preaction system, 402–403
Precooling, 72
Preheat coil, 174
Present value, 441–442
Present-worth method, 444
Pressurization, 68–69, 106, 179, 396–398
Pressurized air toilet, 369
Pressurized stairwell, 396–397
Primary source (of light), 256
Prime mover, 316, 317, 330–331
Products of combustion detector, *see* Ionization detector
Psychrometric chart, 18, 49–52
PTAC (packaged terminal air conditioner), *see* Incremental unit (for heating and cooling)
Public sewer, 355, 370, 374
Pull station, manual fire alarm, 336, 337, 386, 387
Pulse-combustion process, 146
Pumps, 142, 162, 183
Purging, 398, 404
PURPA (Public Utility Regulatory Act), 316, 319
Pyrolysis, 151

Raceway, 292, 303–307
Radiant heating equipment, 30, 189–191, 192
Radiant panel, 189–190
Radiant plate, 384
Radiation (radiant heat transfer), 9–11, 45–46, 140
    comfort aspects of, 6, 9–11, 12, 140
    defined, 5, 9, 41, 55
    and thermal storage mass, 69
Rainwater harvesting, 129, 130–132
Ramp, temperature, 17

Rankine, *see* Temperature scales
Rankine cycle, 156, 157
Rate of return (ROR), 439, 446–447
Receptacle, 292, 307, 308
Recharge basin, 377
Reciprocals, table of (Appendix Table B4.5), 526
Recirculating hot water distribution, 359
Reclaimed water, 129
Recovery rate, 358
Reflectance coefficient, *see* Reflectance (of light)
Reflectance (of light), 254, 255, 256, 258, 260, 271, 273
Reflection factor, *see* Reflectance (of light)
Reflector, light, 256
Refraction of light, 255, 256
Refractor, light, 256
Refrigerant, 163
Refrigeration, 162, 164, 168
Refuge area, 388, 392–393
Refuse-derived fuel (RDF), 151, 152
Register, air, 21, 142, 185
Reheat coil, 174
Reheat system, 175–176
Relative humidity (RH), 18, 32, 73, 74, 75
    defined, 18, 49, 53
Resistance:
    electrical, 290, 312
    thermal (R-value), 42–43, 46, 47, 48, 55
Resistive loads, 293
Resistivity, thermal (r), 42–43
Resonance, room, 419
Return on investment (ROI), *see* Rate of return (ROR)
Reverberation, 411, 418–419, 430
    time, 418–419, 421
Roof:
    drain, 375, 376
    pond, 239, 240, 241–242, 245
Rooftop unit (RTU), 179, 180
Room:
    cavity ratio, 261, 271
    exhaust fan, 69, 197–199
RTT (round trip time), elevator, 343
Run-around heat recovery, 199
Running time, elevator, 343
R-value, *see* Resistance, thermal
R-values:
    of air (Appendix B4.1), 517–518
    of typical building materials (Appendix Table B4.2), 519–524

Sabin, unit of acoustic absorption, 417
Sanitary piping, 370–375
Saturated air, 6, 9, 49–52, 72–73
Saturation line, 49, 72
Seasonal performance factor (SPF), 108, 166
Secondary source (of light), 256

Security systems, 338–340
Seepage pit, 375
Selective surface coating, 155
Sensible heat:
    defined, 5, 49, 51
    and evaporative cooling, 6, 8–9, 53
    gains and losses of the body, 8–9, 12–13, 25–26
    loads, 101–103
    on psychrometric chart, 52
Sensible heat gain, 62–63, 80–81, 83–84
Sensitivity analysis, 444
Sensor, 203–204, 336
Separator, *see* Interceptor
Septic system, 374
Septic tank, 374
Septum, 428
Service disconnect, 292, 297, 299–301
Service entrance, electrical, 291, 292, 296–302, 312
Service sink, 360, 363
Sewage ejector pump, 373
Sewer gases, 365
Shading coefficient (SC), 88–89, 90–93
Shading coefficients:
    for glass (Appendix Table B3.3), 499
        with draperies (Appendix Table B3.6), 501–502
        with exterior shading (Appendix Table B3.5), 500
    for plastic (Appendix Table B3.4), 500
Shadow lengths and widths for building exterior projections
        (Appendix Table B3.9), 505–513
Short circuit, 290, 300
Short-cycling, 295
Siamese connection, 401
Sick building syndrome, *see* Indoor air quality (IAQ)
        problem
Side-lighting, 279
SIL (speech interference level), 415
Silicon solar cell, 326. *See also* Photovoltaic solar cell
Simple payback, 439–440
Single-phase power, 293, 294, 297
Single-step method, 107–109
Single-zone system, 172, 175, 180
Sink, 360, 363
Skin-load-dominated building, 220, 221, 222, 224, 236,
        241
Skylights, 65, 96 224, 279, 280
Sling psychrometer, 18, 19
Slop sink, *see* Service sink
Small-scale hydro, 328, 330
Smoke, 30, 31, 384, 392, 393–394, 395–398
Smoke-control systems, 392, 395–398
Smoke detectors, 336, 337, 385, 386, 391, 398
Smoke management, 395, 398
Smoke shafts, 391, 395, 398
Smoke vents, 391, 395, 398

Soil:
    piping, 370, 371
    stack, 370, 372
Sol-air temperature, 96–97
Sol-air temperatures (Appendix B4.7), 528–529
Solar altitude and azimuth angles, 91, 93, 153
Solar cell, *see* Photovoltaic solar cell
Solar collector, 154–161
    concentrating, 156, 159
    flat plate, 155, 156, 157, 158, 161
Solar constant, 45–46, 87, 153
Solar desiccant cooling, 157
Solar energy system:
    active, 153, 154–161, 235–236
    hybrid, 153
    passive, 153, 236–241
Solar heat gain (SHG), 88–95
Solar heat gain factor (SHGF), 88, 89–90, 92–95
Solar heat gain factors:
    for externally shaded glass, maximum (Appendix Table B3.8),
        504
    for glass block, with and without shading devices (Appendix
        Table B3.7), 503
    for unshaded glass (Appendix B3.1), 480–498
Solar load, 83–84, 88, 93, 95
Solar reflectances of various foreground surfaces (Appendix Table
        B3.2), 498
Solar time, 87
Sone, 415
Sound:
    absorption, 413, 417, 420
    airborne, 422, 423–424
    audibility, 416
    insulation, 413–414, 415, 419, 423–424, 425–428, 429–430
    isolation, 421–430
    leaks, 424, 425
    masking, 187, 416
    pressure level, 413, 414–415, 430
    solid-borne (structure-borne), 422, 423–424, 425–428, 429
    trap (sound attenuator), 181, 430
Specific heat, 5, 69, 70, 94, 156
Specific humidity, *see* Humidity ratio
Specular reflection, 88, 254–255
Speech intelligibility, 416, 419
Split system, air-handling, 180
Sprinkler head, 399–403
Stack effect, 61, 105, 397, 398
Stacks, waste and vent, 370–372
Standard SI units (Appendix Table A2), 455
Standby electrical systems, 308
Standpipe, *see* Fire hose standpipe
Static pressure, 356–357
Steam turbine, 317, 318
Storage-type water heater, *see* Tank-type water heater

Stormwater drainage, 355, 356, 375–378, 379
Storm sewer, 376
Storm window, 66, 67
Street main, 356
Structural discontinuity, 425–428
Subpanel, 292, 307–308
Suction tank, 357
Sunspace, passive solar heating, 239, 240–241
Superinsulation, 68, 228–229
Supply water, 356–360
Surface conductance, 44, 46, 62
Surface effect, 185
Surface raceway, 303, 304
Surface runoff, 375, 377
Surveillance system, *see* Security systems
Sustainable design, *see* Green design
Sweating, 7, 8–9, 12–13, 24, 67, 73
Switchboard, 301
Switchgear, 297, 301

Takeoff water heater, *see* Indirect water heater
Tankless water heater, *see* In-line water heater
Tank-type water heater, 357–358
Task lighting, 257–258
Tax credit, 443–444
Tax deduction, 439, 443–446
Temperature scales, 7, 23, 41
Terminal reheat, 175–176
Theoretical power, 330
Thermal acceptability limit, 15, 16
Thermal bridge, 96, 99
Thermal capacity, 70, 94, 228
Thermal conductivity, *see* Conductivity, thermal
Thermal detector, *see* Heat detector
Thermal energy storage (TES), 200–203, 311
Thermal equilibrium, 40
Thermal lag, 70, 94, 100, 228, 229
Thermal lag factor (TLF), 88, 94–95, 100
Thermal lag factors:
    for glass solar load:
        with interior shading (Appendix Table B3.11), 516
        without interior shading (Appendix Table B3.10), 514–515
    for walls and roof (Appendix Table B4.11), 532–533
Thermal storage mass, 58, 69–72, 94, 228, 238–241
Thermal storage wall, 71, 224, 228, 238, 239–240
Thermodynamics, laws of, 57
Three-phase power, 293, 294, 297
Three-pipe system, 183, 192
Through-the-wall unit, 195–196
Throw, 185
Time lag, *see* Thermal lag
Time-of-use metering, 202, 311
Time value of money, 441–444
TL (transmission loss), 414

Tone, 412
Ton of refrigeration, 162
Top-lighting, 279, 280
Topping cycle, 320
Total energy (TE), 319, 326
Total light wattage, 102
Traction-type elevator, *see* Elevator, electric-type
Transformer, 292, 296, 297–299
Transmission:
    of light, 255, 256
    of sound, 413–414, 423, 425–428, 429
    thermal, 46–47, 63, 81, 83–84, 97, 95–100
Transmission loss (TL), 414
Transmission factor, *see* Luminous transmittance
Transmittance:
    of light, *see* Luminous transmittance
    thermal, *see* U-value
Trap:
    plumbing, 365, 368, 370, 379
    sound, *see* Sound trap
Travel time, elevator, *see* Elevator travel time
Two-pipe system, 183, 192

U-factors:
    conversion table for various wind velocities (Appendix Table B4.6), 527
    for slab doors (Appendix Table B4.4), 526
    for windows and skylights (Appendix Table B4.3), 525
Underfloor raceway, 305, 306
Undershot water wheel, *see* Water wheel
Underslung arrangement, 346–347
Underwriters Laboratories (UL), 266, 292
Uninterruptible power system (UPS), 309
Unitary package unit, 179, 180, 195
Unitary split system, 179, 180
Unit conversions (Appendix Table A1), 453–455
Unit heater, 188–189
Unit substation, 301
Unit ventilator, 194
Upfeed distribution, 357
Urinal, 360, 361, 366–368
Usage charge, 310
Usage factor, *see* Diversity, factor
Useful life, 440
U-value, 46–47, 48, 55, 73, 74

Vacuum breaker, 365
Vapor:
    Barrier, 75
    -compression cycle, 162
    pressure, 18, 53, 55, 73, 74–75
Variable air volume (VAV), 175, 177–179
Ventilation, 33–35, 68–69, 106
    automatic control of, 205

defined, 32, 44
load, 62–63, 106
with various systems, 171, 183, 194, 195
Ventilation requirements (Appendix Tables B6.3 and B6.4),
550–553
Vent stack (vent pipe), 370–372
Vertical landscaping, 127
Vibration isolation mountings, 163, 206, 422, 428, 430
Visual indicators, 335–337
Voice-activated alarm systems, 338, 339
Voice fire alarm, 388–389
Voltage, 289–290, 295

Wall case, 195
Wall solar azimuth, 91, 93
Wash fountain, 363
Waste heat boiler, 200, 145. *See also* Heat, recovery incinerator
Waste piping (waste stack), 370, 371
Water closet (WC), 360, 361, 366–368
Water cooler, 359. *See also* Drinking fountain
Waterflow switch (waterflow detector), 336, 337, 386
Water heaters, 357–359
Water-side economizer cycle, 168
Water turbine, 328, 329–331
Water wheel, 328–331
Watt (W), 6, 63, 291, 292

Wavelength:
of solar radiation, 88, 153, 229–230
of sound waves, 412
of thermal radiation, 45–46
WECS (wind energy conversion system), *see* Wind turbine
Wet-bulb depression, 18
Wet-bulb (WB) temperature, 6, 18, 21–22, 49, 53, 54
Wet column, 372
Wet pipe system, 402
Wind channeling, 58–60, 220
Wind furnace, 325
Wind generator, *see* Wind turbine
Windmill, *see* Wind turbine
Wind sheltering, 219–220, 324
Wind turbine, 323–326
Wire, 302–303. *See also* Conductor, electrical
Wireway, 303–304
Work plane, 256, 257
Work station, 256

Xeriscape, 129, 130

Zones:
elevator, 345
HVAC, 63, 68, 172, 175–179, 183
Smoke-control, 397–398